Jet Web

Dietrich Eckardt

Jet Web

CONNECTIONS in the Development History of Turbojet Engines 1920 - 1950

 Springer

Dietrich Eckardt
München, Germany

ISBN 978-3-658-38533-0   ISBN 978-3-658-38531-6   (eBook)
https://doi.org/10.1007/978-3-658-38531-6

This Springer imprint is published by the registered company Springer Fachmedien Wiesbaden GmbH, part of Springer Nature.
The registered company address is: Abraham-Lincoln-Str. 46, 65189 Wiesbaden, Germany

Foreword

Without doubt, the turbojet engine is one of the greatest technological achievements of the twentieth century—both in technical and societal perspective. Never before had mass transportation flight been possible at such speeds and over transcontinental distances. Modern jet transport with its high speed, safety, and economical impact has fundamentally altered the concepts of travel and mobility. Methods of communication have accordingly changed, as have business operations. A lasting success story: Taking just the past two decades, the global economy, measured in terms of gross domestic product, grew at 2.8% annually, while the world passenger air traffic, expressed in revenue passenger kilometres, increased at an average annual growth rate of 5.0%. Consequently, under changing priorities, the fight to reduce the impact of global air traffic on climate change implies enormous challenges for the jet industry in the envisaged timeframe up to 2050.

Like many preceding and following future engineering developments, the decisive advance of gas turbine propulsion, in the words of historian Edward Constant II *The Turbojet Revolution* (1980), was driven by military needs on the eve of the Second World War. Like nuclear bomb and radar, the turbojet aircrafts and engines belonged to the category of radical innovations that profited from military funded programmes. Its breakthrough, however, was only achieved after the 1960s, when civil air travel began to skyrocket.

According to conventional historiography of technology, already in the late 1920s Frank Whittle pioneered the turbojet development; then in due course, and unaware of the works in England, Hans Joachim Pabst von Ohain followed after 1936 at the Ernst Heinkel Flugzeugwerke. *'Jet Web'* deviates from this established, as we know in the meantime, somewhat constructed *'dual inventor narrative'* of early turbojet engine developments. Correspondingly, the author Dietrich Eckardt, a professional turbojet engineer, and in the meantime awarded engineer-historian for his foregoing book *Gas Turbine Powerhouse* (2014), has considerably expanded the international scope of the main historical actors of this fascinating story, but included also many, so far overlooked, *connections* between them. In addition, he deepened our understanding of the turbojet engine development up to and beyond the end of WW II, by connecting it to parallel scientific developments and technological innovations, such as industrial gas turbines, emerging numerical methods, and turbochargers, to name a few.

The *Turbojet Revolution* can only be fully understood in light of the foregoing—in hindsight rather short—era of reciprocating aero engines. After the first four decades of powered flight, further technical refinement was only possible under increasingly cumbersome endeavour. The technology of that class of aircraft had reached a plateau with little prospect of major improvements in the future. Under these auspices the radical innovation of the jet engine, based on the scientific achievement of aeronautical research of the interwar period and a specific production regime during World War II, allowed for a new level of technological performance: The specific engine power (thrust) more than duplicated—at a stunning flight speed surplus in comparison to the established propeller propulsion systems—and considerably reduced manufacturing cost, time, and demand in fuel quality.

Over four decades from 1920 to 1960, *Jet Web* unfolds a fascinating panorama of often surprising social and technological relations and connections—to the edification of experienced, and the inspiration of future engineers, and—hopefully—the millions of visitors to our global network of museums of science and technology.

Deutsches Museum   Prof. Dr.-Ing. Helmuth Trischler
München, Germany
June 2022

Figures and Copyrights

- This book—containing ~250 figures—observes the Copyright principles:
- Figures in the Public Domain are not designated in general.
- Figures from the ABB Archive—operated by Docuteam, CH-5405 Baden-Dättwil—are marked ©ABB.
- Author and Springer Verlag thank the various copyright owners for the permission of reprint.
- In a few cases the copyrights were not detectable, though diligent good faith search for the copyright holders took place; the figure sources are provided in any case. Potential copyright owners are asked to contact the author.

Contents

About the Author

Dietrich Eckardt studied Chemical Engineering at TU Berlin, before he finished his doctorate in centrifugal compressor research under Profs. W. Traupel, ETH Zurich, and H. Gallus at RWTH Aachen in 1976. He has more than 40 years of professional experience in turbomachinery research at DLR Cologne, in the aero engine industry at MTU Aero Engines, Munich, and finally since 1995 in the area of power generation gas turbines at Alstom/ABB in Baden/Switzerland. Since 1992, he has a honorary professorship for gas turbine development at TU Dresden.

In 2013, he published *Gas Turbine Powerhouse. The Development of the Power Generation Gas Turbine at BBC-ABB-Alstom* about 100 years of power generation gas turbine development, which received the 2017 ASME Engineer-Historian Award.

The present, technical-history book *Jet Web. Connections in the Development History of Turbojet Engines 1920–1950* covers for the first time in detail corresponding parallel developments in Great Britain (USA), Germany, and Switzerland—with many surprising interconnections.

Abbreviations

ABB	Asea Brown Boveri Ltd, CH and D
ABBEC	American Brown Boveri El. Corp., USA
AC	Allis-Chalmers, USA
AC	alternating current
ACA	Advisory Committee for Aeronautics, UK
AD	Anno Domini
ad	adiabatic
AEDC	Arnold Engg. Development Center, TN, USA
AEG	Allgemeine Elektrizitaets-Gesellschaft, D
AIPS	advanced integrated propulsion system
AK	Ackeret–Keller process
ALR	Arbeitsgruppe für Luft- und Raumfahrt (Aerosp. Dev. Group), Zurich, CH
AM	British Air Ministry, London, UK
Ar	Arado Flugzeugwerke, Rostock-Warnemuende, D
ARC	Aeronautical Research Committee
ASM	Armstrong Siddeley Motors, Coventry, UK
ASME	American Society of Mechanical Engineers, USA
ATAR	Atelier Aérodynamique de Rickenbach (French turbojet)
AVA	Aerodynamische Versuchsanstalt Goettingen, D
BAC	British Aircraft Corp., UK
BAe	British Aerospace, UK
BASF	Badische Anilin- und Soda-Fabrik, Ludwigshafen, D
BBC	A.-G. Brown, Boveri & Cie., Baden, CH, and Mannheim, D
BC	Before Christ
BIOS	British Intelligence Objectives Sub-Committee, UK
BL	boundary layer
BLI	boundary layer ingestion
BMFT	Bundes-Ministerium für Forschung u. Technologie (Fed. Min. for Res. & Tech.), D
BMW	Bayerische Motoren Werke, Munich, D
BPR	bypass ratio

BR	British Railways, UK
Bramo	Brandenburgische Motorenwerke AG, Berlin-Spandau, D
BSc	Bachelor of Science
BSEL	Bristol-Siddeley Engines Ltd, UK
BT-H	British Thomson-Houston, UK
c	absolute velocity
CalTech	California Institute of Technology, Pasadena, USA
CCPP	combined cycle power plant
CDA	controlled diffusion aerofoil
C.E.M.	Compagnie Électro-Mécanique, Paris, F
CEO	chief executive officer
CFD	computational fluid dynamics
CFM	CFM International (CF6 + M56 turbofan engines), Paris, F
CH	Switzerland (Corps Helveticum)
CIOS	Combined Intelligence Objectives Sub-Committee, UK
cm	centimetre
CR	counter-rotation
CRISP	counter-rotating integrated shrouded propfan
D	Germany (Deutschland)
D	diameter
DAAD	Deutscher Akad. Austausch-Dienst (German students exchange programme)
DB	Daimler Benz, Stuttgart, D
DC	direct current
DCA	double circular arc aerofoil
DE	Germany (Deutschland)
DED	Department of Engine Development (AM), London, UK
DFG	Deutsche Forschungs-Gemeinschaft, D
DFL	Deutsche Forschungs-Anstalt fuer Luftfahrt, Braunschweig, D
DFS	Deutsche Forschungsstelle fuer Segelflug, Darmstadt/ Ainring, D
DGER	Direction Gen. des Études et Recherches (French secr. service, today DGSE)
DGW	Deutsche Gesellschaft für Windenergie (German Society for Wind Energy)
DLR	Deutsches Zentrum fuer Luft- und Raumfahrt (German Res. Centre for Aeronautics and Space), D
DKW	Motor car and bike manufacturer (Dampf Kraft Wagen), Zschopau, D
DM	Deutsche Mark
DoD	Department of Defense, USA
DVL	Deutsche Versuchsanstalt fuer Luftfahrt, Berlin-Adlershof, D
EADS	European Aeronautic Defence and Space
EC	European Community
EEC	English Electric Co., UK
EHF	Ernst Heinkel Flugzeugwerke A.G., Rostock-Marienehe, D

EMPA	Eidgen. Materialprüfungsanstalt (Swiss Fed. Laboratories for Material Testing), CH
ESC	Engine Sub-Committee, UK
ETH	Eidgen. Techn. Hochschule (Swiss Federal Institute of Technology), CH
ETHZ	Eidgen. Techn. Hochschule (Swiss Fed. Institute of Technology) Zurich, CH
ETF	engine test facility
FEM	finite elements method
FKFS	Forschungsinstitut für Kraftfahrwesen und Fahrzeugmotoren (Res. Institute of Automotive Engineering and Vehicle Engines) Stuttgart, D
FRS	Fellow of the Royal Society, UK
FVD	Flugtechnischer Verein Dresden, D
GALCIT	Guggenheim Aeronautical Laboratory at the California Institute of Technology, since 1961 GAL—stands for Graduate Aeronautical Laboratories, USA
GE	General Electric, Evendale, Oh., USA
GEC	General Electric Company, Coventry, UK
GT	gas turbine
h	total enthalpy, or hour
HAPAG	Hamburg-Amerikanische Packetfahrt-Actien-Gesellschaft, D
He	Ernst Heinkel Flugzeugwerke A.G., Rostock-Marienehe, D
Hermos	SNECMA aero engine repair shop, Casablanca, after Hermann Oestrich
Hermso	BMW project designation, after Hermann Oestrich
HeS	Heinkel Strahltriebwerk (turbojet)
HFI	Hermann Foettinger Institute, TH Berlin, D
hp	high pressure
hp	horse power
HPC	high pressure compressor
HPT	high pressure turbine
HR	human resources
HVA	Heeres-Versuchs-Anstalt, Peenemuende, D
IAI	Israel Aircraft Industry
IC	internal combustion
ICAM	Int'l. Congress of Applied Mechanics
ICE	internal combustion engine
IEEE	Institute of Electrical and Electronics Engineers, USA
IIE	Institute of International Education, USA
IMechE	Institution of Mechanical Engineers, London, UK
in	inch
ip	intermediate pressure
IPK	Institute for Product Engineering (KIT), Karlsruhe, D
IPR	intellectual property right
is	isentropic

JFM	Junkers Flugzeug- und Motorenwerke AG, Dessau, D
Jumo	Junkers Motorenbau GmbH, Dessau, D
Kalag	Kali AG (potash mine), D
kg	kilogram
KIT	Karlsruhe Institute of Technology, D
km	kilometre
KoBü	Konstruktions-Büro (design office), D
KTA	Kriegs-Technische Abteilung, CH
KTL	Kraftfahr-Technische Lehranstalt, Vienna, A
KVA	Kraftfahrtechn. Versuchs-Anstalt (KTL)
LFM	LuftfahrtForschungsanstalt Munich, D
lp	low pressure
LPC	low pressure compressor
LPT	low pressure turbine
m	metre
$\dot{m}$	mass flow
Ma	Mach number
MAN	Maschinenfabrik Augsburg-Nuernberg, D
Me	Messerschmitt AG, Augsburg, D
Metrovick	Metropolitan-Vickers, Manchester, UK
MFO	Maschinenfabrik Oerlikon, Zurich, CH
MIT	Massachusetts Institute of Technology, Cambridge, MA, USA
ML	Motor-Luft (motorjet)
mm	millimetre
MoD	Ministry of Defence, London, UK
MTU	MTU Aero Engines AG, Munich, D (former Motoren- and Turbinen-Union)
MV	Metropolitan-Vickers, Manchester, UK
MWF	Magdeburger Werkzeugmaschinenfabrik, D
NACA	National Advisory Committee of Aeronautics, Wash. D.C., USA
NAE	U.S. National Academy of Engineering, USA
NASA	National Aeronautics and Space Admin., Wash. D.C., USA
NASIC	National Air and Space Intelligence Center, USA
NECIES	North-East Coast Institution of Engineers and Shipbuilders, Newcastle on Tyne, UK
NEIMME	North of England Institute of Mining and Mechanical Engineers, Newcastle on Tyne, UK
NRC	National Research Council, USA
NSDAP	National-Soz. Deutsche Arbeiter-Partei (NS German Workers Party), D
NTM	Norsk Teknisk Museum, Oslo, N
OEM	original equipment manufacturer
OKB	Soviet experimental design office
OKM	Oberkommando der Marine (Naval high command), D

OMW	Otto-Mader Works (Jumo), Dessau, D
ONERA	Office National d'Études et de Recherches Aérospatiales, Meudon, F
P	power
PCF	Parti Communiste Français
pol	polytropic
POW	prisoner of war
PR	pressure ratio
Pr	Prandtl number
PRC	People's Republic of China
PRU	Photo-Reconnaissance Unit, RAF Medmenham, UK
P&W	Pratt & Whitney, East Hartford, CT, USA
q	heat
R	degree of reaction
RAE	Royal Aircraft Establishment, Farnborough, UK
RAeS	Royal Aeronautical Society, London, UK
RAF	British Royal Air Force
RC	remotely controlled
R&D	research and development
RDLI	Reichsverband der Deutsch. Luftfahrt-Ind. (Reich Assoc. of German Aero-Ind.), D
Re	Reynolds number
RfP	request for proposal
RIC	Research Information Committee, USA
RKL	Reichskommissariat für die Luftfahrt (Reich Commissionar for Air Traffic), D
RLM	Reichs-Luftfahrt-Ministerium (Reich air ministry), Berlin, D
RM	Reichs-Mark (German currency)
RPA	Reichs-Patent-Amt (Reich patent office), Berlin, D
RVM	Reichs-Verkehrs-Ministerium (Reich communication/ traffic ministry), Berlin, D
rpm	revolutions per minute
RR	Rolls-Royce plc, Derby, UK
RRHT	Rolls-Royce Heritage Trust
s	entropy
SAW	submerged arc welding
SBZ	Schweizerische Bau-Zeitung (newspaper), Zurich, CH
SCM	supply chain management
SFA	Svenska Flygmotor AB, S
SFr	Swiss Francs
SFR	Swiss Federal Railways
S.L.	sea level

SLM	Schweizerische Lokomotiv- und Maschinenfabrik (Swiss Locomotive and Machine works), Winterthur, CH
SMAD	Soviet Military Administration in East Germany
SNECMA	Société Nationale d'Études et Construction de Moteurs d'Aviation, Paris, F
SOCEMA	SOciété de Constructions et d'Équipements Mécaniques pour l'Aviation, Le Bourget, F
ST	steam turbine
SU	Soviet Union
T	total temperature
TACOM	US Army Tank Automotive Command, USA
TBC	thermal barrier coating
TCh	turbo charging
TEC	Turbo Engineering Corp., USA
TEL	tetra-ethyl lead
TGZ	Technische Gesellschaft (Technical society) Zurich, CH
TH	Technical High School
THB	Technische Hochschule Berlin-Charlottenburg, D
TIBB	Tecnomasio Italiano Brown Boveri
TIG	tungsten inert gas welding
TIT	turbine inlet temperature
TMF	thermal–mechanical fatigue
TO	take-off
TVA	Torpedo-Versuchs-Anstalt (Torpedo test facility), Eckernfoerde, D
TU	Technical University
u	circumferential speed
V	volume flow
V-1	Vergeltungswaffe 1, Fieseler Fi 103, D
V-2	Vergeltungswaffe 2, A4 rocket, D
VDI	Verein Deutscher Ingenieure, Dusseldorf, D
VDIZ	VDI Zeitung (VDI newspaper)
VIGV	variable inlet guide vane
VKI	Von Kármán Institute for Fluid Dynamics, Rhode-St. Genèse, B
VP	Vice President
VR	VerwaltungsRat (supervisory board)
w	relative velocity
W	work
WPC	World Power Conference
WMF	Württembergische Metallwaren-Fabrik, Geislingen, D
WW	World War
ZF	(ZahnradFabrik) ZF Friedrichshafen, D
η	efficiency
ρ	density

Introduction

The writing of history, historiography, is the 'doing of history',
and engineers should make sure that the historiography of
engineering is not left entirely to historians. Recognition of this
activity by the profession is very important.

The invention of the *turbojet engine*, the gas turbine for aircraft propulsion, belongs to the
most important technical achievements of the twentieth century[1]. However, *invention . . .*
thermodynamically, it was more a continuous-flow derivative of the established (instantaneous) piston engine. Nevertheless its introduction was a *'turbojet revolution'*[2] towards the

---

[1] The *motto* at top, issued by Werner Albring, puts the engineer to the centre of engineering
historiography. Prof. Dr. Werner Albring (1914–2007),—see Wikipedia, 'Werner Albring' in German, was head of the Institute of Applied Fluid Mechanics at TU Dresden from 1952 until 1979, and
author of one of the most recommendable textbooks on fluid mechanics—see Albring, Angewandte
Strömungslehre. He wrote several outstanding papers on the history of engineering and science
(Prandtl, Helmholtz, Hagen, dimensional analysis, etc.) and in this context he is known for his credo
that qualified engineers should return to a responsible leadership role in interpreting technological
developments, in fact the decisive encouragements for this book, and the foregoing 'Gas Turbine
Powerhouse' (2017 ASME Engineer-Historian Award),—see Koeltzsch, Was bleiben wird, including four original Albring papers (in German, especially his 2004 Helmholtz Lecture) under www.
albring.info.

[2] See Constant, The Origins of Turbojet Revolution.

end of the Second World War for practical reasons. The turbojet engine—and the later turbofan (by-pass) engine already thought of—provided significant operational advantages in comparison to the reciprocating aero engine, driving a conventional propeller:

- The specific weight of the new turbo engines per unit thrust was just in the order of 50% of that of the replaced, reciprocating propulsion systems,
- turbojet/turbofan avoided the speed-limiting compressibility losses during propeller-driven aircraft cruise, achieved their best performance at higher flight speed (and for the turbofan at an equivalent performance level),
- the new turbo-engines were capable of generating a very large amount of power in a single unit without prohibitive mechanical complexity, and
- war-time limitations aside, the turbo-engines quickly became more reliable than reciprocating engines, were manufactured at lower cost, and operated for thousands of hours without major maintenance work.

Though turbojet engines are now prevalent, a parallel idea of the humble beginnings and the introductory system transition as the key technologies evolved, is still rather rare. In 1980 Edward Constant II laid the groundwork with a first comprehensive description, in which he illustrated the basic complexity of this task, by highlighting four necessary components to understand and evaluate the '*turbojet revolution*':

- the vast majority of turbojet engine histories followed the concept of independent, multiple inventions by very few key '*pioneers*', working largely alone with visionary insight and unshakeable determination. The comprehensive description of the impressive work of Sir Frank Whittle as a '*British inventor hero*', followed this concept over decades, supplemented to a smaller extent by comparable publications of the competitive activities in German turbojet development by Hans Joachim Pabst von Ohain;[3]
- the turbojet gas turbine is heir to two centuries of largely European turbomachinery development, including water pumps and turbines, steam turbines, piston engine turbochargers and internal combustion gas turbines. Nevertheless, the turbojet engine is a unique, new system of unprecedented performance capabilities;
- in fact, the turbojet engine replaced the foregoing, highly successful piston engine technology, by using and at the same time being constrained by the receiving airframe structure, the engine accessory system, aircraft production and organisational technologies, all sharing characteristics with the conventional propulsion system;
- finally, the turbojet engine both depended upon and was developed at the same time along with a few, most remarkable achievements in aeronautical sciences—

---

[3] As outlined in Sect. 12.1, Fig. 12.6, this widespread, readily accepted approach to historic technology writing culminated in the '*independent dual-inventor-narrative*' of the mid-1970s—with the negative consequence that the historic reality, although still available when researching the detail, was outshone by the new storyline, largely overshadowed and its effect distorted.

aerodynamics, aerostructures and material properties, though the precise nature and extent of this dependence has not yet been investigated.

This book by and large follows Constant's described structure, in chronological order over roughly 30 years,—between the 1920s and 1950s—with a few extending narrative strings on both ends. The personnel scope goes considerably beyond the simple Whittle vs. Ohain saga. Four intermediate Chaps. 3, 5, 7 and 9 are reserved for double-portraits of leading *'influencers'* of the turbojet engine history. Constant's topic 2 (above), the associated early stationary fluid- and turbo-machinery history as basis of the aero gas turbine developments has been outlined in detail in the author's foregoing *'Gas Turbine Powerhouse'* (2014).[4] It is used here as part of the condensed chronological *'timeline'* of Chap. 2.

Instead of independent and isolated developments, the book emphasises the concept of information exchange, traces inherent aspects of freely-flowing ideas, definitively up to the late-1930s; and even after the outbreak of war and the introduction of considerable secrecy measures between the combatants, a reliable, supporting *'Jet Web'* carried and accompanied the major development groupings. Of course, the largest and most important part of the historic account is drawn from Great Britain and Germany, but Switzerland— represented by the industrial heavyweight *Brown Boveri & Cie. (BBC)*—kept a decisive position throughout, and for both sides. Brown Boveri not only had a historic *hinge function* as a participating party within the first—still unsuccessful—development phase of internal combustion gas turbines before the First World War (Stolze, 1899; Armengaud and Lemâle, 1906; Holzwarth, 1903) and corresponding component developments (Parsons, 1884; Rateau, 1902),[5] but played a decisive role in the second, success period 1930–1945. Here in important component technology sections as e.g. the axial compressor, Brown Boveri led the earlier industrial gas turbine developments and accompanied the then rapidly catching up and even outpacing aero engine development activities. Consequently, as illustrated in Fig. 1.1, there was the country triangle of Great Britain, Germany and Switzerland which represented, in simplified form, the *'Jet Web'* structure, and corresponding exchange of information, know-how, of personnel and of relevant hardware. Graphically differentiated are three shades of grey—arrows (1) and (5) *light*, arrows (2) and (3) *medium*, and arrow (4) *dark*—corresponding to the roughly assessed *substance* of the covered activities for the turbojet engine developments on the receiving side. In this respect, the transfer of the Whittle engine to the USA in 1941—arrow (4)—exists without parallel in this context.

In short, arrow (1) marks a period of intensive scientific exchange between Great Britain and Germany after WWI, typified by Hermann Glauert and Ludwig Prandtl, Fig. 4.39, followed by the glider comradeship on the Wasserkuppe (Beverly Shenstone and

---

[4] See Eckardt, Gas Turbine Powerhouse. The development of the power generation gas turbine at BBC—ABB—Alstom.

[5] For reference years, - see Chap. 2.

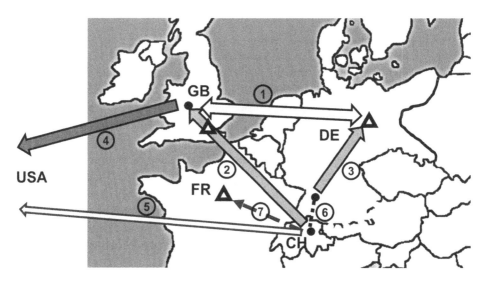

Fig. 1.1 *'Jet Web'*: Main inter-national turbojet technology flows, 1935–1945

Alexander Lippisch), Fig. 4.40, the strange *'period of openness'* between military and industrial camps as part of mutual *'appeasement 1935–1938'* (Roy Fedden and Ernst Udet), Sect. 6.1.3, and the still, largely non-transparent shift of aero know-how carrying, politically-racially prosecuted engineers before 1939 from Nazi-Germany to England (Fritz Heppner and Gustav Lachmann), Chap. 5. Arrow (2) stands for the sustainable axial compressor and gas turbine engineering support between 1935 and 1940 from the Swiss BBC out of Baden, CH, mostly towards RAE Farnborough and Metrovick Manchester, Sect. 6.3.2 f., to some extent balanced by the later axial compressor design support, arrow (3), out of BBC Mannheim for Hermann Oestrich's *BMW 109–003* turbojet engine, and the various *Hermso* derivative study activities by Hermann Reuter's group, Chap. 9. Arrow (5), was mostly executed locally via BBC's U.S. licence partner company Allis-Chalmers, but also by direct influence of the BBC upper management over the years; first by Claude Seippel's collecting U.S. experiences between 1923 and 1928, especially in the consulting engineering office of Earl H. Sherbondy. Here Seippel, later acknowledged as *'father of the axial compressor'*, received his first decisive encouragements in that direction, Chap. 3 and Sect. 4.1.2; then followed the omnipresent Adolf Meyer on various occasions at the *'cradle of US turbojets'*, Sect. 6.4.2. The remainder stands for BBC internal activities; the dotted line (6) indicates some granted exchange up to the beginning of WWII—and perhaps beyond, mostly from Baden to Mannheim on technical issues like the first aero turbochargers, the instantaneous *Holzwarth* gas turbine, the *Velox* boiler and, in due course, various power generation gas turbine and wind tunnel drive projects, 1939–1945. Vice versa, contacts directed from Mannheim to Baden during and immediately after the war were apparently non-existent, so that e.g. the opportunity of an effective in-house transfer of advanced know-how from the experienced group of Hermann Reuter in support of the *'French connection'* of turbojet engine developments, arrow (7), to BBC Baden and their daughter companies C.E.M and SOCEMA did not materialise, Sect. 10.3.2.

Besides these rather systematic *'lines of influence'* within the *'Jet Web'* network, there are numerous unplanned incidences in the time before the outbreak of WWII—without which the turbojet development history might have unfolded differently:

- *1910*, during the mentioned first, still unsuccessful gas turbine development period Brown Boveri's Walter Noack—later renowned for inventing the Velox boiler with the first operational *all-axial turbomachinery* drive—travelled to Paris-St. Denis to meet the then still living Marcel Armengaud,[6] taking his strong recommendation that *'a company like BBC should try to create something new in the* (compressor) *field'*,
- *1911*, Henry Tizard, later influential head of *ARC*, the Aeronautical Research Committee and strong supporter of Whittle's gas turbine project, studied then at Berlin with Professor Walther Nernst, who had just failed to get the *Nernst tube* concept of an all-radial gas turbine rotor patented, which 30 years later Whittle's inventor-competitor Hans von Ohain managed successfully. It is what became his second *'secret patent'*, and his first turbojet engine demonstrator HeS 01 in 1936,[7]
- *1924*, after an international *'scientific scandal'*, Alan A. Griffith decided to give up a promising career in the science of metal fatigue and fracture mechanics, to switch to and continue in advanced turbomachinery design, where he became one of Great Britain's most influential principal scientific officers, and *'one of the greatest philosophers of the science of aircraft propulsion'*,[8]
- *1925*, Max König (32), scientific *'rookie'* from Swiss Brown Boveri gathered his (few) cards together, stood up in front of the honourable assembly of the *North-East Coast Institution of Engineers and Shipbuilders* at Newcastle upon Tyne, and maintained against heavy *'headwind'* from the illustrious audience, *'that in time, the flight gas turbine will entirely replace the piston engine'*,[9]
- *1932*, the beginning of a patent lawsuit between Brown Boveri and the German-speaking turbomachinery industry, led by AVA Aerodynamic test establishment, Goettingen. It lasted until 1937 with Brown Boveri's unrestricted patent success. This might have substantially influenced AVA's, in the end untenable axial compressor design philosophy with 100 percent reaction blading, which *'infected'* both Jumo 004 and BMW 003, the German turbojet engine series production configurations,[10]

---

[6] René and Marcel Armengaud, together with Charles Lemâle built in 1905/1906 an experimental, self-sustained gas turbine of 6–10 kW power output—see Eckardt, Gas Turbine Powerhouse, p. 70f.

[7] See Fig. 6.22, and—see Sect. 12.1.1 for the history of Ohain's *'Secret Turbojet Patents'*.

[8] See Rubbra, Alan Arnold Griffith, and—see Sect. 4.2.1.

[9] See Chap. 3.

[10] See Sect. 4.3.3, Figs. 4.46–4.49.

- *1935*, Adolf Meyer (55), mighty Director of BBC's Thermal Department, and different to Max König, a life-long cautious sceptic about the feasibility of flight-worthy gas turbines, in a series of '*Jet Web*'-type contacts, met, first Helmut Schelp (23) at Stevens Institute of Technology, NY,[11] who would become after four years Germany's most influential turbojet engine programme manager, and thereafter at Long Beach, CA, Vladimir Pavlecka (37), presumed to be America's most daring turbojet engine pioneer and Northrop's chief engineer,[12] ... and finally, at his own historic gas turbine presentation at IMechE London on 24 February 1939 also Frank Whittle (32), with whom developed a well-documented, highly interesting fundamental discussion,[13] just months in advance of the essential breakthrough of the turbojet engine.

The origins of this book project date back nearly 20 years, when a first sketch of the historic development of BBC's advanced gas turbine technology was published in 2002.[14] In a side view to the parallel development of the early aero turbojet, gas turbine developments, the authors stated with some astonishment the fierce and secret race towards the best propulsion gas turbine, specifically compressor concept—especially in Britain and Germany in the mid-1930s, at times, when BBC had already sold a few of their newly developed axial turbomachinery sets. By stipulating certain differences in the relevant technologies between aero and stationary applications, it was obvious to assume at the same time certain '*communication activities*' to bridge apparent gaps in know-how. The focus was put on the triangle Switzerland, Germany and England from 1925 onwards, with some links to other participating countries. This at first rather speculative approach nourished some in-depth research to investigate more carefully possible links and know-how dissemination paths— and the findings in the form of the present '*Jet Web*' were worthwhile. The 2002 ASME paper stated: '*The history of this specific development/flow of ideas and know-how has still to be written; within this context* (of the paper) *only a few notes shall be sketched*'.

In the meantime the relevant new materials grew considerably, both with and without immediate BBC participation, so that a separate publication emerged, now with the focus on the European turbojet development in the timeframe 1920–1950. In the light of the extensive von Ohain/Whittle discussion of recent years on first turbine-powered flight 1939, it appeared beneficial to investigate the role and contributions of potential third players in this context/contest, e.g. to trace the dissemination of key ideas, mainly e.g. the application of aerofoil theory for advanced compressor design.

---

[11] Schelp achieved his MSc in Mech. Engineering at the Stevens Institute during a 10-month stay after Sep. 1935, a period in which Meyer, received an honorary doctorate there—see Chap. 7.

[12] See Sect. 6.4.2, and Fig. 6.52.

[13] See Sect. 6.3.5, and Fig. 6.47.

[14] See Eckardt and Rufli, Advanced Gas Turbine Technology.

At the end of these introductory remarks, special thanks still go to the author's Swiss power engineering community at ABB/ Alstom which laid the groundwork and provided the encouragement over time, to enable the 2012–2014 publications of *'Gas Turbine Powerhouse'*, and now the twin-book *'Jet Web. Connections in the Development History of Turbojet Engines 1920–1950'*. Both books summarise the deduction of historic reminiscences out of an abundant pool of the author's professional experiences in gas turbine development, 1996–2010 in the field of power generation, and 1969–1995 *'back to the roots'* in turbojet engineering.

Again this work was considerably facilitated by a thorough preparation of the relevant literature body for this comprehensive task. The used *'GT History References'* database covers more than 1200 objects (papers, journal articles, books) that have been thoroughly collected, digitised and put into a searchable form by my former Alstom-colleague Robert Marmilic, who thus also prepared the reliable foundation for this book project. Wherever applicable, the great resources of the ABB Historic Archive, Baden-Daetwil, CH, led by Tobias Wildi and the *Docuteam*, have been applied again. Special thanks go to Norbert Lang, for his invaluable, generous support in many special areas, and to Georges Bridel and Marc Immer of ALR Zurich for their Me 262 flight simulation calculations, with modified Jumo 004 engines, Fig. 9.7.

In the area of turbojet developments the author is in debt for the patience of many lengthy discussions, resulting in valuable hints and specific contributions, to his former colleagues of MTU Aero Engines, Munich, foremost to Hanns-Juergen Lichtfuss, to Uwe Schmidt-Eisenlohr, Helmut-Arndt Geidel and Klaus-Peter Rued. Helmut Schubert was ready for problem-solving nearly 24/7, and organised, together with Sabine Hechtl, many rare nuggets of information at short notice, Jessika Wichner thankfully mined over the years the vast resources of the DLR- and former AVA-Archives at Goettingen. Hedwig Sensen used her wide DGLR network for information and contacts, and many employees of the Archives of Deutsches Museum Munich, of TU Dresden, TU Berlin and KIT Karlsruhe, and of the city archives of Goettingen, Brunswick, Duisburg and Dessau helped by generously spending their time for effective researches. Several times Lutz Budrass kindly provided his valuable support as professional historian; correspondingly, thanks go to Helmuth Trischler for contributing the Foreword. Peter Hamel, Ernst Hirschel and Arthur Rizzi contributed valuable insight from the perspective of aircraft aerodynamics. I am especially grateful for valuable personal information about their famous relatives, from Ian Whittle, Chris Heppner. Tom and Michael P. Schelp,—and owe numerous useful input and assistance to the following:

John Ackroyd, Peter Albring, Frank Armstrong, Philipp Aumann, John Bailey, David Bloor, Wolfgang Brix, Neil Chattle, Peter Collins, Nick Cumpsty, Calum Douglas, John Dunham, Hans J. Ebert, Michael Eckert, Soeren Flachowsky, Richard Fuchs, Herbert Gassert, Hermione Giffard, Georg Gyarmathy, Franz Heitmeir, Dietmar Hennecke, Heinz Hoheisel, John Horlock, Hans Hungenberg, Ernst Jenny, Günter Kappler, Wolfgang Keppel, Maria Kissler, Volker Koos, Wolfgang Koschel, Lee Langston, Achim Leutz, Bernd Lukasch, Gero Madelung, Hans-Ulrich Meier, John Moore, Pierre Mouton, Lutz

Mueller, Harald Müller-Berner, Andrew Nahum, Michael J. Neufeld, Gerhard Pappert, Dave Piggott, Juergen Potthoff, Christian Rabl, Jacques Renvier, Hans Rummelsberger, Thomas Sattelmayer, Stefan Schmunk, Arne Seitz, Leroy H. Smith jr., Konrad Vogeler, Hans Wettstein, Heinrich Weyer, Jakob Whitfield and Gert Winterfeld.

Dietrich Eckardt
eckardt@bluewin.ch
Lenzerheide/Munich
June 2022

Cover picture, an *Alamy* vector graph ID KEG6PX of Jeerawut Rityakul '*Spiralling white* (jet) *streaks on black background*' is meant to symbolise the compact whirl of early '*Jet Web'* development string interactions, unfolding to a blossoming global gas turbine industry.

2.1   Gas Turbine/Turbojet Timeline Up to 1930

In the following survey the important developments in the history of air-breathing turbojet engines are highlighted in chronological order from early beginnings.[1] Although the gas turbine powered jet engine was clearly a twentieth century invention, many of the needed advances in theory and technology were made well before this time. In addition, essential elements of gas turbine component technology have been deduced from the parallel development[2] of the power generation gas turbine at Brown Boveri & Cie. in Switzerland (CH), one reason for this company's prominent role also in early turbojet developments.

2.1.1   The Lead-Up to 1900

~ 400 BC   The earliest known account on the jet propulsion principle tells of the construction of a flying dove of wood by Archytas, the founder of theoretical mechanics, living in Tarentum in Southern Italy. Aulus Gellius

---

[1] An adaptation of—see Wikipedia, 'Timeline of jet power' in English.

[2] It has become common practice in technical historiography to consider both the first full-load certification testing of the 4 MW BBC stationary GT power plant at Baden, CH and the subsequent, first flight of the gas turbine powered Heinkel He 178 at Rostock-Marienehe, Germany, D—6 weeks apart in 1939—as corresponding hinge dates.

	in his *Noctes Atticae* gave credence to the idea that the dove flew by means of expanding vapour contained within it.[3]
~ 70 BC	Roman records of pure impulse, paddle-type water wheels for grain grinding mills.
1st. cent. AD	Heron of Alexandria (c.10 AD–c. 70 AD), Greek inventor and geometrician in Roman Egypt describes an *'aeolipile'*, a simple steam-powered pure reaction thermal engine.[4] A closed spherical vessel mounted on a bearing axis is fed with steam from a boiler. The steam discharges tangentially through jet pipes at the vessel's periphery in the opposite direction in a plane perpendicular to the axis of rotation, thus put in rotation by the reactive forces. Centuries passed before the reaction principle was applied again, after which a continuous line of development can be traced up to the present time.
1040	The solid propellant rocket, commonly known as the black powder rocket, and related fire arrows were based on the reaction principle. The earliest reference to this kind of weapons used by the army during the Chinese Sung Dynasty was recorded in the *'Wu Ching Tsung Yao'*, an official publication of that year. It is believed that they were first introduced to Europe during the Mongol invasions.[5]
c. 1500	Leonardo da Vinci (1452–1519) sketches the first known hot gas, axial turbine wheel to drive a rotating spit in a chimney; *Codex Atlanticus fol.5 verso-a*,[6] Fig. 2.1.[7] Interestingly, besides numerous descriptive reports, Montaigne saw one of these devices in practical use on his European tour in 1580 at Baden CH, later the site of Brown Boveri, as a kind of foretelling thermomechanical turbomachinery sign over the centuries.
c. 1543	Blasco de Garay (1500–1552), captain in the Spanish navy during the reign of the Holy Roman Emperor Charles V, used a primitive steam machine to move a ship in the port of Barcelona.[8]

---

[3] See Lancaster, Jet Propulsion Engines, p. 5.

[4] Mother Nature uses the reaction principle according to Newton's third law of motion since +400 million years in the class of marine *'cephalopods'* (e.g. octopus, squid), swimming by a natural built-in reciprocating *hydrojet*. Some cephalopods are able to fly through the air for distances of up to 50 m. While not particularly aerodynamic, they achieve these impressive ranges by jet propulsion; water continues to be expelled from the funnel while the organism is in the air. The animals spread their fins and tentacles to form wings and actively control lift force with body posture. See Wikipedia, 'Cephalopod' in English.

[5] See Lancaster, Jet Propulsion Engines, p. 6.

[6] See Eckardt, Gas Turbine Powerhouse, p. 44.

[7] See Eckardt, Gas Turbine Powerhouse, p. 50; also for Fig. 2.1 (r)—see Ackeret, Euler, who puts this unique piece of hardware with uncertain dating (~1620)—owned by AG Rieter&Co., Winterthur, CH—in context to a corresponding mill type at Basacle (Toulouse), built acc. to Belidor ~1737.

[8] See Wikipedia, 'Steam Turbine History' and 'Blasco de Garay', both in English.

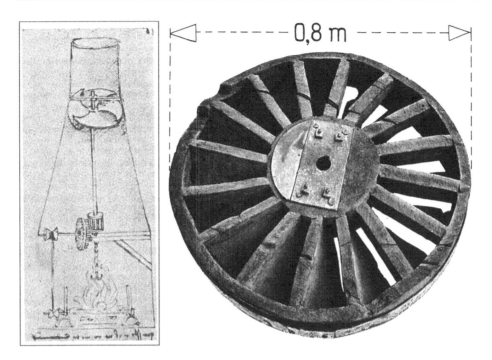

Fig. 2.1 Early Turbo- and Fluid Machines: Leonardo da Vinci—Hot gas axial turbine wheel, ~1500 (l), Reaction turbine water wheel with 3D-curved wooden blading, Southern France, ~1620/1740 (r)

1551	Taqi-al-Din (1526–1585), Ottoman engineer, described a rudimentary impulse steam turbine, again to drive a rotating spit.[9] He may have inspired Giovanni Branca, who proposed similar steam turbines in 1629. Branca was formerly considered to be the originator of the impulse steam turbine.
1556	Agricola, Georg (Bauer/Pawer, Georg 1494–1555), German metallurgist, who wrote *De Re Metallica Libri XII* (12 Books from Mining and Metallurgy) under the Latinised version of his name. The first depiction of a (centrifugal) turbocompressor is found in woodcut prints attached to the sixth book on mining tools and machines.[10]
1629	Giovanni Branca (1571–1640), Italian engineer, proposed a useful working impulse steam turbine for a stamping mill. A jet nozzle directed steam onto a horizontally mounted turbine wheel; the rotation of which was then

---

[9] See Wikipedia, 'Taqi ad-Din Muhammad ibn Ma'ruf' in English.

[10] See Eckardt, Gas Turbine Powerhouse, pp. 45–46.

converted to a stamping action by means of bevel gearing that operated the mill.[11]

1687   (Sir) Isaac Newton (1642–1727), English mathematician, astronomer and physicist, who is widely recognised as one of the most influential scientists of all time, presented his three laws of motion in the *Principia Mathematica Philosophiae Naturalis*. His third law states that for every action (force) in nature there is an equal and opposite reaction. For aircraft and jet engines, the principle of action and reaction is very important. It helps to explain the generation of lift from an aerofoil: the air is deflected downward by the action of the aerofoil, and in reaction the wing is pushed upward. A jet engine also produces thrust through action and reaction: hot exhaust gases flow out the back of the engine; in reaction, a thrusting force is produced in the opposite direction.

1690   Denis Papin (1647–1713), French physicist, mathematician and inventor became known for his pioneering contributions to the development of the heat engine, of a type of pressure cooker with safety valve (London 1679) and a submarine vessel (1691). As a Calvinist he was forced to leave France to England, Italy and Germany, where he invented most of his patents as professor at Philipps University Marburg 1687–1707. In 1690, having observed the mechanical power of atmospheric pressure on his pressure cooker, Papin built and reported on a model of the first effective working one-cylinder piston steam engine and described, in 1705 a series of centrifugal blowers and pumps, from which there is little information on further similar developments until the nineteenth century.[12]

1709*[13]   Abraham Darby I (1678–1717), born to an English Quaker family, established a coke-fired blast furnace to produce cast iron (after as apprentice, he would have seen the use of coke to fuel malting ovens, preventing the sulphur content of coal contaminating the resulting beer, but also avoiding the use of the scarcer charcoal as a fuel). Coke's superior crushing strength allowed blast furnaces to become taller and larger. The

---

[11] See Wilson, Turbomachinery, p. 29.

[12] See Wilson, Turbomachinery, p. 32, and see Wikipedia 'Denis Papin' in English and German.

[13] This and the following, star-marked entries in the timeline correspond with the original 1978 10-episode documentary television series *Connections, Episode 6: 'Thunder in the Skies'* of succeeding prerequisites for technological advancement which implicates the Little Ice Age (ca. 1300–1850) in the invention of the chimney, as well as knitting, buttons, wainscoting, wall tapestries, wall plastering, glass windows (Hardwick Hall [1597] has *'more glass than wall'*), and the practice of privacy for sleeping and sex, before—here specifically for the gas turbine—*the genealogy of the steam engine* is examined—created, written and presented by science historian James Burke (born 1936)—see Wikipedia, 'Connections (TV series)' in English.

ensuing availability of inexpensive iron was one of the factors leading to the *Industrial Revolution*.[14]

1712*   Thomas Newcomen (1664–1729), English inventor, created the first commercial true, atmospheric steam engine using a piston, which was used for pumping flood water from coal and calamine[15] mines. The engine was operated by condensing steam drawn into the cylinder, thereby creating a partial vacuum, thereby allowing the atmospheric pressure to push the piston into the cylinder. It was the first practical device to harness steam to produce mechanical work; 104 units were in use by 1733. Eventually over 2000 of them were installed in Britain and Europe.[16]

1752   John Smeaton (1724–1792), English civil engineer and experimenter, introduced the study of turbomachinery by (small scale water) models. His experiments led him to support Leibniz's early theory of conservation of energy. He also defined the still fundamental thermodynamic concept of power being equivalent to the rate of weight lifting. A lift equation, containing the *Smeaton coefficient*, was used and corrected by the Wright brothers. In modern analysis, the lift coefficient is normalised by the dynamic pressure instead of the Smeaton coefficient.

1754   Leonard Euler (1707–1783), Swiss mathematician, then at the Berlin Academy of Sciences working on a contract by Frederick II of Prussia to improve the *Sanssouci Castle* waterworks, analysed Heron's turbine, the water equivalent of it, Segner's water wheel, and carried out corresponding experiments around 1750. He published his application of Newton's law to turbomachinery, now universally known as *Euler's Turbine Equation* in 1754, and thereby immediately permitted a more scientific approach to design than the previous trial-and-error methods.[17]

1763*–1765   James Watt (1736–1819), Scottish inventor, mechanical engineer and chemist improved on the 1712 Newcomen steam engine, which was

---

[14] See Wikipedia, 'Abraham Darby I' in English.

[15] *Calamine* is a historic name for an ore of zinc, which was used in Europe since the sixteenth century for making cheap brass produced by alloying/'*cementing*' four parts of copper ore with one part of calamine directly, rather than first refining it to metallic zinc. This—and therefore better quality brass—was only possible, when James Smithson (1765–1829), English mineralogist, found out in 1802 that what had been thought to be one ore was actually two distinct minerals—the impure zinc silicate and the nobler zinc carbonate, today known as Smithsonite. Smithson financed in his will the founding of the Smithsonian Institution in Washington, DC, though he had never been in the USA. In 1905, Alexander Graham Bell (1847–1922), Scottish scientist, who is credited with inventing the first practical telephone and regent for the Smithsonian, requested that Smithson's remains be moved from Genoa to the Smithsonian Institution Building. See Wikipedia, 'Calamine (mineral)' and 'James Smithson'.

[16] See Wikipedia, 'Newcomen atmospheric engine' in English.

[17] See Eckardt, Gas Turbine Powerhouse, p. 48.

fundamental to the changes brought by the *Industrial Revolution* in both his native Great Britain and the rest of the world. Asked to repair a model Newcomen engine, Watt demonstrated that about three-quarters of the thermal energy of the steam was being consumed in heating the engine cylinder on every cycle. His critical insight was to cause the steam to condense in a separate chamber distinct from the piston, and to maintain the temperature of the cylinder at the same temperature as the injected steam by surrounding it with a *steam jacket*. Thus very little energy was absorbed by the cylinder on each cycle, making more available to perform useful work.

1767   Jean-Charles, chevalier de Borda (1733–1799), a French mathematician, physicist and sailor introduced stream tube analysis and the study of ideal waterwheels. In fluid dynamics the Borda-Carnot equation is an empirical description of mechanical energy losses of the fluid due to a sudden flow expansion, describing how the total head reduces due to the losses. This is in contrast with Bernoulli's principle for frictionless flow, where the total head is a constant along a streamline.

1773*–1775   John Wilkinson (1728–1808), English industrialist who pioneered the manufacture of cast iron and the use of cast-iron goods during the *Industrial Revolution*—and, in the present context, known as inventor of a precision-boring machine in which the shaft that held the cutting tool extended through the cylinder and was supported on both ends, unlike the cantilevered borers then in use. That could bore cast iron cylinders, such as those used in more efficient steam engines of James Watt, but also for iron guns from a solid piece, rotating the gun barrel rather than the boring-bar. This technique made the guns more accurate and less likely to explode; Wilkinson's boring machine has also been called the first *machine tool*.

1767*–1778   Joseph Priestley (1733–1804), English theologian, philosopher, chemist—and Wilkinson's brother-in-law—gained considerable scientific reputation on his invention of soda water, his writings on electricity, and his discovery of several *airs* (gases). In 1767, his semi-popular 700-page *The History and Present State of Electricity* was published in English and appeared in a French translation in 1771. Early in 1772 Priestley and Alessandro Volta (1745–1827), Italian physicist, chemist, and pioneer of electricity and power, started an intense scientific correspondence and thus expanded this timeline at the end of the *age of enlightenment* to an international level. In the years between 1776 and 1778, Volta—influenced by Priestley—studied also the chemistry of gases. He researched and discovered methane after reading a paper by Benjamin Franklin on *flammable air*. In November 1776, he found methane at Lake

	Maggiore, and by 1778 he managed to isolate methane. He devised

Maggiore, and by 1778 he managed to isolate methane. He devised experiments such as the ignition of methane by an electric spark in a closed vessel, and invented gas detectors and igniters accordingly.

1787   James Rumsey (1743–1792), American mechanical engineer, exhibited a boat propelled by machinery on the Potomac River at Shepherdstown, WV comprising a pump driven by steam power, ejecting a stream of water from the stern of the boat and thereby propelling the boat forward.[18] Along these lines German master-shipbuilder and engineer Alexander Seydell built 1855 at Stettin, D the water-jet propelled turbine steam ship '*Albert*' with 30 hp engine, used on river Oder up to 1859 for regular passenger transport for the 50 km between Stettin–Schwedt and back.[19] Thermal jet propulsion engines did not make their appearance until the twentieth century.

1791   John Barber (1734–1801), English inventor of the basic gas turbine concept, received the British patent #1833 for *A Method for Rising Inflammable Air for the Purposes of Producing Motion and Facilitating Metallurgical Operations*. The patent description clearly follows the thermodynamic sequence that later became known as Joule Cycle (or Brayton Cycle in the USA).[20]

1799   (Sir) George Cayley (1773–1857), English engineer, inventor, and aviator set forth the concept of the modern aeroplane as a fixed-wing flying machine with separate systems for lift, propulsion, and control at Brompton, near Scarborough in Yorkshire. He discovered and identified the four forces which act on a heavier-than-air flying vehicle: weight, lift, drag and thrust. Modern aeroplane design is based on those discoveries and on the importance of cambered wings, also identified by Cayley. He constructed the first flying model aeroplane, and designed the first glider reliably reported to carry a human aloft.[21]

1822   Claude Burdin (1788–1873), French mining engineer, coined the word *turbine,* derived from the Latin *turbo* or vortex in a memoir '*Des turbines hydrauliques ou machines rotatoires à grande vitesse*', submitted to the Académie royale des sciences in Paris. He also called himself '*Teacher of Fourneyron*' after the latter's success in 1827, and is said having experimented on a hot-air multi-stage turbine in 1847.[22]

---

[18] See Wikipedia, 'James Rumsey' in English.

[19] See Friedrich, Vom Wagner/Mueller.

[20] See Eckardt, Gas Turbine Powerhouse, p. 67.

[21] See Wikipedia, 'George Cayley' in English, and Ackroyd, The United Kingdom's contributions, Part 1, p. 15.

[22] See Koenig, Gas Turbines, p. 365.

1824   Nicolas Léonard Sadi Carnot (1796–1832), French physicist and military engineer, gave the first successful theoretical account of heat engines, now known as the *Carnot Cycle*, thereby laying the foundations of the second law of thermodynamics. It is the most efficient cycle capable of converting a given amount of thermal energy, consisting—in this sequence—of isentropic work input (compression), isothermal heat addition (at hot temperature), isentropic work output (expansion) and isothermal heat rejection (at cold temperature). Carnot is often described as the *father of thermodynamics*; his father Lazare N.M. Count Carnot (1753–1823), general and organiser of the French Revolution Army with 1,500,000 conscripts, named him for the Persian poet Saadi of Shiraz (1210–1291/1292), and he was always known by his third given name as Sadi Carnot.[23]

1827   The utilisation of water-power had a long and continuous history in Western Europe. While by the 1820s Great Britain was abruptly ceasing its development and increasingly relying on steam engines, in France to the contrary and with lasting (also negative) effect, there was enormous, steady and meticulous interest in making traditional waterwheels more efficient, since here and there the new machines of the '*Industrial Revolution*' required more power. Benoît Fourneyron (1802–1867), French engineer, developed a radial-outflow *turbine*, a term coined by Burdin already in 1822, consisting of two sets of blades in a horizontal plane, curved in opposite directions to get as much power as possible from the water's motion through this stator/rotor arrangement. Contrary to its forerunners, Fourneyron's turbine used efficient blade angles and ran fully loaded, rather than in the *partial admission* mode in the form of a single incoming jet. In the development of his turbine Fourneyron used the brake produced in 1822 by the Baron Riche de Prony (1755–1839) which permitted much greater accuracy in measuring efficiencies. With the greater reporting accuracy, the efficiencies of the engines also began to increase rapidly.[24]

1851   James Prescott Joule (1818–1889), English physicist, mathematician and brewer, sometimes referred to as the last self-taught person who contributed substantially to science, states in the introduction of his presentation:[25] *It has long been suspected that important advantages might be derived from the substitution of air for steam as a prime mover of machinery. It has been alleged that the air-engine would be safer, lighter, and more economical in the expenditure of fuel than the steam*

---

[23] See Eckardt, Gas Turbine Powerhouse, p. 37 and 48.

[24] See Eckardt, Gas Turbine Powerhouse, p. 49.

[25] See Joule, On the Air-Engine.

engine. The theoretical GT Joule process, in the literature sometimes also named after the Boston engineer George Brayton (1839–1892) who patented a corresponding piston engine in 1872 (without reference to a cycle), is a representation of the properties of a fixed amount of air, as it passes through an operating gas turbine. It is similar, but with losses—by undergoing two isentropic and two isobaric changes of state.

1853   Louis-Marcellin Tournaire (1824–1886), French engineer and mineralo-gist, described and presented the basic concepts of multi-stage axial-flow steam and gas reaction turbomachinery (compressors and turbines) in a memorandum to the French Académie des Sciences.[26] In addition, he discussed the origins of losses due to leakages, vibrations, flow channel friction, circumferential irregularities and by vortices/swirling flows.

1859*–1885*   Edwin Drake (1819–1880), American businessman and the first American to successfully drill for oil at Titusville, PA,[27] laid the ground for the coming global motorisation of land, air and sea traffic, at best illustrated by the overwhelming success of the internal combustion engine, at first for the automotive piston-engine, designed for liquid combustibles. Almost simultaneously, there are the inventions of *automobiles* of Carl Benz (1844–1929),[28]—Gottlieb Daimler (1834–1900) and Wilhelm Maybach (1846–1929), German engine designers and industrialists, who developed light, high-speed internal combustion engines, suitable for mass-production in a rapidly widening range of applications. They also invented (in 1892/3) the carburettor (inspired by medical atomisers with roots back to Priestley's work) and a new ignition system inspired by Volta's *bad air* detection spark gun.

1863   Jean-Charles de Louvrié (1821–1894), French scientist and inventor, was presumably the first to suggest aero jet propulsion in modern times. In a lecture at the French Academy of Sciences he presented a paper on *Locomotion aérienne: l'aéronave* with an engine which materialised in detail in follow-up publications after 1867 as the idea of a *pulso-reacteur*, very similar to what was actually realised some 70 years later in the German *flying bomb V-1*.

13 Dec. 1872   Paul Haenlein (1835–1905), German inventor and airship pioneer, succeeded with the earliest aeronautical application of the internal com-bustion engine in a dirigible-balloon flight at Brno/ Brünn, using a

---

[26] See Tournaire, Sur des appareils à turbines multiples.

[27] See Wikipedia, 'Edwin Drake' in English.

[28] See Wikipedia, 'Carl Benz' in English. There was a controversy about the first automobile: Some UK sources including the 1978 TV series 'Connections' favoured—see Wikipedia, 'Siegfried Marcus' in English.

4-cylinder, 5 hp Lenoir engine, running on coal-gas fuel at 40 rpm. The low efficiency of ~5% allowed for a flight speed of 19 km/h.[29]

1875   Osborne Reynolds (1842–1912), Irish innovator in the understanding of fluid dynamics, filed a patent for the so-called *turbine pump*, which was for the combination of a centrifugal (water) impeller with a *vaned diffuser*. That same year Reynolds also built a multistage axial-flow steam turbine running at 12,000 rpm. In 1883 Reynolds most famously studied the conditions in which the flow of fluid in pipes transitioned from laminar to turbulent flow, which established the foundations of flow similarity for laminar/turbulent transition in channels, based on *dimensional analysis* (Lord Rayleigh, 1872). From these experiments came the dimensionless *Reynolds number* for dynamic similarity—the ratio of inertial forces to viscous forces. The concept was originally introduced by the Irish physicist and mathematician (Sir) George Stokes (1819–1903) in 1851, but the Reynolds number Re was named by the German theoretical physicist Arnold Sommerfeld (1861–1951) in 1908 after Osborne Reynolds, who had popularised its use. In 1885 Reynolds also described mathematically the convergent-divergent nozzle, an item of great relevance for future steam turbines which was independently discovered by De Laval.[30]

1882   Gustaf de Laval (1845–1913), Swedish engineer, introduced the concept of the first practical power delivery impulse steam turbine and in 1887 built a small impulse steam turbine to demonstrate that such devices could be constructed on that scale. In 1890 Laval developed a nozzle to increase the steam jet to supersonic speed, working from the kinetic energy of the steam, rather than its pressure. The nozzle, now known as a *de Laval nozzle*, is used in modern rocket engines and supersonic wind tunnels. De Laval turbines can run at up to 30,000 rpm; the higher speed of the turbine demanded that he also designed new approaches to reduction gearing, which are still in use today. Since the materials available at the time were not strong enough for the immense centrifugal forces, the output from the turbine was limited, and large-scale electric steam generators were dominated by designs using the alternative compound steam turbine approach of Charles Parsons.[31]

1884   (Sir) Charles Algernon Parsons (1854–1931), Anglo-Irish engineer, patented the compound steam turbine. The patent application has an excursion towards the stationary gas turbine, in which he notes that the turbine could be driven *in reverse* to act as a compressor; a comment which

---

[29] See Wikipedia, 'Paul Haenlein' in English.

[30] See Meher-Homji, The historical evolution.

[31] See Wikipedia, 'Gustaf de Laval' in English.

misled early axial compressor development. He immediately utilised the new engine to drive an electrical generator, which he also designed. Parsons's steam turbine made cheap and plentiful electricity possible and revolutionised marine transport and naval warfare.[32] In the same year he patented his first axial-flow compressor, which will be further discussed in the context of the Stolze design in Sect. 4.1.2. In 1899 he made an 80-stage compressor—the number of stages is probably an all-time record—for which an efficiency of 70% was claimed, though the aerodynamic quality was poor.[33]

One can assume that Parsons's axial-flow compressor, typical for the design and manufacturing capabilities of the time, had similar flaws in his axial turbine design of 1895, as depicted in the enlarged view in Fig. 2.2.[34] The circumferential blade staggering—pitch/chord ratio—is apparently too small, the tip thinning on the blade suction side (presumably to gain in through-flow area) causes a sharp, aerodynamically disadvantageous suction side turning. The aerofoil trailing edges are far too thick and the surface quality is generally poor.

| 1887 | Ernst Mach (1838–1916), Austrian physicist and philosopher, after initial studies in the field of experimental physics on the interference, diffraction, polarisation and refraction of light, presented together with physicist-photographer Peter Salcher (1848–1928) their ground-breaking paper on important explorations in the field of supersonic fluid mechanics by means of *Schlieren* photography,[35] which had been invented in 1864 by the German physicist August Toepler (1836–1912). They deduced and experimentally confirmed the existence of a shock wave of conical shape, with the projectile at the apex. The ratio of the speed of a fluid to the local speed of sound, a critical parameter in the description of high-speed fluid movement in aerodynamics and hydrodynamics, is now called *Mach number*. This designation was proposed first by the Swiss aeronautical scientist and engineer Jakob Ackeret.[36] |

| 1891 | Otto Lilienthal (1848–1896), German aviation pioneer, entrepreneur, inventor and author, the first person to make well-documented, repeated, |

---

[32] See Wikipedia, 'Charles Algernon Parsons' in English.

[33] See Wilson, Turbomachinery, p. 32.

[34] Courtesy of John Denton, 19 Aug. 2019,—see Denton, The evolution.

[35] See Mach and Salcher, Photographische Fixierung; German *Schlieren* stands for streaks of varying optical density.

[36] Ackeret (see '1921') suggested the Mach number definition in his ETHZ inauguration lecture on 4 May 1929, entitled *The aerodynamic drag at very high speeds*. See Eckardt, Gas Turbine Powerhouse, p. 128.

Fig. 2.2 Charles A. Parsons's 100 kW steam turbine rotor close-up, 1895

successful flights with gliders. At the beginning, in 1891, Lilienthal succeeded with jumps and flights covering a distance of about 25 m. In 1893, in the Rhinow Hills (45 km west of Berlin), he was able to achieve flight distances as long as 250 m. This record remained unbeaten for him or anyone else at the time of his death on 9 August 1896, when his glider stalled and he was unable to regain control; falling from about 15 m, he broke his neck and died the next day, 10 August 1896. Lilienthal did research in accurately describing the flight of birds, especially storks, and used polar diagrams for describing the aerodynamics of their wings. He made many experiments in an attempt to gather reliable aeronautical profile data.[37]

2 Oct. 1891   Charles E.L. Brown (1863–1924) of Brighton, England and J. Walter D. Boveri (1865–1924) of Bamberg, Germany established a limited partnership under the company name *Brown, Boveri & Cie.* at Baden, Switzerland for the manufacturing of electric machines. Over time the company grew shortly before 1995 up to 220,000 employees world-wide, then under the company name *ABB Asea Brown Boveri*. For more than a century the company was at the forefront of technologically advanced engineering, especially in the area of turbomachinery and steam and gas turbines. Up to the early 1950s, BBC and its international daughter companies in Germany, and in France (C.E.M.) considerably influenced developments in aero engine propulsion, in aero engine turbochargers, engine altitude test chambers and high-speed wind tunnels.

---

[37] See Wikipedia, 'Otto Lilienthal' in English.

3 Nov. 1897	The relatively lightweight and efficient Otto-cycle gasoline engine began with developments in England and Germany in the 1880s, stimulating the early automobile developments and found aircraft applications soon after. The first flight of this engine type was apparently that of a 40 m long, dirigible airship at Berlin-Tempelhof on 3 November 1897, powered by a 12 hp Daimler engine, and 10 months after the death of its designer, the Croatian-Hungarian aviation pioneer David Schwarz (1850–1897).[38]
1899	Franz Stolze (1836–1910), versatile German inventor, photographer, stenographer and writer, in 1899 (after a first rejected application in 1873) patented and, in 1904 built at Berlin-Weissensee a *hot air machine*, in principle the first modern gas turbine with one-shaft, two-bearing rotor arrangement and an all-axial compressor/turbine turbomachinery set.[39] The decisive flaw of Stolze's concept was that compressor and turbine, stator and rotor blading were—in the sense of Parsons (1884)—completely *'mirrored'*; the sheet metal blading had no aerodynamic profiling and the blade turning—though acceptable for the turbine—was way too high for an efficient compressor flow.

2.1.2   The Gas Turbine Preparation Phase 1900–1930

1901/2*	Wilhelm Kress (1836–1913), Austrian piano-maker, aviation pioneer and early aircraft designer unsuccessfully attempted to fly the first seaplane on an Austrian lake using Daimler's new gasoline engine. During the turn of the century he was one of the world-wide contestants for the creation of a breakthrough powered airplane. This last example, taken from James Burke's *Connections*, should illustrate on the one hand the critical interdependence (including failure) of these early protagonists from each other, especially in direct collaboration,[40] and on the other hand

---

[38] See Wikipedia, 'David Schwarz (aviation inventor)' in English.

[39] See Eckardt, Gas Turbine Powerhouse, pp. 68–69. The 'first modern gas turbine' in the context of Stolze refers to his patent application in 1873, which in hindsight appears to be rejected unjustified due to his own fault. Wrongly, he had claimed priority for the multi-stage compressor and turbine concept, rather for the combination of both all-axial components on one shaft.

[40] The engine for Kress's aircraft, a kind of motor-powered triplane-type hang glider with two counter-rotating propellers, had been correctly specified, but the actual delivery from Daimler in Stuttgart was nearly twice as heavy and operational limits on the *Wienerwaldsee*, a shallow and space-limited water reservoir 20 km west of Vienna, caused the final disappointing outcome. See Wikipedia, 'Wilhelm Kress' in English.

the inherent principle of competition as a basis for technological progress, especially in view of the successful Wright brothers shortly thereafter.

1902   Auguste C.E. Rateau (1863–1930), French mining and mechanical engineer-entrepreneur, was—besides Parsons—the other major pioneer working on compressors and steam turbines at that time, who had published a major paper on turboblowers '*Considérations sur les turbomachines*' already in 1892. However, a turbocompressor which he designed for PR 1.5 at 20,000 rpm gave an isentropic efficiency of only 56%. Subsequently, his turbocompressors improved in pressure ratio and mass flow, and gradually gained also in efficiency, but as will be seen most attempts, which in the beginning struggle to produce a working gas turbine failed simply because of poor compressor efficiency.[41]

1903   Aurel Boleslav Stodola (1859–1942), Slovak engineer, scientist, professor for mechanical engineering at ETH Zurich (1902–1929) and global teacher gained highest influence and reputation with his book *Die Dampfturbinen* (The steam turbines). The first publication had 220 pages and was the result of an expanded lecture at VDI Dusseldorf, the Association of German engineers in 1902. Since the author consistently tried to cope with the growing know-how in his field, the sixth and last German edition of then *Dampf- und Gasturbinen* in 1924 comprised 1150 pages, 1200 neatly drawn steel engravings, 13 tables and weighed 3.1 kg. The second edition in 1905 was also translated in English and French. The sixth German edition was translated to become the second English version, appearing in two volumes simultaneously in London and New York in 1927. Three years after Stodola's death there appeared a reprint of the US edition in 1945; moreover, there is an offset facsimile print of the last German edition distributed in China. Stodola devoted most of his gas turbine studies—several thick unpublished volumes—to the Holzwarth explosion turbine (and its contested performance) which at that time was the only cycle that promised acceptable efficiencies. However, Stodola had himself given the first place to continuous combustion turbines in *Steam and Gas Turbines,* where a complete theory was available. He appreciated it, therefore, as the culminating point in his career, when Brown Boveri invited him, though an octogenarian, to perform official tests on their first utility combustion turbine in 1939. These were conducted with a thoroughness and precision which could set a standard for all time.[42]

---

[41] See Wilson, Turbomachinery, p. 32.

[42] See Eckardt, Gas Turbine Powerhouse, p. 51f., and—see Seippel, From Stodola to Modern Turbine Engineering.

1903   Sanford Alexander Moss (1872–1946), American aviation engineer,
            and—later—pioneer of the turbosupercharger at General Electric, wrote
            his thesis at Cornell University on *The Gas Turbine, an Internal Com-
            bustion Prime-Mover*, but his experimental set-up failed to produce any
            positive work.[43]

1903   Jens William Ægidius Elling (1861–1949), Norwegian researcher, inven-
            tor and pioneer of gas turbine, built the first constant-pressure gas turbine
            with 6-stage centrifugal compressor, of which the first three stages were
            of the double-flow type and the diffuser had already adjustable vanes.
            This set-up with steam injection to reduce the turbine entry temperature to
            400 °C was able to produce a net output of 8 kW (in the form of
            compressed air), i.e. more power than needed to run its own components.
            Shortly thereafter Elling improved his design to a net output of 31 kW
            (without steam injection, but with heat exchanger) at approximately
            20,000 rpm and with electric power generator (1906), instead of the
            compressed air production. He understood that if better materials for
            higher temperatures could be found, the gas turbine would be an ideal
            source for airplanes.[44]

1903   Hans Theodor Holzwarth (1877–1953), German inventor-entrepreneur,
            proposed in his first patent an explosion or constant-volume gas turbine
            cycle, somewhat related to the *Otto cycle*, in which the high temperatures
            and pressures of gas or fuel combustion that were obtained in an explo-
            sive, repetitive firing system could compensate for the compression
            efficiency deficits of that time, but which caused also enormous heat
            transfer rates in the water-cooled jackets of the turbine component on
            the negative side. Holzwarth stayed persistently with the concept for
            50 years: his patent collection comprised of nearly 200 granted
            applications on the same subject, a life-long occupation; BBC was
            involved in the manufacturing of the machines from the beginning in
            1909, up to 1938. Holzwarth's effort culminated in the ninth prototype, a
            5 MW single-shaft unit with a turbine inlet temperature of 930 °C for
            Thyssen steelworks at Hamborn, D, built and erected again by BBC
            Mannheim in 1938, and a documented last run in 1943, when it suffered
            heavy bomb damage. Holzwarth's aim was to develop and market his
            turbine for industrial use, although, at least one patent was filed in 1930
            for a railway locomotive drive, and another one showed up in Germany
            only after WW II (with priority from March 1945) for a *low-drag,*

---

[43] See Wikipedia, 'Sanford Alexander Moss' in English.
[44] See Eckardt, Gas Turbine Powerhouse, p. 70.

counter-rotating turboprop version of the explosive gas turbine.[45] In close relation to the explosion gas turbine with mechanically acted valves stand the *pulsejets*, actually there are two main types (valved and valveless), both of which use resonant combustion and harness the expanding combustion products to form a pulsating exhaust jet which produces thrust intermittently. The first working *pulsejet* was patented in 1906 by Russian engineer Victor V. Karavodine, who completed a working model in 1907. The French inventor Georges Marconnet patented his valveless *pulsejet* engine in 1908 and suggested it for aircraft applications, and the Spanish Ramon Casanova patented a *pulsejet* in 1917, having constructed one, beginning in 1913. Robert Goddard (1882–1945) invented a *pulsejet* engine in 1931, and demonstrated it on a jet-propelled bicycle. German engineer Paul Schmidt pioneered a more efficient design based on modification of the intake valves (or flaps), issued his basic patents between 1930–1932, earning him first government support from German ministries for jet propulsion projects (see '1931' for Germany).

17 Dec. 1903   Wright brothers, Orville (1871–1941) and Wilbur (1867–1912), American aviators, engineers and inventors, are generally credited[46] with inventing, building and flying for the first time a 12 hp powered, heavier-than-air airplane in a controlled, sustained flight of 12 s duration, over a distance of 37 m in 3 m average height, at a wind-swept beach site, 4 miles south of Kitty Hawk, NC.[47]

1904   Ludwig Prandtl (1875–1953), German scientist, introduced his *boundary layer theory*,[48] an aerodynamic concept to understand flow diffusion up to flow breakdown against a rising pressure, and thus a precondition to realise the gas turbine principle, i.e. a turbine driving an efficient compressor with sufficient surplus power/turbojet energy. His corresponding 1904 paper[49] raised Prandtl's prestige as an aerodynamicist. He became director of the Institute for Technical Physics at the University of Goettingen later in the year, where he worked with many outstanding

---

[45] See Eckardt, Gas Turbine Powerhouse, pp. 72–76.

[46] In March 2013, *Jane's All the World's Aircraft* published under the motto *Justice delayed is justice denied* an editorial which accepted Bavarian-born Gustav Weißkopf (1874–1927, in the U.S. since 1893, anglicised as Gustave Whitehead) and his claimed flight on 14 Aug. 1901 in his self-built No.21 monoplane at Fairfield, CT, USA as the first manned, powered, controlled flight of a heavier-than-air craft, more than two years ahead of the Wrights. The *Smithsonian Institution* is among those who do not accept that Whitehead flew as reported; but this dispute has no impact on this gas turbine/turbojet timeline. See Wikipedia, 'Gustave Whitehead' in English.

[47] See Wikipedia, 'Wright brothers' in English.

[48] See Wikipedia, 'Boundary layer', and 'Ludwig Prandtl', both in English.

[49] See Prandtl, Über Flüssigkeitsbewegungen.

students, creating the greatest aerodynamics research centre of his time, to which belonged since 1925 the *AVA Aerodynamische Versuchsanstalt* (Aerodynamic Test Establishment), Goettingen. The *Prandtl number Pr* is a dimensionless number, named after him,[50] defined as the ratio of dynamic viscosity to thermal diffusivity.

1905   Alfred Buechi (1879–1959), Swiss engineer, filed his first of many turbocharger patents showing a scheme with 7-stage axial(!) compressor, thus starting this important (though only partially turbojet-related) string of developments. Besides the *motor jet* (see '1908') and the constant-volume GT concept (see '1903'), the turbocharger represents the third essential, accompanying path towards the aero gas turbine. The origins of aero-supercharging cannot be determined unambiguously in view of various claims, for the first turbocompressor as engine-coupled super-charger—or for the first free-wheeling turbocharger as compressor-turbine combination—and also for the first meaningful series production. One source has the US Murray-Willat Co. already in 1910 producing a 90 hp supercharged two-stroke rotating piston engine with integrated radial compressor up to remarkable 5.2 km. The French Auguste Rateau in 1916 took out patents for turbocharging equipment, manufactured and tested in 1917. Sanford A. Moss of General Electric was the first to manufacture turbochargers on a regular basis. A resulting GE-US Army Air Corps programme finally culminated in aircraft like B 17 or P-38—with significant impact on the outcome of WW II. In 1917 the US Army also tested a turbocharger designed by Earl Hazard Sherbondy (1887–1958) that had been built by the De Laval Co. for Fergus Motors.[51]

1906   René Armengaud (1844–1909), French inventor, built at Paris-St. Denis on the basis of a 1902 patent for a combustor/turbine combination from Charles Lemâle the first experimental gas turbine, rated at ~250 kW and consisting of a 25-stage radial compressor and two intercoolers from BBC Baden, CH (Rateau compressor design, PR 4.5) and a water-cooled two-stage Curtis wheel turbine of 0.94 m diam. (again of Rateau design).[52] It achieved self-sustained operation by adding some steam, generated in combustor cooling, and feeding it back to the turbine in a kind of early '*steam injection*' (STIG). The actual GT efficiency should

---

[50] This happened apparently not in full agreement with Prandtl. In his *Führer durch die Strömungslehre*, (Essentials in fluid dynamics), 3rd ed., (in German) 1949, he wrote in a footnote on p. 382: *The number ν/a ... shown by Nusselt already in 1909, has been named later Prandtl number Pr. The author did not want to follow this historic incorrectness and preferred therefore the equally short term ν/a.* Information by courtesy of Michael Eckert, 11 June 2018.

[51] See Sect. 4.1.2.

[52] See Eckardt, Gas Turbine Powerhouse, pp. 71–72.

have been between 2% and 3% only, or 6–10 kW of equivalent power production; consequently, the project was abandoned.

1907   Brown-Boveri-Rateau turboblower impeller[53] for Siemens-Martin steelworks at Rote Erde, Aachen D, a 1.3 m diam. steel casting, pressure ratio (PR) 1.2, mass flow ($\dot{m}$) 14 kg/s, 2600 rpm with 750 hp steam turbine drive, basically the same dual-flow centrifugal compressor configuration as used in Frank Whittle's WU (Whittle Unit) first model experimental engine, first run on 12 April 1937, PR ~ 4, $\dot{m}$ ~ 12 kg/s.[54]

1908   René Lorin (1877–1933), French aerospace inventor, patented a reaction-type *motor jet*, sometimes designated *subsonic ramjet*, in which compressed exhaust gas of a reciprocating piston engine was directly used to form hot jets of thrust; he intended such devices to be attached directly to the wings of airplanes. In due course in a 1913 patent, he conceived also the *supersonic ramjet* which was at first viewed as impractical because it only became efficient at high speeds.

After a corresponding 1917 British *motor jet* patent for H.S. Harris, followed in 1923 by Ch. Lemale's invention of an *Avion sans hélice* (aircraft without propeller), the *motor jet* became popular mainly with the Axis powers before and during WWII in various versions. Typically they combine a reciprocating petrol engine with a thermal jet which has been intensified in combustion chamber/ afterburner stage(s). The official German designation was *ML Motor-Luftstrahl* engines; BMW and its forerunner Bramo tested and built *motor jets* as early as 1938 and experimented—like Heinkel and Junkers—with these alongside their well-known turbojets, thus drawing on limited resources up to 1943.

The Italian Campini-Caproni *C.C.2* jet aircraft[55] flew with a *motor jet* engine in 1940 and, in November 1941 a highly publicised Milan—Rome flight to the *Italian Experimental Establishment* at Guidonia took place with an intermediate re-fuelling stop at Pisa, indicating the afterburner usage. Besides its elliptical wing, the C.C.2 carried an advanced jet-directing swivelling nozzle which however, due to low flight speed of 370 km/h, proved best for ground manoeuvring only.

1908   Still without firm basis, Walther Nernst (1864–1941), German chemist and Nobel prize winner,[56] invented or published a *'Nernst rotor, or turbine'*, a high-speed *'U tube'* arrangement with combustion in the

---

[53] See Eckardt, Gas Turbine Powerhouse, p. 47.

[54] See Golley, Genesis of the Jet, p. 250.

[55] The C.C.2 was considered as the first flying jet aircraft for a while, until news spread about the Heinkel He 178 flight one year earlier. See Wikipedia, 'Caproni Campini N.1' in English.

[56] See Wikipedia, 'Walther Nernst' in English.

U-turn, so that air in the ascending leg was compressed by centrifugal force, and after heating, expanded down the descending leg to exhaust with a greater pressure and temperature than at entry, thus having had its energy augmented.[57] Unfortunately this simple gas generator was impracticable, because of the low compression obtainable with available materials. Nevertheless, the device contains in a nut-shell the concept of combining a radial flow compressor with a radial flow turbine, with which Hans von Ohain experimented in 1936, as his first turbojet engine HeS 01.[58]

24 Feb. 1910   Walter G. Noack (1881–1945)[59] visited the *Cabinet Armengaud Frères* at St. Denis and there met—after René Armengaud's death—the remaining brother Marcel to collect first-hand information about gas turbine test experiences and further prospects, thus opening a series of personal cross-border contacts of gas turbine pioneers typical for the emerging *Jet Web*. M. Armengaud predominantly recommended *to improve the turbocompressor*, considering the circumferential speed of the Rateau design as too low for gas turbine applications: '*A company like Brown Boveri should try to create something new in this field*'. But, as Noack noted later in his memoirs, *there was a lack of heat-resistant materials at that time, and also the turbocomponents had not reached an efficiency level, as required for turbines to drive their own compressor (and more).*[60]

1910   First operative gas turbine according to the *Holzwarth-Principle* (see '1903') at BBC Mannheim; however, the 1000 hp design produced 200 hp only. The principle was successfully demonstrated, but after a short time of operation the set-up deteriorated thermally, even with the wide use of water cooling for all thermally exposed parts. So, measured against its cylindrical size—2.95 m diameter at the bottom, an impressive height of 6.36 m—and an efficiency of approximately only 8%, it was less favourable than available reciprocating engines of the time.

14 Dec. 1917   Alan Arnold Griffith (1893–1963) and (Sir) Geoffrey I. Taylor (1886–1975), both then from *Royal Aircraft Factory*, Farnborough (what became *Royal Aircraft Establishment [RAE]* in 1918) suggested to the *ACA Advisory Committee for Aeronautics* the use of soap films as a

---

[57] The Nernst patent is mentioned in—see Ermenc, p. 106, and described—see Whittle, Gas Turbine Aero-Thermodynamics, p. 103, and Stodola, Dampf- und Gasturbinen, pp. 1010–1011; for details—see Fig. 12.1.

[58] See Wikipedia, 'Heinkel HeS 1' in English, and—see Fig. 6.22.

[59] See Eckardt, Gas Turbine Powerhouse, p. 88f.

[60] See Eckardt, Gas Turbine Powerhouse, p. 71f.

unique experimental technique of studying stress problems, especially for investigating the stress distribution in the root of propeller spars—without naming the original inventor of this *membrane analogy* some 14 years earlier: Ludwig Prandtl.[61] What might have been explainable as a result of hostile tensions during the Great War, became untenable in the European scientific community after the war, especially in view of the relative prominence of both in fracture mechanics.[62] At the next, unavoidable encounter of Prandtl and Taylor/Griffith at the *First International Congress of Applied Mechanics (ICAM)* in the last week of April 1924 at Delft, NL—Griffith took the certainly painful opportunity under considerable pressure to name the correct, earlier authorship of Prandtl in front of a gathering of 200 top-scientists from all parts of Europe and America. Thereafter, Griffith ended his career in metal fatigue and fracture mechanics apparently quite abruptly, only to reappear in 1926 (see below) in an all new field of aerodynamic theory of axial turbomachinery for the benefit of future aero gas turbines.

Sep. 1920   William J. Stern (1890/1–1965) of the Air Ministry Laboratory in South Kensington wrote a report *The Internal Combustion Turbine*[63] in response to an Advisory Committee for Aeronautics (ACA)[64] request about the possibilities of developing a 1000 hp gas turbine engine to drive a propeller. His report was extremely negative: Given the low performance of existing turbocompressors, such an engine appeared to be mechanically inefficient. In addition to poor fuel efficiency and high weight assumptions (cast iron and bronze for combustor and turbine*), ... the weight of a 1000 hp set comes out to something of the order of 10 lb per hp, ...*, [0.1 hp/lb], Fig. 2.3, i.e. 1/5 of the actual values for piston engines of the time. Stern was sceptical that there were materials available that would be suitable for use in the high-heat areas of the

---

[61] See Griffith, The Use of Soap Films, and—see Prandtl, Zur Torsion, and—see Sect. 4.2.1.

[62] See Rossmanith, Fracture Research, and Sect. 4.2.1.

[63] See Stern, The Internal Combustion Turbine and—see Wikipedia 'William Joseph Stern' in English.

[64] Note: ACA, founded on 30 April 1909, became the better-known *ARC Aeronautical Research Committee* in 1919, which was disbanded in 1979 as *Aeronautical Research Council*. See Wikipedia, 'Advisory Committee for Aeronautics' in English.

turbine. Stern's paper proved to be so convincing there ceased to be any official interest in gas turbine engines in Britain for several years.[65]

Correspondingly in 1924, Edgar Buckingham (1867–1940), American physicist at the United States National Bureau of Standards, published a report on *Jet propulsion for airplanes*, coming to the same conclusion as W.J. Stern, that the turbine engine is not efficient enough: *The relative fuel consumption and weight of machinery for the jet, decrease as the flying speed increases; but at 250 miles per hour the jet would still take about four times as much fuel per thrust horsepower-hour as the air screw, and the power plant would be heavier and much more complicated. Propulsion by the reaction of a simple jet cannot compete, in any respect, with air screw propulsion at such flying speeds as are now in prospect.*[66]

25 Jan. 1921   Hermann Glauert (1892–1934) and Ronald McKinnon Wood (1892–1967), two researchers from RAE, visited Prandtl's laboratory in Goettingen, thus resurrecting a personal scientific relationship between Britain and Germany after the hostilities of the Great War. McKinnon Wood was deputy director of the *Aerodynamics Department* at Farnborough and worked on propellers and the experimental side. Glauert, born in Sheffield to a family of German origin, had been appointed to the staff of the Royal Aircraft Factory in 1916. The first one-day-contact with Prandtl was exclusively on technical issues of wind tunnel tests at Goettingen, which were particularly important for a comparative assessment of international trials.[67]

1921   Jakob Ackeret (1898–1981), Swiss scientist and aeronautical engineer, received his Diploma in mechanical engineering from ETH Zurich (ETHZ) in 1920 under the supervision of Aurel Stodola. From 1921 to 1927 he worked with Ludwig Prandtl at the *Aerodynamische Versuchsanstalt* in Goettingen, where a planned one-year stay actually

---

[65] Interestingly, W.J. Stern reappeared 25 years later in Sir Roy Fedden's *Mission to Germany* (see Wikipedia, 'Fedden Mission' in English, which lasted two weeks in June and one week in July 1945)—in the role as *translator*. Some of his universal language background is explained in his obituary in Vacuum, Nov. 1965. Fedden even paid written tribute to the cooperation with W.J. Stern [*Flight*, 13 Sept. 1945], and a few weeks later also H.E. Wimperis—in 1920 superintendent of the South Kensington Labs—felt obliged to join that chorus of praise [*Flight*, 4 Oct. 1945], though it remains open, if these accolades were meant to compensate the foregoing severe criticism.

[66] See Buckingham, Jet propulsion for airplanes.

[67] See Bloor, Enigma, p. 314, and—see Wikipedia, 'Hermann Glauert' in English. In June 1929 Glauert attended—presumably invited by Th. von Kármán—an aerodynamic conference at RWTH Aachen,—see Gilles, Vorträge, with a lecture '*The force and moment of an oscillating aerofoil*', pp. 88–95, where he met a.o. L. Prandtl, A. Betz, A. Busemann, and E. Trefftz, names to reappear frequently in the following. Information, courtesy of Michael Eckert, 23 Feb. 2022.

lasted nearly seven years. He thus witnessed a legendary period in the development of modern fluid dynamics, there and at the end of his term, in close cooperation with Albert Betz (1885–1968), laid the ground for the AVA axial compressor design methodology. After completing his PhD at ETHZ in 1927, Ackeret worked at Escher Wyss AG in Zurich as chief engineer of hydraulics, where he applied, with great success, modern aerodynamics to the design of water turbines and variable-pitch propellers for ships and airplanes. In 1931 he became professor of aerodynamics at ETHZ in 1931, where Wernher von Braun was one of his students during the summer term 1931.[68] Ackeret's Mach 2 closed-circuit supersonic wind tunnel with 0.4×0.4 m test section at ETH Zurich in 1935—built by BBC—allowed for the first time an independent Ma, Re setting via pressure variation.

1921   Maxime Guillaume (1888–xx), French engineer and inventor, still gets credit for patenting[69] the first all-axial aero gas turbine by combining an axial compressor and a multi-stage turbine, though at least his patent drawings still lack aerodynamic profiling, in this respect related to the failing Parsons's and Stolze's turbomachinery designs some 20 years earlier.

1922   Walther Bauersfeld (1879–1959), German engineer, lifelong employed as a leading manager of *Carl Zeiss* Jena/Oberkochem, D[70] produced as a kind of mathematical *finger exercise* outside of his normal activities a nine-page key paper[71] on the application of aerofoil theory to the design of turbomachinery. It had enormous impact on further turbomachinery development, especially by the Brown Boveri axial-flow compressor success 10 years later, due to its radical change in design perspective, away from established stream tube and flow channel theory towards the lift/ drag principle, deduced from comprehensive single aerofoil

---

[68] See Wikipedia, 'Jakob Ackeret' in English.

[69] Apparently the British patent office for Whittle's turbojet patent of 1930, as well as the German *'Reichspatentamt'* for what became Hans von Ohain's secret turbojet patent of 1935 did not detect this French patent FR534'801 with priority on 3 May 1921, and other patented turbojet forerunners in time, so that both patents (in the agreed form) might have been formally unjustified, as Hans von Ohain stated rather bemused in 1996,—see Sect. 12.1.1, and—see Mattingly, Elements, p. xxiii, and Wikipedia, 'Maxime Guillaume' in French.

[70] Bauersfeld is most renowned for a series of planetariums erected between WW I and II at the *German Museum Munich* (1923, initiated by Oskar von Miller), Wuppertal-Barmen, Leipzig, Jena, Dresden, Berlin, Dusseldorf, Rome, Paris, Chicago, Los Angeles and New York,—see Wikipedia, 'Walther Bauersfeld' in English.

[71] See Bauersfeld, Die Grundlagen, and—see Sect. 4.1.3.

measurement data from AVA Goettingen, published by L. Prandtl in 1921.[72] This paper was probably not responsible for the subsequent renewal of interest in the axial compressor after the early works of Parsons, Rateau and Stolze, but its existence provided the basis for early practical designs of such machines up to Keller's widely-spread doctoral thesis in 1934.

1922   Rudolph Birmann (1899–1968), Swiss engineer and inventor, wrote his diploma thesis as a future mechanical engineer from ETH Zurich before his emigration to the USA, on the subject of a radial-inflow turbine used for a gas turbine, a turbine type which was later used by Hans von Ohain in his early Heinkel centrifugal gas turbines.[73] He believed that the centripetal turbine type would be superior to the ordinary axial-flow turbine by better efficiency, be capable of handling larger pressure ratios, operate at higher speed for a given flow, and was easier to be cooled. Birmann became an extraordinarily versatile engineer who influenced turbomachinery development significantly in many areas, expressed by his more than 75 relevant patents. Beginning 1923, he worked for 45 years for the De Laval Steam Turbine Co, Trenton NJ, and since 1937 for TEC Turbo Engineering Corp., a De Laval offshoot to exploit Birmann's patents in other than steam applications. New in the United States, he designed and built at his own expense a small turbocharger using both the centripetal turbine and another component of novel design: a compressor impeller of the mixed- or diagonal-flow type. Also the turbine components were often more mixed-flow than radial inflow, i.e. the flow had a significant axial component at rotor entry; an early design achieved a peak efficiency of 85% at an expansion ratio of 4:1. Birmann's resulting turbocharger efficiency was considerably superior as recognised from early on by Brown Boveri Baden, leading to a collaboration in this area till the 1960s, when Birmann's *pulse converter* ideas were used in thousands of BBC/ Sulzer products.[74] With all likelihood Helmut Schelp, then still a student from TH Dresden at Stevens Institute

---

[72] See Prandtl, Ergebnisse.

[73] See Schlaifer, The Development, p. 454. The gas generator of Birmann's study engine was a *free-piston engine,* an idea brought at that time to Stodola's ETHZ institute by the Swiss engineer and inventor Robert Huber (1901–1995), later becoming *'Mr. Free Piston'* in his decade-long career as Technical Director of the *Bureau Technique Pescara*, Paris, since 1939 *S.E.M.E.* The engine concept was basically the same as the Pratt & Whitney PT-1 of the 1940s, the development of which gave P&W engineers valuable experience in designing gas turbines that it would later apply to projects that resulted in the JT3 turbojet engine. Huber was also a pioneer in the development of *common rail* fuel injection in the 1960s; – see Chap. 7, and Fig. 7.5.

[74] See Jenny, The BBC Turbocharger, p. 105.

of Technology, Hoboken NJ, met Birmann in 1935/1936 and carried his diagonal compressor idea back to Germany, where he made sure this, as one of his personal *'pet projects'*, was adopted for the Heinkel HeS 011 turbojet engine in 1942. Personally, Birmann comes back into focus, when in 1941 the first US government support for a systematic theoretical investigation of all the possibilities involved in aircraft gas turbines came from the US Navy in the form of a contract for Turbo Engineering Corporation.[75] In due course, the first aero gas turbine which the Navy contracted with TEC in 1942 was a small booster turbojet, aimed at 500 kp of sea-level static thrust.[76]

31 Aug. 1923   Claude Seippel (1900–1986)[77] finished his work as *Stodola's number cruncher*—on lease basis from BBC to solve severe (steam turbine) disk vibration problems—to set sails for what became a 5-year stay in the United States, where he worked from December 1924 to August 1926 at E.H. Sherbondy's five-man design office at Cleveland, OH, marking what *was one of the most fruitful working periods of my life.*[78]

26 April 1924   Georges Darrieus (1888–1979), French chief engineer of *C.E.M. Compagnie Électro-Mécanique* at Paris, the French subsidiary of Brown Boveri, started multi-stage axial compressor considerations, resulting in his first surviving axial compressor concept sketches from July 1926. In the years 1925–1931 C.E.M. built under his supervision a series of wind turbines, ranging from 8–20 m wheel diameter. Correspondingly, Darrieus recommended the application of the aerofoil theory also for axial turbomachinery at BBC Baden CH for the first time.[79]

1924   A special free-wheeling turbocharger configuration was suggested by Christian Lorenzen in 1924, typically combining the dual-flow centrifugal blower with a one or two-stage axial turbine on top in an integral wheel with hollow turbine blades; the radial outflow of the compressed air at the same time cools the turbine blading. The compressor pressure ratio was in the order of a rather modest PR 1.25 and—similar to the stacked (axial) turbine above compressor blading arrangement by Griffith and Heppner in England—the major drawback of this concept was also the compressed air short-cut flow directly to the turbine duct. One feature of

---

[75] It may be coincidental, that Schelp carried out the same study as his first job in the RLM Reich Air Ministry in 1937.

[76] See Schlaifer, The Development, p. 457.

[77] See Eckardt, Gas Turbine Powerhouse, p. 92f.

[78] See Chap. 3.

[79] See Sect. 4.1.3, and Figs. 4.18 and 4.21.

Lorenzen's work was beneficial, though, as Edward W. Constant[80] states correctly, *his work on air-cooled blades proved invaluable to the Germans* (jet engine developments) *during the Second World War.*[81]

1924   From 1920 onwards, annual gliding competitions were held on *Wasserkuppe,*[82] leading to a regular meeting point of European glider enthusiasts. Records being set and broken for height, distance and duration of unpowered flight. After 1924 the Wasserkuppe had a gliding school, workshops for building gliders and a funded research facility. On 27 April 1924 Fritz Heppner (1904–1982), student in the third semester of Technical Physics at TH Dresden, and later German-English turbojet engine designer, files as co-inventor his first patent together with E. Pohorille about *Differential gears with ball bearings, especially for automotive (model) cars.* Between January–March 1925 he participated in a glider training course which he finished with a 400 m straight, 30 s proof flight on 24 March 1925, and an official Wasserkuppe glider certificate. Alexander Lippisch was appointed as the managing director of the new glider Rhön-Rossitten Society on Wasserkuppe, maintained there a *(Delta)* aircraft design office to which belonged also the Canadian Beverley Shenstone, later known for his elliptic *Spitfire* wing.[83]

6 June 1924   Test date of turbocharger VT402,[84] the world's first heavy-duty exhaust gas turbocharger, leaving still within this week the responsible steam turbine department of BBC's Baden works for SLM Swiss Locomotive and Machine Works at Winterthur, CH to power boosting one of their experimental two-stroke diesel engines under development. The two-stage centrifugal compressor provided PR 1.35 at a volume flow of 1.83 m^3/s.[85] Shortly thereafter, this advanced turbomachinery design laid the ground for a rather prophetic prognosis on the future of aero gas turbines by Max Koenig, a former assistant to the BBC board of directors.

24 April 1925 Max Koenig (1893–1975), Swiss engineer-entrepreneur, presented a lecture on *Gas Turbines* to *the North-East Coast Institution of Engineers*

[80] See Constant, Turbojet revolution, p. 148.

[81] Lorenzen's investigations were based on a license for a 1920 Brown Boveri patent,—see Eckardt, Gas Turbine Powerhouse, pp. 399–401.

[82] See Wikipedia, 'Wasserkuppe' in English.

[83] See Chap. 5 (Heppner), and—see Sect. 4.3.3 (Shenstone), Figs. 4.39–4.41.

[84] The letters VT in VT402 stand for a Verdichter-Turbine (compressor-turbine) set-up, 400 is the impeller diameter in mm and the final 2 indicates a two-stage configuration. The BBC tests were carried out with a steam-driven turbine,—see Chap. 3, and Fig. 3.4.

[85] See Jenny, The BBC Turbocharger, p. 46. Actually the exhaust gas turbocharger fed into the existing scavenging Roots blower; this turbocharger is exhibited today at *Technorama,* Winterthur CH.

and Shipbuilders at Newcastle on Tyne, GB, in which he expressed strong belief in the aero gas turbine replacing the piston engine *entirely in time to come*, even against strong conservative academic head-wind from the floor, arguing that *he is too optimistic ... that the gas turbine may ... replace the piston engine for aerial navigation. The very exhaustive investigation carried out by Stern for the Air Ministry has shown that, compared with the highly efficient and light high-speed petrol engine, the gas turbine alone is an impracticable proposition.*[86] (see 'Sep. 1920')

22 May 1926   The *Paris Air Agreement*[87] was signed. It cancelled all restrictions of the *Treaty of Versailles* on the quality of German commercial aircraft, lifted all of the technological restrictions and allowed the Germans to construct dirigibles. In return, Germany agreed to stop subsidising sport flying. The effect of the new treaty was to restore a large measure of Germany's air sovereignty. Although the *Paris Air Agreement* did not lift the ban on German military aviation, it did allow some in the military to take up sport flying; by chipping away at the restrictions, the stage was thus set for dual-use developments and for the rapid military build-up that later occurred during the 1930s.

7 July 1926   Alan Arnold Griffith filed his ground-breaking, though internal *RAE Royal Aircraft Establishment, Farnborough* report on *An Aerodynamic Theory of Turbine Design* with appendix and a supplement,[88] containing the scheme for a small turbo-compressor test rig designed to verify his theoretical approach,—and presented it on 14 October 1926 at RAE to members of the Aeronautical Research Committee (ARC). This report is considered as a basic milestone in suggesting a gas turbine as an aircraft powerplant. He had worked out a method for the aerodynamic design of multi-stage axial compressors and turbines which was based on Prandtl's circulation theory. He showed that designs up to this point had been flying *stalled*, and that by giving the compressor blades an aerofoil-shaped cross-section their efficiency could be dramatically improved. The paper went on to describe how the increased efficiency of these sorts of compressors and turbines would allow an axial jet engine to be produced, although he felt the idea was impractical, and instead suggested using the power as a turbo-prop. As a result of his proposal and unanimous ARC recommendation, the first British cascade experiments were begun and

---

[86] See Koenig, Gas Turbines, and—see Chap. 3 for details.

[87] *'Agreement on aerial navigation with a view to the application of article 198 of the Treaty of Versailles, with protocol of May 7, 1926'*, concluded at Paris between Germany, and Belgium, British Empire, France, Italy and Japan.

[88] See Sect. 4.2, and—see Griffith, An aerodynamic theory.

Griffith himself designed a single-stage compressor and single-stage turbine on the same shaft for testing. The blading was of free-vortex type with 50% reaction and tests on the unit (with 4 in/10.2 cm outer annulus diameter, pressure ratio 1.16, max. speed 27,150 rpm, mass flow 0.56 kg/s) run in 1929, showing compressor efficiencies according to Griffith's data of 88.3%. In November 1929 Griffith summarised *The present position of the internal combustion engine as a power plant for aircraft*;[89] in this report he reversed Stern's negative view on the aero gas turbine of Sep. 1920 completely: *The turbine is superior to existing service engines and to projected compression ignition engines in every respect examined. Efficiency is higher and weight and bulk less. No external cooling is required. ... Any liquid fuel of suitable chemical composition may be used without reference to anti-knock value and volatility.* He used for confirmation the work of Betz at AVA Goettingen, who had run an axial test compressor with an aerodynamic efficiency of approximately 0.85.[90] But he suggested also the contraflow engine concept with integral turbine/compressor blades combined to *prevent stalling of the blading*, which occupied part of the British engine developments for 15 years, leading to nothing. Nevertheless, Griffith played an important part in aero gas turbine development, but his well-known early rejection of Whittle's concept in 1929 delayed government assistance at a most critical juncture. Unfortunately at the same time, proposals to continue and extend Griffith's own work were not carried out, and a lapse of six years followed in British government research efforts (see 'April 1930').

7 Oct. 1926   Date of the first preserved document in the history of BBC's axial compressor development, the *Test Programme* for a 4-stage axial compressor test rig with 540 mm max. tip diameter and a max. speed of 5000 rpm, which had been built as test vehicle for *the application of new aerodynamic theories to compressors to improve efficiency and simplify construction,* issued by Jean von Freudenreich (1888–1959), head of the Baden test laboratory, even before Seippel's return from the United States in April 1928. However, the *jump start* with a 4-stage set-up failed partially, due to unexplainable test data. First the originally untwisted blades were replaced by a swirl-adapted design—with moderate success, then a radical step back to a single-stage configuration helped to confirm fundamental fluid dynamic assumptions on the basis of a specific *double conus* compressor test rig in 1930. The rig created clean and undistorted

---

[89] See Griffith, The present position.

[90] See Howell, Griffith's early ideas.

flow conditions for the isolated rotor with 1.19 m tip diameter up to 3600 rpm. These tests irrefutably confirmed the polar plot characteristics of the Goettingen aerofoils used. The corresponding BBC *double-cone ventilator* test data fitted perfectly into the expected range of single aerofoil measurements, thus allowing also for a sound extrapolation towards a multi-stage configuration.[91]

11 July 1927   One year after the Versailles aero limitations had been finally removed, a target specification of the newly emerging German *Luftwaffe* demanded *pre-compression equipment and variable-pitch propellers for high altitude flights above 10,000 m.*[92] In due course, as one of the first concrete steps Friedrich Schmidt-Ott (1860–1956), former Prussian minister of education and personal, long-term friend of Emperor Wilhelm II, invited on behalf of the *Notgemeinschaft der Deutschen Wissenschaft* (Rescue fond for the German science) on 11 July 1927, 11 am to a meeting at the Berliner Schloß (residence palace) to discuss *Problems of the high-altitude aircraft to investigate the upper air space* officially; unofficially very soon strange, obviously abiding WW I ideas of dropping poison gas from un-armed, high altitude aircraft have been ventilated in this context. This became the starting point of the Junkers Ju-49 high-altitude research aircraft, and in due course of a multi-stage axial turbocharger development project at AVA Goettingen, the nucleus of German turbojet developments some 10 years later.[93]

1927   Martin Schrenk (1896–1934), German aviation engineer, aircraft designer and assistant professor at TH Berlin-Charlottenburg (THB), presented as test engineer of *DVL Deutsche Versuchsanstalt für Luftfahrt* (German test institute for aviation) at Berlin-Adlershof a lecture and paper[94] about *The problems of altitude flight.* This inspired Prof. Herbert Wagner (1900–1982), Austrian aviation scientist, and co-workers (M.A. Mueller, R. Friedrich) at his THB *Flugtechnisches Institut* (Institute for aeronautics) to analyse high-speed flight problems at high altitudes in 1935. Schrenk had suggested a kind of turbocharged *motor jet* piston engine, with backward-curved exhaust nozzles. Wagner and his team simplified Schrenk's concept by replacing the combustion cylinder by a continuously operating combustion chamber, which resulted in 1936 and 1938 respectively, in corresponding turboprop and turbojet patents.

---

[91] See Sect. 4.1.3, and—see Eckardt, Gas Turbine Powerhouse, pp. 102–106.

[92] See Gersdorff, Flugmotoren, p. 42.

[93] See Sect. 4.3.2, and—see Fig. 4.32.

[94] See Sect. 4.3.2 and Fig. 4.33, and—see Schrenk, Probleme.

1928   Anselm Franz (1900–1994), Austrian jet engine engineer, later known for
                       his responsibility in the first mass-produced *Junkers Jumo 004* turbojet
                       development programme since 1939, stayed at the *Eidgenoessische
                       Technische Hochschule Zurich* (Swiss Federal Institute of Technology)
                       in search of a doctorate, after first finishing his mechanical engineering
                       studies at *Graz University of Technology* in 1924. Following this he
                       moved on to Berlin to work as a design engineer,—for hydraulic torque
                       converters at *Berliner Maschinenbau AG*, former *L. Schwartzkopff,
                       Berlin*, predominantly a railway locomotive manufacturer, which started
                       the building of diesel locomotives in 1924. Presumably, in 1934 the Franz
                       area became part of *AEG Allgemeine Elektrizitäts-Gesellschaft* (General
                       Electric Soc.) Berlin. In 1936, he joined *Junkers Flugmotoren* Dessau, D
                       and during much of the 1930s thereafter, he was in charge of supercharger
                       and turbocharger development. In 1940, he received a doctorate in aero-
                       nautical engineering about turbojet/exhaust gas propulsion from the TH
                       Berlin-Charlottenburg.[95]

16 Jan. 1930   Frank Whittle (1907–1996), at that time *Flying Instructor* in the RAF,
                       filed the first ever patent for a gas turbine to propel an aircraft directly by
                       its exhaust gas (GB347,206: *Improvements relating to the Propulsion of
                       Aircraft and other Vehicles*). This contained all the main features of his
                       later designs, but was allowed to lapse after five years through lack of
                       financial support. The path to this successful patent application had
                       already started in 1928, when Whittle, in officer training wrote a thesis
                       on potential aircraft design developments, notably flight at high altitudes
                       and speeds over 800 km/h. In *Future Developments in Aircraft Design* he
                       showed that incremental improvements in existing propeller engines were
                       unlikely to make such flight routine. Instead, he described what has been
                       mentioned already above (see '1908') as a *motor jet*; a motor using a
                       conventional piston engine to provide compressed air to a combustion
                       chamber whose exhaust was used directly for thrust—essentially an
                       afterburner attached to a propeller engine. Of course, the idea was not
                       new and had been talked about for some time, but Whittle's aim was to
                       demonstrate that at increased altitudes the lower outside air pressure
                       would increase the design's efficiency. For long-range flight, using an
                       Atlantic-crossing mail-plane as his example, the engine would spend
                       most of its time at high altitude and thus could outperform a conventional
                       powerplant. Whittle continued working on the *motor jet* principle after his
                       thesis work, but eventually abandoned it, when further calculations
                       showed it would weigh as much as a conventional engine of the same

---

[95] See Fig. 6.27, and—see Boyne, Air Warfare, p. 234, and Hirschel, Aeronautical Research in
Germany, p. 232.

thrust. Finally, he came up with the apparently simple, though revolutionary patent idea, instead of using a piston engine to provide the compressed air for the burner, a turbine could be used to extract some power from the exhaust and drive a similar compressor to those used for superchargers, and the remaining exhaust thrust would power the aircraft.[96] In late 1929 Whittle sent his concept to the *Air Ministry* which passed it over to Griffith for evaluation. He appears to have been convinced that Whittle's *simple* design could never achieve the sort of efficiencies needed for a practical engine, that the centrifugal compressor design would be too large for aircraft use and that using the jet directly for power would be rather inefficient. In due course the RAF returned these comments to Whittle, referring to the design as being *impracticable*. Nevertheless, friends encouraged Whittle to go for the patent which was finally granted on 16 April 1931.

April 1930   A panel of the *Engine Sub-Committee* of the Aeronautical Research Committee had studied the Griffith report from 1929 and related experimental results available between January and April 1930. Chairman of the panel was (Sir) Henry T. Tizard (1885–1959), chemist[97]—specialising in aeronautics and then Rector of Imperial College, London. Other members included Professor of Engineering Arnold H. Gibson (1878–1959), University of Manchester; Hermann Glauert (1892–1934), Principal Scientific Officer of the Royal Aircraft Establishment, Farnborough; Professor (Sir) David R. Pye (1886–1960), then deputy-director in the directorate of scientific research, established at the Air Ministry under Dr. H.-E. Wimperis; (Sir) Thomas E. Stanton (1865–1931), Superintendent of the NPL's (National Physical Laboratory) Engineering Department, London. Summarising, the panel stated: *The Panel consider that, at the present state of the knowledge, the superiority of the* (gas) *turbine with respect to the reciprocating engine cannot be predicted, and they have no intention of advocating the large expenditure that probably would be involved in any attempted development of a turbine power plant by the Air Ministry.* This first paragraph of the Panel's report, and the general tenor of the discussions at the Panel meetings, must have discouraged the Air Ministry from investing any effort in the gas turbine. It certainly discouraged Griffith who, although becoming Head of the Engine Department at RAE in 1931, gave up working on the gas turbine, and consequently—except for the persistent Whittle—British work on the gas turbine aircraft was discontinued for some six years.[98]

---

[96] See Wikipedia, 'Frank Whittle' in English, and—see Fig. 6.31.

[97] See Fig. 6.31, and—see Footnote 141.

[98] See Hawthorne, The Early History.

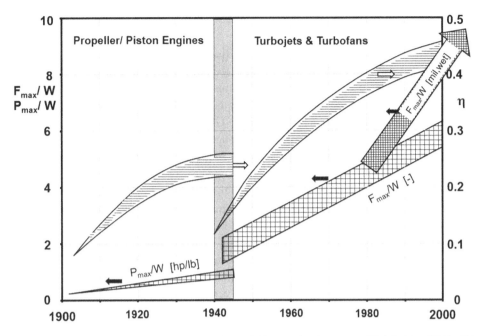

Fig. 2.3 Aero Propulsion Systems from 1900 to 2000: Trends of Power to Weight P_{max}/W [hp/lb] resp. Thrust to Weight F_{max}/W [−] Ratio and Overall Engine Efficiency η

Though the actual transition from piston aero engines to turbojet propulsion began only 10 years after Whittle's far-sighted analysis and his corresponding patent, it is worthwhile to visualise this *'hinge period'* and the corresponding radical changes in characteristic aero propulsion parameters already now, Fig. 2.3:[99] The view on the twentieth century is clearly split in the first, propeller and piston-engine dominated phase up to ca. 1940 and the subsequent, successful era of turbojets and turbofan engines. The *Power-to-Weight* trend for piston engines starts typically with a value of 0.07 hp per lb engine weight for the *Wright Flyer I* as early as 1903 with a 12 hp straight-4 water-cooled piston engine of 170 lbs.[100]—and levels out at the end of WW II with an average value, in its own right impressive of $P_{max}/W \sim 0.8$ hp/lb ranging e.g. from 0.79 hp/lb for the *DB 603A*[101] up to 0.96 for the *RR Merlin 61*.[102] This great improvement was achieved by engine design structures and materials, advanced fuel injection, advanced aerodynamic shapes of the propeller blades, variable-pitch propellers, and engine superchargers. The overall

---

[99] A classic representation, presumably shown for the first time by Hans von Ohain in Boyne, The Jet Age, 1979, and again in Mattingly, Elements, 1996.

[100] See Wikipedia, 'Wright Flyer' in English.

[101] See Wikipedia, 'Daimler-Benz DB 603' in English.

[102] See Wikipedia, 'Rolls-Royce Merlin' in English.

efficiency (engine and propeller) reached about 28%, while the power output of the largest engine amounted to about 5000 hp.

The turbojet trend starts in the first half of the 1940s with a clearly visible *jump* in comparison to the piston engines, all in a sudden the power density is duplicated; expressed in typical numbers—in Great Britain F_{max}/W ratios rose during wartime from 1.47 for the later version of *Whittle's W1 engine* to 1.68 for the *Rolls-Royce B.23 Welland*, the mass-produced version of the Power Jets W.2B for the *Gloster Meteor I* aircraft, introduced into service in mid-1944, while German production engines showed F_{max}/W 1.40 for the *BMW 109–003 A-1* and 1.25 for the *Jumo 109-004 B-1*, and unfinished C-versions of both engines—in parts with improved BBC axial compressor designs—would have been upgraded to 1.39–1.48. Besides the initial jump the turbojet/turbofan trend for the further decades shows a more than duplicated power gradient in comparison to the foregoing propeller/piston engines, thus illustrating the enormous development potential of the new engine concept. Present-day civil turbofan engines have typically $F_{max}/W \sim 6$, as specified e.g. for the *RR Trent 1000* family, while military turbofans with afterburners broke off the civil development pace around 1980, and reach in the meantime ~9, e.g. for the *Eurojet EJ 200* fighter engine.

In addition to the discussed specific *power/thrust* trends, Fig. 2.3 shows in reference to the right ordinate the *overall engine efficiency* curves—with clearly flattening tendencies at the end of the compared development periods. The initial drawback in fuel consumption at the time of the introduction for the early turbojets becomes clearly visible, which was, however, already fully compensated for the early 1950s, and advanced turbofans surpassed the η ~0.4 level in the meantime. As can be seen, progress was rapid. In half a century, the power/weight ratio increased about five-fold, and the overall efficiency exceeded that of a diesel propulsion system. The power output of today's largest gas turbine engines surpasses 100,000 equivalent hp. Stronger and lighter structures and greater aerodynamic quality of aircrafts combined with—as described—greatly advanced overall efficiency and enormously increased power output/weight ratios in aero-propulsion systems which had a tremendous impact upon flight performance, such as flight range, economy, manoeuvrability, flight speed, and altitude.

With the timeline of the *gas turbine preparation phase* passing the '1930' marker, it is appropriate for the timeframe 1930–1945 to go into more detail and to break down the further developments in the *Jet Web* for the country-triangle Switzerland—Great Britain—Germany separately and to look at the respective actions of their corresponding countrymen abroad.

2.2   Jet Web Timeline in Parallel 1931–1945

2.2.1   Turbomachinery and Aero-Related Activities Up to 1935

Switzerland and Swiss Abroad

18 Feb. 1932   BBC received an order from the *Soc. Métallurgique de Normandie* for two 11-stage axial compressors as commercial product with PR 3.4 as part of the first *Velox* steam-generating boiler to burn blast furnace gas for the Mondeville ironworks near Caen, F and practically at the time 13-stage axial compressors were ordered for installation in high-speed, closed-circuit wind tunnels (acc. to Ackeret's design) at ETH Zurich, *Institute of Aerodynamics* (Ma 2, PR 3, 3800 rpm) and shortly thereafter for the Italian Research Centre *Città dell' Aria* at Rome-Guidonia. These compressors were designed for a typical degree of reaction R 0.55. The two Mondeville axial compressors were of similar size, one for air and one for blast furnace gas. Both were equipped with a row of inlet guide vanes adjustable before starting, so that the optimum air/ gas ratio could be set for operation.[103]

15 Oct. 1932   Walter Noack, summarising recent achievements with the Holzwarth gas turbine at BBC test labs wrote: *Dealing with gas turbine problems— indeed a difficult problem!—has brought us (besides the Holzwarth pro-totype) a number of challenges and inspirations, which resulted in a new product, the Brown Boveri Velox boiler, a steam generator with turbo-charged combustor.* Having the extraordinary high heat transfer rates of the Holzwarth turbine in mind, here was the opportunity to change an obvious disadvantage of the unloved Holzwarth project into a striking success for BBC. And, just-in-time, the required high mass flows of pressurised air could be ideally provided by the newly developed axial compressor.[104] The first public demonstration of this *Velox steam genera-tor* was at the *Scientific Conference* of *VDI Verein Deutscher Ingenieure* (Association of German Engineers), Berlin, 15 October 1932 and the first Velox paper was printed[105] accordingly, carrying the term *Gas Turbine* in the title. At last now it must have become obvious that the all-axial compressor (10/11 stages, PR 2.4) and turbine set for commercial Velox boilers became aerodynamically so efficient that, in combination with a high-intensity combustor, the turbine drive generated a surplus/net power

---

[103] See Sect. 4.1.4, Figs. 4.25 and 4.26, and—see Eckardt, Gas Turbine Powerhouse, p. 108 f. and p. 128 f.

[104] See Eckardt, Gas Turbine Powerhouse, p. 158 f.

[105] See Noack, Druckfeuerung von Dampfkesseln.

output. The gas turbine principle was at hand, first time to be exploited commercially in the *Houdry oil cracking process* (see Swiss '1936' below).

1933   The *British Admiralty* ordered a Velox boiler of 50–55 t/h, typical for what was known as a *Torpedo Boat Destroyer (TBD)*. Due to military secrecy nothing is known about the further execution and use of that reference order for the UK. But Adolf Meyer stated in his speech at ASME Cincinnati in June 1935 (see below) in this context: *By taking full advantage of the technical means offered by the Velox boiler, war vessels with previously unseen striking power can be built. For obvious reasons no information about Velox plants in warships can be provided except the fact published in British papers that one unit has been built for the British Navy by Yarrow* (& Co., Glasgow) *and* (the British licensee) *Richardson Westgarth* (& Co., Hartlepool)*, the former building the boiler proper, the latter the gas-turbo blower and other auxiliaries.* A conventional water tube boiler produced 29 kg (steam)/m^2 while the Velox targeted for impressive space-saving 510 kg/m^2.[106]

1934   Curt Keller (1904–1984), Swiss engineer and assistant to Profs. Stodola and Ackeret at ETH Zurich, and head of the *Hydraulic and Caloric Laboratories for Turbomachinery* at *Escher-Wyss Zurich,* designed the two identical blowers for the subsonic wind tunnel of Ackeret's newly founded Institute of Aerodynamics (IfA). This resulted in Keller's widely spread doctoral thesis *Axial Blowers in View of Aerofoil Theory* of 1934, detailing advanced axial compressor design know-how—originating via Ackeret also from Goettingen—to a broader engineering public for the first time. In due course the thesis was translated into English and published in shortened form in 1937.[107] The compressor for the closed supersonic wind tunnel came entirely from Brown Boveri, Baden, CH.[108]

---

[106] See Eckardt, Gas Turbine Powerhouse, p. 164 f. In addition to these considerations, one can speculate that potential Velox applications for the Navy had become partially obsolete due to the introduction of the Admiralty 3-drum boiler, especially in its small-tube version, also called '*express boiler*', as destroyer standard equipment since 1934. Possibly, the order was put on hold and renewed in the context of a 1.6 MW GT delivery,—see 11–16 Aug. 1939 in the Swiss section.

[107] See Keller, Axialgeblaese.

[108] See Sect. 4.1.3 and Fig. 4.26.

Great Britain and British Abroad

13 Sep. 1931   The British team secured the *Schneider Trophy*[109] for the UK permanently with the 1931 uncontested win, after winning 1927 at Venice, and 1929 at Calshot Spit, the same location as in 1931. The winning Schneider flight was performed by a *Supermarine S.6B*, piloted by Flt. Lt. John Boothman, having attained a recorded top speed of 547.19 km/h, and flown seven perfect laps of the triangular course over the *Solent*, the strait between the Isle of Wight and the British mainland.

1932   While British turbojet engine developments were at a standstill, it was— besides Glauert—a young Canadian Beverley S. Shenstone,[110] who developed at that time the closest aeronautical working relations to German institutions like the just forming *Glider scene* at the *Wasserkuppe*, to the Junkers Flugzeugwerke at Dessau, and as a professional intermediary to Prandtl and the Goettingen-influenced aero research community. In the same year he was hired back to England, to *Supermarine Aviation Works* on River Itchen close to Woolston, Southampton, a subsidiary of Vickers-Armstrong, where he became famous for his ingenious and meticulous aerodynamic design of the *Spitfire* fighter aircraft, especially its *'elliptic wing'*, under Reginald J. Mitchell, Supermarine's brilliant chief engineer.[111]

Germany and Germans Abroad

Oct. 1929–   Helmut Schelp (1912–1994) and Hellmut Weinrich (1909–1988), two
March 1931   strong *influencers* of Germany's coming turbojet engine activities in the 1930s, Schelp at *RLM Reichsluftfahrtministerium* (Reich Air Ministry) and Weinrich, with some likelihood together with Fritz Heppner, inventor of the BMW 109-002 counter-rotating turbojet engine, studied jointly mechanical engineering and fluid machinery at the Technical College Chemnitz.

1931   Paul Schmidt (1898–1976), an independent engineer and inventor at Munich, designed and patented as a primary propulsion system for aircraft a *pulsejet* engine (see '1903' Holzwarth) which was used in the coming war as the powerplant of the V-1 flying bomb. He applied for support to Adolf Baeumker, then head of the Research Division of the *RVM Reichsverkehrsministerium* (Ministry of Communications) which

---

[109] See Wikipedia, 'Schneider Trophy' in English.

[110] See Wikipedia 'Beverley Shenstone' in English.

[111] See Sect. 4.3.3 and Fig. 4.40, and—see Wikipedia, 'R.J. Mitchell' in English.

was the predecessor of the *RLM*. Schmidt's request was granted and his work was continuously financed by first the Verkehrs- and then from 1935 the Reichsluftfahrtministerium until the end of the war.[112]

15 April 1932 Beginning German '*patent confusion*', which lasted up to 1936,[113] mainly between AVA Goettingen, its turbomachinery protagonists Albert Betz (1885–1968) and Walter Encke (1888–1982), and numerous supporting industrial companies, and on the other side Brown Boveri, Mannheim and Baden CH, after BBC filed under this priority date a patent application of Claude Seippel, later '*father of the axial compressor*', on a 3+ stage axial turbocharger, and a corresponding, supplementary strategic move of BBC's French daughter company C.E.M., also possibly considered as a threat to AVA's preferred axial compressor design practise, to apply for a (short time) patent for *degree of reaction* R 0.5 designs.[114]

1934 Hans Joachim Pabst von Ohain (1911–1998), student of physics at *Georgia Augusta University* at Goettingen, began in that year (perhaps already in autumn 1933), annoyed by the strong vibrations and noise of conventional aircraft piston engines, to think about alternative ways of aircraft propulsion, which led him independently to the continuous radial gas turbine concept—with high mass flow and low vibrations and the prospects for considerably improved power to weight ratios, especially suited for high speed flight at high altitudes. Starting from the concept of a '*Nernst Turbine*' (see '1908') he suggested a centrifugal compressor rotor, back to back arranged to a radial-inflow turbine and combustion chambers within the casing, which allowed fuel combustion at low through-flow Mach numbers close to stagnation pressures. The main advantage of this concept was its extreme simplicity which minimised the development risks. The fuel surplus energy as far as not used up in the turbine to drive the compressor wheel is transferred to kinetic thrust energy in the exhaust nozzle.[115]

---

[112] See Schlaifer, The Development, p. 381. Since 1939 Schmidt's *pulsejet* fell in the responsibility of *Sondertriebwerke* under Hans Mauch, who—most remarkably—left his desk job at RLM in Sep. 1939 to be followed by H. Schelp, only to show up in 1943/4 in a critical phase of V-1 development as a finally successful hands-on manager and project coordinator, see German 'Sep. 1943'.

[113] Finally decided in favour of Brown Boveri on 26 Feb. 1936 in a patent law suit at Reichspatentgericht (Reich patent court),—see Sect. 4.3.3, *In the Patent Arena*.

[114] G. Darrieus patent FR781,182: '*Compresseur à couronnes d'aubes alternativement fixes et mobiles juxtaposées* (Compressor with alternatively fixed and rotating blade rows)' with priority 3 Feb. 1934 in the name of Brown Boveri, which was obviously cancelled ~1936 to prevent a conflict with that established design principle of the friendly English designers community.

[115] See Sect. 6.2.2 and Fig. 6.13.

2.2.2   Early Turbojet Developments 1935–1939

Switzerland and Swiss Abroad

1935	Sanford Moss (63, see '1903' Moss), GE turbocharger legend and then near his retirement age,[116] toured Europe for turbocharger *news* and wrote an intriguing eight-page travel report (amongst other about visits at RAE and Rolls-Royce in England as well as AEG Berlin).[117] With reference to BBC, Moss wrote: *Buchi's scheme ... is now in use in many exhaust superchargers for diesel engines made by the Brown Bovery (sic!) Company for motor ships and rail cars. I tried to visit the Brown Boveri Company but could not obtain permission. However, I was informed that the Brown Boveri Co. have never made exhaust superchargers for aviation engines. ...* It cannot be ruled out that this unfriendly, untypical gesture in view of Swiss general welcome culture was a reaction to the General Electric vs. Sherbondy US turbocharger conflict 18 years before,[118] based on Seippel's reminiscences (see '1905' and '31 Aug. 1923'). Actually, about that time Rolls-Royce and BBC had started secret turbocharger collaboration which comprised advanced single-stage centrifugal as well as seven-stage axial compressor deliveries from Switzerland, which brought amongst other and 25 years before similar aero engine designs became known, first film-cooled turbine blading.[119] Following the Moss report, nothing of these activities leaked out at Derby—and also Swiss BBC had lifted the draw-bridge in time.
1935	Adolf Meyer (1880–1965),[120] Director of the *BBC Thermal Department* from 1923 to 1946—with comprehensive responsibility for all turbomachinery, steam and gas turbines as well as turbocharger activities—propagated the advantages of the newly developed *Velox boiler* in comparison to conventional steam generation in his comprehensive survey lecture *The Velox Steam Generator—its Possibilities as Applied to Land and Sea*[121] which he delivered, on invitation, as the *second Calvin W. Rice Memorial Lecture* during the Semi-Annual Meeting, Cincinnati OH,

---

[116] See Fig. 4.13.

[117] The report has been found at the National Air and Space Museum, Washington DC and was reproduced in 2009 in the Derby branch magazine of the RR Heritage Trust (by courtesy of Dave Piggott, RRHT).

[118] See Sect. 4.1.2.

[119] See Eckardt, Gas Turbine Powerhouse, p. 397 f.

[120] See Eckardt, Gas Turbine Powerhouse, p. 87 f.

[121] See Meyer, The Velox Steam Generator.

17–21 June 1935 of *The American Society of Mechanical Engineers (ASME)*. He describes the Velox development steps as part of the Holzwarth's explosion gas turbine in 1926, then the switch to continuous combustion with the application of a two-stage centrifugal compressor and finally the successful replacement of the latter by an all-axial turbo-set with high mass flow in 1933. For illustration the 10-stage *Mondeville* axial *blower* was shown, Fig. 4.25, presumably the first presentation of an axial compressor in North America, designed on the basis of aerodynamic aerofoil theory.

1936   The Velox operational principle of combustion at raised pressures was also used in chemical engineering for the first time. Demand for hot compressed air in connection with the *Houdry*[122] *oil cracking process* created the first field application for the newly emerging constant-pressure gas turbine. In many chemical processes industrial supercharging had been impossible for a long time due to the high cost of compressing the air, but if gases from the process could be expanded in a turbine with enough power to drive the compressor, then supercharging became economical. The sheer size of the required *Houdry turbo-set* was a challenge for BBC, with a volume flow of nearly 20 m³/s at a pressure ratio of PR 4.2. The compressor unit with 20 stages was a considerable step forward in comparison to the foregoing Mondeville and Zurich units. The excellent performance of the first delivered turbomachinery (axial compressor and turbine) set for the *Marcus Hook* oil refinery of *Sun Oil Co.* Philadelphia,[123] meant the 4.4 MW compressor driving power was more than compensated by the 5.3 MW turbine output, so that the difference could be used up in a gearbox-coupled generator on the turboshaft to produce electricity.[124] The stationary gas turbine was in sight. In 1939 Sun Oil already had 10 Houdry plants in operation, based on further turbomachinery deliveries from Baden and the first, licence-produced equipment from *Allis-Chalmers*, which became—based on their BBC compressor know-how—also part of early US turbojet developments. First since 1941 as a member of the NACA *Special Committee on Jet Propulsion* together

---

[122] Eugène Jules Houdry (1892–1962) was a French, later naturalised American, mechanical engineer who invented catalytic cracking of petroleum feed stocks,—see Wikipedia, 'Eugene Houdry' in English.

[123] Houdry unit *Eleven Four* came on line there on 31 March 1937.

[124] See Eckardt, Gas Turbine Powerhouse, p. 169 f.

with *Lockheed* and *Northrop*, thereafter selected for licence production of the *de Havilland Halford H.1B Goblin* centrifugal-flow turbojet.[125]

1 Oct. 1938   Foundation of a specific *Gas Turbine Department* in the BBC Baden Thermal Design Office with Claude Seippel as head of this new *Dept. GT* with the responsibility for Velox and electro boilers as well as *Gas turbine groups for power generation and other* (aero) *purposes*.

24 Feb. 1939   Adolf[126] Meyer gave the first comprehensive presentation *The Combustion Gas Turbine: Its History, Development and Prospects*[127] of BBC's gas turbine achievements, including the new 4-MW power generation plant at Neuenburg CH, the gas turbine-electric locomotive, GT-powered, compact and lightweight (destroyer) ship propulsion and an excursus on the benefit of a *combined gas turbine and steam plant* which also boosted the vision for possibilities of successful aero gas turbine developments considerably. The presentation was given at an extra general meeting of the *IMechE Institution of Mechanical Engineers* in its traditional headquarter site since 1897 at London-Westminster, One Birdcage Walk[128] with attendance and discussion contributions of

- David M. Smith from *MV Metropolitan-Vickers* Manchester, together with MV's chief mechanical engineer of Swiss origin Karl Baumann (1884–1971) responsible for the Metropolitan-Vickers *F.2* early turbojet engine in close cooperation with RAE, also the first British design to be based on an axial-flow compressor, and, then,
- Squadron Leader Frank Whittle, RAF Rugby, who remarked at the end of the discussion, . . . *so far little has been said about the gas turbine in relation to aircraft, which I regard as a most hopeful field for it, for, as the gas turbine is taken up into the air the reduced atmospheric temperature makes possible very much higher efficiencies than could be attained on the ground.* Herewith, Whittle pointed out that the gas turbine was especially attractive from the aircraft point of view and that there was a very good reason why it should be more successful in this

---

[125] See Sect. 6.4.3, and—see Wikipedia, 'de Havilland Goblin'. The *Halford H.1B*, in the US known as *J36*, experienced lengthy delays in 1944.

[126] An amusing aside: Adolph (with -ph) Meyer is also the name of a German agent character in Sir Arthur Conan Doyle's *The Adventure of the Bruce-Partington* (submarine) *Plans*, published in 1908; this may have become an issue during AM's frequent contact with Britain.

[127] See Sect. 6.3.5, and—see Meyer, The Combustion Gas Turbine.

[128] See Eckardt, Gas Turbine Powerhouse, p. 87 f. The building, erected in the Queen Anne, *streaky bacon*, style in red brick and Portland stone, was remodelled in 1933 by expanding the library and introducing electric lighting. After Meyer's lecture the building would go on to host the first public presentation of Frank Whittle's jet engine in 1945.

field than in any other. He was, of course, careful to avoid any reference to its application as a jet propulsion device, or to give any indication of the work he was doing. Meyer, apparently informed, wrote in his response: ... *Squadron Leader Whittle's remarks on the possibilities of the gas turbine as a prime mover for aircraft seemed rather optimistic even to himself—and he was not very bashful—but Squadron Leader Whittle no doubt knew much more about this field of application, so that one might hope that his forecast would soon come true. ...* F. Whittle, M.L. Bramson and L. Cheshire attended this lecture together.[129] One can speculate with some certainty that local engineering *grandees* A.A. Griffith, H. Constant, W.S Farren, H.T. Tizard, H. Roxbee Cox might have attended that historic meeting as well, but thorough search in the IMechE Archive produced no invitation/attendance list yet; O. Zweifel[130] (1911–2004), Swiss aerodynamicist, took part for BBC, staying in England at that time on an industrial visiting leave.

At the IMechE lecture we see Meyer in his role as a (sceptical) *Jet Web* networker. In 1935 he had received a Doctor honoris causa from the *Stevens Institute of Technology* in Hoboken, NJ, exactly during the period when the young German Helmut Schelp (1912–1991)—three years later responsible head of German turbojet developments in *RLM Reichsluftfahrtministerium* Berlin (Reich Air Ministry)—studied there for a Master Degree and there, in all likelihood, came into contact with Meyer. Similarly, there are links of Meyer to US turbojet developments at *Northrop Aircraft Corporation* in Hawthorne, CA, and their Chief of Research Vladimir H. Pavlecka (1901–1980). This was presumably in 1939 when Northrop's *Turbodyne* propeller-turbine engine was still in its

---

[129] See Golley, Whittle—The True Story, p. 122; in 1935, Whittle had met Mogens Louis Bramson, a well-known independent consulting aeronautical engineer, who wrote the positive, historically famous *Bramson Report* in Nov. 1935, that paved the way to Whittle's further financial support. At the time of the IMechE lecture Leslie Cheshire was still employed by *BTH British Thomson-Houston Co.,* Rugby; in Sep. 1939 he joined Whittle's company *Power Jets* on loan from BTH, becoming one of Whittle's closest collaborators.

[130] Since 1936, BBC used an aerodynamic loading coefficient for design calculations and cascade test evaluation, suggested by Zweifel. After official publication in 1945 up to today, the *Zweifel number* is well-known and still useful for compressor and turbine preliminary design in an adapted form. See Eckardt, Gas Turbine Powerhouse, p. 117.

infancy.[131] In general, Meyer was perhaps (too) pessimistic at that time that the low GT *power/ weight ratios*, which he knew from BBC's power generation experiences could be radically improved short term—measured against the first flight of a GT-powered Heinkel He 178 only six months later. The close ties between Swiss engineers and IMechE were never extinguished: Adolf Meyer received the IMechE *George Stevenson Medal* for *best achievement of the year* in 1939 and 1943, Karl Baumann, Metrovick's chief mechanical engineer, was IMechE *Honorary Fellow* of 1954. One can assume the early acquaintance of Baumann/Meyer out of Stodola's camp of assistants back in 1906/7 was instrumental, even in view of a principle competition between both companies, to establish direct contacts between the RAE axial compressor designers and BBC Baden from 1937/1938 onwards and to prepare the platform for Meyer's seminal IMechE lecture.

28 June 1939   Claude Seippel, BBC Baden, files a turboprop engine patent with priority in Germany (US2,326,072) on a *Gas Turbine Plant, of the type including an axial flow compressor for supplying combustion air and cooling air, a combustion chamber or chambers in which fuel is burned under constant pressure, and a turbine operated by the combustion gases to drive the compressor and develop useful power. The turbine assemblies contemplated by the invention may be used for various purposes but are particularly adapted for use on aeroplanes or other high-speed craft.*[132]

July 1939   Gyoergy Jendrassik (1898–1954), Hungarian mechanical engineer and inventor, commenced the development of what is considered the first bench-tested turboprop engine in 1940 at Budapest. The turboprop project, designated the Cs-1, featured a 15-stage axial compressor, an annual reverse-flow combustion chamber, an 11-stage axial turbine (with extended-root turbine blades to reduce heat flow to the turbine discs) and a fixed-area exhaust nozzle. It was aimed at an initial rating of 1000 shp at 13,500 rpm. Its rotor was mounted on only two bearings and there was an airscrew reduction gearbox in front of the annular air intake. Jendrassik's confidence in the turboprop project relied to a large extent on his foregoing positive experiences in developing a small scale, stationary 100 hp gas turbine with heat exchanger, which achieved on the basis of independent, official measurements on 7 Jan. 1939 the remarkably high

---

[131] See Chap. 7 and Sect. 6.4.2, and—see Fig. 6.52.
[132] See Sect. 8.3.

overall efficiency of 21.2%.[133] Between 1937 to 1939 Jendrassik was a regular visitor to Professor Ackeret at ETH Zurich and at BBC Baden;[134] therefore, the proximity of Seippel's turboprop patent (priority date of 28 June 1939, above) and Jendrassik's immediately following turboprop development activities may not be completely accidental. During 1940 the Cs-1 engine bench tests were hampered by considerable combustion problems, which limited the output to 400 shp only and unfortunately, the further engine development was halted after the signing of a German-Hungarian Mutual Armament Programme in June 1941. Strangely, the Germans apparently never showed any interest in this unique, advanced development and Hungarian Air Force went to war with German conventional piston-engine planes.

7 July 1939   The acceptance test for the world's first industrial gas turbine set from *Brown Boveri* took place that Friday between 10.10 and 11.10 h at the BBC Baden test laboratory—commissioned to the *nestor* (doyen) in the field of thermal machines, the 80-year-old ETH professor Aurel Stodola. This decision was certainly also made in view of Stodola's international reputation as an independent expert, not least in view of some unpleasant discussions which had accompanied similar performance assessments of Holzwarth's explosion gas turbine a few years ago.[135] Stodola himself wrote about the event: ... *In spite of the simplicity of the arrangement the test showed the efficiency referred to the heat contained in the fuel, and the heat equivalent of the electrical output of the generator to be 17.38%,*[136] *representing an efficiency which, together with the many constructional advantages of the set, renders it in many cases competitive. An especially noteworthy feature is the fact that the plant requires no cooling water. ...* This gas turbine set with 4 MW output power had been ordered for the *Municipal Power Station* in Neuenburg/ Neuchâtel, CH as a standby-unit

---

[133] See Sect. 6.4.4, and—see Wikipedia, 'György Jendrassik' in English, and—see Kay, Turbojet Vol. II, p. 252 f.

[134] Personal communication with Franz Farkas on 31 Aug. 2018, who himself contributed considerably to BBC/ABB/Alstom gas turbine developments between 1961 and 1997, predominantly in the area of advanced turbomachinery,–see Eckardt, Gas Turbine Powerhouse, p. 285 f. G. Jendrassik was maternal uncle to Prof. G. Gyarmathy (1933–2009), former BBC director and successor to Stodola's chair at ETH Zurich (1983–1998); see Eckardt, Gas Turbine Powerhouse, p. 2.

[135] See Eckardt, Gas Turbine Powerhouse, p. 183 f.

[136] This value has to be put into the perspective of today's efficiency range for simple cycle gas turbines. From the beginning the gas turbine was characterised by performance deficits in comparison to contemporary steam turbines, which already achieved 25% at that time. This excluded the gas turbine from the base load power generation for nearly 50 years, and the situation only changed when both concepts were successfully merged into *CCPP Combined-Cycle Power Plants,* which in the meantime claim record efficiencies of 63+ percent. See Eckardt, Gas Turbine Powerhouse, p. 347 f.

	and was used continuously from 1940 onward for that purpose for 62 years after commissioning.[137]
11–16 Aug. 1939	Acceptance tests at BBC Baden for a 1.6 MW gas turbine set on order of the *British Air Ministry* fulfilled all guarantee values—with 20-stage axial compressor and 5-stage turbine, in principle and with parameters very similar to the series of *Houdry gas turbines* in production: Output of turbine 6050 kW, blower rating 4400 kW, pressure ratio 4.1, gas inlet temperature 560 °C, speed 5180 rpm. Accordingly, first test runs of this unit at RAE took place between 3 and 6 May 1940. The extraordinary short delivery time of less than 10 months is a further indication that the order was a BBC standard product at that time and could be delivered almost straight from the shelf.[138]
1939	In 1936, rumours spread in Switzerland about the gas turbine-related activities at BBC. Professor Jakob Ackeret, ETH Zurich and Curt Keller, then head of the Escher Wyss Zurich research department, decided on a different approach. Their basic, patented idea was to operate the machine in a closed air cycle, the so-called *AK process,* to which the heat is added from outside the closed loop by means of an air heater (heat exchanger). A first test set-up became operable at Escher Wyss during 1939 and the acceptance test showed a promising efficiency of 32% in January 1945.[139]

Great Britain and British Abroad

16 March 1937	The *ARC Engine Sub-Committee* corrected its fateful decision of April 1930 in an important meeting, again chaired by Henri Tizard. Other members included now Professor (Sir) Leonard Bairstow (1880–1963),[140] then Britain's leading aerodynamicist, who held the

---

[137] In 1988 the ASME awarded the Neuenburg gas turbine the status of a *Historic Mechanical Landmark*, reserved for milestones of outstanding technical development. At that time, this powerplant was #135 listed, the eighth outside the United States. In the meantime (Feb. 2022) the list grew up to 277 items. https://www.asme.org/about-asme/engineering-history/landmarks In 2005 under the leadership of Walter Graenicher, then President of *Power Service and of Alstom (Switzerland) Ltd,* was decided to relocate, restore and display the machine at the power generation development and production facility in Birr, 14 km south-west of Baden, CH. Visits and guided tours can be arranged via *ABB Birr*, Portier 1, Tel +41 (0)584466156,—see also Lee Langston's informative travel report of 2010: Langston, Visiting the museum.

[138] See Sect. 6.3.4 and Fig. 6.44.

[139] See Eckardt, Gas Turbine Powerhouse, p. 287 f.

[140] Bairstow plays a curious role in a clash of political philosophy and technology. In 1909 V.I. Lenin had published his philosophical book *Materialism and Empirio-Criticism* which found strong opposition in Russia. In the preface of the 2nd edition of 2 Sep. 1920, he called these Russian

Zaharoff Chair of Aviation at *Imperial College*, (Sir) Harry R. Ricardo (1885–1974), one of the foremost engine designers and researchers in the early years of the development of the internal combustion engine, who represented then his consulting and research firm of *Ricardo & Co.* at Shoreham, (Sir) Alfred Ch.G. Egerton (1886–1959),[141] chemist and successor to H. Tizard at *Clarendon Lab* at *Oxford University* in 1923; in 1936, he assumed the chair of *Chemical Technology* at the *Imperial College of Science*, and Dr. Alan A. Griffith. Triggered by a highly persuasive report from Hayne Constant (1904–1968),[142] the ARC authorized the RAE to proceed with Griffith's turboprop scheme, thus finally opening the door to look again into the business of adopting the gas turbine for aeronautical use. In May 1937 the Engine Subcommittee of the ARC considered Constant's paper and the newest experimental results from Whittle together (see below) and made a favourable report accordingly. The Directorate of Scientific Research, influenced in part by this report but even more by information that *Brown Boveri was offering for sale gas turbines with guaranteed performance,*[143] decided to support research on gas turbines, even though it still believed that they would have no practical application for a long time to come.

12 April   First run of Whittle's WU engine which consisted of a centrifugal com-
1937   pressor with a double-sided impeller to obtain the greatest airflow from a

---

opponents *Machists* after the philosopher-physicist Ernst Mach (obviously without Mach's personal involvement) with the consequence, when Ackeret propagated the term *Mach number* after 1930, that the Soviet Union refused acceptance for a while and used a *Bairstow number* instead, e.g. in 1969 the Tu-144 SST achieved in 19 km altitude speeds up to *Bairstow 2.05*.

[141] Egerton and Tizard had presumably already studied together since 1913 at the *Friedrich Wilhelm University* (today *Humboldt University) Berlin* in the laboratory of Professor Walther H. Nernst (1864–1941), a German chemist, whose contribution to the 3rd law of thermodynamics won him the 1920 *Nobel Prize in Chemistry*. Nernst's own study and laboratory always presented aspects of extreme chaos which his co-workers termed appropriately *the state of maximum entropy*. Already in 1911 Nernst and Tizard had published a joint theoretical chemistry paper. At this time Frederick A. Lindemann (1st Viscount Cherwell, 1886–1957), British physicist and influential scientific advisor to Winston Churchill from the early 1940s to the early 1950s was also there for his doctorate with Nernst in physical chemistry, and the three became friends at first, until during WW II when a severe disagreement broke out between Lindemann (in support of R.V. Jones, RAE) and Tizard about German electronic warfare. See Wikipedia, 'Battle of the Beams' in English. Interestingly Hans von Ohain's first patent application in May 1935 (what became secret patent #318/38), see Sect. 12.1.1, was based on a *Nernst Tube*, originally a rotating, U-tube-shaped chemical reactor on two bearings with gas in- and outflow along the axis of rotation which also resembled Ohain's first hydrogen-fuelled demonstrator HeS 1;—see Prisell, The Beginning.

[142] Tizard had persuaded Constant to return to the RAE to assist Griffith with gas turbine research after a period spent lecturing at Imperial College.

[143] See Schlaifer, The Development, p. 349.

given size and to keep the inlet Mach number at the tips of the axial inducer sufficiently low. The impeller was given the greatest number of blades (30) that manufacturing limitations allowed, to reduce loading and avoid stalling at impeller inlet. The compressor unit was driven by a single stage axial flow free-vortex turbine without nozzle guide vanes, made like the compressor from a single forged disc.[144]

13 April 1937	RAE design office started preliminary design work on various general arrangements of gas turbine engines, in preparation of a planned, first contact to Metropolitan Vickers Ltd on 3 June 1937, whose name had been suggested by the ARC sub-committee as a firm who would be willing to co-operate. The investigated GA spectrum shrank to an all-axial lp/hp turbomachinery arrangement only, after news from Brown Boveri and their successful operation of a gas turbine with axial compressor/ turbine spread in 1938.[145]
Summer 1937 –May 1938	RAE delegations visited BBC Baden, (Alan A. Griffith, 1937 and Hayne Constant and (Sir) William S. Farren (1892–1970) in 1938), negotiated also about a license agreement, but came to the conclusion that *exclusivity on the BBC axial compressor design would not be granted.* In due course and recommended by MAP Ministry of Aircraft Production, RAE slowly started negotiations with Metropolitan-Vickers (Metrovick), a steam turbine manufacturer like Brown-Boveri, to consider collaboration on the development of the F.2, a simple axial turbojet. In July 1940 actual detailed design and development of this engine had been entrusted to Metrovick and in the early part of 1941 the collaboration partners were instructed to devote all available energies to the development of this jet engine, while turboprop activities were completely abandoned.[146]
12 Oct. 1938	(Sir) Roy Fedden (1885–1973) receives the *Lilienthal Ring* at Berlin, and John B. Bucher, Armstrong Siddeley Motors, is attending as part of a presumed MI5 mission to clarify some of Fritz Heppner's reported German turbojet engine developments.[147]
30 June 1939	After a period of indifference, a successful demonstration of the *Power Jets WU* at full power in a continuous 20-minute run was made before a delegation of the Air Ministry led by David R. Pye (1886–1960), so that

---

[144] See Sect. 6.3.1, and Figs. 6.31 and 6.32, and—see Golley, Whittle–The True Story, p. 249.

[145] See Sect. 6.3.1, and—see Hawthorne, The Early History.

[146] See Sect. 6.3.3, and—see Fig. 6.40, and—see Schlaifer, The Development, p. 356.

[147] See Sect. 6.1.3, and—see Figs. 6.6, 6.8 and 6.9.

the Ministry quickly arranged to buy the engine to give Power Jets working capital. In addition, a contract was placed for a flight engine, the *W.1*. Finally, in August 1939, the Gloster Aircraft Company was also given a contract for the design and construction of two airframes, the *E.28/39*, intended primarily as experimental vehicles for flight testing the Whittle engines.[148]

Germany and Germans Abroad

Mid-1935–
Feb. 1936

Already in 1934 Hans von Ohain discussed his ideas with Max Hahn (1904–1961), car mechanic at the garage of Bartels & Becker, Goettingen, who built a ~ 1 m diam. demonstrator from sheet metal during 1935. In the time-frame mid-1935 up to February 1936, a demo test of the so-called '*Garage model*' was accomplished in the backyard of the Institute of Physics, where Hans von Ohain stayed after finishing his doctorate with Professor Robert W. Pohl (1884–1976). The test, which already required permission from RLM, failed to achieve self-sustained operation of the apparatus with less than 200 m/s tip speed, but motivated Pohl to contact in a letter, dated 3rd March 1936, and successfully interest Ernst Heinkel (1888–1958), aircraft designer and owner of *EHF Ernst Heinkel Flugzeugwerke AG*[149] at Rostock for von Ohain's ideas. H.J. von Ohain started his job at Rostock together with Max Hahn immediately after the Easter break on 14 April 1936, followed already on 15 April 1936 by an (again failing) test repetition of the *Garage model* in a wind tunnel set-up. In due course a first engine design HeS 01 was on Wilhelm Gundermann's design board[150] by mid-1936.

5 Sep. 1935–
26 June 1936

As part of the newly launched '*Flugbaumeister*' education programme Helmut Schelp studied mechanical engineering at Stevens Institute of Technology, Hoboken, NY, finishing with a prize winning master thesis. In addition, he should have come in contact with the Swiss inventor-engineer Rudolph Birmann (see '1922') and Birmann's *diagonal compressor* concept, to be realised in the Heinkel HeS 011.[151]

---

[148] See Sect. 6.3.1, and Fig. 6.31.

[149] The number of EHF employees grew from ~1000 in 1932 to approx. 9000 in mid-1939 up to ~16,000 at the end of 1944.

[150] Dipl.-Ing Wilhelm Gundermann (1904–1997) was turbojet chief engineer at Heinkel Rostock.

[151] See Chap. 7.

31 Sep.– 5 Oct. 1935	Fifth International Volta Congress at Rome under the motto 'Le Alte Velocità in Aviazione (High-speed in aviation)' had certainly been influenced by Italy's international speed world record[152] in the previous year. An international bunch of hand-picked, leading aerodynamicists attended: L. Crocco, I; G.I. Taylor and H.R. Ricardo, UK; Th. von Kármán and E.N. Jacobs, USA; Jakob Ackeret CH. L. Prandtl led the German delegation, A. Busemann (1901–1986), then Associated Professor at TH Dresden, presented—rather unnoticed—his *'Aerodynamischer Auftrieb bei Ueberschallgeschwindigkeit (Aerodynamic lift at supersonic speed)'*, introducing wing sweep as an elegant method to reduce the *effectively felt Mach number*, an essential precondition to bring the turbojet to its full speed potential after the war. Congress participants were taken to see the just inaugurated complex of test facilities *Città dell'Aria* at Guidonia,[153] east of Rome. Highlight of the tour was the Mach 2 supersonic wind tunnel with 0.4×0.4 m test section, to a large extent a 1:1 copy of Ackeret's Zurich wind tunnel, except for the drive power which was more than doubled to 2100 kW, thus considerably expanding the researchable Re number range. The wind tunnel installation came from BBC's Italian daughter company TIBB (Tecnomasio Italiano Brown Boveri) and was finished in 1936.[154]
10 Nov. 1935	Hans-Joachim Pabst von Ohain receives for his invention of an all-radial jet engine configuration the German secret patent #317/38 *Stroemungserzeuger fuer gasfoermige Mittel* (Flow generator for gaseous means), under this priority date. Von Ohain's second patent of a Nernst tube type gas turbine engine configuration has been granted as German secret patent #318/38 on 05 June 1936.[155]
8 Feb. 1936 14 Aug. 1938	German patent priority dates for Herbert Wagner's all-axial, geared turboprop engine [GB495,469] and—more than two years later—for an all-axial turbojet [DE724,091]. On 1 April 1936 Wagner had been

---

[152] 23 Oct. 1934, Francesco Agello, 709,209 km/h with Macchi-Castoldi M.C.72 seaplane with 3100 hp, 24 cyl. Fiat V-engine, counter-rotating propellers, over Lake Garda. The aircraft was too late to participate in the 1931 Schneider Trophy competition; in due course three pilots were killed until Agello was successful.

[153] Named after Alessandro Guidoni (1880–1928), general of the *Italian Royal Airforce*, killed during parachute testing at (later) *Guidonia*.

[154] See Eckardt, Gas Turbine Powerhouse, p. 128 f.

[155] For details,—see Sect. 12.1.1 Shortly before publication deadline, the Author detected patent ES137,729, with priority 29 March 1935 for—see Wikipedia, 'Virgilio Leret Ruiz' in English, which resembles the CR radial Ljungstroem turbomachinery of Fig. 5.9 (l). Eventually, this invention would fall between 16 Jan. 1930 (Whittle's first GT patent) and this 10 Nov. 1935. With complimentary thanks to Antón Urquidi and Günter Kappler, 4 Sep. 2022.

installed as new Junkers aircraft development chief engineer at Dessau. Part of the agreement was to continue the activities towards an all-axial aero gas turbine at the Junkers manufacturing branch Magdeburg, led by Max Adolf Mueller (1901–1962), former senior engineer at Wagner's THB institute, and assisted by Rudolf Friedrich (1909–1998), who had the responsibility for axial compressor design and corresponding manufacturing tasks e.g. the contour milling of twisted aerofoils. The decision for the selected axial compressor design, remarkable for its use of a degree of reaction R 0.5 blading (so that both rotors and stators contributed equally to the compressor pressure rise), came from Wagner with marine steam turbine background (where R 0.5 represented the classic standard), and who *'knew the industrial gas turbine'*, a clear reference to BBC's *Velox turboset* with a likewise R 0.5 axial compressor design, on sale since 1932.

The Magdeburg group carried out systematic compressor tests with 1-, 5-, 12- and 14-stage(s). A first test—called *RT0 Rueckstoss (repulsion) Turbine engine Ohne Leistungsabgabe an Propeller* (without power output to propeller)—with 14-stage compressor, annular combustor and 2-stage axial turbine was set-up at Magdeburg in early 1939, at a time when Herbert Wagner had already left Junkers to Henschel Berlin. However, these tests failed to achieve self-sustained operation; further testing with propane gas was interrupted, when the RLM insisted to transfer the activities from Magdeburg to the Junkers Motor Works at Dessau.[156]

Mid-Jan. 1937– Mid-Sep. 1937	Hans von Ohain is forced to demonstrate short-term feasibility of the gas turbine principle, by giving up the *Nernst tube* principle, and introducing a row of diffuser vanes instead to decelerate the flow at combustor inlet and—decisive—to switch for the demonstration purpose to hydrogen combustion. This equalised the flow temperatures at turbine entry and allowed for the first self-sustained demonstrator operation.[157]
23 July 1937– 1 Aug. 1937	Fourth International Flight Meeting at Zurich-Duebendorf with—first time—international demonstration of six, newly developed *Messerschmitt Me 109* fighter aircraft. This event and the exhibited Me 109 speed performance is important, since it roughly fixes also the starting date of Schelp's first study task for his *Flugbaumeister* qualification training, in which he simply was asked to investigate the new speed limits and *'what would be required for aircraft* (propulsion), *if*

---

[156] See Sect. 6.2.1.

[157] See Sect. 6.2.2 and Fig. 6.22, and—see Koos, Heinkel, p. 49.

	you wanted to double their speed. This was the only specification I was given'.[158]
August 1937	Helmut Schelp joined the RLM Technical Department's section LC1, their short-lived pure-research arm. Neither LC1 nor DVL shared his enthusiasm for the jet engine, so without any backing nearly one year passed without any real turbojet progress. But when the RLM was re-organised in 1938, he found himself in the LC8 division for *special propulsion engines* development. Here he found an ally in the assertive Hans Mauch, in charge of rocket and pulsejet development within LC8 since 15 April 1938.[159]
11 Nov. 1937	Dr.-Ing. Hermann Wurster (1907–1985) flew his modified Me 109 with a tuned, 1660 hp DB 601 injection engine along the straight railway track Augsburg–Buchloe to a top speed of 610.95 km/h, and thus breaking the foregoing speed record of Howard Hughes of 567.1 km/h of 13 September 1935.
15 April 1938	Hans Mauch (1906–1984) takes over the *Sondertriebwerke* (Rocket development office) within RLM *Development Department*.[160] He expands the charter of his office and starts a massive jet development project under Helmut Schelp, who joined Mauch's group in August 1938, becoming responsible for turbojet development co-ordination. Mauch spurns Heinkel and Junkers, concentrating only on the *big four* engine companies BMW, Bramo, Daimler-Benz and Jumo Junkers Flugmotoren. Mauch and Schelp visit all four over the next few months, and find them initially lukewarm, interested only in the turbojet engine concept.
9–13 June 1938	The Heinkel He178 V1 turbojet aircraft is tested in the large AVA wind tunnel (7 × 4.7 m nozzle) at Goettingen under extreme secrecy; test pilot Erich Warsitz familiarises himself with the aircraft characteristics in the low speed range up to 32 m/s.[161] The completed testbed is now awaiting an engine.
Nov. 1938	A small team at BMW Berlin-Spandau led by Hermann Oestrich (1903–1973) builds and flies a simple piston-engine powered *motor jet* (BMW project P.3301); at that time Oestrich believed that the

---

[158] See Chap. 7. On Fri 23 July 1937, Oberstdivisionaer Hans Bandi, head of the Swiss Air Force, opened the event with a gala reception at the Zurich Hotel Baur au Lac and, Luftwaffe General Erhard Milch at his side.

[159] See Chap. 7.

[160] See Sect. 6.2.3; formally, RLM *Research* and *Development* were closely connected, both being divisions of the Technical Office; in fact, however, the connection was rather loose.

[161] See Koos, Heinkel, p. 30.

efficiency of a gas turbine would be too low to be practical. Before the end of 1938 and after the first encouraging RLM contacts on turbojet propulsion, however, the programme was broadened to include theoretical studies of gas turbines and actual experiments with corresponding combustion systems. A counter-rotation turbojet engine had been integrated on intervention of Hans Mauch as project P.3304, in spring 1939. This *statorless* P.3304 CR-project is linked to the names of Hellmut Weinrich and Fritz Heppner, a Jewish emigrant from Germany, who got this concept patented in the UK in 1941. In parallel, a rather conventional, single rotation (SR) turbojet engine (BMW Project P.3302) with 6-, later 7-stage axial compressor was launched as supplementary test vehicle. The CR concept was actively prosecuted as BMW 109-002 turbojet engine up to prototype testing in 1942, until it was abandoned and the SR-concept became the *BMW 109-003*, with first test runs during August 1940, developed and produced up to War's end in a small scale series production for *Arado Ar 234* bomber and *Heinkel He 162* fighter aircraft.[162]

Dec. 1938– Jan. 1939	Schelp's RLM technical officer colleague (also qualified as *Flugbaumeister*) Hans M. Antz (1909–1981)[163] issued a contract to *Messerschmitt* with Schelp's engine figures of *thrust-specific weight and dimensions* as a kind of specification, but without prescribing the actual aircraft configuration,—just the basic demand for 1 h flight endurance and 850 km/h flight speed was set. In due course, Messerschmitt starts the preliminary design of a twin-engine fighter under the direction of Woldemar Voigt (1907–1980), becoming *Messerschmitt Me 262*, the world's first turbojet fighter aircraft in series production; as a follow-on three Me 262 prototype aircraft were ordered on 1st March 1940.
Late autumn 1938	RLM/ Industry discussions about a turbojet development project based on a H. Schelp/H. Wagner script of an all-axial engine concept with afterburner—with A. Franz, M.A. Mueller and O. Mader from Junkers attending.[164]
Spring 1939	RLM/Industry contract for turbojet developments with a) Junkers Flugmotoren (Jumo) Dessau, represented by Otto Mader (1880–1944), Jumo director, Dessau and Anselm Franz as responsible

---

[162] See Sect. 6.2.4.

[163] See Sect. 6.2.3; like Schelp, Antz had also been a *Flugbaumeister* exchange student to the MIT Cambridge, MA and GALCIT Pasadena, CA, between mid-1933 to mid-1935, obtaining his Master's degree at the California Institute of Technology in Dec. 1936. See Kay, Turbojet, Vol. 1, p. 175.

[164] See Sect. 6.2.3, and Figs. 6.25 and 6.26.

project leader of what later became the Jumo 109-004 turbojet engine, and b) BMW Flugmotoren, Berlin-Spandau, represented by Bruno W. Bruckmann (1902–1997), the engineering manager, and Hermann Oestrich as project leader of what became the BMW 109-003. Based on a suggestion from H. Schelp the partners agreed to use 6-, (7-) and 8-stage axial compressors from AVA Goettingen, designed and tested by Walter Encke; first experimental axial engine tests started at BMW in August 1940, followed by A. Franz's prototype testing of Jumo 109-004 A on 11 October 1940.

26 Apr. 1939   Fritz Wendel (1915–1975), Messerschmitt chief pilot and world record holder up to 1969, flew the then fastest aircraft with piston engine *Messerschmitt Me 209 V-1* along a 3 km railway track section Augsburg - Buchloe, achieving a horizontal speed of 755.14 km/h, powered by a 12 cyl. 33.6 l DB601 Re V of 2770 hp.

June 1939   The Heinkel *HeS F3b flight quality engine*, the first truly usable jet engine, is tested in several 1 h and 10 h runs, achieving 500 kp thrust, until the engine-aircraft integration with the He178 V1 was carried out end of June 1939. On 3 July 1939 the fully equipped He178 was part of an advanced aero technology presentation at the Rechlin Luftwaffe test site—with a disinterested Adolf Hitler attending.[165]

27 Aug. 1939   First flight of a turbojet-powered *Heinkel He178*, piloted by Erich Warsitz,[166] with Hans von Ohain's centrifugal engine *HeS 3b* in the early hours of a sunny Sunday morning at the Heinkel test site in Rostock-Marienehe, thus ushering in the jet era, five days before Germany began the Second World War. Officially the event was noted—rather half-heartedly[167]—by a visiting RLM delegation in Rostock-Marienehe on 1 November 1939 by repeating the He178 demonstration flight in front of Luftwaffe Air Inspector General Erhard Milch (1892–1972), Director-General of Equipment for the Luftwaffe

---

[165] See Koos, Heinkel, p. 55. Later, Ohain found out from AH's adjutant, *'it was too early in the morning'*.

[166] Erich Warsitz (1906–1983), at that time chief test pilot at the Luftwaffe test site at Peenemuende-West had also flown the first rocket-powered aircraft He176 from there on 20 June 1939 with a Walter liquid-fuelled (hydrogen peroxide monopropellant) rocket engine. See Wikipedia, 'Erich Warsitz' in English.

[167] Negative impressions of the visit and of the visitors' attitude towards advanced propulsion systems have been reported by Ernst Heinkel and Hans von Ohain. However, already before the visit, Heinkel received on 17 Oct. 1939 a preliminary construction order for the He 280 fighter aircrafts, in fact a competitive development to the equally ordered Messerschmitt *Project P.1065*, the beginning of the Me 262;—see Schabel, Die Illusion der Wunderwaffen, p. 54.

Ernst Udet (1896–1941) and Luftwaffe General-Engineer Roluf Lucht (1901–1945).

1 Sep. 1939   Max Adolf Mueller and his former Junkers engine team (18) re-started work at Heinkel Rostock-Marienehe on their axial-flow engine design, now known as the Heinkel HeS 30. First contacts dated back to spring 1939, when Mueller decided not to work directly for Otto Mader, head of Jumo engine department without intermediate protection from Herbert Wagner, head of Junkers aircraft development between February 1938–May 1939. Wagner had resigned after an internal management conflict, which well could have been connected with the failing high-altitude research aircraft project *Junkers EF 61* under his responsibility. Wagner officially left Junkers for Henschel Flugzeugbau, Berlin-Schoenefeld at the end of April 1940.[168]

2.2.3   The War Years 1940–1945

Switzerland and Swiss Abroad

25 Nov. 1940   Testing of the world's first locomotive gas turbine started at BBC Baden. The commercial arrangements demanded for the GT locomotive 92 t total weight, a maximum drive power of 1600 kW/ 2200 hp for the selected 6-axle configuration with four driven axles; 110 km/h maximum speed and a one-year operational test were required before SFR Swiss Federal Railways would accept the delivery. For BBC, the new GT locomotive was the highlight of the company's 50th anniversary celebrations on 29/ 30 September 1941, touring between Zurich-Baden-Olten, and back.[169]

9 Dec. 1940   BBC patent application CH221,503 for a *motor jet* - type aircraft propulsion unit with two (or four) propeller units, relatively lightweight gas generators and axial turbine drives distributed on the aircraft wings, and a central, multi-stage axial compressor with turbocharged piston engine drive in the main axis of the aircraft fuselage.[170]

1940   The development of a gas turbine with semi-closed cycle started at Sulzer Ltd Winterthur, CH, with their special interest in ship propulsion units. The advantages of this cycle were seen in the high peak- and part-load efficiencies, the high power concentration and the relatively

---

[168] See Sect. 6.2.1.

[169] See Eckardt, Gas Turbine Powerhouse, p.196 f.

[170] See Pfenninger, Die Gasturbinenabteilung, p. 685, and Eckardt, Gas Turbine Powerhouse, p. 209.

small intake and exhaust volume flows, especially beneficial for that space-limited application. Tests of a gas turbine plant with a conventional open cycle followed there in 1942.[171]

29–31 July 1941   Claude Seippel travelled to Lyon and Grenoble (then in the unoccupied *Vichy Zone* of France) where he met the chief engineers of C.E.M. Georges Darrieus and Paul Destival from *SOCEMA Société de Constructions et d'Équipements Mécaniques pour l'Aviation,* C.E.M.'s daughter company, to discuss and support aero engine projects, unknown to the German occupants.

14–19 Sep. 1941   A mixed Swiss university/ industry delegation consisting of Professor J. Ackeret (ETH Zurich), Cl. Seippel (BBC Baden), C. Keller (Escher Wyss, Zurich, participation unconfirmed) and the Ackeret co-workers P. de Haller and G. Daetwyler met on invitation of the RLM Research Department (Adolf Baeumker) the Professors Ludwig Prandtl, Albert Betz, Adolf Busemann and Otto Walchner at AVA Goettingen, thereby following up several foregoing visits from German delegations. Purpose was to discuss the possibility of delivering two of the Ackeret-designed, Guidonia-type Mach 4 high-speed wind tunnels from BBC Baden, CH—one for installation at the AVA site at Reyershausen near Goettingen and the other at *LFM Luftfahrt-Forschungsanstalt Muenchen,* an all-new aeronautical research and test establishment, under planning in Baeumker's responsibility. Most of the LFM research institutes and numerous engine test facilities were to be erected on the main site at Ottobrunn on the southern outskirts of Munich, except one huge high-speed wind tunnel in the Oetztal,[172] approximately 170 km south of LFM headquarters and 50 km west of Innsbruck. Presumably Walter Encke, responsible for the AVA axial compressor developments within the German turbojet programme, joined part of the obligatory AVA facilities tour, and the subsequent compressor expert discussions.

---

[171] See Eckardt, Gas Turbine Powerplant, p.290 f.

[172] The Oetztal wind tunnel was supposed to have an 8 m diameter nozzle with operation up to high subsonic Mach numbers at atmospheric conditions. This—then biggest wind tunnel in the world—was designed for 76 MW driving power for high-speed aerodynamic tests on complete aircraft models at significantly higher Re numbers than any other existing facility, on full-scale component parts of aircraft and on full-scale nacelles with operational propulsion systems. After the war, then in the French occupation zone in Austria, it was dismantled between December 1945 and June 1946 and moved to Modane-Avrieux in the French Alps, where the set-up was reconstructed and finished, and has been operational since 1952 as the *ONERA wind tunnel S1 MA.*

Great Britain/USA and British Abroad

30 May 1940– 18 March 1941	Internment of the German-Jewish family of Fritz Heppner on the Isle of Man; with some likelihood first contacts to the upper management of ASM which led to Heppner's firm employment at *ASM Armstrong-Siddeley Motors*, Coventry on 8 December 1941.[173]
June 1940	A *Committee on gas-turbines* set up by the American National Academy of Science, amongst its prominent members also Th. von Kármán, reported[174]—that the promise of gas turbines for ship propulsion was very good, that development should be begun at once and that ship GT operation at a turbine inlet temperature (TIT) of 800 °C might soon be possible, even without resort to turbine cooling. With a compressor and turbine of the same efficiencies as those of Brown Boveri's Neuchâtel GT plant, an increase in TIT from 650 °C to 800 °C would raise the overall efficiency from 17% to 26% or to about 2/3 of the efficiency of the much heavier and more expensive diesel engine. The report added, however, '*In its present state . . . the gas turbine engine could hardly be considered a feasible application to airplanes mainly because of the difficulty in complying with stringent weight requirements imposed by aeronautics*'. The gas turbines would provide only 0.078 hp/lb, Fig. 2.3, against little more than 1 hp/lb for conventional aircraft engines. The error was due largely to the fact that no member of the Committee had any knowledge of aircraft engine practise, either as regards lightweight construction or as regards the reduced durability, acceptable in this field.
September 1940	The *Tizard Mission*, officially the British Technical and Scientific Mission, under the lead of (Sir) Henry Tizard (see 'April 1930') travelled to the then still neutral United States during the *Battle of Britain* to convey a number of technical innovations to the USA in order to secure assistance in maintaining the war effort. The shared technology included *radar* (in particular the greatly improved cavity magnetron), the design for the VT variable-time proximity fuse, the first technical exposition of an airborne atomic weapon, written by expatriate German physicists Otto R. Frisch (nephew of Lise Meitner) and Rudolf Peierls in March 1940 at the University of Birmingham,—and details of Frank Whittle's jet engine.

---

[173] See Chap. 5.

[174] See U.S. Navy Dept., An Investigation.

26 March 1941	Fritz Heppner, ASM Coventry, files patent US2,360,130 on a *High speed propulsion plant,* the first design concept of a counter-rotating turbojet, in the meantime after his leave from Germany realised since 1938 as BMW 109-002. On 28 October 1941 he applied for what became US2,404,767 of a corresponding counter-rotating turboprop configuration, which Weinrich had offered in vain to RLM and *OKM Oberkommando der Marine* (Naval High Command) between 1936 and 1938, before Hans Mauch, RLM established Weinrich's working contact to Bramo/ BMW at Berlin-Spandau.[175]
15 May 1941	*Gloster E.28/39* maiden flight with Whittle W.1 centrifugal turbojet engine from RAF Cranwell at 19.45 h, piloted by Gloster's chief Flight Lt. Gerry Sayer (1905–1942) for 17 min, achieving a top speed of more than 560 km/h.[176]
3 October 1941	Strategic break at *ASM Armstrong Siddeley Motors*, Coventry in agreement with the *Ministry of Aircraft Production* to stop the manufacturing of piston engines in favour of turbojet engines. In due course Stewart S. Tresilian (1904–1962), ASM Chief Engineer and designer of legendary automobiles and their motors was replaced by Fritz/ *Fred* Heppner on 19 January 1942.[177]
Mid-August 1942	Based on a government initiative ASM starts the development of its first turbojet engine, of *simple design* and 1270 kp TO thrust, with design support from RAE for the 14-stage axial compressor (PR 5)—without Heppner's involvement. First test run of that engine *ASX* already on 22 April 1943, what became finally the commercially successful ASP *Python* turboprop of 3656 shp (+ 500 kp thrust), certified on 20 April 1945. As the result of an additional government order started in August 1943 the development of the *ASH* (H—Heppner) turbofan engine in the upper thrust class of 5000 kp, apparently based on Heppner's adaptation of a design concept related to Griffith's counter-rotating *C.R.1 engine*.[178]
2 October 1942	*Bell XP-59A*, later in 1943 designated as *P-59 Airacomet*, made its first US turbojet flight, powered by two *GE I-16* radial turbojet engines (later *J31*) of 725 kp TO thrust (derived from the *Whittle W.2B/23 engine*), piloted by Bell test pilot Robert Stanley at the controls from Muroc Army Air Field (today, Edwards Air Force Base) in California.

---

[175] See Chap. 5 and Sect. 8.2.4.

[176] See Sect. 8.2.1.

[177] See Chap. 5.

[178] See Sects. 8.2.2–8.2.4, and Figs. 8.27, 8.28 and 8.31.

After a visit to England, General Henry H. Arnold was so impressed by news of flight demonstrations of the *Gloster E.28/39 jet aircraft*, that he arranged for the *Power Jets W.1X turbojet engine* to be shipped by air to the USA, along with drawings for the more powerful *W.2B/23* engine, so that the U.S. could develop its own jet engine.

5 March 1943   The fifth prototype of the *Gloster F9/40 Meteor,* powered by two substituted *de Havilland Halford H.1* centrifugal *engines Goblin,* owing to problems with the intended *Whittle W.2 engines,* became the first British turbojet aircraft, airborne at RAF Cranwell and piloted by Michael Daunt. The first aircraft with Whittle engines flew on 12 June 1943 (later crashing during take-off on 27 April 1944). The first flight of a Meteor, powered with the all-axial *Metrovick F.2 engine,* based on a RAE design of H. Constant, took place on 13 November 1943. This prototype was lost in an accident on 4 January 1944, the cause believed to have been an engine compressor failure due to overspeed.[179]

27 April 1944   Anthony *Tony* F. Martindale (19xx–1959), RAE test pilot, RAF Wing Commander and later Chief Development Engineer of *Rolls-Royce Motor Dívísion*, accelerated his *Supermarine Spitfire PR Mk XI (EN409)* with laminar wing and *Rotol* variable pitch propeller from an altitude of ~8 km into a daring 45 deg. dive at Farnborough, up to estimated 975 km/h (606 m.p.h.), powered by a *Rolls-Royce Merlin 61* of ~1600 hp. At ~4 km flight height the propeller/ gearbox unit disintegrated and suddenly tail-loaded, the aircraft shot up to ~12 km (40,000 ft) altitude, where he slowly regained orientation and glided home, landing safely after a 32 km detour.

21 January 1944   Westinghouse's all-axial J30 (19A) engine with 550 kp TO thrust made its maiden flight under a Vought FG-1 Corsair; the US Navy was impressed by the progress made by Westinghouse, unaided by the British, and considered that the J30/ 19A engine was superior in many respects to British engines (but said nothing about engine life).[180]

January–May 1945   Fritz Heppner (1904–1982) at his final professional stage at *Rolls-Royce, Derby*—before being life-long hospitalised with *schizophre-nia* in August 1945; his move from ASM to RR presumably had been supported by Alan A. Griffith, who had joined Rolls-Royce

---

[179] See Sect. 8.2.2.
[180] See Kay, Turbojet, Vol. 2, p. 106.

from RAE in 1939, and was working there until 1960, when he retired from his post as the company's Chief Scientist.

7 November 1945   Hugh J. *Willie* Wilson (1908–1990), test pilot and RAF Group Captain flew a *Gloster Meteor Britannia F 4* at Herne Bay, UK to the first, officially registered FAI record of a turbojet-powered aircraft with a horizontal speed of 975,675 km/h (606.26 m.p.h.), powered by two *Rolls-Royce Derwent Mk V* radial turbojet engines of 1590 kp take-off thrust.

Germany and Germans Abroad

11 October 1940   The first prototype of *Junkers Jumo 109–004 A* with 8-stage axial compressor (AVA), six separate combustion chambers and a single-stage turbine, was tested at Junkers Dessau, still without an exhaust nozzle, giving 430 kp TO thrust at 9000 rpm by the end of January 1941.

30 March 1941   *Heinkel He 280 V-2*, first demonstration flight (3 min) of the world's first 5 April 1941 turbojet-powered fighter aircraft by Heinkel test pilot Fritz Schäfer at Rostock-Marienehe, and successfully repeated on 5 April, piloted by Paul Bader from the Luftwaffe test site Rechlin. This was a precondition for the promised acquisition of engine manufacturer *Hirth Motoren GmbH* at Stuttgart-Zuffenhausen by Heinkel, which was formally accomplished still during April 1941. The axial/centrifugal turbojet engine *HeS 8A*, designed by Hans von Ohain, achieved only 500 kp thrust compared to the specified 700 kp. Further problems with *HeS 8* development caused a switch in the propulsion system to gain time for high speed testing of the fuselage. Consequently, the first *He 280-V1* prototype was re-equipped with *Argus pulsejets* and towed aloft for transfer from the Luftwaffe test base Rechlin to the Laerz airfield nearby on 13 January 1943.[181] Bad weather hampered visibility and presumably also caused aircraft icing to block controls, so Argus test pilot Rudolf Schenk became the first person to put an ejection seat to use in 2000 m altitude. In due course, the *Jumo 004* turbojet was installed. However, the corresponding *He 280* flights showed speed deficits, general weaknesses of the fuselage and limits in the tank volume, so that the

---

[181] Several Internet sources in English claim erroneously the 13 Jan. 1942 for the first successful ejection seat usage in He 280.

	He 280 programme was stopped—not the least by EHF's own request—on 27 March 1943 by Luftwaffe General Erhard Milch.[182]
May 1941– January 1942	Weinrich's BMW 109-002 (P.3304) counter-rotating turbojet engine, foreseen to power the *Lippisch Li P.01-115* (a forerunner of the *Me163* high-speed interceptor aircraft) in 1942, was apparently never tested in full, but a corresponding test rig was prepared.[183] Instead a 100 hp marine unit was used to demonstrate the general feasibility of the CR concept; after news of a failure of that rig spread during running combustor tests for BMW 002, the end of project P.3304 came in sight; information about the actual project end vary, as indicated, between May 1941 and January 1942.[184]
April 1941	*BBC Brown, Boveri & Cie.* Mannheim founded their *Dept. TLUK/ Ve*[185] under Hermann Reuter (30), mainly to carry out RLM initiated design activities to replace the AVA compressor design by a degree of reaction R 0.5 modification in the *Hermso I–III* studies, applicable for *BMW 003C* and *003D* turbojet engines, providing superior 7- and 10-stage axial compressors with top performance.[186]
2 October 1941	Heini Dittmar (1911–1960), flight captain, Messerschmitt test pilot and glider world champion 1937, flew the rocket aircraft Messerschmitt *Me163A V-4* at the Luftwaffe test site Peenemuende-West, achieving a new (unofficial) speed record in horizontal flight of 1003.67 km/h, powered by a *Walter rocket engine HWK RII 203* of 7.35 kN thrust.
18 July 1942	First (all-jet) flight of third prototype of *Messerschmitt Me 262 V3* fighter aircraft of 12 min length, achieving 600 km/h, from Leipheim airstrip, piloted by Messerschmitt chief test pilot Fritz Wendel. The power provided two *Junkers Jumo 004 A-0* turbojet engines of 840 kp take-off thrust. In due course Messerschmitt received an order to increase the pre-production series from 15 to 30 units. First flight of Me 262 V2 followed on 1 October 1942, and this month saw also the first successful 50 h demonstration run.[187]
25 May 1943	Decision on the mass production of Jumo 004B jet engine in a RLM meeting of Luftwaffe Air Inspector General Erhard Milch and his

---

[182] See Sect. 8.1.3 and Fig. 8.12, and—see Koos, Heinkel Raketen- und Strahlflugzeuge, p. 150.

[183] The shipment papers of that CR rig from BMW Spandau to Weinrich's shop at Chemnitz carry a 1942 date.

[184] See Chap. 5, and—see Masters, German Jet Genesis, p. 111.

[185] TLUK/Ve, standing for '*Turbine-Luftfahrt-Konstruktion/ Verdichter* (Turbine-aviation-construction/compressor)'.

[186] See Chap. 9 and Fig. 9.4.

[187] See Sect. 8.1.1 and—see Kay, German Jet Engine, p. 60.

leading department heads. In addition, the meeting decided which type of fuel was to be used in the turbojet. Following a proposal from Helmut Schelp, it was agreed to switch two of the hydrogenation plants reserved for the Luftwaffe to diesel fuel, so that the powerful new jet also entered into service with a safer fuel.[188]

Sep. 1943   After leaving RLM in late 1939, Hans A. Mauch as *assault patrol leader* at *Volkswagen* got involved in the task of accelerating the V-1 *buzz bomb* production since May 1941. In September 1943 Mauch and his *Arbeitsstab* (task force) *FZG 76* (*Flak-Ziel-Gerät,* V-1 code name) was put in charge of the production coordination. Mauch's task force had 70 members and met regularly once per month in the large RLM conference room. The shooting against London from 64 starting catapults, which had been erected in Northern France between Calais and Rouen began one week after the Allied landing in the Normandy in the night of 12/13 June 1944. On 20 July 1944 Hans Mauch received the *Knight Cross to the Merit Cross with Swords* from the German Air Ministry, then the highest non-political decoration for civilians.[189]

1 January 1944   Capture of former Heinkel (dept. von Ohain) employee, now Wehrmacht soldier Franz Warmbrünn near Vitebsk, Belarus, provided the Soviet Union with first time, immediate insight to the German turbojet engine development work, not only at Heinkel-Hirth, but also about the Messerschmitt jet aircraft activities.

5 June 1944   Feldwebel (Sgt.) Karl-Heinz Herlitzius flied a Messerschmitt pre-series version *Me 262 S-2/V11* (with heavy serial canopy) in a 35 deg. dive over the Messerschmitt test site Leipheim, in the west of Augsburg up to a new unofficial record speed of 1004 km/h, powered by two *Jumo 004 B-1* engines of 900 kp take-off thrust.[190]

6 July 1944   Messerschmitt *Me 262 S-3/ V12,* equipped with lighter *racing canopy,* achieves, with unknown pilot, at Leipheim a horizontal speed of 998.5 km/h, powered by two modified *Jumo 004 C* engines with 1000 kp take-off thrust. On the same day at Lager Lechfeld, south of Augsburg, Heini Dittmar pushed his own speed record for a rocket aircraft *Messerschmitt Me 163B V-18* in horizontal flight up to 1130 km/h, powered by a *Walter rocket engine HWK109-509* of 1600 kp thrust.[191]

---

[188] See Sect. 8.1.1, and—see Budrass, Review, p. 184.

[189] See Sect. 7.4.

[190] See Messerschmitt Test Report No. 262 08 L44, dated 28 June 1944; copy, courtesy of Hans J. Ebert, Munich, April 2016.

[191] See Green, Warplanes, p. 625.

26 July 1944	A first, documented air fight of a *Me 262 A-1a S-12* turbojet and a piston powered *DH.98 Mosquito* PR Mk16 reconnaissance aircraft (544 Sqn RAF) cruising in 28,000 ft over Munich, was reported by the Me-pilot Lt. Alfred *Bubi* Schreiber[192] as a successful kill of his adversary, while the British crew F/Lt. Albert E. *Bertie* Wall (pilot) and P/O Albert S. *Jock* Lobban (nav.) managed their—though damaged—near escape to the Italian target airfield Fermo near Ancona, where the Mosquito was presumably destroyed by crash landing.[193]
August 1944	Substantial series production of *Jumo 004 B-1* engine of 900 kp take-off thrust was taken up at the Junkers Koethen and Muldenstein plants, and with less emphasis at Berlin-Spandau, then Berlin-Zühlsdorf (near Oranienburg, 30 km north-west of Berlin, beginning October 1944), before a rudimentary production line was installed together with H. Oestrich's development team underground at Neu-Stassfurt for the *BMW 003 A-1* engine of 800 kp take-off thrust. Up to May 1945 the *Jumo 004* production achieved 6010 units, mainly to power the 1800 built Messerschmitt *Me 262* fighter aircraft, whereas the ~450 *BMW 003* engines were mostly installed on the *Arado Ar 234* bomber and the *Heinkel He 162 'Volksjäger'* fighter aircraft.[194]
7 August 1945	The *Nakajima Kikka Kai* (Orange blossom), Japan's first turbojet aircraft, flew one day after Hiroshima had been nuclear-bombed, from Kisarazu Naval Airfield (Tokyo Bay) for the first time, in a 20 min test, piloted by Lt. Cdr. Susuma Takaoka. The aircraft was constructed as a simplified Me 262 design, powered by two *Ishikawajima Ne-20* axial turbojet engines of 475 kp TO thrust, on the basis of one *BMW 109-003 A-0 GA* drawing.[195]

2.3   Gas Turbine/Turbojet Timeline for the Post-War Period 1946–1950

7 Nov. 1945	*Gloster Meteor Britannia F 4* aircraft with two turbojet engines *RR Derwent Mk V* (1590 kp TO thrust), piloted by Hugh J. *Willie* Wilson

---

[192] Lt. Schreiber (1923–1944) was credited for further four aerial victories, making him one of the first *jet aces* in history. On 26 Nov. 1944 he was killed in a crash landing at *Lager Lechfeld* airfield, flying the same Me 262 A-1a, *WNr.130,017*, when his wheels caught the lip of a slit trench.

[193] See https://aviation-safety.net/wikibase/177214, this source provides no hint to the final, otherwise reported crash landing. The aircraft had started at RAF Benson, Oxfordshire to a planned transfer recce flight to Fermo.

[194] See Sect. 9.2, and—see Kay, German Jet Engine, p. 119.

[195] See Kay, Turbojet, Vol. 2, p. 166 f. and—see Wikipedia, 'Ishikawajima Ne-20' in English.

	(1908–1990), test pilot and RAF Group Cpt., starting from Herne Bay UK, achieved the first FAI-registered world record of a turbojet-powered aircraft with a horizontal speed of 975,675 km/h.
1 April 1946	Claude Seippel succeeds Adolf Meyer as Director of the *Thermal Department* at BBC Baden, CH.
22 Oct. 1946	Deportation of approximately 650 engine and aircraft specialists from Junkers/ BMW, led by the Austrian chief engineer Ferdinand Brandner (1903–1986) Dessau and Stassfurt (in total approx. 2000)[196] to contribute to turbojet/ turboprop developments at the *N.D. Kuznetsov Design Bureau* at Uprawlentsheski near Kuybyshev[197]/Samara, some 860 km east of Moscow. Engineering activities resulted in a know-how transfer and continuous development of the *Jumo 022* turboprop engine towards the *NK-12* turboprop engine—with max. 15,000 shp, still the most powerful turboprop engine—and the *Jumo 012* turbojet to the *Pirna 014* with 3300 kp take-off thrust. Developed in East Germany in the 1950s for the *Baade 152*, 48–72 seater passenger jet, which was to be developed and constructed as pre-production aircraft at *VEB Flugzeugwerke Dresden*, it was abandoned in 1961 after a prototype crash, and the result was the halt of all aviation industry activities in East Germany in due course, so that a large extent of the engineering development teams flew to the West, just before the building of the *Berlin Wall* on 13 August 1961.
10 April 1947	Ivan E. Fedorov (1912 ~ 1987), test pilot and SU air force colonel, and Mark L. Gallai (1914–1998), TsAGI test pilot, engineer-writer and SU air force colonel, shared the success of making the first afterburner test flights up to a top speed of 950 km/h in a Lavochkin La-150 M fighter aircraft, powered by a *Klimov RD-10F* of 1240 kp take-off thrust, an SU derivative engine of the *Jumo 004B*, of which in total 1280 copies have been manufactured by the Klimov design bureau in the *Plant #26* at Ufa. On 24 June 1947, Fedorov improved the horizontal top speed to

---

[196]*Action Ossawakim* or *Operation Ossoawiachim* (russ. for *Society for the Advancement of Defence, Aviation and Chemistry*) was a secret operation of the *Soviet Military Administration in Germany (SMAD)* in the *Soviet Occupied Zone (SBZ)* and the Soviet sector of Berlin, beginning in the early morning hours of TUE 22 Oct. 1946, when more than 5000 selected German specialists (scientists, engineers and technicians), were deported—in most cases against their will—for a forced labour stay of 5+ years in the Soviet Union.

[197]The town of Kuybyshev, named between 1935 and 1990 after W.W. Kuybyshev (1888–1935), one of Stalin's mighty deputies, had a certain strategic importance between 1941 and 1943, when the town was chosen to be the alternative capital of the Soviet Union, should Moscow fall to the invading Germans. Government and military administrations, the complete diplomatic corps with 20 embassies, as well as the ensemble of the *Bolshoi Theatre* and numerous artists were moved there, outside of the reach of German bombers,—see Wikipedia, 'Samara' in English.

	1050 km/h, piloting a re-designed Lavochkin La-160 *Strelka* (Arrow), now with a 35 deg. swept wing with boundary layer fences (which later-on became standard in the *MiG-15*), but otherwise unchanged, powered by the *RD-10F*.

1947   British Motor Gun Boat *MGB 2009* was fitted with a Metrovick *Gatric* gas turbine, using the F.2/3 engine as gas generator, driving a new 4-stage power turbine. This was the world's first gas turbine powered naval vessel, another engineering first of the English-Swiss Karl Baumann.

15 August 1948   First flight from Hucknall of a Rolls-Royce *Avon*, RR's first all-axial turbojet engine, originally known as a A.A. Griffith design of 1945, *AJ 65* (Axial Jet, 6500 lbf), and nearly 15 years after the Air Ministry stated that the jet could not be a *'serious competitor to the airscrew engine combination'*.[198]

27 July 1949   The world's first commercial jet airliner, the *de Havilland DH 106 Comet (I)*, took off from the company's Hatfield Aerodrome in Hertfordshire. It featured an aerodynamically clean design with four de Havilland *Ghost 50* turbojet engines of 2240 kp TO thrust, buried in the wing roots, a pressurised cabin, and large square windows. For the era, it offered a relatively quiet, comfortable passenger cabin and was commercially promising at its debut in 1952. On 2 May, 1952, the *British Overseas Aircraft Corporation (BOAC)* began the world's first commercial jet service with the 44-seat *Comet 1A*, flying paying passengers from London to Johannesburg. The *Comet* was capable of traveling 770 km/h, a record speed at the time. However, the initial commercial service was short-lived, and due to a series of fatal crashes in 1953 and 1954, the entire fleet was grounded. Investigators eventually determined that the planes had experienced metal fatigue, resulting from the need to repeatedly pressurise and depressurise.

20 June 1951   Jean *Skip* Ziegler (1920–1953), Bell chief test pilot, achieved with the *Bell X-5* experimental aircraft, an adaptation of the unfinished Messerschmitt *P.1101* swept wing prototype aircraft a horizontal speed of 1134 km/h.

---

[198] See John Mortimer, 'Dr. Griffith and Sir Frank Whittle', IMechE Professional Engg. Newsletter, Readers Letters, 30 April 2012, and—see Wikipedia, 'Rolls-Royce Avon' in English.

Excursion I: Max Koenig (1893–1975) and Claude Seippel (1900–1986)

First signs of technical developments at Brown Boveri Switzerland which gained considerable importance for future axial turbojet engines could be observed in the early 1920s and will be discussed here in some detail at the example of two, very different key personalities in engineering. Both were BBC employees, but while Max Koenig's most remarkable contribution can be narrowed to just one day, the 24 April 1925—Claude Seippel, often called the *'Father of Axial Compressor'* influenced gas turbine engineering for more than four decades.

Max Koenig was born in Rumania, visited the schools at Canton Zurich and studied mechanical engineering at the Swiss Federal Technical University at Zurich (ETHZ) between 1912 and 1917. Then the diploma engineer worked for one year as Technical Assistant to Professor Aurel Stodola, ETHZ, until he moved on as Assistant to the Technical Directorate of Brown Boveri & Cie., Baden, CH. In 1921 he became Chief Assistant Engineer in the engineering staff of EEC English Electric Co. Ltd, London,[1]—however, still associated to his ETH Institute of Mechanical Engineering, led by his

---

[1] The English Electric Company Ltd was a British industrial manufacturer formed after World War I by amalgamating five businesses which, during the war, had been making munitions, armaments and aeroplanes. It initially specialised in industrial electric motors and transformers, railway locomotives and traction equipment, diesel motors and steam turbines. Two English Electric aircraft designs became landmarks in British aeronautical engineering; the *Canberra* and the *Lightning*. In 1960, English Electric Aircraft (40%) merged with Vickers (40%) and Bristol (20%) to form British Aircraft Corporation. In 1968 English Electric's operations were merged with General Electric Co.'s, the combined business employing more than 250,000 people.

D. Eckardt, *Jet Web*,
https://doi.org/10.1007/978-3-658-38531-6_3

doctoral thesis supervisor Professor Ernst Meissner.[2] Approximately at the time of his historic lecture about *'Gas Turbines'* on invitation of the *North-East Coast Institution of Engineers and Shipbuilders NECIES*, Newcastle-upon-Tyne on Friday 24 April 1925, Koenig was granted leave from EEC, so that he could finish his *'Dr.sc.techn.'* doctoral thesis at ETH Zurich. Entitled *'An approximate method for determining the vibration modes of profiled circular disks'*, his thesis with E. Meissner as first and A. Stodola as second supervisor could be presented in print in 1927.

Still in the same year the engineering office *Drs. Honnegger & Koenig* was started at Zurich, specialised in aluminium alloy/sheet metal construction. From 1932 to 1963 the *'Dr. Ing. Koenig AG'*—now run by Max Koenig alone—had its offices at the renowned Zurich-Paradeplatz, from where Max Koenig steadily developed his business from sales representation and commodity import towards a modern sheet-metal mounting and fixation service centre. At the end of the 1950s a special highlight was the conversion of the alpine-going yellow fleet of *'PostBus Switzerland'* from steel to Al-sheet-metal car bodies.[3] Besides his professional activities, Max Koenig led for 20 years as President the *TGZ Technische Gesellschaft Zürich* (Technical Society Zurich), founded in 1825 to support the technical education at Zurich and to inform about advanced technical developments abroad, in historical perspective mostly from England. The TGZ and its activities contributed also considerably and laid ground for the foundation of the ETH Zurich (Swiss Federal Technical University) in 1912 and its forerunner, the *'Polytechnikum'* since 1855.

Similar to TGZ, but some 60 years younger, the North-East Coast Institution of Engineers and Shipbuilders NECIES held its inaugural meeting at Newcastle-on-Tyne on 28 November, 1884, and from then[4] had been one of the most active bodies concerned with the advancement of the sciences of engineering and shipbuilding not only in Great Britain, but also throughout the world. Its main aims were to organise the discussion and publica-tion of technical papers and to facilitate the exchange of ideas and information among engineers and shipbuilders. The need for such an organisation was widely felt and by the end of its first session the Institution had already 454 members. The Institution's facilities soon included a Library and in 1889 a Graduate Section was established. Shortly after its establishment, the Institution rented offices opposite Newcastle Cathedral, before in 1910

---

[2] Prof. Dr. Ernst Meissner (1883–1939) was a Swiss mathematician, who was trained at the Canton School Aarau by the same math teacher—Prof. Dr. Heinrich Ganter (1848–1915)—as Albert Einstein (1879-1955), who became Meissner's colleague at ETHZ for a short while in 1912/1913. In 1926 Meissner organised the second ICAM Intern. Congress of Applied Mechanics, Zurich—after the first ICAM 1924 at Delft, NL which played a salient role in the early scientific life of A.A. Griffith and his subsequent engagement in British turbojet engine developments,—see Sect. 4.2.1.

[3] See Wikipedia, 'PostBus Switzerland'in English. In the mid-1960s the company moved from Zurich to Dietikon, AG and changed the name to KVT—Koenig Verbindungstechnik AG, https://www.kvt-fastening.ch/en/company/history/

[4] Up to 1992, when NECIES was transformed to the North East Coast Engineering Trust.

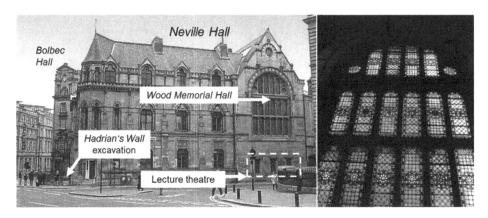

Fig. 3.1 Site of M. Koenig's lecture on 24 April 1925 at Newcastle: *Neville Hall* and *Bolbec Hall*, Westgate Rd./Orchard St. (l), *Wood Memorial Hall* stain glass window (r)

the Institution moved to more spacious premises in *Bolbec Hall*,[5] Westgate Road, in the immediate neighbourhood to the *North of England Institute of Mining and Mechanical Engineers*,[6] in short *'The Mining Institute'*, founded in 1852 and residing in the connected *Neville Hall*, Fig. 3.1.

This is in fact historical ground: along the east-west stretching Westgate Road, on the left of the Figure, run since 122 AD *Hadrian's Wall*, of which a short section has been excavated under the pavement outside Neville Hall. In addition, the Figure illustrates the large stain glass window of the Wood Memorial Hall (Library), covering the western façade of Neville Hall and below, the location of the lecture theatre where with all likelihood[7] Max Koenig presented his *'Gas Turbines'* paper to a NECIES audience on Friday, 24 April 1925, Fig. 3.2.

The Figure shows also the self-confident, then presumably 27-years old Swiss diploma engineer at the beginning of his BBC career.[8]

---

[5] Named after the Bolbec family, coming originally from Bolbec, Normandy, 25 km east of Le Havre, north of the mouth of river Seine. The Newcastle branch of the family owned land at Hexam, some 35 km in the west of Bolbec Hall.

[6] See Wikipedia, 'North of England Institute of Mining and Mechanical Engineers'; NEIMME—the Mining Institute developed one of the largest collections of mining information in the world. Its library, named after the first President Nicholas Wood contains more than twenty thousand volumes of technical literature. In 1902 the lecture theatre was opened, modelled on that at the Royal Institution in London. It features a steep rake of seating constructed from Cuban mahogany and the walls display portraits of all the Mining Institute's Presidents since 1852.

[7] The Neville lecture theatre has a capacity of 200. In addition, the *Lit & Phil* building, between Neville and Bolbec Hall, had its own lecture theatre of 500 seats between 1860-1960, but—as checked—it had not been leased to the 'Shipbuilders' in 1925.

[8] Thanks to Norbert Lang, former head of BBC/ABB Archives, who remembered on 1 June 2018 a blue, linen-clad photo album of the early 1920 with BBC key personnel, 4 cm thick. The album was found on 6 June 2018 by the present Archive head Tobias Wildi, containing the picture of <König TF

Fig. 3.2 Max Koenig, ~1920, and *Neville Hall* Lecturing Theatre for the *North of England Mining Institute*, Newcastle, 1902

Though many had used and thought about using gas turbines before 1925, few considered this type of powerplant suitable for aircraft propulsion. Stern in 1920 had examined the possibility of employing aero gas turbines—with negative conclusions. Maxime Guillaume deserves the credit for the first aero gas turbine patent for reactive jet propulsion in 1921,[9] while Brown Boveri's Max Koenig was obviously the first who put future developments in an optimistic, though cautious wording; his paper concludes on p. 397 of the Transactions with the rather prophetic lines: *'The gas turbine has, furthermore, already established itself in the field of aerial navigation as an exhaust turbine. In time to come, it may possibly replace the piston engine for this purpose entirely, being certainly at a much greater advantage with regard to weight.'*[10]

Dir> (Directorate of Turbine Factory); the album cover was black. Koenig's Newcastle presentation is meticulously documented in the Transactions of the North-East Coast Institute of Engineers and Shipbuilders, Vol. XLI—1924/1925 (41 years after NECIES foundation), pp. 347–399 with 16 additional pages of discussions attached.

[9] See Chap. 2, 'Survey' for <1921>; FR534,801 *'Propulseur par réaction sur l'air'* from 3 May 1921. Guillaume expresses the new reactive principle as *'locomotion aérienne'* and continues by describing the apparatus lifting off vertically(!) and *'as slow as the pilot wishes'* before it tends to become horizontal under the effects of a pitch elevator and carrying wing surfaces to take up speed.

[10] Though Koenig's brave assessment is remarkable for that time, given his youth and in view of an early contribution from Brown Boveri to the area of aero propulsion, there is another, surprisingly early recommendation for aero turbomachinery applications,—see Armstrong, Farnborough, p. 18. The visionary comments came from Jeremiah Head (1835–1899), former IMechE president; in Oct. 1888 in discussing a presentation by Charles Parsons, he made the far-sighted suggestion that the high-speed, smooth running characteristics of a turbine drive might in the future *'. . . render it suitable for aerial navigation . . .'* in which application *'. . . burning petroleum instead of coal . . .'* would be appropriate, *'. . . with direct internal combustion to create hot gas from atmospheric air, rather than using steam.'*

Two RAeS papers of 1976[11] helped to preserve the memory of Max Koenig; Frank Armstrong, one of the authors, wrote in a mail on 8 Nov. 2018 that the last comment '... *shows that Koenig recognises the particular attraction of low weight in an aircraft engine, provided that the challenges of achieving a practical and efficient gas turbine can be overcome. It is clear from the paper that Koenig also understands both the nature and magnitude of these challenges—in fact the paper presents a very good survey of the gas turbine situation as it was in 1925, covering both the theoretical thermodynamic and the practical engineering aspects. He expresses very well his balanced view in the last paragraph of his introductory abstract of the paper, saying* "The author does not in any way share the undue optimism of some engineers on the future prospects of the gas turbine", *but* "On the other hand, he is equally opposed to those sceptics who state that the gas turbine is impossible". *However, Koenig's 1925 paper came perhaps a little too early for him to identify a specific approach to gaining the big improvement in the efficiency of aerodynamic compressors which he (and others) had shown to be needed for the gas turbine to become a serious competitor in the heat engine field.'*

Scant proposals of gas turbines as primary aircraft power plants after WW I were mostly turned down by a common reaction similar to the one, expressed in a now famous report by W. J. Stern of the British's Air Ministry's South Kensington laboratory and issued by the Aeronautical Research Committee in 1920.[12] The engine which Stern considered the best possible would use a 7-stage, PR 10 centrifugal compressor and a 2-stage Curtiss impulse turbine at 475 °C and a tip speed of 240 m/s. While Stern's component efficiencies were quite optimistic, in estimating engine weight and size, however, he was far off and simply followed conventional industrial practice. Thus Stern arrived at a total weight of 2700 kg for a 1000 hp engine or 0.167 hp/lb, Fig. 2.3. Since contemporary aircraft piston engines weighed only about 0.4 hp/lb and since the fuel consumption of his gas turbine would be twice as great, Stern concluded finally: *'In its present state of development, the internal combustion turbine is unsuitable for aircraft, on account of weight and fuel consumption, the main difficulty in the case of aircraft being the design of a light, compact and efficient compressor'.*[13]

An unexpected gas turbine proposal is shown in Koenig's paper on p. 363 for a locomotive application which at first look—besides the use of centrifugal turbomachinery, but with elements like regeneration and turbo-electric drive—shows a high similarity to BBC's first, actually-built GT locomotive in 1940. Koenig gives no details of origin, except

---

[11] See Howell, Griffith's Early Ideas, which lists Koenig's 'Gas Turbines' paper as Ref. 11 and—see Armstrong, The aero engine.

[12] See Chap. 2, 'Survey', <Sep. 1920>.

[13] It is remarkable that not only Stern and Buckingham were this far off in the early 1920s, but that some 20 years later, in 1940, industrial-turbine engineers in the USA were making estimates of the minimum relative weight of gas turbines which were twice as high as Stern's, despite all the progress in basic knowledge which had intervened; the U.S. turbojet developments as far as relevant in this context will be addressed in Sect. 6.4.

the usage of a '*N.E.R.* (North Eastern Railway, UK) *wheel arrangement and frame*'. The main feature of this proposal with which he apparently was confronted as EEC expert, is a partially sub-atmospheric cycle. A multi-stage centrifugal compressor driven directly by a 2-stage axial turbine supplies air to the combustion chamber at approximately 3.2 bar, and after combustion at constant pressure, the gases expand to ~0.2 bar (abs.), from where the exhaust gases being re-compressed up to atmospheric pressure by a centrifugal exhauster on the same shaft. Steam is raised by the cooling of these exhaust gases, admitted in a separate nozzle group of the gas turbine and then condensed in a regenerator.

By far the greatest value of Koenig's lecture, expressed by broad appreciation in the discussion contributions, was the most recent information including detailed photographs of the assembly process of the '*Explosion type turbine*', in short also named after its inventor, the '*Holzwarth gas turbine*'. The difficulty to design and build an aerodynamic satisfying i.e. axial compressor for large volume flows and in due course an efficient gas turbine was insurmountable during the first decades of the twentieth century. There were in principle three major hurdles: compressor and turbine (component) efficiencies were too low and the turbine blade materials set too low a limit on the gas temperatures. Consequently, an intermittently operating, discontinuous working process was selected, where the required pressure rise was achieved in a—for a short-time—hermetically sealed combustion chamber by self-ignition of an explosive atmosphere. First operational gas turbines according to this '*constant-volume*' Holzwarth-principle were introduced in the mid-1920s (after early beginnings in 1906) with steadily increasing output power up to 5 MW. In this turbine, the fuel (oil, blast furnace gas or pulverised coal) was fed to this closed combustion chamber filled with compressed air, and the exploding mixture caused the pressure to rise approximately 4½ times of its original value. The combustion chamber, nozzles, impeller, and blades were water-cooled. The power consumption of the compressor was only a fraction of that required for the combustion turbine and consequently, a poor efficiency of the compressor had no longer such a disastrous effect; only a small amount of excess air was necessary for combustion (since water was used for cooling) and the air needed only be compressed to just one quarter of the final explosion pressure. These straightforward advantages were, however, accompanied by considerable complication and increase in cost of the plant. BBC got involved in building several of those Holzwarth turbines, unit no. 2 as early as 1909–1913 and still, in a second approach in 1928. Brown Boveri proposed for it an improved two-chamber, two-stroke cycle. This unit was installed in a Thyssen steel plant at Hamborn, Germany, where it has been several years in operation with blast furnace gas.[14]

For a short while the Holzwarth gas turbine was more efficient than the continuously operated, '*constant-pressure*' combustion type in the 1920s and early 1930s, but it became hopelessly complex, especially after the axial compressor design breakthrough was

---

[14] For details—see Eckardt, Gas Turbine Powerhouse, 'The Holzwarth Gas Turbine', pp. 72–76.

achieved by BBC's Claude Seippel. This was especially true in comparison to the resulting, finally successful power generation gas turbine, which Brown Boveri brought to market in 1939, after its derivative all-axial forerunners—the Velox boiler and the Houdry refinery process—had been produced already as early as 1932.[15]

Edward W. Constant II, author of the ground-breaking *'Origins of Turbojet Revolution'* of 1980 appeared to formulate with his first sentence also the motto over Koenig's historic lecturing session on a late Friday afternoon on 24 April 1925 at Newcastle: *'Time, not reason, separates real from absurd. Nothing is so certain as what is, nothing quite so unsure as what might be. What today is, is yesterday's possibility, a selection out of what-might-have-been's.'* And continuing *'Presumptive anomaly occurs in technology, not when the conventional system fails in any absolute or objective sense, but when assumptions derived from science indicate either that under some future conditions the conventional system will fail or function badly or that a radically different system will do a much better job. No functional failure exists; an anomaly is presumed to exist; hence presumptive anomaly.'* And, rounding it off *'Because presumptive anomaly is science-based, it is most likely to be recognised initially, and in some cases, singularly, by those very close to the intellectual foundations of their respective fields. This circumstance explains why young outsiders, those who have been recently exposed to the leading edge of relevant science and who are not yet fully committed to the conventional technology, tend to be the instigators of presumptive-anomaly-based technological revolution.'*[16]

Does this scenario apply to young Koenig, in front of a phalanx of representatives of the *'conventional system'?* Out of a group of seven persons identifiable by name in the paper due to participating in discussion and correspondence, three are depicted in Fig. 3.3: Tom Westgarth (73), Gerald Stoney (62) and William Goudie (57), compared to Koenig (32) their average age (64) is just twice as old, certainly adorned by life-long experience, but also endowed by some sceptic conservatism and even acid criticism (Goudie).

Tom Westgarth, who was President of the North-East Coast Institution of Engineers and Shipbuilders during the session 1924/1925, was then also chairman of Messrs. Richardsons, Westgarth & Company,[17] marine engine builders including steam reciprocating, steam turbine and diesel engines with then head office at Hartlepool and additional works also at Middlesbrough and Sunderland; the total workforce was typically some 3500 (1914). In 1901, Tom Westgarth had toured American and Continental iron, steel and engineering works; upon his return, he warned that foreign competitors were gaining on British manufacturers. He called upon British workers to lose less time, take fewer holidays and to be more adaptable to changing conditions in order to ensure that indigenous industry remained competitive. Richardsons Westgarth had never built an

---

[15] See Eckardt, Gas Turbine Powerhouse, 'The Early BBC Gas Turbines', pp. 158–209.

[16] See Constant, The Origins—in the order of quotation, p. 1, p. 15 and p. 17.

[17] See Wikipedia, 'Richardsons Westgarth & Company' in English.

Fig. 3.3 In the audience of Max Koenig's lecture: Tom Westgarth (1852–1934) President NECIES (l), Prof. G. Gerald Stoney (1863–1942), Univ. of Manchester (m), Prof. William J. Goudie (1868–1945), Univ. of Glasgow (r)

engine for the Admiralty, and at the beginning of the First World War, orders were slack. Consequently, the Company wrote a letter to the government advertising its services, and war orders began from 1915. Between that time and the end of 1920, the firm engined 202 vessels, including 59 for the Admiralty, 57 for the Ministry of Shipping and 86 for the Mercantile Marine, with a total horsepower of 685,000. 51 ships were equipped with engines in 52 weeks in 1917 alone. Richardsons Westgarth built its first turbine engines during this period and also 28 turbines for generating electric power onshore. At the request of the Admiralty, Richardsons Westgarth opened a shell manufacturing plant at Middlesbrough in 1915. Tom Westgarth supervised the project, and eventually, four, six and eight inch shells were being produced at the rate of 1000 a week. Richardsons Westgarth produced the largest number of marine engines in Britain in 1920, with a total horsepower of 96,000. Worldwide, the company ranked sixth among marine engine builders, behind five American firms. However, the profitability of the marine engines business had declined substantially since the pre-War period.

The Company was exclusive licensee of Brown Boveri product in the UK, especially for BBC turbo-alternators since 1927, and for several multi-purpose Velox boiler deliveries up to 1940. From 1933 onwards several marines launched orders or at least started experimental tests, for example the British Admiralty who ordered Velox boilers.[18] In this respect it is thinkable that Koenig's lecturing at Newcastle was also part of a diplomatic mission between Meyer and Westgarth to strengthen ties between the BBC *mother house* and the influential British licensee.

---

[18] See Chap. 2., 'Survey', <Switzerland 1933> and Eckardt, Gas Turbine Powerhouse, pp. 164–166. Highlight of the cooperation is the delivery of the first all-axial 1.6 MW gas turbine set on British ground via Richardsons Westgarth for testing to RAE Farnborough in 1939, see Sect. 6.3.

Dr. George G. Stoney was a mechanical engineer, who stood in life-long connection with Sir Charles A. Parsons and C.A. Parsons & Co.; already his father George Johnstone Stoney,[19] a physicist, had been employed as an astronomical observer for Parsons's father, the Earl of Rosse, Ireland. In 1889, he founded C.A. Parsons & Co. together with Parsons, working initially as a fitter. After several patents together with Parsons on the reaction turbine, he became chief designer of the steam turbine department in 1895. During the period 1894–1897 the steam turbine propulsion for the S.Y. *'Turbinia'* was developed, and Stoney was one of the original crew, being the last survivor at the time of his death. In 1909 Stoney published his seminal paper *'The tension of metallic films deposited by electrolysis'*. In that paper he derived an expression for the curvature of a steel strip due to the stress in a metallic coating applied to one side of the strip. This expression, like all subsequent versions of what has become known as the *Stoney equation* is still in use today to designate e.g. the film stress in single crystal silicon wafers. In 1911 he was elected FRS Fellow of the Royal Society. In WW I, he was a member of the Admiralty board of invention and research, and on the anti-submarine scientific research committee. In 1917 he went to Manchester as Professor of Mechanical Engineering at the College of Technology, where he remained during the next nine years, before he returned at the invitation of Sir Charles Parsons as Director of Research of the famous *Heaton Works*, Newcastle. In private life Stoney was said being kindly and sympathetic, which is also reflected in his generous discussion of Koenig's lecture where he made at the end of his contribution the essential point*: 'One thing that must be remembered in connexion with gas turbines is that most materials creep when under stress at high temperatures and this may have serious effects in gas turbines'.*

Professor William J. Goudie, born in Kilmarnock, Scotland, was trained as a mechanical engineer before matriculating to study at the University of Glasgow in 1891. He graduated BSc in 1895, worked for a marine engineer's consultancy and went to London in 1907 as Assistant Professor and Reader at University College. Returning to Glasgow University he was James Watt Professor of Theory and Practice of Heat Engines (what became Mechanical Engineering) from 1921 until 1938. Goudie was a leading authority on steam engines, best known for his 1917 text book *'Steam Turbines'* with a much enlarged second edition of 1929, and in 1932 a completely rewritten edition of William Ripper's *'Steam Engine Theory and Practice'*, which was published originally in 1908.

In the mid-1920s Holzwarth had to defend his concept of an *'Explosion or constant-volume turbine'* against accusations of over-optimistic or even manipulated efficiency values which were indeed rather complicated to determine due to considerable heat losses in the water cooling jackets. Finally, his reputation was stabilised when he managed to get the renowned Professors A. Stodola and François L. Schuele (1860–1925, since 1901 Director of EMPA, Swiss Federal Material and Research Establishment, Zurich) involved

---

[19] G.J. Stoney introduced the name *'Electron'* for the elementary electric charge, discovered in 1897 by J.J. Thompson, who had called it originally *'a corpuscle'*.

in the complicated thermodynamic analysis of the engine. In 1927, Stodola had been called upon to test a 500 kW oil-fired Holzwarth gas turbine in the Thyssen shops at Muelheim/ Ruhr. From these results, Stodola predicted in comprehensive, paid studies that an optimally designed Holzwarth GT should have the potential for efficiencies up to 30 percent, a number fundamentally disputed by BBC.

In 1925, in preparation for his *Gas Turbines* lecture at Newcastle, Koenig alone was absolutely overcharged with this task. Expectably his thermodynamic analysis revealed flaws and lacked of convincing clarity, which Goudie discloses and attacks relentlessly in his attached, lengthy correspondence on pp. 407–415: *'I find . . . the opening section of the Paper, in which the Author proceeds to outline the thermodynamic theory of the possible types of internal combustion turbine* (constant-pressure vs. constant-volume) *somewhat disappointing. . . . Unless a writer on this subject clearly defines the efficiency terms he uses the inevitable result is confusion and mistake on the part of the readers.'* And so on, seven pages—but the trap is set . . .

As his most advanced turbomachinery example of the time, Koenig illustrated the state-of-the-art by a *'Diesel exhaust gas turbine of Brown, Boveri'* in his final Figs. 33 and 34. In 1923, Swiss Locomotive and Machine Works (SLM) were testing a two-stroke experimental diesel engine that needed bringing up to a higher power level with better fuel consumption. Brown Boveri recommended using an exhaust gas turbocharger that would feed into the scavenging blowers. It consisted of a 2-stage centrifugal compressor and a single-stage action turbine. The turbo group was already mounted on roller bearings—with a combined *'Radiax'* bearing to cope with the axial thrust. The resulting axial load was small though, since it was nearly completely compensated by a rotating balance piston on the compressor side. SLM subsequently placed an order for such a machine and in June 1924, turbocharger VT402, the world's first heavy duty exhaust gas turbocharger, left the Baden works of Brown Boveri, Fig. 3.4. The resulting power increase of the turbo-charged diesel engine was ad hoc nearly 40%.

Goudie comments this laconically *'As a waste-heat engine or exhaust gas turbine, used in conjunction with the* (aero) *petrol engine for supercharging, it is, of course, in quite a different category* (than the other land-based examples)*, and its successful adaptation for this purpose, through the enterprise of Prof. Rateau is now a matter of history.'* . . . (However,) *in his final summary the Author has stated the case very well for the land development of the gas turbine, but I think that he is too optimistic in the suggestion that the gas turbine may possibly replace the piston engine for aerial navigation. The very exhaustive investigation carried out by Stern for the Air Ministry has shown that, compared with the highly efficient and light high-speed petrol engine, the gas turbine alone is an* impracticable *proposition.'*

True, in 1925 Koenig offers here not much more than his gut instincts, but 20 years later he was confirmed right . . . while the Sterns and Goudies were plainly rebutted; not bad for a *'scientific rookie'.*

The two engineers which are spotlighted here have more in common than just the BBC employment. Presumably without knowledge from each other they were part of an

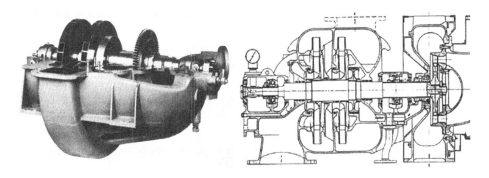

Fig. 3.4 BBC's first 2-stage turbocharger VT 402 for a 500 hp SLM diesel engine, 1924—the last slides in Koenig's presentation

emergency taskforce which asked for *'all hands on deck'* in Brown Boveri's economically most important steam turbine business. In 1921 serious events rocked the Company—a Brown Boveri turbine disc exploded at Copenhagen, at least four turbines were wrecked in the United States and a couple in Great Britain.[20] Disc vibrations were recognised to be the cause and Brown Boveri called on Stodola as the man, if any, who could deal with this situation. Under the action of periodical forces of various frequencies, a disc may vibrate in the axial direction, the rim assuming the shape of a wave and one or a number of nodal diameters being in neutral position. On a revolving disc periodical stresses may derive from stationary forces, fixed in space, which give an impulse revolving relatively to the disc. The wave deformation travels with respect to the disc and subjects it to fatigue.[21] On short notice Stodola, his and neighbouring ETH institutes like the one of Prof. Ernst Meissner organised an analysis and calculation taskforce, to which belonged for the first eight months of 1923 also the newly employed Claude Seippel and indirectly with his relevant doctoral thesis also Max Koenig up to 1927, when all related problems were finally solved.

Seippel describes the decisive project phase in his Parsons Lecture—with a lasting insight: *'There was a major interest in finding ways to calculate the resonant frequencies with various numbers of nodes, for any disc (shape), accounting for the influence of centrifugal forces which stiffen the disc and of heat or shrinkage stresses which depending on their distribution may raise or lower the frequency and were difficult or even impossible to determine by experiment: Stodola selected Rayleigh's method*[22] *(of dimensional analysis,* a conceptual tool used in engineering, physics and chemistry, expressing a functional relationship of some variables in the form of an exponential equation.)

---

[20] Further details could be found in *Engineering,* 20 May 1921.

[21] See Seippel, From Stodola to Modern Turbine Engineering. This Paper was delivered by Claude Seippel as the 17th Parsons Memorial Lecture before NECIES on 31 Oct. 1952 at Newcastle's Bolbec Hall, 27 years after Max Koenig's *Gas Turbines* Lecture.

[22] See Wikipedia, 'Rayleigh's method of dimensional analysis'.

Numerous discs were calculated with no other tool than a 2ft. slide rule and the Author was engaged under Stodola's direction to cover endless pages with figures. Some of the discs were checked by test at Brown Boveri's. The excellent coincidence of measured and calculated frequencies up to 5 or 6 nodal diameters caused no little surprise. Today such calculations would doubtless be performed on electric computers. Working numerically 'by hand' conferred, however, an insight into the problem and a sense of proportion which, after some practice enabled the work to be carried out very quickly. It is to be hoped that 'mechanised calculation' will not result in this advantage being lost.

I can remember Stodola exclaiming with joy over the success of Rayleigh's method, not its elegance—there is no algebraic solution—but its clearness, its call for assiduity which leads to port under all circumstances. 'Truly British' he called it.'

Claude P. Seippel had broad impact on a large variety of BBC engineering fields, but by far his greatest influence was in axial turbomachinery as brilliant scientist, engineer and inventor. Moreover, he had significant influence on the Swiss education scene and on public life in general. He was born on 14 June 1900 in Zurich. His father Paul Seippel (1858–1926), a renowned writer and journalist from Geneva, was professor for the French language and literature at the Swiss Federal Institute of Technology in Zurich (ETHZ) since 1898, so that Claude Seippel received certainly strong impressions of the literate and philosophical atmosphere at home. He went to the Canton school Zurich, studied at ETHZ and received his diploma in theoretical electrical engineering from Prof. Karl Kuhlmann (1877–1963), after passing the practical stage at the Ateliers des Charmilles S.A. in Geneva. On 1 July 1922 he started at the Brown Boveri transformer lab at Baden, CH, but was transferred effectively to the steam engine Dept. D already on 2 January 1923, Fig. 3.5 (l), to support—as described before—Prof. Stodola at ETHZ in solving disk vibration problems[23] which had been detected in the context of the severe Copenhagen steam plant damage. After accomplishing this task, he did not return immediately to BBC Baden, but left the Company on 31 August 1923 for what became a nearly 5-year stay in the United States[24] of which for his further training in the emerging axial compressor design technology the stay at Earl H. Sherbondy,[25] Consulting Engineers at Cleveland, Oh. from 1 December 1924 onward for the next 20 months had the most sustainable effect.

---

[23] Seippel referred to his role in this period in hindsight jokingly as *'Stodola's Rechenknecht'* (number cruncher), but the mathematical training helped him during his following stay in the USA e.g. to handle complex mechanical problems of automotive gear-trains by applying a set of Lagrange's equations in a competent manner.

[24] For all U.S. stations—see Eckardt, Gas Turbine Powerhouse, 'Claude Seippel', pp. 92–96.

[25] For Earl Sherbondy (1887–1958) and his role in U.S. turbocharger development during WW I—see Sect. 4.1.2 and Fig. 4.12 f. On 1 Dec.1924 Seippel started in Sherbondy's five-men-office at what *'was one of the most fruitful working periods of my life'*. The working conditions must have been rather weird; Seippel reports of his windowless office space—and of his colleague Glen Smith, a former professional catcher, now the responsible draftsman of the team, who *'tirait ses dessins avec*

Fig. 3.5 Dr. h.c. Claude P. Seippel (1900–1986), Brown Boveri & Cie. ~ 1923 (l), ~ 1955 (m), ~ 1970 (r) ©ETHZ (l), ABB

Main project at the Sherbondy office was a newly designed, hydrodynamic speed converter—along the Foettinger concept—to be tested in combination with an existing 300 hp marine motor of Sterling Engine Co., Buffalo. The test failed, presumably due to then unknown cavitation effects of the speed converter—and caused at the same time, the sudden death of Sherbondy's small engineering office. Before, the collapse however, Sherbondy issued the idea of an axial-flow compressor design to charge large automotive engines, unavailable since the (failed) trials of Parsons at the beginning of the century.[26] And, Sherbondy directed young Seippel to prepare a basic compressor design theory on the principles of aircraft wing lift and drag by organising for him the wind tunnel test results from Gustave Eiffel (1832–1923), and the corresponding *'NACA design tables'*.[27]

---

finesse et précision' (drew his designs with finesse and precision), as Claude Seippel describes the episode in his U.S. diary.

[26] The *'addiction'* of Sherbondy, then 38, to high-powered cars had a long history, dating back to his early youth as 16-year-old boy. During his education at the University School of Cleveland, he discovered his apparently natural talent for designing internal combustion engines; he made a single-cylinder two-cycle engine—with claimed 16½ hp, followed by his first 4 cyl. air-cooled 38 hp engine in 1905. In 1908 together with his elder brother they joined forces with A.F. May, owner of a local drug store, to form the short-lived *Derain Motor Co.* (1908–1911), known for a 7-seater luxury car with a 3.19 m wheelbase. The name *Derain* is ominous in this context, except for the French fauvist-painter André Derain (1880–1954), also called the painter of the *'premier jet'* and known for his exquisite collection of 14(!) high-powered Bugattis, which he considered—perhaps also in the sense of Sherbondy—*'to be more beautiful than any work of art'*,—see Wikipedia, 'André Derain' in English.

[27] See Eiffel, La résistance, and—see Munk, Elements. Seippel's reference to the *'NACA design tables'* is not easy to verify, since all standard NACA airfoil reports in this respect have apparently not been published before 1930 and were actually summarised in the famous NACA Report No. 460 in 1935 only, except for Max Munk's NACA-TR-191 from 1924, as used as reference here. On Munk's

From 1929 to 1940 he managed the steady step-by-step design, development and testing of high-performance, multi-stage axial compressors, from early single-stage efforts[28] to the evolution of the first 4 MW power generation gas turbine for the City of Neuchâtel, CH in 1939, and the foregoing compressor implementations for the first supersonic wind tunnel at Prof. Ackeret's Aerodynamic Institute at ETH Zurich as of 1934/5, see Sect. 4.1.4, as well as the axially turbo-charged Velox boilers and the refinery process equipment for the Houdry process of that time, see Sect. 6.4.3.[29]

Claude P. Seippel was a brilliant engineer, inventor, scientist and manager in one; even as Managing Director of Brown Boveri's Thermal Department, Fig. 3.5 (m), he took his time to analyse technical problems, solving numerical problems by breaking down complex mathematics into engineering-wise simplified sub-tasks. Seippel dedicated his whole professional career to power generation and turbomachinery design, but beginning already in the mid-1930s towards axial aero turbocharging and more and more also in support of British turbojet developments. During the Second World War he took several times the opportunity—even under the threat of the German occupation—to contact Brown Boveri's French daughter companies to discuss and prepare the possibilities of future Company aero engine activities after the war.[30]

Claude Seippel could not have played his role as 'wizard in engineering', and especially could not have maintained the addressed ties to the Allied countries in the area of axial turbojet development without consistent support and encouragement by his superior (and forerunner as Director of the BBC Thermal Department) Dr. h.c. Adolf Meyer (1880–1965).[31] Adolf Meyer's exceptional role for Brown Boveri is not only described by his responsible leadership for steam turbines as well as for the emerging gas turbine business in decisive years, but by his multi-facetted personality which combined the convincing salesman with the sensitive team-builder with broad personal experience as an outstanding engineer of international calibre. Twice he was honoured as doctor honoris causa: 1935 from the Stevens Institute of Technology in Hoboken, NJ, USA[32] and 1941 from his 'alma mater', the ETH Zurich. He held lifelong close contact to IMechE, the

---

background at Goettingen—see Sect. 4.1.3, and in the United States—see Fig. 4.39 in Sect. 4.3.3 and the corresponding entries to Edward P. Warner—and especially to Jerome C. Hunsacker.

[28] See Fig. 4.21 f. and Sect. 4.1.3.

[29] The decisive question of the optimum *Degree of Reaction* for axial compressor design has been discussed at length in the context of Fig. 4.48 f.

[30] Brown Boveri's aero engine collaboration with British institutions and here explicitly with Rolls-Royce will be mainly dealt with in Sect. 6.3, while the corresponding French affairs together with C.E.M. and SOCEMA will be discussed in Sect. 10.3.

[31] See Eckardt, Gas Turbine Powerhouse, p. 87 f.

[32] Exactly then or shortly thereafter, also the young German engineering student Helmut Schelp was at Stevens Institute of Technology, Hoboken NJ, who received there his Master's degree with recognition (best SAE paper) after a study year in June 1936, before he returned back to Germany and his new post at RLM, becoming within years the most influential single person on the German turbojet development programme,—see Chap. 7.

Institution of Mechanical Engineers in London, probably at best documented in the first comprehensive presentation of the BBC gas turbine development at London on 24 February 1939, including already the Neuenburg plant, the gas turbine-electric locomotive, GT-powered (destroyer) ship propulsion and an excursion about the benefit of a '*Combined gas turbine and steam plant*'.[33] He received the IMechE George Stephenson Medal for '*best achievement of the year*' in 1939 and 1943. He was Member of the ASME American Society of Mechanical Engineers since 1946, Hon. Member of the American Academy of Science and Arts and of the TGZ Technische Gesellschaft Zurich since 1950. Meyer and the Swiss-born, long-time Chief Engineer and—since 1927—Director of Metropolitan Vickers, Manchester, Karl Baumann (1884–1971) represented the basis of a reliable engineering exchange network between Brown Boveri and British institutions on axial compressor design and corresponding turbojet applications, as will be outlined in detail in Sect. 6.3.

Seippel's greatest impact on engineering came from his activities in almost every important field of turbomachinery. His inventions led to the granting of 38 major patents, and he wrote some 50 significant papers and fundamental scientific essays. In addition to his early patents on axial compressors/ turbocharging, the substantial implications of which on the German high-altitude piston engine developments since 1932 will be outlined in Sect. 4.3.3, *In the Patent Arena;* Seippel's patents covered gas turbine governing as well as pressure wave generators, which for a short while replaced under the brand name '*Comprex*' conventional automotive turbocharging in the 1970s.

In the second period of Seippel's career from 1941 to 1960[34] he continued development work on the gas turbine, but was also responsible for significant improvements in steam turbine blading design and for the preparation of the first successful gas/steam turbine combined cycle powerplants. This period became an especially productive one, when his patent for an exhaust turbocharger in 1942 led to a new market, which after the war grew to Brown Boveri's most important business. Nevertheless, his scientific mind was not preoccupied that he forgot the coming generations of young engineers, required for the development of a more peaceful, post-war society. The (later) motto '*Fordern und Fördern*' (Demanding and promoting) virtually guided his thinking already in 1953 when one of his publications outlined what industry offered, wanted and expected from the future

---

[33] The wording to imply a '*Combined Cycle Power-Plant*' is there, but thorough reading reveals that the GT-attached exhaust gas boiler was actually foreseen by Meyer still for steam production only. Carnot's early idea of a CCPP was apparently still not in reach—and had to wait until BBC realised the first large 75 MW CCPP at Korneuburg, A in 1961. Meyer's IMechE presentation attended also then RAF Squadron Leader Frank Whittle, delivering an interesting discussion contribution,—see Sect. 6.3.5.

[34] In that period Seippel discovered and announced also an important limitation of the second law of thermodynamics; the importance of his discovery was only recognised years later when his definition of '*exergy*' found general acceptance. His discovery grew steadily in importance over time in line with the engineering trend to energy conservation and environmental protection.

generation of engineering academics. The Swiss Federal Institute of Techology (ETHZ), which had appointed him to the board of trustees *('Schulrat')* already in 1947, elected him vice-chairman of the board in 1957 and awarded Claude Seippel its *'Doctor honoris causa'* in 1959 for his contributions in the field of turbomachinery. Finally in 1965, he was named Chairman of the Institute's board, and he served in that capacity for many years. The American Society of Mechanical Engineers elected Seippel an honorary member in 1982 and the U.S. National Academy of Engineering (NAE) co-opted him as *'foreign associate'* in 1987, Fig. 3.5 (r), with the lines: *'Dr. Seippel's many contributions to the development of thermal power have earned him a lasting place in the history of technological development. The world is a better place because of his brilliance, foresight, perseverance and dedication.'*[35]

---

[35] See Roe, NAE Memorial Tributes, pp. 311–314.

Turbomachinery and Aero-Related Activities at BBC and in International Context up to 1935

4

Why to start with BBC? In 1867 the German Werner von Siemens presented the first *'dynamo'* after having discovered the principle of electrodynamics. In 1879 Thomas A. Edison invented the light bulb, thus laying ground to create the powerful General Electric (1895). In 1891 Charles E. L. Brown succeeded in transmitting 220 kW of power the 175 km from Lauffen/Neckar to Frankfurt/Main, Germany. From this moment on, driving power no longer had to be generated and consumed at the same site. An electrical cable could now link the source of energy with the place at which it was utilised; centuries of limiting mechanical transmission equipment were over.

As will be clear in the following, many companies which started originally with the generation and distribution of electricity added later-on turbo-generators and turbomachinery in general, after steam also gas turbines, and in due course even aero gas turbines to their product portfolio: besides BBC Brown Boveri in Switzerland/Germany/France, there were GE General Electric and Westinghouse for the USA, MetroVick Metropolitan Vickers and BTH British Thomson-Houston—a GE subsidiary—in England, and Siemens, AEG Allgemeine Elektrizitäts-Gesellschaft for Germany on this path, just to name the most prominent. All this happened in a decade-long continuous preparation and technological maturation phase for which the company history of BBC is an early, well-documented example, based on an unique archival situation.

In quintessence the whole excursion refers to the ambiguous fact that advanced axial compressor designs became finally key to the successful and sustainable realisation of the aero gas turbine—before and after 1939/1940. All this happening on the one hand in great secrecy, reflecting the fact that jet-powered flight was considered as and actually played a salient role in the war machinery of both sides. However, in strange contrast to this later development stands the fact that BBC openly sold advanced axial compressor hardware unrestricted to everybody—as early as 1931. Given the later importance, this in parts covert, nearly 10-year-long dissemination period of axial compressor technology is rather astounding—not to mention todays short-term technology cycles.

4.1   Brown Boveri & Cie

4.1.1   The Founders and Their Works

The successful development of the Swiss machine industry started mainly as a spin-off of the then superior British mechanical industry to the Continent in the mid of the nineteenth century.[1] Johann Jakob Sulzer-Hirzel (1806–1883), one of the later-on famous Sulzer Bros., who had founded in 1834 together with their father the Sulzer foundry, recruited Charles Brown Sr. (1827–1905), Fig. 4.1 (l), a self-made engineer, native from Uxbridge (now a suburb in the west of London) in 1851 from England to Winterthur, Switzerland for the building of steam engines. Here he fulfilled this job until 1871, when the firm decided against entering the locomotive market, whereupon Brown Sr. left the Sulzer shop with then ~100 employees to set up SLM Schweizerische Lokomotiv- und Maschinenfabrik on an adjacent site, a company especially noted for its rack locomotives. The figure shows him in his function as SLM technical director in ~1875, besides the next generation of founders of Brown Boveri & Cie. in 1891, his son Charles E.L. Brown and Walter Boveri.

The *'System Brown'* indirect drive that was to be a feature of the smaller SLM locomotives for some thirty years, Fig. 4.2, was originally devised to facilitate the construction of tramway locomotives, but found also its application in the famous System Brown-SLM *Gotthardbahn*, 1882. These, being subject to regulations on overall width and complete enclosure of moving parts, set Brown something of a problem which he solved by placing the cylinders above the footplate and driving through a rocking lever, thus reducing the overall width relative to the gauge and putting all the vulnerable parts of the motion where they could be inspected and lubricated from inside the locomotive without lifting the covers over the wheels.[2]

In 1880 SLM commenced under Brown's direction the building of electric locomotives. In 1884–1886 he joined MFO Maschinenfabrik Oerlikon at Zurich, where he implemented an electric department which very soon was transferred in the responsibility of his then 24-year old son Charles E. L. Brown, while Brown Sr. restlessly moved on to Newcastle for the construction of an ammunition factory and to Pozzuoli near Naples where the upgrading of the marine shops fell in his responsibility. He retired officially in 1891, in the same year when his son together with Walter Boveri founded BBC Brown, Boveri & Cie.

---

[1] https://www.gracesguide.co.uk/Expatriate_British_Engineers_in_the_Industrial_Revolution   has the names of some 100 expatriate British engineers up to the 1850s, of which nearly 30% went to France, to USA 27% and Germany 10%; Charles Brown is the only one listed for Switzerland.

[2] Examples of Brown tramway locomotives are presented in Switzerland at the Lucerne Transport Museum and in the SBB Swiss Federal Railways shed at Glarus, Switzerland.

Fig. 4.1   Charles Brown Sr. (1827–1905) as Technical Director of SLM ~ 1875, Charles E. L. Brown (1863–1924) and Walter Boveri (1865–1924) as of ~1891, the year of BBC's foundation, left-to-right

at Baden, Switzerland. Charles Brown Sr. accompanied and supported the young company with all his experience up to his death in 1905 at Basle.

Charles Eugene Lancelot Brown (1863–1924) was born in Winterthur as one of six children; he attended the Engineering School at Winterthur, before he worked in his father's factory and became the director of the MFO electrical department in 1887. He designed—in parts together with his father—a variety of direct and alternating current machines for MF Oerlikon and was awarded a grand prize for a dynamo design at the Paris exhibition in 1889. A biographical sketch of Brown published in a technical periodical in 1891 described him as being *'one of the brightest and best known of the continental electricians [with] a reputation of international importance.'*

The practical development of the polyphase motor was realised during the year 1890. Simultaneously with Michail von Dolivo-Dobrowolski (1861–1919), AEG's chief designer, he originated the polyphase winding with rectangular coils in slots, a design practically unchanged till today. The year 1891 brought the opening of the International Electrical Exhibition at Frankfurt/M. and, beginning on 24 August 1891, the celebrated power transmission from Lauffen to Frankfurt over 175 km with three-phase current at 25,000 volts, this effectively demonstrating the technical possibility and commercial feasibility of the transmission of power over great distances by electricity. C.E.L. Brown designed for this power transmission the 40-pole generators and the first oil-cooled transformers. This demonstration provided convincing evidence of the economic and technical feasibility of supplying power generated at remote locations to industrial centres—and convinced the City of Frankfurt to adopt alternating current for its municipal power plant from BBC which began operation in 1894 and also influenced the adoption of alternating current at the large hydroelectric plant at Niagara Falls, New York in the following year. American versions of the Lauffen type alternator

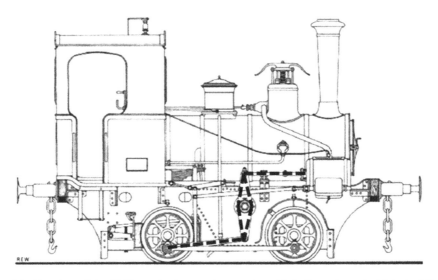

Fig. 4.2 Drawing of the SLM standard gauge locomotive with indirect drive (bold dashed line), derived from material in *'The Engineer'*, 31 May 1878

were introduced by General Electric in 1897, and later by the Westinghouse Company and Allis-Chalmers.[3]

On 2 October, 1891 the following entry was made in the commercial register of the Swiss Canton of Aargau: *'Charles E. L. Brown of Brighton, England and Walter Boveri of Bamberg, Germany, both resident in Baden, have established a limited partnership under the company name Brown, Boveri & Cie., Baden. The nature of the business: Fabrication of electrical machines'.* Brown Sr. and the two founders complemented each other perfectly: the Browns influenced the early technical developments and the German Walter Boveri (1865–1924) was the dynamic and visionary businessman, Fig. 4.1.

Switzerland offered the BBC founders what they needed to get off to an excellent start, for the Alps had huge, as yet untapped resources for hydroelectric power. The orders of the young company included equipment for power stations and subsequently electric railroads. In the electric railway field the Lugano tramway, installed in 1894, was the first traction system for which three-phase motors were used, and this was followed by many mountain and other polyphase railways, of which most are in successful operation at the present day. The *Jungfraubahn*, a railroad electrified by BBC, carried its first tourists to the base of the

[3] IEEE, the Institute of Electrical and Electronics Engineers, designated 1991 as the centennial year for the industrial use of alternating current power. Among the reasons for this selection was the successful and well-publicised transmission of polyphase power beginning 24 August, 1891 from Lauffen on River Neckar, Germany (90 km south-east of Heidelberg), to the site of the international electrical exhibition in Frankfurt/ Main over a distance of about 175 km. IEEE was formed in 1963; as of 2018 it is the world's largest association of technical professionals with more than 423,000 members in over 160 countries around the world.

Eiger Glacier in 1898 and by 1912 was extended all the way up to the *Jungfraujoch*, 3454 m above sea level. The locomotives were considered technological wonders of the world at the time. In the 1900s Brown Boveri took the lead in electrification of Europe's railway network by electrifying, at its own risk, a 20-km line on behalf of the Swiss Federal Railways through the *Simplon tunnel* using 3-phase electrification. At this time the number of employees in comparison to the start nine years before had increased by a factor of thirty.

Soon BBC became an international company, essentially shaped by one outstanding product. In 1900 the company made the courageous and momentous decision to include steam turbines in its range of products. Watt's steam piston engines had triggered the first industrial revolution in the early nineteenth century; a hundred years later steam turbines, coupled with generators, were to play a role of similar importance. Rotating turbo-engines subject to constant impingement by jets of steam replaced the venerable piston steam engine. The decisive move to high-speed turbomachinery was initiated at BBC still by Charles Brown Sr., who had established reliable personal contacts to Charles A. Parsons and which made BBC the first European company to license Parsons's steam turbine patents in due course in 1900. In a kind of travel report[4] he wrote in 1896 to his nephew Eric Brown (1866–1942), since 1900 the first director of the BBC turbine factory at Baden:

> ... I have now seen three central stations run entirely with Parsons turbos ... At all three stations I got the most favourable accounts of the working of these engines as being easy to manage, requiring hardly any attention, being entirely free from breakdowns, and effecting a very enormous saving in oil! ... I feel convinced that for electric work the steam turbine is far superior to the common engine type: they govern so much more easily! The light is as steady as from an accumulator battery ... they may use a trifle more steam, but this does not amount to the interest and amortisation of the extra cost for buildings and machinery and the extra expense for wages and oil!—without taking into account the advantages from an electrical point of view.

BBC's acquisition of manufacturing licences under the Parsons steam-turbine patents turned C.E.L. Brown immediately to some of the special problems due to the introduction of high-speed machinery, and he soon recognised that for turbo-alternators of large outputs and high peripheral speeds the construction with projecting field-poles could not be satisfactory. This led to the successfully patented creation of the 3000 rpm *'turbo-generator'* rotor in the form of a cylinder with radial or parallel slots for carrying the excitation winding, Fig. 4.3, which has proved to be the only possible constructive solution and has been universally adopted.

Consequently, the production of turbine generators soon became a major line of business at BBC. The fast-rotating alternating current generator, a stroke of genius on the part of Charles Brown Jr., led to the breakthrough of turbine generators at the turn of the century, and to an influx of orders for BBC from around the world. By 1902, BBC had delivered 17 steam turbines, one of them with an output of 3 MW. By 1905 the product was accounting already for half of the total company sales. Illustrating the successful

---

[4] See Bolter, Sir Charles Parsons, p. 159.

development, Brown Boveri supplied the world's largest steam turbine (40,000 hp) already
in 1914. Steam turbines became the company's largest and most important product which
set up subsidiaries all over the world but with the main operations remaining in Switzerland
and Germany.

After the conversion of Brown, Boveri and Cie. into a limited company in 1900,
C.E.L. Brown became chairman of the board of directors and held that position until
1911. The rapid growth of the company demanded a more strategic management at the
upper level, which was not in line with Brown's creative talents and interests. The mutual
support of both founders diverged, and after a severe conflict with Boveri, Brown retired
abruptly altogether from his business and technical activities. In 1912 he was awarded the
honorary degree of Doctor by the Technical College of Karlsruhe as a kind of last
engineering honour, before he left Baden completely to his family retreat near Lugano,
CH. Charles E.L. Brown died there in May 1924 from a heart attack much too early in the
age of 61, only to be followed by his company partner Walter Boveri just six months later,
who died from a car accident in the Netherlands.

C.E.L. Brown and the BBC board of directors wrote in 1903 a kind of prophetic legacy,[5]
after they observed a certain business consolidation: '. . . *Today the position of the steam
turbine appears to be firmly established and—unless we are very much mistaken—it or its
derivative, the gas turbine, will become the engine of the 20th century, while the piston
steam engine has passed its climax with the outgoing 19th century.*'

It appears to be a curious twist in Brown's biography that his widespread, presumably
distracting interests occupied him even under the severest business pressure. As his
biograph Nobert Lang puts it, '. . . *news in science and technology caused his ardent
interest*'.

He was evenly fascinated by the first automobiles as well as from news of early flight
tests. After he learnt of Lilienthal's first successful trials, he contacted him immediately by
several visits at Berlin to acquire a copy of the Lilienthal delta glider—and in fact Brown

---

[5] See Anon., Die industrielle und kommerzielle Schweiz.

received No.1 of what in hindsight became the first effectively brand-marketed and series-produced aeroplane in spring 1894.[6] Otto Lilienthal (1848–1896) sold between 1894–1896 nine of these so-called *'Normalsegelapparate'*.[7] Fig. 4.4 shows Otto Lilienthal with a slightly improved version of his glider in 1895—with which he flew distances of up to 250 m; his flights inspired the Wright Brothers to pursue the challenge of flight. This 1894 glider, one of only six Lilienthal gliders still in existence, looks and works remarkably like its direct descendent, the modern hang glider. After Lilienthal crashed in a flight test out of 15 m altitude and died from his injuries on the following 10 August 1896, not much is known about Brown's aero experiences thereafter.[8] In December 1904 he offered his Lilienthal glider successfully to the Deutsches Museum Munich for free: *'Though the apparatus has been slightly damaged during my trials, I hope it will suffice for your purposes.*'[9]

In 1910 the Swiss aviation pioneer Martin Hug (1883–1944) took the initiative to kindle the idea of a Swiss aviation industry which finally failed when the apparently wealthy and known aero enthusiast Charles Brown Jr. denied his support. In a return letter to Hug's request from 12 December 1910, he answered just one day later somewhat swaggeringly *'... personally I can claim of being experienced in all aviation-related question, since I followed all developments in this area from early on with great interest—20* (sic!) *years ago I possessed already a Lilienthal flying apparatus,'*[10] and after describing his broad experience in aviation technology, gained by attending two flight meetings at Reims and the corresponding Paris exhibition, in addition to numerous visits at prime aero manufacturers, he concludes: *'Therefore, we feel (at present) not being intrigued to start the production of flight apparatuses.'*

This harsh reaction might have been additionally influenced by experiences of Brown's former company MFO Maschinenfabrik Oerlikon, which brought to market the first Swiss 4 cyl., 50/60 hp, *'boxer'* flight engine, which caused with an exceptional power weight of 1 hp/kg considerable attention at the Paris Air Show, 22 October—10 November 1910; however, after some failing attempts with a higher powered 8 cyl. configuration MFO stopped this exercise in 1913.[11] But the conditional *'at present'* in Brown's letter was

---

[6] See Lang, Brown and Boveri, p. 44.

[7] See Wikipedia, 'Normalsegelapparat' in German—with the names of the other eight customers after C.E.L. Brown: H. Seiler, Ch. de Lambert, A. Wolfmueller, K. Frank, T.J. Bennett, G.F. Fitzgerald, W.R. Hearst, N.J. Joukovski.

[8] See Tilgenkamp, Schweizer Luftfahrt II, who allocates Brown's glider tests to the slopes below Schartenfels Castle at Baden, CH.

[9] In fact the Museum sent Brown's glider to various aero exhibitions up to 1917, but thereafter the set-up was so torn down that it had to be removed from the exhibition to be scrapped in 1922. However, the Museum has remains of two other original Lilienthal *'Normalsegelapparate'* which are presently under renovation.

[10] The letter is reproduced in—see Tilgenkamp, Schweizer Luftfahrt II, p. 145.

[11] See Tilgenkamp, Schweizer Luftfahrt II, p. 145.

Fig. 4.4 Otto Lilienthal and his *'Normalsegelapparat'* w. enlarged vertical tail fin and four added profile stabilizing rails, wing span 6.7 m, weight 20 kg, 1895 © Deutsches Museum Munich

meaningful. The aero gene was in BBC's company DNA from the beginning and re-appeared several times over the coming decades up to an energetic final effort immediately after World War II to gain access to a promising aero engine market by means of its French subsidiaries.

Since the second half of the nineteenth century more and more Swiss enterprises decided to produce abroad due to the limited home-market, difficulties to organise material supplies, a sufficient workforce and the hinderance of customs and valuta regulations. In addition to supplying the flourishing domestic market, BBC was export-oriented even in the first years of business. But high tariffs in other countries were an obstacle to expansive exportation. With the aim to penetrate new markets and avoid long-distance shipping, the company had to grant either licenses or to establish local subsidiaries. The transformation into a joint stock company in 1900 was both an occasion for and the consequence of the internationalisation already underway and helped to procure capital for the cost-intensive manufacture of steam turbines.

Since 1894 the French C.E.M. Compagnie Électro-Mécanique, which built and operated electric power plants as early as 1885, took the right to produce and distribute BBC engines in France and became BBC's full subsidiary *BBC France* in 1901. On 15 June 1900 operations of the German daughter company started at Mannheim-Kaefertal to satisfy the growing market in Germany for steam turbines, generators, electric motors and transformers, supplemented since 1908 by the development and production of electric railways. By the outbreak of the 1st World War, BBC had established a foothold in the key industrialised countries of Europe.

In retrospect the early years must not be viewed solely as a time of technical innovation and success; they were also rife with intense labour and social disputes. Shareholders, for their part, suffered major disappointments in the 1920s and 1930s; economic difficulties have been as much a part of BBC's history as grand triumphs. From 1903 to 1914 the German AEG Allgemeine Elektrizitäts-Gesellschaft (General Electric Company) held a large part of the BBC shares. After WW I the BBC Group had no choice but to join forces for a short time with the powerful British Vickers Ltd; in 1919 the companies entered into a licensing agreement which gave the British firm the right to manufacture and sell Brown Boveri products throughout the British Empire and in some parts of Europe. The agreement gave Brown Boveri a significant amount of money and the promise of substantial annual revenue, and also helped the company expand into foreign markets at a time when protectionist policies inhibited international expansion.

In 1916 British Westinghouse, following the impression that American control of the company hindered its performance during World War I, began the transition to a British owned company. Metropolitan Carriage Wagon Company bought a controlling interest in the group in 1917. The same year the board accepted collaboration with Vickers, which took complete control in 1919 and the name Metropolitan-Vickers was chosen. In due course by the early 1920s and supported by the newly established connection to Vickers, the remainder of the European Westinghouse Companies had been sold to Brown, Boveri and Co. As most important part Westinghouse's locomotive plant some 50 km west of Genoa was integrated to TIBB Tecnomasio Italiano Brown Boveri, since 1903 BBC's Italian daughter company in full.

In view of a stagnant European market BBC went even further and declared its intent to invest in the United States and become a rival to General Electric and Westinghouse on their home market. As a first step BBC acquired the facilities of the New York Shipbuilding Corporation in August 1925 and changed its local name to ABBEC American Brown Boveri Electric Corporation, with a somewhat fateful connotation in view of the end in 1987, when the name Brown Boveri was merged to ABB Asea Brown Boveri.

The principal planned market for BBC's U.S. output was to be public utilities which—assumed—were spending 1 billion USD annually on equipment. In addition, American Brown Boveri would also continue in shipbuilding, where at the time of the acquisition of the prestigious Camden shipyard the aircraft carrier USS *Saratoga* had been just launched, Fig. 4.5, and which was commissioned into a later-on famous service on 16 November 1927.[12]

---

[12] See Wikipedia 'USS *Saratoga* (CV-3)' and '*Lexington*-class aircraft carrier', both in English. USS *Saratoga* served throughout the war, despite being torpedoed twice, notably participating in the Battle of the Eastern Solomons in mid-1942. She supported Allied operations in the Indian Ocean and South West Pacific Areas till the end of the war to protect U.S. forces during the *Battle of Iwo Jima* in early 1945, but was badly damaged by *kamikazes*. The continued growth in the size and weight of carrier aircraft made her obsolete by the end of the war. In mid-1946, the ship was purposefully sunk during nuclear weapon tests. Technically unique was USS *Saratoga*'s turbo-electric propulsion unit, which

Fig. 4.5 USS *Saratoga (CV-3)* aircraft carrier at ABBEC shipyard, Camden NJ, ~1926, displacement 37,000 t, length 271 m, power 180,000 shp © ABB

However, in view of the overall business performance in the U.S. market, it soon became clear that BBC—its efforts notwithstanding—failed to achieve its ambitious goals. Already in the summer of 1927 ABBEC's management recognised that it could not proceed as anticipated, since its *'existing financial resources'* did not justify further expansion.[13] Its optimistic foray had dissipated in a general economic downturn and an increasingly protectionist environment with extraordinary rapidity. By 1929 the two diversified, U.S. owned giants in the U.S. market General Electric and Westinghouse prevailed, American business had complete hegemony in the large and dynamic U.S. electric industry, while Brown Boveri was on full retreat, divesting its shipbuilding facilities at Camden and just retaining its manufacturing subsidiary. In 1931, Allis-Chalmers Manufacturing Company acquired from American Brown Boveri all the latter's

---

actually came from General Electric, but mistakenly was also claimed as part of Brown Boveri's shipbuilding supplies after the acquisition,—see Catrina, BBC, p.48. The sheer power of nominal 180,000 shp for the USS *Saratoga* propulsion unit was clearly in the reach of Brown Boveri's capabilities after their *Hell Gate* steam plant for United Electric Light and Power Co. in New York,—see Eckardt, Gas Turbine Powerhouse, p. 80 f., held for a short while in 1926 the global record of installed power,—160 MW, i.e. the equivalent of 215,000 shp. But the turbo-electric drive principle had been chosen already in 1920 for the originally planned battlecruisers in view of certain operational advantages but also because American companies struggled to produce the very large geared steam turbines necessary for such big ships and was retained when they were converted into aircraft carriers in 1922. BBC started its marine steam turbine business in 1899, in parallel to their land-based power generation activities. Hallmarks of BBC Mannheim deliveries to the German Navy were the geared steam turbine propulsion units for—see Wikipedia, *Scharnhorst*-Klasse (1936) in German, 150,000 shp, entry into service 1938;—see Wikipedia, German battleship *Tirpitz*, 161,000 shp, 1941, and—see Wikipedia, German aircraft carrier *Graf Zeppelin*, 200,000 shp, unfinished, both in English.

[13] See Wilkins, The history of foreign investment, p. 260 f.

assets related to the electrical business including the corresponding patent rights. Allis-Chalmers became the U.S. licensee of BBC, most effectively in the joint production of *Houdry* refinery plants from 1936 onwards, which brought BBC's original axial flow compressor technology for the first time to the United States—with direct impact on U.S. aero gas turbine developments from the beginning 1940s. After the 2nd World War BBC's U.S. business was relaunched in 1946, after cancelling the license agreement with Allis-Chalmers and founding the Brown Boveri Corporation, New York. It required then another 20 years until BBC was at the top again: In 1967 TVA Tennessee Valley Authority, the largest U.S. electricity provider, ordered a steam turbo group of unheard of 1300 MW which was manufactured at the newly opened BBC-plant at Birr, approximately 10 km in the west of the BBC headquarter at Baden, CH. This unit covered the electricity demand of a city with 1.3 million inhabitants—with many of these *'mega machines'* to follow in the coming years.[14]

Curiously in 1927, when Brown Boveri was tied up in the heavy shipbuilding business, it was also close to become a significant player in the U.S. aero (engine) market, but only close so. In the 1920s BBC owned a large share in the Swiss magneto producer *'Scintilla'* from Solothurn, known for their high-quality products. The engineers in Switzerland had developed an efficient magneto that provided a fat spark at high speeds firing the sparkplugs in aircraft engines. On 21 May 1927 the doors to the U.S. aircraft market were wide open, Charles Lindbergh landed at Paris after crossing successfully the Atlantic non-stop with his *'Spirit of St. Louis'*. Its 9 cyl. Wright J-5C Whirlwind radial engine was fired—by Scintilla magnetos. Ironically however, BBC had sold already again its Scintilla shares at the time of the historic event—with considerable losses: a life-long *'writing on the wall'* to the quality-minded BBC, not to forget profit in the pursuit of engineering perfection. In 1929 the Bendix Aviation Corp. purchased Scintilla, adding this division to their group operations.

Over the years BBC's U.S. business developed always in the shadow of a special relationship to its main competitor, the power branch of the mighty GE General Electric Company, formed in 1892 just one year after Brown Boveri through the merger of Edison General Electric Company of Schenectady, New York, and Thomson-Houston Electric Company of Lynn, MA. The BBC records contain a first contact as early as 1904, when GE asked for information about conditions to manufacture BBC machinery.

Paying out a dividend to Brown Boveri shareholders was out of the question from 1921 to 1924 and from 1931 to 1938, but—in parallel—the technical, innovative progress in the newly founded company appears having continued nearly not affected. By the end of the 1920s, BBC's American competitor, General Electric, had managed to acquire 20% of the company shares, and the Brown Boveri families did not hold the majority anymore. In

---

[14] See Eckardt, Gas Turbine Powerhouse, p. 8 f., and Catrina, BBC, p. 94.

order to avoid a growing influence of the American company on BBC, or a hostile takeover, the General Assembly of the board decided in 1929 to introduce registered shares, whose value was five time weaker than the bearer shares, but which had an equal voting right. These registered shares were of course restricted to the Swiss shareholders, and the company remained in the hands of the Boveri family; General Electric's interest in taking over BBC was defended by this energetic effort of the family owners of the time and Swiss banks.[15] By 1933, all the leading European electrical manufacturers were bound with General Electric and/or Westinghouse in cartel accords that effectively precluded their making direct investments in operations in the United States. In this context in 1933 GE also requested for a seat in Brown Boveri's supervisory board based on holding a sufficiently large share in BBC stocks, but the BBC board turned down categorically also this second approach of getting influence.

The whole Brown Boveri structuring process of course did not follow a straight and consequent course. There were drawbacks and detours. Economic and political crises and management disappointments alternated with unforeseen successes over longer periods, all of this interrupted by the two world wars. The personnel numbers over time reflect these unpredictable events, nevertheless with a surprisingly stable growth trend over several decades. Figure 4.6 illustrates this on a global scale and from Swiss perspective, roughly in the scope of this book. While overall the BBC personnel grew rather steadily to a total of 50,000+ until the mid-1950s, the corresponding employment figures for Switzerland alone grew rather modestly—starting from a BBC workforce of approximately 800 only in 1900. Nevertheless, the Swiss employment followed the global trend in general, so that up to the 1960s the employment at the Swiss mother company consistently represented 25% of the total.

---

[15] After nearly 100 years of unbroken interest by General Electric, its Power branch was finally successful in taking over a substantial part of the former BBC/ABB (service) business in 2014 for what some consider an *'overprice'* of 13.5 billion USD, GE's largest-ever industrial acquisition, when Alstom decided to leave this market. The acquisition allowance was given by the EU Commission under certain restrictive terms of competitive business control—the gas turbine development team at Baden, CH was not part of the deal, but came to the Italian *Ansaldo Energia* in due course. Alstom and GE had a long history and in fact, Alstom was formed after GE's former self, Thomson-Houston Electric Company, merged with Société Alsacienne de Constructions Mécaniques to form Alstom (originally called Alsthom). According to GE, the two firms had many ties with factories side by side, and even share a cafeteria. The long history between the two has definitely attributed to the strong preference from GE to acquire the Alstom assets. However, after a short while it became clear that GE had apparently outdone itself by this deal at the worst possible time.

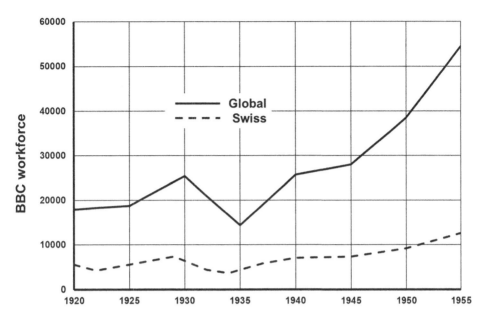

Fig. 4.6 BBC personnel development 1920–1955

4.1.2 Early Turbomachinery and Turbochargers

The history of the first stand-alone, net power generating GT gas turbine power plants has been described in detail in 'Gas Turbine Powerhouse',[16] so that it is sufficient to repeat here only a few key milestones of importance for the understanding of the coming aero gas turbines.

Since gas turbines were and are still limited in process temperatures, the demand for a positive work output requested low losses, especially on the compression side. Consequently, for decades many inventors produced machines that either never ran self-driven at all or the output was so small, that the overall effort was not justified in view of other competitive concepts. The first category will be addressed in the following mainly by reviewing the works of Franz Stolze (1836–1910) with an all-axial engine configuration—with some remarks to the comparable gas turbine activities of Charles Parsons (1854–1931), while the title of the first gas turbine with positive though small, measurable and continuous power output goes to the Norwegian inventor J.W. Ægidius Elling (1864–1949) for the years 1903/1904.[17]

---

[16] See Eckardt, Gas Turbine Powerhouse, p. 66 f.

[17] The subsequent phase of the *Holzwarth gas turbine* with explosion or constant-volume cycle is of no relevance for the coming aero gas turbine and is skipped here; interested readers should—see Eckardt, Gas Turbine Powerhouse, p. 72 f.

Fig. 4.7 The Gas Turbine Creators: John Barber (1734–1801), Franz Stolze (1836–1910) and Charles A. Parsons (1854–1931), l-t-r

Undisputed, John Barber of Nuneaton, Warwickshire, England was the first man to describe in detail the principle of the gas turbine, Fig. 4.7. He patented several inventions between 1766 and 1792, of which the most remarkable was one for a gas turbine in 1791. Planned as a method of propelling a *horseless carriage*, Barber's design included a chain-driven, reciprocating gas compressor, a combustion chamber, and a turbine; unfortunately, nothing practical came out of this patent. The first gas turbine hardware produced Charles A. Parsons from Newcastle upon Tyne, mainly known for his steam turbine invention[18] in 1884, and Franz Stolze from Berlin, both portrayed also in Fig. 4.7. Patent-wise Parsons had the lead, who received British patent No.6735 for a combination of a multi-stage axial-flow air compressor, a combustion chamber for gaseous, liquid and solid fuels and a multi-stage axial turbine on 23 April 1884. Stolze's patent DE101,959 dates from 21 Aug 1897, the illustrating Fig. 4.8 is part of his corresponding Swiss patent No. CH18,721 for a *Hot-Air Machine* of 1899.

However, there are clear indications that Stolze tried to patent his gas turbine idea already some 24 years earlier based on the same configuration by an application on 19 March 1873—but with the obviously false claim of being the first to get a patent for a multi-stage axial turbine in a thermal engine, instead of asking for a patent on the first combination of multi-stage axial *compressor and expansion turbines* on one shaft, i.e. the turbo-components abbreviated K for *Kompression* and E for *Expansion* in Fig. 4.8. Both ends of the shaft belt drove alternators that were rated at 150 kW. The axial flow compressor comprised 10 stages, target pressure ratio PR 2.5, and the reaction turbine

---

[18] See Wikipedia 'Charles Algernon Parsons' in English. The turbine Parsons invented in 1884 utilised several stages in series; in each stage the expansion of the steam was restricted to the extent that allowed the greatest extraction of kinetic energy without causing the turbine blades to overspeed. Parsons' turbine was fitted with a condenser in 1891 for use in electric generating stations, and in 1897 it was successfully applied to marine propulsion in the *Turbinia*, a ship that attained a speed of $34^1/_2$ knots, extraordinary for the time. The turbine was soon used by warships and other steamers.

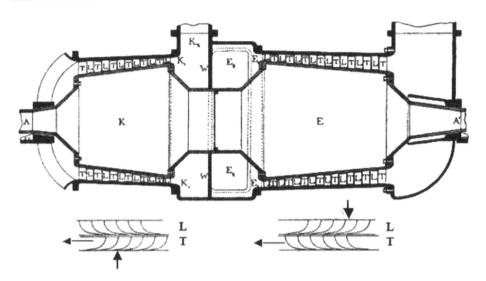

Fig. 4.8 Franz Stolze: All-axial *'hot air machine'* on two bearings, 1899 K—compressor, E—turbine, L—stator, T—rotor

had 15 stages, as he incorrectly thought necessary due to the higher energy head acting in the turbine. The machine operated at a turbine inlet temperature of 400 °C, which Stolze considered the allowable maximum. Consequently, the compressor efficiency would have to be over 70% for the unit to be self-sustaining, a value completely out of the aerodynamic reach.[19]

Franz Stolze can even be considered the inventor of the annular combustion chamber, and thus basically for the aero gas turbine in its present form. While his standard concept of a *fire turbine* showed a vertical combustion chamber—as seen rudimentarily also in Fig. 4.8—with compressor air passing upward through a perforated grate and a kind of fluidised bed of burning anthracite, his British patent no. 7398, applied for in 1898, clarifies that *'If gaseous or liquid fuel be made use of—the combustion may take place directly within the cylindrical mantle or casing, ..., thereby obviating the* use *of a separate furnace.'*

The weak point of these early concepts can be seen at the bottom of Fig. 4.8, where compressor and turbine, stator and rotor blading are completely *'mirrored'*. The expected flow turning in the compressor blading is apparently too large, and the sheet metal blading has no signs of aerodynamic profiling with thickness variation. An especially puzzling observation, since Otto Lilienthal carried out his glider experiments in Berlin since 1894 just 20 km away from Stolze's shop, Fig. 4.4, using already polar diagrams for describing the lift/drag aerodynamic performance based on profiled (stork) wing studies. Moreover, as

---

[19] See Meher-Homji, The Historical Evolution, p. 291.

is known today, the strong aerofoil curvatures (camber) would work for expanding turbine flow, but would cause boundary layer separation, low lift values and high losses for the *'diffusing'*, separation-prone compressor flow. This insensitivity for the aerodynamic peculiarities of diffusing flows with a pressure rise is typical for the time, and also caused Parsons to fail in his struggle for sufficiently high compressor efficiency. In his case the compressor blade and vane profiles of 1884 had little or no camber. In 1907, C.A. Parsons & Co. had made or on order 41 axial flow compressors,[20] in some cases even with a blade design based on propeller theory. But all were plagued by aerodynamic problems; some were reported operating partially stalled with hot-red casings—and low efficiencies in the range of 50–60%. Consequently, this first period of axial flow compressors ended rather soon, Parsons ceased production in 1908 and operationally resilient centrifugal compressor configurations dominated the gas turbine developments for the next 30 years. The category of operational gas turbines with small scale power output begins with the Norwegian inventor J.W. Ægidius Elling (1864–1949), who realised the first constant-pressure gas turbine to produce a net output of 11 hp in the form of compressed air in 1903. It had in a complex set-up an advanced six-stage centrifugal compressor with angle-adaptable diffuser vanes and water-injection between the stages, of which Fig. 4.9 (l), shows a centrifugal wheel with axial inducer. A heat exchanger produced steam to be mixed with the combustion gases before the nozzles of a driving centripetal turbine. In 1904, Elling built the first regenerative gas turbine in the modern sense, in which the turbine outlet gases heated the compressor delivery air. This turbine raised the turbine inlet temperature to 500 °C (from originally 400 °C) and the power output to 44 hp, this time directly to an electric generator.[21] In a patent from 1923 (US1,766,886), Elling described a three-shaft engine with intercooling and reheat and an independent power turbine, the hardware of which existed already in 1912 according to NTM Norsk Teknisk Museum Oslo (Norwegian Museum of Science and Technology) and which was improved up to 1932, when the engine reached a power output of approximately 75 hp.[22] Figure 4.9 (r), shows the entry shaft of this arrangement—for which Elling uses the term *'sub-aggregate'*—with dual-flow centrifugal compressor wheel and integral reaction turbine, both with 0.45 m rotor diameter as exhibited at NTM.[23]

---

[20] See Meher-Homji, The Historical Evolution, p. 314.

[21] See Johnson, Ægidius Elling.

[22] See Bakken et al., Centenary.

[23] See Wikipedia 'Ægidius Elling' in English, especially the quotation *'. . . Many years later, Sir Frank Whittle, building on the early work of Elling, managed to build a practical gas turbine engine for an airplane, the jet engine. . .'* Though there appears to be a stunning coincidence of Elling's isolated shaft scheme of 1924 with Whittle's later turbojet engine configuration of 1937, see Fig. 6.32, there were no contact or information exchange between both inventors, as confirmed by a letter from F. Whittle to K. Weedon, Blommenholm, N, dated 6 Oct. 1968,—courtesy of Lee Langston, 14 June 2018. In the past 50 years NTM and a bunch of Norwegian authors (D. Johnson, J. Mowill, L.E. Bakken et al.) managed successfully to push the memory of their long-time

Fig. 4.9 Ægidius Elling: Gas turbine rotors, centrifugal wheel of 1906 (l), and shaft of 1912 (r, dual-flow centrifugal impeller and reaction turbine) © NTM

Elling's unit was shortly followed by another stand-alone, net power generating GT power plant by the French René Armengaud (1844–1909) and Charles Lemâle for their company Société Anonyme des Turbomoteurs at Paris St. Denis in 1905/6, for which Brown Boveri provided in 1906 a 25-stage radial compressor (System Rateau, distributed to three casings), then—as illustrated in principle for one sub-unit in Fig. 4.10—one of the largest turbo-compressors in the world.[24] The three compressors with a power consumption of nearly 300 kW produced in series together a PR 4.5 at an adiabatic efficiency of 65–70%, had an operational speed of 4250 rpm with a volume flow of 3600 m³/h. The machine, driven by a water-cooled two-stage Curtis wheel turbine of 0.94 m diam. (again a Rateau design), achieved self-sustained operation by adding large quantities of steam generated in combustor cooling and feeding it back to the turbine in a kind of early *steam injection process*. Since there were combustion temperatures up to 1800 °C claimed for the carborundum-lined combustion chamber, a considerable amount of water was necessary to cool the gas to the 450–470 °C, permissible for the intermediate turbine stator row. The actual GT efficiency should have been between 2 and 3% only, or 6–10 kW of equivalent power produced.[25]

After the unsuccessful efforts of Stolze, the first industrially meaningful pressure ratios resulted from radial turbo-compressor tests which Auguste Rateau (1863–1930) carried out between 1900–1904 at Paris. A single-stage test blower of 0.25 m impeller diameter was operated between 8000–20,000 rpm, producing PR ~1.6 at an adiabatic efficiency of

---

overlooked champion of early gas turbine history to the attention of the turbomachinery community. However, present descriptions of Elling's work from that sources appear somewhat isolated and out of context to foregoing and parallel developments in Norway and abroad. Further research along these directions might be appropriate to illustrate the flow of influential ideas and interdependencies.

[24] See Eckardt, Gas Turbine Powerhouse, p. 71.

[25] See Stodola, Dampf- und Gasturbinen, p.1025.

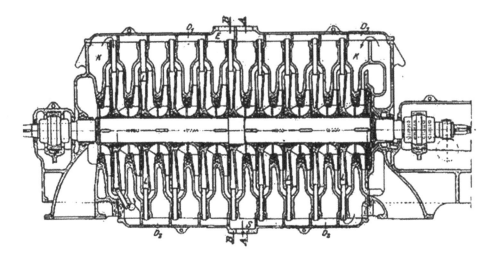

Fig. 4.10 Brown Boveri 10-stage centrifugal compressor (similar to the Armengaud-Lemâle gas turbine set-up) with K—cooling water pockets and horizontally split casing, ~ 1905

~60%. This positive result caused Rateau to put several wheels together, which led in parallel to the first high pressure compressor demonstration of PR 6 at Sautter, Harlé & Co., Paris (with A. Rateau as consultant) in 1904 and subsequently, Brown Boveri taking a licence for the production of multi-stage centrifugal compressors of the *System Rateau*.[26]

The so-called *'2D shrouded impeller'* of this design, although now manufactured in a different way, is still the workhorse of today's process centrifugal compressors in chemical industries, refineries and oil/gas fields—with a cover disk and backward curved blades. Since the water cooling as shown in Fig. 4.10 was prone to leakages, separate intercoolers were introduced soon instead.

The *turbocharger development history* has been outlined for industrial applications before, so that the following will address aero engine needs only.[27] Since the description of the engineering life of Claude Seippel, Brown Boveri's *Father of the axial compressor* was touched in the foregoing chapter also in view of experiences he collected in the U.S., it may be worthwhile to address the U.S. turbocharger competition at the end of WW I in more detail. The turbocharger is an intriguing piece of early turbomachinery history in the context of this book since it is the vehicle to get air-borne.

---

[26] After BBC in 1907, other companies took licences for Rateau's turbo-compressors between 1908–1910—in Germany GHH Gutehoffnungshuette Oberhausen, KKK A.-G. Kuehnle, Kopp & Kausch Frankenthal, Jaeger & Co. Leipzig, Masch.-A.-G. Pokorny & Wittekind Frankfurt a.M., A.E.G. Allgemeine Elektrizitaets-Gesellschaft Berlin and—in Switzerland, in Brown Boveri's immediate neighbourhood, Escher Wyss & Co. Zurich and Sulzer Bros. Winterthur.

[27] See Eckardt, Gas Turbine Powerhouse, pp. 53–65.

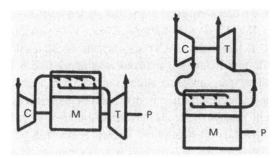

Fig. 4.11 Turbocharger alternatives: turbo-compound motor with geared/gearless supercharger (l) and conventional, free-wheeling turbocharging concept (r), M internal combustion engine, C compressor, T turbine, P actual net output

The Swiss engineer Alfred Buechi (1879–1959) today is considered the inventor of the turbocharger technology in general, though this does not hold true in this broad sense. He certainly was the first to address the subject of increasing the performance and power output of internal combustion engines with the application of turbo-components. Instead of the gas turbine—for which it was still too early—it was the turbocharger which assisted the combustion (piston) engine to undreamt of improvement. Though the *'System Buechi'* with its tuned, heavy piping found no aero applications, it was interestingly Buechi in his first patented turbocharger scheme of 1905 which showed an eight-stage *axial* compressor and thus—rather intuitively—anticipated similar aero turbomachinery configurations 20 years ahead of practical realisation.[28] The term *'supercharger'* or *'turbo-compound'* are used, whenever the drive power of the charging compressor reduces the engine output, i.e. when there is a mechanical, geared link between piston engine and compressor. Alternatively, the free-wheeling *'turbocharger'* exploits the waste energy of the motor in a turbine which directly drives the compressor, Fig. 4.11.[29] Effective high-altitude flight with piston engines beyond 10 km could only be realised by turbocharging—with the inherent drawback of a characteristic turbocharger acceleration lag.

The origins of aero supercharging cannot be determined unambiguously and as a result of its importance, there are several fathers of the application for the first turbo-compressor as supercharger, for the first free-wheeling turbocharger (compressor-turbine combination) and for the first meaningful series production. According to one source,[30] the American *'Murray-Willat Company'* in 1910 already produced a 90 hp super-charged two-stroke rotating piston engine with integrated radial compressor for applications up to remarkable 5.2 km. Professor Auguste C.E. Rateau in 1916 took out patents for turbocharging equipment that was manufactured and subjected to practical trials in 1917. The military use of the airplane in the Great War of 1914–1918 resulted in considerable research being undertaken in Europe to compensate for the loss of engine power output with increase in

---

[28] See Eckardt, Gas Turbine Powerhouse, with Fig. 3.6 there, illustrating Buechi's Swiss patent CH35,259.

[29] See Jenny, The BBC Turbocharger.

[30] See Schwager, Hoehenflugmotor.

altitude. During 1915–1916 experimental work was started on the exhaust driven turbo-supercharger by the Royal Aircraft Factory (since 1918 Establishment, RAE) at Farnborough, England and in France, by Auguste Rateau in conjunction with the Farman Company. In 1917, the first Rateau supercharger was applied on a 175 hp engine and maintained the ground pressure up to an altitude of ~5.5 km. Rateau's decisive break-through to the turbosupercharger freed him from the necessity of heavy gearing, at the same time allowing for high rotational speeds for the compressor impeller with sufficient pressure out of a single-stage configuration. Rateau began tests in early 1917;[31] the presently popular narrative of Rateau's turbocharger technology transfer to the United States can be traced back to a 1948 book: '. . . *Dr. Durand, at that time chairman of the National Advisory Committee for Aeronautics, came in contact with the work of Prof. Rateau through the Interallied Commission and when he came back to the United States interested Dr. S.A. Moss of the General Electric Company in this French development* . . .'[32] But revealing, recent research provided more detail on Durand's itinerary in this time frame.[33] The United States started very early to share scientific information with their Allies, and the First World War can be seen as the beginning of the intense relationship between the U.S. and Great Britain in terms of security and defence science. As early as March 1915, the National Advisory Committee for Aeronautics (NACA) was founded as a U.S. federal agency, which was to become NASA in 1958. The National Research Council (NRC) was founded in September 1916 with a first official scientific mission to Britain and France still in 1916. The NRC was also responsible for establishing the Research Information Committee (RIC) in December 1917, whose task was to provide the linkages with the offices of Military and Naval Intelligence and to coordinate their international scientific activities. Science attachés were appointed in *January 1918* and sent to Europe: engineer William Frederick Durand (formerly with NACA) was sent to Paris together with his deputy Karl T. Compton (later MIT president, 1930–1948) to London. The offices in London, Paris and Rome sent reports on a weekly basis, and the RIC was to receive, index

---

[31] Rateau's first turbo-supercharger had a 0.24 m diam. centrifugal impeller with 380 m/s tip speed (30,000 rpm), producing a stage pressure ratio of PR 2 and a turbine inlet temperature of approx. 750 °C. The RAE turbosupercharger, designed by James E. Ellor in cooperation with Metropolitan Vickers Company before the end of the war, was later considerably modified under the influence of Rateau's machines. The National Archives, Kew hold under the subseries *'Engines used with RAE and other aircraft'* an arrangement drawing of a Rateau turbo-compressor, ref. AVIA 14/57/19/2.

[32] See Vincent, Supercharging. E.H. Constant propagated this story with reference to Vincent in his 1980 book, see Constant, The origins, p. 123—with additional details '. . . *Durand had been a professor at Cornell and "remembered the black smoke from the [Moss] gas-turbine research and knew of the extensive business that the General Electric Company then had in steam turbines and centrifugal compressors." When he returned to the United States, Durand asked Sanford Moss to undertake development of a turbocharger for the U.S. Army Air Corps. Drawings of Rateau's turbocharger were provided, but G.E. apparently did its own detail design. . .',*—see Moss, Gas Turbines, 1944.

[33] See Lehmann, Science, p. 19.

and distribute these records to the American military agencies. On the other hand, it is known that Rateau tested at least one of his turbochargers in a *SPAD XIII* (serial no. S-706), a successful French fighter aircraft—known as the *Spitfire of WW I*—which appeared at the Western front on 4 April 1917. Given the staccato of events and decisions in this decisive period of the Great War, the indicated gap of nine months between the first aero appearance of the Rateau turbocharger and Durand's arrival in Europe causes suspicion about his effective role in this context.

Later in Sect. 6.4.1, reference is made to the *'Tizard Mission'* with the objective to cooperate in science and technology with the U.S., which was neutral and, in many quarters, unwilling to become involved in the war. The U.S. had greater resources for development and production, which Britain desperately wanted to use. The shared technology included amongst other *Radar*, in particular the greatly improved cavity magnetron and Frank Whittle's jet engine, which after a few years put the receiving General Electric automatically—though with considerable own efforts—in the pool position of a developing military and civil turbojet engine market.

Interestingly, a similar situation occurred on 24 May 1917, just six weeks after the United States had declared war against the German Empire, when the French Premier Alexandre Ribot sent U.S. President Woodrow Wilson a cablegram with a detailed French aeronautical programme request with unheard of figures, asking for an immediate contribution of 4500 aircraft, 5000 pilots, and 50,000 mechanics to the war effort. Of course, these numbers proved to be absurd because U.S. aircraft production capability was in no way prepared for such a requirement. Moreover, the aviation section of the U.S. Signal Corps, the aerial warfare service of the United States from 1914 to 1918, and a direct statutory ancestor of the United States Air Force, was relatively unprepared for the task, not being equipped with any aircraft suitable for combat operations. However, considerable optimism and energy was put into addressing this identified need, leading to the mobilisation of American industry to set about the production of contemporary combat aircraft. As there were no suitable aircraft domestically, a technical commission, known as the *Bolling Commission,*[34] was dispatched to Europe to seek out the best available combat aircraft and to make arrangements to enable their production to be established in the United States.

For production in the United States, the Bolling Mission—which toured aircraft production sites in England, France and Italy during July 1917—selected the British De Havilland DH-4 for observation and day bombing first with some preference;[35] thereafter,

---

[34] The Commission was headed by Raynal C. Bolling (1877–1918) of United States Steel. A corporate lawyer by vocation, he became an early Army aviator and the organiser of both of the first units in what ultimately became the Air National Guard and the Air Force Reserve Command. After the mission he remained in France instead of returning home and was the first high-ranking officer of the United States Army to be killed in combat in World War I on 26 March 1918.

[35] The De Havilland was partly chosen because the British government granted free use of its licence for the aircraft rather than because of any superiority to French aircraft, which required payment of a royalty to produce; an aspect to be taken into account also in view of the Rateau turbocharger deal.

also British Bristol and French *SPAD* as fighters, and the Italian *Caproni* tri-motor as a long-range night-bomber. However, there were unforeseen teething problems; in April 1918, a full year after the war began for the United States, the total production was 15 of the modified British De Havilland DH-4, out of a contracted order backlog of 10,000.

In the light of an almost endless variety of aircraft and engine types,[36] the decision of the U.S. Aircraft Production Board in May 1917 to focus all effort at least just on one standardised, modular engine concept was revolutionary. The resulting *Liberty L-12*[37] was a 27 l, water-cooled 45° V-12 aircraft engine of 400 hp, designed for a high power-to-weight ratio and ease of mass production—to the benefit of the united U.S. automotive industry. Insofar as the *Liberty* was turned out in quantity—over 15,000 by November 1918—the engine was a success in production; moreover, since the *L-12* was modified with progressively increasing horsepower output from 330 to 440 hp, the engine was a success in design, capable of expanding its margin of performance to maintain superiority in combat.[38]

The French *SPAD S.XIII* biplane fighter,[39] powered with a new 200 hp *Hispano-Suiza 8B*, which also had been selected by the Bolling Mission, arrived in the United States in September 1917, where the sample SPAD was shipped directly to the Curtiss factory in Buffalo, New York. The manufacturer had no sooner received an order for 3000 SPADs

---

[36] For the aircraft ratio of 171(!) types on the Allied vs. 28 on the German side,—see Holley, Ideas and Weapons, p. 124, while the engine ratio of 37 Allied types vs. 7 Germans has been deduced from the certainly limited scope of Sherbondy's book,—see Sherbondy, Textbook, part II.

[37] On 29 May 1917 the Aircraft Production Board brought two top engine designers, Jesse G. Vincent (1880–1962), Packard Motor Car Co. of Detroit and Elbert J. Hall (1882–1955), Hall-Scott Motor Co. of Berkeley Ca. together at the luxury *Willard* Hotel, Washington (two blocks east of the White House), where the two were asked to stay until they produced a set of basic drawings. After just five days, Vincent and Hall left the *Willard* with a completed design for the new engine, which had adopted, almost unchanged, the single overhead camshaft and *rocker arm valvetrain* design of the *Mercedes D.IIIa* engines of 1917/8. In July 1917, an eight-cylinder *L-8* prototype assembled by Packard's Detroit plant arrived in Washington for testing, and in August, the 12-cylinder version was tested and approved. See Wikipedia 'Liberty L-12' and Sherbondy, Textbook, p. 311 f.: *'The 180 hp Mercedes, ... taken from the captured German Albatros D Va (G. 97)'* and p. 318 f. *'The 260 hp Mercedes', ... collected from a captured three-seater 'Gotha' biplane of the pusher type'.*

[38] See Holley, Ideas and Weapons, p. 124.

[39] The *SPAD S.XIII* was developed by *Société Pour L'Aviation et ses Dérivés*,—see Wikipedia, 'SPAD S.XIII' in English, and—see Sherbondy, Textbook, p. 123 f. 'The Hispano Suiza' (already categorised here as an *American Type*). The Hispano-Suiza 8 was a water-cooled V8 90 deg aero engine with single overhead camshaft (SOHC), designed by Hispano-Suiza's chief engineer, the Swiss Marc Birkigt (1878–1953). His basic contribution to engine design was the en-bloc cylinder construction with a cast-Al water jacket containing steel cylinder barrels, with enclosed and lubricated valves and valve gear. This engine started a revolution in liquid-cooled engine design, culminating in the *Rolls-Royce Kestrel* and *Merlin* and as prototype for the Mercedes and Junkers engines of the German *Luftwaffe* in WW II. It was with 21,000 units the most commonly used, non-rotating liquid-cooled engine in the aircraft of the Entente Powers during the First World War.

and begun to prepare for production, than a cancellation arrived and the whole project stopped. A number of factors probably entered into the decision, but the concern that it would be impossible to convert the SPAD to carry a Liberty engine, might have been key.[40] Though there is no explicit proof, the SPAD delivery was with some likelihood accompanied with substantial information about the *Rateau* turbocharger (hardware and/or drawings) and after the manufacturing stop the package presumably was forwarded to McCook Field, one mile north of downtown Dayton OH, where the Army Signal Corps Engineering Department, originally at Washington D.C., had established its *Foreign Data Section* to evaluate foreign scientific and aircraft information in 1917.[41] The Section's work centred on collecting information on foreign aircraft capabilities, editorial work, exhibits, photography, and statistical recognition; maintaining a library of technical works and history; and even handling public relations. The function exists still, since 1961 as NASIC—the National Air and Space Intelligence Center, now headquartered at Wright-Patterson AFB in a considerably enlarged capacity as the source of air and space intelligence for the U.S. Department of Defense (DoD) and produces integrated, predictive air, space and specialised intelligence to enable military operations, force modernisation and policymaking. In its own mission statement *'NASIC is a global intelligence enterprise which fulfills the needs of today's and tomorrow's warfighter, aids in shaping national and defense policy and guides the development of future weapons systems. NASIC products and services play a key role in ensuring that United States forces avoid technological surprise and can counter existing and evolving foreign air and space threats.'*

A November 1917 organisation chart shows Earl Hazard Sherbondy (1887–1958) serving as the head of the Foreign Data Section of the *Airplane Engineering Department* and consequently, NASIC listed him in this tradition in a survey, covering 1917–2014, as first commander for the year 1917. Here starts for the years 1917/1918 a confused, in parts contradictory story—possibly influenced by certain interests or blurred in historic hindsight—from which reality and truth may be deduced only cautiously from various sources.

The United States Air Service started work on the Rateau type turbocharger in the last year and a half of WW I—embodied in two separate programmes, one by said Sherbondy and the other by Sanford Moss and the General Electric Company. Both of these turbos appear to have depended on the French Rateau example, yet all three differed in major details. Sherbondy entered the field first when in June 1917 he was—apparently delegated by the Army—to investigate the potentialities of the aircraft supercharger. Initially he concentrated on the gear driven type but soon settled on the turbine driven configuration. E.H. Sherbondy worked in conjunction with the *Rateau-Bateau-Smoot Co.*, which handled

---

[40] See Holley, Ideas and Weapons, p. 126.

[41] The Engineering Department of the Army Signal Corps moved as a whole to McCook Field, which officially opened on 4 December 1917.

the Rateau patents in the U.S. By May of 1918 preliminary testing was begun at the *De Laval Steam Turbine Co.* of Trenton, NJ., continued by a slightly different version in June, both tested on steam only.[42] The next month Sherbondy's No. 1 and 2 turbo-superchargers were taken to McCook Field where they were installed on a wind tunnel-mounted Liberty engine. Considerable trouble was encountered due to overheating of the exhaust-driven turbine and even the use of a special heat-resisting metal in this part did not overcome the problem.[43] In addition the diaphragm which divided the inlet gas from the exhaust gas warped out of shape, changing the nozzle shape and allowing gas to flow from nozzle to nozzle. Much of the trouble was due to the lack of construction materials which were able to withstand the 800 °C exhaust heat of the engine.[44] Additional problems encountered were seizing of the turbine rotor because of the failure of the rear turbine bearing. This occurred several times and it was decided in June/July 1918—that air cooling of that bearing was not adequate and water cooling would be needed. Sherbondy's No. 3 turbo-supercharger was redesigned accordingly to eliminate the problems; it was tested with steam in the same manner as No. 1 and 2 and then shipped to McCook Field, so that the third version was ready for Liberty testing there in December.[45] Fig. 4.12 shows a detailed sectional drawing, rotor hardware and engine mounting of Sherbondy's turbo-supercharger No. 3. In comparison to the foregoing, nearly identical versions No. 1 and 2, the whole arrangement was axially considerably squeezed e.g. by putting both exhaust bypass valves on top, and the rear bearing on the turbine side received water cooling from the Liberty cooling system, which completely surrounded the turbocharger lubricating oil chamber. Nevertheless, later analysis showed that this modification was insufficient for the required water flow. The turbine wheel with 72 blades had 150 mm diam., while the centrifugal impeller with 10 radial blades measured 229 mm, corresponding to a maximum tip speed of 378 m/s at 31,500 rpm, producing a total pressure ratio PR ~ 2. Stodola calculated for this set-up an approximate effective compression work of 22 hp.[46]

It is now time to switch to the second, finally winning team of Sanford Moss and the General Electric Co. Sanford A. Moss (1872–1946), Fig. 4.13, received his Ph.D. from Cornell University where he built his first gas turbine engine[47]—unsuccessful as others at that time. In 1903 after graduation, Moss became an engineer for General Electric's Steam Turbine Department in Lynn, MA, and up to 1917 nothing of Moss' work and of General

---

[42] See Constant, The origins, p. 124. Constant states that the Sherbondy design was built by the De Laval Steam Turbine Company at Trenton, NJ for Fergus Motors, without outlining the role of the latter in detail. It could stand for Fergus Motors of America, Inc. from Newark, NJ as an early producer of the Liberty engine or parts thereof.

[43] See Hallett, Superchargers.

[44] See Neal, A Technical & Operational History, p. 327.

[45] See Olmsted, Turbo-Supercharger Development, also confirmed by—see Hallett, Superchargers, supposedly an especially reliable source from 1920.

[46] See Stodola, Dampf- und Gas-Turbinen, p. 1039.

[47] See Moss, The Gas Turbine.

Fig. 4.12 Sherbondy turbosupercharger No. 3: sectional drawing (l), rotor shaft with compressor and turbine wheel (m), mounting on Liberty L-12 (r)

Electric in general pointed towards aviation, and even the decisive first steps in this direction have not been conveyed unambiguously as the following quotations demonstrate.

Constant[48] writes *'W.F. Durand,*[49] *chairman of the U.S. National Advisory Committee for Aeronautics (NACA), was in Europe during 1917 and learned of Rateau's work through the Interallied Commission [ref. to Vincent, Supercharging pp. 1–2, followed by the foregoing quotation of FN31].* Schlaifer[50] kept it shorter *'...In 1917 a Rateau turbo or the design of one was sent to the American Army. The Army was told of Moss's early work on gas turbines at GE and persuaded GE to develop a turbo under Moss's direction.'*

Neither Constant nor Schlaifer address Sherbondy's development lead nor the sensitive fact how a concurrent design was phased in by the U.S. Army in competition to an earlier contracted and presumed Army member Sherbondy. Our only contemporary witness George Hallett,[51] Fig. 4.13, who must have had his own special relationship with Sherbondy, as described below, writes in his 1920 report *'...Soon after Mr. Sherbondy began work on the turbo-compressor, Dr. S.A. Moss, chief of turbine research department of the General Electric Co., asked permission to carry on some work on the same general type. He built one turbo-compressor which was also a modification of the Rateau type but differed considerably from Mr. Sherbondy's machine.'* Finally Neal, who provides most details on these parallel developments, writes *'... By the fall of 1917 the General Electric*

[48] See Constant, E.W., The origins, pp. 123–124.

[49] See Wikipedia 'William F. Durand' (1859–1958), Fig. 4.13, as outlined before Durand's trip to Europe was rather later in January 1918—with consequences to the further GE activities.

[50] See Schlaifer, The development XII, p. 327.

[51] See Hallett, Superchargers, p. 221.

Fig. 4.13 Men around Earl H. Sherbondy (1887–1958): William F. Durand (1859–1958, l), Sanford A. Moss (1872–1946, m), George E.A. Hallett (1890–1982, r)

Company and their principal turbine engineer, Dr. Sanford A. Moss, had been requested by Dr. William F. Durand of the National Advisory Committee for Aeronautics to also pursue the turbo-supercharger' and *'Moss began work on his supercharger design late in November of 1917, some five-months behind the start of the Sherbondy effort.'* Apparently, details matter in this context, Robert Neal knows *'In spite of the later start, the machine had been designed, constructed, steam-tested at GE's Lynn River Works and was delivered to McCook Field on June 4, 1918, about the same time the first Sherbondy arrived.'*[52] If this timing is correct, the Lynn team would have foreseen potentially emerging difficulties with over-heating of the critical turbine bearing well in advance. Factually, the water-cooled turbine bearing was already a feature of the one and only GE turbo-supercharger before the decisive motor tests started which suggests that the Lynn design team somehow benefitted from corresponding information either deduced from foregoing Rateau and/or Sherbondy No. 1 and 2 test experiences.

After this flying start the extraordinarily smooth testing of the Moss design, operationally fitted on a Lincoln Liberty engine, continued from 17 June to 7 July, 1918 and showed minimal problems.[53] At that point it was decided that the design should be tested at altitude, also to circumnavigate the disturbing backfiring of the turbo-charged piston engine at sea level. An improved version of the truck-mounted test stand used on Pikes Peak, Colorado for Liberty engine tests a year earlier was ordered, left McCook Field on 1 Sep 1918 and was driven to Pikes Peak summit, elevation 4300 m, where testing ensued from 10 September through 7 October 1918. Altitude tests went reasonably satisfactory and

---

[52] See Neal, A Technical & Operational History, p. 243 and p. 331.

[53] See Neal, A Technical & Operational History, p. 331, stating about the Moss turbosupercharger *'The rear bearing was water-cooled and proved to be reasonably trouble free.'*

showed that it was possible to raise the mean effective pressure of the Liberty to practically the same value obtained at sea level.[54]

The further U.S. turbosupercharger development was significantly influenced by the armistice, ending WW I on 11 November 1918. In due course all supercharger development work was stopped—or nearly so, if the cited sources are correct. After the successful test campaign of the GE turbosupercharger only one more 4-h endurance test was appointed for the Sherbondy No. 3 during the week of 7–14 December 1918 which became—nearly predictably, as a suspicious observer may want to state—in sum a complete disaster. As documented in a report, officially dated 16 December 1918, numerous problems occurred: Bearings were burned-out in spite of the water cooling. After about three hours into the 4-h endurance run the motor was stopped to correct problems caused by loosened oil and water connections. After sitting some 15 min and cooling, it was found that the turbine wheel could not be moved because the rear bearing had melted. In addition, the bypass valves were stuck in the closed position. According to Neal, the immediate conclusion out of this situation was '*Although the Sherbondy turbosupercharger had a number of novel features and certainly appeared to be a well-made and quality finished unit, its design did not overcome the problems presented by the materials then available* '[55] which in reverse conclusion was then in the reach of GE's design capability.

In stark contrast to Neal's positive record of 2009, the near-time witness Hallett knew in 1920 '*The Moss supercharger, as first built, was of rather crude construction, and much mechanical trouble was encountered with all parts except the rotating element*', and continued about the further course of decisions '*When the engineering division of the Air Service took over McCook Field and started to plan peace-time development, the supercharger situation was carefully considered. It was decided that it was important to continue development work along this line. It then became necessary to decide whether work should be continued on both the Sherbondy and the Moss machines, and, if not, which one should be developed. It was noted that although Dr. Moss' machine was comparatively crude, it contained some inherent advantages over the Sherbondy type, and no way* (sic) *was seen to overcome the faults of the Sherbondy machine. Therefore, the latter was dropped and the General Electric Co. was given a contract to rebuild the old supercharger designed by Dr. Moss. The new device is now being tested in actual flight and giving very interesting results.* '[56] By late December 1918 the Sherbondy project had been cancelled—and Moss and his team at GE Lynn works and the Army Engineering Department had the coming 15–20 years to steadily test and improve the once crude design to a superb, strategically decisive quality product. The most difficult problem was the construction of

---

[54] See Kimble D. McCutcheon 'The first turbosupercharged U.S. aircraft engine', https://www.enginehistory.org/superchargers.shtml

[55] See Neal, A Technical & Operational History, p. 329.

[56] See Hallett, Superchargers, p. 222—the wording of this statement is especially noteworthy in view of Hallett's role as Sherbondy's de-facto successor at McCook Field after the war, where he was placed in charge of the power plant branch of the Army Engineering Department.

the turbine wheel which could withstand the exhaust temperature of a gasoline engine, and here in the field of metallurgy is where the *turbo* had a lasting impact to the development of the gas turbine.

From 1919 through the Second World War the Army paid the entire costs of continuous development, headed until 1937 by Sanford Moss.[57] It was not until World War II that the full benefits of these high-altitude test flights were realised, when planes such as the B-17, B-24, B-29, P-38 and P-47 were so successful in carrying the war to the enemy at high altitude.[58] They all used General Electric turbo-superchargers based upon those primitive models tested and constantly improved in the 1920s and into the 1930s. Notwithstanding the long-term success of the GE turbo-supercharger the circumstances of the competitive decision between a three-times failing Sherbondy against a one-time successful Moss camp create doubts. Neal provides a direct hint in the direction of data manipulation, when he refers to a historical documentation of 1926, referencing the Sherbondy design as 'Form A-1'. He concludes that this designation was generated *'after the fact'* and that the Sherbondy was never known by that name when it was an active project, while the term *Form* was exclusively used by GE. *'Logically the Sherbondy would have been called the "A" and the Moss would have been the "A-1" since the Sherbondy actually preceded the Moss.'*[59] Besides the poor quality of Moss' first product as mentioned by Hallett, it is noticeable that apparently until today[60] no original design drawing of the Moss turbo-supercharger is in the public domain, while all three Sherbondy designs are neatly documented by general arrangements similar to the one reproduced in Fig. 4.12. Somewhat surprisingly, given GE's superior turbomachinery capabilities at that time, the first GE turbo-supercharger used an impulse turbine provided by the De Laval Steam Turbine Company, i.e. the same source as of Sherbondy's machine; in this context another source

---

[57] See Schlaifer, The development XII, p. 328.

[58] In a typical air raid profile over Germany a Boeing B-17 *Flying Fortress* had to climb to ~8+ km to cross the German Flak defence belt on its way back and forth, which challenged the endurance of turbo-superchargers to hot-red material temperatures for hours, before the carpet bombing level of ~4.5 km was reached over the target zone. See Wikipedia 'Kammhuber Line' in English.

Schlaifer states in this context '... *with the beginning of carpet bombing over Germany by the 8th U.S. fleet, the GE turbosupercharger on board of B-17/B-24 made these the outmost strategic weapon for winning the battle in the European theatre.*' See Schlaifer, The development XII, p. 329 All American WW II combat aircraft equipped with a turbo also had an internal, gear-driven supercharger, i.e. had two superchargers per engine. The turbo's function was to maintain sea level pressure at the engine inlet as the airplane climbed, and the internal supercharger then compressed the air further before delivery to the engine cylinders. See Olmsted, Turbo-Supercharger Development, p. 99.

[59] See Neal, A Technical & Operational History, p. 331; it remains unclear if this 1926 documentation was issued by the Army and/or GE.

[60] See Norris, GE at 100.

of potential information leaks between the two competing camps.[61] And finally, to round up this list of strange coincidences, it is striking that apparently nothing is known about a GE-Rateau patent or licence relationship, though the origins back to Durand's pick-up of Rateau drawings and the design similarity have been independently stated several times. All this, in obvious contrast to the Sherbondy variant which was inescapably *'hooked'* to Rateau's patent agent, the Rateau-Bateau-Smoot Co. (see above) from early beginnings in 1917. As stated in Footnote 35 above on the preferred *'for free'* selection of the British government-supported De Havilland DH-4 in comparison to licence-charged French and Italian aircraft designs, one can assume that similarly on the engine side potential licensing demands might have been the real *red rag* for the Army Engineering Department in a precarious financial situation, and thus might have considerably contributed to the decision against the Sherbondy concept.

Sherbondy, certainly frustrated after this development, left McCook Field in 1919, presumably being followed after the Great War by George E.A. Hallett (1890–1982) in a position responsible for the Army powerplant branch.[62] There, Hallett established an engineering department and an engine test and development facility, which led to a number of innovations, particularly with engine superchargers—and which explains his familiarity with the foregoing Sherbondy and Moss activities. In 1920 Sherbondy together with G. Douglas Wardrop, managing editor of *'Aerial Age Weekly'*, published a 363 p. *'Textbook of Aero Engines'* which summarised Sherbondy's experiences at McCook Field from an Allied view in a comprehensive description of 14 U.S., 10 British, 9 French, 7 German and 4 Italian contemporary flight piston engines, however without any reference to the turbocharger issue.

Claude Seippel, Brown Boveri's axial compressor pioneer worked at Earl H. Sherbondy's engineering office at Cleveland, OH, between December 1924–August 1926, as outlined in the foregoing Chap. 3. Sherbondy, according to Seippel, *'a genius without success'*, had no wish to join GE after he had lost the U.S. turbocharger

---

[61] See Constant, E.W., The origins of the turbojet revolution, p. 124, and—see Eckardt, Gas Turbine Powerhouse, p. 54.

[62] See Wikipedia 'George E.A. Hallett' in English. In 1911 Hallett had been a member of the small Curtiss group who succeeded in making the first successful flights off the water. In June 1914 he and John Cyril Porte planned the first transatlantic flight with the flying boat *America*, commissioned by the department store magnate Rodman Wanamaker, but were prevented by the successful flight of the British aviators John Alcock and Arthur Brown on 14–15 June 1919, and the start of WW I. Late in 1914 Hallett left Curtiss and accepted a government position as the Army's aviation mechanic, supervising engine overhaul and serving as a consultant on engine matters. He then developed a course of study for engine mechanics and developed a new method for investigating engine problems and failures. This led—similar to Sherbondy—to his preparation of a book on the topic,—see Hallett, Airplane Motors, which became a standard of aviation ground school instruction. In 1922, Hallett resigned from the Army to become a research engineer and section head of the General Motors Research Laboratory in Detroit, where his work laid the foundation for GM's very successful diesel engine programme.

competition. He continued as an independent inventor, and his only financial resources were provided by friends; sadly, he finally became mentally deranged.

Since after even thorough search no picture of this tragic figure (and NASIC's first commander) could be found, it remains only to remember at him by his contemporaries, and in parts presumed adversaries, as grouped in Fig. 4.13, and by a line from Bertolt Brecht's *'The Threepenny Opera'*:[63]

There are some who are in darkness, and the others are in light
And you see the ones in brightness, those in darkness drop from sight.

After this excursion to the early U.S. aero turbocharger developments with only indirect, nevertheless substantial implications with Brown Boveri, it is now worthwhile to repeat in short the involvement of BBC Mannheim in corresponding key developments on the German side near the end of WW I. German researchers began work on supercharging in 1915, and quickly settled on the gear-driven centrifugal compressor. It was the increasing effectiveness of anti-aircraft defence, especially against the relatively slow *'giant aircraft'*, replacing the Zeppelins over the battlefield in 1916/1917, which urgently demanded to fly at greater altitudes. The German Walter G. Noack (1881–1945) from the Swiss Brown Boveri at Baden, CH had to report to arms in 1915 and in 1917, being too old for pilot training, achieved the technical responsibility for the machinery equipment of the 'giant aircraft' fleet at *Rea Riesenflugzeug-Ersatz-Abteilung* (giant aircraft reserve department) at the *Flugzeugmeisterei der Inspektion der Fliegertruppen* at Berlin-Charlottenburg.[64]

---

[63] *'Denn die einen sind im Dunkeln, Und die anderen sind im Licht, Und man sieht nur die im Lichte, Die im Dunkeln sieht man nicht.'* The rarely heard final stanza, not included in the original play, but added by Brecht for the 1931 movie, expresses the theme and compares the glittering world of the rich and powerful with the dark world of the poor.

[64] See Eckardt, Gas Turbine Powerhouse, p. 88 f. The WW I flying test engineer Noack became later one of the prominent Brown Boveri engineering personalities at Baden, CH in the context of turbochargers, high-pressure steam turbines, *Velox* boilers and constant-volume as well as constant-pressure industrial gas turbines.

Part of the *Flugzeugmeisterei der Fliegertruppen* (Directorate of Aircraft Production of the Aviation Troops), after 8 Oct. 1916 called *Luftstreitkräfte* (Air Force) was the *Inspektion der Fliegertruppen,* abbreviated *Idflieg,* which was responsible for all training units in the interior, home front as well as the procurement dealing with aircraft production and replacements. Subordinate to Idflieg were several important Departments, amongst these the Flugzeugmeisterei der Inspektion der Fliegertruppen. The corresponding Flugzeugmeisterei office in Charlottenburg, in the west of Berlin was responsible for all aeronautical material, i.e. through this office was purchased aircraft, engines and accessories; associated were the Commands of Training Squadrons (Fea Fliegerersatzabteilung) and of Giant Airplane Sections (Rfa Riesenflugzeugabteilungen)—the latter with the Subgroup Rea. A subordinate section of Flugzeugmeisterei was also *Prüfanstalt und Werft,* (P.u.W., Testing Section and Workshops), located at Adlershof, attached to the Johannisthal Flughafen, about 11 km SSE of Berlin; it was here where the military testing was done on all aircraft, engines, etc. On 22 Dec. 1916 a *Wissenschaftliche Auskunftei fuer Flugwesen* (WAF, Scientific Information Office for Aviation) was established, the task of which was primarily to publish new

In total, there were 37 R.VI giants built by several manufacturers, amongst these 18 of the Zeppelin-Staaken type, the rest by Siemens-Schuckert and Gotha.[65] The supercharged Zeppelin-Staaken R.VI combined the best of German compressor technology with the world's largest series-produced wooden aircraft, Fig. 4.14. With a wingspan of 42.2 m— more than 10 m greater than the B-17—and a take-off weight of 11,848 kg, the bi-plane monster aircraft required an 18-wheel landing gear. Entering production in late 1916, the bomber was initially posted to the Eastern Front, then transferred to the West where it participated in raids on the English capital. An impressive flying machine, powered by 2×2 Mercedes D.IVa 6-cylinder, water-cooled inline with 4 valves per cylinder, engine tandem (pusher/tractor) arrangement with 2×2 propellers of 4 m diam., and of 260 hp take-off power/motor. R.VI serial number R30/16 was used as the test bed for supercharger experiments. A Brown Boveri four-stage compressor, Fig. 4.15, powered by a separate 120 hp Mercedes D.II SC (single overhead camshaft) engine was installed in the central fuselage immediately behind the pilot, and by putting the pressurised air piping to the two tandem engine nacelles and to the blower motor. The speed of this auxiliary motor was hand-controlled, thus omitting the inherent disadvantage of over-powering of directly coupled turbochargers at lower altitudes. Moreover, it provided a failsafe approach to supercharging, since failure of the central supercharger would do no more than reduce power to the drive engines. The speed difference between drive motor (1450 rpm) and BBC blower (6000 rpm) was bridged by a carefully manufactured ZF gearbox. In its normal configuration the R.VI had a service ceiling of ~4 km (reached within 2½ h) and a flight speed of 135 km/h, which improved charged to ~6 km in 1½ h and 160 km/h.

First aero engine altitude test facilities had been used by the engine company *Luftfahrzeug-Motorenbau GmbH*[66] on Mt. Wendelstein in the Bavarian Alps at an altitude of 1.800 m (which were relatively easily accessible by a cog railway since 1912) and by Benz—as a mobile facility—on 1.600 m high Schneeberg, Lower Austria as early as 1916/ 17. They served as test beds for oversized and super-compressed altitude engines like the specially designed 250 hp Maybach MB.IVa, which up to an altitude of 2.000 m had to be throttled in flight. Figure 4.16 shows Karl Maybach's altitude test campaign in 1917;

scientific results from government research contracts in a *'secret journal'*, the *'Technische Berichte der Flugzeugmeisterei der Inspektion der Fliegertruppen, Charlottenburg'* (abbreviated TB), to which Ludwig Prandtl's *MVA Modellversuchsanstalt* (Model Test Institute) Goettingen contributed a total of 24 communications since 1917,—see Eckert, Ludwig Prandtl, p. 96 f. Among these, 10 were authored or co-authored by Max Munk (1890–1986), Prandtl's assistant from 1916 and 1918, and together with Albert Betz (1885–1968) Prandtl's closest research collaborators in the development of the *Aerofoil Theory*.

[65] Eighteen R.IVs were built, serialised 'R25' to 'R39' and 'R52' to 'R54'; exceptional, the 'R30' was used exclusively as a supercharged engine test bed between April 1918 and June 1919, when the Treaty of Versailles stopped corresponding test activities. See Eckardt, Gas Turbine Powerhouse, p. 57 f.

[66] A joint venture of Wilhelm Maybach and Ferdinand Graf von Zeppelin 1909–1918, which became *Maybach-Motorenbau GmbH* thereafter.

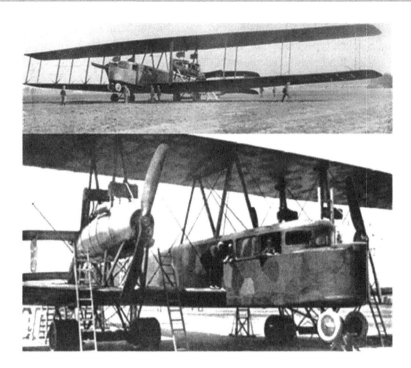

Fig. 4.14 German Zeppelin-Staaken R.VI (*R Riesen*, giant) aircraft with separate BBC supercharger in central fuselage, crew 7, flight endurance 7–8 h, 1917/1918

however, these mountain test beds proved somewhat impractical due to the limited altitude, weather dependence and the distance from the home plant, so that Zeppelin decided to build a low-pressure engine test cell at Friedrichshafen on Lake Constance as early as 1916. Figure 4.17,[67] a photo rarity from the Zeppelin Archive, shows the test personnel with oxygen breathing pipes within the altitude chamber,[68] led by the head of the Zeppelin test department, engineer Leitmann in the middle. It became an invaluable tool for assessing the altitude performance of flight piston engines. It was constructed by Zeppelin's renowned chief designer Ludwig Dürr[69] and used by Walter G. Noack in a comprehensive test campaign between November 1917 and February 1918 on a Daimler D-IVa engine[70]

---

[67] Motor identification, thanks to Michael J. Jung, Mercedes-Benz Classic, Stuttgart, 26 Aug. 2019. The Benz Bz IIIa had 185 hp at sea level with 5.65:1 compression ratio and weighed 276 kg, i.e. 43% of the empty weight of its application in a Roland Rol.D.VIb aircraft. The Benz Bz IIIa engine made the Rol.D.VIb an equal to the more famous Fokker Fok.D.VIIF with the BMW IIIa engine. It was as fast as the D.VII and had a slightly better rate of climb.

[68] Noack in his test report (see below) describes the use of oxygen masks, but the difference could not be dissolved.

[69] See Wikipedia 'Ludwig Dürr' in English.

[70] See Noack, Tests of the Daimler D-IVa engine, translated by the NACA Paris Office from Technische Berichte III, Sec.1 as NACA TN-15, Oct. 1920.

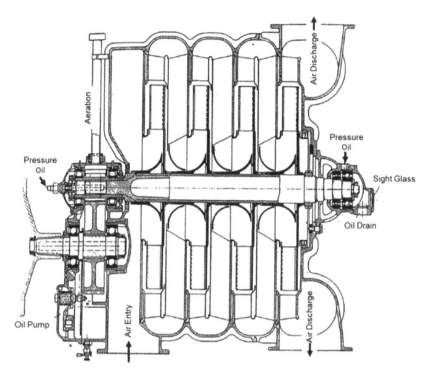

Fig. 4.15 BBC four-stage radial compressor for 1200 hp R aircraft engine equipment, 6000 rpm, 0.47 m impeller diam., air mass flow 4600 kg/h, PR 1.75

with and without the separately driven BBC four-stage radial compressor of Fig. 4.15. The building of this unit had been initiated by Noack at BBC Mannheim[71] as application in a Zeppelin-Staaken R.VI giant aircraft. According to one source the facility was *replicated* in similar form by the U.S. Bureau of Standards till the end of 1918 for Liberty engine altitude testing, though details of a corresponding information transfer are lacking.[72] The Zeppelin vacuum chamber consisted of a ferro-concrete building with a ground area of 8.5×4 m and 3.4 m at its highest pitch. The vacuum was generated by an electric-powered, positive displacement blower of the company C. Enke, Schkeuditz near Leipzig. Unpressurised engine tests revealed a nearly linear performance/power decrease with altitude up to simulated 6.000 m, down to ~50% of the nominal 260 hp of the Daimler D-IVa engine.

---

[71] See Noack, Airplane Superchargers. A paper which discloses Noack as incessant advocate of aero turbocharging with early German examples—besides those from Brown Boveri Mannheim—from companies Schwade, AEG and Siemens-Schuckert. The four-stage, separately powered BBC turbocharger existed also as directly powered two- and three-stage versions.

[72] See Dempsey, P.: 'Notes on WW I German Superchargers', https://www.enginehistory.org/Piston/Before1925/WW1SC/WW1SC.shtml

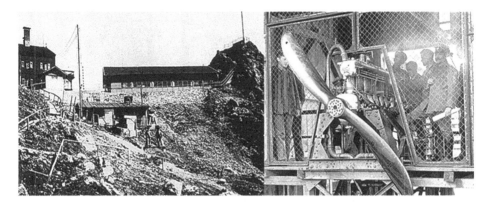

Fig. 4.16 Altitude test cabin of oversized/super-pressurised Maybach MB.IVa aero engine on Mt. Wendelstein, 1.800 m, 1916/17 © MTU-F

The turbocharger tests—though not unambiguously interpreted—revealed successfully the possibilities to compensate decreasing motor power by supercharging in a low pressure test chamber, an early reference to BBC's later unique experience in building high altitude test facilities, Sect. 8.3.3. They were also preparatory to the practical installation of a turbocharging BBC compressor in a R.VI giant aircraft for the purpose of maintaining constant power at high altitudes. Correspondingly, the first flight of the experimental giant aircraft R 30/16 with a supercharged motor plant took place on 2 April 1918. The supercharger produced with its four stages a pressure ratio PR 1.75, lifting the operational ceiling from 3800 to 5900 m.

4.1.3   The Path from Aerofoil to First Axial Compressor

The successful realisation of the known gas turbine concept depended—as has been outlined before—on an efficient axial compressor, this was the most critical component. Brown Boveri at Baden, CH, succeeded in the development of first commercial axial compressors in 1931—and of the first utility gas turbine in 1939, just weeks before the first turbojet flight based on aero gas turbine propulsion—a precondition for superior high-speed flight, what was thought to become one of the most essential issues in WW II warfare, 1939–1945. Apart from first axial compressor development activities at BBC in Switzerland, beginning after 1926 under the lead of Claude Seippel, independent axial-flow compressor development also started in England apparently at the same time at RAE Farnborough on the initiative of Alan A. Griffith, Sect. 4.2, followed at *AVA*[73] Goettingen

---

[73]*AVA Aerodynamische Versuchsanstalt Goettingen* (Aerodynamic Test Establishment) resulted in 1919 from the *MVA Modellversuchsanstalt für Aerodynamik,* which changed its name in 1915 from *Modellversuchsanstalt der Motorluftschiff-Studiengesellschaft* (Society for Motor Airship Studies),

Fig. 4.17 Zeppelin Works Friedrichshafen engine altitude test cell, ~ Sept.1917, 6 cyl. water-cooled Benz Bz IIIa aero engine on test (rear ctr.), fuel supply (rear l.), water brake (front ctr.) © Archive Luftschiffbau Zeppelin GmbH

shortly thereafter in ~1927/8 by Albert Betz and Walter Encke (Sect. 4.3). Jakob Ackeret, returning to Switzerland from his stay at Prandtl's Institute at Goettingen (1921–1927), commenced first at Escher Wyss Zurich, later in 1930 together with his assistant Curt Keller at ETH[74] Zurich a second Swiss attempt[75] of scientifically based axial-flow compressor design methodology. This was with all likelihood, due to Ackeret's close cooperation with Betz at the end of his Goettingen stay, a copy or close to the Goettingen design system, which in itself, though not documented in detail, exploited the lift/drag characteristics of some investigated aerofoils out of the comprehensive Goettingen

---

founded in 1907 by Ludwig Prandtl, and which played an essential role in the development of scientific aerodynamic principles and thus—as described in this chapter—also in early axial turbomachinery.

[74] *ETH Eidgenoessische Technische Hochschule* (Swiss Federal Institute of Technology).

[75] Historian Edward Constant, otherwise quite reliable, is wrong in ignoring Brown Boveri's stand-alone lead role in deducing practical axial compressor design principles derived from Prandtl's aerofoil experimental results in combination with lift/drag assessments in the wake of early *Aerofoil Theory* applications, – see Constant, E.W., The Origins, p. 110. Notwithstanding the spatial vicinity of only some 20 km between Zurich and Baden, the two design camps cultivated no in-depth information exchange and especially BBC protected its design know-how advantage on the basis of its earlier and comprehensive experimental start. Typical peculiarities will be outlined in Sect. 4.3.3 'Degree of Reaction' in the context of R preferences—R 0.5 at BBC and R 1 at AVA/ ETH.

catalogue of nearly 350 different profiles. The described transparency of widely independent axial compressor design origins can be upheld at best still between 1926–1933, then in 1934 appeared Curt Keller's doctoral dissertation *'Axialgebläse—vom Standpunkt der Tragflügeltheorie'*, followed in 1937 by the immediately translated and highly disseminated English version *'The Theory and Performance of Axial-Flow Fans'*.[76]

The present topic from *'Aerofoil to First Axial Compressor'* has been described already at length in the context of BBC's development of the power generation gas turbine,[77] so that only some highlights and newly appeared aspects will be reviewed in the following. Brown Boveri in Switzerland developed its axial compressor design methodology steadily between 1926 and 1932, later with the advantage of the early beginner, but during this process confronted with enormous uncertainties about the transferability of single aerofoil wind tunnel results into the turbomachinery environment. As will be seen at various occasions, communication in the information triangle Switzerland—Germany—England worked surprisingly well however, so that even short-time followers appreciated the benefit of omitting some of the critical and elaborate proof-of-concept efforts. Of course, BBC experienced also valuable support and could rely on priceless preparatory work especially from Prandtl and co-workers at Goettingen, who had developed and—noteworthy to mention—*published* their basic part of *Aerofoil Theory* between 1904 and 1921.[78]

Figure 4.18 illustrates three of the most important and influential contributors to Brown Boveri's finally successful breakthrough to an effective axial compressor. The professional versatility of these early engineers is remarkable. The role of an early intermediary between the emerging *'aerofoil theory'* at Goettingen and possible design applications for *'fluid*

---

[76] See Keller, Axialgebläse, and in this context the situation in the U.S. before and after the reception of Keller's book, Sect. 6.4. Curt Keller (1904–1984) studied Mechanical Engineering at ETH Zurich (1923–1927) with internships at BBC Baden and Escher Wyss. Afterwards he became according to his thesis CV a private assistant to Professor Stodola, before he joined Escher Wyss where he led the Thermal Test Dept. from 1931. Returning from Goettingen, J. Ackeret was chief scientist and head of the Escher Wyss Hydraulics and Fluid Machinery Lab (1927–1931), before he became an ETH professor and founded the renowned Institute of Aerodynamics, accompanied by Keller. In the last two years at Goettingen, Ackeret had participated in the development of independent AVA compressor technology in close cooperation with Prandtl's deputy Albert Betz. Keller, working closely with Ackeret, specialised in axial compressor aerodynamics when compressors for two wind tunnels were required for the new Institute's equipment. The two blowers for the large, subsonic open wind tunnel were of identical Keller design, while the compressor for the closed supersonic wind tunnel, Sect. 4.1.4 and Fig. 4.26 came entirely from BBC,—see Wikipedia 'Jakob Ackeret' in English. Keller mentions in his thesis that Ackeret addressed fundamentals of axial compressor design in his *'Samstagsvorlesungen'* (Saturday lectures), an institution as part of a *'Studium Generale'*, comprising law, economy, music, arts, history, ethics, of which every engineering student had to attend at least two during his ETHZ study time; saturdays were picked to allow engineers in industry to attend as well. (Information, courtesy to the late Prof. Dr. W.R. Gundlach, Lodz, Pl, 11 Sep. 2002).

[77] See Eckardt, Gas Turbine Powerhouse, pp. 98–121.

[78] See Eckert, The Dawn, p. 55 f., Prandtl, Ergebnisse, I. Lieferung,1921 and the three publications of Munk/ Hueckel, 1917/1918.

Fig. 4.18 Axial compressor 'influencers': Erich Hueckel (1896–1980, l), ~1936, Walther Bauersfeld (1879–1959, m), ~ 1925, Georges Darrieus (1888–1979, r) ~1940

machinery' played Walther Bauersfeld (1879–1959)[79] with his 1922 VDI paper *'Die Grundlagen zur Berechnung schnelllaufender Kreiselräder'* (Basic evaluation of high-speed impellers). Typically, the paper title does not even reveal the novelty of the addressed subject and its applicability for axial-flow turbomachinery. Bauersfeld, often concerned about cavitation when writing the paper, had rather an axial-flow Kaplan turbine-type impeller in mind, but his loss and efficiency assessments covered also air flow, and he reassures the reader several times that the given data base was sufficiently reliable for the intended system analogy between a single static aerofoil and a rotating aerofoil cascade. He owed this subject to a suggestion of his doctoral thesis supervisor Ernst Reichel (1857–1934),[80] Professor for Mechanical Engineering and Water Power Machinery at TH Berlin-Charlottenburg, where he finished his thesis *'On the automatic regulation of* (water) *turbines'* in 1904. The Bauersfeld design approach assumes relatively few rotor blades, with the general assumption that the circumferential blade spacing t has at least two- to three-times the blade chord length b. The flow is considered to occur on cylindrical, radially stacked sections. Deviating rotor configurations can be shaped by *conformal mapping*. The *effective work* is determined by means of the *lift force A* and the *drag force W*, with A normal to the *mean relative velocity* w_∞ and W in the direction of w_∞— and in due course follows the *rotor efficiency*. With the number of blades undetermined to

---

[79] Walther Bauersfeld (1879–1959) was a versatile, both theoretically and practically talented German engineer, lifelong employed as a leading manager of the optical company Carl Zeiss Jena/Oberkochem, D, however interrupted in 1907/8 by a research award for the investigation of airscrews (sic). He is most renowned for a series of planetariums/*orreries* erected between WW I and II at the German Museum Munich (1923, initiated by Oskar von Miller), Wuppertal-Barmen, Leipzig, Jena, Dresden, Berlin, Dusseldorf, Rome, Paris, Chicago, Los Angeles and New York. https://en.wikipedia.org/wiki/Walther_Bauersfeld, https://www.zendome.de/en/company/history and https://www.ips-planetarium.org/page/a_chartrand1973?&hhsearchterms=%22chartrand%22

[80] See Hager, Hydraulicians, Reichel.

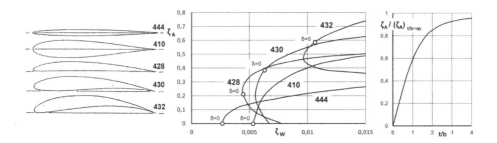

Fig. 4.19 Bauersfeld design tools: typical Gö aerofoils (l), related lift/drag curves (m), spacing dependent lift reduction (r)

the very end, the blade dimensions are deduced from the annulus cross section area. The calculus applies for h, infinite wing width or blade height respectively. The lift coefficient ζ_A and drag coefficient ζ_W—experimentally determined for model wing chord/width ratio b/h 1:5[81] are calculated as well for infinite wing width (blade height); similarly the incidence angles at the profile leading edge are adjusted. As outlined in the following, the best profile form (from the profile catalogue) and the design operation point are picked so that the ratio of lift/drag coefficient becomes a maximum, corresponding to the best rotor efficiency. Key points of the Bauersfeld design methodology are combined in Fig. 4.19: five extremely different profile shapes have been selected out of the Goettingen profile pool to assist in the profile selection process. The number designations #444, 410, etc. are identical with those in the Goettingen reports. The measured lift/drag characteristics of these selected profiles are shown as $\zeta_A = f(\zeta_W)$ on the middle graph, already adjusted—as required—to infinite wing width (blade height). The incidence angle δ has to be referred to the straight line of each of the given five profiles. Zero incidence $\delta = 0$ is designated on each curve. The corresponding angles of other ζ_A are easily calculated.

In line with Bauersfeld's discussion we observe that profile 444 achieves its best efficiency at $\zeta_A = 0.12$ for ζ_W: $\zeta_A = 1$: 28, and that small ζ_W values can be only achieved with slender and straight profiles, which on the other hand show obvious disadvantages when high loadings would be required. Profile thickening on the basis of an unchanged symmetrical profile shape increases the drag resistance as illustrated by curve 410. Quite

---

[81] Calculation and design of BBC's first axial compressor products, ordered within months in 1932—the ETHZ supersonic wind tunnel facility, Fig. 4.26, and the first steam-generating Velox boiler to burn blast furnace gas for the Mondeville ironworks near Caen, France—were based on Gö265 aerofoils which had been delivered in TB Vol.II for a b/h 1/6; later experience showed that thicker profiles have beneficially a less sudden separation behaviour and more bending strength, so that Gö384 from TB Vol.I with b/h 1/5 was preferred. The Bauersfeld design method is best suited for compressor bladings with pitch/chord ratios t/s > 3, i.e. with small mutual blade interference, see Traupel, Thermische Turbomaschinen.

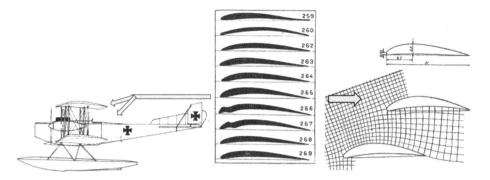

Fig. 4.20 Goettingen wing profile # 265 for *Friedrichshafen FF41* hydroplane (1917, l), 10 Gö wings of TB II (1917, m) and conformal mapping of BBC's first axial compressor (1931, r)

attractive is the situation for profile 428 with a flat pressure side, which has for ζ_A 0.33 a drag/lift ratio $\zeta_{W:}\ \zeta_A = 1:66$.

The bowed profiles 430 and 432 are close to these best efficient drag/lift ratios, but achieve much higher loadings in comparison. In addition to these lift/drag considerations, Bauersfeld takes also the diminishing effect of circumferential blade spacing on the actual cascade lift into account. The right graph in Fig. 4.19 provides a first guide derived from a Kutta flat plate analogy. It illustrates that the lift of a single plate (aerofoil) achieves only 85% for t/b 2, and shrinks further to 59%, if the circumferential blade spacing comes in the order of the plate (aerofoil) chord length t/b 1, always in comparison to the undisturbed plate lift for t/b ∞. Herewith the initial design tool kit was complete and it lasted until 1936 before Brown Boveri replaced this simplified approach for blade count calculation by an— even today—well-known and still useful aerodynamic loading evaluation for compressors and turbines, the *Zweifel number*.[82]

After Bauersfeld's convincing argumentation in favour of the bowed profiles 430 and 432 in Fig. 4.19 in view of loading capability and efficiency, Brown Boveri's subsequent decision for profile 265 is a logic consequence, and it appears additionally intriguing to learn more about the historic origin of this design, which is shown in Fig. 4.20 as part of ten profiles reproduced in the Technical Berichte, Vol.II of 1917.[83]

The AVA Goettingen (since 1919) and the precursor institution MVA Modellversuchsanstalt grew considerably during the world war periods, mostly in military importance, but while WW II saw impressive numbers of up to 800 employees, the humble beginnings during WW I culminated in just 50 members. In 1918 the scientific staff including Ludwig Prandtl (43) and his deputies Albert Betz (33) and Max Munk (28) comprised 9 persons, the workshop had 17 workers and mechanics, while the largest group of scientific assistants and supporters had 23, recruited from local scholars and

---

[82] See Eckardt, Gas Turbine Powerhouse, p. 117.

[83] See Munk and Hueckel, Weitere Goettinger Fluegelprofiluntersuchungen, Table 153.

students.[84] The enormous load of carrying through contracted measurement orders for nearly 300 aircraft wing profiles, received in Goettingen after 1916, in line with the stupendous gain in importance and prestige of aeronautics in general, and the corresponding, in parts weekly reporting efforts, which added up at war's end to three volumes of the Technische Berichte with more than 1000 pages point to lasting merits of Max Munk and Erich Hueckel in this context.[85] Munk's biography, who was officially at MVA from March 1915 to April 1918 (thereafter at the naval aircraft test facility at Warne-muende) and who continued his career after the war at NACA Washington and—with mixed success—at Langley, is well-known,[86] while the not less remarkable biography of co-author Erich Hueckel, Fig. 4.18, went rather unnoticed in aerodynamic circles, while he apparently missed several times the expected Nobel Prize in chemistry, when he was a Professor for Theoretical Physics at Marburg, D.[87] Hueckel, born at Goettingen and in summer 1916 a student in physics and mathematics from Goettingen University, joined—freed from military service on Prandtl's demand—the MVA between August 1916 and May 1918, before he followed Munk for the remaining war period to Warnemuende. While Munk in the critical period 1917/1918 naturally must have been focussed on the preparation of his two doctoral theses at Hannover and Goettingen, Hueckel apparently bore the responsibility for the wind tunnel experiments and their meticulous documentation.

As documented in the quoted TB report, Footnote 80, aerofoil #265 was foreseen for the *Friedrichshafen* hydroplane *FF.41* wings, as illustrated in Fig. 4.20.[88] The aircraft, alone through its small scale of production, played no prominent role, except for one event by which it became the first German aircraft to sink a surface warship on 22/23 August 1917.

---

[84] Prandtl praised Munk's merits in organising a *Kriegshilfsdienst* (war auxiliary service) in the *Geschichtliche Vorbemerkung* (historic preliminary remark) of—see Prandtl, Ergebnisse, p. 3 f. There were groups for measurement preparation, measurement execution and post-measurement evaluation. Therefore, early in 1917 a small barrack with four office rooms was erected and made operational.

[85] See Munk and Hueckel, three reports for *TB der Flugzeugmeisterei*, Vol. I and II, 1917/1918.

[86] See Wikipedia 'Michael Max Munk' in English. On the claimed secrecy of the TBs, Munk mentioned in his memoirs *'All of my Goettingen work was published in secret reports, by the German Army, for this was during the First World War. Nevertheless, they were translated in England a week after appearance and distributed there and in the US.'*,—see Munk, Thoughts and Memories. After leaving NACA in 1927, he worked at several stations, apparently one year (1928) also at ABBEC, Camden NJ, Fig. 4.5.

[87] See Wikipedia 'Erich Hückel'in English. Hueckel's exceptional scientific contributions lay in the just emerging field of *quantum chemistry*—covering theoretical problems between physics and chemistry, and therefore were 'overlooked' at times.

[88] See Wikipedia *'Friedrichshafen FF.41'* in English; up to nine examples of FF.41 were supplied from February 1916 onwards. Powered by twin 150 hp Benz Bz III engines, it was intended for torpedo-carrying duties with a crew of 2–3: Span 22 m, length 13.3 m, speed 125 km/h, range 575 km. https://en.wikipedia.org/wiki/Friedrichshafen_FF.41

On the second day Wolfram Eisenlohr (1893–1991) directed his FF.41 hydroplane some 60 km to the north from his base on the Baltic shores, for a finishing torpedo attack;[89] 25 years later he was General-Engineer and Director of the Department for Engines and Accessories in the Technical Office (01/1938-01/1944) of the RLM Reichsluftfahrtministerium (Reich Air Ministry)—and thus immediately responsible for all turbojet engine developments on the German side (Sects. 6.2 and 8.1). In a kind of visionary *fork in the road* that moment combined present and future technical developments, as indicated in Fig. 4.20, where the dual-use of Gö 265 out of a group of *'ten further profiles'* by Munk/Hueckel on the FF.41 wings in 1917, and as a kind of lead profile in BBC's first axial compressor product in 1931 has been illustrated.

First ideas for a scientifically based axial compressor design were obviously introduced to Brown Boveri in Baden, CH, via its French daughter Compagnie Électro-Mécanique C.E. M. of Paris, belonging to BBC since 1901. C.E.M. built a series of wind turbines under the supervision of their chief engineer Georges Darrieus (1888–1979),[90] Fig. 4.18 (r), ranging from 8–20 m in wheel diameter, Fig. 4.21 (l).[91] Different from what is known today as a

---

[89] Eisenlohr, shortly thereafter promoted to Lieutenant at Sea of the Reserve (and at that time, in addition to his flying duties the unit's carrier pigeon officer) and his pilot Petty Officer second Class Gruber left their naval flyer base at Windau/Ventspils, Latvia to sink the 350 t Russian torpedo-boat destroyer *Stroini* (1907–1917) which had been running aground while protecting mine operations off southwest Sörve Peninsula on Oesel Island, Estonia already the day before on 21 August 1917. See Prommersberger, Seeschlachten, Kap. 5 *Seekrieg 1917*.

[90] Georges Darrieus, son of a vice-admiral of the French Navy, worked continuously for C.E.M. from 1912 to 1958 as a versatile engineer and scientist in areas such as ballistics, fluid- and thermodynamics, turbomachinery and electrical engineering. Today, he is mostly known for his pioneering work in wind energy, comparable to Albert Betz, AVA Goettingen. The mentioned 20 m diam. rotor with horizontal axis produced 35 kW of electric power at a wind velocity of 29 km/h, but was destroyed in a storm. Seippel describes these wind turbines *'as designed according to modern aerodynamic principles, thus supporting the impulse towards further developments along these lines.'* Personal appreciation between Darrieus and the 12 years younger, French-speaking Seippel (with a Geneva family background) certainly helped the early axial compressor developments at BBC.

[91] Independently, both Darrieus and Seippel gave credit to Walther Bauersfeld and his 1922 VDI paper as highly influential on their own design learning curves, mentioning his radical change in cascade design perspective away from established stream tube and flow channel theory towards the lift/drag principle deduced from single aerofoil measurements. Strangely Betz, Ackeret and others might have overlooked Bauersfeld's 1922 VDI paper, since they discuss and welcome the comparable method of Schilhansl in 1927 as rather unique,—see Schilhansl, Naeherungsweise Berechnung. A similar observation may apply for the book of—see Eck, Ventilatoren, which mentions Bauersfeld's 1922 VDI paper for the first time in the second edition of 1952 (on p. 189), while the information is missing in the first edition of 1937. Given the importance of Edward W. Constant's 1980 book on 'The Origins of the Turbojet Revolution' for the aero gas turbine history in general, it is worthwhile, to highlight a few of his inconsistencies and questionable conclusions in this context, in light of recent in-depth research. On p.110 he states: *Prandtl's aerofoil theory, and other insights derived from it, by 1926 led four men—A.A.Griffith in England, Jakob Ackeret in Switzerland, and Albert Betz and W. Encke in Germany—to investigations of axial compressors and turbines which would prove of*

Fig. 4.21 Fluid machinery applications of Prandtl's *Aerofoil Theory*, ~1929: Darrieus' wind turbine (l) and Brown Boveri's axial-flow fans for generator cooling (r), both designed with Bauersfeld's method

Darrieus Rotor (with vertical axis, thus independent of wind direction, but not self-starting), it has a look similar to a conventional three-bladed axial rotor, but enlarged and in combination with the patent drawing of US 1,820,529 (with French priority dating back to 1927), it reveals an uncambered lancet profile in a dual-spar girder construction, apparently to overcome mechanical integrity deficits compared to present day single long-spar composite turbine blade design.

Work on axial compressors at BBC Baden started quite unspectacularly; the increasingly larger quantities of air necessary for the cooling of the electrical generators of high output determined the need for this new type of high-flow capacity blower. A series of single- and two-stage axial fans were developed directly driven by the generator and especially in view of the constricted space demands, Fig. 4.21 (r). Since it was still unknown how the aerodynamic behaviour of a rotating blade would deviate from an isolated aerofoil, a simple *double-conus wind tunnel* was created in which the generator fans worked under ideal upstream and downstream conditions, so that the measured lift/drag curves of the ventilator blading could be compared with the single aerofoil measurements from the AVA Goettingen deliveries, Fig. 4.22. The figure shows the BBC double cone compressor rig with a rotor tip diameter of 1.19 m, a corresponding rotor hub diameter of 0.83 m and the entry spinner/rotor hardware—with speeds up to 3600 rpm. Data sheets from these tests allowed to deduce for every required ventilator volume flow and pressure the resulting wheel and blade dimensions as well as efficiency. The tests irrefutably confirmed the polar plot characteristics of the Goettingen aerofoil

---

tremendous moment for the turbojet revolution. It appears that BBC's contribution as early as 1926 has been omitted or falsely mixed up with Ackeret's role.

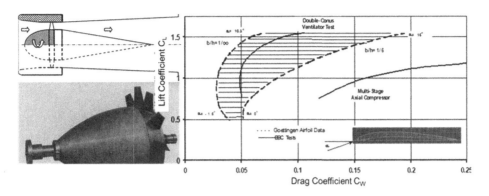

Fig. 4.22 Aerofoil Gö265 lift/drag polar plots in comparison of single aerofoil wind tunnel and rotating compressor rig tests

used, as illustrated for Gö265 in Fig. 4.22, where the lift coefficient C_L has been plotted over the drag coefficient C_W. Both coefficients have no dimensions: in the first case the lift force of an aerofoil is referred to the dynamic flow pressure and a reference planform area, while the drag coefficient of the investigated wing section is defined as the ratio of its drag force divided again by the dynamic flow pressure and a reference area. The Goettingen single aerofoil data for the profile Gö265 show measurement results for an aerofoil chord/width ratio b/h 1:6, which was then recalculated as an infinitely long wingspan ratio b/h $1/\infty$ to eliminate the induced wing tip drag by simulating a ducted rotor with zero tip clearance. Various angles of attack between $\alpha = -1.5° \ldots + 15°$ are shown along the Goettingen aerofoil measurements (hatched area). The corresponding BBC double-cone ventilator test fits perfectly into the expected range of single aerofoil test data. In addition, a multi-stage axial compressor curve has been drawn, considerably shifted to the right in the area of higher losses and with less average loading capability. *'Radial equilibrium'* designs—with the radially acting forces on the fluid in equilibrium—were verified at BBC as early as 1929.[92] However, tests showed that the effect of additional corrective radial blade twist had only small influence on the compressor efficiency, and was therefore omitted in follow-on designs. This considerably facilitated the task of the designers and allowed for simpler and cheaper blade forms.

Besides these preliminary, though important preparatory basic aerodynamic investigations, Brown Boveri's first axial compressor test rig was built at Baden, CH in 1926 and started in due course a five year long test campaign.[93] Even before Claude Seippel's return from the United States in April 1928,[94] the head of BBC's test department

---

[92] For further details, − see Eckardt, Gas Turbine Powerhouse, p. 116 f.

[93] See Eckardt, Gas Turbine Powerhouse, p. 102 f.

[94] See the foregoing Chap. 3

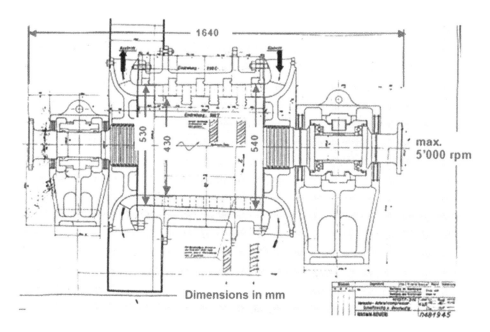

Fig. 4.23   BBC four-stage axial compressor test rig, drawing 18 Sep. 1926

Jean von Freudenreich initiated axial compressor activities by building a four-stage test rig, Fig. 4.23. Several times modified and adapted, the first blading had four identical stages with untwisted, symmetrical blades and vanes on a cylindrical hub of 0.43 m diameter and conical outer contour variation from 0.54 to 0.53 m diameter. At the entry, an inlet guide vane prepared the flow conditions for the following symmetrical stages. An exit guide vane was planned for axial outflow re-direction, but was removed after initial testing.

The preserved test programme for that rig dates from 7 October 1926, stating in the words of Jean von Freudenreich: *'The purpose is the application of new aerodynamic theories to compressors to improve efficiency and simplify construction. . . .The compressor wheel consists of an endless row of wings, which interact. Airplane wing profiles are therefore not necessarily optimum for compressors and turbines. Possibly compressor aerofoils should be more cambered than the original aircraft wings; this to investigate is a prime test target. . . .'*

Immediately after surge and instability problems had been overcome, as described in the following Section, and the design rules for optimum blade/vane staggering had been investigated together with the selection of optimum degree of reaction for stage design,[95] BBC successfully resolved and implemented in the mid-1930s another fundamental cascade principle—with the advantage of exclusive usage for more than a decade. Suggested

---

[95] This important design feature will be discussed in the context of the AVA Goettingen design practice in Sect. 4.3.3.

by their young employee Otto Zweifel (1911–2004), BBC has used an aerodynamic loading coefficient Ψ for blade number design calculations and cascade test evaluations, which today is still well-known and useful for preliminary compressor and turbine design in an adapted form: the *Zweifel number*. Surprisingly, it was shown[96] that optima for small and large turnings occurred almost always at values of $\Psi = 0.8 \ldots 1.0$. Consequently, the designers achieved much faster, more acceptable blade shapes and forms of flow channels. At Brown Boveri, Zweifel was awarded the task of designing the turbine for the first power generation gas turbine at Neuchâtel, CH, in 1938. He kept Ψ 0.8 for all rows from root to tip and produced the best turbine efficiency of all BBC designs up until then.

4.1.4   First Axial Turbo-Products and Wind Tunnel Blowers

After the realisation of the first axial compressor at Brown Boveri in the early 1930s, the following decade saw a manifold of successful compressor reference projects:

- the first blast furnace gas compressor system,
- compressor power for the first supersonic wind tunnels,
- the steam-generating *Velox* boilers with axial turbo-sets,
- turbo-blowers for the *Houdry* cracking process,
- the first power generation gas turbine, and finally
- the first locomotive gas turbines.

The development of these various products has been described already in detail,[97] so that a general overview on the basis of sixteen machines might be sufficient, illustrated in Fig. 4.24 as the development of axial compressor stage pressure ratio over time, before parallel aero gas turbine activities come into focus.

The sixteen industrial projects,[98] selected for this visualisation of technical progress, belong to three main categories:

---

[96] For details of Zweifel number deduction, – see Eckardt, Gas Turbine Powerhouse, p. 117 f.

[97] See Eckardt, Gas Turbine Powerhouse, pp. 123–135, pp. 158–190 and 196–203

[98] Of these 16 listed objects, #8—the first power generation gas turbine of 4 MW for the former Municipal Power Station at Neuchâtel, NE, CH (1939) is preserved since 1988 as the ASME 135th International Mechanical Engineering Landmark, https://www.asme.org/about-asme/engineering-history/landmarks, presently on exhibition at the General Electric plant, Birr AG in Switzerland and under the auspices of IndustrieWelt Baden, CH, an association for the preservation of industrial culture in the Baden area https://www.industrieweltbaden.ch/ (in German); the historic turbo-set of a 2200 hp BBC locomotive gas turbine is still on storage there. Regretfully, no leftovers of the first ETHZ supersonic wind tunnel (#5, compressor and measuring section) exist any longer, after these historically invaluable parts had been already stored at the *'Swiss Scientific Center Technorama'*, Winterthur; they were scrapped in 1990 without any further notice. Information, with courtesy to Georges Bridel, 25 June 2016.

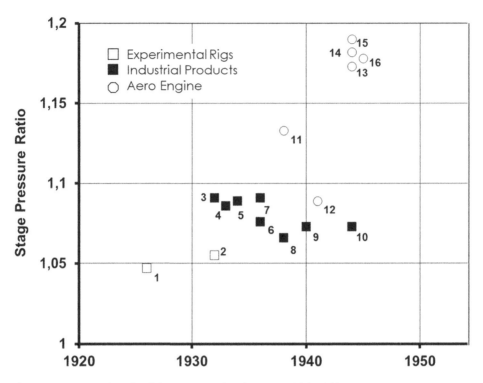

Fig. 4.24 History of BBC axial compressor development, 1926–1945

A. Experimental rigs
1. 4-stage test rig, 1926, Fig. 4.23
2. 4-stage test rig, 1932
B. Industrial products
3. 11-stage Mondeville compressor, 1933, Fig. 4.25
4. 13-stage 'typical product compressor', 1933
5. 13-stage ETHZ supersonic wind tunnel blower, 1934, Fig. 4.26
6. 20-stage Houdry compressor, 1936
7. 11-stage Velox compressor, 1936
8. 23-stage Neuchâtel GT compressor, 1938
9. 21/20-stage RAE compressor, 1940, Fig. 6.44
10. 19-stage wind tunnel compressor, 1944
C. Aero engine projects
11. 7-stage aero turbocharger for RR Crecy, 1938, Fig. 6.38
12. 15-stage SOCEMA/C.E.M. TGA -1 compressor, 1941, Fig. 10.20
13. 10-stage BBC compressor for BMW—003D, 1944, Fig. 9.4
14. 9-stage BBC compressor for tank GT 102, 1944
15. 7-stage BBC compressor for BMW—003C, 1944
16. 8-stage SOCEMA/C.E.M. TGAR-1008 compressor, 1945, Fig. 10.21

Fig. 4.25 Mondeville 10-stage test compressor rotor, PR 2.4, 10,000 rpm, $\eta_{is} = 0.83$, degree of reaction R 0.5, #3 in Fig. 4.24, ~ January/February 1933 © ABB Archive

The stage pressure ratio for the two *experimental rigs* starts with values of ~1.05, while the field of *industrial products* jumps immediately to 1.09—from where it flattens out over time conservatively to approximately 1.07; clearly separated is the area of *aero engine projects*[99] in the 1937–1945 time frame, where peak values of up to 1.19 were achieved, mostly as a result of considerably higher circumferential speeds (and in due course higher aerodynamic loading capability) in comparison to the industrial projects.

Comparable to Griffith's experiences in the UK,[100] compressor *stability* represented the largest obstacle for the Swiss developments. It was known relatively early how to achieve stability on two stages from the foregoing generator blower activities, but there was a great deal of apprehension as to how 10 or more stages in a row would perform. Theoretical studies had already revealed that stability is not only a function of the flow conditions inside the compressor, but depends largely on the whole system volume connected to it. Considerable flow and combustion instabilities overshadowed the first test period, until it was realised that the addition of one to two stages brought great improvements, avoided all instabilities and saved the compressor from being scrapped due to heavy, surge-induced vibrations. Already the first commercially sold axial compressors for the Mondeville plant had to be *boosted* by an 11th stage to achieve stable design data for 12,000 rpm with PR 3.3, at a volume flow of 3.5 m^3/s and a temperature-measured adiabatic efficiency of 83%; the corresponding test rotor with still 10 stages is shown in Fig. 4.25. This turbo-blower had to supply compressed air to blast furnaces. The furnace gases were burned in a combustion chamber and the exhaust gas drove a turbine which in turn powered the air compressor.

A prominent point in the history of Brown Boveri's axial compressor development is the propulsion system for the world's first supersonic wind tunnels, realised between 1932 and 1952 in six cases for Mach numbers 2–4, and for the Kochel, D/Tullahoma,TN facility with serious planning up to Ma 10. Though most of these were part of military research programmes, e.g. for ballistics and missile aerodynamics, there is generally more project

---

[99] All these six aero projects of Brown Boveri and partners will be reviewed in detail in various sections of this book: #11 in Sect. 6.3.2 (Fig. 6.38), #12 and #16 in Sect. 10.3.2, #13 (Fig. 9.4) and #15 in Sect. 9.1, #14 in Sect. 8.3.4.

[100] See Sect. 4.2

Fig. 4.26 13-stage BBC axial compressor for ETHZ supersonic wind tunnel facility, front view (l) and installation set-up (r), 1934 © ETHZ (l) and ABB Archive

information available via the scientific community than for comparable industrial projects at that time and consequently, most of these projects were also internationally interrelated.[101]

Besides lengthy preparatory engineering work for the first supersonic wind tunnel drive, Fig. 4.26, the success of this project for Prof. Ackeret's Institute of Aerodynamics at ETH Zurich was not the least determined by a psychological component. Seippel praised Prof. Ackeret several times for his courageous step of issuing an order without significant experience from BBC at that time. The order intake document[102] confirmed an average total pressure ratio PR 3 at 3800 rpm, a maximum intake volume flow of 50–55 m³/s and for the design point a typical pressure rise from 0.125 to 0.275 bar at 45 °C, 3400 rpm and a compressor efficiency of 75%.

The actually measured compressor map[103] confirms these guarantee values with excellent agreement. This outstanding result for the first-ever design of a multi-stage axial compressor impressively illustrates how well Claude Seippel's design team prepared for such a success. Apparently not only the aerofoil aerodynamics fulfilled expectations, but also the complete ducting system both up- and downstream of the compressor were carefully taken into account. The peak performance of η_{ad} 0.75 appears unimpressive at first glance, but the copy of this compressor, immediately produced afterwards for the

---

[101] See Eckardt, Gas Turbine Powerhouse, p.126 f.

[102] Officially ordered in 1933, Ackeret's ETH Institute received two wind tunnels, a large subsonic wind tunnel with a 2 × 3 m exhaust nozzle, manufactured by Escher-Wyss with two single stage axial blowers in parallel, designed by Ackeret's assistant Curt Keller and the first supersonic wind tunnel with 0.4 × 0.4 m measuring section, built in the manner of Prandtl's closed circuit concept by Brown Boveri, including design and manufacturing of the 13 stage axial compressor. The statement of—see Constant, E.W., The origins, p.115—*that Ackeret designed the Zurich blower would seem to indicate that his understanding of the axial compressor was somewhat more advanced than Brown Boveri's*— is fundamentally wrong.

[103] See Eckardt, Gas Turbine Powerhouse, p. 127.

Guidonia wind tunnel[104] revealed the impact of low Reynolds numbers for the ETHZ case. At Guidonia, the increase in driving power to 2100 kW allowed for a raise in inlet pressure and for a Reynolds number increase by a factor of 3, the significant benefits of which were shown in the compressor efficiency by an addition of six to eight percentage points.

The Velox boiler, first built by Brown Boveri at Baden, CH, in 1931–1932, is another example of a proto-gas turbine. Per definition a high pressure, forced circulation water-tube boiler, the *Velox* combines the use of combustion under pressure, with very high flue gas velocities and with a gas turbine, whereby the pressure in the combustion chamber, the high gas velocities and the pressure drop of the gases for production of work are created by means of a compressor driven by the gas turbine, the latter being actuated by the products of combustion. The great significance of the Velox idea lies in the fact that by means of the increase in pressure and velocities, the components are greatly reduced in size, and the processes are accelerated without impairing the efficiency. The corresponding BBC sales slogan was that the Velox provided steam in a certain number of minutes, for which conventional steam generation required the same amount in hours. The whole arrangement is thus a gas turbine, used not as a source of power, but as a source of heat. This GT application rendered essential the creation of a compressor set having a high efficiency. The problem was solved at Brown Boveri as early as 1932 by the development of a four- or five-stage reaction turbine and a ten- to twelve-stage axial compressor with the design taking into account the results of the then latest research in the field of aerodynamics. By the end of the 1930s and after steady axial compressor performance improvements, the turbine not only ran the compressor, but delivered a certain excess power which was used for electricity generation.

The Velox operational principle of combustion at raised pressures was also used in chemical engineering for the first time in 1936, when demand for hot compressed air in connection with the catalytic *Houdry oil cracking process*[105] created the first field application at the Marcus Hook refinery of Sun Oil Co., Philadelphia, for the newly emerging BBC constant-pressure gas turbine. The primary purpose of these machines, as of the foregoing blast-furnace applications, was to supply compressed air, but the Houdry turbines have no combustion chamber of their own, the hot by-product gases of the cracking process being used directly to drive the turbine. The sheer size of the required Houdry turbo-set was a challenge for BBC, with a volume flow of nearly 20 m³/s at a pressure ratio PR 4.2. The compressor unit with 20 stages was a considerable step forward in comparison to the previous Mondeville and ETH Zurich units. Like the later Velox boilers, the Houdry turbo-blowers had a net power output. The excellent performance of the delivered turbomachinery meant the 4.4 MW

---

[104] See Eckardt, Gas Turbine Powerhouse, p. 126 f.

[105] After the French mechanical engineer Eugène J. Houdry (1892–1962), who invented catalytic cracking for petroleum feed stocks.

compressor driving power was more than compensated by the 5.3 MW turbine output, and a gearbox-coupled generator produced electricity accordingly.

By the end of the 1930s the efficiencies of the Brown Boveri axial-flow compressors and turbines were remarkably high, each being close to 85%. It was the substitution of axial for centrifugal compressors which brought a net power output from the Velox boiler, and the overall efficiency of the Houdry gas turbines was as low as it was—not much over 15%— chiefly because the used temperatures were very low. The turbine inlet temperature was never higher than 550 °C and usually closer to 480 °C, because at higher temperatures the turbine life was too short to be economical.

Even with efficiencies only a little over 15%, gas turbines could be used as independent engines in special circumstances: where low first cost outweighed high operating cost as in a stand-by plant, or where the engine replaced was particularly inefficient, as in a railroad locomotive.[106] As mentioned before, on 7 July 1939 Prof. Aurel Stodola, then already 80 years old, supervised the acceptance test of the first power generation gas turbine, manufactured at BBC Baden for a 4 MW stand-by electric plant for the City of Neuchâtel, and by 1940 it had built four more engines of the Neuchâtel type—with at design point—an adiabatic efficiency for the 23-stage axial compressor of 84.9% (total pressure ratio PR 4.39), an adiabatic efficiency for the 7-stage axial turbine of 88.4% (expansion ratio 4.27) and 550 °C inlet temperature. At a total mass flow of 62.2 kg/s, the overall thermal total efficiency was 17.38%.[107]

Summarising his view of 'The Second Internal Combustion Revolution', Edward Constant II[108] states: '. . . by 1940 a very few projects had brought the gas turbine to the verge of general acceptance, and had in fact secured its wide adoption for special purposes. The main burden of this incipient revolution fell to Brown Boveri, although other experimenters also made significant contributions. Brown Boveri was ideally situated to create a revolution in internal combustion turbines. The company had an excellent background in steam turbine manufacture; it had constructed the multistage centrifugal compressor for the Armengaud-Lemale turbine (1906) and had subsequently marketed centrifugal compressors; it had constructed a number of Holzwarth intermittent combustion (explosion) turbines, and had also manufactured centrifugal turbosuperchargers for diesel engines and geared centrifugal superchargers for aero-engines during the First World War. And, as noted above (in this chapter), . . . it was in the forefront of scientific investigation of axial compressors. . . . by 1935–36 the turbines and compressors (of the Velox and Houdry plants) were so highly developed that the turbine not only delivered enough power to run the compressor but also could run a small

---

[106]The first 2200 hp gas turbine-powered locomotive was delivered for the Swiss Federal Railways by Brown Boveri, Baden, CH in 1941,—see Eckardt, Gas Turbine Powerhouse, p. 196 f.

[107]See Sect. 4.3.3 The World's First Utility Gas Turbine Set at Neuchâtel (1939) in Eckardt, Gas Turbine Powerhouse, p. 181 f.

[108]See Constant, E.W., The Origins, p. 145.

auxiliary generator.' The door to the power generation gas turbine (1939)—and at the same time to the turbojet gas turbine engine—stood widely open.

4.2   A.A. Griffith and Early British Activities Towards Aero Gas Turbines

4.2.1   A 'Scandal' To Start With

Alan Arnold Griffith (1893–1963) was a versatile scientist, best known for his work on stress and fracture in metals—now known as metal fatigue—at the beginning of his career, as well as after a rather abrupt change in subject in the mid-1920s as one of the first to develop a strong theoretical basis for the jet engine. Griffith's advanced axial-flow turbojet engine designs were integral in the creation of Britain's first operational axial-flow jet engine, the Metropolitan-Vickers F.2 which first ran successfully in 1941.[109]

Griffith took a first in mechanical engineering, followed by a master's degree and a Doctorate from the University of Liverpool. In 1915 he was accepted by the Royal Aircraft Factory as a trainee, before joining the Physics and Instrument Department the following year in what was soon be renamed as the Royal Aircraft Establishment (RAE).

During this period, in December 1917, a joint paper by A.A. Griffith, M.Eng., and G.I. Taylor,[110] M.A., entitled *'The use of soap films in solving torsion problems'* was read before the Institution of Mechanical Engineers in London with an attendance of 71 members and 36 visitors. Timoshenko in the 1953 edition of his book *History of Strength of Materials*[111] remarks that *the practical importance of Prandtl's soap film analogy was recognised by (Professor) Geoffrey I. Taylor and his student A.A. Griffith, who used this technique for the determination of torsional rigidities of rods of various cross-sections*, which contradicts somewhat with the actual submission and reception of the paper. The written discussion to the paper begins with a statement of the IMechE President confirming that *'. . . the paper was both original and very interesting'*. However, although mention is made of *'. . . a few isolated experiments . . . in this country and in Germany'* on page 792 of the Griffith-Taylor paper, there was no reference to the original soap bubble analogy work of Ludwig Prandtl[112] and his students. Considering the fact, that especially

---

[109] For details on 'Early Turbojet Developments in Great Britain',—see Sect. 6.3 together with a picture of A.A. Griffith, Fig. 6.39 (l).

[110] See Wikipedia 'G.I. Taylor' in English. (Sir) Geoffrey Ingram Taylor (1886–1975), scholar of Trinity College Cambridge and *'one of the most notable scientists of the twentieth century'*, was sent to the Royal Aircraft Factory at Farnborough at the outbreak of WW I to apply his knowledge to aircraft design, working amongst other, on the stress on propeller shafts, where he met Griffith.

[111] See Timoshenko, History of Strength.

[112] See Wikipedia 'Ludwig Prandtl'in English. Ludwig Prandtl (1875–1953) received his education at the then Royal-Bavarian Technical High School Munich (now Technical University Munich) where,

Taylor should have been very well aware of the newest developments in plasticity and materials testing in Europe, it seems at first glance very strange that a technique, widely known for 14 years, had to be reinvented in the UK in 1917.[113] But it is apparently difficult to name the true incitements in this context. 1917, the culmination of the war-time struggle with lots of anti-enemy excesses[114] on both sides, also in the scientific work of our two, then 24- and 31-year-old protagonists?

Griffith continued to specialise in *Fracture Mechanics*; his work *'The phenomena of rupture and flow in solids',*[115] published in 1921, analysed the mechanics of crack formation and resulted in a new awareness in many industries. The *'hardening'* of materials due to processses such as *cold rolling* was no longer mysterious. Aircraft designers were better able to understand why their designs had failed even though they were built much stronger than was thought necessary at the time, and soon turned to polishing their metals to remove cracks.

Griffith's troubles started in the early 1920s, not only internally at RAE when his nickname *'soap bubble'* spread, but also on the floor of the international community of mechanical engineers, when Theodore von Kármán,[116] encouraged by his multi-lingual sister Josephine took the initiative to expand his science beyond national boundaries and to

---

after graduating with a Ph.D. in 1900, he assisted his thesis supervisor (and future father-in-law) August Foeppl (1854–1924), renowned Professor for Technical Mechanics and Graphical Statics at the Technical University of Munich, to reorganise the materials testing laboratory. Working at the Hannover Polytechnic Institute in 1903, Prandtl published his famous membrane analogy—see Wikipedia, 'Membrane analogy' in English—of the torsion problem (—see Prandtl, Zur Torsion) where he shows that by using a soap film all the information on stress distribution in torsion can be obtained experimentally, —see Timoshenko, History of Strength. The mentioned Wikipedia article on 'Membrane analogy' clearly names L. Prandtl as original inventor in 1903, in the contrary to the present Wikipedia article (Nov. 2019) in English about 'Alan Arnold Griffith' which maintains unshakeable still: *'Some of Griffith's earlier works remain in widespread use today. In 1917 he and G.I. Taylor suggested the use of soap films as a way of studying stress problems.'*

[113] See Rossmanith, Fracture Research in Retrospect, p. 47.

[114] Another example of an anti-German, nationalistic atmosphere of war in 1917 will be told in Sect. 6.3 for Metropolitan-Vickers's mother company British Westinghouse Electric and Manufacturing Co. which forced its German-born well-beloved General manager 'Papa' Philip A. Lange (1856–1937) to leave the company when war tensions could apparently not be left out of company dealings any longer.

[115] See Griffith, The phenomena.

[116] See Wikipedia 'Theodore von Kármán' in English. Th. von Kármán (1881–1963) was a Hungarian-American mathematician, aerospace engineer and physicist who was active primarily in the fields of aeronautics and astronautics. He is regarded as the outstanding aerodynamic theoretician of the twentieth century. After graduating at Budapest in 1902, he joined Ludwig Prandtl at the University of Goettingen, where he received his doctorate in 1908, teaching at Goettingen for another four years. In 1912 he accepted a position as Director of the Aeronautical Institute at RWTH Aachen University where he stayed—interrupted by his service in the Austro-Hungarian Army during WW I—until 1930, before he accepted the directorship of the Guggenheim Aeronautical Laboratory at the California Institute of Technology.

establish a series of international conferences in mechanics, of which the first one took place rather informally in September 1922 at Innsbruck.[117] Already two years later the first ICAM International Congress of Applied Mechanics took place at Delft, Nl in April 1924. Despite the still mixed feelings about Germany, especially from the French, the First International Congress for Applied Mechanics was a success and a remarkable demonstration of international scientific cooperation only six years after the war; 214 scientists from 21 countries gathered there and presented some fifty papers, the French for one exception were absent. The congress committee under Dutch leadership of the mechanical engineer Cornelis B. Biezeno (1888–1975) and the physicist Jan Burgers (1895–1984), at that time both professors at TU Delft, bore the main load of conference preparations, supported amongst others from the English side by A.A. Griffith and G.I. Taylor,—Th. von Kármán, L. Prandtl and Richard von Mises[118] for Germany and Ernst Meissner,[119] representing Switzerland.

Though nowhere enunciated explicitly, the *Proceedings* of the First International Congress for Applied Mechanics[120] at Delft, 22–26 April 1924, illustrate that Alan Griffith must have experienced considerable headwind about the *soap bubble affair* in the context with Prandtl, especially from the circle of the congress committee, which arranged the first session amongst themselves in a quite unusual style. On Wed 23 April 1924, 9 am, Prof. G.I. Taylor, Cambridge had the conference chair and opened the lectures at the General Meeting, which saw A.A. Griffith's (first) presentation as number 3 in the row of presenters, after the conference organiser C.B. Biezeno and the head of the English delegation E.G. Coker.[121] Griffith's paper was entitled *'The Use of Soap Films in solving Stress Problems'*, that means it was word-by-word nearly identical to the incriminated 1917 paper together with G.I. Taylor, which had referred to *'Torsion Problems'*. Instead of 54 pages then, the Delft paper, now authored by A.A. Griffith alone had just three pages— and after a few introductory remarks stands its essence: *'This analogy was first pointed out by PRANDTL'*, the name in capitals, ending with three references, to Prandtl's 1903 soap film publication and two subsequent Griffith-Taylor papers to the same subject of 1917 and 1918, where Prandtl's foregoing contribution had been omitted.

As number 4 in the row of presenters is listed L. Prandtl with his lecture in German *'Spannungsverteilung in plastischen Körpern'* (Stress distribution in plastic bodies, 12 p.)

---

[117] See Eckert, The Dawn of Fluid Dynamics, p. 96 f.

[118] Richard von Mises (1883–1953), Professor of Applied Mathematics at Berlin University,—see Wikipedia, 'Richard von Mises'in English.

[119] See Wikipedia 'Ernst Meissner'in German. Ernst Meissner (1883–1939), mathematician and since 1910 Professor of Technical Mechanics at ETH Zurich, supervisor of the doctoral thesis of Max Koenig in 1927, see Chap. 3.

[120] See Biezeno, Proceedings.

[121] See Wikipedia, 'Ernest George Coker', (1869–1946) in English, for the British mathematician and engineer, specialised in stress analysis and photo-elasticity.

followed again by A.A. Griffith's (second) paper *'The Theory of Rupture'*, 9 p., and then further down contributions by G.I. Taylor, Th. von Kármán and J.M. Burgers.

In addition to this *walk on eggshells*, it is striking that both Griffith's conference papers had to bear critical *Notes of the editors* at the end which put apparently some of Griffith's references for the reader under the impression of being in need of a kind of corrective and improving perspective, as for a notorious second offender.[122]

One could conclude that this *Delft showdown* was more or less the end of Griffith's career in fracture mechanics, at least there appears to be no further publication of him in this field thereafter. While G.I. Taylor developed not before long a very relaxed and friendly relation with Prandtl, Griffith became—at least in international science—rather isolated with a tendency for separation and somewhat exaggerated concealment. Already in 1925 Taylor became a corresponding Member of the Goettingen Academy of Sciences, followed on Taylor's initiative, Prandtl being invited to London to the prestigious *'Wilbur Wright Memorial Lecture'* in May 1927 of the Royal Aeronautical Society.[123] And as if the lecture were not sufficient an honour, the occasion was also used to award to him the *'Gold Medal of the Royal Aeronautical Society'*.

Though most Griffith's biographers leave his exceptional break in study subjects at RAE in 1926 uncommented, referring to a rather natural development,[124] there is at least one other reference[125]—presumably based on Griffith's own account—which indicates

---

[122] At the end of Griffith's 3 p. *Stress paper* quite formally: *'The editors take liberty to draw attention to the following paper: F.A. Vening Meinesz, de Ingenieur 1911, No. 3, p.180, in which the soap film method was treated for the first time in relation with the problem of finding the shearing stress in a beam of uniform section fixed at one end and bent by a single load.'* And at the end of *'The theory of rupture'*: *'Dr. Griffith's paper mentioned . . . also . . . some considerations on the molecular theory of strength phenomena. The calculation of cohesional* (sic) *forces . . . has in the last years been made the subject of a series of papers, amongst which those of Max Born (Goettingen) and his collaborators may be mentioned.'* Again a sensitive blow against Griffith's supposed attitude of ignoring professional expertise abroad; in 1954 Born received the Nobel Prize in physics.

[123] On this occasion the correspondence with Prandtl on behalf of RAeS was led by the then RAeS chairman William F. Forbes-Sempill, 19th Lord Sempill (1893–1965), a Scottish peer and record-breaking air pioneer who was later shown to have passed secret information to the Imperial Japanese military before WW II. See Wikipedia 'William Forbes-Sempill, 19th Lord Sempill' in English. Sempill and Prandtl met again 11 years later on the occasion of another RAeS Gold Medal award ceremony to the German Zeppelin pioneer Hugo Eckener, this time on the eve of the Annual International Conference of the Lilienthal Society at Berlin, 12–15 Oct. 1938. An event which could be also called *'The spy summit'* in view of corresponding activities to learn more about the secret German turbojet engine developments; see Sect. 6.1.3.

[124] For example,—see Rubbra, Alan Arnold Griffith, p. 120, knows *'Griffith occupied the position of Senior Scientific Officer for eight years; Sir Ben Lockspeiser was one of his colleagues for part of this time. [Lockspeiser will reappear after WW II,—see Sect. 10.1 and Fig. 10.1]. During this period his work on* (torsional stress in) *airscrews led Griffith to the study of the gas turbine and to one of his greatest contributions to the science of aircraft propulsion.'*

[125] See Nixon, Aircraft Engine Developments, p. 165.

unusual circumstances here. An accidental fire caused by a glass-melting torch is said to have led to inquiries and accordingly, Griffith being instructed to undertake other work, since aircraft faced *'no fatigue problems'*. Whatever the real motives, Griffith's head start in his new preferred research subject of axial turbomachinery design justifiably raised high expectations for further progress in aero gas turbine development.

4.2.2   Griffith's Secret Reports of 1926 and 1929

There is a—in the meantime classical—narrative to *tell the story of the early history of the aircraft gas turbine in Britain*[126] as two independent *'strands'*: the first one, initiated by Griffith in 1926, led to a practical design methodology for axial turbomachinery—in combination with a turboprop engine application, the second represented by Whittle after 1930, accomplished finally a technological breakthrough by the realisation of the turbojet propulsion concept. As said already in the *Introduction* to this book, its main purpose is not to repeat well-known technical history, but rather to close still existing gaps in the reported tradition, especially in view of international information flows. Under this aspect it is predominantly the first 'strand' which will be addressed here from Griffith's early beginnings to the restart of RAE axial compressor and turbine activities in the mid-1930s to further developments at Metropolitan-Vickers and Armstrong Siddeley, which stood in considerable information and know-how exchange with both Switzerland and Germany.

In hindsight, it appears that competitive rivalry accompanied and influenced mutually the turbojet developments in England and Germany from the beginning, more so when the military and strategic implications of these revolutionary new engine concepts became broadly visible. The developments of the essential axial compressors started on both sides approximately at the same time in 1926, at RAE and AVA (see the following Sect. 4.3) and in parallel to the first BBC activities in Switzerland, as discussed already in Sect. 4.1.

In July 1926 A.A. Griffith at RAE had outlined his aerofoil theory of compressor and turbine design, based on Ludwig Prandtl's general lift and drag aerofoil theory of 1918.[127] At the end of his report, Griffith suggested the use of a single-shaft turbine engine with a multi-stage axial compressor to drive an airscrew via reduction gearing.[128] In continuation, according to E.W. Constant II, *Griffith—after receiving the first positive test result—continued to develop his axial-flow compressor ideas, ..., and in November 1929*

---

[126] See Hawthorne, The early history; similar approaches by Bailey, The early development, Dunham, A.R. Howell, etc. Dunham's perception however, is wrong when he states in a side comment on p. 2 of the addressed 2000 ASME paper, describing parallel design activities at the Swiss BBC: *'Meanwhile, Brown Boveri tested a successful axial compressor about 1938'*,—by being 5–6 years too late.

[127] See Griffith, An Aerodynamic Theory (RAE report H.1111, 1926).

[128] In addition, Griffith predicted there correctly 200–500 hp as the lower power limit of useful gas turbine applications.

submitted a memorandum[129] *(unpublished and secret) to the ARC* [Aeronautical Research Council, chaired by H. Tizard] *containing a design study for a very complex contrarotating, contraflow, 500 hp turboprop engine.'*[130] A decisive, and in hindsight fatal change of mind.

Brown Boveri, as described in the foregoing Sect. 4.1.3, had started its axial compressor test and design evaluation activities in Baden, CH, towards the end of 1926 by means of a 4-stage compressor rig, Fig. 4.23. In 1927 Betz and Encke began experiments on the behaviour of axial-flow compressors at AVA Goettingen, Germany. Their investigations— as outlined in Sect. 4.3 were based upon aerofoil theory as well, and were conducted using a specially constructed turbine-driven test rig, Fig. 4.43; their explicit objective was to gain the information necessary to design an axial-flow piston-engine supercharger. By 1929, Betz had in operation a six-stage test compressor built by Junkers that gave a 2:1 pressure ratio with 85% efficiency[131] at 37,000 rpm.

On the basis of the new theory, Griffith believed that he could achieve an axial compressor efficiency high enough to make a gas turbine a practical turboprop engine, and in his 1926 report H.1111 he proposed an axial gas turbine engine to be developed as an aircraft prime mover. In due course the ARC approved *preliminary experiments to verify the theory*. Griffith's documented formulation of an axial turbomachinery design method in combination with a practical aero application makes him unique in comparison to the German/Swiss approaches, where at the time similar design tools—e.g. based on the Bauersfeld (1922) or Schilhansl (1927) papers must have existed—but none of these implied design methods were actually conveyed in a comparable report or accompanied by revolutionary aero propulsion recommendations.

The paper H.1111 gives—with one exemption—no detailed references so that at first glance the impression of an unique stand-alone design method could occur. After outlining the then conventional turbomachinery *'flow channel'* theory with the inherent disadvantage of most blading running stalled, Griffith describes for the new approach the application of the circulation theory of aerodynamics for which the blade turning angle *'is limited by the magnitude of the maximum attainable lift coefficient'*. Here *'the blades, instead of being regarded as the walls of channels, whose shape determines the velocity and pressure changes taking place in the fluid, are to be regarded as aerofoils and the changes in velocity and pressure are to be calculated from the blade reactions'*. In the *'Analysis of a two-dimensional case'*, he considers *'the case of a row of an infinite number of parallel aerofoils of infinite length'*, in this respect using the same start as Bauersfeld[132] in 1922. Similarly, Griffith 1926 states *'The simplest way of attacking this problem is to employ*

---

[129] See Griffith, The Present Position (AML report 1050A, 1929).

[130] See Constant, E.W., The Origins, p. 111.

[131] This value was known to Griffith shortly thereafter, see Postan, Design... of Weapons, p. 188.

[132] See Bauersfeld, Die Grundlagen.

Prandtl's circulation theory'—with the explaining hint in a footnote[133] that *'The solution of the ideal circulation problem here given is due to H. Glauert',* without further reference. The required aerofoil energy loss coefficient σ, as the ratio of drag/lift coefficient, is deduced from NPL wind tunnel experiments. For theory proof, both Bauersfeld 1922 and Griffith 1926 refer to *'a special type of water turbine, known as the "propeller turbine"* (that) *'has recently come into use, especially in Germany, for dealing with low heads of water',* all of the foregoing in italics from Griffith H.1111. Bauersfeld provides here already the established name for axial water turbines with radially positioned *'blade wings'*—the *Kaplan Turbine.* Griffith knows *'With this turbine the maximum overall efficiency obtainable is rather more than 0.9. The blade efficiency, of course, must be still higher. It therefore appears that the foregoing theoretical figures of 0.960 to 0.966 may be regarded as quite a reasonable estimate of the maximum blade efficiencies attainable with known sections'.*[134] In the contrary Bauersfeld's prediction of the rotor efficiency of a Kaplan turbine appeared rather conservative with 90.4%. It was the fact that stage efficiencies were worked out that made the Griffith theory a practical one; the efficiencies were simply too high, because *'secondary flow losses'* had been omitted.

Vortex flows were not explicitly considered, but it became clear very early that the theoretical assumption of irrotational flow and of constant circulation along the blades meant in fact a *'free vortex blading'.*[135] As a first step, experiments were carried out at the RAE in 1927 under Griffith's supervision on compressor and turbine blade designs, using cascades of aerofoils. The development of this piece of experimental equipment enabled major advances to be made and culminated in the establishment of a successful design method for axial flow compressors for aircraft gas turbines. However, speculation by Bailey[136] that Harris and Fairthorne[137] used this equipment, inspired by Griffith, as the

---

[133] See Griffith, An Aerodynamic Theory, p. 4. Griffith changes this superficial quotation style only once in H.1111, p. 17, when he refers in detail to a paper in English of 'Dr. Ing. Bruno Eck: The Hydrodynamical Theory of Turbines and Centrifugal Pumps', Engineering, Vol. CVVI, Jan. 22 and 29, 1926, pp. 98–101 and pp. 125–127, who—important for Griffith's intended priority claim— *'deduced no expression for efficiency'.* There are no indications that Griffith knew, or would have understood Bauersfeld's 1922 paper in German.

[134] See Howell, Griffith's Early Ideas—concluding, that Griffith's estimates of efficiencies would be generally some 4% too high.

[135] According to Howell, Griffith's Early Ideas, similar concepts on circulation and vortex flow were expressed at the same time by—see Darrieus, Contribution a tracé, which implies that this design philosophy was also at hand for Brown Boveri latest in 1929 (but—as not significant—was not necessarily used in their standard 50% reaction blading for reasons of cost and design complexity); the same source indicates already as early as 1910 an interest of Stodola in radial changes in the degree of reaction due to *'centrifugal forces'.* In England, Whittle appeared to be the first to push the *'vortex flow'* concept for his axial turbines, followed by H. Constant, who referred to *'free vortex'* blading from 1936 onwards.

[136] See Bailey, The Early Development, p. 4.

[137] See Harris and Fairthorne, Wind tunnel experiments.

world-wide first cascade wind tunnel are presumably misleading, since the first Goettingen cascade wind tunnel was of importance for the thesis of Karl Christiani,[138] who had his doctorate examination at Goettingen University on 27 July 1926, with L. Prandtl as first and A. Betz as second examiner.[139] This and later cascade wind tunnels became extremely important to the development of axial compressors and turbines in Great Britain, especially in the period immediately after war's end. Independent of the early start of this measuring technique in Germany, cascade wind tunnels never grew up here to a similar importance in the initial phase of axial compressor development.

The data gained from wind tunnel tests of a large number of these cascades helps the designer model the performance of the blades in an operating compressor or turbine. Griffith used his cascade data to find the blade angles which would yield the desired air flow. Thus, Griffith supplemented the theoretical approach with empirical data. Neither body of knowledge had been sufficient by itself, but the combination of the two proved useful.[140]

Based on Griffith's 1926 paper and the additional cascade data, ARC authorised at RAE the construction of a small single-stage axial compressor/turbine compound set-up with free vortex blading on a common shaft, Fig. 4.27.[141] Tests of aerodynamic efficiency could be made by forcing air through this simple gadget of just 260 mm axial length; speeding it up to approximately 15,000 rpm yielded stage efficiencies of—claimed—better than 90%.

After the wind tunnel cascade experiments and the model rig investigations Griffith must have been very satisfied that the results of 1928/9 fully verified his theoretical ideas for the axial blading of the aircraft gas turbine components. In November 1929 he issued as a further report, the already cited AML report 1050A, which covered his recent assessment in his *Conclusion* in nearly euphoric words:

> The (axial aero gas) turbine is superior to existing Service engines and to projected compression-ignition engines in every respect examined. The efficiency is higher and the weight and bulk less. No external cooling is required. At high altitude there is an inherent supercharging effect, coupled with a substantial decrease in the specific consumption. The use of a variable airscrew is unnecessary. Starting presents no difficulty and control is simpler than in the case of existing engines. Any liquid fuel of suitable chemical composition may be used, without reference to anti-knock value or volatility.

---

[138] See Christiani, Experimentelle Untersuchung. A last trace of Karl Christiani (1900–) has been found so far in 1945 as Assistant at the TH Berlin-Charlottenburg, Institute for Applied Mathematics (Prof. A. Timpe).

[139] Courtesy of Goettingen University, 19 Sep. 2019; the thesis was supervised by Albert Betz.

[140] See Nichelson, Early Jet Engines, p. 120.

[141] The drawing from Kay, Turbojet, Vol. 1, p. 13 represents the implemented hardware. Originally Griffith in H.1111 had suggested in flow direction a R-S-S-R aerofoil sequence with the advantage of undisturbed compressor entry flow, see Howell, Griffith's early ideas, Fig. 3.

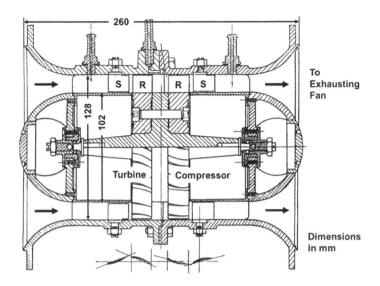

Fig. 4.27 RAE 1929 model turbine-compressor

Griffith gained the certainty in his judgement about the possibility of a superior aero GT design in comparison to the established piston engine not only from the exploitation of his own resources, but drew also conclusions from foreign information. In the context of interpreting his own efficiency data in the order of 0.88 to 0.90, he states laconically in the *Introduction* of his 1929 report—again without further reference:

> On the aerodynamic side of the problem, Betz in Germany has made a bladed compressor having an aerodynamic efficiency of approximately 0.85.

The subject of information exchange by visitors—mostly from England in a private, informal way to the increasingly famous research institutions of Ludwig Prandtl at Goettingen between the early 1920s up to 1933, the year of Hitler's seizure of power in Germany, will be dealt with in the Sect. 4.3.3 '*The AVA Goettingen and the axial-flow compressor*'—and for the time till the outbreak of WW II, then only on '*official*' level in Sect. 6.1 under the header '*International information exchange/transfer*'.

The early *AVA Gaestebücher* (guest books)[142] have a relevant entry for 1928 which could explain the detailed information flow to Griffith about the then started AVA axial compressor rig investigations, which will be also addressed in Sect. 4.3.3. For *27 April 1928,* there is a hand-written entry of *W.S. Farren, Cambridge,* who became officially RAE Director after 1941. William S. Farren (1892–1970)[143] was from 1920–1937 a lecturer in

---

[142] Accessible in the DLR Archives Goettingen, approximately for the period 1920 to 1933.

[143] See Wikipedia 'William Farren (engineer)'.

Engineering and Aeronautics at the University of Cambridge (under the famous Sir Melvill Jones, author of the seminal *The Streamline Airplane*[144] of 1929), being a Fellow of Trinity College, Cambridge. Whilst at Cambridge, he sat also on the Aeronautical Research Committee; with all likelihood Farren in his ARC function was also aware of the various committee decisions on Griffith's proposals.

Griffith's 1926 and 1929 Reports had been classified as *'secret'* and as such not officially published,[145] inaccessible e.g. for the Germans which on the other side maintained up to the 1930s quite an *'open house'* policy, supporting especially from Prandtl's side the idea of international scientific exchange.[146] As will be seen later, Griffith and his intermediaries as well as the contemporary Swiss compressor community at Brown Boveri must have learnt relatively early a *'peculiarity'*, if not fault of the Goettingen axial compressor design system with regard to the selected *Degree of Reaction* which—if openly discussed according to the standards of scientific discourse, e.g. in the style of the communication between Ludwig Prandtl and Hermann Glauert[147]—might have triggered second thoughts and even a redirection of the Goettingen design approach. On the other hand the strange ideas—seemingly directly out of the engineering *'ivory tower'*—which Griffith laid down in his 1929 Report, and which misled the British engine development in parts for more than a decade, might have been stopped presumably earlier, if discussed for example with a hands-on engineer like Frank Whittle.[148]

Griffith, meanwhile promoted to principal scientific officer at the Air Ministry Laboratory, South Kensington, continued to develop his axial-flow compressor ideas and in November 1929 submitted his second report to the ARC as AML1050A (again 'secret' and officially unpublished), containing a design study for a very complex contra-rotating, counter-flow 500 hp turboprop engine. While at the end of his 1926 report Griffith had still suggested quite reasonably a single shaft engine with 16-stage axial-

---

[144] Jones discovered that then-current airplanes wasted up to 75 percent of their power in overcoming unnecessary turbulent drag. After Jones's report the aeronautical community began to realise how much power was actually being wasted by contemporary aircraft or, alternatively, what increases in speed were theoretically possible.

[145] Both Reports were *'declassified'* in 1963 only.

[146] It cannot be decided, if at that time and in this context still certain hard feelings existed towards *Goettingen* as remnants of Griffith's *'soap bubble affair'*. Professor G.I. Taylor, Griffith's co-author of the 1917 paper tried to harmonise his relationship with Prandtl continuously, which finally led to Prandtl's stay in Taylor's home, when invited to Cambridge in 1934 and 1936,–see Eckert, Ludwig Prandtl, p. 217.

[147] See Bloor, The Enigma, p. 312 f.: *Postwar Contact with Göttingen.*

[148] E.W. Constant's statements are somewhat ambiguous in this context; see Constant, E.W., The origins, on p.184, he confirms that Whittle at the end of 1929 neither knew the Stern report nor Griffith's work on axial compressors, while on p.186 is stated for approx. 1930, *Whittle first learned on Griffith's work on axial compressors when someone at the Air Ministry handed him a copy of Griffith's report to read while he was waiting for an appointment.* This latter information being based on a letter from Sir Frank Whittle to Constant, E.W., dated 20 May 1970.

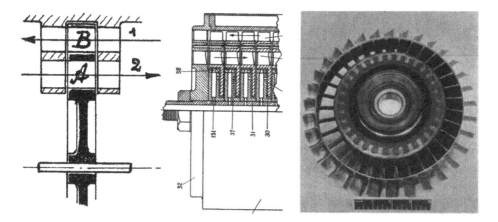

Fig. 4.28 Contra-flow counter-rotating turbomachinery patents: F. Stolze and R. Barkow, Berlin, DE181,147, 1905 (l); A.A. Griffith, Derby, US2,391,779 with priority in 1929 (m);RAE CR9 test wheel, 1942 (r)

flow compressor[149] with a total pressure ratio PR 11 to compete with the piston engine, now here, all in a sudden there was a break. In 1929, satisfied, that high blading performance was obtainable at the design point, he now devoted much attention to the multi-stage compressor stability problem. This resulted in the extraordinarily complex contra-flow, counter-rotating arrangement, where actually each compressor stage was driven by its own turbine row, mounted on the ends of the compressor blades outboard of a shroud ring. The turbine gas flow annulus thus surrounded the axial compressor flow and compressor/turbine compound wheels were independently mounted on a common shaft—with adjacent wheels counter-rotating (CR). In 1942 when a nine-stage CR compressor rig according to Griffith 1929 ideas had been actually tested,[150] Fig. 4.28 (r), the authors concluded:

'In the contra-flow unit the leakage between the shrouds separating the compressor and turbine annuli is a special problem. Owing to the departure from design conditions and the intake air boost the leakage observed on the unit was at times as much as 50 per cent of the entering air. The leakage likely to be obtained in a unit operating under designed conditions is estimated at 4 per cent.' ... And, still with some reverence to the great Griffith, then already with Rolls-Royce Derby: ... *'It was concluded that, in spite of the poor aerodynamic performance, there was no fundamental reason why similar units should not operate efficiently and why a good mechanical performance should not be obtained.'*

---

[149] See Howell, Griffith's early ideas, Fig. 2: reconstruction of the 16-stage compressor.

[150] See Baxter and Smith, Contra-flow turbo-compressor tests.

In hindsight, after several, comparably conventional turbojet generations, one can state: While ingenious in some respects, the Griffith CR concept certainly does not take advantage of the fact that it is aerodynamically much easier to extract work from the flow than adding work to it. A smaller number of turbine stages is able to drive a larger number of compressor stages,—with handsome weight savings for airborne applications. This was exemplified by the success of Whittle in producing an efficient only one-stage turbine to drive the PR 4 centrifugal compressor of his jet engine. Quite unusually, Griffith had ended his 1929 Report with a Provisional (Patent) Specification[151] of his turbine-compressor compound ideas, accompanied by a drawing of a typical 14-stage counter-rotating contra-flow arrangement, which can be seen in parts in Fig. 4.28 (m)—and in full in Fig. 8.28 (l).

Some readers may not be surprised after the foregoing description of some oddities in Griffith's scientific work, that also this rather uncommon concept had again already its patented forerunners,[152] though there are no indications that neither Griffith nor the reviewers at the British Patent Office had been aware of either German patent DE181,147 to the axial turbomachinery pioneer Franz Stolze, Fig. 4.7, and his co-inventor Rudolf Barkow at Berlin-Charlottenburg, issued for a *Gasturbine* on 5 March 1905, Fig. 4.28 (l)—or French patent FR411,473 to E.B. Mérigoux for a similar contra-flow counter-rotating *turbo-compresseur,* with invention priority on 11 January 1910.

As already addressed before, Griffith's counter-flow concept was mainly hampered by aerodynamic, thermodynamic and mechanical problems—due to the sharp 180 deg. bend from compressor to turbine and leakage of hot gas from the turbine channel to the compressor air.

Though Griffith states in his '*preliminary patent specification*' somewhat self-reliant '... *both the thrust bearings and the journal bearings are of the normal air-lubricated type*', sufficient lubrication of the disc bearings would have been difficult in any case.

A special panel of the ARC's engine sub-committee examined Griffith's submission in early 1930. Contrary to Griffith's assessment they concluded that the superiority of the gas turbine over the piston engine was not proven, so that the development of an aero gas turbine was not recommended. However, a multi-stage test rig should be built, incorporating Griffith's aerofoil blading to check the theory—and certainly also some questionable features of his suggested mechanical design. In addition, the Committee noted the problems of high intensity combustion and the lack of suitable material for the turbine.[153]

---

[151] See Griffith, An Aerodynamic Theory, Appendix II.

[152] Griffith was not alone here, Hans von Ohain's second secret patent of a Nernst tube reactor would have been prevented as well by another feature of the Stolze-Barkow patent, if it had been observed by the Reichspatentamt, see Sect. 12.1.1 for further details.

[153] See Bailey, The Early Development, p. 5.

4.2.3   Griffith and Whittle: A Patent Conflict?

When Griffith wrote his 1929 Report on aircraft gas turbines he was *Superintendent* of the Air Ministry's South Kensington Laboratory. It was in October[154] of the same year that he was asked by the Air Ministry to meet Whittle, the RAF pilot with an unique proposal for aero propulsion. Instead of thrust from a propeller, in line with Griffith's established considerations, the gas turbine would provide a high velocity jet at exhaust. According to Golley in the same month,[155] and with all probability *before* the meeting with Griffith, Whittle realised that the gas turbine could be even *substituted* for the piston engine, because the exhaust would propel the aircraft, making the propeller unnecessary. He realised that this would require the compressor to have a much higher pressure ratio than the one he had foreseen for his earlier piston engine scheme. But once thought to an end, he was puzzled that he had taken so long to arrive at a concept which seemed so very obvious in hindsight.

The circumstances of the meeting between Griffith and Whittle cannot be reconstructed after such a long time and innumerable, mostly speculative accounts. One of the earliest reports comes from Whittle himself in his documented 1945 Clayton Lecture,[156] where— after an illustrious frontispiece quotation by Francis Bacon:[157]

> ... I do not endeavour either by triumphs of confutation or pleadings of antiquity, or assumption of authority, or even by the veil of obscurity, to invest these inventions of mine with any majesty ...

—he continued in a rather conciliatory tone: '*At the time I started thinking about the subject* (of a constant pressure gas turbine*), i.e. in 1928–9, the many failures had led to a general belief in the engineering world that it had no future. I now know that, apart from myself, there were others who refused to conform to the prevalent view and believed that the problems of the gas turbine were not insuperable. They included Dr. A.A. Griffith, Mr. H. Constant and certain others at the Royal Aircraft Establishment, and engineers of the Brown-Boveri Company led by Dr. Meyer.*'

Actually, according to E.W. Constant,[158] '*Griffith, although a believer in the gas turbine for propeller drive, thought Whittle's assumptions overly optimistic; and the Air*

---

[154]Confirmed by Ian Whittle on 28 Sep. 2019.

[155]See Golley, Whittle, p. 33 and p. 63. Whittle had initially become aware of the promising qualities of the gas turbine whilst looking into Stodola's book '*Dampf- und Gasturbinen*' at the Leamington reference library, a foresight to intensifying turbomachinery relations between England and Switzerland some 15 years later. In 1928 at Cranwell (RAF) College, his calculations showed that the gas turbine had the potential to become the prime mover for aero-propulsion. The memory of his early schoolboy experiences was renewed in 1936 when he received the now two volumes of Stodola's book as a college prize at Cambridge.

[156]See Whittle, The Early History.

[157]Francis Bacon (1561–1626), English philosopher and statesman,—see Wikipedia 'Francis Bacon' in English; for quotation,—see Bacon, The Works, Preface p.19.

[158]See Constant, E.W., The Origins, p. 184.

Ministry, probably reflecting Griffith's opinion, informed Whittle that since his design was basically a gas turbine, its development was considered impracticable because of limitations on materials, temperatures and stresses.' At this point, Golley,[159] presumably with the agreement of Whittle, condemned—now rather fiercely—Griffith's attitude to Whittle's ideas in his 1987 Airlife book: *'Griffith was a highly qualified scientist with a growing reputation in the academic world. He certainly had the knowledge to do a quick design study of Whittle's turbojet proposals. Had he done so with intellectual honesty he would inevitably have come to the conclusion that, with easily foreseeing improvements in materials and component efficiencies, a revolutionary (in every sense of the word) aircraft propulsion engine was at least within reach.'*

Though strangely, Griffith was unsupportive to Frank Whittle's suggestion of the turbojet engine concept, describing his assumptions as over-optimistic and finding fault with his calculations, the accusation of a lack of *intellectual honesty* on Griffith's side appears unjustified. Putting the circumstances of this historic meeting into the perspective of age, one realises that the then 22-years-old pilot officer Whittle met the 36-years-old ministerial expert Griffith, both ambitious and convinced that their own inventions could lead to a nearterm breakthrough in aircraft propulsion. And while Whittle's proposals and especially his most recent idea of a turbojet must have been screened between them in detail, with all likelihood Griffith on the other side played with hidden cards, not revealing his own activities of just one month ago. But instead of speculating about the psychological side of this encounter, it is perhaps more helpful to sort the known facts over the short period of several months:

- On 5 Sep. 1929[160] Griffith filed his patent ideas for a contra-flow counter-rotating *Turbocompressor* of individually rotating turbine-compressor disks—without any kind of combustion, and thus a turbomachinery arrangement, but definitively no new ICE Internal Combustion Engine concept.
- In Oct. 1929 there are two decisive milestones, (a) Whittle—according to his own statement—had his *'eureka moment'* for the revolutionary, considerably simplified *Turbojet* concept, where simply just one turbine stage would do to drive the compressor and (b) clearly thereafter, presumably in the second half of October, both had their fateful meeting at the Air Ministry's South Kensington Laboratory with the known outcome.
- In Nov. 1929 Griffith's AML Report 1050A appears with the 'preliminary (patent) specification' as Appendix II, mainly in line with the actual patent application two months earlier, but now with combustion and for a geared turboprop configuration only.

---

[159] See Golley, Whittle. p. 35.

[160] This original patent priority date stems from <*espacenet*>, the European Patent Office database for US2,391,779, the unchanged U.S. patent application of Griffith's *Turbocompressor* of 5 Oct. 1942, which was granted on 25 Dec. 1945. This was Griffith's second patent application after a *Flowmeter* measuring device in 1928.

- On 8 Dec. 1929 William L. Tweedie of the Air Ministry's Directorate of Engine Development, who had arranged the meeting between Whittle and Griffith in October, wrote to Whittle[161] *'After careful study of all the papers … no reason has been found for any alteration of the preliminary criticism advanced against your various assumptions. … However, it has been of real interest to investigate in detail your scheme and I can assure you that any suggestion submitted by people in the Service is always welcome. Your calculations are returned herewith.'*
- Finally, on 16 Jan. 1930 Whittle files what became GB347,206 *'Improvements relating to the Propulsion of Aircraft and other Vehicles'*, and thus the first *Turbojet* patent issued with an unequivocal purpose visible in the header, addressing the revolutionnary turbojet principle.

It may be worthwhile to highlight here the fact that the comprehensive Whittle/Griffith literature nowhere traces the origin of their conflict in October 1929 back to an immediately foregoing patent application (by Griffith). In comparing Griffith's patent text of the September 1929 filing, and his version of a *'preliminary* (patent) *specification'* in the AML Report of November 1929, the different emphasis on combustion, and thus the definition of a complete ICE in the later Report is striking. Given the sequence of events one must conclude that this reflects a certain impact of Whittle's argumentation on Griffith, who must suddenly have realised the decisive flaw of his own inconclusive engine approach in September. Griffith must have been also troubled by the simplicity of the Whittle idea. He was certainly not amused being told by a young RAF pilot that the next generation of aircraft should not use only gas turbines but that Griffith's favourite—the propeller—should be replaced by a high-speed propulsive jet. And he must have guessed even that, if he were to admit feasibility, the Air Ministry would possibly make him work on it and he would then have to turn his attention away from his own project.

On the other hand, Whittle must have felt the threat after the encounter with Griffith of being out-manoeuvred before the winning post. Pat Johnson, one of his flight instructors, who had had experience as a patent agent and who helped him with his patent submission, said that they should hurry to get the application submitted as soon as possible, before others, who had seen details already, could possibly present their own application without delay.[162]

Whatever reason Griffith and in his wake the Air Ministry had for the negative assessment of Whittle's proposal, it had missed a chance to make a significant technological breakthrough and created a considerable delay of 6–8 years for British turbojet developments. If Griffith had hoped to eliminate an inconvenient competitor, his own progress in getting government support was likewise disappointing. During the economic

---

[161] See Bailey, The Early Development, p. 6.
[162] Communication with Ian Whittle, 28 Sep. 2019.

crisis in the early 1930s the Air Ministry withdrew the funding for his *turbocompressor* project. They did not resurrect the aero gas turbine work until they became uneasy when Whittle's Power Jets Ltd was formed in 1936, instructing Griffith and his RAE team to get back to work. Hence the F.1 turboprop project at the RAE began, converted to the F.2 turbojet in 1939.[163]

4.3   Up To High Altitude: The German New Frontier

On Monday, 11 November 1918 the *New York Times* appeared with the front page header:

ARMISTICE SIGNED, END OF THE WAR!
BERLIN SEIZED BY REVOLUTIONISTS;
NEW CHANCELLOR BEGS FOR ORDER;
OUSTED KAISER FLEES TO HOLLAND

Though this Chap. 4 deals mainly with early development activities up to 1935 on the way to the first aero gas turbines in England (supported from Switzerland) and—here in Sect. 4.3—Germany, the attentive reader has in any case without saying to accomplish the *'full picture'* by adding political, economic and social historic background and implications of the times. And there were many.

The unofficial historical designation for the German state from 1918 to 1933 is *Weimar Republic,* derived from the city of Weimar,[164] where its constitutional assembly first took place. In its fourteen years, the Weimar Republic faced numerous problems, including hyperinflation, political extremism (with contending paramilitaries) as well as contentious relationships with the victors of the First World War. Resentment in Germany towards the *Treaty of Versailles* was strong especially on the political right, where there was great anger towards those who had signed and submitted to the treaty. The Weimar Republic fulfilled most of the requirements of the Treaty of Versailles, although it never completely met its disarmament requirements and eventually paid only a small portion of the war reparations. From 1930 onwards, the *Great Depression,* exacerbated by the national policy of deflation, led to a surge of unemployment. In 1933, Adolf Hitler was appointed Chancellor with his Nazi Party being part of a coalition government. Within months, the *Reichstag Fire Decree* and the *Enabling Act of 1933* brought about a state of emergency, wiping out constitutional governance and civil liberties. These events brought the republic to an end—as democracy

---

[163] A story with considerable involvement from Switzerland, described in Sect. 6.3.

[164] The *Weimar Classic* goes with the names of German poets Johann Wolfgang von Goethe (1749–1832) and Friedrich Schiller (1759–1805) for the period between 1786 (Goethe's first Italian travel) and 1832 (Goethe's death); the Junkers Aircraft and Engine Works, Dessau are located 100 km away in the north-east of Weimar, the Buchenwald Concentration Camp—to name two thinkable extremes in this context—just 7 km off, in the north-west of Weimar.

collapsed, the founding of a single-party state began the warmongering dictatorship of the Nazi era.

As seen *Versailles* had an immediately felt impact on German political and economic life, not the least on post-war aeronautical developments. The *Treaty,* signed on 28 June 1919, effective since 10 January 1920, prohibited the production and import of aircraft of all kinds in and into Germany. At war's end 4000 aero engines had to be surrendered to the *Allied Forces,* more than 25,000 were scrapped. A few manufacturers could retail some 2500 engines to neighbouring neutral countries. The Dutch aerospace industrialist Antony Fokker managed in 1919 to send in an adventurous shift 60 railway wagons with semi-finished aircraft and 400 engines from his Schwerin plant across the border to Holland.[165]

BMW Bavarian Motor Works at Munich found a relatively elegant way-out of these difficulties, when they had to cease production of aircraft engines. To remain in business, the company began producing small industrial engines (along with farm equipment, household items and railway brakes). In 1920, a flat-twin petrol engine was released; despite being designed as a portable industrial engine, the *M2B15* was also used for motorcycle drive.[166]

Only with implementing the *'Versailles aero definitions'* of 5 May 1922, it was again possible to develop and build sporting and light commercial aircraft, however limited to a maximum speed of 170 km/h, flying altitudes up to 4000 m and a payload of 600 kg. Engine production was allowed only up to 60 hp for single-seat aircraft; engines for commercial aircraft were limited to 200 hp (in 1923 expanded to 300 hp).[167]

Under these circumstances the sudden strive for new flight altitude records appears as a kind of subconscious psychological *outbreak* of energy. Franz-Zeno Diemer,[168] Fig. 4.29, was the first to start an attempt to go for a German altitude record. On Tue 17 June 1919, he flew a *DFW Deutsche Flugzeugwerke F37*[169] powered by the newly developed BMW IV engine[170] to an unofficial world record height of 9760 m from

---

[165] See Von Gersdorff, Flugmotoren, p. 42.

[166] One application were the *BFW Bayerische Flugzeugwerke Helios* motorcycles. When BMW merged with BFW in 1922, the *Helios* became the starting point for the first BMW motorcycle. Released in 1923, the *BMW R 32* used a 486 cc flat-twin petrol engine, which was transversely mounted to eliminate cooling problems. This 8.5 hp engine resulted in a top speed of 95 to 100 km/h. At a time when many motorcycle manufacturers used total-loss oiling systems, the new BMW engine featured a recirculating wet sump oiling system with a drip feed to roller bearings; a design which BMW used until 1969. The *R 32* also started the tradition of direct shaft drive, which was used on all BMW motorcycles until 1994.

[167] See Hirschel, Aeronautical Research in Germany, p. 204.

[168] See Wikipedia 'Franz-Zeno Diemer'in English; his grandmother Wilhelmine von Hillern (1836–1916) wrote the Bavarian folklore-novel *Die Geier-Wally (The Vulture Maiden)*, 1875.

[169] *DFW F37*, belonging to a family of German reconnaissance biplane aircraft, was first used in 1916 in WW I—with in total 3250 units, the German aircraft type with the highest production numbers.

[170] *BMW IV*, a 22.9 l, 6 cyl., water-cooled inline aircraft engine in the 250 hp range, here with special altitude carburettor, for details see BMW IV handbook, 1918 (in German): https://bmw-grouparchiv. de/research/detail/index.xhtml?id=3052440 After 1925 the motor was also licence-manufactured as *Junkers L5*.

Fig. 4.29 Enthusiastic crowds on Munich-Oberwiesenfeld follow the ascent of Franz-Zeno Diemer to the world record altitude of 9760 m on 17 June 1919

Oberwiesenfeld,[171] reaching that altitude in 87 min. The BMW IV was an oversized high-compression engine by the famous engine designer Max Friz,[172] at that time in principle competition to the supercharging approach of Paul Daimler.

Later Diemer announced that the engine still had reserves in hand, but that he himself had reached the limits of his capacity. After all, in his open pilot's seat he not only had to contend with temperatures as low as -50 °C, but also with the low oxygen supplies from an additionally installed oxygen bottle, which took their toll on him physically. Diemer's record flight broke the foregoing, just 3-days-old record of the French fighter ace Jean Casale, who had achieved 9650 m on a *Nieuport 29C1*—and lasted till 27 February 1920, when Major Rudolph Schroeder, then U.S. Army Signal Corps' chief test pilot at McCook Field, pushed his *Lepere Lusac-11* biplane, powered by a 400 hp turbocharged Liberty L-12 engine, Sect. 4.1.2, beyond the 10,000 m marker up to the new record height of 10,093 m.

Hugo Junkers (1859–1935), famous German aircraft entrepreneur and engineer,[173] commissioned his first, very advanced commercial product, the *F13 Herta*—named after

---

[171] The *Oberwiesenfeld*—known since 1792 as *Wiesenfeldt* (meadow field)—is an area in the north of Munich, near the BMW plant and what today is the 1972 *Olympics Park*, a former artillery range, military parade ground and barracks area. From 1909 onwards, it was also used as an airfield, so for French Premier Edouard Daladier and British Prime Minister Neville Chamberlain during the *Munich Conference*, 29–30 Sep. 1938. The name of the *Herbitus* engine altitude test facility is a camouflaging Latin expression of Wiese/ Oberwiesenfeld, where it was located on the BMW range at Munich-Milbertshofen, after mid-1944 also used for BMW 003 and Jumo 004 turbojet testing, Fig. 8.46; today operated at the AEDC Arnold Engineering Development Complex, Tullahoma TN, see Sect. 8.3.3.

[172] See Wikipedia, 'Max Friz'in English.

[173] See Wikipedia, 'Hugo Junkers'in English.

his eldest daughter—on 25 June 1919, three days before the signing of the Treaty of Versailles and piloted by Emil Monz.[174] The ground-breaking aircraft, a stroke-of-genius of the then Junkers' chief engineer Otto Reuter,[175] was the first aerodynamically clean, cantilever all-metal plane made of strong yet light *duralumin*, an aluminium alloy, and it was produced at the Junkers factory in Dessau, D, until 1933.[176]

Already on Sat, 13 September 1919, the F13 left the Junkers airfield at Dessau to an— again unofficial—altitude record flight[177] of 84 min with seven passengers up to 6750 m; the aircraft was powered by a BMW IIIa of 185 hp.[178] Figure 4.30 shows the record aircraft *'Annelise'*—named after Junkers's second-eldest daughter—immediately after the safe landing at 8.32 am. The aircraft was steered again by Junkers chief pilot Emil Monz (1), responsibility for barometric altitude measurements had the Swiss Robert Gsell[179] (2). Georg Madelung[180] (3) took part as then part-time student from TH Berlin-Charlottenburg,

---

[174] Emil Monz (before 1900–1922), became Junkers'chief pilot in Jan. 1919, after he flew reconnaissance missions during WW I. In 1920 he demonstrated the F13 capabilities in a selling campaign in the USA, where he established also several records, e.g. he set up a speed record with 210 km/h between Atlantic City and Philadelphia. He died on 18 Feb. 1922 when his F13 crashed under unexplained circumstances near Lauenburg/Lebork, Eastern Pomerania—due to fog, snow storm or, not unlikely, suicide.

[175] Otto Reuter (1886–1922) had started at Junkers as engine designer in Nov. 1915; he modified the Junkers diesel MO2 to the first gasoline direct injection, opposed piston engine FO2, 6 cyl., 17.1 l, 475 hp with advanced aluminium crankcase and pistons.

[176] Ninety years after the first flight a German-Swiss project to build a reconstruction of the F13 was launched in 2009; the aircraft first lifted off in Sep. 2016 and is now a highlight of air shows and fly-ins. The reconstruction is equipped with radio and a transponder, and uses a 1930s Pratt & Whitney *R-985 Wasp Junior* motor, but is otherwise as close as possible to the original.

[177] Both record flights of Diemer and Monz were not officially registered by FAI Fédération Aéronautique Internationale, the World Air Sports Federation; although Germany was a FAI founder member in 1905, it had been excluded as a result of the First World War.

[178] *BMW IIIa* was an inline 6-cylinder valvetrain (with single overhead camshaft), water-cooled aircraft engine, the first-ever product from BMW GmbH which laid the foundation to future BMW success. It is best known as the powerplant of the *Fokker D.VIIF,* which is said to have outperformed most allied aircraft during WW I.

[179] Robert Gsell (1889–1946), Swiss aeronautical pioneer, who flew the first all-metal aircraft, the *Reissner Ente* (duck) at Aachen-Brand on 1 June 1912. As pilot of a *Flugzeugwerke Friedrichshafen FF-1* he achieved with three passengers an endurance world record of 3 h, 11 min on 2 Sep. 1913. At the time of the F13 record flight he led the *DVL Deutsche Versuchsanstalt für Luftfahrt* (German Research Institute for Aviation), Department of Instrumentation. In 1927 he joined ETH Zurich, where he became Professor for Aeronautical Technology in 1939, – till his death in 1946.

[180] See Wikipedia, 'Georg Hans Madelung' in English; the statement there under *'WW I service'*: *'He was involved with fellow engineer Fritz Haber's deployment of poison gas* is obviously wrong, mixed up with his elder brother Erwin Madelung (1881–1972), renowned theoretical physicist, specialised in atomic physics and quantum mechanics,—see Wikipedia 'Erwin Madelung' in English and—see Szoelloesi-Janze, Haber, p. 328. There is stated, that a special troop was formed for gas warfare (Pioneer Regiments 35 and 36) with Haber as advisor. Haber actively recruited physicists, chemists,

Fig. 4.30 Junkers *F13* *'Annelise'*after altitude record flight up to 6750 m on 13 Sep. 1919: pilot E. Monz 1, R. Gsell 2, G. Madelung 3, E. Brandenburg 4 (l), and as crop duster ~1925, × forest marking flags (r)

where he had calculated the strength of a F13 wing in his diploma thesis, and Ernst Brandenburg[181] (4) participated presumably as independent observers of the newly founded *RVM Reichsverkehrsministerium* (Reich Ministry of Transportation); both will reappear in the following Sects. 4.3.2 (EB) and 4.3.3 (GM).

---

and other scientists to be transferred to the unit. Future Nobel laureates James Franck, Gustav Hertz, and Otto Hahn as well as physicists W. Westphal, E. Madelung and H. Geiger served as gas troops in Haber's unit. Georg Madelung (1889–1972), a versatile German engineer, professor and aircraft designer, was founding member of *WGL Wissenschaftliche Gesellschaft für Flugtechnik*, Germany's first aeronautical society in 1912, his doctoral thesis at TH Hannover in 1921 dealt with propeller theory, in the same year he designed the single-beam *Vampyr* sailplane, which launched the breakthrough to thermal flight. From 1922–1925 he was—recommended by L. Prandtl—employed at Glenn L. Martin Co., Cleveland, OH., adding to this company's prestigious list of one-time employed aircraft designers (D.W. Douglas, L. Bell, J.S. McDonnell, W.E. Boeing); in April 1922 the U.S. Navy ordered 36 observation monoplanes MO-1 which Martin had developed, assisted by Madelung. Due to his foregoing experiences at Junkers, the MO-1 had some design features in common with the Junkers J4 to J10 aircraft. Between 1925 and 1929 he was head of the Aircraft Department of DVL, scientific member of the DVL board and associate professor at TH Berlin-Charlottenburg; he married the sister of Willy Messerschmitt (1898–1978), leading German aircraft entrepreneur and designer in coming years. Since 1929 Professor for Aeronautical Engineering at TH Stuttgart, Georg Madelung founded there *FIST Flugtechnisches Institut Stuttgart* and in 1940 *FGZ Forschungsanstalt* (research institute*) Graf Zeppelin*.

[181] Ernst Brandenburg (1883–1952), Commodore of *Bogohl 3* [B̲ombergeschwader der O̲bersten H̲eeresleitung] led the first German air raid to London by means of 17 'giant aircraft' on 13 June 1917, for which he was—heavily wounded—decorated by the *Pour le Mérite*, the highest German bravery medal. From 1924 to 1942 he held high and influential posts in the Ministry of Transport. In 1926 he belonged to the founding members of *Luft Hansa*; in 1942 he was disposed of all functions as member of the anti-Nazi *Confession Church*.

4.3.1   Thinking the Unthinkable: Poison Gas Bombing

The use of poison gas by all major belligerents throughout World War I constituted war crimes as its use violated the *1899 Hague Declaration* concerning *Asphyxiating Gases* and the *1907 Hague Convention on Land Warfare*, which prohibited the use of *poison or poisoned weapons* in warfare.[182] Widespread horror and public revulsion at the use of gas and its consequences led to far less use of chemical weapons by combatants during World War II; however, poison gas *Zyklon B*, the trade name of a cyanide-based pesticide invented in Germany in the early 1920s, is infamous for its use by Nazi Germany during the *Holocaust* to murder approximately one million people in gas chambers installed at Auschwitz-Birkenau, Majdanek, and other extermination camps.[183] Unexpectedly the topic emerged during investigations about the early beginnings of axial turbomachinery for German aero gas turbines—and shall consequently be reported here in short.

Saying goes, that the original German impetus for putting poison gas on stock was driven by fear of conventional explosives and ammunition shortages due to the Chilean saltpetre monopoly, which actually threatened, enforced by the Entente's naval blockade after the outbreak of war in 1914. But magically, it was Fritz Haber[184] (1868–1934), Fig. 4.32, and his *Haber-Bosch process* of synthesising ammonia from nitrogen and hydrogen gas which yielded German independence for large-scale production of explosives (and fertilisers) just in time,—and Haber was among the first to think about poison gas usage early in the war, especially to break up static warfare along the French frontier, thus shortening the war and indirectly saving lives. But after the war was lost, Fritz Haber feared arrest as a war criminal for his poison gas research and usage, so that he preferred to retreat to neutral Switzerland for a while. He regained very soon his characteristic self-confidence when rumours spread in early 1919, that he was awarded the 1918 Nobel Prize in Chemistry[185] for his *Synthesis of ammonia from its elements*. From 1919 to 1923 Haber continued to be involved in Germany's secret development of chemical weapons, working with Hugo Stoltzenberg,[186] and helping both Spain and Russia in the development of chemical gases. During the *Rif War* in Spanish Morocco between 1921 and

---

[182] See Wikipedia, 'Chemical weapons in World War I'.

[183] A thought, originally expressed by Michael J. Neufeld, Smithsonian Inst. in private communication, 19 Nov. 2015.

[184] See Wikipedia, 'Fritz Haber' in English.

[185] Fritz Haber received his Nobel Prize for 1918 officially 1 year later, in 1919, since in 1918 none of the candidates fulfilled the criteria; the award ceremony took actually place on 2 June 1920 at Stockholm—with his Nobel Lecture https://www.nobelprize.org/uploads/2018/06/haber-lecture-1.pdf
A statement in the book,—see Stoltzenberg, Haber, that Haber had defended gas warfare against accusations also in his Nobel Lecture, that it was inhumane, saying that death was death, by whatever means it was inflicted,—cannot be confirmed by the text of Haber's lecture.

[186] See Wikipedia, 'Hugo Stoltzenberg'in English.

1927, the Spanish *Army of Africa* dropped chemical warfare agents in an attempt to put down the rebellion led by Berber guerrilla leader Abd el-Krim. These attacks in 1924 marked the first confirmed, large-scale use of mustard gas being dropped from *Farman F60 'Goliath'* airplanes.[187] On invitation the *Reichswehr*[188] had sent observers on the spot—in civilian clothes—and consequently, in 1925, first corresponding activities could be observed at *Junkers Flugzeugwerke* at Dessau.[189]

It all started one day with a funny éclat, when an angrily protesting cyclist came to the hangars—splattered all over with red paint. He described an aircraft spraying a reddish cloud, which the wind drove over to his country road, and the result could be seen. Finally, the man was calmed down—and after some in-house investigation the background was found out as some secret Reichswehr activities—in agreement with the company owner Hugo Junkers—on the neighbouring old Junkers airfield at Mosigkau. As reported, a certain *Marquard*[190] had experimented there, simulating poison gas spraying.

After the occupation of the *Ruhr* by French troops in 1923 the German defencelessness was very obvious; saying is that the remaining stock of artillery ammunition was 170 shots. One way out—especially after an evaluation of the fresh chemical Rif warfare experience—was seen in a strong air force in combination with chemistry. Chemical agents could be either dropped as bombs or sprayed; the German preference, especially after the foregoing *'successful'* Spanish raids, was directed to *Lost*[191] or Mustard gas, as it was called due to its characteristic smell. In April 1925 Marquard's first spray simulations by means of a F13 were extraordinarily primitive. He used glycol with physical features similar to Lost which he colourised red by all available *Eosin*[192] from the Dessau pharmacies. Then he covered parts of the air field with nettle cloth to evaluate the pour-spray effect of the red glycol solutions from various altitudes.

However, the incident on the country road had shown the restricted possibilities at a place like Dessau, a single cyclist could jeopardise the required secrecy. Therefore in June

---

[187] See Wikipedia, 'Chemical weapons in the Rif War' in English.

[188] German Armed Forces, 1921–1935, according to the regulations of the Treaty of Versailles limited to 100,000 men.

[189] Described after—see Sietz, *'Es riecht nach Senf!'* ('It smells of mustard!')

[190] Identified as Dipl.-Ing. Ernst Marquard (1897–1980), who entered the service of *RWM Reichswehrministerium* (Reich Defence Ministry) on 1 Oct. 1924 (up to 14 June 1926) as *Managing Director and Test-Director of Combat Material Equipment and Bombs* with the firm *Junkers Luftverkehr*, Dessau (presumably a self-definition), see http://www.oocities.org/~orion47/WEHRMACHT/LUFTWAFFE/Ingenieur/MARQUARD_ERNST.htm and,—see Forsyth, Luftwaffe, about Marquard's development of a Mach 20 hollow-charge warhead in 1943, hoisted under a Ju-88 (1.8 m long, 1700 kg of high explosive), designed to penetrate armoured battleship steel or thick concrete.

[191] Mustard gas was originally assigned the name *LOST*, after the scientists Wilhelm LOmmel and Wilhelm STeinkopf, who developed large-scale production for the Imperial German Army in 1916.

[192] *Eosin*—today mostly used for the colouring of HE Haemalaun-Eosin histo-pathologic sections.

1925, Marquard and his small team moved on to Germany's unsettled *'far-east'* of East Prussia to the dunes north of Rossitten,[193] well-known from the *RRG Rhoen-Rossitten-Gesellschaft* (RR Society) sailplane activities of the early 1920s with 30,000 trained scholars. But even here the secret tests remained not unnoticed: *'It smells of mustard !'* shouted a lonely *wanderer* along the beach to the flying chemists. Again, the final way-out was a shift to the Soviet Union; between 1928 and 1933 the Reichswehr *Inspection of Artillery (In 4)* practised development of and training in chemical warfare together with the *Red Army* at the secret *Tomka* gas test site off Volsk/Saratov, some 1400 km south-east of Moscow on the Volga river.[194] A special target of these activities was the investigation of *winter-LOST* and its application for territorial barrage at cold temperatures.

Figure 4.31 illustrates details of the Dessau location with its remarkable concentration of technological-cultural places and not nameable *no-places*: Just some 3 km in the west of Dessau's city centre and the famous *'Bauhaus'*[195] was the Junkers' new main airport since 1924—with a unique, heated 1000 m concrete runway. From there approximately 2 km to the south lay Junkers's first airfield with a grass runway, in use between 1916–1925, where besides many other firsts the F13 altitude record flight had taken place on 13 September 1919, and also Marquard's poison gas spray simulations. Further back in direction to the city centre lay the huge areal of office buildings and test facilities for *Jumo Junkers Motorenbau GmbH* (Engine Works) since 1923 and *Jfa Junkers Flugzeugwerk AG* since 1919, both were merged to *JFM Junkers Flugzeug- and Motorenwerke* AG in 1936. Nearly in the heart of Dessau lay the production site of *Dessauer Zuckerraffinerie AG* (Dessau Sugar Refinery), where since 1924 on order from *Degesch Deutsche Gesellschaft fuer Schaedlingsbekaempfung m.b.H.* (German Corporation for Pest Control), originally launched in 1919 by Fritz Haber as co-founder and patent owning company since 1924, a cyanide-based pesticide was produced under the trade name *Zyklon B*[196]—also during the 2nd World War, where approximately 2/3 of the quantities for the *Holocaust* gas chambers came from Dessau, from a small 20-men-shed production *no-place* high above the roofs of the conventional sugar refinery. This landmark of unnameable horror is today surrounded

---

[193] Rossitten, today Rybatschi in the Russian oblast Kaliningrad (Koenigsberg) is located on the UNESCO World Natural Heritage site *'Curonian Spit'(Kurische Nehrung)*. Glider starts took place from the 60 m high *Gora Chornaya (Schwarzer Berg)* towards the *Courland Lagoon (Kurisches Haff)*; 5 km in the north, beyond the present Russian/ Lithuanian border at Nida (Nidden) Thomas Mann (1875–1950), German novelist and 1929 Nobel Prize literature laureate built from the prize money a cottage (from the locals called *'Uncle Tom's Hut'*), where he spent the summers 1930–1932,—see Wikipedia, 'Thomas Mann' in English.

[194] See Wikipedia, 'Tomka gas test site' in English. Today the place is known as Shikhany, Saratov Oblast, where in the early 1970s the Novichok agent as the deadliest of new binary chemical weapons was developed. It came to public attention after being used to poison opponents of the Russian regime, including the Skripals and two others in Amesbury, UK (2018) and Alexei Navalny (2020).

[195] See Wikipedia, 'Bauhaus'in English.

[196] See Wikipedia, 'Zyklon B' in English; the *Dessauer Zuckerraffinerie* was nearly completely destroyed by an Allied air raid on 7 March 1945(!).

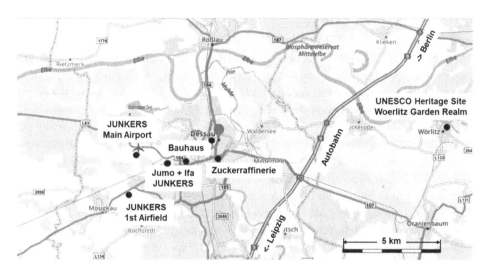

Fig. 4.31 Dessau locations, based on © *Michelin* map

by cultural and natural UNESCO World Heritage sites, the *Bauhaus Dessau* since 1996 and the *Woerlitz Garden Realm* since 2000,[197] one of the first and largest English parks in Germany and continental Europe, erected under the influence of *Enlightenment* between 1765 and 1800.

Due to the *Versailles* aircraft production ban, the company Junkers like other German aircraft manufacturers was forced to a number of uncommon decisions in the early 1920s to secure its economic survival. As part of an agreed Russian-German cooperation in the *Treaty of Rapallo*, *Junkers Flugzeugwerke* founded additional operations in 1922 at Fili (7 km due west of Moscow centre) and in 1924 in Sweden. However, the Soviet-Russian venture, though planned for 30 years, had to be closed already again in 1926 on request of the USSR government. Within Germany Junkers tried to diversify the aircraft business immediately after the Great War. Already in 1919 the *Junkers Luftbild* (aerial photo) was founded, in 1921 followed by *Junkers Luftverkehr* (air transport) with a widely stretched flight route network towards Turkey, Persia and South America. It was estimated that in 1925 nearly 40% of all aerial traffic were generated by using Junkers aircraft. Finally in 1925, the *Junkers Schaedlingsbekaempfung* (pest control) was founded, immediately after Marquard had finished his spray tests at Dessau-Mosigkau; Fig. 4.30 (r) illustrates a corresponding F13 sortie as crop duster still in 1925. First flights were organised over large forest areas in Pomerania and Upper Palatinate/Bavaria against *pine moth* and *pine looper*, using the stomach insecticide *Esturmit,* a sticking calcium-arsenic powder developed for winegrowing applications by *E. Merck, Darmstadt.*[198] Contrary to the spatial

---

[197] See Wikipedia, 'Bauhaus Dessau' and 'Dessau-Wörlitz Garden Realm', both in English.

[198] See Wikipedia, *'Merck Group'* in English; founded in 1668 E. Merck is the world's oldest operating chemical and pharmaceutical company, as well with about 56,000 employees and at present in 66 countries one of the largest pharmaceutical companies in the world.

vicinity at Dessau, nothing is known about a cooperation at any time between Junkers and the Dessauer Zuckerraffinerie or related companies.

The aforementioned Treaty of Rapallo was an agreement signed on 16 April 1922 between the German Republic and Russian Soviet Federative Socialist Republic (RSFSR) under which each renounced all territorial and financial claims against the other after World War I.

The two governments also agreed to normalise their diplomatic relations and to 'co-operate in a spirit of mutual goodwill in meeting the economic needs of both countries'. Secretly the two sides established elaborate military cooperation, while publicly denying it. The cost for these activities were carried by the *Ministry of the Reichswehr,* while the official responsibilities were with private individuals to prevent any implications with the *Versailles* restrictions. In addition to the already mentioned Junkers aircraft plant at Fili and the Tomka gas test site, there was a military training ground for tank equipment and corresponding staffs at Kazan (800 km east of Moscow) and—here of special interest— the first and most important of the Russo-German training centres established under the military agreement of August 1923: the air base at Lipezk, a small provincial town some 400 km south-east of Moscow. Flight testing started there in June 1925, basically with 50 newly developed Fokker D.XIII, of which hundred had been ordered after the Belgian-French occupation of the *Ruhr* in January 1923, and seven *Heinkel D17* two-seater fighter aircraft. But the start-up was delayed after the *Manchester Guardian* had revealed the secret military cooperation in December 1926 in general. This caused a fundamental cabinet crisis at Berlin, so that the facility was back to normal only in the 1928–1933 timeframe.

The slow start was also due to the complicated set-up in principle. A number of dangerous frontiers lay between the two theatres of *Reichswehr* activity in Germany and Russia. Since the Russians only supplied the basic building materials, such as wood and stone, for the numerous hangars and facilities that had to be built at Lipezk, everything else, down to the last nail had to be shipped from Germany. Neither men nor materials could be sent across Poland, which was the most direct route, for fear of detection. The safest railroad was—after crossing the *Polish Corridor*[199] in sealed trains—circuitous via Koenigsberg/Kaliningrad — Kaunas/Kovno (Lithunia)—Duenaburg/Daugavpils (Latvia)—Smolensk and on to Moscow. This trip involved six border and customs inspections, all of which were best avoided. The airplanes themselves and valuable precision instruments, such as bomb-sights, were flown to Russia, and then always without intermediate landings. Most of the materials were sent by the Baltic Sea, usually via the Free Port of Stettin/Szczecin to Leningrad/St. Petersburg.[200]

---

[199] See Wikipedia 'Polish Corridor'in English.
[200] See Freund, Unholy Alliance.

The German *Bundesarchiv* (Federal Archive) published a significant document about the Lipezk undertaking, a travel report[201] from the then Hauptmann a.D. Kurt Student about a trip from Berlin to Moscow and Lipezk between 15 August to ~ end of August 1926:

- He arrived at Moscow after 2½ days which corresponds to the described six-border-railroad tour.
- Student is not specific in writing about his field of activity, but with all likelihood his first stop was at Podosinki,[202] 60 km in the north of Moscow, where the Russians had offered the possibility of a *sharp* (poison gas) sprinkling test in a populated area, where the nearest living quarters were just 3 km away. There he signed a related contract order and the tests were apparently accomplished thereafter still in 1926 with '*satisfactory results*'.
- After visiting the *Junkers* plant at Fili in the morning of 21 August 1926, he continued in the evening of the same day by train to L.(ipezk), where he arrived on 22 August afternoon.
- In his *Summary* Student concludes: '... *Certain test with air warfare agents can be carried out only in R.(ussia) Recent progress in this field requires now the establishment of a specific test area latest till spring 1927, where the effect of aircraft weaponry, bombs and means for Schaedlingsbekaempfung* (pest control) *can be investigated.* [Hand-written marginal note: *Gut!* (Good)] ... *For sharp tests we require a site in less than 50 km distance from L.(ipezk), comprising an area of at least 4×4 km. .... For the specific test department we need trained personnel and separate own equipment.*'

Nothing is known about the actual usage of poison gas at or near Lipezk from the German command, but Student's travel report with his camouflaging, dual-use terminology *Schaedlingsbekaempfung* is unequivocally the clearest evidence so far about serious

---

[201] BArch RH 2/2213, *Geheime Kommandosache* (Secret Command Document) *"Z"*, dated 10 Sep. 1926; *"Z"* stands for *"Zentrale"*, the Reichswehr coordination command at Berlin, internally *'Department T3'*, to be differentiated from *"Z(Mo)"*, the corresponding coordination subsidiary at Moscow. See Speidel, Reichswehr und Rote Armee, p. 19 and—see Wikipedia, 'Wilhelm Speidel (General)' in German. K. Student (1890–1978) had been a fighter pilot in WW I, from 1922 to 1928 as Captain (off duty) in the *Heereswaffenamt* (Army Ordnance Office), Inspection for Weapons and Equipment, he was responsible for aircraft developments. In May 1941 Student, in the meantime head of German paratroopers, directed the airborne invasion of Crete. See Wikipedia, 'Kurt Student' in English.

[202] Other sources name a place *Ukhtomsky* (today Rayon Kosino-Ukhtomsky, 18 km east of Moscow city centre) which later led to the otherwise meaningless designation *Tomka* for the transferred gas test site near Volsk/Shikhany, Saratov Oblast. And here at Tomka, there were also four aircraft stationed since 1927, so that what had started two years before at Dessau by Marquard's simulated gas spraying could be accomplished now in a sharp/realistic test.

considerations on the German side to prepare in 1927 for the use of poison gas agents in future aerial warfare.

But Student's opinion was not an isolated one. As early as February 1924 Joachim von Stuelpnagel (1880–1968), then head of the Army Department within the *Truppenamt*[203] developed—as result of the small 100,000 men-army—ideas for a kind of future provoked guerrilla war against France or Poland on German territory (with assumed British neutrality) which should be *'channelised by chemical contamination of certain territorial strips'*. He assumed with the onset of hostilities *'that large aerial squadrons will attack our cities—also Berlin—and important railroad junctions day and night with explosive and gas ammunition'*. He therefore asked for protective counter-measures. . . and strong own aerial forces *'to give back what us was given'*. In view of the present own weakness, there was to him only to demand from the convened young officers to tool aeronautics *'eagerly with phantasy'*.[204]

In fact, the weakness of the conventional arm of the German military increased the relative strategic importance of poison gas and its possibilities. The special machinery set-up for the production of *Phosgen* and *Lost* at Tomka had a capacity of one million gas grenades, which would have added more than 200% to the allowed Reichswehr stock of conventional artillery ammunition. Since in the 1920s France and Great Britain reduced their chemical warfare activities nearly to zero as a result of the world-wide proscription, this suddenly opened up a chance for Germany to compensate for its conventional inferiority.

In this situation even weird strategic considerations found a positive military resonance. The recognised military theoriser Adolf Caspary forestalled the *neutron bomb* when he stated: *'While attacking, cities (and industrial agglomerations) are much too precious to be destroyed. And—quite frankly—that is not necessary at all: Poison gas is effective against human beings, but preserves material values.'*[205]

4.3.2 The Altitude Aircraft Developments: From *Ju 49* To *Ju 86R*

On 25 May 1926 the *Treaty of Paris* removed all still existing restrictions on civil aviation in Germany as leftovers of the Treaty of Versailles (military developments were still interdicted till the end of 1930). The *'Notgemeinschaft der Deutschen Wissenschaft'* (Emergency Association of German Science),[206] an industrial/government fond to alleviate

---

[203] The *Truppenamt* or 'Troop Office' was the cover organisation for the German General Staff from 1919 through until 1935, when the General Staff of the German Army (*Heer*) was re-created. This subterfuge was deemed necessary in order for Germany to be seen to meet the requirements of the Versailles Treaty.

[204] See Deist, Die Reichswehr, p. 85.

[205] See Mueller, Gaskriegsvorbereitungen and—see Caspary, Wirtschaftsstrategie, p. 158.

[206] See Wikipedia 'Notgemeinschaft der deutschen Wissenschaft'in English.

Fig. 4.32 The *'Notgemeinschaft'* of German Science 1925: President F. Schmidt-Ott (l), NG premises at Berlin Castle (m), Vice-President F. Haber (r)

economic stresses to maintain a minimum of scientific research especially in the wake of the hyperinflation 1922/3 in Germany, had been founded already on 30 October1920. The financial endowment was rather low at first, with small contributions from the Reich Ministry of Finance and an industrial association for the promotion of science, represented by Carl Friedrich von Siemens and his deputy Hugo Stinnes. The *'Notgemeinschaft'* was led from 1920–1934 by its President, the lawyer Friedrich Schmidt-Ott (1860–1956),[207] actually the last Royal-Prussian Minister for Cultural Affairs 1917–1918 and—as his deputy—Fritz Haber (1868–1934), a chemist and—as discussed before in Sect. 4.3.1—a *'pioneer'* of chemical warfare in WW I, who nevertheless had received the Nobel Prize in Chemistry in 1919, Fig. 4.32.

Atmospheric high-altitude research had been suggested as a joint research topic as early as October 1925—and in due course, the project of a high-altitude aircraft was agreed by the *Notgemeinschaft* on 16 July 1927.[208] The selected research subject combined nearly ideally the interests of the various parties involved. It could address the pure science of the atmosphere, especially of the *stratosphere* beyond 12 km, by closing still existing gaps

---

[207] Schmidt-Ott's career had been highly interwoven with the imperial *Hohenzollern* family and the last German Emperor Wilhelm II (1859–1941) with whom he kept a lifelong friendship. Born at Potsdam, young Fritz (13) belonged already to a small, hand-picked group of school mates to Crown Prince Willy, and he shared also with him the spartan years at the *Friedrichsgymnasium* at Kassel, 1873–1878. Consequently, the premises of the *Notgemeinschaft* at the second floor of the Berlin Castle, Fig. 4.32, indicated a certain continuity in this relationship, while the exiled *Kaiser* enjoyed his pastime with wood chopping at Doorn, Nl,—see Wikipedia 'Friedrich Schmidt-Ott' in English.

[208] See Flachowsky, Das größte Geheimnis, p. 9, referring to a memorandum of Hugo Hergesell (1859–1936), Director of the *Lindenberg Aeronautical Observatory* and one of the influential key consultants to the *Notgemeinschaft,* in which he suggested as a future research topic *'the development of instruments which will allow the determination of vertical atmospheric movements with sufficient accuracy, at the ground, but mostly at larger altitudes'.*

after intensive ballooning, but also the applied science of high-altitude flight. Already in the last year of the Great War, the Krupp-built *Paris Gun* with its range of up to 130 km had dramatically illustrated the advantages of high-altitude flight at low atmospheric density.[209] And of course, there were the military interests of the time, which meant—after the foregoing introduction nearly self-evident—*poison gas*.

The involvement of Fritz Haber led continuously to speculations about assumed intentions for the promoted high-altitude aircraft project. In fact, taking all the negative experiences with poison gas usage, beginning with the exposure of Haber's troops to their own gas due to moody winds, up to endangering anti-aircraft fire against low flying gas bombers in the *Rif Mountains*, the prospects of an unarmed, at 15 km nearly invisible and inaudible, high flying gas bombing or reconnaissance aircraft must have developed a certain fascination—at least for a short while.

In the mid-1920s Schmidt-Ott was in regular contact with Karl Becker (1879–1940),[210] since 1926 head of Sect. 1 *'Ballistics and ammunitions'* of Army Ordnance's Testing Division, then especially interested in solid-fuel rockets as a means of launching poison gas against enemy troops on the battlefield.[211] The fact that *Versailles* omitted any mention of rocket development might have kindled Becker's interest in the technology. But of more significance was the *Versailles* ban on heavy artillery, an important class of weapons that was Becker's speciality. Provided that rockets or alternatively the high-altitude aircraft could be made sufficiently powerful, they could replace not only short-range battlefield weapons like the *Nebelwerfer* (smoke mortar) and the *Livens Projector*,[212] but also long-range heavy guns.

The inspired German 'Hoehenflugzeug' (high-altitude aircraft) project started with a design flop on the basis of a *Siemens SSW D.IV* combat aircraft—even before the official kick-off meeting, to which Schmidt-Ott invited to the *Notgemeinschaft* premises in the Berlin Castle on FRI 15 July 1927, 11 am, Fig. 4.32, using the antiquated form of address *'Ew. Hochwohlgeboren (Revered Excellency)'*.

One single copy of that WW I fighter skirted the *Versailles* demolition, and was handed over to *DVL* Berlin-Adlershof with permission of the Allied authorities for future civil

---

[209] The *Paris Gun* fired its 21 cm grenades at a steep angle of 55 deg, so that the ballistic trajectory culminated 40 km above ground in the upper *stratosphere*. Conventionally fired at a rather shallow 35–45 deg angle the range would have decreased to 30–40 km only.

[210] See Szoelloesi-Janze, Fritz Haber, p. 652 and 677 f.

[211] See Neufeld, The Rocket and the Reich, Chap. 1 'The Birth of the Missile', pp. 5–9. Neufeld (same source, footnote 25) is also certain that Becker and his subordinates discussed the possibility of using liquid-fuel missiles for chemical warfare against civilians—based on a corresponding note in the war diary of General Franz Halder, Chief of the Army General Staff (1938–1942). Halder records a tour of a chemical weapons plant with Becker on 26 Sep. 1939: Poison gas and the use of the long-range rocket against London are mentioned, but not clearly linked. There is at least one further hint that the *Notgemeinschaft* of 1926 dealt also with the ingredients for chemical agent production, see http://www.spiegel.de/spiegel/print/d-45234195.html in German.

[212] See Wikipedia, 'Nebelwerfer' and 'Livens Projector' in English. In the first World War the simple mortar-like weapons could throw large drums filled with flammable or toxic chemicals and thus became the standard means of delivering gas attacks on both sides of the frontline. It remained in the arsenals—partially upgraded with rocket support—until the Second World War.

aviation research purposes. In 1926, *DVL* directed the *Albatros* Berlin-Johannisthal Works, to build the high-altitude research aircraft based on the *SSW D.IV* airframe. Martin Schrenk,[213] then head of the *DVL Hoehenflugstelle* (altitude flight office), carried out the necessary design changes, comprising the enlargement of the wings to a significant span, its reinforcement with two pairs of struts, changing of the empennage outline and the installation of a specially designed, high-altitude propeller; after reconstruction, the aircraft received the *Albatros H-1* marking, Fig. 4.33 (l),[214] but apparently flight testing had to be stopped very soon due to dangerous wing flutter.

Main data of German high-altitude research aircrafts

	Albatros H-1	Junkers Ju 49
Year of construction	1926	1929/1930
First flight	1927	02 Oct 1931
Crew	1	2
Cockpit	open	pressurised cabin
Flight altitude	10,000 m (plan)	12,500 m
Engine	Siemens Sh IIIa, 160 hp	Jumo L88a, 800 hp
Supercharger	none	2-st. radial, intercooled
Propeller diameter	2-blade, 3.68 m	4-blade, 5.6 m
Aircraft take-off weight	900 kg	4250 kg
Aircraft wing span	biplane, 12.56 m	mono-pl., 28.26 m

(continued)

---

[213] Martin Schrenk (1896–1934), German pilot, aviation engineer and designer, professor at TH Berlin-Charlottenburg,—see Wikipedia 'Martin Schrenk'in German. Also in 1926, DVL chairman Prof. Wilhelm Hoff followed a recommendation to install Schrenk's study friend Wunibald Kamm (1893–1966) as head of the DVL aircraft engine department, a position, which he kept till 1930, before he became first Professor of Automotive Engineering and Vehicle Engines at TH Stuttgart, establishing in addition the aero-influential, privately organised Research Institute of Automotive Engineering and Vehicle Engines Stuttgart (FKFS), growing in the coming years up to 650 employees.

Martin Schrenk died on 13 May 1934 together with meteorologist Victor Masuch, presumably suffocating during high-altitude research beyond 10,000 m on board the largest German gas balloon *Bartsch von Sigsfeld* (9900 m^3, see picture at https://hannsklemm.wordpress.com/wichtige-personen/martin-schrenk/), starting from Bitterfeld, 30 km in the north of Leipzig, and crashing after a 1300 km, 15 h long uncontrolled drift on to Russian territory near Duenaburg/ Daugavpils (Latvia). After thorough investigations, the use of oxygen mouthpieces—similar to those shown in Fig. 4.17—were identified as main cause for the accident, so that thereafter only oxygen masks were allowed.

[214] The restored *Albatros H-1* is on exhibition at the Polish Aviation Museum, Kraków, Pl. Originally the *Albatros H-1* had been in the 1930s part of the *Deutsche Luftfahrtsammlung* (German aviation collection) at Berlin-Moabit. With the onset of heavy Allied air raids on Berlin in 1943, the unique air collection was partially destroyed in the night of 22/23 Nov. 1943. The rest of approx. 30 aircraft was brought to Czarnikau/Czarnkow, 35 km south of Schneidemuehl/Pila in Western Prussia, east of the River Oder, an area that became Polish after WW II. Consequently, this and other pieces of early aircraft and engine history are now on exhibition at the Polish Aviation Museum, Cracków since 1963, and still in Polish custody.

Fig. 4.33 German high-altitude research aircrafts: *Albatros H-1* restored, upper wing indicated, 1926 (l), *Junkers Ju 49* assembly 1930 (r) © *Pol.Av.Museum* and *Junkers*

	Albatros H-1	Junkers Ju 49
Aircraft wing area	14.4 m^2	98 m^2
Aircraft length	5.70 m	17.20 m
Aircraft speed	190 km/h	146 km/h

Comment on H-1 1st flight date above: Some sources—see e.g. Wikipedia 'Albatros H.I' in German—claim that an observed lack of stiffness in the extended wing structure prevented from flying at all, but official DVL records,—see Hoff, Jahrbuch, p. XIII—confirm that the aircraft was flight tested during 1927

In due course after project start, the Junkers design department presented four alternative proposals on 16 May 1928, which were decided in favour of a cantilever, low wing mono-plane of 28+ m wing span with corrugated-sheet metal planking. The engine was an 800 hp supercharged *Junkers L88a*[215] which drove a single 5.6 m diam. geared four-blade propeller at the front end (which required a high-legged landing gear) with the prospect of a 13 km flight ceiling. The pressure-proof, double-skin, temperature-insulated high-altitude chamber, *body-shaped* for a two-member crew, was integrated into the fuselage as an independent unit, Fig. 4.34 (m).

Adolf Baeumker (1891–1976), later presumably Germany's first and most powerful *State Science Manager,* stated in spring 1926 as representative of The Department of Aerial Defence within *Truppenamt*, officially a member of RVM Reichsverkehrsministerium (Ministry of Transportation)—with a hidden military agenda: *'We demand for the highly loaded wartime aero engine—in comparison to the civil aircraft engine—a climb*

---

[215]See Wikipedia, 'Junkers L88' in English.

Fig. 4.34 Junkers high-altitude research aircraft *Ju 49*, 1930: Front view with 5.6 m diam. propeller (l), 2-seater altitude pressure cabin (m), interior view of pressure cabin (r) © Junkers

capability up to highest levels in shortest imaginable time lapses, in combination with extraordinary overload margin during critical combat situations. Military target is a large high-altitude engine of 1000 to 1500 hp, water-cooled—and an air-cooled fighter engine of 500 hp. Besides engine developments, flight range, bomb payload and flight speed also the cruising altitude is of utmost importance. Fighting for best ceiling height is key for achieving tactical air superiority in aerial warfare'.[216]

Indeed, as the *Ju 49* development showed, the two problems of building a reliable pressurised cabin, Fig. 4.34 (m), and of a high-altitude engine with efficient, high-capacity blowers represented the most important technical challenges, which had to be solved before stratospheric flight, also from a military point-of-view. And, in view of the demanded power split, also resulting engine design implications became clear. Beyond the basic differentiation in small, air-cooled fighter engines and heavy water-cooled bomber drives, it was presumably also Martin Schrenk, DVL who pointed out the consequences on supercharging for the first time: smaller fighter engines would have to rely on the—in the meantime broadly established—radial/centrifugal super- or turbo-chargers, eventually in one to three multi-stage configuration; the multi-cylinder engine variants for heavy bomber applications demanded considerably more compressed air, which only could be provided by applying the aerodynamically demanding, high-flow axial charger design

---

[216] See Flachowsky, Das größte Geheimnis, p. 15.

concept, which urgently had to be developed.[217] The following Sect. 4.3.3 *'The AVA Goettingen and the axial-flow compressor'* will basically address these developments, which eventually also paved the ground for subsequent axial-flow turbojet developments.

The mentioned military aspects were also discussed in the context of the newly developed high-altitude research aircraft. Consequently, apparently in July 1927 Schmidt-Ott achieved agreement with the leading RVM *Ministerialdirektor* (assistant interior secretary) Ernst Brandenburg, seen already eight years before on Fig. 4.30, who was now in charge of the Ministry of Transportation's camouflaged military investment budget; in the end Schmidt-Ott had Brandenburg's full support for the *Notgemeinschaft* and their Ju 49 planning.

In March 1929 the Junkers development project received the official designation *Ju 49*[218]*/EF 29*, the latter standing for Entwicklungs-Flugzeug (development aircraft), Fig. 4.34 (l). As a valuable experience backup for the ongoing high-altitude aircraft development at Junkers, company pilot Willy Neuenhofen (1897–1936) achieved on 26 May 1929 a new single-seater world altitude record with 12,739 m, the second German entry in this category after Franz-Zeno Diemer in June 1919, Figs. 4.29 and 4.35. The *Junkers W34*, a follow-on, five-passenger-design to the F13, achieved the record ceiling after 45 min, starting still from Junkers's first (grass) airfield, Fig. 4.31, and powered by Roy Fedden's[219] *Jupiter VII* engine, a 400 hp supercharged, nine-cylinder single-row piston radial engine built by the *Bristol Aeroplane Company*. Like the F13 before, the W34 had still an open cockpit so that Neuenhofen's sole protection against cold was a thick fur coat. He used a *Draeger breathing apparatus;* since he had to fly alone for weight reasons, Junkers installed—presumably for the first time—a *'dead man's knob'* at the W34 steering horn, which prevented in case of a pilot unconsciousness further climbing of aircraft; this in fact saved Neuenhofen's life in a foregoing exercise flight test.

After the first flight on 2nd October 1931, the flight tests lasted nearly two years until September 1933, when the aircraft was transferred to DVL, where during 1935 finally flight altitudes nearly up to the aspired 13,000 m were achieved. The aircraft showed its practical merits especially for the development of a high-altitude (bomber) engine, which led to a number of secret developments from the 'Reichswehr' (German Armed Forces of that time), first for the 12-cylinder, 800 hp altitude motor Junkers L 88a with 2-stage, individually adjustable centrifugal blowers and corresponding charge-air cooler.

Since DVL—from the early beginnings of Martin Schrenk—had considered the high-altitude aircraft as its own, special sphere of interest, it was entrusted from both, RVM and the *Notgemeinschaft* with the task of an independent technical project controller, positioned as competent mediator between the official principals and the contracted company Junkers.

---

[217] See Schrenk, Probleme.

[218] See Wikipedia, 'Junkers Ju 49' in English.

[219] See Wikipedia, 'Roy Fedden' in English, and Sect. 6.1.3.

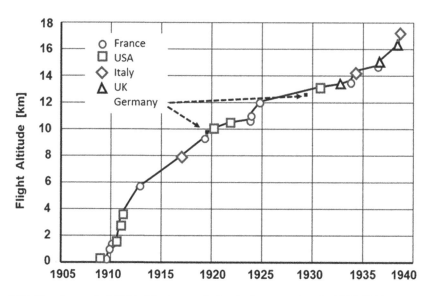

Fig. 4.35 Development of flight altitude records for single-seater aircraft, 1905–1940

Consequently, on 24 October 1929, it was DVL which placed the order for one experimental Ju 49/EF 29 (development, design and manufacturing with Jumo L88a) to Junkers Flugzeugwerk (aircraft work) AG and Junkers Motorenbau (engine production) GmbH, Dessau. The DVL purchasing contract fixed a total sum of 574,000,- RM (Reichsmark), of which the Notgemeinschaft carried a share of 200,000,- RM, while the RVM incurred the rest.

In general the Ju 49 development was considered as a success, though there remained fuselage structure and engine problems which prevented finally e.g. the installation of a three-stage blower for planned altitudes up to 16 km.[220] Ju 49 testing was finished in 1936, since the experiences should be transferred to the development of a new twin-engine, high-altitude aircraft Junkers EF 61 with pressurised cabin, as the basis for a planned high altitude, long range standard (bomber) aircraft.[221] The aircraft was powered by

---

[220]The non-existence of this high-pressure, three-stage supercharger prevented after 1932 the achievement of new, firmly timed altitude records for the German side; after a considerable delay of eight years, the device was finally ready for installation in the *Ju 86P*. On 5 Feb. 1940, one day after his 65th birthday (and 3 months before the German campaign in France) L. Prandtl thanked for *'your recognition as a stimulus for my continued work'* in return letters to Goering and Milch, adding the strong recommendation towards the latter *'We must urgently increase the ceiling height of our long-range reconnaissance aircraft. ..This can be achieved short-term, since a correspondingly designed DVL blower is now ready for installation in the existing reconnaissance aircraft. . . . Acc. to information from DVL, we could gain here 1–2 kilometres in ceiling height.'* See Eckert, Ludwig Prandtl, p. 244.

[221]See Wikipedia, 'Junkers EF 61' in English.

2× *Daimler-Benz DB 600A*, V-12 inverted, liquid-cooled, direct fuel injection piston engines of 900 hp each, since the originally planned *Junkers Jumo 211*[222] of 1500 hp was not ready for installation then. Project leader at Junkers became Herbert Wagner,[223] who failed—not the least in the development of a reliable, wide-scope front canopy for the EF 61, Fig. 4.36 (l). He had filed a corresponding patent[224] shortly after his start at Junkers with considerable momentum: After describing the inherent disadvantages of a conventional multi-pane/rib construction, especially for terrestrial observation and aiming, the basic idea of the new invention was a continuous, self-supporting, transparent, relatively thick one-piece inner shell body -2- and a thinner, outer wrap -3-, so that both parts could thermally freely expand and the intermediate space could be heated/de-iced by means of a 20 cm diam. tube -4- to distribute motor exhaust gas in-between. After a thorough evaluation together with specialists from *IG Farben*,[225] it was decided to go for an organic non-inflammable (cellulose tri-acetate) glass which would allow shooting penetration— without a complete collapse under the existing inner cabin pressure (which prevented an acrylic glass solution).

Without the perfect panorama canopy, flight testing—unchanged in comparison to Ju 49, Fig. 4.34 (r),—became a tremendous challenge after the prototype EF 61-V1 first flew on 7 March 1937. Ground visibility through one large front bull-eye was considerably hampered, so that meeting the landing cross at Dessau relied every time on a nerve-racking communication interplay between pilot/observer in the cabin and the ground crew. Though

---

[222] Mass produced between 1937–1944, there were 68,248 Jumo 211 built in total, which thus became #1 in the German engine inventory,–see Wikipedia 'Junkers Jumo 211' in English. High-altitude vers-ions of this engine would have been equipped with a 5-stage axial turbocharger, developed at AVA Goettingen, the axial compressor nucleus of the coming *Jumo 004* and *BMW 003* turbojet developments after 1938, see Sect. 4.3.3. The *Jumo 211* was a scaled-up version of the preceding 730 hp *Jumo 210* which was considered the German equivalent to the *Rolls-Royce Kestrel* of the early 1930s. There is evidence that the *Jumo 210* has been used as test vehicle for the AVA 5-stage axial turbo-charger ~1934, but limitations in small high-speed bearings for >30,000 rpm prevented a series production.

[223] Herbert A. Wagner (1900–1982) was a versatile Austrian scientist who developed numerous innovations in the fields of aerodynamics, aircraft structures and guided weapons. After a professorship at the TH Berlin-Charlottenburg, Institute for Aviation Technology, he joined Junkers Dessau in Sep. 1935, where he became a board member in the succession of Hugo Junkers (who had been forced out of his Company still in 1933) with responsibility for all-Junkers aircraft design. In this period, he filed also two ground-breaking patents for all-axial turboprop and turbojet engines, which will be dealt with in Sect. 6.2.1.

[224] Patent DE693,159 with German priority from 29 Dec.1936 for H. Wagner and J. Muttray on a *'Self-supporting, transparent shell body for aircraft'*. Justus Muttray had the pressurised cabin design responsibility, beginning from the Ju 49 in 1927 till the post-war East-German/GDR *Baade B-152* 72-passenger aircraft in ~1960.

[225] *Interessengemeinschaft Farbenindustrie AG* (German for 'Dye industry syndicate corporation'), commonly known as *IG Farben*, was a German chemical and pharmaceutical conglomerate, formed in 1925 from a merger of six chemical companies with headquarter at Frankfurt/M.

Fig. 4.36 H. Wagner et al. patent for all-plastic glass canopy 1937 (l), *Junkers Ju 86P* reconnaissance aircraft 1940—with two-seater glass pane canopy (m), crewman climbing aboard to take-off (r) © *Junkers,* and Luftkrieg.net

the EF 61 integrated the lessons-learnt from Ju 49, in addition to numerous firsts e.g. the first-time usage of *Fowler flaps* in Germany, Wagner's reputation was considerably tarnished after the greatly announced, patented ideas of a plastic canopy did not materialise in time.[226] However, only the fact that the two aircraft prototypes crashed already in 1937 within three months[227] developed absolutely disastrous, and ended also Wagner's career at Junkers prematurely in 1939. In hindsight, this moment would have been suited to reflect about the actual achievements in the German strive for air superiority in the high-altitude flying domain during the past decade. Figure 4.35 illustrates the development of altitude records for single-seater aircraft in the time frame 1910–1940, differentiated for the main four contributing countries France, USA, Italy and UK. There are 27 entries in total, of which the French share represents 44%, followed by 30% from the USA and 11% each for Italy and the UK. The two, German (marked) entries of Franz-Zeno Diemer in 1919 and Willy Neuenhofen in 1929 illustrate that German technology was competitive then for a short while, but far from developing the planned dominant role.

---

[226]The task was nearly perfectly realised at the same time by Kurt Matthaes (1900–1999), up to May 1937 Hans von Ohain's superior at the Heinkel test department, for the He 176 rocket-powered aircraft,—see Koos, Heinkel Raketen- und Strahlflugzeuge, p. 51, and—see Wikipedia, 'Heinkel He 176′ in English.

[227]The first prototype crashed on 19 Sep. 1937 at the Junkers Dessau airfield, when a wing configuration with modified Fowler flaps disintegrated during speed flying at 3500 m altitude, while the second prototype EF 61 V-2 crashed only three months later in Dec. 1937, even before high-altitude testing had started, when the two-men-crew observed commencing flutter of the double-fin tail—and decided to bail out. Surprisingly, the flutter disappeared then and the pilotless aircraft gravely circled down—10 years after the poison-gas rain simulation tests had started, exactly there.

This was basically the result of a lack in supercharging power, while the development of advanced aircraft piston engines appeared to be competitive by and large; especially, the salient Junkers diesel approach, manifested not only by being the first, but for more than half a century represented also the only successful generation of aviation diesel powerplants.[228]

Simultaneously, approximately around mid-1936 the altitude-directed German bombing strategy must have been given up—as obviously in view of limited resources without any reasonable chance of strategic success—and consequently being replaced in a kind of last-minute compensation rather by a speed-oriented approach, which naturally supported the further turbojet developments.[229] Wagner in 1939, after also the axial turbojet developments at Junkers Magdeburg had been stopped, moved on to Henschel Flugzeug-Werke AG at Berlin-Schoenefeld[230]—certainly also by realising that the gas turbine-powered turbojet would be the better altitude engine concept.[231]

Somewhat typical for a confused and confusing administration of the RLM Reich Air Ministry of that time, the EF 61 development drawback triggered immediately parallel high-altitude aircraft developments at *Henschel* (*Hs 128* and *Hs 130*) and *Dornier* (*Do 217P*), which led to nowhere and consequently could be forgotten, if there had not been an interesting detail in their turbocharging configurations. Both the high-altitude variants *Henschel Hs 130E* and the *Dornier Do 217P* were equipped with a *Daimler-Benz HZ-Anlage* (*Hoehenlader-Zentrale*, central altitude charger) i.e. a *DB 605 T*, V12 aircraft piston engine with ~1000 hp at altitude, mounted separately inside the aircraft fuselage, to drive a large two-stage centrifugal turbo-compressor for pressurised air supply of all three engines. In principle, this arrangement revived a WW I idea, when Brown Boveri (BBC) had placed their four-stage radial altitude blower/Mercedes D.II combination inside the

---

[228] See Wikipedia, 'Junkers Jumo 204' and 'Junkers Jumo 205' in English, the latter covering also the variants Jumo 206/207/208/223.

[229] This apparent decision milestones coincides with the death of General Walther Wever (1887–1936), whose self-piloted *Heinkel He 70 Blitz* crashed during start from Dresden-Klotzsche on 3 June 1936. An investigation revealed, that aileron gust locks were forgotten to be removed before a hurried start. Wever, then Chief of Staff of the *Luftwaffe*, was an early proponent of the theory of strategic bombing and supported correspondingly developments of heavy four-engine bombers,—see Wikipedia, 'Ural bomber' in English. This initiative was stopped with far-reaching consequences shortly thereafter, mainly due to economic restrictions, in favour of lighter, twin-engine (dive) bombers.

[230] At Henschel's special department F, Wagner was responsible for the development of the *Hs 293* radio-controlled anti-ship glide bomb with rocket booster,—see Wikipedia, 'Henschel Hs 293' in English, of which more than 1000 have been built, thus preserving the war-time existence of Henschel's Berlin facilities. Mid-1941 he detected and kept,—see Wikipedia 'Konrad Zuse' in English, amongst his Henschel employees, which paced the development of modern computer technology. Wagner's interests went also to the building of a super-cyclotron as essential pre-condition for German nuclear weapons,—see Beisel, Bomben-Stimmung in German.

[231] H. Wagner—to the members of the *German Academy for Aviation Research* on 28 Oct. 1937: *'After very thorough considerations I believe, that the gas turbine is the right altitude engine.'*

central fuselage of a *'R 30' Giant* aircraft.[232] Now BBC got this idea patented for both axial and centrifugal central compressor arrangements inside of the fuselage—with priority of 9 Dec. 1940,[233] sufficiently in time to the practical prototype testing of these aircrafts. This coincidence in timing may be an indicator of a small, though decisive information leak across the border during wartime, either on the German side between Daimler Benz Stuttgart to BBC Mannheim, and further to BBC Baden, CH, and/or simply across Lake Constance between the Dornier plants at Friedrichshafen and Altenrhein, CH, and from there again transported further to BBC Baden, CH.

In the meantime, Junkers recovered quickly and expanded its universal *Ju 86* aircraft series, of which between 1936 and 1945 more than 900 units have been manufactured— also by international licensees, steadily also in a successful high-altitude version *Ju 86P*. Figure 4.36 (m) shows the pressurised two-seater cabin with multi-pane glass canopy, isolated and removable from the aircraft main nacelle, and the characteristic, vertically stretched Jumo 207 diesel engine nacelle with two-stage turbocharger and intercooler; Fig. 4.36 (r) illustrates the direct, sealed access to the pressurised cabin bottom, which was also used when the crew had to bail out during a high-altitude mission.

The Ju 86P could fly at heights of 12,000 m and higher on occasion, where it was felt to be safe from Allied fighters. Satisfied with the trials of the new Ju 86P prototype in January 1940, the Luftwaffe ordered that some 40 older-model bombers be converted to Ju 86 P-1 high-altitude bombers and Ju 86 P-2 photo reconnaissance aircraft. Those operated suc- cessfully for some years over Britain, the Soviet Union and North Africa. Finally, the long- time idea of a nearly unarmed high-altitude aircraft materialised in the reconnaissance versions *Junkers* Ju 86P and R[234]—with a wing-span finally stretched up to 32 m and demonstrated flight ceilings of nearly 16 km—unmatched at least for a short while between January 1941 and August 1942,[235] before the jet-powered *Arado Ar 234 Blitz* took over, a few months before war's end in August 1944.

---

[232] See Eckardt, 'Gas Turbine Powerhouse', p. 58 f. and Fig. 4.14.

[233] See BBC patent CH221,503 *'Installation of aircraft propulsion system'*, patent grant 31 May 1942.

[234] The Ju 86R was further improved by Jumo 207C turbo-Diesel engines of 1000 hp take-off power, methanol-water injection during take-off and a *'GM1'* cryo-N_2O altitude boost injection, which was combined with intensified charge air cooling for better combustion (with the result of an excessively rising fuel consumption).

[235] For Ju 86P and R, see Wikipedia, 'Junkers Ju 86' in English. The belief in *'unarmed air superiority'* ended in 1942 after special *Spitfires* had been adapted to high-altitude fight, see P. Lehmann *'The Luftwaffe's High-Flying Diesel-Powered Bomber'*

https://www.historynet.com/luftwaffes-high-flying-diesel.htm—before it was resurrected in the mid-1950s with the *Lockheed U-2* at altitudes beyond 20 km.

4.3.3 The AVA Goettingen and The Axial-Flow Compressor

The foregoing Sect. 4.3.2 went close to the end of WW II with the German high-altitude aircrafts Ju 86 P and R—exceptionally beyond the set *hinge year 1935* of Chap. 4, since the technical account on that subject should be concluded in a continuing flow.

The present section about the AVA Goettingen, Ludwig Prandtl's Aerodynamische Versuchsanstalt (Aerodynamic Test Establishment) will return first to the beginnings of axial-flow compressor research in the mid-1920s,—with an introduction about the directing, slowly contour-gaining *Government Aviation Organisation*. Thereafter, *AVA— the Institution* and its turbomachinery key personnel will be addressed—also in exchange with a considerable number of curious international visitors. Then follows, what could be deduced from a rather lean archival situation *On Axial Blowers—Axial Superchargers for High-Altitude Piston Engines*, i.e. about first contracted development orders of turbo-charging high-altitude axial blowers (and their potential engine applications)—as the forerunners of coming axial-flow compressors for aero gas turbines—initiated from industry, or the Notgemeinschaft and the government, the latter two mostly coordinated by DVL Deutsche Versuchsanstalt fuer Luftfahrt (German Research Institute for Aviation) as their executing arm.

Then follows a necessary technical excursion about *The Degree of Reaction—The Weak Point of AVA Compressor Designs*, a design peculiarity which overshadows the German turbojet developments in full—and with significant implications. Finally and unexpectedly follows *In the Patent Arena—BBC Challenges the AVA Compressor Design*. Here the international *Jet Web* flares up in a hefty patent conflict in the years 1932–1936 between *Brown Boveri* and *AVA* for the German side on axial-flow compressors design features in the sensitive area of altitude turbocharging a battle which finally ended in the court room of the *Reichspatentamt Berlin* (Reich Patent Office), illustrating the subliminal stress in context with the background build-up of the emerging Luftwaffe.

Government Aviation Organisation From the very beginning, structure, personalities and budget of the (civil) *Aviation Department* within the Reich Ministry for Transportation were completely dominated by the Reichswehr, the German Armed Forces of that time.[236] Between 1924 and 1933 the department was led by the former professional officer Ernst Brandenburg, whose role during the definition of the high-altitude aircraft project has already been described in the foregoing Sect. 4.3.2, and also in the context of Fig. 4.30. Brandenburg negotiated successfully the removal of the *Versailles* civil aviation restrictions in the '*Paris Aviation Agreement*'[237] of 21 May 1926 and—nearly parallel on

---

[236] See Budrass, Flugzeugindustrie, pp. 164–166.

[237] German military activities remained restricted and the 'private' flying sport of Reichswehr members was controversially discussed. The original German demand of 200—a number close to the Lipezk piloting trainees, Sect. 4.3.1—was finally decreased to 72. See Budrass, Flugzeugindustrie, p. 217.

6 April 1926—contributed to the founding merger of *Deutsche Aero Lloyd* and *Junkers Luftverkehr* to the *Deutsche Luft Hansa (Lufthansa) Aktiengesellschaft,*[238] the then largest aviation enterprise worldwide.

The Reichswehrministerium (Ministry of the Reichswehr) and the corresponding departments of the *HWA Heereswaffenamt* (Army Ordnance Office), Inspection for Weapons and Equipment started already in 1925 and commenced in 1927 with substantial Luftwaffe rearmament planning in full. The officials in charge of engine planning and development were Helmuth Sachse[239] and Wolfram Eisenlohr,[240] who laid down a detailed *'Aufgabenstellung'* (development task list) in May 1927 on water- and air-cooled flight engines of low specific weight (below 0.5 kg/hp) up to 2000 hp, and on the accessory side *'Vorverdichtungsanlagen und Verstellpropeller'* (supercharging devices and variable-pitch propellers) *'for enabling high-altitude flights above 10,000 m'*. In addition, an early organisational chart of the HWA aviation department of October 1929 has as heads of Dept. Ib *'Engines'* besides Helmuth Sachse also Franz Mahnke.[241] At the same time and at

---

[238] The name *Lufthansa* was chosen in reference to the *Hanseatic League/Hanse*, a commercial and defensive confederation of merchant guilds and market towns in North-Western and Central-European coastal regions between 1200 and 1450.

[239] Helmuth Sachse (1900–1971) finished his studies in mechanical engineering in 1925. In 1926 he was appointed head of engine development at the Rechlin test site. Besides his focus on engine flight testing, he developed one of the first engine altitude test stands, which became a reference case for following, standardised designs from BBC Mannheim. In 1933 Sachse moved on to the newly founded RLM at Berlin, where he took the responsibility for the *'Engines'* development department up to 1937. Thereafter, he was appointed Technical Director of *BMW Flugmotorenbau* (aircraft engines) Munich. During this period emerged the *BMW 801*, as a successful 2000 hp mass product and the 18 cyl. double-radial engine *BMW 802* of 3200 hp passed successfully the test phase. In April 1942 he founded his own company *Helmuth Sachse KG* at Kempten (approx. 100 km in the south-west of Munich) which specialised in close cooperation with BMW on the production of automated propeller and engine control devices, the development of which was due to his initiative.

[240] Dipl.-Ing. Wolfram Eisenlohr (1893–1991) has been addressed before already in Sect. 4.1.3 in context of his WW I activities, Fig. 4.20. Between 1926 and 1933, he was officially employed by the Navy/Naval Command, in charge of the reconstruction of Naval-Luftwaffe, 1933/4 transfer to Luftwaffe as RLM consultant, 1934–1937 Group-Director for aircraft engines at Luftwaffe test site Rechlin, 1938–1944 Director of the Department for Engines and Accessories in the RLM Technical Office, and in this position superior to Helmut Schelp, promoted to General-Engineer in Nov. 1941. In 1943 after armament problems became visible, Eisenlohr and several other engineers were accused (presumably after 1 Oct. 1943, due to a passport dating, signed by Eisenlohr still in charge) at the *Reichskriegsgericht* (war court) for not fulfilling their job, but after thorough enquiries by General-Judge (Luftwaffe) Alexander Kraell (1894–1964), then in the position of highest ranking military prosecutor, it was recognised that the accusations were unfounded and Kraell *'managed'* to cease the case. Eisenlohr was sent into premature retirement in April 1944 and became officially Technical Director of the propeller manufacturer *VDM*, Frankfurt/M.

[241] Dipl.-Ing. Franz Mahnke (1900–1975) studied 1920–1926 after his military service in WW I mechanical engineering at the TH Berlin-Charlottenburg, before he joined HWA 1927–1933. On

Fig. 4.37 AVA pioneers of axial flow compressor development: Albert Betz (1885–1964) l, Jakob Ackeret (1898–1981) m, Walter Encke (1888–1982) r © Archives DLR and ETHZ (m)

the same HWA organisational level Ernst Marquard,[242] who had carried out the first poison gas spray simulations at Junkers Dessau in 1925, Sect. 4.3.1, led here Dept. IIb *'Bombs'*. In the newly founded RLM Reich Aviation Ministry (1933) three already existing organisations were merged: *Dept. Aviation* of the Reich Ministry of Transportation and the two organisational departments of the Ministry of the Reichswehr *'Luftdepartement'* and *'Inspection I'*.

The supervision of the *Versailles* restrictions by the *Military Inter-Allied Commission of Control* ended officially on 1 February 1927, after Germany had joined the League of Nations which also took responsibility for arms control. At the same time the secret German rearmament programmes were publicly revealed for now on parliamentary discussion and control.

AVA: The Institution   The humble beginnings of AVA Aerodynamische Versuchsanstalt Goettingen (Aerodynamic Test Establishment)—officially under this name since 1919— have already been outlined in Sect. 4.1.3. The scientific and applied axial turbomachinery research lasted since the early 1920s predominantly on the shoulders of three men: Albert Betz—Jakob Ackeret—Walter Encke, Fig. 4.37.

April 1933–1936 transferred to Luftwaffe, as RLM official in Dept. LC III 2, where he was promoted to Group-Director 1936–1938, thereafter till 1941 Group-Director and Temporary Department-Chief at RLM LC 8 and LC 3, and then at the time of Eisenlohr's military tribunal RLM Office-Group Chief GL/C B-3 *'Engines'*, accordingly in Dec. 1942 he became General-Engineer as well.

[242] Marquard was promoted to General-Engineer in Sep. 1941, two months before Eisenlohr, and more than one year ahead of Mahnke.

In February 1911 Ludwig Prandtl described in a memorandum the future working areas of a research institute for aero- and hydrodynamics at Goettingen: I. Air drag measurements of all kinds, II. flow studies of air and water in channels, III. investigations on bladed propellers and blower wheels, IV. development of measuring equipment for air flows and its precision calibration, V. studies of natural wind and its impact on air resistance. Albert Betz joined Ludwig Prandtl's aerodynamic research institute at Goettingen in 1911, after he had finished his naval engineering studies at TH Berlin-Charlottenburg; he became Prandtl's closest research associate over a period of more than 40 years, predominantly in the aforementioned areas I, III and V—and thus also Prandtl's successor in various administrative functions, most importantly in 1937 as head of the AVA. The Aerodynamische VersuchsAnstalt were then already 10-times grown to an organisation of approximately 600 employees; the administration of this research establishment automatically represented the end of Betz's scientific career. The comprehensive aerofoil lift/drag measurements of Max Munk together with Erich Hueckel under Betz's supervision during the Great War and typical for *Area I* research, has already been outlined in Sect. 4.1.3. Betz received his doctorate from the Goettingen University in 1919 for his work on *'ship propellers with minimum loss of energy'*, and qualified three years later as an (extraordinary) Assistant Professor of Physics. Betz's basic wind research activities of *Area V* fell in the years 1918–1926. Betz's Law, first published in his 1920 paper about the theoretical limit for best wind utilisation,[243] states that, independent of the design of a wind turbine, only 16/27 (or 59%) of the kinetic energy of the wind can be converted to mechanical energy. Finally, Betz's relatively short period of active (axial) turbomachinery research *(Area III)* can be allocated in the time frame 1919–1929, first alone, then since 1924 in a fruitful collaboration with Jakob Ackeret up to Ackeret's leave back to Switzerland in early 1927; thereafter with the newly engaged Walter Encke, who prepared the axial blower test facility and launched the corresponding test activities, which will be reviewed in the following Sub-Section *On Axial Blowers.*[244] Some 30 years later Betz summarised these turbomachinery experiences in a seminal book *'Introduction to the Theory of Flow Machines'*, which appeared both in German (1959)[245] and English (1966, 2014).

In autumn 1921 Ackeret came to Goettingen, when he had just received his Diploma degree in Mechanical Engineering from ETH Zurich in 1920 under the supervision of Professor Aurel Stodola; he expected to stay for a year. But then in 1922 opened up the opportunity for him to become one of Prandtl's assistants in succession to Carl Wieselsberger (1887–1941), who had supported Prandtl in the foregoing ten years—and

---

[243] See Betz, Das Maximum.

[244] In the early 1930s, after thoroughly supervising Karl Christiani's doctoral thesis on first compressor cascade experiments in a wind tunnel, Sect. 4.2.2, Betz had also learned to fly a glider and qualified—then already 49—for his licence in 1934; a very practical way of coming to grips with the subject of his research.

[245] See Betz, Einführung.

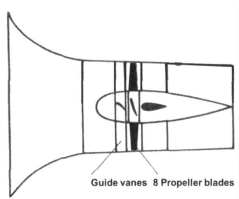

Guide vanes 8 Propeller blades

Fig. 4.38 Siemens-Betz axial blower, 1929 and north portal of Kaiser-Wilhelm-Tunnel, Cochem, 1937, with ten 19 kW axial blowers, 480 rpm, 1.7 m diam., 10×40 m^3/s © A. Leutz

Ackeret remained at Goettingen till 1927, an excellent preparation for his subsequent scientific career in Switzerland.

Betz's axial blower activities can be easily related to two patents, which he filed in 1918 and 1923. The first, DE319,413 provides a ruling for axial blower aerofoil geometry, which should be—deviant from a conventional airplane wing—over-cambered in the blade root section up to >18% of the local blade width, while vice versa at the blade tip the optimum camber should be reduced to <12% of the local blade width—to get best overall lift/drag ratios. The basic idea was inspired by test observations that accumulated boundary layer material was centrifuged out of the higher loaded hub region—so that the flow stabilised there, while the natural accumulation of flow losses at the casing required reduced turning in the separation-prone blade tip region. In continuation of this thought, Betz tried to achieve a well-balanced radial loading of axial blower blades in DE399,062 by inlet guide vanes (IGV), upfront of the rotor, and correspondingly induced co- or counter-rotating inlet swirl over blade height, which led to then attractive efficiencies of up to 72%. This single-stage perspective of flow directing IGV in combination with a pressure-rising rotor, followed either by a vaneless diffuser as shown in Fig. 4.38 (l), or by an aerodynamically simple, just flow-turning, non-diffusing row of vanes determined the unofficial *Goettingen School of turbo-machinery design* up to 1945, with—as will be seen later—rather negative consequences.

In the mid-1920s Betz was forced to transfer these patent rights for economic reasons to the *SSW Siemens-Schuckert Works*, Berlin, which in due course developed a range of very successful industrial products, the *Siemens-Betz Schraubengeblaese* (axial blowers)—the first time exhibited at a foundry trade show at Dusseldorf in 1929, Fig. 4.38. These blowers found in 1937/8 a salient application for the ventilation of Germany's longest (till 1988) railway tunnel, the 4205 m long *Kaiser-Wilhelm-Tunnel*[246] with its northern access at

---

[246] See Wikipedia, 'Kaiser-Wilhelm-Tunnel'in German, named after the German Emperor Wilhelm I (1797–1888) and—see Leutz, Foettinger.

Cochem at the River Mosel. Satisfactory exhaust ventilation of this tunnel caused ongoing problems from the very beginning of intensive steam railway usage after the tunnel construction in 1874–1877. A final solution was only reached, when Professor Hermann Foettinger (1877–1945), head of the Fluid-Dynamics Institute at the TH Berlin-Charlottenburg suggested on the basis of comprehensive flow model tests as of 1929 the installation of the new *Siemens-Betz* high-performance blowers.

Figure 4.38 illustrates the blower installation with—for aerodynamic reasons—an inclination of 14 deg. towards the tunnel axis. Normally, the operation of four blowers was sufficient, but with the maximum capacity of all ten blowers running up to 400 m^3/s could be pushed with 8 m/s through the tunnel tube, so that a complete air exchange over the 4.2 km length was achieved within 9 min. The necessary tests for the Siemens-Betz blower development were carried out to a large extent also by the newly hired Ackeret. Moreover, Betz praised his valuable help in the development of unique wind tunnel propeller model drives—for which high-speed electric motors of up to 30,000 rpm were built. In addition, both practised jointly on inventive lift improvements by means of the *Magnus effect* for the *Rotor Ship*[247] propulsion as well as the effect of boundary layer suction on high-lift aircraft wing configurations. In this context a joint patent DE513,116 *'Drag reduction for a body in liquids and gases'* was filed in 1923, which will be further discussed in Sect. 8.2.4 and in the context of Fig. 12.7 together with the presently revived, fuel-saving *BLI Boundary-Layer-Ingestion* technique at the aft end of large aircraft fuselages, Fig. 8.34.

In November 1923 a currency reform gave also Prandtl's research institutions again a reliable financial basis. Early in 1924 an expansion to the *Institute for Hydrodynamics* was started, in which the new *Department of Fluid Research* was headed by Jakob Ackeret; theoretical work started with Walter Tollmien, who had taken up his research work when still a student with Prandtl. In 1925, the department was joined, among others, by Adolf Busemann, Oskar Tietjens and Johann Nikuradse; names, which later gave the Institution its world-wide reputation.

After completing his PhD at ETH Zurich in 1927, Ackeret worked at Escher Wyss AG in Zurich as Chief Engineer of Hydraulics, where he applied, with great success, modern aerodynamics to the design of turbines; thereafter, he became a Professor of Aerodynamics at ETH Zurich in 1931. During this period 1928–1933 Ackeret's personal dedication for axial-flow turbomachinery research grew to full blossom—in the context of Curt Keller's generally renowned doctoral thesis[248] *'Axial blowers in view of aerofoil theory'*. With this later-on seminal work Ackeret and Keller launched a second Swiss attempt of a kind of scientific axial-flow turbomachinery design methodology at ETH Zurich in the wake of the

---

[247] A *Rotor* or *Flettner Ship* is designed to use the *Magnus effect* for propulsion, a force acting on a spinning body in a moving airstream, which acts perpendicular to both the direction of the airstream and of the rotor axis.

[248] See Keller, Axialgeblaese.

Goettingen School, and fully independent of the then already established industrial design process for axial compressors at Brown Boveri, some five years earlier.

Keller, working closely with Ackeret, specialised in axial compressor aerodynamics when the equipment for Ackeret's new *IfA Institute for Aerodynamics* was prepared. As said before, the two axial fans for the large, open subsonic wind tunnel were of an identical Keller design, while the 13-stage axial compressor for the closed supersonic wind tunnel, Fig. 4.26, came entirely from BBC. As mentioned in the *Introduction* of his thesis, Keller thanked for Ackeret's suggestion to investigate especially the single-stage, axial blowers, developed by Betz at Goettingen, Fig. 4.38, theoretically and experimentally from the perspective of aerofoil theory.[249]

Most important for the understanding of the AVA axial compressor activities over the next nearly twenty years (1927–1945), the first phase of which till 1935 will be discussed in the following sub-section, is Walter Encke (1898–1982), who joined the Institution as *'Assistant'* on 1 Feb. 1927, in time with Ackeret's departure towards Switzerland. The personalities are incomparable, the decision for hiring the then 29 year old Dipl.-Ing. without any scientific background or perceptible ambition in that direction cannot be reconstructed. Correspondingly difficult are the assessment of motivation, inspiration, drive of these activities in the times to come. The only existing portrait photo shows Encke presumably already after the war at the age of ~60, Fig. 4.37; similarly meagre are the available biographic data.[250] Born on 19 August 1898 at Magdeburg and, presumably delayed by the aftermath of WW I, he commenced studies in mechanical engineering at the TH Technical University of Hannover in 1920, finishing as Dipl.-Ing. in 1923. In due course he worked from May 1924 till 31 January 1927 as mechanical designer for centrifugal pumps at the Machine Factory *Oddesse*[251] at Oschersleben near Magdeburg, approximately 100 km in the east of Goettingen, his future area of living. As originally planned Walter Encke might have fulfilled satisfactorily his assistant role in dependence to the scientifically dominant Albert Betz in the period of roughly 1927–1932. Thereafter, one can assume that Betz was more and more charged with the administrative tasks of

---

[249] Somewhat jokingly, Keller's colleagues at the ETHZ-Institute put his thesis success in perspective *'No wonder, large parts of it wrote the chef himself..'* Another possible source for the mentioned investigations of single-stage axial fans at the AVA Goettingen is the doctoral thesis of Paul Ruden (1903–1970), who was at the AVA between 1930 and 1934 and used the same axial compressor test rig as will be described in the next sub-section for the investigations of W. Encke. Regretfully, apparently no further details exist about Ruden's cooperation with Betz/Encke. See Wikipedia, 'Paul Ruden' in German, and—see Ruden, Untersuchungen.

[250] By courtesy of Dr. Jessika Wichner, DLR Archive Goettingen, Jan. 2018, and of Dr. Sigrid Dahmen, Stadtarchiv Goettingen, 14 Oct. 2021. Between the end of 1950 up to 1964, Encke worked for DEMAG Duisburg, before he returned to Goettingen, where he died on 4 March 1982; his grave is at the Goettingen-Junkerberg cemetery.

[251] *Oddesse* combines the names of the English engineer Philip Francis Oddie (1862–1940) and the German engineer/merchantman Gustav Hesse, founded in 1897 as a steam pump manufacturer, the company produces still at the same site—in the meantime, submersible motors and motor pumps.

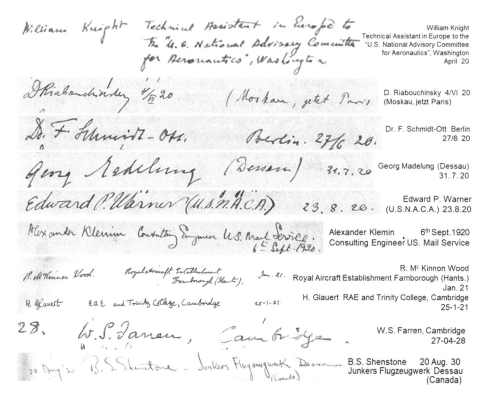

William Knight
Technical Assistant in Europe to the
"U.S. National Advisory Committee
for Aeronautics", Washington
April 20

D. Riabouchinsky 4/VI 20
(Moskau, jetzt Paris)

Dr. F. Schmidt-Ott Berlin
27/6 20

Georg Madelung (Dessau)
31. 7. 20

Edward P. Warner
(U.S.N.A.C.A.) 23.8.20

Alexander Klemin   6th Sept.1920
Consulting Engineer US. Mail Service

R. Mc Kinnon Wood
Royal Aircraft Establishment Farnborough (Hants.)
Jan. 21

H. Glauert RAE and Trinity College, Cambridge
25-1-21

W.S. Farren, Cambridge
27-04-28

B.S. Shenstone   20 Aug. 30
Junkers Flugzeugwerk Dessau
(Canada)

Fig. 4.39 AVA Goettingen Guest book entries, selection 1920–1930 © Archive DLR

managing the ever-growing AVA, so that Encke and his limited capacity stood largely on his own in front of the challenges of the developing axial compressor and aero gas turbine technology.

Prandtl and his Goettingen Institutions—the *Institute for Technical Physics* at the University of Goettingen since 1904, which was transformed to the *Kaiser Wilhelm Institute for Flow Research* in 1925 and the AVA Aerodynamic Test Establishment since 1919 as follow-on organisation of the MVA Model Test Institute—gained considerably in international reputation in the years after World War I. Primarily due to the quality of the scientific research but also by the openness towards visitors with which especially Prandtl himself tried to overcome German isolation and to promote the idea of international scientific collaboration. The *AVA (Visitor) Guest Books* of the period between the Wars have luckily survived in the present day DLR Archive at Goettingen and represent a unique information source in this context. A few selected excerpts have been collected in Fig. 4.39, mostly for personalities which play a salient role in this presentation and will be discussed top-down in a short review. The scientific exchange process started at first between the German side and U.S. American initiatives, not the least due to the fact that the Americans themselves recognised a certain backwardness in aeronautical science and technology after the war-time experiences. Scientifically the rapid rapprochement between former adversaries has been discussed under headers like *'strategic internationalism'* and *'transfer of technical knowledge'*.[252] The early beginnings of these activities date back—as already described in short in Sect. 4.1.2 under *'Turbocharger development history'*—still to 1916 for the United States *NRC National Research Council* and their Paris liaison office, which was led since January 1918 by the *'Scientific attaché'* Frederick F. Durand, then an ex-Professor of Mechanical Engineering at Stanford University. Durand himself a pioneer in aeronautics in the United States, had served as chair of NACA before he went to Paris, and after his return he reported on the state of the field in Europe to the executive committee, fuelling NACA's determination to catch up.

William Knight,[253] a first lieutenant in the U.S. Army Air Service, had himself suggested sending a representative abroad to gather scientific intelligence, was finally picked by NACA to lead the post-war Paris office, which became operational during July 1919. In November 1919, Knight introduced himself to Prandtl by letter as *'Technical Assistant in Europe to the U.S. National Advisory Committee for Aeronautics'*, a term which he used exactly also for his entry in the *AVA Guest Book*, when he finally visited Goettingen in April 1920, Fig. 4.39. This first visit must have been very impressive to him, since he finished his Guest book entry emphatically: *'I have been greatly impressed with the enormous contribution to the progress of Aerodynamics as brought about by Prof. Prandtl and I wish to express to him my respectful admiration.'*

The next selected Guest book entry is from *Dimitri P. Riabouchinsky (1882–1962)*, the scientific assistant to Nikolay Y. Zhukovsky[254] (1847–1921), one of the Russian founders

---

[252] For a specific study with reference to the field of Aerodynamics in the relation between the United States and Germany, see Eckert, Strategic Internationalism.

[253] For a more comprehensive survey of Knight's activities as NACA representative for Europe between July 1919 and autumn 1920, when he was released,—see Eckert, Ludwig Prandtl, pp. 119–122. Knight's further life can be reconstructed in parts. On 5 July 1924 the Junkers Corporation of America was founded in New York, after Junkers's first U.S. foundation of 1914, the American Junkers Company, came to an end in 1917 and had to be dissolved. The new *Jucoram* in the Canadian Pacific Building at New York's 342 Madison Ave., led by William Knight, was a representative organisation for all Junkers companies in Germany; it existed for several years, however, the entry into the U.S. aviation market was unsuccessful. The late William Knight, one time associate to German-born Charles P. Steinmetz (1865–1923), early U.S. socialist and founder of the GE Research Lab, was member of Technocracy Inc., wearing in the mid-1930s a such a uniform consisting of a *'well-tailored, double-breasted suit, grey shirt and blue necktie, with a monad insignia on the lapel'*. See Wikipedia, 'Technocracy movement' in English.

[254] See Wikipedia, 'Nikolay Zhukovsky (scientist)'in English. In 1904, the world's first institute for aerodynamic research was created, based on Zhukovsky's laboratory, practically on the family *dacha* site of the rich Riabouchinsky (banker) family at Koutchino, located 40 km south-east of Moscow. In 1917 the Riabouchinskies had to flee Russia during the Revolution; in 1918 the laboratory became the nucleus of TsAGI Central AeroHydrodynamics Institute under the leadership of N.Y. Zhukovsky and in 1935 the place became Stakhanovo (after A. Stakhanov, Soviet record miner), renamed Zhukovsky in 1947.

of modern aero- and hydrodynamics. Riabouchinsky himself extended in 1932 considerably the *'shallow water-flow analogy'* for 3D compressible flows, a powerful tool to visualise and investigate transonic and supersonic turbomachinery cascade flows.[255]

The following two Guest book entries as depicted in Fig. 4.39 are for *Friedrich Schmidt-Ott (1860–1956)*, whom we met before in the context of Fig. 4.32 and for *Georg Madelung (1889–1972)*, who appeared the first time in Fig. 4.30—and who will be addressed in more detail in one of the coming Sub-Sections: *In the Patent Arena*. The early contact with Prandtl and the AVA in mid-1920 could indicate that *Goettingen* was part of the joint discussions about future German aviation concepts (civil and military) from the very beginning. The next two listed visitors came from the United States, indicating that William Knight's *ice-breaking mission* four months earlier showed first results.

Edward P. Warner (1894–1954)[256] represented then on 23 August 1920 the U.S.N.A.C. A. *under construction* and was thus in a short time the second high-ranking visitor from that organisation after *Jerome C. Hunsacker (1886–1984)* at AVA Goettingen in July 1920, a visit only noted afterwards in the Guest book apparently by an inhouse secretary. At that time Warner had been just appointed as NACA's Chief Physicist in charge of aerodynamic research at Langley Field; after WW II he became first President of ICAO International Civil Aviation Organisation up to his retirement in 1957. In July 1920 Hunsacker, later influential NACA Chairman between 1941 and 1956, negotiated the transfer from Max Munk from Goettingen to Washington and shortly thereafter to Langley still in the same year (with explicit consent from U.S. President Woodrow Wilson), where Munk proposed building the new *VDP Variable Density Tunnel*, which went into operation in 1922 for more accurate testing of small-scale models than could be obtained with atmospheric wind tunnels.

The signature of the second here addressed U.S. visitor *Alexander Klemin (1888–1950)* belongs to a fascinating, originally English-born personality, who came to the United States in 1912. By 1917 he had become the head of the *MIT Aeronautics Department*, an U.S. citizen, and a First Lieutenant and the Officer-in-Charge of the Army Air Service Research Department at McCook Field in Dayton, Ohio—an area already addressed at length in Sect. 4.1.2 and in the context of the *Turbocharger development history*. While at Goettingen Klemin was a consultant to the U.S. Mail Service and taught in parallel the first Aeronautics course at New York University, 1919–1925. Later-on he became Director and Professor of the Guggenheim School of Aeronautics at New York University from 1925 up to his retirement in 1945.

---

[255] See Eckardt, Gas Turbine Powerhouse, pp.386–387.

[256] See Wikipedia, 'Edward Pearson Warner' and 'Jerome Clarke Hunsacker', both in English.

Another four months after this group of U.S. visitors[257] appeared the first British scientific two-men team[258] from RAE Royal Aircraft Establishment, Farnborough at Goettingen on 25 January 1921:[259] *Ronald McKinnon Wood*[260](1892–1967) and *Hermann Glauert (1892–1934)*, Fig. 4.39. The names of Prandtl and Glauert are tied together for a number of aerodynamic terms—of which the *Prandtl-Glauert rule*[261] is most prominent—though they were not the result of immediate collaboration. Glauert's major achievements are all the more astonishing when it is realised that they occurred over the mere 14-year-period following his acquisition of Goettingen's aerodynamic theory, before his tragic death in 1934. Hermann Glauert was born at Sheffield; his father, a cutlery manufacturer had immigrated as young man from Germany. He joined the RAF Royal Aircraft Factory

---

[257]Not shown is the entry of Milton C. Baumann (1895–1976) from *Dayton Wright Co.* on 19 Oct. 1920, who came from France, where shortly before on 28 Sept.1920 his *RB-1 Racer* design had participated in the 1920 *Gordon Bennett Cup*,–see Wikipedia, 'Dayton-Wright Racer'in English.

[258]An opportunity to throw a broader—here political view on historic British relations to Goettingen. Although the British Royal Family had been predominantly German since 1714 when King George I mounted the throne, Ministers of the British Crown never learned to speak German. The one exception was Richard B. Haldane (1856–1928), Secretary of State for War 1905–1912, who spoke fluently German; he had persuaded his father to permit a course of study in philosophy at the University of Goettingen. In 1873, Haldane first arrived here, where he remembered later his professors, looking *'as if they had seen more books than soap or tailors' shops. Most of them are men of about sixty, wearing colored spectacles, broad Tyrolean hats, with dirty, badly shaven faces and their clothes almost tumbling off. They lecture, sometimes in Latin, sometimes in German.'* Quoted from—see Massy, Dreadnought, p. 805.

[259]The whole visit was arranged rather spontaneously and uncomplicated, after Glauert had sent a hand-written note, dated Fri 21 Jan.1921 to Prandtl in German from Hotel Hessler (renowned for the in-house pastry shop) at Kantstrasse, Berlin-Charlottenburg, the visit at Goettingen took already place four days later on Tue 25 Jan. 1921, —see Bloor, Enigma, chapter *'We have nothing to learn from the hun'*, p. 315. In general, an excellently investigated, comprehensive account about the positive effects from this first contact, which followed several more, with a considerably improved understanding of Prandtl's work in Great Britain. For a comprehensive survey of Glauert's life and achievements together with a substantial list of references, —see Ackroyd, Glauert.

[260]Ronald McKinnon Wood worked at RAE from 1914 to 1934. After a short period in politics he was attached to the Ministry of Aircraft Production between 1940 and 1946. After the war he became a member and 1957–1958 chairman of the London County Council. McKinnon Wood returned to Goettingen in 1945 in a mission similar, but more formal and grimmer in tone, to the one he had undertaken with Glauert in 1921. He was then working for the Combined Intelligence Objectives Sub-Committee (CIOS) and questioned Dietrich Kuechemann (1911–1976) about the latter's work on swept wings and the flow over engine ducts and fairings, Figs. 4.41 and 8.17, before Kuechemann came to Britain in the same year and eventually took British citizenship. In 1963 he was elected a fellow of the Royal Society and in 1966 became the head of the Aerodynamics Dept. at Farnborough. See Bloor, The Enigma, p. 437.

[261]The *Prandtl-Glauert rule* states a compressibility correction, that the pressure coefficient at any point in the subsonic flow of a fluid about a slender body is equal to the pressure coefficient at that point in the corresponding incompressible fluid flow, divided by $\sqrt{1 - M^2}$, where M is the Mach number far from the body. (Acc. to McGraw-Hill Dictionary of Scientific & Technical Terms, 2003).

Farnborough in 1916 on recommendation of a friend from his undergraduate days, _William Scott Farren (1892–1970),_ Fig. 6.39 (whose Guest book entry follows next), and became over time its principal scientific officer. Glauert spoke fluent German and was sent after the end of WW I to Goettingen to study the work of Prandtl and AVA. The preserved AVA Guest book has the entries of Glauert (for RAE and _Trinity College Cambridge)_ and McKinnon Wood (RAE) on 25 January 1921; seven years later on 27 April 1928 W.S. Farren,[262] Cambridge and future RAE Director after 1941 visited AVA at the time of Betz's and Encke's first successful axial compressor experiments and—as indicated already in Sect. 4.2.2—with all likelihood carried this news to Griffith. Tragically in August 1934, Glauert was killed during a walk with his family by a chance fragment of a tree stub which was blown up by sapper troops on Aldershot Common, Farnborough. His premature departure from the aeronautical scene was not only a blow to the developing British-German scientific contacts, but had also a far-reaching effect in slowing the post-war progress of the British aeronautical industry.[263]

Besides Glauert, it was a young Canadian—Beverley S. Shenstone[264]—who developed at that time the closest aeronautical working relations to German institutions like the just forming _Glider scene_ at the _Wasserkuppe,_ to the Junkers Flugzeugwerke at Dessau, and as a professional intermediary to Prandtl and the Goettingen-influenced aero research community. Shenstone became famous for his ingenious and meticulous aerodynamic design of the _Spitfire_[265] fighter aircraft family under Reginald J. Mitchell (1895–1937), Supermarine's brilliant chief engineer; the Spitfire was designed as a short-range, high-performance interceptor aircraft at _Supermarine Aviation Works,_ which operated as a subsidiary of _Vickers-Armstrong_ from 1928. Mitchell pushed the Spitfire's distinctive, aerodynamically sophisticated _'elliptical wing',_ designed by Beverley Shenstone with all likelihood on the basis of earlier Prandtl's textbook publications[266] to have the thinnest

---

[262] Farren will re-appear 10 years later in another axial compressor scouting mission—this time to Brown Boveri in Baden, Switzerland, see Sect. 6.3.

[263] See Sect. 10.1 _'Great Britain—an Industrial Head Start, Still Going Radial'_ for the post-war period 1946–1955.

[264] See Wikipedia, 'Beverley Shenstone' in English.

[265] See Wikipedia, 'Supermarine Spitfire' in English; it is said, the name _Spitfire_ was suggested by Sir Robert McLean, director of Vickers-Armstrong at the time, who called his spirited elder daughter Annie Penrose _'a little spitfire'._ The word dates from Elizabethan times and refers to a fiery, ferocious type of person; at the time it usually meant a girl or woman of that temperament.

[266] See Ackroyd, The Spitfire Wing Planform. Max Munk, Prandtl's doctoral student, deduced that the induced drag is a minimum when the wing has an elliptically distributed lift loading, which vice versa most simply is achieved by an elliptic planform wing. Surprisingly, see Prandtl, Tragfluegeltheorie—the idea is illustrated as shown in Fig. 4.41 (l) by two semi-ellipses which is the selected Spitfire wing; in Prandtl's words _'The form of an aerofoil of this type results when it is established that the profile form and the effective angle of attack α' are to be constant, when bounded by two half ellipses'._ For the further information flow to the U.S./GB—see Ackroyd, The Spitfire Wing Planform, p. 125.

possible cross-section, helping give the aircraft optimum lift at a higher top speed than several contemporary fighter aircrafts. In an era when most aircraft were still fabric-covered wooden biplanes, Shenstone realised that metal monoplanes were the future and that German industry was leading that field. He applied for a job with Dornier in Friedrichshafen,[267] but was unsuccessful. Perseverance and useful contacts paid off, with the assistance of the British Air Attaché at Berlin Group Captain M. G. Christie,[268] he got a position with Junkers in November 1929 at Dessau. According to his own statements, he worked for a year at Junkers learning metal-working techniques such as panel beating and riveting. He worked in technical departments such as the engine workshop and also studied the Junkers flying wing concepts. In the summer of 1930, Shenstone learnt to glide at the Wasserkuppe, the premier gliding centre in Europe at the time. There he met Geoffrey Hill[269] and Alexander Lippisch,[270] both pioneers of all-wing aircraft. At this time Lippisch was leading the technical branch of the *RRG Rhoen-Rossitten Gesellschaft*. Shenstone spent the winter of 1930/1931 working with Lippisch and his team developing tailless gliders, as shown in Fig. 4.40[271] for the Delta I.

While in Germany, _Beverley Shenstone (1906–1979)_ travelled to Goettingen and met presumably also Ludwig Prandtl at AVA on 20 August 1930, as indicated by his entry at the bottom of the Guest book list, Fig. 4.39, still as employee of Junkers Flugzeugwerke Dessau with (Canada) in parentheses. It was also at the Wasserkuppe that he met Air Commodore John Adrian Chamier[272] and acted as his translator. Chamier suggested that

---

[267] On the northern shore of Lake Constance, originally known as home of the *Zeppelins*.

[268] M. Graham Christie (1881–1971) is another remarkable personality at the British Embassy in Berlin, two more will be introduced in Sect. 6.1. Himself a wealthy businessman, he had lived for 9 years in Germany before 1914, obtaining the degree of a *Doktor-Ingenieur* at Aachen University. This background and his status as a veteran pilot from the Western Front, helped Christie make contacts in German political, military, aviation and industrial circles (Goering, Milch, Junkers, R. Bosch). After 1933, Sir R. Vansittart (1881–1957), senior British diplomat in the period 1929–1941, came to regard Christie as Britain's best source on the inner workings of the Nazi state. He will reappear in Sect. 6.1 now in private mission after 1935. See Wikipedia, 'Graham Christie' in English and—see Ferris, Intelligence, pp. 62–65.

[269] See Wikipedia, 'Geoffrey T.R. Hill' in English.

[270] See Wikipedia, 'Alexander Lippisch' in English; the speed record flight with his Messerschmitt Me 163 Komet rocket-powered delta-interceptor took place on 2 Oct. 1941. The rather rare Fig. 4.40 (r) from the DLR Archives Goettingen (courtesy to Mrs. Jessika Wichner, 4 May 2015) was taken in winter 1930/31 at the Ursinus House on Wasserkuppe, when the design for Lippisch's Delta I was in progress—the world's first tailless delta wing aircraft, to fly motor-powered in 1931 at Berlin-Tempelhof. Besides Shenstone and Lippisch, there is known Guenther Groenhoff (1908–1932), Lippisch's test pilot, who deadly crashed there in 1932 and Hans Jacobs (1907–1994) glider designer and head of *DFS Institute for Aircraft Design* 1933–1939.

[271] See Wikipedia, 'Hans Jacobs' in English.

[272] John Adrian Chamier (1883–1974), British RAF officer and known as *'The Founding Father of the ATC'* for his role in the foundation of the *Air Training Corps* for young pilots. He had been

Fig. 4.40 Beverley S. Shenstone (l) at *RRG* design office, *Ursinus House,* Wasserkuppe, 1930 (r): B.S. Shenstone, G. Groenhoff, F. Kraemer, H. Jacobs (rear), A. Lippisch, W. Hubert, l-t-r © Archive DLR

Shenstone should come to work for the British at Vickers-Armstrong of which Chamier was a Director. After some drawbacks he finally got through Chamier an interview in 1932 with Reginald Mitchell at Supermarine which was part of Vickers-Armstrong. While disappointed with Shenstone's knowledge of monoplane-wing construction, Mitchell was impressed by his expertise in aerodynamic theory and gave him a full-time position, following a 2-months trial.

As well as providing technical support, Mitchell charged Shenstone with bringing an external perspective to Supermarine's designs. As a result, Shenstone travelled with Ernest Hives[273] (1886–1965) of Rolls-Royce again to Germany in early 1934 and later that year to the U.S. where he visited NACA and a number of aircraft manufacturers.

Summarising the situation here, *'The Germans and their glider design bureaus invented the new thin-winged, sleek monoplane form. Only the blinkered could refuse to see the potential of the transfer of such knowledge to powered aircraft. . . . Ironically, perhaps the British Spitfire was the first powered aircraft to incorporate aspects of such German learning into its design.'*[274] After John Ackroyd[275] had traced and updated the origins of

---

Technical Director at *Vickers Supermarine* from April to Dec.1927 and thereafter till 1931 he was Director at the board of Vickers (Aviation) Ltd;–see Wikipedia, 'John Adrian Chamier' in English.

[273] See Wikipedia, 'Ernest Hives, first Baron Hives' in English. Notable engines were developed under Hives leading to the *'R' series*, which powered the Supermarine S.6 seaplanes that won the *Schneider Trophy* in 1929 and 1931 for Rolls-Royce, and most importantly the famous *Merlin* engine. It was thanks to Hives that a total of a hundred and sixty thousand Merlins were produced by 1945. In 1941 Hives quickly decided *'to go all out for the gas turbine'*, ensuring the company's outstanding role in developing jet engines for civil and military aviation.

[274] See Cole, Secret Wings, p. 65.

[275] See Ackroyd, The Aerodynamics of the Spitfire.

Fig. 4.41 Prandtl's wing planform 1918 (l) and *Supermarine Spitfire*—with possibly German-influenced design features © quoro.com, mod

the Spitfire's *elliptic wing* planform, Fig. 4.41, he mentioned also his former speculation that the appearance of wing root fillets in the Spitfire's design development during the spring of 1935 would also have been favoured by Shenstone—after his visit to the United States. There these features had been adopted by, for example, the *Northrop* and *Douglas* companies. However, it appears as Ackroyd states in his more recent investigation that it was not Shenstone who instigated this.

At the time of the Spitfire's design, the published data on fillets came from Muttray in Germany and Klein, working with von Kármán[276] in California; both favoured fairly extensive fillets, beginning at the wing leading edge.

---

[276] According to another variant of origin, it was von Kármán himself, who mentioned the positive effect of fillets in a Paris presentation first in 1932,–see von Kármán, Aerodynamics (Information, by courtesy of Ernst H. Hirschel, 8 Jan. 2020). However, as indicated in—see Muttray 1934, Die aerodynamische Zusammenfuegung, there was an earlier reference publication—see Muttray 1928, Untersuchungen, comprising the same subject. Further research revealed that the mentioned work on the Junkers pressurised cabins, Sect. 4.3.2, had belonged to Georg Justus Muttray, while the fillet results were from Horst Wilhelm Muttray, both siblings, born on 30 Aug.1898 at Groeditz, Saxony, Germany. After studying Aircraft Design and graduating as Dipl.-Ing. at TH Dresden, Justus specialised at Junkers Dessau in the problems of stratosphere flight, and was continuously responsible for all pressurised cabins from Ju 49 to Ju (Alexejew) 150, while Horst joined the Kaiser-Wilhelm-Institut fuer Stroemungsforschung (AVA) at Goettingen, where he published a.o. the fillet studies; in Nov.1937 he moved on to DFS,–see Wikipedia, 'Deutsche Forschungsanstalt fuer Segelflug (German Res. Inst. for Sailplane Flight), Darmstadt as department head, which became DFS Ainring after 1940. There the name 'Muttray' is associated with some obscure publications about *'flying disks/ saucers'*. After the end of WW II both brothers were deported on 22 Oct.1946 from Dessau (Soviet Occup. Zone) to the Soviet Union as part of 'Operation Ossawakim/Ossoawiachim'—in total approx. 2000 scientists, with family members 10,000–15,000. After returning to Dessau in 1950, both were

Besides the details of his Wikipedia CV, it is worthwhile to mention that Shenstone became in 1948 *BEA British European Airways'* chief engineer, was raised to that Board of Directors in 1960 and switched in 1964 as Technical Director to *BOAC British Overseas Airways Corporation.* In the meantime Beverley Shenstone held the post of President of RAeS Royal Aeronautical Society for 1962–1963 and rounded off an unconventional life by making the first proposals that man-powered flight presented a challenge in aeronautical design and human achievement.

On Axial Blowers *Axial Superchargers for High-Altitude Piston Engines* The beginning of axial compressor research at AVA Goettingen can be reconstructed on the basis of meeting minutes[277] of the Board of Trustees (*Kuratoriumssitzung*) of the *Kaiser-Wilhelm-Institute for Fluid Mechanics together with the Aerodynamic Test Establishment Goettingen* on Sat 11 December 1926 at the Berlin Castle in the premises of the Notgemeinschaft, Fig. 4.32. The meeting was led by *Staatsminister Dr.Dr. Schmidt-Ott,* who opened the session with Prussian punctuality at 10.15 h, welcoming the other 10 participants, amongst these *Professors Dr. Prandtl and Dr. Betz.* In the *Report of the Director* (Prandtl) is already the first statement *'Thanks to the great interest of the Reichsverkehrsministerium* (Reich Ministry of Transportation*), the whole Institute experiences a very encouraging development. Especially the Aerodynamic Test Establisment, for which Herr Prof. Betz will provide details, could considerably expand its field of activities.'* And in the following *Report of the Deputy Director,* Betz explains that after the Inflation ended and the Aviation Restrictions were relaxed, the demand for services of the Establishment have grown considerably. *'In this context we welcome the additional financial resources for aviation research from the RVM which stabilise our future operations. . . . For the next year we plan the preparation of a larger wind tunnel and, the development of a turbo-compressor for high-altitude aircrafts.'* This statement implies that principle decisions in this context must have been made already during 1926, so that the aforementioned financial agreement between Schmidt-Ott and Ernst Brandenburg, RVM in July 1927, Section 4.3.2, was just the confirming cornerstone of the Ju 49 project. And, given the coordinating role of the DVL Altitude Flight Office, one can take for granted that its head Martin Schrenk, Fig. 4.33, contributed his idea to develop a multi-stage axial supercharger for high power aircraft engine applications to the AVA

---

again arrested on 18/19 Sept.1951, presumably for violating secrecy constraints about their former SU stay. Justus Muttray was sentenced to forced labour in SU till 1956. His brother Horst Muttray was sentenced to death at Berlin-Lichtenberg on 1 Feb.1952 by a Soviet Military Tribunal, mercy petition rejected on 12 April 1952, executed at Moscow on 18 April 1952 and buried at Moscow-Donskoje cemetery—with memorial for foreign victims of Stalinism, rehabilitated on 5 May 1997,—see Boening, Ich wundere mich nur . . . and Roginskij, Erschossen, about the 1000 German victims of Stalinism 1950–1953.

[277]DLR Archive Goettingen registration no. GOAR 2713-J.

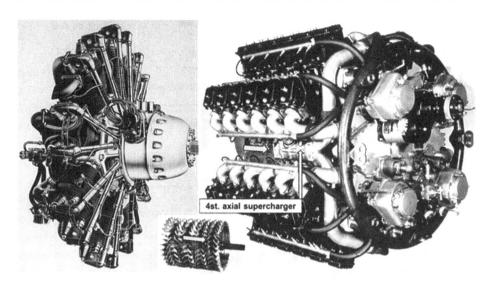

Fig. 4.42 AVA axial supercharger developments for *Siemens Sh 20*, 600 hp, 9 cyl. radial engine, ~1930 (l) and *FKFS* Gruppen-Motor A, 2000 hp, 48 cyl., ~ 1940 (r) with four-st. axial supercharger (inset) © Siemens Arch./FKFS

tasks. Actually Betz's *Annual AVA Activity Report 1928*[278] confirmed the installation of special compressor test facilities for high-altitude aircraft research during that year, for which already *'first tests could be accomplished.'*

After Ackeret's departure in January 1927, Albert Betz and Walter Encke, newly employed at AVA on 1 Feb. 1927, started their contracted work for the RVM/Notgemeinschaft on the behaviour of multi-stage axial-flow compressors. Their investigations were based on aerofoil theory and were conducted using specially constructed test rigs at the AVA. Their explicit objective was to gain the information necessary to design an axial-flow piston engine supercharger. In 1940 Encke published a relatively detailed measurement report[279] of 28 pages, which appeared already in April 1947 as NACA TM 1123 in English. The paper was mostly in view of the then recent applications of the studied experimental wheels for the German seven- and eight-stage axial turbojet compressor developments, (and will therefore be discussed in more detail in

---

[278] GOAR 1290, p. 3. The AVA spending of RVM money for a turbo-compressor test stand in 1927/8 adds up to 15,000 RM incl. 6600 RM for Encke's annual salary—GOAR 2713-A, while the cost projection for 1928/9, GOAR 1289 issued on 8 March 1928, specifies for corresponding tests on RVM/Notgemeinschaft account in this context—under rather camouflaging titles *'Origin of wind tunnel noise, blower tip leakage losses'* and, on the small blower rig *'Dispersion of a rotating jet'*—for 13,500 RM.

[279] See Encke, Untersuchungen. The corresponding NACA TM1123 translation dates the German original paper wrongly to April 1944, it should be April 1940(!) instead, based on reliable quotation in—see Eckert, Axialkompressoren.

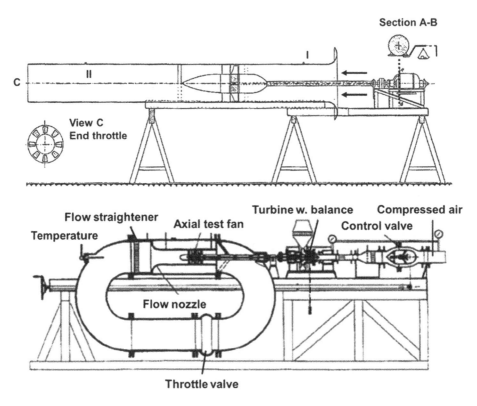

Fig. 4.43 AVA axial flow compressor test beds, 1927: Top: Single-stage tests, 0.3 m diam., 3000 rpm with DC e-drive, Bottom: Multi-stage tests with variable density, 0.15–0.2 m diam., 33,000 rpm, 70 kW (95 hp) air turbine drive © Archive DLR, mod

Sect. 6.2), but provided also some introductory remarks in reference to the *'old'* measurements *'in early 1930'*, in the following directly quoted from the NACA TM: 'These *led to the designing . . . of a purely axial supercharger for use at an altitude of 5 kilometers; this design was planned as a five-stage axial compressor with a rotor speed of 30,000 rpm but due to the bearing difficulties then existing it was not completed'*. In this context an Internet source claimed to know that this 15 cm diameter axial blower was successfully tested as 4-stage configuration in combination with a liquid-cooled, gasoline, inverted-V12 Jumo 210[280] of 730 hp take off power, which was after its first run in 1932 in continuous production between 1934–1938 up to 6500 units.

Over the years the Junkers/RLM files at the Archive of the Deutsches Museum Munich contain a number of *Minutes* which reveal the non-existence of the planned axial superchargers as a considerable deficit.[281] The indicated ground demo test of Jumo

---

[280]See Wikipedia, 'Junkers Jumo 210' in English.

[281]By courtesy of Peter Pietschmann, Dec. 2019/ Jan. 2020.

210 with attached AVA four-stage axial supercharger was announced in a meeting at
Junkers Dessau with the RLM-representatives Helmuth Sachse, LC II/2 and Franz
Mahnke, LC III/2 on 20 August 1936[282]—for the coming month. Apparently, the pressure
for axial supercharging capacity grew over time. In a meeting at Junkers Dessau on 27 Feb.
1940 with the high-ranking attendance of then still Fl.-Stabsing. (flight staff engineer,
General-Engineer since Nov.1941) Wolfram Eisenlohr, RLM and Professor[283] Dr. Otto
Mader, Jumo, an improvised solution was suggested by adding an axial front stage
(inducer) to the established two- and three-stage radial superchargers in use, for
applications like the 12 cyl., 35 l liquid-cooled with direct fuel injection Jumo 211F
(1340 hp) and the derivative Jumo 213 (2000 hp). Since there exist no pictures of AVA-
related axial supercharger installations, Fig. 4.42 shows the first high-power altitude engine
Sh 20 from Siemens ~1930 of which according to subsequently discussed information from
the DLR Archive at Goettingen a diesel derivative *R 20*[284] was prepared for an AVA five-
stage axial blower and in addition, ten years later, the 2000 hp *Gruppen-Motor A*[285] from
*FKFS Forschungsinstitut fuer Kraftfahrwesen und Fahrzeugmotoren Stuttgart (Research
Institute of Automotive Engineering and Vehicle Engines Stuttgart,* Prof. W. Kamm) with a
four-stage axial supercharger as shown on the inset. The data of this remarkable device,

---

[282] In the same meeting there was also reported that Jumo had contacted recently Brown Boveri about
their axial supercharger project (without further specification)—with the information that
corresponding activities at BBC were in the early stages only, so that further checks would be
required. The only axial turbocharger project with 7 stages which BBC obviously had already offered
or was planning to offer for Rolls-Royce at that time, will be described in Sect. 6.3.

[283] Mader (1880–1944) received the title officially from his *Alma mater* TH Muenchen, just for his
short period as Professor for Technical Mechanics between 1 April 1927–1 Aug. 1928, but the title
was apparently continued to be used as a reverence, as here in official minutes. Mader habilitated
i.e. qualified for a professor already in 1912 at TH Aachen in *Measuring Technology;* thereafter in
1915, he became scientific head of the Junkers Research Institute at Dessau, before he received
overall technical responsibility for Jumo Junkers Motorenbau in 1923.

[284] In 1929 Siemens started the development of Diesel aero engines, followed by BMW in 1931, in
the hope to achieve longer flight distances by the lower fuel consumption at higher altitudes,
unhampered by the trouble-free self-igniters. In 1932 an air-cooled 9 cyl. 2-stroke Diesel radial
engine R20 with 31.5 l engine displacement and fuel injection was planned, deduced from the SH
20, since 1926—when Siemens had acquired via Gnome & Rhône a production licence of Fedden's
reliable 9 cyl. Bristol-Jupiter—the first and with 500–600 hp strongest German radial piston engine.
Simultaneously to these AVA/ Siemens contacts, Ju 49 altitude flight tests were carried out, powered
by Jumo L88a engines which were equipped with conventional radial superchargers,–see Wikipedia,
'Junkers L88' in English.

[285] FKFS *'Group Motors A, B and C'* belong to a series of high-power, multi-cylinder, multi-shaft
piston engines which were started on RLM initiative in 1938/1939 at Prof. Kamm's Institute
Stuttgart,—see Potthoff, Kamm, under the header—see Wikipedia, 'Amerikabomber'in English,
and which included also the development of axial superchargers for that application in Eckert's
FKFS group 'Fluid machinery'; for the BBC Mhm./ FKFS cooperation between 1941–1945,—see
Sect. 8.3.

designed by Bruno Eckert, Fig. 9.3 (l), then head of Turbomachinery Design at that Institute 1940–1945, are as follows:[286]

German high-altitude axial superchargers	AVA ~1930	FKFS ~1940
Cylindrical outer diameter D	150 mm	170 mm
Radial aerofoil height at entry h_1	37.5 mm	25 mm
Radial aerofoil height at exit h_2	37.5 mm	16 mm
Stage count	5	4
Rotor blade count (front to rear)	12/row	27 + 27 + 33 + 33
Overall length L	455 mm	200 mm
Speed n	33,000 min^{-1}	30,800 min^{-1}
Volume flow V	1.0 m^3/s	1.0 m^3/s
Total pressure ratio PR	2.19	1.8
Ad. efficiency η	0.85	0.85
Degree of reaction R	1.0	0.5

Nothing is known about the *'bearing difficulties'* of the four-stage AVA axial super-charger as indicated by Encke before. They could have originated from the small size in combination with high speed, but could be also the inherent result of the AVA-typical design with a degree of reaction R 1 and thus considerably higher axial thrust loading, subject to further discussion in the following sub-section—and in reference to the *Jumo 004* turbojet in Sect. 8.1.1. Another negative side effect of the selected degree of reaction is clearly revealed by comparing the foregoing tabulated data. The AVA design is with L 450 mm more than twice as long as the otherwise comparable FKFS axial supercharger, mainly due to the required, excessive axial extension of the guide vanes. A drawback with which AVA designers struggled continuously, either by considerable performance reduction of an overall axial cut-back or the extra complexity e.g. by vaneless counter-rotating axial compressor arrangements, which AVA investigated seriously in the mid-1930s.[287] The AVA axial compressor test equipment has been described of consisting since 1927 basically of two rigs,[288] Fig. 4.43, one for low speed, single stage testing with 0.3 m diameter up to 3000 rpm test speed, powered by a DC electro motor and the other for high speed testing of one- or two-stage configurations with max. 0.2 m diameter up to

---

[286] See Eckert, Axialkompressoren, pp. 171–172, with measured compressor map, detailed blading plan and measuring set-up including e-drive motor.

[287] See Betz, Axiallader, p. II 185, where reducing the axial vane length by 35% resulted in an efficiency decrease of 3 points, due to Reynolds number effects. However, nothing is known about a potential connection between these AVA CR compressor tests and Heppner/ Weinrich's CR compressor proposal as realised in the BMW 109-002 turbojet engine, and discussed in Chap. 5.

[288] See Encke, Untersuchungen. The upper single stage test arrangement in Fig. 4.43 is from—see Ruden, Untersuchungen; Encke's corresponding arrangement had the bell mouth entry on the left-hand side with a wind-shielded e-motor at the backend.

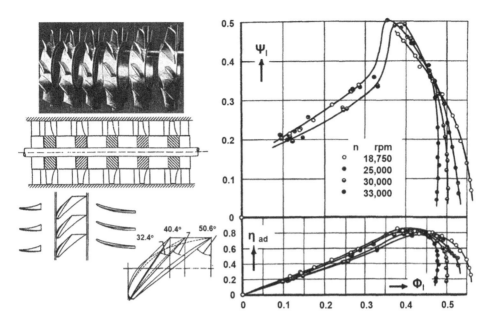

Fig. 4.44 AVA axial supercharger hardware: 6-stage rotor (top, l), annulus drawing of 5-st. version (mid,l), 1-st. test version (bottom, l) and measured 1-stage compressor map, 12 blades, 150 mm diam

~37,000 rpm by means of an 70 kW air turbine drive (since high speed e-drives were not available then). The latter arrangement has the Goettingen-typical closed loop (wind tunnel) set-up, but tests at 30,000 rpm with throttled inlet pressure from 1 bar down to 0.25 bar for a 150 mm diam. single stage (with 11 inlet guide vanes and 13 exit vanes, see also Fig. 4.44, bottom left) showed no significant variation of the pressure/performance distributions and corresponding Re number sensitivity. The speed dependent results for this set-up are illustrated on Fig. 4.44 (right)—with variations from 18,750 to 33,000 rpm.

The measurements are presented in non-dimensional form; the flow coefficient Φ_I is the ratio of the axial-flow velocity c of the intake air to the peripheral velocity of the rotor u, while the pressure coefficient Ψ_I is the ratio of the pressure rise Δp produced by the compressor to the dynamic pressure $\rho_I/2 \, u^2$. The change in the shape of the performance curve with a change of the Mach number may be clearly seen. As the rotor speed increases, the curve becomes more nearly vertical and the maximum value of the flow coefficient is lowered; then at the lower flow coefficient values, higher pressure ratios are obtained at the same flow ratio as the Mach number increases. At the given rotor speeds, the Mach numbers computed on the basis of the relative flow velocity at the blade tip in the range of best efficiencies are 0.45, 0.60, 0.72, and 0.80. Both the DC motor and the air turbine were flexible mounted on a cradle to determine the reaction torque, and thus together with the rotor speed the applied driving power. The efficiency of the high-speed rig was additionally controlled by temperature measurements ahead of and behind the compressor. Whereas the low-speed test rig could be operated rather unlimited, the test time of the high-speed test bed was restricted both due to the insufficient storage capacity of the air supply tank and also in view of bearing over-loading. Consequently, all instrument readings (18 pressure tubes, manometer, tachometer and dynamometer) were taken simultaneously by photo-recording.

Unique specialities of the Goettingen axial compressor design, which will be re-addressed in the following sub-sections in detail, are illustrated on the left-hand side of Fig. 4.44. Following the foregoing, decade-long in-house experiences with axial blower single-stages, it appears that this thinking determined also the axial compressor multi-stage design approach. Though today completely outdated, a design for higher pressures by simply, subsequently adding 'repetitive' (identical) stages was considered acceptable at AVA then, while it was early on superseded by contemporary, individual stage-to-stage axial compressor designs—in industrial practise (BBC), at other scientific institutes (FKFS, Fig. 4.42) as well as abroad, with examples from Ackeret, Switzerland to Griffith, Great Britain. Besides the repetitive sequence of identical rotor wheels as shown in Fig. 4.44, upper left, for an indicated six-stage axial supercharger hardware with 12-bladed rotor, Fig. 4.44 middle left is the principle sketch of the corresponding five-stage axial blower on the basis of the same, identical wheels with 0.15 m diam. Figure 4.44, bottom, illustrates the other peculiarity of Goettingen axial designs, the never changed usage of a degree of reaction R 1, which implies in short that all work done and all pressure rise takes place within the rotors, while the stator blading fulfils just the required flow turning and repetitive re-directing for the following next standard stage.

According to Edward W. Constant II,[289] *'by 1929 Betz had in operation a six-stage test compressor for Junkers that gave a pressure ratio of PR 2 with 85% efficiency at 37,000 rpm.'*[290] In fact, as mentioned already in Sect. 4.2.2, Griffith heard after Farren's visit at Goettingen of Betz's favourable results and cited them in his 1929 proposal. Betz meanwhile had by 1931 through the thorough investigations of Karl Christiani fully investigated the use of aerofoil cascades as an experimental aid in compressor design. Consequently, again following E.W. Constant, *'by the end of 1930, Betz and Encke could get the same PR 2 out of five wheels instead of six, and by about 1936 needed only four wheels, giving a stage pressure ratio of just 1.2. Betz and Encke also experimented with*

---

[289] See Constant, E.W., The origins, p. 115.

[290] While—see Encke, Untersuchungen of 1940 has only single- and two-stage investigations up to 33,000 rpm for a 150 mm diam. rotor, mentioning a five-stage axial supercharger application only once in the introduction,—see Betz, Axiallader of 1938 refers to a six-stage configuration, built by Junkers acc. to AVA design and tested at AVA, with 125 mm rotor diam. and 37,000 rpm design speed (actual tests at 35,400 rpm). This Junkers axial supercharger is quoted with best values of PR 2.14 at V 0.7 m^3/s and η_{ad} 0.83. And also Encke's former five-stage results are cited by Betz as PR 2.19, corresponding to a stage pressure ratio of 1.17. In 1938 he believes that PR 2.14 should be possible within four stages, equivalent to a stage pressure ratio of 1.21. In view of the following discussion, Constant's timing is doubtful. Betz in 1929 apparently had a certain confidence to achieve the four- to six-stage design targets based on the then one- to two-stage test results, while the actual operation of a Junkers six-stage configuration was rather in the 1935–1938 time frame.

various configurations for the axial compressors. They considered not only straight-through units, but also counter-rotating units (as separately proposed by Griffith). Betz and Encke encountered mechanical difficulties not only in the counter-rotating sets with more than two stages', but also with their small four- to six-stage high-speed axial superchargers—one reason, why the original idea of an axial-flow piston-engine super-charger did not materialise so often as originally planned, and was finally superseded by the axial-flow turbojet development. Up to this point one can/must follow Constant's statements—not the least due to a lack of any kind of complimentary information. However, his following conclusions have to be reviewed carefully. *'Betz and Encke were probably more advanced than Griffith or Ackeret[291] even in 1930, and their experiments went on uninterrupted right through the Second World War. By 1936–37, the critical period for the turbojet in Germany, Betz and Encke at the AVA had presumably a uniquely comprehensive knowledge of axial compressors* (which has to be discussed in the follow-ing). *Although the 1927 test rig was driven by a compressed-air turbine (fed from a tank), and although Betz must surely have been aware of his works value to possible gas turbine design, there is no evidence that he sought to exploit his knowledge in that direction. Unlike Griffith, who pushed for the turboprop from 1926 on, Betz and Encke would wait for others to make use of the insight they had generated.'*

Constant's timing and the mentioned AVA-Junkers connection at that time have to be reviewed on the basis of archived AVA materials; clearly, three phases can be differentiated within the scope of this Section up to 1935:

---

[291] Independently, both Darrieus and Seippel gave credit to Walther Bauersfeld and his 1922 VDI paper as highly influential on their own design learning curves, mentioning his radical change in cascade design perspective away from established stream tube and flow channel theory towards the lift/ drag principle deduced from single aerofoil measurements. Strangely Betz, Ackeret and others might have overlooked Bauersfeld's 1922 VDI paper, since they discuss and welcome the compara-ble method of Schilhansl in 1927 as rather unique,—see Schilhansl, Naeherungsweise Berechnung. A similar observation may apply for the book of—see Eck, Ventilatoren, which mentions Bauersfeld's 1922 VDI paper for the first time in the second edition of 1952 (on p. 189), while the information is missing in the first edition of 1937. Given the importance of Edward W. Constant's 1980 book on 'The Origins of the Turbojet Revolution' for the aero gas turbine history in general, it is worthwhile, to highlight a few of his inconsistencies and questionable conclusions in this context, in light of recent in-depth research. On p.110 he states: *Prandtl's aerofoil theory, and other insights derived from it, by 1926 led four men—A.A.Griffith in England, Jakob Ackeret in Switzerland, and Albert Betz and W. Encke in Germany—to investigations of axial compressors and turbines which would prove of tremendous moment for the turbojet revolution.* It appears that BBC's contribution as early as 1926 has been omitted or falsely mixed up with Ackeret's role and indicated in a footnote 101, Constant apparently has difficulties to differentiate between *'Ackeret'* and *BBC* and their independent performances at that time.

(I) 1927–1929, AVA contract developments for RVM/DVL,
   • design and construction of two axial compressor test rigs, Fig. 4.43,
   • low- and high-speed tests of single- and two-stage configurations,
   • two test reports, covering the period up to October 1928, and for October 1928–May 1929.[292]

(II) 1930–1932, a bundle of in total 11 letters[293] illustrates the AVA/Siemens (DVL) cooperation in this context, comprising two options for the axial blower development: (a) a high-altitude supercharger for the *Sh 20/R 20*, Fig. 4.42, with PR 2—to be achieved by a five-stage AVA axial compressor configuration, (b) a purging blower with PR 1.3 for the flight diesel engines under development.[294] Betz's axial compressor patents as discussed already in Sect. 4.3.3 belonged to Siemens in the meantime, which explains a certain preferential treatment.[295]

(III) 1935–1938, Junkers Motorenbau Dessau asked AVA W. Encke for axial blower design support for spark ignition and diesel engines of mass flows of 0.5–1.0 kg/s. Three of the remaining four Junkers letters from that period at the AVA Archive were signed by Prof. O. Mader and Dr. J. Gasterstädt,[296] the last one by Jumo 004 turbojet designer Anselm Franz, who at that time was still interested in an axial-radial supercharger.

Consistently in the Phases I–III, the five- to six-stage axial compressor supercharger was addressed. Typical for Phase I the ~1928 test set-up for the two-stage closed circuit arrangement with 'one shot' camera recording of U-tube pressure readings is shown in

---

[292] A. Betz and W. Encke: 'Versuche zur Ausbildung von achsialen Kompressoren für Höhenflugmotor (Tests for designing axial compressors for high-altitude motor), GOAR 2852, AK-3121(1928) and AK-3122(1929).

[293] Nine of these conveyed documents from Dipl.-Ing. Harald Wolff (1885—~1970), Director of Siemens-Halske Flugmotorenwerk, Berlin-Spandau to Prof. A. Betz/ W. Encke, AVA, two vice versa from Encke to Wolff. In a letter from 11 Aug. 1932 Encke mentions also the design of a piston engine cooling propeller of 1.75 outer diam., an idea which finally materialised in the cooling fan of the 41.8 l, 1500–2000 hp, 14 cyl. radial aircraft engine BMW 801 in the early 1940s, of which 61,000 units have been built,—see Wikipedia, 'BMW 801' in English. The versatile Wolff, designer of Germany's best WW I fighter aircraft Siemens-Schuckert D IV, will show up due to his professional stations from Siemens to Bramo to BMW to Heinkel to BMW several times, most explicitly in Sect. 8.1.3 in the context of a court-martial investigation in 1944 against him and his then protégé at Heinkel-Hirth, Stuttgart, Hans-Joachim Pabst von Ohain. Besides Wolff, Siemens flight engines were represented in the early 1930s in the AVA cooperation by Dr. Fritz Gosslau (1898–1965), who after some preparatory work in the late 1930s contributed after 1942 decisively to the development of the Fieseler FI-103, better known as—see Wikipedia, 'V-1 flying bomb'.

[294] This three-stage axial blower had already a light-weight 'Electron' casing (90% Mg, 10% Al).

[295] For details,—see Von Gersdorff, Flugmotoren, p. 51 f.

[296] Dr. Johannes Gasterstädt (1888–1937) had the Junkers diesel engine design responsibility after he had brought the Jumo 204 to production readiness in 1932, from which the smaller and lighter Jumo 205, better suited for aircraft installation was deduced.

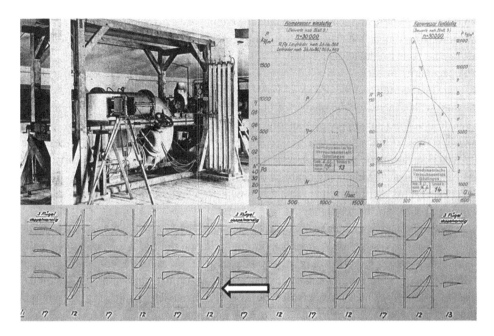

Fig. 4.45 AVA axial superchargers 1928–1936: two-stage closed circuit test set-up (upper left, then clockwise), measured single stage characteristics, calculated five-stage characteristics, axial blading of Junkers measured six-stage arrangement

Fig. 4.45. The Betz-Encke test report of 1929 contains—further in clockwise direction in Fig. 4.45—the measured single stage performance characteristics for a 12 blade rotor #866 of 0.15 m outer diam. at 30,000 rpm—with a peak pressure of 1700 kg/m^2, for which the five-stage performance characteristics for 0.15 m outer diam. at 30,000 rpm—with a peak pressure of 12,200 kg/m^2—were correspondingly calculated and finally for Phase III at the bottom—in through flow direction from right to left—the axial blading of a six-stage axial supercharger design from Junkers with 0.125 m outer diam. which was successfully tested at AVA Goettingen.

Not shown is what remained test intent during the Siemens cooperation (Phase II), when AVA had to give up test plans for a five-stage supercharger configuration with 0.15 m outer diam. due to the lack of high-speed bearings for the 30,000+ rpm operation range. The potential application of this—in various design versions—four- to six-stage axial supercharger is not known, but it could be certainly brought in the context of Junkers's most ambitious piston engine project—the Jumo 222; for which the design work started in 1937. The engine consisted of six engine blocks with four cylinders each, arranged around the central crankshaft. The engine looked like a radial due to the arrangement, but the internal workings were more like a V-engine, and it was liquid cooled. The cylinders were arranged such that neighbouring banks had exhaust and intakes beside each other, resulting in simpler piping from the rear-mounted supercharger position and having three sets of exhausts. With a bore and stroke of 135 mm, the engine had a displacement of 46.5 litres. It was forced to run at a fairly low 6.5 compression ratio, the best possible given the low-octane fuels available in Germany.[297]

A peculiarity of early AVA axial compressor design is visible in the blade count of the six-stage configuration at the bottom of Fig. 4.45: consistently all six rotors have 12 blades, a feature which increases the danger of blade excitation, and in fact both the eight-stage AVA-designed axial compressor—and the single stage turbine—of the Jumo 004 turbojet as well as the six- and seven-stage axial compressor versions for the later BMW-designed BMW 003 had to cope with compressor vibrations in the front stages in the early 1942/1943 test periods,[298] which only could be overcome by consequently introducing ratios of blade/vane (struts and combustor cans) ratios without a common denominator—or better by using prime numbers throughout.

In hindsight it appears that the 1927 design directions by Martin Schrenk, DVL towards a two- and a ~ five-stage axial supercharger configuration put the AVA activities for the contracted RVM/DVL work from the very beginning in the right direction, so that after 1930 the emerging industrial demands from Siemens and Junkers could be fulfiled on the basis of established measurement procedures and proven test results. As an obvious drawback one can state that Walter Encke and his small team were intermediately not sufficiently qualified to gain in design flexibility, giving up the rigid, aerodynamically fixed design concept towards more practically oriented investigations—as at the same time carried out by Claude Seippel and his team at Brown Boveri Baden, CH. An accordingly designed and tested seven-stage axial supercharger from BBC (presumably) for the Rolls-Royce *Crecy*, a 2500 hp two-stroke V12 liquid cooled aero engine, will be presented in Sect. 6.3.2. As an isolated, stand-alone effort from FKFS Forschungsinstitut fuer Kraftfahrwesen und Fahrzeugmotoren Stuttgart (Prof. W. Kamm) exist two comprehensive design reports for six- and seven-stage axial superchargers,[299] based on a 1941 BMW research contract. With all likelihood these axial superchargers were planned for applications on the powerful BMW 801, 41.8 l air-cooled, 14-cyl. radial aircraft engine with max. 2000 hp; however, nothing is known about a corresponding practical realisation.

---

[297] See Wikipedia, 'Junkers Jumo 222' in English. Designed by Ferdinand Brandner (1903–1986), the 222 delivered finally 2500 hp (1850 kW) at take-off by increasing the speed of the engine to 3200 rpm; the dry weight was 1088 kg. Between 1946 and 1953 Brandner and a team of Junkers/BMW engineers developed during their internment at Kuibyshev, Soviet Union at the Kuznetsov design bureau the *NK-12*, with 12,000 hp take-off power still in the forefront of the most powerful turbo-prop engines and—presumably, one of the aircraft engine in longest, continuous production. See Wikipedia 'Ferdinand Brandner' and 'Kuznetsov NK-12' in English.

[298] The AVA axial compressor design for the Jumo 004 turbojet had 27 blades in the two front rows and 6-times 38 blades in the remaining stages; after a correcting design change by BMW for the 003 the thin AVA aerofoils were replaced by more robust NACA profiles.

[299] See Eckert, Berechnung, FKFS reports No. 395 and 399.

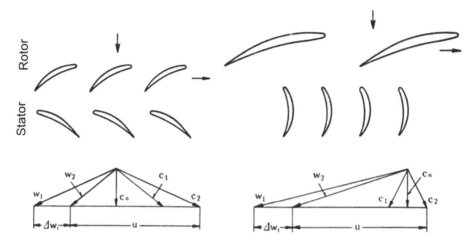

Fig. 4.46 BBC axial compressor test configurations for differing degrees of reaction: R 0.5 (l) vs. R 1.0 (r), 1933, c—abs. velocity, w—rel. velocity, u—circumf. speed, 1—blade row inlet, 2—blade row exit

Degree of Reaction *The Weak Point of AVA Compressor Designs* Probably the widest difference in axial compressor designs arises from the different degree of reaction R, i.e. the ratio of the static temperature or pressure rise over the rotor to that for the whole stage. Accordingly, a stage with an *'impulse stator'* and all the pressure rise in the rotor is 100% reaction, R 1,—while most compressors tend towards 50% reaction, R 0.5, splitting the pressure rise half-way between rotor and stator. The physical appearance of the blading is largely dependent on this parameter, as can be seen from Fig. 4.46. This compares 50% reaction (sometimes also called *'symmetrical blading'* for the possibility to use the same profiles for rotor and stator) with 100% reaction, where the stator just fulfils the task of flow turning and re-adjustment at a constant pressure level. Some authors[300] suggested the former being typically adopted for British and most American compressor designs, while the latter should have been representative for Continental design practice. But this does not hold true in general, especially not for the period under consideration between 1900–1940:

In chronological order the earliest example of a R 1 axial compressor design goes back to *Parsons*, who however did not follow the self-evident thought to just copy and reverse the flow direction of his foregoing steam turbine design, Fig. 4.47 (r). Besides a considerable lack of aerodynamic know-how with a tendency to compressor stalling, the profile hardware quality itself—as illustrated already in Fig. 2.2 for Parsons's turbine blading—must have contributed significantly to the poor performance so that Parsons axial

---

[300] See Roxbee-Cox, Gas Turbine Principles, Sect. 5: 'The Axial Compressor' by A.D.S. Carter and—see Kruschik, Die Gasturbine, 2nd ed., p. 143.

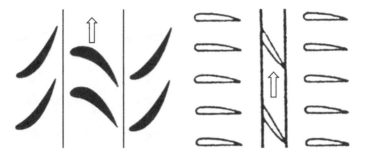

Fig. 4.47 Degree of Reaction design ancestors: High pressure steam turbine, 1924 for R 0.5 (l) and Parsons axial test compressor, 1901 for R 1 (r)

compressor initiative was stopped already after a short while in favour of multi-stage centrifugal compressors.

In due course, a reverse (high-pressure steam) turbine design approach is reported for the early Brown Boveri axial compressor considerations and their R 0.5 design track, Fig. 4.47 (l). In June 1924 two company representatives—for BBC Jean von Freudenreich (1888–1959), head of the Baden test laboratory and—for BBC's French daughter company C.E.M. Georges Darrieus (1888–1979) strolled jointly through the World Power Conference exhibition at London, Wembley and according to Darrieus' later account:[301] *'In front of an open Parsons turbine with typical 45° blading and a degree of reaction of 0.5 Freudenreich developed the idea that this symmetrical blading might be suited to run this engine in reverse as a compressor by changing the direction of rotation. We ignored the fact that Parsons had already tried to realise an axial compressor himself in 1901, giving up this idea due to unsatisfactory result.'* For Brown Boveri and their foregoing, decade-long reaction steam turbine design and manufacturing experience this was clearly an obvious decision. To the contrary nothing is known explicitly about *Alan Griffith's* motives in 1926 at this fork in the road, but a look at the velocity triangles of Fig. 4.46 may have sufficed for him to avoid the considerably higher relative velocities and the implied Mach number losses for the R 1 configuration in the upper subsonic regime.

As straightforward as the choice of R 0.5 might appear for somebody with Brown Boveri's steam turbine design background in combination with a self-evident aerodynamic interpretation of the flow velocity triangles, this was apparently not sufficient reason for *Claude Seippel*, when he took over BBC's emerging Velox development group after his return from the United States in 1928. Under extraordinary development pressure when both the first commercial 13-stage axial compressor for Ackeret's supersonic wind tunnel at ETH Zurich as well as the first all-axial furnace gas-fired Velox boiler with two 11-stage

---

[301] See Eckardt, Gas Turbine Powerhouse, p. 69 f.

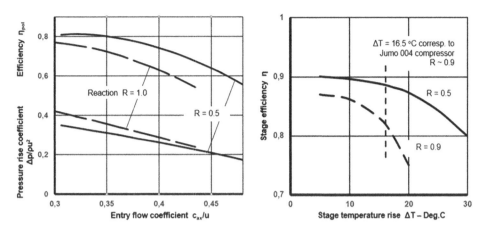

Fig. 4.48 BBC experimental characteristics of 11-stage axial compressor for R 0.5 vs. 1.0, 1933 (l), dependence of calculated stage efficiency on stage temperature rise for R 0.5 vs. 0.9, Howell 1945 (r)

axial compressors for combustion air and furnace gas compression[302] for the Société Métallurgique de Normandie steel works at Mondeville/Caen awaited delivery at the end of 1932, he decided to equip and investigate one of these 11-stage compressors with four different blade/vane settings, corresponding to a systematic—up till now unique—degree of reaction variation from R 0.5, R < 1.0, R 1.0 and R 1.25, Figs. 4.25 and 4.48.[303]

In Fig. 4.46 the blade staggering of rotor blades (above) in relation to stator vanes (below) for the variants R 0.5 and R 1 are shown together with the velocity triangles for equal work transfer u Δc_u. It appears that the principle drawback of R 1 blading becomes evident just by analysing the velocity triangles, given the much more balanced velocity distribution between rotor and stator for R 0.5 and the especially striking difference in the relative velocity vector lengths. The obvious advantages of an R 0.5 design are the relatively low velocity levels even at circumferential speeds in the supersonic regime. In a corresponding R 1 arrangement, shock losses and related separation losses can occur; early aerofoil aerodynamics typically showed a significant high-speed sensitivity beyond Mach number Ma 0.6–0.7.

The corresponding BBC test results[304] have been replotted in Fig. 4.48 (l). It shows two settings for the 11-stage Mondeville test compressor bladings, a) with the classic '*symmetrical*' degree of reaction R 0.5 (solid lines) and b) for R 1 with pressure rise in rotor only (dashed lines). The diagram combines polytropic efficiency curves (top) and pressure rise

---

[302] Both axial compressors had a row of variable guide vanes at entry to adjust the combustion gas mixture. The air compressor had a design volume flow of 3.17 m³/s, providing in 11 stages a PR 2.6 for ambient entry conditions.

[303] See Eckardt, Gas Turbine Powerhouse, p. 111 f.

[304] See BBC, Axialgeblaese fuer Veloxkessel, 1932–1933.

characteristics $\Delta p/(\rho u^2)$ with the entry through-flow coefficient c_{ax}/u. While the pressure data fulfil the expectation of higher stage pressure ratio for R ~ 1 design and thus overall in principle a lower stage count, the results confirm BBC's long-term preference for the R 0.5–0.6 aerofoil setting due to the impressive measured efficiency advantages of 8–12(!) percentage points in comparison.

The 1933 BBC measurement results on the performance impact of degree of reaction on axial compressor design were kept absolutely company-confidential for the next 25 years, before they appeared for the first time in John Horlock's book on *Axial Flow Compressors*.[305] A. Raymond Howell was obviously the first to publish a detailed assessment of compressor cascade losses on a theoretical basis,[306] though delayed until 1945, when the war against Germany was decided. He makes the point explicitly in a figure reproduced here as Fig. 4.48 (r), where the efficiency drawbacks due to profile losses of an R 0.9 design—exemplified by the German Jumo 004 compressor with a characteristic average stage temperature rise of 16.5 °C—are compared with a clearly superior R 0.5 design. Consequently, the 100% reaction compressor is much more bulky than the 50% reaction design and was therefore never expected to be used in aircraft gas turbine engines—what consequently caused considerable astonishment after the war when exactly this design principle was identified in the German Jumo 004 and in most of the BMW 003 turbojet engines. With his theoretical approach Howell also confirmed the BBC efficiency differences and highlighted for the first time a basic weakness of the AVA axial compressor design concept in the responsibility of *Albert Betz* and *Walter Encke*.[307]

In hindsight, the Allied explanations for this German axial compressor design peculiarity circled around the corresponding low-cost simplicity of built sheet-metal vanes, the ease of sealing of just turning, constant pressure vanes and the lack of strategic materials at war's end in Germany, but the fundamental decision in that direction—if any[308]—must have been much earlier, long before the war-triggered restraints.

---

[305] See Horlock, Axial Flow Compressors, pp. 92–93, there Figs. 4.11 and 4.12—with the explicit comment 'Courtesy Cl. Seippel, Brown Boveri Ltd., unpublished data, March 1957'. In private correspondence with the author, Sir John replied on 1 Dec. 2008: *'With regard to the reaction data I quoted in my compressor book [fifty years old this year!] it was given to me personally by Claude Seippel when I visited BBC as a consultant for CEGB* [Central Electricity Generating Board] *UK in about 1955 or so. He learned I was writing a book on compressors and encouraged me in this venture and to use the data unpublished, I think. He was very kind to me and I had much respect for him—a gentleman engineer of the old school'.*

[306] See Howell, Design of Axial Compressors; Howell's results were reproduced in German in—see Kruschik, Die Gasturbine, 2nd ed. 1960, p. 144.

[307] Howell's figure was first reproduced in German literature in 1960 in—see Kruschik, Die Gasturbine, 2nd ed., p.144, hinting at the resulting Messerschmitt Me 262/Jumo 004B flight performance drawbacks which will be further discussed together with corresponding mechanical design restrictions in Sect. 9.1 and Fig. 9.7.

[308] Even after intensive search the term *Degree of Reaction* (Reaktionsgrad) could not be found in any of the Betz and Encke publications from early on up to and including—see Betz, Einfuehrung

One possible cause might have been hidden in Betz's and Encke's patent history, which especially for Betz represented also an important economic aspect in his 1920s. All three patents under consideration—Betz's DE319,413 *'Screw ventilator and screw pump'* with priority of 31 Jan. 1918, his DE399,062 *'Axial blower or pump'* of 28 March 1923 and Encke's DE707,013 *'Multi-stage axial compressor for high rotating speeds'* of 27 May 1936 put the single axial compressor rotor in the focus, so that a multi-stage arrangement could at best be achieved by adding geometrically nearly equal or highly similar individual rotors together, Fig. 4.44. Betz's first patent relied on the observation that due to centrifugal and secondary flow transportation of boundary layer material out of the rotor hub/suction side region, the energy transfer/loading can be increased in this separation-prone region while the rotor tip region should be unloaded in comparison. Betz's second patent dealt with inlet guide vanes upfront of the rotor blading to achieve a balanced/radially optimum flow distribution by redressing the inflow swirl accordingly; an idea basically implemented in the concept of the Siemens-Betz axial blower, Fig. 4.38. And finally, Encke's 1936 patent followed the basic idea to keep a pre-selected ratio u/a of circumferential speed u to speed of sound a, close to loss optimum by varying the rotor diameter of the rear stages in line with the pressure/temperature rise so that fundamental rotor design parameters remain unchanged (as highlighted before by a largely constant stage-by-stage blade count).

From a theoretical perspective—for which the scientifically trained AVA designers were certainly best prepared—the radial distribution of the degree of reaction implies however a considerable effect. Under the assumption of a radially constant total pressure distribution, there is only the radially constant R 1 case characterised by constant axial velocities in meridional direction c_{m2} at the vane exit of a rotor/stator stage arrangement, while in comparison e.g. a radially constant R 0.5 design shows an axial overspeed of Δc_{m2} + 30%—compared to the radial average—at the hub and correspondingly a velocity deficit of the same magnitude at the casing.[309]

Interestingly the only basic description of the AVA axial compressor design methodology is from Hans von Ohain, who attended a few lectures presented at Goettingen by Walter Encke, and who subsequently invited Encke as axial compressor consultant to Heinkel Rostock after November 1938:[310]

(Introduction to the theory of flow machines) of 1959. Immediately after the war the AVA provided the 'Goettingen Monographs', a kind of 5000 pages summary report about their wartime activities to the British Ministry of Supply, which contained also a Chapter J3 *'Turbine and Jet Engines'* with *3.1 Compressors* by W. Encke and R. Foell (19 p. in German), in which basically Encke's foregoing publications were repeated, again without any specific discussion of the 'degree of reaction riddle'. A single exemption is—see Eck, Ventilatoren, 1st ed. 1937, p. 143 which discusses the degree of reaction for single stage axial blowers with preferably inlet guide vanes and R 1 as nearly design optimum, while for the case with exit guide vanes the difficulty of low-loss flow diffusion is emphasised,—and Mach number effects for the relative rotor flow are completely ignored.

[309] For typical meridional velocity distributions of a R 0.5 axial compressor stage,—see Eckert, Axialkompressoren, p. 170, Abb. 190.

[310] See Von Ohain, Turbostrahltriebwerke, p. 23.

The work transfer to the first, rotating row of an axial compressor shall be maintained radially constant. The following stator row shall (only) redress the resulting circumferential momentum from the foregoing rotor back to zero, so that there is again co-axial flow at the entry to the second rotor—and this process continuing further, stage-by-stage. Ideally this implies only a stagewise adaptation of flow density changes by adjusting the annulus height, while the rest of the blading geometry would be unchanged. For these conditions the circulation between rotating and stationary rows remains constant, resulting in a radially constant axial flow distribution—without considering tip leakages and boundary layer interaction at the non-rotating annulus borders. Consequently, the static pressure rise happens nearly completely in the rotating blading, the degree of reaction is close to 100 percent.

Given Goettingen's preference to create as described a multi-stage axial compressor arrangement at best by adding up in a repetitive manner largely unchanged optimum single stages, the selection of an R 1 basic design suggests—especially under time pressure—the least problems. Clearly from the very beginning the R 0.5 approach asks already for a more time-consuming, individual stage-by-stage design effort, typical for modern axial compressor design methodology.[311]

Another possible explanation for an R 1 design could have been exaggerated fear of flow instabilities, so that rows of non-diffusing, just flow-turning vanes might have been considered as buffers especially in view of fighter aircraft gas turbine applications.[312] But, real reasons remain in the dark; seen from today, the period between 1933 and 1938, i.e. between the end of small-scale axial supercharger developments and the actual beginning of axial compressor design for German turbojet engines might look ideally suited for *AVA Goettingen* to review their axial design principles on last minute in a decisive phase. In fact Encke in this period was mostly on his own, since Betz was more and more absorbed by administrative Institute management tasks. Encke in his logic addressed then the most fundamental problem of the Goettingen design approach (in agreement with Betz)—the excessive axial length, which he tried to improve by jumping to a compact, counter-rotating and vaneless axial compressor design—without a clue like other followers, how to

---

[311] Gratefully following a suggestion of Uwe Schmidt-Eisenlohr, 7 Sep. 2019.

[312] This argumentation is indirectly underlined by—see Seippel, The evolution, where he revealed that the R 1.25 variant (with even flow acceleration in the vane sections) in the Brown Boveri test set-up had been suggested by Curt Keller—together with J. Ackeret a member of the AVA camp—to improve flow instability. And in fact, as is confessed somewhat hidden in an internal BBC report—see Seippel, Die Entstehungsgeschichte, p. 8, a few of the early BBC axial compressors for Velox boiler applications were actually built with an R 1 blading for this reason. However, though the direct comparison of test results for R 0.5 and 1.25—both for an 11-stage axial compressor configuration—showed considerably higher through-flow coefficients for the latter case, no significant enlargement of the stable operation range $\Delta V/V_o$ could be observed—and nothing is known about a potential information feedback to Goettingen in this context. Of course, the superior performance advantage of R 0.5 was kept strictly company-confidential.

overcome the problems of mechanical complexity.[313] So, when end of 1938 the urgent RLM call for a reliable multi-stage axial compressor design ready for industrial transfer reached AVA, the jump to the repetitive R 1 stage arrangement promised—somewhat short-sightedly—the fastest result with least complications.

Besides being dragged from their own patent background, Betz and Encke could have been also somewhat forced in the chosen design direction by an external patent conflict which will be outlined in the next sub-section. After mentioning the degree of reaction usage for axial compressors from *Parsons* and *Griffith* to *Seippel* and *Betz/Encke*, there remain only two more names for completion. *Curt Keller*, Ackeret's doctorate student at ETH Zurich in 1934 is his master's voice in full, no mention of the term *Reaktionsgrad* just on one of the 190 pages of his praised thesis, but all his drawings represent the Goettingen R 1 design philosophy. Unaware of the curtain of secrecy closing around him on the critical degree of reaction issue by Griffith and Brown Boveri, Encke might have found additional confidence in his position with Ackeret/Keller as well-known partners.

Herbert Wagner (1900–1982) appeared in the foregoing in the context of the Junkers high-altitude aircraft activities. In 1934, still as professor for aviation technology at TH Berlin-Charlottenburg he personally[314] designed (and patented) an all-axial gas turbine engine for a geared turboprop drive. He took the large General Arrangement drawing of that engine with him to his next professional station at Junkers Dessau in September 1935;[315] the GA was characterised by a multi-stage R 0.5 axial compressor which a small team of engineers—of which the group head Max Adolf Mueller (1901–1962) and the axial compressor specialist Rudolf Friedrich (1909–1998) accompanied Wagner from TH Berlin to Junkers—tried to develop an all-axial turbojet demonstrator engine in the coming years. Especially Friedrich, who came to Wagner as young engineer (24) in 1933 from TH Hannover (and who left the development team at Heinkel in August 1941) insinuated after the war his responsibility for the R 0.5 approach on the German side,

---

[313] As revealed by—see Betz, Axial Superchargers, the reduction of the axial chord length resulted in a considerable performance decrease, a direct sign that especially the small AVA axial superchargers with 0.125–0.15 m diam. must have been close to being hampered by Reynolds number effects. An observation which was repeated—without clear identification of cause of origin—during altitude testing of the Jumo 004B turbojet engine shortly before war's end.

[314] In his short memoirs—see Wagner, Meine Arbeiten, reads the first sentence *'My lecturing* (at TH Berlin) *comprised also aircraft propulsion systems and the concept of the stationary gas turbine was known to me. In 1934 I applied this to aircraft propulsion … I executed then without any support from my Institute for Aviation Technology a detailed drawing of such a* (turboprop) *engine'.* Given Wagner's background as Austrian steam turbine engineer, the hint towards *'the stationary gas turbine'* points with high likelihood to Brown Boveri and their all-axial Velox boiler gas turbine concept up for sale since 1932. Wagner's turboprop patent made no reference to the axial compressor degree of reaction.

[315] See Von Gersdorff, Flugmotoren, p. 299; for further details on Wagner's engineering team and their axial turbojet developments at Junkers Magdeburg (and later Heinkel)—see Sect. 6.2.

which in interaction with the dominant Wagner and in view of the latter's explicit statements in this context is not very likely.

This interpretation is further corroborated by the actual turbojet development history in the early 1940s, when the RLM under the responsible technical officer for turbojet development Helmut Schelp (1912–1994) must have realised that his original whole-hearted support for the AVA axial compressor design philosophy in 1938/1939 implied certain substantial performance flaws and operational drawbacks which had to be urgently corrected. Though the Junkers Jumo 004 engine development was considered then as already too far ahead, a special axial compressor technology programme was launched to correct the earlier, wrong decision at least for the BMW turbojet family, disposed to Brown Boveri's experienced German turbomachinery branch at Mannheim and its newly founded aero gas turbine design group under the leadership of Hermann Reuter (1911–1981),—see Sect. 9.1.[316]

In the Patent Arena *BBC Challenges the AVA Compressor Design* In the middle of the German high-altitude aircraft and engine development activities, BBC Baden, CH, applied on 15 April 1932 for a German patent on *'Compressor unit for pressurised combustion in steam generators or supercharging of combustion engines'*[317] which caused in due course considerable turmoil in related German/Swiss industrial and scientific communities, but was finally positively decided pro BBC against all objections (interestingly with a certain impetus from the Swiss Escher-Wyss, Zurich) by the Reich Patent Office, Berlin on 26 Feb. 1936 and published as granted patent DE655,698 on 30 Dec. 1937. With all likelihood the BBC-inventor was Claude Seippel, though not explicitly named in the German application and one can assume, that Walter Noack, after his salient role in the development of aircraft turbocharging at the end of WW I and with certainly still existing excellent connections to the very Reichswehr units, had strongly encouraged the patent project to re-establish BBC's once lead role in this area. The excitement about a potentially negative effect on German aero engine developments was especially surprising, since the patent itself makes no reference whatsoever to aero turbocharging and flight applications. The actual, translated patent claim is stunningly short and simple: *'Said compressor unit, in which a blower is driven by an exhaust-gas turbine with variable, load-dependent speed, characterised by a multi-stage axial compressor i.e. with more than three stages and with wing-type aerofoils'.*

---

[316] BBC's aero engine development activities in the newly founded development department TLUK/ VE since April 1941—culminating in the RLM-issued *'Hermso I–III compressor technology programmes'* for the advanced engine versions BMW 003 C/D—are dealt with in detail in Sect. 8.1—and corresponding tank gas turbine activities in Sect. 8.3.4.

[317] Original BBC patent title in German*: 'Verdichteranlage fuer Druckfeuerung von Dampferzeugern oder Aufladung von Brennkraftmaschinen'.*

In hindsight, it is difficult to understand why the principle of a free-wheeling turbo-charger—which has nothing else but a self-regulating *'variable, load-dependent speed'* and consequently air flow delivery—should be patentable at all at that time, independent of the actually used radial or axial blower concept. On top, the patent text provides no specific clue about the *'more than three stages'* so that one can only speculate, if the filing simply wanted to prevent interference with existing plans of German turbocharger development, since here—as seen before—the maximum imaginable complexity was a combination of three stages. Another interpretation is that BBC somehow had learnt of the AVA 1928–1930 tests which according to a statement in a letter of Encke towards the RLM patent attorney comprised actually three test stages maximum, while project drawings of an axial blower for the Siemens Sh 20 radial piston engine, Fig. 4.42, showed five stages. The design data of this compressor with 15 cm cylinder diameter, 455 mm length were—as already provided in tabulated form in Sect. 4.3.3, Sub-Section *'On Axial Blowers'*: mass flow ~1 kg/s, pressure ratio PR 2.19, efficiency ~85%, speed 33,000 rpm.

The basic BBC patent ideas are summarised in Fig. 4.49 from DE655,698. The actual, innovative idea follows a different path via the interpretation of the characteristic, different shapes of the p-v pressure-volume flow compressor speed lines—*'steep/narrow'* for the axial with more than three stages, *Fig. 4*, vs. *'flat/broad'* for the centrifugal compressor, *Fig. 5*. System (pressure) instabilities which can occur in both systems are rather damped for the suggested all-axial set-up, while the centrifugal system supports tendencies to react on pressure fluctuations Δp by wide Δv excursions—and with potentially resulting combustion instabilities. The usage of the referred, wing-type ('tragdeckaehnlich') blading is beneficial, since it generates exactly these type of steep p-v compressor characteristics. A typical Velox turbo-set with a 12-stage axial compressor and an axial multi-stage drive turbine is shown as *Fig. 1*, while the corresponding *Fig. 2* illustrates a typical 50% reaction or *'symmetrical'* blading—*'having the same profiles in rotating and stationary sections, with the same angles and blade pitches'*, but this is no absolute necessity and *'the blading can be different in this respect as well'*; finally, the *'wing-type'* blading is illustrated by *Fig. 3*.

The reactions on the German side after the detection of BBC's very patent application take in parts burlesque forms—but in the end the outcome of all counter-initiatives is in vain, actually the battle appears to be lost before being even started. The patent application was published on 31 January 1935 so that according to standard German practice the period of possible objections ran out after three months latest on 30 April 1935. As can be reconstructed with AVA archive materials, only the Swiss company Escher-Wyss Zurich had objected in time—somewhat helplessly and without meaningful reference to the BBC patent application—with reference to an AVA publication of 1925[318] and to Charles Parsons 1901 axial compressor patent *'Turbine pump'*, DE132,884.

---

[318]L. Prandtl, Ergebnisse der Aerodynamischen Versuchsanstalt zu Goettingen, 1. Lieferung, 3. Aufl. 1925, p. 73, aerofoils #364, #365 and #366.

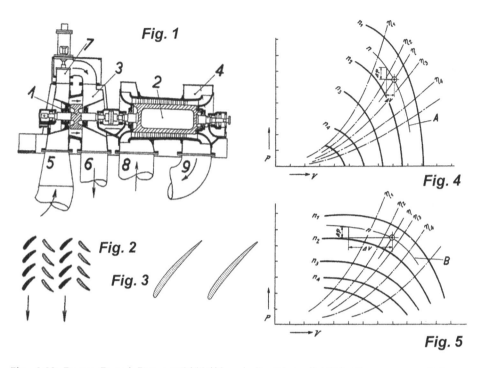

Fig. 4.49 Brown Boveri Patent DE655,698, priority 15 April 1932: *Compressor unit for . . .*
supercharging of combustion engines

AVA realised the situation only on 22 August 1935, when Professor Betz on his way to his holiday resort[319] in the south stopped to visit TH Stuttgart and was informed by his colleague Georg Madelung about the patent calamity in view of the German *'Hoehenflugzeug'* (high altitude aircraft) project. The general mood is expressed in a letter from Madelung to Encke only two days later: '. . . *The BBC patent application can cause problems. As far as I know the RLM and Herr Baeumker,*[320] *it will not be tolerated, if a patent of a foreign company would obstruct the further development of the supercharger blower. With a German salute—Heil Hitler! Georg Madelung.'* AVA's industrial network

---

[319] Stuttgart is some 80 km away from Baden-Baden and in the vicinity of traditional *Black Forest* resorts. In principle, Prandtl's newly built house at Mittelberg, Kleinwalsertal, just 10 km across the German-Austrian border and 220 km south of Stuttgart,—see Vogel-Prandtl, Ludwig Prandtl, p.134 (information, courtesy to Michael Eckert, 3 March 2020)—would have been an option, but since Prandtl himself inaugurated the house only with the family winter vacations 1935/1936, a kind of advance summer rent for *'dry dwelling'* https://educalingo.com/en/dic-de/trockenwohner by the thrifty Betz can be rather excluded.

[320] Baeumker's role as coming *State Science Manager* in the early stages of the high-altitude aircraft project has already been described in Sect. 4.3.2. For more details—see Hirschel, Aeronautical Research in Germany, p. 71 f. and—see Wikipedia, 'Adolf Baeumker' in German.

at that time was represented by existing relics of the former *'Zoelly Syndicate'* (1904–1929) for impulse and action turbines, then still led by Director Heinrich Zoelly (1862–1937) of Escher-Wyss Zurich, CH,[321] and Ravensburg, Germany with the member companies Siemens Schuckert Werke, Siemens & Halske, Brueckner & Kanis Dresden, Krupp, Thyssen, MAN, Borsig AG. In the following months a substantial correspondence grew amongst AVA and these partners to find relevant objection materials against the BBC application, in the end in vain, because (a) the timing was too late anyhow, (b) the Escher-Wyss materials were discarded as not appropriate.

The RPA *'Reichspatentamt'* (Reich patent office) at Berlin decided on 26 February 1936[322] against the Escher-Wyss objection, mainly due to the argument that the usage of an axial compressor as turbocharger was new and confirmed the BBC patent grant officially since 15 April 1932. Escher-Wyss objected immediately against this decision, again in vain. Finally, the industrial network wanted the RLM to launch a patent revocation action (*'Nichtigkeitklage'*) which was denied already by the RLM lawyers as not very promising. At last it was tried to gain at least the right of a *'joint use'* for the *'Reichsverwaltung'* (Reich administration). The documented dispute between the RLM and BBC went till April 1941, when the patent office of BBC Mannheim insisted cautiously, nevertheless tough-minded and cunningly to gain proof, that the AVA compressor rigs had actually been turbine-driven during test operation, as patented; but, of course as visible in Fig. 4.43, the AVA test compressors had been powered either by electro motor as common practice in test facilities everywhere, or—though turbine-powered—had not a patent-like self-regulating compressor-turbine-arrangement where the compressor mass flow determined the turbine power.

As said before, Brown Boveri must have been surprised about the resonance of their 1932 axial turbocharger patent application, but in short they realised the unique AVA—considered as the leading German aerodynamic institution—position and from BBC perspective AVA's backwardness with respect to the preferred degree of reaction close to R 1.0 for axial turbomachinery. In due course already on 13 Nov. 1934 the Swiss Brown Boveri applied for a French patent with Swiss priority on 3 February 1934 and granted as FR781,182, for a *'Compressor with neighboured, alternating rotating and non-rotating blading'*[323] with again a very simple, straightforward and comprehensive patent claim, which might have found also the interest of the Goettingen aerodynamicists: *'The compressor with alternatively fixed and mobile blading similar to a steam turbine design is to a large extent equipped with both fixed and mobile rows with similarly profiled,*

---

[321] See Eckardt, Gas Turbine Powerhouse, p. 287 and—see Wikipedia 'Heinrich Zoelly' in English.

[322] The judgement was rendered by Dr. phil. Dipl.-Ing. Bruno Czolbe, who was established as Oberregierungsrat already in 1921 by Reich President Ebert and confirmed as President of the RPA Senate in 1941 by Adolf Hitler.

[323] Original BBC patent title in French: 'Compresseur à couronnes d'aubes alternativement fixes et mobiles juxtaposées'.

symmetrically oriented blading, where the pressure rise is achieved in both fixed and mobile rows and the resulting degree of reaction is therefore close to 0.5.'

Nothing is known about the resulting resonance for this patent, but apparently Brown Boveri themselves must have developed *'second thoughts'* in this context—possibly after some contacts with British institutions as e.g. RAE and/or Rolls-Royce, where R 0.5 had been firmly established latest since 1926 when A.A. Griffith had put down his axial compressor design principles. Consequently, Brown Boveri liquidated the unpromising endeavour to keep and defend an universal 50% reaction design patent—by abandoning the already granted patent after some 18 months. Perhaps, this move followed the insight that the loss of an unsustainable patent might diminish their chances in the main patent conflict, which the Reich patent office decided finally in favour of Brown Boveri on 26 February 1936. After this intermezzo the AVA compressor design team felt presumably highly unsure—and tried for further axial compressor design activities not to leave their established, firm ground of a degree of reaction of R 1. Thereafter, the Swiss Brown Boveri must have been seen critical from German perspective which vice versa apparently favoured BBC's inclination to seek for business partnership in Great Britain.

Publicly however, Albert Betz showed AVA and himself unshaken when he stated on 28 February 1938 in a presentation to the Lilienthal-Society at Berlin: *'The required compression that must be supplied by the supercharger increases with the altitude. A supercharger with constant speed will, at low altitudes, give either too high pressures or deliver too large air quantities. With regard to economy, therefore, it will be necessary to allow the speed to increase with the altitude. In the case of mechanical drive from the engine shaft this will naturally raise considerable difficulties. If the drive is by means of an exhaust turbine, (sic!) however, which is also desirable for reasons of economy, then it is possible to vary the speed with the altitude since the pressure drop available in the exhaust gases increases with the altitude.*[324] With less words, this is in fact the description of a freewheeling, self-regulating turbocharger, Fig. 4.11 (r)—a concept which at least for civil applications was, as Betz knew definitively since more than two years, protected back till 1932 by Brown Boveri patent DE655,698. In the documented, secret discussion of this Lilienthal turbocharger specialist meeting Walter Encke had his coming out in favour of the patented BBC advantages of the 'steep' axial compressor characteristics—in view of a thus considerably reduced effort to balance turbocharger-turbine instabilities. However he states, *'Goettingen is not convinced that solely the axial compressor could provide the required steep pressure-volume characteristics and therefore, this compressor concept has to be* (automatically) *considered superior to the centrifugal compressor. . . . We believe, we*

---

[324] Quoted according to—see Betz, Axial Superchargers, NACA TM 1073, p. 3; the Encke statement towards the advantages of the BBC patent in the corresponding discussion of the Betz 'Axiallader' paper in German.

could also design a steep centrifugal compressor characteristic, if required'. A somewhat strange, stand-alone position without any second thought to the impact of the chosen degree of reaction in comparison to the BBC approach and its implied performance differences, approximately just one year ahead of the general German all-axial turbojet development programme start on the initiative of Hans Mauch and Helmut Schelp, RLM, which relied also considerably on the foregoing axial compressor development activities at AVA Goettingen. On page 54 of the mentioned secret report follows a one-page-summary about the general meeting topics:

- axial supercharger,
- blade shape of centrifugal chargers and
- questions of mechanical design.

'The opinions about the prospects of axial superchargers are highly differing. Some research results indicate certain advantages. ... One is the high efficiency of the axial supercharger, which however is diminished by the small operation range; this hampers aero applications which ask for flat pressure-volume characteristics for a wider operation range. etc.'

The list of in total 68 conference participants reads like a *'Who's who?'* of German aero engine specialists; in view of the immediately forthcoming axial turbojet/aero gas turbine developments there are only a few names missing like the already mentioned Hans Mauch from RLM Technical Office (but from RLM—LC1 Research Office Dr. Hermann Lorenz[325] and—presumably still for DVL—Helmut Schelp instead) and Herbert Wagner/Max Adolf Mueller from Junkers (here represented by Dipl.-Ing. Ferdinand Brandner and Dipl.-Ing. Anselm Franz instead). Nowhere is the slightest indication of a potential use of axial compressors for flight gas turbines. Brown Boveri Mannheim, which sold multi-stage axial compressors in the forerunners of industrial gas turbine configurations (Velox boiler, Houdry plant) since 1932 was represented by Dipl.-Ing. Koeckritz and—top of the tops—EHF Heinkel Flugzeugwerke Rostock sent their quiet *'observers'* Dipl.-Ing. Wilhelm Gundermann and a person named *H.-J. v. Oheim,* which with all likelihood is the mis-spelled name of Hans-Joachim (Pabst) von Ohain. If nobody else in this *'Group of 68'* had seen or heard of Frank Whittle's published aero gas turbine patent of 1930, one can at least and at best assume that Gundermann/von Ohain stayed intentionally mum. At that time in spring 1938 they were preparing after foregoing successful combustor tests their first flight-worthy turbojet engine HeS 3.

---

[325] In 1942 as 'Fliegeroberstabsingenieur' successor to A. Baeumker, who then took over the responsibility for the huge, newly founded LFM Luftfahrtforschungsanstalt Muenchen (Aviation Resarch Establishment Munich).

Excursion II: Fritz Heppner (1904–1982) and Hellmut Weinrich (1909–1988)

'Fritz Albert Max Heppner's name presumably appeared for the first time in English print in the 1940s.[1] He is introduced there as 'German domiciled in England', *and his patent record actually confirms that he presumably left (or was forced to leave) Germany and his home-town Berlin-Charlottenburg as part of the persecution of Jews before or latest during 1936,* Fig. 5.1(l). *Thereafter, his patents indicate several living stations in England, among these a presumed stay in an internment camp on the Isle of Man in 1940,* Fig. 5.7(l). *His first identifiable patent application in the name of his new employer Armstrong Siddeley Motors Ltd—US2,360,130—has a priority date 26 March 1941 and shows in principle, surprisingly, the counter-rotating (CR) BMW 002 concept, which in German technical literature is associated with the name of Hellmut Weinrich,* Figs. 5.1 (r) and Fig. 5.11 (top).

Besides the patent, the essential engine general arrangement was also published for the first time in Geoffrey G. Smith's wide-spread book 'Gas Turbines and Jet Propulsion for Aircraft', *3rd edition, April 1944, p. 30, Fig. 20 and* Fig. 5.11 *(bottom, l). The actual acquaintance of Weinrich and Heppner could still not be confirmed, but the patent portfolio of both has the same 90/10 automotive/turbomachinery structure and there is one post-war witness*[2] *account that the Weinrichs had Jewish friends before the war. Heppner's special background accelerated obviously his rapid career in the aircraft/ automotive company Hawker/Armstrong Siddeley Motors (ASM).*[3]

---

[1] See Smith, Gas Turbines, 1st ed., p. 30 and 78.

[2] Courtesy of Dr. H. Griepentrog, 2012, colleague to H. Weinrich at GHH Oberhausen, D up to 1980.

[3] Richard Hodgson's website www.wolfhound.org.uk has details in this context: *'F.A.M. Heppner was a German refugee retained by Armstrong Siddeley during the war to work on highly complex gas-turbine schemes that would have been profitable had they worked. Heppner had the unshakeable*

© The Author(s), under exclusive license to Springer Fachmedien Wiesbaden GmbH, part of Springer Nature 2022
D. Eckardt, *Jet Web*,
https://doi.org/10.1007/978-3-658-38531-6_5

Fig. 5.1 Fritz A.M. Heppner
~1937 (l), Hellmut Weinrich
~1955 (r) © priv

The success of Heppner's CR engine concept at Armstrong Siddeley may be in parts explainable by the fact that there had been manufactured and tested in 1939–1940 before his arrival an experimental engine based on a layout suggested by the esteemed Dr. Griffith in 1926, then designed by RAE Royal Aircraft Establishment at Farnborough. Experts from the Rolls-Royce Heritage Trust came to the conclusion that the designation of the ASH Mark III engine in their archives, started after his arrival in 1942, stands presumably for Armstrong-Siddeley-Heppner; on top, he had adapted AS-internally his German-sounding first name 'Fritz' to the more neutral 'Fred'. The patent records see him at Canada since 1949/1950; he and/or his wife were last heard of in Montreal, Canada in about 1959.

Further in-depth studies on Fred/Fritz Heppner and his potential role of a technology and information carrier between Germany and England appear to be worthwhile. In the moment one can only speculate what he actually knew about early German turbojet developments and what, if any, advantages the English side might have gained thereof. His rapid rise within ASM and the involvement of that company in the further all-axial turbojet development in England appears to provide some hints.'

The foregoing text stub represents the Author's collected background knowledge about Fritz Heppner approximately at the end of 2013.[4] As German immigrant to England in decisive times for jet engine developments, he represented potentially also an interesting candidate for immediate insight into a special kind of technology transfer. It appeared to be worthwhile—though with little hope—to intensify the search effort. Contacts to the Berlin registry office and Jewish organisations came to nothing. Similarly, the response of Max Amichai Heppner, author of the moving refugee story *'I Live in a Chickenhouse'* was disappointing. Finally, there was contact to—and response from—a young *Klezmer*[5] duo at

---

backing of both Sopwith and Spriggs (the directors of Armstrong Siddeley and its parent, Hawker Siddeley). Senior engineers, including the chief engineer Stewart Tresilian, who openly questioned the viability of Heppner's schemes, were quite simply required to leave in early 1942.'

[4] Most of this information could be confirmed later-on, except for the (indicated) fact that Fritz Heppner might have left England towards Canada after WW II; he stayed continuously in the UK up to his death in 1982.

[5] See Wikipedia, 'Klezmer'in English.

Solothurn, CH, Liora Heppner, violin/viola and Micha Hornung, accordion: *'We have no idea. But there is a kind of Heppner clan historian—living in England, perhaps ...?'* Michael Heppner answered on 25 Jan. 2014:

> Dear Mr. Eckardt
> Very belatedly I am replying to your email of 9 July 2013!
> Your email got lost in my system and I have just come across it.
> Fritz Heppner was my father's first cousin. His father Hermann Heppner was the eldest brother of my grandfather ... His son Christopher Heppner emigrated to Canada where he became a professor at McGill University. On his retirement he moved to British Columbia ... here is his email address.

Thankfully, most of the following information about the private life and fate of Fritz Heppner came over the past years from his son Chris Heppner, born at Berlin in 1933, up to ~2000 Associate Professor of English literature at McGill University, Montreal, and a leading expert and author[6] on William Blake (1757–1827), English poet, painter and printmaker, and closely connected to the Swiss painter Johann Heinrich Fuessli, engl. Henry Fuseli (1741–1825). Moreover, some technical gaps could be closed by visiting the archives of the Rolls-Royce Heritage Trust, Derby, where a lecture on turbojet engine history was presented as the 2017 President Evening Lecture[7] on 12 April 2017. The following outline of the *'parallel lives'* of Fritz Heppner and Hellmut Weinrich covers by and large their common period up to the beginning of WW II, while Heppner's time at Armstrong Siddely Motors, Coventry, and thereafter will be dealt with in the special Sect. 8.2.4.

Fritz Albert Max Heppner was born in 1904 at Posen/Poznań, then West-Prussia, Germany to Hermann Heppner and his wife Jenny; Fritz had a two-year-elder sister Ilse. As result of WW I and the reconstitution of an independent Polish state approximately 50,000–60,000 German inhabitants left the area after 1919, amongst these the Heppner family some 300 km to the west from Posen to Dresden, where H. Heppner's profession could still be found in a post-war address book as *'Baker and confectioner, cookies and gingerbread production'*. In March 1922, Fritz finished the local gymnasium Dresden-Johannstadt, Marschnerstr. 18/Pillnitzer Str., Fig. 5.2 (l), called *'Oberrealschule'* for emphasising natural sciences, with his *'Abitur'*, the university-entrance diploma.[8]

One year later Fritz Heppner commenced officially his studies in Mechanical Engineering at the local *'Sächsische Technische Hochschule Dresden'* (Saxonian Technical

---

[6] See Heppner, Reading Blake's Designs.

[7] With special thanks to Paul Stein, RR Derby and Alan Newby, RRHT,—and for the valuable archive support from Peter Collins and Dave Piggott.

[8] On 15 Dec. 2021, courtesy of Mrs. Angela Buchwald, TU Dresden Universitätsarchiv, Ilse Heppner was also successfully identified as a THD-student of chemistry between 1919 and 1922 (StudA A 4871).

Fig. 5.2 Oberrealschule 1921/2 at Dresden-Johannstadt (l), '*Vordiplom*' certificate (m), TH Dresden main building 1875 (r) © TUD UA (m)

Highschool, eq. to university),[9] Fig. 5.2 (r), where he passed the intermediate '*Vordiplom*', now in Theoretical Physics, the equivalent of a Bachelor degree, on 22 June 1929 with the third-best examination grade '*good—passed*', Fig. 5.2 (m).[10] This more than six years period[11]—for what normally should have lasted only 2–2.5 years—indicates a study interruption of some three years, for which Michael Heppner remembered in 2017 a presumed employment at the German car and motor-cycle manufacturers *DKW* at Zschopau, or the *Horch/Audi* works at Zwickau, Saxony. This could explain Heppner's expertise on automotive (partially *automatic/'hydrokinetic'*) gearboxes and torque transmitters, documented also by corresponding patents.[12] One should keep in mind that Heppner when he started studying was just 19, so he apparently had time. Being at Dresden, he was automatically at the centre of an emerging youth movement in Germany—after Versailles—the upturn of the motorless glider aircraft movement. *FVD Flugtechnischer*

[9] The THD main building of 1875 at Bismarckplatz (Architect R. Heyn) was destroyed during 1945 bombing, and the ruins removed thereafter.

[10] Information and certificate, courtesy of Dr. M. Lienert, TUD Universitätsarchiv, 31 Jan. 2014.

[11] Acc. to the THD students' file '*Fritz Heppner, born at Posen (Germany)*' studied between Easter 1923 and end of SS 1926 in the THD Mech. Department; info. courtesy of TUD Archive, 15 Dec. 1921.

[12] Heppner's potential CR axial turbojet partner/ contact Hellmut Weinrich visited the Oberrealschule Greiz (20 km west of Zwickau) up to Easter 1925, thereafter the Technical Academy (College) Chemnitz up to March 1931. Chemnitz is also only some 35 km north-west of Zwickau. For Heppner's known 21 patent ideas, the first five with priorities between 1924 and 1940 in his name were exclusively automotive gearbox-related, while the remainder of 16 inventions between 1941 and 1945 was solely in the connection with aero gas turbines. A contract between Fritz Heppner and Armstrong Siddeley Motors of 11 May 1943 assigned all his, then 35 ASM patents to the Company, in exchange for a three percent royalty up to a maximum of 10,000 pounds per annum—corresponding today roughly to a limit of 500,000 EUR/a. Some of Heppner's patents will be reviewed in detail also in Sect. 8.2.4, Figs. 8.30–8.37.

Fig. 5.3 Fritz Heppner at the *Wasserkuppe* glider camp, Glider ID #101 (l), training camp group Jan.-March 1925 (r): Heppner, right and Tracinski, left,—standing

Verein Dresden (Flight Technical Club) had the organising lead in Germany and initiated as early as 1920 first national glider flight meetings at the *Wasserkuppe*,[13] a unique mountain plateau (950 m above SL) in the German state of Hesse, approximately 110 km north-east of Frankfurt/M., from where great advances in sailplane development took place during the interwar period. Between January and March 1925, Heppner participated in a glider training course which he finished with a 400 m straight, 30 s proof flight on 24 March 1925, Fig. 5.3 (l). The glider ID #101A is signed by the meteorologist, then head of the Hamburg weather station Prof. Dr. Walter Georgii (1888–1968), and founder of German glider research, who became head of DFS Deutsche Forschungsanstalt fuer Segelflug (glider research establishment), in the 1930/40s a natural camouflage name for high speed research e.g. on *V-1* derivatives, and who rose in 1942 to one of four members of the influential *RLM Forschungsfuehrung* (research leadership), an organisation to coordinate all aeronautical research activities during wartime. The photograph, Fig. 5.3 (r), shows Heppner to the far right of his schooling group, the assumed glider teacher, sitting, and on the left, standing, a young man named Richard Tracinski from Breslau, Silesia.

On top of Wasserkuppe stands a historic, nationalistic anti-Versailles *'Flyer Monument'* which was inaugurated on Thu 30 August 1923 with—claimed—100,000 spectators amongst these some prominence of the *Old Reich*: the *'Red Baron's'* mother, Freifrau Kunigunde von Richthofen; Prince Heinrich von Prussia, the *'Kaiser's'* younger brother; Admiral Alfred von Tirpitz; General Erich Ludendorff, the *'Hero of Tannenberg 1914'*, and the *'Sea devil'* (and British blockade breaker) Felix Count von Luckner, whose later speciality was to tear thick Berlin phone books apart. In this spirit the inauguration festivities were combined with a glider demonstration air show, though the weather conditions with reported wind gusts of up to 130(!) km/h must have been rather unpleasant, so that some of the daring starters had to pay a tribute. Max Standfuss, a WW I pilot,

---

[13] See Wikipedia, 'Wasserkuppe'in English.

Fig. 5.4 Conditioning for scientific aeronautics, Dresden ~1927: E. Pohorille (l), FVD model glider group (m), Prof. E. Trefftz (r) © TUD UA

crashed and died in hospital thereafter, the third fatal accident on Wasserkuppe. In addition, three more pilots were injured on the same day, when their planes went down. Amongst these was Richard Tracinski, who lost the control of his single decker *'Galgenvogel'* (rogue/gallows bird) over Abtsroda, approx. 1 km north-west of the starting point, receiving head injuries and a concussion.

As a fixpoint in Heppner's otherwise shaky early CV data, there is his first patent DE415,115 about a *'Differential gearbox with bearing balls'*, priority 27 April 1924, together with Emil Pohorille,[14] Fig. 5.4 (l),[15] about his age and since April 1923 one of his fellow students at the TH Dresden (THD). It appears that both, before getting more and more involved in flying, practised along the second FVD focus—*car modelling*, and the patent actually solved the basic curve driving problem, as in reality, by a cheap differential gearbox.[16]

As a student, Fritz Heppner must have been very intelligent with an extraordinary quick perception capability, spontaneous, enthusiastic and outgoing. The most recent element of Heppner's life mosaic is a photograph, Fig. 5.4 (m),[17] which shows a group of nine Dresden model glider enthusiasts, presumably 1927 in a wide heather area *'On the Heller'*,[18] a former WW I military drill ground, in the vicinity of where today is the local Dresden-Klotzsche airport. The FVD group proudly lays out in front of them their self-built

---

[14] Besides Pohorille (and Heppner), the German-Jewish *'glider hall of fame'* comprises the names of Wolfgang B. Klemperer (1893–1965), assistant to Th. von Kármán,—see Wikipedia, 'Wolfgang Klemperer' and—see Wikipedia 'Robert Kronfeld', both in English.

[15] Courtesy TU Dresden, Universitätsarchiv No. 8566.

[16] According to one found source, the present RC model scene dates the introduction of model car differentials just to the mid-1970s, an obvious gap.

[17] Picture by courtesy of Lutz Mueller, Dresden on 2 April 2020.

[18] See Wikipedia, 'Heller (Dresden)' in German; the article mentions also the foregoing, 1926-1935 airfield 'Dresden-Heller', and a forced (Jewish) labour camp, the inhabitants of which were deported to Auschwitz in 1943.

glider models—from 0.4 to 1.5 m wing span. The youngest shown might be 16; the picture originates from a photo album, owned by the later aircraft carpenter Hans Schreier—a. Next to him stand with some likelihood, Emil Pohorille—b, and Fritz Heppner—d, then perhaps 23—is in the rear. In the front—c, there is the man with the inapt bow tie—which could be identified as Professor Erich Trefftz (1888–1937),[19] Fig. 5.4 (r), then head of the local *Akaflieg* academic model and sailplane glider club—and since 1922 full professor in the faculty for Mechanical Engineering at TH Dresden. Trefftz, with a British grandmother, spoke English fluently. Before coming to Dresden, he had stations at Goettingen (Hilbert, Prandtl) and Aachen, interrupted in 1909/1910 by a stay at Columbia University, NY on the initiative of his uncle, mathematician Carl Runge. At Dresden, Trefftz had responsibility for teaching and research on technical mechanics, strength of materials, theory of elasticity, hydrodynamics, aerodynamics and aircraft technology, all followed with mathematical rigour, but his special inclination always belonged to aerodynamics. Just a thought, in such a situation even without being officially qualified with the required bachelor exam, was attending Trefftz's THD lectures not a must for Heppner?[20]

Trefftz was considered an excellent and inspired teacher, joining his deep theoretical and practical understanding of mechanics with thorough empathy with his students and their further professional career. In a responding letter to a former student H. Boettcher[21] (then at DFS Darmstadt), dated 19 January 1934, who proudly had mentioned his successful flight instructor exam before, Trefftz insisted not to neglect finishing his studies formally. And, continued about *'Fritz'* (Heppner), who has to be brought back to *'orderly circumstances'* (Heppner then unmarried, his wife a second time pregnant,—and presumably unemployed), and that he, Trefftz, therefore had directed already several job requests

---

[19] See Wikipedia, 'Erich Trefftz' in German, and *'An appreciation of Erich Trefftz'* in English, https:// shellbuckling.com/cv/trefftz.pdf, where especially the following remark in view of Trefftz's personality is noteworthy with reference to Richard von Mises, co-founder together with L. Prandtl of the renowned *GAMM Ges. f. Angew. Mathematik und Mechanik* (Soc. f. Appl. Mathematics and Mechanics) in 1922 and first editor of their Zeitschrift/ journal ZAMM: *'Trefftz had a life–long friendly relationship with von Mises who, being a Jew, had to leave Germany in 1933. Trefftz felt and showed outgoing helpful solidarity and friendship to von Mises, and he clearly was in expressed distance of the Hitler regime until he died at a young age in 1937.'* His early death was the result of a renal infection, and the non-availability of antibiotics. Trefftz successor at THD became W. Tollmien, known for the *Tollmien-Schlichting waves* during laminar/ turbulent boundary layer transition,—see Wikipedia, 'Walter Tollmien' in English.

[20] Heppner and Schelp received their bachelor levels (*'Vordiplom'*) with the overall qualification 'good—passed' from TH Dresden, Heppner as *Technical Physicist* in June 1929, Schelp in *Mechanical Engineering* in June 1935; the latter with examinations '1b Very Good' in *Technical Mechanics* and '2a Good' in *Strength of Materials* by Professor E. Trefftz.

[21] Letter, courtesy to Lutz Müller, 18 April 2020, from TUD-Universitaetsarchiv <Trefftz Files>. Later, Boettcher apparently followed Trefftz's strong recommendations; the internet has *Harry* Boettcher, born 24 July 1909, as ‚Flugbauführer' (head of aircraft construction) at DVL Berlin-Adlershof.

to Blohm & Voss,[22] Hamburg, where the family of Heppner's wife lived. Erich Trefftz is not only important as mentor and *encourager* of his students towards a frank personality, along the path of scientific glider aerodynamics and design education, but in view of the international aero gas turbine *Jet Web* network, he is also an important and recurrent reference point and helping hand, irrespective and open-minded beyond any political or cultural boundaries,—here for the struggling Jew with presumably communist leanings Fritz Heppner, and surprisingly also in the coming Chap. 7 for the young Helmut Schelp, early half-orphan and nearly consequently, enthusiastic Nazi party member with 19, the later creator and maker of the German turbojet programme, again in a very decisive period of his personal development.

Heppner's fellow student Pohorille migrated in 1934 to England and further to Palestine, where he became—living in *Kibuz Afikim*—under the alias name *Emil Poran*, founder and senior chief engineer in 1948 of what—later—grew to *IAI* Israel's leading Aircraft Industry. The internet has traces of *Gestapo* records, in which in 1934 a *'Jewish girl'* Pohorille distributed communist propaganda material near Pirna, Saxonia, very similar to Fritz Heppner's wife Grete at Berlin at approximately the same time. Consequently, one can assume with some likelihood that Heppner and Pohorille were in 1924 not only engineering- and glider-, but also political-friends. It is very likely that Pohorille might have been also the driving force to bring Fritz Heppner in contact with *Wasserkuppe* and early aircraft aerodynamics in the mid-1920s. The glider training school on the Wasserkuppe passed between 150 and 200 students annually between 1925 and 1927; in this respect Heppner's glider license #101 assigns him to the *'early birds'*.[23] Pohorille became known amongst the early glider community on the Wasserkuppe since 1925 as one of Alexander Lippisch's designers/draftsmen, besides Hans Jacobs, Fig. 4.40. Their joint masterpiece was the perfect aerodynamic detail design of the new world record sailplane *'Wien'* in 1929, Fig. 5.5 (r), for which Alexander Lippisch had increased the wing span from so far 16 m to unheard-of 19.15 m. The elliptical cross-section of the sailplane body was tailor-made to the small figure of the Austrian glider champion Robert Kronfeld (1904–1948),[24] Fig. 5.5 (l), who was the first to accomplish the transition from *hill soaring* to *cloud flying,* and thus the unchallenged #1 in 1929/1930.

On 15 May 1929, Kronfeld set up a new glider record for distance flight beyond 100 km, which earned him also a newspaper prize of 5000 RM (Reichsmark), open for challenge since 1927. To achieve the exact 102.2 km on a straight point-to-point line, Kronfeld had to fly actually 145 km in 5+ hours, meandering in the *Teutoburger Wald* ridge lift in south-

---

[22] *HFB Hamburger Flugzeugbau GmbH* (Hamburg aircraft construction) was founded in June 1933 as a spin-off of the Blohm & Voss shipyard, and was apparently in Trefftz's mind a potential future employer for Fritz Heppner.

[23] See Lance Cole, Secret Wings, Ch.4 *'Wasserkuppe'*, p. 58 f. (online as Google book) has the further outlook *'over 50,000 young Germans (including women) became qualified glider pilots before 1936. Neither Britain nor America could boast such a cadre of pilots in reserve.'*

[24] See Wikipedia, 'Robert Kronfeld' in English.

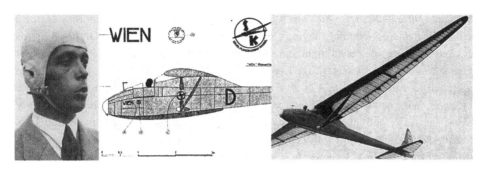

Fig. 5.5 R. Kronfeld (l), flying record sailplane *'Wien'*, 19 m wingspan, 1929 (r) front side view drwg.—cockpit with Gö 549 mod. wing profile, manufacturer *Kegel Flugzeugbau Kassel* (m)

easterly direction from hill top to hill top, roughly between Ibbenbueren and Detmold, with the speciality of a final 360 deg circle around the *'Hermannsdenkmal'* (monument for *'Hermann, the German'*) near Detmold.[25] Up to 30 July 1929, he broke his own world records twice, setting the distance record to 150 km and the altitude record to 2589 m. On 20 June 1931, Kronfeld was the first pilot to fly a glider across the English Channel,[26] making a return flight the same day; for this he won 1000 £ sponsored from the *Daily Mail*. In 1933, Kronfeld fled Germany first to Austria, later for the United Kingdom. In 1939 he became a British citizen and during WW II he served as RAF Squadron Leader. On 12 Februar 1948, as Chief Test Pilot for *General Aircraft Ltd*, he was killed in a crash of an experimental flying wing glider.

The astonishing row of Jewish sailplane pioneers is rounded up by Dr.-Ing. Wolfgang B. Klemperer (1893–1965), who was born at Dresden to Austrian parents. Following his service as reconnaissance pilot for the Austrian-Hungarian Army at the Alpine front, he returned to Dresden to complete his studies in mechanical engineering with honours in 1920. On 24 March 1920, still from Dresden, he sent out a call for the first *Rhoen Glider Contest*, before in May, he moved on to the Aerodynamic Institute[27] of TH Aachen as assistant of the newly appointed Professor Theodore von Kármán, whom he had met at Vienna during the war years. Together with von Kármán, he founded the FVA Flugwissenschaftliche Vereinigung Aachen 1920 e.V. (flight research association), designed and built sailplanes and won the first contest on the Wasserkuppe in the FVA-1

---

[25] See Wikipedia, *'Hermannsdenkmal'* in English. The monument was constructed between 1838 and 1875 to commemorate the *Cherusci* war chief *Arminius* (in German, *Hermann*) and his victory over *Rome* at the Battle of the Teutoburg Forest in 9 AD.

[26] After a towplane had brought him up to altitude over his starting marks. For an impression of Kronfeld's cross-channel glider,—see https://www.youtube.com/watch?v=_1YFhgfHLR8

[27] There W. Klemperer also met E. Trefftz, then Prof. for Mathematics at TH Aachen since 1919, who had already published together with Th. von Kármán two seminal papers—on aircraft vibrations and stability (1914/5) and aerofoil theory (1918).

'Schwatze Düvel' (Black Devil).[28] In 1921, soaring an improved version, the *'Blue Mouse'*, he flew the first cross-country, staying aloft for 13 minutes, thus exceeding Orville Wright's previous 9 minute 45 second flight and earning Soaring Certificate #1. In 1922, Klemperer became test department head of the Zeppelin Works Friedrichshafen, Fig. 4.17, before he emigrated to America in 1924 as research manager, together with a group of Zeppelin engineers, to the newly founded Goodyear-Zeppelin Corp. at Akron, OH. There in the early 1930s, the rigid, helium-filled airships USS 'Akron' (ZRS-4, 239 m length) and USS 'Macon' (ZRS-5) were built with Zeppelin technology, both used for reconnaissance and as *'flying aircraft carriers'*.[29] In 1936, Klemperer moved to Southern California as head of special research for Douglas Aircraft Company,[30] where he eventually became Chief of Missiles Research, up to his retirement in 1958.

The next move in Heppner's vita is again to be deduced from a patent, now his second patent[31] CH172,804, priority 19 Dec. 1932, together with O. Kantorowicz,[32] both at Berlin-Charlottenburg, for a *'Change gear with continuously variable transmission*

---

[28] The name resulted from the waterproof impregnated, black 'voile' textile, with which the wing sections were covered.

[29] See Wikipedia, 'USS *Akron*' in English.

[30] At his time at Douglas, W. Klemperer should have met also Vl. Pavlecka, Sect. 6.4.2 and Fig. 6.52, and therefore, should have come in contact with first emerging ideas for US turbojet propulsion.

[31] It appears that Fritz Heppner for a period of 1932–1940, Otto Kantorowicz for an unknown period and Hellmut Weinrich mostly all his adult life (1930–1989) lived as a kind of independent patenting privateer/ entrepreneur (with in Weinrich's case, his wife administrating and invoicing), generating income by propagating their patented intellectual property via license agreements in typically, in all three cases identical fields of gas turbines/ turbomachinery and automotive hydraulic transmissions. One can imagine that they visited regularly the reading room in the Berlin Reichspatentamt (Reich Patent Office), Gitschiner Str./corner Lindenstr., a habit which Heppner might have continued at the London Patent Oficce up to 1939.

[32] Dr. Otto Kantorowicz (1906 Berlin–1973, Ulverston, GB) wrote his Ph.D. thesis—see Kantorowicz, Zur Leitfähigkeit (On electric conductivity of metal powders) at Berlin University in 1932 under the 1920 Nobel Prize winner W. Nernst, who had amongst his students also Frederick Lindemann, technical consultant to Winston Churchill in WW II,—see Wikipedia, 'Frederick Lindemann, first Viscount Cherwell' and Sir Henry Tizard, Rector of Imperial College London 1929-1942 and since 1932 head of the influential ARC Aeronautical Research Committee,—see Wikipedia, 'Henry Tizard' in English;—see also Chap. 2, 'April 1930'. Otto Kantorowicz was the eldest child of Hermann U. Kantorowicz (1877–1940), who in an expertise on the *'Question of war responsibility 1914'* for a Reichstag parliamentary investigation committee in 1923 brought up strong evidence against the German Reich,—with predictable consequences after the Nazi rise to power in 1933,—see Wikipedia, 'Hermann Kantorowicz' in English. O.Kantorowicz came already in 1933 to the UK, and stayed there in life-long contact with Ilse Heppner. Up to 1936 he worked—possibly ministered by Tizard at the Imperial College London, thereafter at the GEC Research Labs. Other than Heppner, he had the dubious privilege of being listed on the 1940 SS *'Special search list GB'*,— see Wikipedia 'The Black Book', which and their prominent entries will be further addressed in Sect. 6.1.

Fig. 5.6 Fritz Heppner, second from left, 1929/30 in the vicinity of C. Lorenz Radio Works, Berlin-Tempelhof; sign of steel wholesaler *'DELLSCHAU'* in the rear (white dashed box) used for localisation

ratio'. In addition to this patent his stay in Berlin is confirmed by a photograph of 1929/30 together with white-coated physicists (apparently during a lunch break) in the vicinity of C. Lorenz Radio Works,[33] Berlin-Tempelhof, where he might have gained some experience as technical physicist in a longer internship, Fig. 5.6.

Presumably between 1930 and 1932 Fritz Heppner studied Mechanical Engineering at TH Berlin-Charlottenburg and graduated there with the degree of *'Diplom-Ingenieur'*, as documented in Heppner's emigration passport, and on the corresponding Austrian patent version AT140,809, while all student records of that period at TH Berlin-Charlottenburg have been destroyed during the war.

On 23 July 1934 Fritz Heppner married Margarete Siems, actress at Berlin, whom he had met there in the left scene in 1931. Before the antisemitic and racist *Nuremberg Laws* were issued in September 1935, which forbade marriages etc. between Jews and Germans, Fritz Heppner gave way to the increasing Nazi persecution pressure and left Germany. He arrived on 10 January 1935 at Harwich, presumably by train from Berlin to Hoek van Holland, later the route of the *'Kindertransporte'*. Letters to his wife indicate first prospects of a job in August 1935, before his wife Grete and the children Ruth(1) and Christopher (2 ½) followed from their intermediate stay at the home of Heppner's parents[34] in Dresden

---

[33] C. Lorenz with then 20,000 employees is mostly known for their *Enigma* coding machine add-on's—the Lorenz *SZ40 etc. rotor stream cipher machines,* used by the German Army since 1940,—see Wikipedia, 'Lorenz cipher'in English.

[34] Heppner's parents managed to flee Germany as late as March 1939, where the family together with sister Ilse was re-united for a short while, already in an own house at Harrow, north-western London.

to London, Golders Green,[35] where Heppner's 1937 portrait, Fig. 5.1 (l), was taken in the garden of a Mrs. Kensall's private house; the family occupied two rooms there, one was Fritz's study.

Not much is known, how Fritz Heppner made a living between his arrival in England in January 1935 and the internment of the family on Isle of Man, beginning 30 May 1940. There is a short hint on Hodgson's website[36] that Fritz Heppner was *'initially retained by Roditi Agencies'*, which with all likelihood stands for *'The Roditi International Corp. Ltd, London'*. The Company was incorporated—in agreement with Heppner's aforementioned, first job announcement to his wife—on 29 August 1935, and rapidly developed into a successful trading company, exporting synthetic textiles, Chinese silk fabrics and commissary goods to British embassies and other international organisations throughout the world. They operated also in the Interpreting and Technical interpreting industries.[37] During his early years in England up to the internment, Heppner filed four patents (two on hydraulic torque transmissions, and two for an industrial gas turbine, Fig. 8.30), which also allow to trace the family's relocations from Golders Green (1935–1936), to Enfield (1937–1938), where the family occupied its first three-room flat in England, and finally the acquisition of a house for the complete family in the London Burrough of Harrow in 1939, also for Fritz Heppner's parents, who had managed to leave Dresden as late as March 1939.

There are indications that Fritz Heppner might have tried to continue in London his—at Berlin enforced 'undercover'—life as independent patent privateer, as a regular visitor of the Patent Office's reading room, then at 25, Southampton Buildings (nearly 1 km east of British Museum), which was also known for its very extensive collection of technical and scientific publications; it might have been here in the 1937–1939 period where he laid ground to his extraordinary broad knowledge of most recent scientific research results, in England and Germany, and the corresponding patenting status.[38]

With others, Heppner tried to solve the problem of devising a hydrokinetic power transmitter capable of performing effectively the functions both of a clutch and of a change speed gear. His patents on *'hydraulic torque transmissions'* may have been the actual technical reason for bringing Heppner in contact with the car manufacturer Armstrong

---

[35] In Golders Green the Heppners lived near and made friends with the Jewish Moholy-Nagy family, also coming from Berlin to London in 1935. See Wikipedia, 'László Moholy-Nagy' (1895–1946), cousin of Sir Georg Solti, was a Hungarian painter and photographer as well as professor in the Bauhaus school, Weimar and Dessau, 1923–1928. Walter Gropius and Moholy-Nagy planned to establish an English version of the Bauhaus, but could not secure backing. In 1937, Moholy-Nagy moved to Chicago to become the director of the New Bauhaus.

[36] See http://www.designchambers.com/wolfhound/

[37] Heppner's sister Ilse, who worked for GEC patenting at London-Wembley, spoke several languages including Spanish, so possibly she had introduced her brother at Roditi International. Information, courtesy of Chris Heppner, 26 Oct. 2021.

[38] See his personal *'turbojet development gap'* in Fig. 8.29.

Siddeley Motors at Parkside, Coventry.[39] As a kind of industrial standard of the time, the product areas automotive cars and engines, aero piston engines and the just emerging turbojet engines were led by the same upper management under the owner and chairman Thomas O.M. Sopwith, Fig. 5.14 (l), and his long-time deputy Frank S. Spriggs. In 1938 ASM struggled with front suspension problems on a '20 hp car' under development, and the suspension specialist Henry S. Rowell (1885–1952) was hired as employed consultant, becoming ASM Director and General Manager in 1939, and since mid-1940 direct superior to Fritz Heppner at ASM.[40]

Following the declaration of war on 3 September 1939, some 70,000 German and Austrian residents in the UK became classed as 'enemy aliens'. In due course Fritz Heppner with wife and children were interned on the Isle of Man from early June 1940 to mid-March 1941. After arriving together with 1200 other Germans at the internment Port Rushen, Castletown, the families were separated: after a 19 km drive to the north-east Fritz with other men to a fenced compound at Onchan,[41] Fig. 5.7 (l), while women and children were brought to a neighboured camp without fences at Port Erin, 6 km in westerly direction; a train could be used to unite the families at Onchan every four weeks. The inhabitants of Onchan learned that the internees included 121 artists[42] and literary workers, 113 scientists and teachers, 68 lawyers, 67 graduate engineers—including potentially Gustav Lachmann, 38 physicists—among them Fritz Heppner, 22 graduate chemical engineers, 19 clergymen, and 12 dentists. At the other end of the social scale were 103 agricultural workers.

On 6 December 1941, Fritz Heppner and his wife were invited to be interviewed at ASM Armstrong Siddeley Motors, Coventry. The ground was apparently prepared, they passed successfully, resulting in his immediate and firm employment. The outside success became

---

[39] The 1929 Armstrong Siddeley 5 l 'Thirty' was one of the first cars to use a Wilson—see Wikipedia, 'Preselector gearbox' in English. After the war preselector and hydrokinetic transmission (Heppner's patents) were combined to 'automatic transmissions'. The 1942–1945 German Tiger I armoured tank used a form of pre-selective gearbox, built by Maybach; the gearbox experts for analysing a captured Tiger I in 1943 came from ASM.

[40] Data of 'Henry Snowden Rowell' acc. to 'Grace's Guide to British Industrial History'; though actually a chemist, he had first entered the car and lorry industry in the 1920's, and made a name for himself, including his work on various industry wide committees. In 1910–1911, after post-graduate studies at NPL London, Rowell had studied at the Universities of Goettingen and Berlin.

[41] Another rather prominent German-Austrian co-internee with aeronautical background was—see Wikipedia, 'Gustav Lachmann' in English, co-inventor of the wing leading edge slot and finally, head of research at aircraft manufacturer Handley Page. With respect to the described surveillance by MI5, mail interception and espionage screening, there may not have been much difference between the Lachmann and Heppner cases. And, internment or prison (Lachmann) was no hinderance for exploitation, or more friendly 'collaboration' with official institutions like MI5, which in the case of Heppner might have started as early as 1937/8.

[42] After Heppner's term at Onchan Camp, June 1940–18 March 1941, the German artist Kurt Schwitters (1887–1948), known for his installations and collages,—see Wikipedia, 'Kurt Schwitters' in English, was interned in the 2 km distant, 'Artists Camp' at Douglas, 17 July–21 Nov. 1941,—see Wikipedia, 'Hutchinson Internment Camp' in English.

Fig. 5.7 Fritz Heppner's social rise: Onchan (Isle of Man), male internment camp, 1940 (l), family home at Leamington Spa after July 1942 (r)

visible already several months later, when the family bought a house at 5 Woodcote Rd., Leamington Spa, Fig. 5.7 (r), some 13 km south of ASM headquarters at Parkside, Coventry.[43]

At a time when Heppner still struggled to organise his own professional life and that of his family in the new environment, a man of his age and a brilliant automotive designer Stewart S. Tresilian (1904–1962),[44] who in May 1939 was installed as chief engineer at Armstrong Siddeley Motors at Coventry and—as turned out later—as Heppner's predecessor, was about to coin the hallmark of his career. Since the 1920s he worked at Rolls-Royce on aero-engines, in the early 1930s as the chief assistant to Arthur Rowledge (1876–1957) at Rolls-Royce on the *'R engine'*, especially constructed for air racing purposes, and thereafter, on the *RR Merlin*. End of April 1935 Walter O. Bentley (1888–1971), founder of Bentley Motors Ltd. and famous car and aircraft engine designer, finished his intermediate stay at Rolls-Royce and moved to the car manufacturer *Lagonda Ltd*[45] at Staines-upon-Thames (Surrey) with the majority of Rolls-Royce's racing department staff. Bentley made Tresilian chief designer for the new *'Lagonda V12'* project, and the engine of 1937 displaced 4480 cc, delivered 180 hp and was said to be capable of going from 10 to 170 km/h in top gear and revving to 5000 rpm. The chassis was also new and features independent torsion bar front suspension and live rear axle with hypoid final drive; the braking system is *Lockheed* hydraulic. The engine is connected to a four-speed gearbox

---

[43] Picture, courtesy to Chris Heppner, 23 July 2021.The site of the house appears strategically picked—only 100 m to the next Coventry bus stop on Kenilworth Rd.—so that Heppner might have reached his ASM office within approx. 1 h.

[44] See Wikipedia, 'Stewart Tresilian' in English.

[45] *Lagonda* was founded in 1901 by the US-American Wilbur Adams Gunn (1859–1920) from Springfield, Ohio—on the banks of the Lagonda Creek, presumably named after the *Shawnee* Indian term meaning buck's horn.

Fig. 5.8 Stewart S. Tresilian's masterpieces: Lagonda V12 Drophead Coupé 1938 (top), Mod. Lagonda V12 *'Le Mans, #3/ #4, 1939'* (bottom)

with centrally mounted change lever. Only 189 cars were produced between 1938 and 1940; coachwork could be by *Lagonda* or a number of independent coachbuilders and to suit various body designs, a wheelbase of 3.150 m to 3.505 m could be specified, as *'tourer, saloon, coupé or limousine'*; even with a *saloon* body the car could reach 160 km/h. Two modified *V12s* with four carburettor engines boosted up to 206 hp at 5500 rpm were entered for the 1939 *24 Hours of Le Mans* where they finished third[46] and fourth on 7 June 1939, Fig. 5.8.

Viewed from a historic, luxury car perspective, the success of the aero gas turbine and the Second World War meant also the end of some fascinating car and corresponding engine developments. And, there may be more than one nostalgic old-timer enthusiast who blamed—if not Adolf Hitler, so at least Fritz Heppner for the premature career end of the exceptionally gifted car designer Stewart Tresilian.

Fritz Heppner's further ASM activities will be discussed at length in the context of *'British Turbojet Engine Developments'* in Sect. 8.2.4. Here, it is worthwhile to step back to the mid-1920s and have a look on young Hellmut Weinrich, Heppner's potential partner or contact in the realisation of the contra-rotation turbojet idea in Germany. Weinrich was five years younger than Heppner, which potentially might have given the latter a natural authority in this relationship, underlined in addition by Heppner's rather theoretical, higher

---

[46] The Lagonda works team on #3, Dobson/ Brackenbury, drove in 24 h, 3227.7 km or 239 laps, on an average speed of 133.9 km/h.

university training in comparison to Weinrich's more practical college education in combination with a full journeyman fitter training.

Hellmut Adolf Weinrich (1909–1988) was born at Greiz, Thuringia, as son of Adolf Weinrich, a personal coachman to the local Prince Reuss. From 1919 to 1925 he attended the local *'Oberrealschule'*, from where he continued his studies from October 1927 to March 1931 in electrical engineering and fluid machinery at the Technical College Chemnitz,[47] where he received also his *Abitur/A* levels qualification at the same time. From 1931 to 1945 Weinrich maintained a backyard workshop at Palmstr. 31, Chemnitz— an enlarged fitter's shop with attached wind tunnel(!)—which allowed him to build model and required test rigs as well as carry out first principle tests on his own, and as part of acquired R&D contracts from the German Navy and Luftwaffe. The official technical history in Germany assumes that Weinrich—at first relatively unsuccessful—tried to sell *'his concept'* of a compact turboprop engine already before 1936,[48] supported by a small-scale CR contra-rotation compressor demonstration model,[49] to RLM and later to *OKM Oberkommando der Kriegsmarine* (Naval High Command) before Hans Mauch, RLM[50] established Weinrich's working contact to Bramo Brandenburgische Motorenwerke (Brandenburg Motor Works), Berlin-Spandau.[51]

It was only in spring 1939 that Mauch first heard about a Hellmut Weinrich of Chemnitz, developing a gas turbine for the Navy. On investigation it appeared that Weinrich had already in 1936 sent to the RLM plans for a turboprop with a counter-rotating compressor and turbine, but no attention had been put on the proposal and it had been completely forgotten in the Air Ministry. After receiving no answer, Weinrich had contacted the newly established OKM at Berlin-Tiergarten. In contact with the

---

[47] Helmut Schelp (1912–1994), after 1938 responsible at RLM for the German turbojet development programme, studied there mechanical engineering between Oct. 1929 and March 1933, and could have met H. Weinrich (and F. Heppner?) already then as one of his fellow students.

[48] See Kay, German Jet Engine, p. 212. This turboprop engine concept might have had some similarity with Heppner's later turboprop patent, Fig. 5.13 (bottom, l).

[49] Presumably identical with hardware for a 100 hp marine gas turbine with which the Weinrich developments for the German Navy had started; at least these early beginnings, based on preliminary tests at Chemnitz, could have taken place still before Heppner's migration to England.

[50] For Hans Mauch (1906-1984) see Sect. 6.2.3 and Fig. 6.13.

[51] Acc. to—see Kay, Turbojet, Vol.1, p. 198, in April 1936 Weinrich had submitted to RLM plans for a turboprop engine with counter-rotating compressor and turbine, which he had followed up with own experiments, but these proposals were officially ignored. Another significant indicator of Weinrich's early occupation with CR design is his patent DE663,935 with priority 18 May 1931(!) for a method of introducing fuel at the front face of co-rotating shafts. In due course, layout and design of the BMW 109-002 was the responsibility of the engineering consultant H. Weinrich, Chemnitz.

open-minded, then Marinebaurat Gustav Menz (1898–1986)[52] was agreed to redesign the engine as an auxiliary power plant for a '*Schnellboot*' (PT Patrol Torpedo boat), and it was this which was currently developing with Navy support. It was intended to produce about 100 hp, and an experimental engine was actually developing about 50 hp on test, when Weinrich's activities were re-directed to aero propulsion.

Bramo and consequently the Weinrich concept was integrated in 1939 to BMW Flugmotorenwerke Brandenburg GmbH under the lead of chief engineer Hermann Oestrich.[53] The united, single BMW turbojet programme at Berlin comprised then two projects, Weinrich's counter-rotation, promising high-performing concept in the main focus—and certainly supported by the RLM-'*mastermind*' H. Schelp—under the internal project designation BMW P.3304 (officially BMW 109-002), Fig. 5.11 (top), and then more as a simplified, preliminary test vehicle the single rotor BMW P.3302 (BMW 109-003); both shown for comparison in Fig. 8.8.

The P.3304 had four axial compressor stages on a rotating drum which was connected with another one, mounting three axial turbine stages. This compressor/turbine arrangement was surrounded by a shell rotor that carried five compressor and four turbine stages, the whole rotating in the opposite direction to the inner drum rotor. The rotating annular combustor arrangement was carefully labyrinth-sealed at the inner diameter to prevent hot gas backflow to the compressor rotor inner space. Cooling air was drawn in through a hollow fairing at intake centre that also held the oil cooling radiator. The cooling air passed through the inner compressor drum, cooled the turbine root sections and left through the fixed exhaust cone. Fuel supply and auxiliaries were in the space upfront of the compressor.

Interestingly, the BMW 109-002 was coupled since July 1941 to a similarly advanced aircraft concept, Alexander *Lippisch's Li P.01-115* single-seat fighter, Fig. 5.11 (bottom, r), as part of a number of tailless concepts similar to the Me 163 rocket aircraft.[54] The fuselage held armament, fuel and powerplants on the basis of a unique dual-power concept: The compact BMW P.3304/109-002 turbojet was located in the upper rear, fed by a dorsal air intake on top of the fuselage. In addition, a 1500 kp HWK 109-509 (A-0)[55] liquid fuel,

---

[52] Menz could be identified with the kind assistance of Prof. H.-J. Lichtfuss in Sept. 2020. Menz maintained marine gas turbine developments at the company Brueckner-Kanis, Dresden, where both Rudolf Friedrich (from Heinkel) and Hellmut Weinrich (from BMW) met after 1942. As OKM department head, Menz achieved responsibility for all German Navy engine projects. Dr.-Ing. G. Menz regained this position—now promoted to Ministerialdirigent (Director-General)—in Sept. 1954 for the Bundesmarine (Federal Navy) at Bonn. Before his retirement in 1963, he fulfilled apparently an engineer's dream by launching a 16 MW pump-jet propulsion unit for a 37 kn, 700 t gunboat on the basis of a Rolls-Royce *Olympus TM1a* gas turbine, described in some detail,—see Eckardt, Gas Turbine Powerhouse, p. 339 f.

[53] See also Chap. 9.

[54] See Masters, German Jet Genesis, p. 111.

[55] For '*Hellmuth Walter Kommanditgesellschaft*', a Ltd located at Kiel,—see Wikipedia, 'Walter HWK 109-509' in English.

bipropellant rocket engine was mounted beneath the turbojet and provided take-off and perhaps emergency power. Overall aircraft dimensions were 6.75 m length at 9 m wing span. The wings were swept back at 27 deg. and only a single vertical tail and rudder was to be fitted, landing was accomplished on a retractable skid. As with other designs in the Lippisch P.01 series, this design went no further than the drawing board stage; all further work was later devoted to the Me 163 and its variants.

Both engine projects—BMW 109-002 and -003—were designed for a take-off thrust of 600 kp. In addition, it was known that Weinrich had built a small-scale 100 hp marine-GT for advance concept testing, which gave some guidance for the preparation of the planned BMW 002 investigations. By 1940 about twenty designers and engineers worked on the project at Spandau. First test pieces for the contra-rotating combustor had been manufactured, but the corresponding, difficult combustor tests were either not carried out or not successful. At the same time spread news about a severe failure of the marine device, so that the BMW 002 development was finally cancelled as too complicated as early as May 1941, and Oestrich focussed the further activities to the former backup and support solution, the BMW 003.[56]

Hellmut Weinrich, however, moved on as designer consultant to the next project with contra-rotation turbomachinery at Daimler Benz Stuttgart, the first German turbofan *'class II'* project DB 007 with 1000–1400 kp TO thrust[57]; in addition, he developed at Brueckner-Kanis Dresden a 10,000 hp *'marine booster'* (aero jet expander, 1.8 m diam.) with a typical contra-rotation, high-pressure compressor/turbine section—of which Antony Kay wrote '... *Used as a turbojet, the original purpose, the Brueckner-Kanis contra-rotating gas turbine would in all probability have been superior to all other German turbojets of similar size, assuming, of course, successful development. Given this probability, it is hard to understand why the RLM failed to back it for aircraft use, especially since it was to be built (in modified form) for marine use ... Certainly, Hellmut Weinrich, who was behind this particular work, was a very forward-thinking engineer.'*[58] H. Weinrich was arrested in May 1945 and sentenced for unknown reasons to five years in the Soviet internment camp

---

[56] See Hagen, Zur Geschichte. Hermann Hagen was Oestrich's deputy in the BMW Flugmotoren development team. His account is reliable by and large; with respect to the timing of the 109-002 programme end, there is a slight discrepancy between his statement <1942>, supported also by—see Kay, German Jet Engine, p. 97, and a number of available BMW documents which indicate rather the earlier <May 1941> date. The fabricated 109-002 prototype and other hardware was sent from BMW Spandau to Weinrich's Chemnitz address; in Aug. 1943, he accepted the cancellation of his BMW employment contract.

[57] The original CR compressor design for the DB 007 was put at AVA Goettingen, W. Encke—and after significant delays alternatively to B. Eckert, FKFS design and H. Marcinowski, J.M. Voith, Heidenheim for test rig and hardware delivery.

[58] For further details and the quotation,—see Kay, German Jet Engine, p.191.

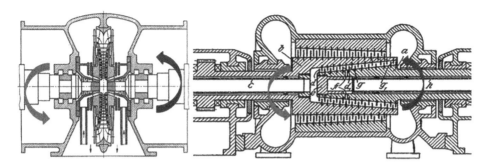

Fig. 5.9 Counter-rotating GT axial turbomachinery predecessors: radial Ljungstroem turbine ~1900 (l), axial Roeder turbine 1926 (r)

Buchenwald (the former *KZ 'concentration camp'* Buchenwald) near Weimar.[59] After his release in February 1950 he lived with his family for a short time again in Chemnitz, GDR, before they resettled in West Germany, where he continued working and patenting as consultant, predominantly for the Voith Getriebe KG, Heidenheim in the area of hydraulic gearboxes and for GHH Gutehoffnungshuette, Oberhausen-Sterkrade in the area of axial turbomachinery.

At this point it appears worthwhile to spend a cursory look backwards on the technical origins of contra-rotating turbomachinery solutions, Fig. 5.9. The earliest shown example is a *Ljungstroem turbine*, a compact, radial inward-to-outward flowing steam turbine, invented and developed by Swedish brothers Birger and Frederick Ljungstroem[60] at the beginning of the twentieth century, Fig. 5.9 (l). In this type of turbine, all of the blading revolves (rather than only half the blades as in most turbines)—one half revolves in one direction and the other half in the other, so that each blade of one turbine half represents equally a vane of the other half. The relative speed of the two sets of blading is doubled and achieves a closer approximation to the correct relation of the speed of steam and blading than is possible in the conventional type. A necessary result of this arrangement is that a Ljungstroem turbine has two output shafts, one for each set of blades which in the case of turbine generators incurs the cost of two generators. During WW II, FKFS used the concept as basis for a unique centrifugal compressor configuration since 1943.[61]

---

[59] From Weinrich's stay in the Soviet internment at Buchenwald survived the story, how he reconstructed/built his most important engineering calculation tool—the slide rule, out of cardboard stripes, on which he marked the scale—from common logarithms which he knew sufficiently precise by heart.

[60] https://www.gracesguide.co.uk/Frederick_Ljungstrom

[61] See Kay, German Jet Engine, p. 212.

As a further step in the interesting direction of axial-flow arrangements with contra-rotating shafts Professor Karl Roeder (1881–1965), TH Hannover[62] invented, as an axial equivalent to the radial Ljungstroem turbine, patents US1,759,817 and CH131,659 with 1926 priorities, *'Steam and gas turbine with contra-rotating blade rows'*, of which Fig. 5.9 (r) shows a typical example. During 1919–1945, i.e. also during the study period of Hellmut Weinrich and Helmut Schelp, Professor Constantin Zietemann (1883–1978) was responsible at the Technical College Chemnitz for the disciplines *'heat engines including steam turbines, thermodynamics and mechanical design'*. After WW II, when Zietemann still lived at Chemnitz, GDR, he published his own 1932 standard work *'Steam turbines, theory and design for studies and practice'*, now together with Karl Roeder at Springer Publ. Berlin, so one can assume, that Roeder's concept of the contra-rotation axial steam and gas turbine will have been presented with a certain emphasis in Zietemann's lectures at Chemnitz.[63]

Consequently, it is known that Weinrich started his metal works business at Chemnitz in 1931 with first own inventions in the direction of contra-rotating steam turbines, eventually for the renowned Turbine Factory Brueckner, Kanis & Co. at Dresden. If not already during Heppner's bachelor studies with excursions to the Saxonian automotive industry near Chemnitz in the 1920s, the potentially assumed, closer collaboration period between Heppner and Weinrich could have been also between 1932, the end of Heppner's Dipl.-Ing. studies at TH Berlin-Charlottenburg, and January 1935, when Heppner went to England.

To understand the importance of the contra-rotation turbomachine principle—both in view of space requirement and power density,—a simplified analytical experiment[64] may be helpful.

In Fig. 5.10 the velocity triangles for a Rotor—Stator combination are compared with a corresponding contra-rotating Rotor 1—Rotor 2 configuration. The assessment on the basis of incompressibility, with *P shaft power* and *ṁ mass flow*, applies for the illustrated cases:

---

[62] Roeder was THH-Professor for *'Steam Engines and Boilers'* since 1926; to his students belonged with some likelihood also Walter Encke, who started at AVA Goettingen in 1927, carried out first fundamental CR compressor tests in 1935/6, as documented in—see Betz, Axiallader, before he received in 1942/3—in the end unsuccessful, the design responsibility for the CR compressor elements of the Daimler Benz turbofan project DB 109-007. Strangely enough, though he discussed after 1945 nothing in the context of his realised Jumo 004 and BMW 003 axial compressors (and their significant *degree of reaction* design flaw), he discussed at length the mechanical feasibility of the CR compressor design concept in view of the DB 007 in his contribution to the 1947 Goettingen Monographs; however, the fact that he knows nothing about the DB 007 CR starting behaviour may indicate that he was at the time of the engine tests already *'out'*.

[63] In fact, also Heppner in his 1943 *ASH engine* patent, Fig. 8.31, makes reference to Roeder's CR steam turbines, see Sect. 8.2.4, which also indicates early Weinrich-Heppner connections.

[64] Courtesy of Prof. Hanns-Juergen Lichtfuss, 22 March 2018.

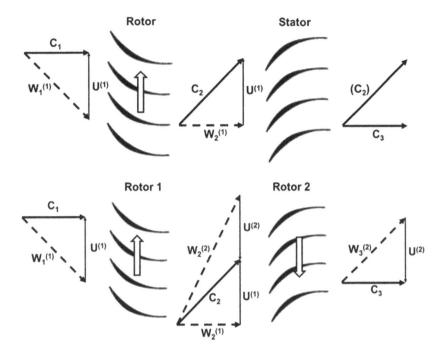

Fig. 5.10 Principle sketch of axial compressor velocity triangles: Single stage (top) vs. contra-rotating stage (bottom)

ROTOR—STATOR

1. C_1, $W_2^{(1)}$ and C_3 are axial - and equal $U^{(1)}$;
2. From Euler's *Turbine Equation*[65] $P/\dot{m} = U_2\, C_{u2} - U_1\, C_{u1}$ follows simplified
   $$\mathbf{P/\dot{m}} = U\,(C_{u2} - C_{u1}) = \mathbf{U}^{\,2} = (P/\dot{m})_{\,1}$$

ROTOR 1—ROTOR 2

3. C_1, $W_2^{(1)}$ and C_3 are axial - and equal $U^{(1)}$; moreover: $U^{(1)} = U^{(2)}$;
4. $W_1^{(1)} = C_2 = W_3^{(2)} = \text{sqrt}\,(2)\ U^{(1)}$ and $W_2^{(2)} = \text{sqrt}\,(5)\ U^{(1)}$;
5. $(P/\dot{m})_{\,2} = U^{(2)}\,(C_{u3} - C_{u2}) = -\,U^{(2)}\,C_{u2} = +\,(U^{(2)})^{\,2} = U^2$
6. $\mathbf{(P/\dot{m})_{GES}} = (P/\dot{m})_{\,1} + (P/\dot{m})_{\,2} = \mathbf{2\ U^2}$

This simplified survey illustrates the principle superiority of contra-rotation vs. a conventional rotor-stator arrangement to enable twice as high a power transfer, and thus roughly to duplicate the achievable total pressure ratio PR. However, even these simplified considerations indicate by the significant rise of the relative velocity level $W_2^{(2)}$ at rotor 2 entry (line 4. above) the extraordinary aerodynamic challenge of this design approach. Nevertheless, all this gives a feeling for the stunning effect of Heppner's presentation to an enthusiastic

---

[65] See Wikipedia, 'Euler's pump and turbine equation' in English.

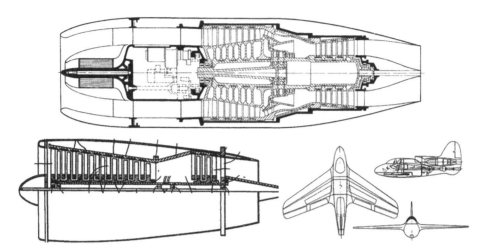

Fig. 5.11 Counter-rotating axial turbojets: H. Weinrich, BMW 109-002, CR design, 1939–1943 (top), F. Heppner CR patent US2,360,130, 1941 (bottom, l), *Li P.01-115* fighter project with BMW 002, 1941 (bottom, r)

Armstrong-Siddeley board of his contra-rotation set-up with a target PR 8(!), Fig. 5.11, in early 1941,—while best British test practice at that time reached rather PR 3.5 only.[66]

When Heppner's CR turbojet patent was presented and discussed in the 3rd edition, April 1944, of G.G. Smith's *'Gas Turbines and Jet Propulsion for Aircraft'*, pp. 29/30 a few special features—in the following in *Italics*—appeared remarkable to be criticised, some understandable, others coloured by special circumstances of the time (technology or know-how limitations):

- *Comparatively modest individual speeds of rotation may be employed as, due to contra-rotation, the combined operating linear speed at the blading will be very high. It may well exceed the velocity of sound in air, so blading having only a small coefficient of deflection can be employed,*
- ignoring bearing friction, *no torque is put into or taken out of the system, and the gas/air propulsive stream will leave without torque,*
- the central exhaust cone *is secured to the rear end of the hollow rotor and projects beyond the end of the nozzle. Why this member should rotate and why the rear bearing, in a high temperature zone, should be arranged to run at the extremely high combined speed is not clear. To mount both these components on the stationary rear bearer would appear to be a more desirable constructional arrangement,*
- *the delivery of fuel to the rotating combustion chamber presents some difficulties.*

---

[66] An outlook to Sect. 8.2, where the British turbojet developments, especially with Heppner's involvement at Armstrong Siddeley Motors in and after 1940, will be dealt with in detail. For the described ASM board reaction (and a correspondingly rather hesitant/ negative mood at Ministry level)—see Giffard, Making Jet Engines, p. 107.

Weinrich has—comparable with Heppner—a still more impressive patenting record of approximately 50+ grants, of which about 90 percent address automotive subjects (hydrodynamic transmissions, torque converters, etc.),[67] while the remainder deals with turbomachinery and gas turbine issues. The general, marked split of the patent portfolio in the two areas of hydraulic fluid machinery and turbomachinery/ gas turbines—for Weinrich with some preponderance for the automotive side, while Heppner's first four patents up to 1940 addressed transmissions only, and the rest was GT related—is typical for the German technical university environment of the 1920/1930s, where both areas flourished at countless Institutes for fluid machinery. The salient example is Professor Heinrich Foettinger (1877–1945) of TH Berlin-Charlottenburg, where he became head of the Institute for fluid mechanics and turbomachinery since 1924. The 'Föttinger transformer' came into prominence at a time when the steam turbine had just achieved its first marine success, and shortly before the war it was looked upon as one of the most important developments in transmission and reduction machinery. Foettinger's prime invention of 1905 had a hydraulicly coupled pump wheel—attached to the drive shaft—and a turbine wheel, connected with the output shaft, later with the addition of a fixed diffuser as guide device. In 1929 three former Foettinger assistants invented a considerable extension, the 'Trilok system', in which in addition to the pump and turbine wheel, the diffuser is also designed as a movable wheel and that all three wheels can be connected to other components either temporarily or permanently via couplings. Heppner's two patents in Great Britain on 'Hydrokinetic torque transmissions'[68] show a familiarity with Foettinger's ideas, with whom Heppner must have been in contact during his THB studies or thereafter.

Amongst the presently known Weinrich patents is no intellectual property protection of the counter-rotation turbomachinery concept whatsoever. As a last explanation one could ventilate the idea of a—so far undetected—secret patent, though with low probability, since then should exist comparable examples from the same or other inventors in this field.[69] It appears that the possibility of a 'deal' between Heppner and Weinrich cannot be excluded: exclusive benefit for the German wartime developments in this context in favour of

---

[67] Foettinger's former assistant (V.?) Speiser moved to DVL, Berlin-Adlershof and joined K. Leist during his short stay at Daimler Benz, Stuttgart and the early DB 007 development period, which might represent another line of influence between Heppner's and Leist's CR compressor designs.

[68] US2,216,411 (1936) and US2,349,725 (1940), of which especially Heppner's 1936 invention with reference to its 'overdrive capability' appears to be a Trilok derivative; information, courtesy of Prof. J. Falkenstein, Uni Rostock, on 5 Nov. 2021.

[69] In hindsight the classification and filing of Hans von Ohain's first two turbojet-related inventions of 1935 as unpublished 'secret patents' resulted rather on his initiative and was rare otherwise in Nazi Germany; for further discussions—see Sect. 12.1 'On aero propulsion patents 1930–1950'.

Fig. 5.12 Daimler Benz Stuttgart, DB 109-007 turbofan engine 1400 kp TO thrust; total PR 8; lp compressor speed 6200 rpm, hp—12,600 rpm; lp airflow 19.9 kg/s, hp—8.2 kg/s, BPR 2.43; length 4.725 m, diam. 1.625 m

Weinrich, against Heppner's exclusive international patent rights then—and in a far-away post-war future.[70]

It is intriguing to continue on this path of thought of an assumed *'quiet agreement'* between Heppner and Weinrich also for the further developments, which—as mentioned— brought Weinrich to Daimler Benz as consultant for the DB 109-007 CR technology, Figs. 5.12 and 5.13 (top), Germany's first turbofan engine. DB 007 chief designer was Karl Leist (1901–1960), who succeeded in 1935 at the TH Berlin-Charlottenburg Prof. Karl Krainer,[71] for thirty years influential head of the Institute for Marine Engineering and Steam Engines, and who gained in parallel up to 1939 responsibility for turbochargers at DVL Berlin-Adlershof, the German test establishment for aviation. Similar to Armstrong Siddeley Motors, Daimler Benz Stuttgart—an established aero piston engine manufacturer—wanted

---

[70] On the other hand, it is striking that Heppner actually had no coverage of the CR combustion problems, which played a significant role in Weinrich's patent portfolio.

[71] Krainer's son in law was Prof. Herbert Wagner (1900–1982), who invented still at TH Berlin-Charlottenburg a gas turbine turboprop GB495,469 (priority 8 Feb. 1936), before he moved on to Junkers Dessau, where he initiated the building and first—still unsuccessful—tests of an axial aero gas turbine, see Sect. 6.2.

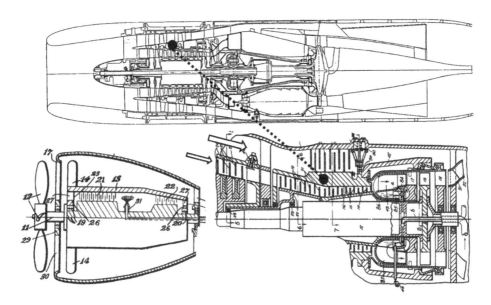

Fig. 5.13 Counter-rotating axial turbofans: H. Weinrich/K. Leist DB 109-007, CR compressor, 1942–1945 (top); F. Heppner et al. CR turbofan patents (bottom), 1941 (l), 1942 (r)

somewhat desperately to participate in the coming turbojet engine business, and hired Karl Leist for a relatively short stint 1939–1941, before he returned to turbocharger (cooling) research at TH Braunschweig. The DB 007 is Germany's first realised turbofan engine with 1300 kg prototype weight, to be reduced in production down to 1100 kg and a calculated specific fuel consumption sfc 1.45 kg/(kph) at sea level to 1.05 kg/(kph) at 12 km altitude— which was tested for the first time in the test cell at Stuttgart-Untertuerkheim on 27 May 1943. Leist's role in DB 109-007 concept definition was not unique; in principle, the turbofan idea was apparently more broadly investigated at that time.[72] However, the DB 007 turbofan configuration is covered by Leist's DE767,704 with priority 30 May 1940[73] and before he left, he managed—somewhat against the (coming) trend of German turbine air cooling—to implement earlier turbocharger cooling ideas[74] in oversized scale also in the DB 007 general

---

[72] As an example, Hans von Ohain filed apparently the first *'Geared Turbofan'* for Ernst Heinkel Stuttgart in a patent DE767,258 with a priority as early as 12 Sept. 1939.

[73] In this context there is an interesting hint that Prof. R. Friedrich since 1964 Professor for Mechanical Engineering at TH Karlsruhe credited his former superior Herbert Wagner, institute head at TH Berlin-Charlottenburg up to 1936 and thereafter leading engineering manager at Junkers, Dessau also with strong influences on the DB 007 design concept (courtesy Günter Kappler, 20 April 2020), see also the following Sect. 6.2 for details.

[74] Karl Leist's cooling patents for circumferentially partially loaded turbochargers, similar to the Moss/GE turbocharger cooling principle are DE767,078 (priority 1934) and DE925,984 (1941).

arrangement, visible in Fig. 5.13 (top) by the non-symmetric combustor arrangement, where the turbine wheel was exposed for a partial sector of ~120 deg. to the cold bypass air stream.

At the same time Fritz Heppner in his early high-time at ASM Coventry filed two—in this context relevant patents—alone, for a *'Jet propulsion plant'*, US 2,404,767 with priority 28 October 1941, Fig. 5.13 (bottom, l), and together with the ASM employees John D. Voce and David R. Evans for *'Torque balancing of jet propulsion turbine powerplant'*, US 2,416,389 with priority 17 June 1942, Fig. 5.13 (bottom, r). The latter was—besides a British application GB586,277—also filed in Germany(!) as a *'Brennkraftturbine fuer Strahlantrieb'* (Combustion turbine for jet propulsion) DE818,277 which might indicate a considered relevance of these patent ideas from ASM perspective in view of ongoing German turbofan developments. In detail the accompanying patent text to the turboprop/ turbofan concept, Fig. 5.13 (bottom, l) illustrates that Heppner clearly saw then also already the option of a *'geared turbofan'*,[75] however, was so convinced of his own approach, that he offered the flexible CR shaft system as a lighter and cheaper alternative in comparison to a planetary gearbox.

The 1942 follow-on patent as illustrated in Fig. 5.13 (bottom, r) shows—as highlighted by the dotted line—the rotating *compressor outer shell* as the decisive common element between the DB 007 configuration of 1943 and the earlier Heppner's turbofan patent. The DB 007 concept has a two-stage lp low-pressure compressor in the bypass duct and a counter-rotating hp high-pressure compressor with nine stages along the inner drum and eight-CR-stages attached to the compressor outer shell.[76] In addition, there is half-way between CR hp compressor and its SR (single rotation) drive turbine—an integrated planetary gearbox to achieve also sufficient operational stability between the shafts. In comparison, the Heppner design has no such gearbox as in the aforementioned Leist patent, but also a counter-rotating turbine with four rotors, of which two (including the rotating combustion chamber with integrated turbine blading) drive the inner hp compressor drum, while the other two turbine rotors drive the compressor outer shell, where both the inner rows of the counter-rotating hp compressor are attached as well as the lp turbofan blading in the bypass duct. Since Heppner was—other than the DB 007 designer—apparently concerned with the engine starting process, he provided and patented also a compressor outer shell *locking mechanism*, so that the engine start-up was sequentially alleviated. Other than for the foregoing *turbojet* patent, Fig. 5.11 (bottom, l), this later *turbofan* patent

---

[75] Today at best known for the—see Wikipedia, 'Pratt & Whitney PW1000G' in English, together with MTU Aero Engines on the way to expand the fuel-saving concept to BPR ~17, a fundamental shift in turbofan technology, to which the Author in the 1990s had the team privilege to contribute a few technological elements—see Geidel, Gearless CRISP, Eckardt, Future Engine Design, and Fig. 8.33.

[76] Acc. to—see Gersdorff, Flugmotoren, p. 291, the demanding pressure ratio of PR 8 was achieved on the basis of a CR compressor proposal from AVA Goettingen.

shows also a fuel transfer solution in detail, Fig. 5.13 (bottom, r).[77] In addition to what is shown here, Heppner's turbofan patent includes also other turbomachinery blading configurations, amongst these Griffith's typical turbine-compressor compound blading, of which Heppner probably knew the foregoing, patented Stolze-Barkow variant, Fig. 4.28. In this respect Armstrong Siddeley Motors and their newly employed designer with German background sent with this turbofan patent not only signals back to Germany, but also to the neighbouring Rolls-Royce, Derby and their newly employed Arnold A. Griffith.

By reviewing the foregoing pages, the author must confess that Fritz Heppner's early life remains partially a riddle, has still year-long gaps which somehow have to be filled with assumed corresponding studies, because otherwise his technical versatility could not be explained. There are mainly three, highly diversified scientific areas where he contributed innovative ideas:

- mechanical and hydraulic automotive transmissions,
- high-speed turbomachinery and
- advanced flight mechanics.

Before stepping back to 1935 in the turbojet development chronology, it is worthwhile to have a closer look to Heppner's turbulent life before the outbreak of war, the internment of his family, and finally his official employment at Armstrong Siddeley Motors, Coventry in December 1941. As indicated by the name of *'Gustav Lachmann'* in the foregoing (Footnote 41) and necessary for the understanding of the next Sect. 6.1 *'International information exchange/ transfer'* it has to be considered as *'normal'* that latest since 1937/ 1938 Heppner had been brought in close contact to the British MI5 secret service and military intelligence, not the least in view of expected background information about German turbojet developments.

Deliberately one could choose the 6 February 1940, the priority date of Heppner's fifth gearbox patent (over a time span of more than 15 years) for a *'Hydrokinetic torque transmitting device'* and still filed in his own name, as the last day of his 'old engineering life'.

Already three months earlier, in November 1939 he sent the engine specification of his counter-rotation turbojet, Fig. 5.11 (bottom, l), for review to the British Air Ministry (AM), directorate for scientific research, and specifically here to Major George P. Bulman (1892–1979), who between 1928–1944 continuously led the Ministry's DED Department of Engine Development.[78] Giffard insinuates that this date was *'well before the invitation from Armstrong Siddeley (to) Heppner'*, which of course would potentially not exclude a

---

[77] An idea, evidently shared with H. Weinrich, his close German engineering counterpart, and thus underlining their special relationship, as illustrated in a corresponding patent of 1951, Fig. 8.37 (r).

[78] Bill Gunston—see Gunston, Jet—identified Bulman as major scape-goat for Whittle's development problems with the Government, which acc. to—see Giffard, Making Jet Engines, p. 220— Bulman denied in his memoirs, written in 1960 and published in 2002. However, again acc. to Giffard, Bulman had a high opinion of Heppner. The mentioned *'engine specification'* could have

kind of *'encouragement'* for such a proposal from the Ministry or MI5, especially in view of a long term AM/ MAP strategy to train and toughen up ASM for their coming role as major turbojet manufacturer besides Whittle's Power Jets. The question has to remain unanswered, if Heppner was finally picked by the ASM upper management (Sopwith, Spriggs) for the new, jet-oriented ASM—or rather sent/ recommended to them as part of the planned bigger package. In any case and with all sympathy, Bulman required for the evaluation of Heppner's proposal the expertise/ agreement of Hayne Constant (1904–1968), RAE, who turned it down for being overly complex.[79]

Between early June 1940—18 March 1941 followed the described internment of the Heppner family on Isle of Man. On 14 November 1940, a German bombing raid destroyed the Coventry Cathedral and damaged severely, 600 m thereof in the south, the actual target—the Armstrong Siddeley Aero Development shop, destroying several Deerhound engines. Another raid on 8 April 1941 further damaged the shop severely, and set engine development back even more, so that it did not recover. ASM *'dog'*-engine development continued at a slow pace until 3rd October 1941, when the British Ministry of Aircraft Production (MAP) cancelled further work. The ASM dog-engines would not be ready in time to be of any use in the war, and the MAP wanted the company to focus on turbine engines.

26 March 1941 is the patent priority for US2,360,130, Heppner's innovative CR engine idea, in discussion with the Ministry since November 1939, now his first turbomachinery patent and already filed in the name of Armstrong Siddeley Motors, apparently as result of a flat transcription of Heppner's turbo-patents to ASM in May 1943. In addition, Hodgson remarks that the upper ASM management *'Spriggs and Sopwith presented . . . Heppner* as early as *August 1941, instructing the then chief engineer Tresilian that he had to give every assistance and top priority to Heppner'.*[80]

This allows to introduce a short portrait of (Sir) Thomas O.M. Sopwith (1888–1989), an English aviation pioneer, business tycoon and yachtsman—and eighth child and only son of Thomas Sopwith,[81] a civil engineer and managing director of the Spanish Lead Mines Company, Linares, Spain. In June 1912, Sopwith et al. set up the *Sopwith Aviation Company*, which got first military orders till the end of that year, consequently moving

---

been only the given patent no. US2,360,130; Giffard's GB577,950 in this context refers to another, unrelated Heppner patent. Bulman will be further addressed in Sects. 6.1.3, 8.2.4 and 10.1.

[79] See Giffard, Making Jet Engines, Ch. 2, FN 146: NA AVIA 5 Dec.1939, Interdepartmental letter Roxbee Cox and H. Constant. Giffard's exact quotation appears more reliable here than Hodgson's vague reference to 1941, see Hodgson, Armstrong Siddeley, p. 14.

[80] See Hodgson, Armstrong Siddeley, p. 10. The delay between this date and the beginning of Heppner's contractually confirmed ASM-employment on 6 Dec. 1941 is presumably part of the *'employment in making'* on the basis of a kind of informal gentlemen's agreement.

[81] On 30 July 1898 whilst on a family holiday on the Isle of Lismore in Scotland, a gun lying across young Thomas's knee went off, killing his father; this accident haunted Sopwith for the rest of his life,—see Wikipedia, 'Thomas Sopwith' in English.

Fig. 5.14 T.O.M. Sopwith (60, l) and his 1934 America's Cup J Class sloop *'Endeavour'* vs. King George's V *'Britannia'* off Isle of White (r) © R. Naughton, Yachting World

on to larger premises. The company produced more than 18,000 WW I aircraft for the Allied forces, including 5747 of the *'Sopwith Camel'*[82] single-seat fighter. Bankrupted after the war, he re-entered the aviation business in 1920 as chairman of a new company Hawker Aircraft, named after his chief engineer and test pilot. Now financially independent, Sopwith detected in the early 1930s his talents for sailing and industrial management, Fig. 5.14 (r),[83] in that order,—by bringing other aircraft concerns under the influence of his business leadership. Firms—as he said—which were 'long on technical skills but short on the business side'.

In 1934, Sopwith challenged the America's Cup with his J-Class yachts, *'Endeavour'*, and with *'Endeavour II'* in 1937, for which he funded, organised and helmed the yachts. The 1934 trial[84] against the defender Harold Vanderbilt's *'Rainbow'* took place off Newport RI, Block Island and Martha's Vineyard, MA between 15 September and 31 October 1934. *'Endeavour'* measured 39.56 m overall, at a mast height of 46.47 m; Sopwith experimented with new running backstay strain gauges, which controlled the trim of the mast and used electronic windspeed and direction indicators. He did not win the Cup,

---

[82] A metal fairing over the gun breeches, intended to protect the guns from freezing at altitude, created a *'hump'* that led pilots to call the aircraft *'Camel'*, although this name was never used officially. The fact that the aircraft was involved in bringing down the *'Red Baron'* Manfred von Richthofen (1892–21 April 1918) increased its high profile, and that of its manufacturer,—see Wikipedia, 'Manfred von Richthofen' in English.

[83] The *'Britannia'* was scuttled after King George's V death in 1935 south off the Isle of Wight, according to his will.

[84] See *'The America's Cup (1934)'*, 1:28 min video: https://www.youtube.com/watch?v=OS8EnfXkMuI

but he became a legend by nearly winning it in a very close 2:3 contest decision which newspapers later commented *'Britannia rules the waves and America waives the rules'*.

His increasing popularity brought him thereafter also industrial-political attention and support.[85] First he bought one of his competitors, the Gloster Aircraft Company. Then he arranged in 1935 a merger with the Armstrong Siddeley Development Group which embraced both the Armstrong Whitworth and the Avro aircraft firms as well as Armstrong Siddeley Motors. Sopwith became chairman of this Hawker Siddeley Aircraft Corporation Ltd together with (Sir) Frank S. Spriggs (1895–1969), whom he had met during WW I and kept as lifelong managing director of his many businesses. During the WW II, the Hawker Siddeley Group employed 100,000 people and turned out 40,000 aircraft including Hurricanes, Lancasters and the Gloster Meteor jets, of which the *E.28/39* flew for the first time on 15 May 1941, powered by a *Power Jets W.1* with 420 kp TO thrust. In 1977, Hawker Siddeley merged with British Aircraft Corporation (BAC) and Scottish Aviation, thus becoming a founding component of the nationalised British Aerospace (BAe)—with Sopwith as official consultant. Sopwith survived Spriggs by 20 years; his 100th birthday was marked by a flypast of military aircraft over his 900 ha home, Compton Manor in King's Somborne, Hampshire, where he died several days after he had accomplished his 101th birthday.

Back to the Armstrong Siddeley chronology before Heppner's official start, end of 1941:

On 26 June 1941 Heppner—apparently still as independent inventor—repeats a reminder about his counter-rotating engine proposal, now to the newly established MAP Ministry of Aircraft Production which RAE (H. Constant) again denies. Like other counter-rotating schemes—meant is Griffith's C.R.1 scheme[86]—Heppner's proposal was judged to have great potential but to be a *'more remote development than the engines at present under construction'*. (sic, DE).

Hidden in the last sentence is the fact that Constant, asked as independent expert-consultant for the Ministry was—in the same manner as Griffith in his fateful meeting with Whittle in late October 1929 at the Air Ministry's South Kensington Lab, Sect. 4.2.3—not really *'independent'*. The all-or-none, decisive test of the all-axial F.2 turbojet with the nine-stage 'Freda' compressor (PR 3.2) from Constant/ RAE, was about to be tested at Metrovick in December 1941. This does not imply that Constant's decision towards the Heppner concept was wrong, but it may explain somewhat the following, strengthened Heppner backing by the mighty ASM upper management, which understood the context with all facets of course completely.

---

[85] Profile *'Sir Thomas Sopwith: The Hustler who always keeps calm'*, New Scientist, 31 Jan. 1957, pp. 25–26 (see Internet).

[86] Griffith had left RAE to RR in mid-1939, where his C.R.1 rig was—built by ASM—prepared for testing, which was actually carried out, beginning 3rd March 1942. Date acc. to—see Giffard, Making Jet Engines, Ch. 2, FN 147, NA AVIA 13/911, 26 June 1941, H. Constant for (RAE Director) Perring to (MAP Dep. Director) Roxbee Cox.

In the aforementioned historic meeting on <u>3rd October 1941</u>, ASM agreed with MAP that it would abandon its current piston engine development programme (*'Deerhound'*) in favour of further 100 percent turbojet/ turbofan activities. And as a kind of confirming sign, Heppner filed with priority of <u>28 October 1941</u> his turbofan concept which has been presented in Fig. 5.13 (bottom, r)—with associated discussions in view of the Daimler Benz 109-007 turbofan engine, but also with implications towards Griffith and Rolls-Royce.

To end this chronology preliminarily, Heppner became contractually fully employed at ASM on <u>6 December 1941</u>. Only 6 days later on <u>12 December 1941</u> in a meeting at Coventry, Heppner repeated his proposal to the MAP, a third time and now officially in the name of ASM. The proposal had a similar structure as in July, MAP unshaken asked RAE to comment and again the proposed engine was rejected as impractical for the firm's first turbojet. The engine concept was considered as being far too complex, and Constant expressed doubts about Heppner's performance assumptions. They exceeded by far anything achieved in contemporary practice in Britain, including his own. Heppner had claimed to achieve a total pressure ratio of PR 8 for his 5/4 stage counter-rotating compressor, while the highest that had yet been achieved in Britain was PR 3.5.

'Connections' and Early Turbo-Jet Developments 1935–1939

6

6.1 International Information Exchange/Transfer

The described period 1935–1939 in this Section is unique—instead of dealing with national or isolated company-related efforts, it reflects mostly a somewhat strange bias for transnational cooperation and communication like a last effort of goodwill—before the outbreak of another Great War, of course always paired with mistrust and coverage, from sounding out to technical espionage. It starts where the foregoing Chapter ended—at Armstrong Siddeley Motors, Coventry, but in a kind of *'hopping procession of Echternach'*[1] by jumping back some five years.

6.1.1 The Dawning of a New Technical Age

Armstrong Siddeley[2] manufactured luxury cars, aircraft engines, and later, aircraft. In 1935, the founder of 1902 John D. Siddeley (1866–1953) purchased his interests for £2 million by aviation pioneer Thomas O.M. Sopwith (1888–1989), owner of Hawker Aircraft, to form—along with the Gloster Aircraft Company and Air Training Services—Hawker Siddeley, a famous name in British aircraft production. Armstrong Whitworth Aircraft and Armstrong Siddeley Motors (ASM) became subsidiaries of Hawker Siddeley, with Sopwith himself the new chairman of Armstrong Siddeley Motors, and in his wake Frank S. Spriggs (1895–1969) as general manager.

---

[1] See Wikipedia, 'Dancing procession of Echternach' in English.
[2] See Wikipedia, 'Armstrong Siddeley' in English.

© The Author(s), under exclusive license to Springer Fachmedien Wiesbaden GmbH,   247
part of Springer Nature 2022
D. Eckardt, *Jet Web*,
https://doi.org/10.1007/978-3-658-38531-6_6

ASM's list of aero radial piston engines comprises over the years some fifteen different products, all named after big cats, followed in 1935 by an all-new family of big piston engines, the *fateful 'dog' series*, of which the *Deerhound*,[3] Fig. 6.1(l), got most development attention till the official programme cancellation in October 1941. The *Deerhound I*, which first ran in 1935, was a triple-row, 21-cylinder, air-cooled radial engine design with the unusual feature of inline cylinder banks. *Fateful* in this context means an interrelation of several causes and influences. The new dynamic and decision-friendly owner-management certainly had short term success expectations which the slow progress under the then chief engineer—without actual engineering education—Lt.-Colonel Louis F.R. Fell (1892–1977), fulfilled apparently only partially; his term ended already in January 1939, after he had joined ASM—coming from Rolls-Royce—only in 1934. Fell was replaced end of May 1939 by the 35-year old Stewart S. Tresilian (1904–1962), who had designed fabulous high-power car engines in his foregoing stations at Rolls-Royce, Lagonda and Templewood.[4] However, Tresilian—though never officially contracted—had not only to clear up the appalling mess of the *Mark II Deerhound* 1500 hp air-cooled (7 × 3) aero-engine[5] to an overly ambitious new 24 cylindre 4 × 6 engine *Wolfhound* of about 61 l to produce 2600–2800 hp, he had also to carry the extra-load of coming aero gas turbine developments. In the end unsuccessful, when a new chief engineer Dr. Henry S. Rowell (1885–1951)[6] was installed in 1941, and in his wake Fritz Heppner, strongly supported by Rowell and the other ASM upper management with comprehensive responsibility for aero gas turbine developments,—as already outlined at the end of the foregoing Heppner-Weinrich personal excursion, Chap. 5.

In 1940, Tresilian went to work on the *Deerhound III* to create an engine free from the issues experienced with the original Deerhound and the Deerhound II. The Deerhound III possessed the same bore, stroke, displacement, and 1.12 m diameter as the *Deerhound II*. However, the engine was essentially redesigned for now 1800 hp and was first run in late 1940. Ironically, a German bombing raid, as described under the date of 14 November 1940 in the foregoing Chap. 5, accelerated the end of Tresilian's stage at ASM—and opened the doors for the Jewish-German refugee Fritz Albert Max Heppner instead.

The *'new age'* at ASM began by hesitant sounding in 1936, when its upper management—with all likelihood Frank S. Spriggs—was approached by the German steam boiler manufacturer Vorkauf-La Mont, Berlin/London about licensing their aero steam engine

---

[3] Besides *Deerhound I–IV*, an increased capacity variant known as *Boarhound* was never flown, and a related, much larger, design known as the *Wolfhound* existed on paper only.

[4] Tresilian's car developments were already addressed in Chap. 5, Fig. 5.8; the following Heppner aero gas turbine activities will be outlined in more detail in Sect. 8.2.4.

[5] See Hodgson, Armstrong Siddeley, p. 9.

[6] H.S. Rowell, originally a (fuel) chemist, published already in 1922 on *'Principles of Vehicle Suspension'* and came as employed, suspension and engine consultant apparently in contact with ASM as early as spring 1938.

Fig. 6.1 Dawning of a new age at Armstrong Siddeley Motors: Tresilian's *'Deerhound III'*, 1940 (l), Vorkauf steam eng. project 1936—as illustrated by the Besler brothers' steam-driven aircraft, July 1933 (r)

with revolving boiler.[7] Instead of going into too much—in the end fruitless—complex design details, it is sufficient to requote what could be read on 16 April 1934 in *The Daily Telegraph*:

> Details are now available of the steam-driven aeroplane which has been under secret construction on the outskirts of Berlin for many months. The inventor is Herr Huettner, chief engineer of the Klingenberg electricity works. The machine is not yet finished, but the plans, according to the "Berliner Tagblatt" have been submitted to experts and found to be theoretically satisfactory. They have been elaborated down to the smallest detail, and give rise to the following expectations:
>
> Range: 60–70 h non-stop flight,
> Speed: 230 m.p.h. on starting, rising to 260 m.p.h. max.,

---

[7] See August-Th. Herpen and Heinrich Vorkauf, Berlin, patents 'Steam generator', DE643,143, priority 21 May 1931 and 'Steam generator with rotating heating surfaces', DE612,627, priority 10 Jan.1932, and—see Kay, German Jet Engine, 'Vorkauf's Drehkessel (rotating boiler) gas turbine unit', p. 176 f. Besides this *'unsolicited proposal'* from Vorkauf to ASM, there is also reported, that ASM's interest stemmed from the request the aircraft manufacturer *Short Brothers* had made for a high-power engine for their new seaplane,—see Lawton, Parkside, p. 66. While the British-German negotiations were always closely accompanied and finally stopped by the Air Ministry, nothing is known about the set-up on the German side with respect to government involvement which of course can be taken for granted as well. In this context it is worthwhile to mention that in due course not only an U.S.-GE daughter company got seriously involved in steam aero engine developments (as discussed in the following), but to a much more comprehensive extent the Russian plan economy in which between 1932–1939 eleven major aero engine projects were installed from 11 research institutions, comprising 34 project years with in the end a significant waste of resources,—see Harrison, The Political Economy of a Soviet Military R&D Failure.

Max. height: 43,000 ft,
Load: One ton for a non-stop flight of 60 h,
Eng. Power: 2500 hp,
Length: 6 ft.

The secret of these claims is said to lie in the fact that for the first time Herr Huettner has succeeded in solving the problem of satisfactory ratio of weight to power. His solution consists of a revolving boiler combined with a steam turbine … In March last the Daily Telegraph Prague Correspondent reported that an article in the Czechoslovak newspaper "Prager Tagblatt", giving details of Herr Huettner's invention, has led to the arrest of the Berlin correspondent of that paper.

Daily Telegraph's 'Huettner engine' and the *'Vorkauf boiler'* offered to ASM were obviously closely related: for the Huettner design the integral steam turbine was of the counter-rotation type and drove both the boiler and the power shaft,[8] whereas Vorkauf proposed an integral steam turbine of the single rotation type for driving the rotating boiler only.[9] For the latter, Hodgson has the following, equally astounding data:

The Vorkauf unit had 4–6 centrifugal compressor stages, burner, gas turbine, steam turbine, and rotary boiler all in one unit. The condenser was separate. The shaft output was 3000 hp.

Overall compressor PR: 6 : 1,
Overall compr. efficiency: 75%,
Combustor exit temperature: 2350 °C,
GT inlet temperature: 1400 °C,
Turbine inlet pressure: 5.8 bar,
Rotary boiler speed: 3000 rpm,
Steam pressure: 80 bar,
ST inlet temperature: 450 °C,
Total plant efficiency: 28%.

Relative power without condenser was about 0.55 hp/lb—at first sight low by comparison with then piston engines with typically 0.75 hp/lb in 1935, Fig. 2.3,—but roughly equivalent due to the claimed efficiency, i.e. attractive in view of total mission weight (engine + fuel weight).[10]

It can be assumed that besides the mystifying effect of the *Daily Telegraph* article, the successful flights of the world's first steam-driven aircraft increased the attention for this new type of rotating aircraft engine considerably. On 12 April 1933 at Oakland, CA, William J. Besler made the first flight with his *Besler/Doble* steam engine installed in a *Travel Air 2000* aircraft, Fig. 6.1(r). The engine was a two-cylinder V-type engine that

---

[8] See Fritz Huettner's patent 'Steam plant with rotating steam generator', Berlin-Karlshorst, DE640,558C, priority 27 Feb. 1931.

[9] See B.I.O.S., Vorkauf Rotating Boiler, p. 3.

[10] See Hodgson, Armstrong Siddeley, p. 3.

generated 150 hp. The engine weighed 180 lb and the boilers and condensers added an additional 300 lb (0.31 hp/lb). The engine could be reversed instantly both in flight and on the ground to steepen landing approaches and shorten the landing run after touchdown.[11] Most important, especially from todays environmental aspects, are the original reports about the relative quietness of flight: *'This blue machine ... sped down the runway and climbed into the air without a sound except the low whine of the propeller and the hum of wind through the wires. Swinging back over the field at 200 feet, the pilot shouted «Hello!» and heard the answering calls from spectators below. Conversation in the craft, ... was as easy as conversation in an open automobile.'*[12]

Interestingly 3000 hp steam engines for *'giant aircraft'* were not only discussed at Armstrong Siddeley Motors as a future production option at times of uncertainty, but also at the settled power generation company Brown Boveri & Cie., Baden, Switzerland in 1939, when the Company was about to launch the first 4 MW power generation gas turbine plant at Neuchâtel, CH. On first February 1939 the topic is suggested for further investigations in the minutes of a board meeting[13]—in reference to a then recent article in *'Génie civil'*.[14] Dr. Adolf Meyer (1880–1965),[15] director of BBC's Thermal Department summarised the Company's position as follows: *'This is a somewhat antiquated story which since 1917/8 has been repeatedly brought up. Difficult are not so much the steam turbines, but rather the boilers and primarily the condensation. There were numerous projects, but none of these has been actually realised. Things become interesting when steam turbine powers of 3000 hp or more are required. As known, England studies at present the gas turbine for aircraft propulsion* (sic! DE) *and we also—if there is time— carry out corresponding studies. It is not impossible that something in this direction can be done. Of course, this will be then special designs which may become interesting, since in case, this will not be single units but with all likelihood series products.'*

With some certainty it can be assumed that Colonel Fell's thoughts went in the same direction, but different to Meyer he required competent help for the further proceeding in 1935. In principle he took the Vorkauf offer seriously and met one of the owners Dr. Herten in Germany, but technically the topic was beyond his reach, so he wanted to appoint an engineer who (1) was sufficiently competent to comprehend this complex matter, (2) spoke at best fluently English and German, and to be prepared for future challenges, (3) knew something about gas turbines. And—where there is a will there is a way—this man was

---

[11] See Besler promotion video https://www.youtube.com/watch?v=2TtHOkgwrk8

[12] Quotation from *'Popular Science Monthly'*, July 1933.

[13] Courtesy to BBC historian Norbert Lang, 22 July 2017. Meyer's board statement was given just weeks in advance of his historic presentation on the status of the *'combustion gas turbine'* at the Institution of Mechanical Engineers, London on 24 Feb. 1939, which will be reviewed at length in Sect. 6.3.

[14] See Génie civil, Les perspectives de la propulsion à vapeur des avions géants.

[15] See Eckardt, Gas Turbine Powerhouse, p. 87 f.

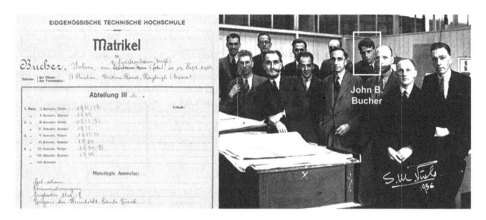

Fig. 6.2 John B. Bucher (1912–1989), student register ETH Zurich 1931–1935 (l), Armstrong Siddeley Motors design office, 12 October 1936, (r) © ETHZ Archive and RRHT Coventry Branch

found in August 1936 as John Bennett Bucher (1912–1989)—with Swiss father and English mother, who had attended at first the secondary Humboldt-Schule Zurich, then an English public school up to A level, and thereafter studied mechanical engineering at ETH Zurich from October 1931 to July 1935. Figure 6.2 (l) shows John Bucher's student register with two places of origin: Twickenham (Engl.), a suburb in the south-west of London, and Schleinikon,[16] a small rural village of ca. 500 inhabitants, some 25 km in the north-west of Zurich. In addition, there is his birthday on 19 September 1912 and the address of his parents A. Bucher at Rayleigh, Essex, England. The photograph on the right-hand side shows part of the Armstrong Siddeley Motors design office[17] with the 24 year-old Bucher in the rear, just two months in his first job. According to Richard Hodgson *'Bucher was made* "head of investigations", *and dealt with all technical queries raised by the Air Ministry and the RAE ... The Vorkauf project was handled in a very secretive way. Given the optimistic specification, Bucher put up a good fight against a somewhat negative report from Dr. Griffith of the RAE who described the Vorkauf plant* as "an internal combustion turbine with auxiliary steam circuit, required primarily for cooling and used incidentally (sic, DE) to increase the power output and efficiency."'

---

[16] Surprisingly in April 2020 a contact to one of the several Bucher families at Schleinikon revealed further details about their ETH student of the 1930s, known to them as *'Jack'*. His father Albert, who came to the family as infant from an unwedded mother in about 1890, grew up in the village, getting training as apprentice in a local Bucher machine shop and toured Europe as a young man, where he came via France to England still before the outbreak of WW I, to be found in the mid-1930s at London-Islington as an agricultural machinery dealer. At the same time his son John B./*Jack* had finished his mechanical engineering studies at ETH Zurich.

[17] The complete picture with names can be found on one of Richard Hodgson's websites https://www.designchambers.com/wolfhound/wolfhoundCutler.htm

Between November 1936 and June 1937, correspondence and technical reports were exchanged between the company, the Ministry and RAE …The Ministry and RAE final view was that the unit was not practical … The project ended on 27 June 1937 when the Ministry's G.P. Bulman told Fell, who had kept pressing him for Ministry funds, that unless Bucher could discover more substantial grounds for counter attack of Griffith's objections based on actual experimental data, there was no point taking the matter further.' Nevertheless, ASM thought enough of Bucher's abilities to ask him to set up their stress office at *Parkside*, Coventry, eventually marking the company's initiation into the world of gas turbine engines.[18]

In the meantime it is known that the *Great Lakes Aircraft Corporation of America,* Cleveland, OH, took interest in the project apparently already in 1933, took finally a license from La Mont after Armstrong Siddeley failed to raise funding—and developed in cooperation with General Electric a 2300 hp prototype—with a claimed efficiency of 23%.[19] In Germany Vorkauf's *'Drehkessel (rotating boiler)'* gas turbine concept was followed further since 1942 at the company Brueckner-Kanis, Dresden for 5000 and 10,000 hp marine versions to accelerate *Schnellboote* (speedboats) beyond their basic diesel-powered speed of max. 42 knots.[20]

In summary, because the steam engine was heavier, less efficient, and more complex than conventional internal combustion engines, it could finally not successfully compete with them and thus was never put into production.

6.1.2   Too Good to Be True

Synchronised with the appointment of Adolf Hitler (1889–1945) as German Reich Chancellor on 30 January 1933, Hermann Goering (1893–1946) became *'Reichskommissar fuer den Luftverkehr (Reich Commissioner for Air Traffic)'* on the same day. This started the clock for the next 6½ years for a furious armament race to make the German *Luftwaffe* fit for war. This happened mostly in secrecy and the correct assessment of the speed from outside was difficult, especially for those responsible on the English side. The speed was high:

- Already on 2nd February 1933 the new institution was renamed into *'Reichskommissariat fuer die Luftfahrt (RKL, Reich Commissariat for Aviation)'* which meant the integration of the *Department of Aviation* from the Reich Ministry of Transport (RVM)

---

[18] See Lawton, Parkside, p. 66.

[19] See Knoernschild, Dampftriebwerke, a surprisingly detailed survey of 1941, comparing company products of Besler, Great Lakes, Brobeck, Huettner, Béchard, Aero-Turbines, Vorkauf and Wagner; alternatively—see Smith, Gas Turbines, 3rd ed., pp. 53–57.

[20] See short reference in Sect. 8.1.

and the *Department of Aerial Defence* from the Reich Ministry of the Interior (RMI). From early on a considerable advantage for the camouflaged build-up of the new *Luftwaffe* was the joint responsibility of that organisation for military and civil aviation issues.

- On first April 1933 RKL issued a letter to the *'Reichsverband der Deutschen Luftfahrt-Industrie (RDLI, Reich Association of German Aero-Industries)'* which was officially read during the RDLI general assembly on 11 April 1933.[21] RKL demanded the procurement of nine specified aero engines from abroad (USA 3, France 3, GB 2, Italy 1)[22] for which RDLI distributed the task of getting near-term offers to their members. The acquisition of one Rolls-Royce *Kestrel* motor, a water-cooled 700 hp class V-12, 22 l aircraft engine, was put on *Arado*,[23] while *Heinkel* took similarly the acquisition task for a RR *Buzzard*, actually a 1.2 scaled *Kestrel* with 800 hp and 36.7 l capacity.

- On 27 April 1933 the RLM Reichsluftfahrtministerium (Reich Ministry of Aviation) was formed from RKL, which had been established just 2½ months earlier with Reichs-marschall (Reich Marshal) Hermann Goering at its head. In this early phase the Ministry was little more than Goering's personal staff. Goering's deputy from 1933 to 1945 was State Secretary Erhard Milch (1892–1972), until Albert Speer (1905–1981) took comprehensive production responsibility in 1943.[24] The RLM triumvirate was accomplished by Ernst Udet (1896–1941), who started as RLM Director of Research and Development and rose further in 1939 to the post of Director-General for Equipment, until after his suicide this responsibility went back to Milch. One of the first, typically brutal actions of this *team* was to requisition control of all patents and companies of Hugo Junkers, the dominant German aeronautical engineer and entrepreneur—unwilling to cooperate with the new regime. The total raid on this private venture included the rights to the Junkers *Ju 52* aircraft, foreseen as backbone of the new Luftwaffe, but implied also considerable financial spin-offs for local NS party officials.

---

[21] The transaction was thus documented on p. 31 of the RDLI Minutes of that meeting, as gratefully pointed out to the Author on 27 April, 2020 by Lutz Budrass, Ruhr-Uni Bochum.

[22] Nothing is known about the planned deals with the other countries except GB; however, 6 years later the situation changed, when France asked now for buying German engines through her emissary Charles Lindbergh in contact with State Secretary Milch on 20 Dec. 1938 and 16 Jan. 1939. The German side offered the supply of Daimler Benz 1250 hp engines, payment in hard currency and secrecy about the deal, i.e. no mentioning of engines but of the purchase of a (Fieseler) *Storch* airplane instead. The *'Storch'* request letter from French Air Minister Guy La Chambre (1898–1975) reached Milch on 28 Jan.1939, thereafter this communication line went dead.—See Lindbergh, Wartime Journals.

[23] The sites of Arado Flugzeugwerke GmbH and the local plant of EHF Ernst Heinkel Flugzeugwerke GmbH lay in immediate vicinity at Rostock-Warnemünde on the Baltic Sea.

[24] See Wikipedia, 'Hermann Göring',—see Wikipedia, 'Erhard Milch,—see Wikipedia, 'Albert Speer' and—see Wikipedia, 'Ernst Udet', all in English.

- On 26 February, 1935, Adolf Hitler signed a secret decree authorising the founding of the Reich Luftwaffe as a third German military service to join the Reich Army and Navy—a major step in the German rearmament programme. The Luftwaffe was to be uncamouflaged step-by-step so as not to alarm foreign governments, and the size and composition of Luftwaffe units were to remain secret as before. However, in March 1935, Britain announced it was strengthening its Royal Air Force (RAF), and Hitler, not to be outdone, revealed his Luftwaffe British cabinet ministers Sir John Simon and Sir Anthony Eden, the most senior delegation that had yet come to Berlin since the Nazi seizure of power, he claimed bluntly—but falsely—that in the mean-time Germany had reached parity in air strength with Britain. Independent of the truth content, the wrong message had the considerable side-effect, that the British side duplicated its air intelligence efforts.
- On 18 June 1935 the Anglo-German Naval Agreement was signed between the United Kingdom and Germany in line with the *Zeitgeist* to regulate the size of the *Kriegsmarine* in relation to the *Royal Navy*. This was an ambitious attempt on both Parties to reach better relations, but ultimately it had to founder because of conflicting expectations. For Germany, the *Agreement* was intended to mark the beginning of an Anglo-German alliance against France and the Soviet-Union,[25] whereas for the UK, it should be part of a series of arms limitation agreements. After accepting already the unveiling of the Luftwaffe, the British side agreed now within months a second time in the violation of *Versailles*—without consulting France and thus sending dubios signals.
- On 12 October 1935 was the topping-out ceremony for the new office buiding of the personally rapidly grown RLM on Wilhelmstrasse 97, designed by Professor Ernst Sagebiehl (1892–1970),[26] Fig. 6.3. This five- to seven-storey steel-skeleton building has a useful floor space of 56,000 m^2 with 2100 rooms and a corridor length of 6.8 km. Though the Ministry of Aviation was amongst the most essential buildings for the German war effort, it survived the Allied bombing raids nearly undamaged.

Up till then nothing was heard about the German engine purchasing activities 2½ years ago, but undercover an agreement must have been reached in German-British industrial contacts—certainly with official consent—to switch from purchasing to bartering. A deal was reached between Rolls-Royce and Heinkel to deliver a He 70G (export version) monoplane for high speed engine testing[27] in compensation to four, then advanced

---

[25] An idea which Hitler had already outlined in '*Mein Kampf*', Vol.II, in December 1926.

[26] See Wikipedia, 'Ernst Sagebiel'in English; Sagebiehl's other most renowned large-scale project was the *Flughafen Berlin-Tempelhof* with a building period 1936–1941. At the time of its completion it was for 2 years the area-wise largest building in the world, before the *Pentagon* at Arlington took over the lead position.

[27] This aircraft with elliptical wing, a design of Siegfried and Walter Guenter, re-engined at Heinkel Warnemuende with a Kestrel engine and delivered to Hucknell in the spring of 1936, was instrumental in testing the Rolls-Royce PV-12, later to become the Rolls-Royce Merlin. Both Supermarine's

Fig. 6.3 Reichsluftfahrtministerium (Ministry of Aviation), Berlin, Wilhelmstr.97, 1935/36, Architect Prof. Ernst Sagebiel

22.2 l, V-12 water-cooled RR Kestrel VI of ca. 700 hp.[28] The bartering trade was accomplished between mid-1935, when the engines were distributed to Heinkel and Messerschmitt to power the German fighter aircraft prototypes in competition mainly between He 112 vs. Bf 109,[29]—and 27 March 1936, when the Heinkel *'Blitz'* (Lightning)

R.J. Mitchell and B.S. Shenstone denied any He 70 influence on their parallel *Spitfire* design, pointing out however, that they had of course very well observed the foregoing usage of an elliptical wing planform for the Guenter twins' 1928 Baeumer B II *'Sausewind'* sports plane,—see Wikipedia, 'Siegfried and Walter Günter' in English. In addition, Shenstone said that the He 70's influence on the Spitfire design was limited to use as a criterion for aerodynamic smoothness.

[28] See Wikipedia, 'Messerschmitt Bf 109', *'1.2 Prototypes'* in English; and see Wikipedia, 'Heinkel He 112', *'The Contest'* in English, both in agreement with respect to the bartering issue, based on— see Forsgren, Messerschmitt Bf 109.

[29] Since the originally foreseen German engines Jumo 210A and DB 600 were not ready then. Power for the prototype Bf 109 V1 first flight at Augsburg-Haunstetten with 27-year-old Hans-Dietrich *'Bubi'* Knoetzsch at the controls on 28 May 1935 was provided by the 695 hp RR Kestrel engine, but the V2 that followed four months later had a 610 hp Junkers Jumo 210A. On 15 October the Bf 109 was flown by Knoetzsch to the Rechlin test facility for the final evaluation contest. Upon arrival over Rechlin, Knoetzsch engaged in aerobatic manoeuvering to impress the assembled crowd. However, the Bf 109 prototype was damaged on landing and ended up in a somewhat embarrassing pose of being perched tail-high on the nose, the one intact undercarriage leg, and a wing tip. Apparently, what the gathered pilots considered *'a perfect three-pointer-landing'*. Quoted from—

He 70G, registered as G-ADZF, was flown from Rostock to Hucknall jointly by the RR test pilot Captain R.T. 'Ronnie' Shepherd and Luftwaffe test pilot Otto Cuno from the Rechlin test centre,[30] *'where it became henceforth a flying test-bed for the* Merlin *of the utmost value'*.[31]

As observed in the foregoing, the actual origins of the bartering or purchasing idea of strategic, foreign aviation supplies could not be resolved. Lutz Budrass[32] presumed that there were elements of a former German Reichswehr and *Voelkerbund* (League of Nations) policy to offer by a network of cross-border license agreements and technology exchanges a kind of voluntary re-armaments control. Apparently in the same category fell the acquisition of two advanced *Boeing B 247* passenger aircraft by *Lufthansa* in 1934,[33] considered the first such aircraft to fully incorporate such as all-metal construction, a fully cantilevered wing, retractable landing gear as well as auto pilot and deicing equipment. Finally, the RLM bought in early 1934 two *Curtiss F11C-2 Goshawk* biplane fighter aircraft for the exclusive use of Ernst Udet,[34] which started the German *Ju 87 Stuka* dive bomber developments. Very soon after the *Heinkel He 70* had been flown to England in spring 1936, it was decided to rely the now fully cranked up rearmament programme solely on own products, developed and built in Germany.

---

see Forsgren, Messerschmitt Bf 109; thereafter, Knoetzsch was replaced by Hermann Wurster (1907–1985), who achieved on 11 Nov. 1937 with Bf109 V13, powered by a 1660 hp DB 601 engine, a new speed record with 610.95 km/h.

[30] The joint pilotage is an assumption, since—see Nockolds, The Magic of a Name, has Shepherd only, while Cuno's preserved flight book—courtesy to V. Koos, 26 Nov. 2020—has the detail of three flight legs on 27 March 1936: Rostock-Warnemuende to Schiphol 12.10–13.50 h, Schiphol to Croydon 14.20–15.55 h, Croydon to Hucknell 16.30–17.10 h, and—see Koos, Heinkel He 70.

[31] See Nockolds, The Magic of a Name; deviating from the described *'bartering trade'*: 1× He 70 aircraft in exchange to 4x RR Kestrel VI engines, the acclaimed British author confirms only in his famous, 10-times reprinted Rolls-Royce saga, 1st ed. Nov.1938(!), that Rolls-Royce *bought* a He 70 for £ 13,000. In a Nov.1959 reprint (which the author received on 23 Sept. 2020, courtesy to A. Nahum) in Ch. VI, p. 161, he states: *'A … Kestrel was sent to Germany to be installed at the Rostock factory, and it was only to be expected that the Germans would study the engine in detail first; they did, even going so far as to install it in two other machines and make some flight tests before putting it in the He 70.'* RR information in this context is apparently only accessible via the National Archives, Kew.

[32] Personal communication 17 April 2020.

[33] D-AKIN was used immediately since Aug. 1934 at the *Rechlin* flight test centre for experimental yaw axis control, but crashed on 13 Aug. 1937 at Hannover-Vahrenwald. D-AGAR was demolished even earlier on 24 March 1935 at Nuremberg in collision with an Air France aircraft, and scrapped thereafter.

[34] D-IRIS crashed already on 20 June 1934 during a demonstration flight over the Berlin-Tempelhof air field, of which Udet successfully bailed out. The second aircraft D-IRIK, after Udet had used it in aerobatic exhibitions during the 1936 Summer Olympics, became part of the collections of the German Aviation Museum, Berlin, survived the war on east-German territory and is today on exhibition at the Polish Aviation Museum, Cracov.

Unnoticed in Britain it was in another He 70, his personal transport, that General Walther Wever (1887–1936), the main protagonist of establishing a German strategic bomber force since 1934, was killed on 3 June 1936 at Dresden-Klotzsche. Leaving in a hurry, his He 70 *Blitz* had not been properly examined during preflight checks. His crash became in international records the second documented case where aileron gust locks were not removed.[35] Wever supported Junkers and Dornier in their respective projects to manufacture the Ju 89 and Do 19 competitors for the *'Ural (America) Bomber'* contract competition. After his death other strategists, mainly Ernst Udet, favoured smaller aircraft as they did not expend as much material and manpower. They were proponents of the dive-bomber concept and a preferred doctrine of close air support and destruction of the opposing air forces on the *'battle ground'* rather than through attacking enemy industries (not to mention unspecified *'carpet bombing'*). As a result, the medium-size, high speed *'Schnellbomber'* concept represented by the Heinkel He 111, Dornier Do 17 and Junkers Ju 88 came in vogue; a misjudgement which in combination with the failed *'Battle of Britain'* might have considerably contributed to Udet's suicide on 17 November 1941.

Hitler's *'parity claim'* of March 1935 had at least one positive side effect—to the British IIC Industrial Intelligence Centre, continuously led by Churchill's confidant Major (Sir) Desmond Morton.[36] The Luftwaffe was now officially recognised; contacts between the British and German services began and opportunities for intelligence collection automatically grew. The new *'openness'* was not only in line with German visions of a future alliance, but was certainly also generally supported by the British intelligence camp. The relaxation in German security paid already unexpected dividends in early 1936, when M. Graham(e) Christie,[37] a retired, in the meantime wealthy group captain who resided in Germany, started reporting about his excellent contacts with Goering, his deputy Erhard Milch, and the RLM as well as a whole host of men in German politics. He established himself as an unusually well-informed source on events in Germany, not as a conventional spy because he was neither paid nor controlled by anyone. He privately sent his reports to his official principal after 1933, the Permanent Under-Secretary at the Foreign Office Sir Robert Vansittart, eluding German security measures by writing from abroad, often expensive hotels outside Germany or from his house beyond the Dutch-German border.

In addition, the IIC benefitted from the reporting from a variety of technical missions and businessmen. W. Lawrence Tweedie, an RAF engineer—who in October 1929, in the Air Ministry's Directorate of Engine Development had arranged the historic meeting between Whittle and Griffith, Sect. 4.2.3—brought back in November 1935 useful information following a tour of German aeroengine plants. He judged the German status 2–3

---

[35] See Wikipedia, 'Gust lock' in English.

[36] See Wikipedia, 'Desmond Morton (civil servant)' in English.

[37] M.G. Christie (1881–1971) had already been addressed in Sect. 4.3.3, AVA—the institution, as instrumental for placing B.S. Shenstone at Junkers Dessau, with further background information.

years behind the British in development, but at the same time was also full of praise of what he had seen at Junkers Dessau—*'undoubtedly the finest factory it has ever been my privilege to see'.*[38] After the Tweedie visit a full Air Ministry team was sent on an inspection tour of Germany in the summer of the 1936 Olympics. Bill Willoughby Lappin (1888–1974),[39] described on his retirement as *'the salesman who never sold an engine'* and the best ambassador for Rolls-Royce, showed then up in Germany as *'personal assistant'* to Ernest W. Hives (1886–1965),[40] at that time general works manager of the RR factory and father of the *'Merlin'* engine (later head of the Rolls-Royce Aero Engine division and chairman of Rolls-Royce Ltd), gave the Air Ministry a detailed account of the Heinkel factory at Oranienburg near Berlin.

But according to Wesley K. Wark *'undoubtedly the most important unofficial source was the well-connected chief engineer of Bristol Aircraft, Roy Fedden.'* After two subsequent visits to Germany on 7–11 June and 2–12 September 1937,[41] he produced a report of 110 pages that improved the accuracy of IIC figures considerably. Not only were his technical assessments useful, Fedden was also a convincing observer on the modernity proper of the German aero industry: *'The aircraft production factories that were visited in Germany were extensive, modern, highly organised plants, of which we have nothing comparable in this country, and, with one or two notable exceptions, the corresponding layout of the British industry can only be described as obsolete and inadequate.'* And *'The first and most important deduction to be made from these two visits is that the declared British policy of having an Air Force which is on the basis of parity with Germany by April 1939 is out of the question.'*[42] Fedden's reports gave the intelligence authorities the advantage of his unsurpassable special knowledge of the German aircraft industry. As pointed out by Wark, he had trained German engineer apprentices in the 1920s and kept in touch with them on holiday visits to Germany during the 1930s; consequently, the head of the German section of IIC regarded Fedden's reports as *'immensely factual about matters almost impossible otherwise to know for certain'* and superior to secret agent reports, which were *'notoriously lacking in higher technical knowledge; they could not be expected*

---

[38] See Wark, The ultimate enemy.

[39] See Pugh, The Magic of a Name.

[40] A joint trip of Hives and Shenstone to Germany in early 1934 has already been described together with some background information in Sect. 4.3.3, AVA—the institution.

[41] Acc. to Gunston, Fedden was invited by General Milch in return to the German visit at Bristol in 1936, Fig. 6.4 (r). Officially Fedden's mission was declared as the opening of sales negotiations on Bristol aero engines, underlined by the participation of Ken Bartlett, Bristol's continental sales manager and Baron William de Ropp, the company's German agent. See Wikipedia,'William de Ropp' in English, ranking him *'as one of the most mysterious and influential clandestine operators'* of the era. In addition, Captain Ken Bartlett shows up in two recent books—see Matthaeus, The Political Diary of Alfred Rosenberg, and—see Griffiths, Fellow Travellers, in highly political, completely contradictory roles of which Gunston in his Fedden biography apparently knew nothing.

[42] Quoted after—see Wark, The ultimate enemy.

to have the insights of a master designer nor contact with the higher administrators of production.'

Technically Fedden and his Bristol engines were in the 1930s nearly unchallenged. Willy Neuenhofen's altitude record of 1929, Fig. 4.35, was already broken on 4 June 1930 by Lt. A. Soucek, US Navy, on a Wright Apache, powered by a 450 hp *Pratt & Whitney R1340 B*, but thereafter began a series of another seven altitude records, all of these either directly powered with Bristol sleeve-valve engines or license-produced derivatives. A clear sign—also noticed in the German RLM and industry—of the supreme position of Fedden's engine designs in this field up to the Second World War.

What Fedden might have recognised only slowly then that the German side was looking for qualified, influential individuals in Britain, France and the United States to whom they could show off their newly gained military and industrial might. Nowhere was this more impressive than in the aircraft industry which by 1936, at least in terms of output, had clearly become No.1 in the world. As Bill Gunston put it in Fedden's biography *'Disclosing its true strength was not a security oversight. On the contrary, the Germans hoped that, by demonstrating to potential enemies that there was no hope of Germany being defeated by force, those nations might be persuaded to stand peacefully aside and not interfere with ... Hitler's plan for the future, which involved taking over almost all of Europe.*[43] The type of man Goering and especially his deputy General Milch were looking for had to be technically highly competent, skilled at assessing industrial potential and at the same time loyal to his own government, so that his assessments in return would be wholly convincing and credible. Clearly the first name on the German list of potential candidates was Roy Fedden (1885–1973),[44]—and with some curtailments Charles Lindbergh (1902–1974) and General Joseph Vuillemin (1883–1963), who became Chief of Staff of the French Air Force in 1938.[45]

In 1936 Fedden did not visit Germany, but he himself and officially his Bristol Aeroplane Company was visited—apparently in response to a foregoing visit of British Vice-Air Marshals Courtney and Evill, end of 1935, in Germany with detailed presentations[46]—by a group of the most senior German aviation men, led by Generals Milch and Udet, Fig. 6.4. By this time, the company had a payroll of 4200, mostly in Fedden's engine factory, and was well positioned to take advantage of the huge

---

[43] See Gunston, Fedden, p.204

[44] See Wikipedia, 'Roy Fedden' in English.

[45] Strikingly in this context, Irving's Milch biography mentions *'Sir Roy Fedden'* solely vaguely in the foreword with gratitude for *'granting an interview or writing a letter'*, but uses his name in the main text just one time indirectly by *'the chief designer of Bristol Aircraft Company was shown round German aircraft factories and opened negotiations for the sale of Bristol aero-engines to Germany'*.

[46] See Irving, The Rise and Fall, p. 59.

Fig. 6.4 Bristol Blenheim 142 M, *'Schnellbomber'* prototype, 1935 (l), Luftwaffe visiting delegation at BAC Museum Bristol, ~1935/6 (r): W. von Richthofen, E. Milch, Mrs. D.M. White, BAC heiress; R. Fedden (?, in her shadow), unknown (with moustache), l-t-r © BAC Museum Bristol (r)

re-armament ordered by the British Government in May 1935. Bristol's most important contribution to the expansion of the RAF at this time was the Blenheim light bomber. Interestingly, the *'Blenheim'*[47] had its origins in some private foresight of Lord Rothermere (1868–1940).[48]

The aircraft was developed as *Type 142*, a civil airliner, in response to a challenge on his initiative to produce the fastest commercial aircraft in Europe. The *Type 142* first flew in April 1935, powered by two 560 hp Bristol Mercury 4 radials, and the performance proved so good that Lord Rothermere presented the aircraft as a gift to the nation. The Air Ministry, evenly impressed, ordered a modified design, *Type 142 M* as a RAF bomber. This first RAF all metal, stressed-skin construction, low-wing monoplane with retractable undercarriage, flaps, a powered gun turret and variable pitch propellers was an immediate

---

[47] Blindheim is today a 1700 inhabitant community, half-way between Stuttgart and Ingolstadt, where on 13 Aug. 1704 the *'Battle of Blenheim'*—in German the *'2nd Battle of Höchstädt'*—took place, a major battle of the War of the Spanish Succession. The overwhelming Allied victory by John Churchill's, the Duke of Marlborough's army, supported by Prince Eugene's troups, ensured the safety of Vienna from the Franco-Bavarian army, thus preventing the collapse of the anti-French *'Grand Alliance'*, formed 1689 between England, the Dutch Republic and the Archduchy of Austria. Of the 108,000 soldiers on the battlefield, nearly 40,000 were killed or wounded. See Wikipedia, 'Battle of Blenheim' in English.

[48] Harold S. Harmsworth, 1st Viscount Rothermere was a leading British newspaper proprietor and pioneer of popular journalism. Two of Rothermere's three sons were killed in action during WW I; consequently, during the 1930s he advocated peaceful relations between Germany and the United Kingdom, using his media influence to this end—up to open support for fascism, praise for Nazism and Oswald Mosley's British Union of Fascists of 1932. See Wikipedia, 'Harold Harmsworth, 1st Viscount Rothermere' in English.

success, having a speed of 500 km/h at 4 km altitude and thus faster than most fighters in the late 1930s.

Figure 6.4 (r), is a rare document[49] of this visit at BAC Bristol-Filton, showing to the left of the examined *Blenheim* bomber model and in front of a larger German delegation, Wolfram von Richthofen[50] and Erhard Milch and on the opposite side Mrs. Daisy May White,[51] daughter of the founder of the Bristol Aeroplane Company in 1910 Sir George White (1854–1916), presumably accompanied by (Sir) Roy Fedden. It appears that this Bristol visit was in the hightime of the *'Schnellbomber'* idea, perhaps inspecting the *'Blenheim 142 M'* might have been even a prime target on the German visitors's list to reaffirm their own changing positions on this issue. The concept developed in the 1930s when it was believed that a very fast bomber could simply outrun its enemies. After the unarmed high-altitude (poison gas) bomber concept of the 1920s[52] had been given up, omitting defensive armament should allow now for significant reduction in drag as well as weight, resulting in improved performance. Adolf Hitler was a staunch supporter of the *Schnellbomber*[53] idea and directed after General Wever's death in 1936, Germany's prime proponent of strategic bombing, to only develop medium bombers, while the USA and British developed both twin-engine medium bombers and four-engine heavy bombers. At the end of July, just weeks before the 1936 England tour, at the fourth International Air Meeting at Zurich-Duebendorf, Milch led the German team, Udet presented a special Me109 in flight, but the principal event was the bomber competition—without RAF participation: Also here, the Dornier Do 17, the then latest of the German Schnellbombers,[54] proved to be

---

[49] Courtesy to Stefanie Vincent, Collection Manager at Bristol Aerospace Museum. The dating of the photo is speculative, but it is suggested that this visit of a high-ranking delegation took place before Nov. 1936, when W. von Richthofen was appointed *chief-of-staff* of the Luftwaffe's Condor Legion, supporting the Nationalist side in the Spanish Civil War; —see Neufeld, Rocket Aircraft, p. 215.

[50] Wolfram Freiherr von Richthofen (1895–1945) was then Lt.-Col. and RLM department head for aircraft development, rising in his further military career up to Field Marshal in 1944. He was a fighter pilot during WW I. During his first mission together with his famous cousin Manfred, the *'Red Baron'* on 21 April 1918, Manfred was killed. After the war WvR studied mechanical engineering at TH Hannover 1920–1923, thereafter he entered the Reichswehr (armed forces) and wrote a doctoral thesis *'The influence of aircraft design on the supply process, especially from military perspective'*, leading to a Dr.-Ing. degree from TH Bln.-Charlottenburg in 1929. On 26 April 1937 German aircraft destroyed under von Richthofen's command the basque town of Guernica, origin of Picasso's 1937 iconic anti-war painting. See Wikipedia, 'Wolfram Freiherr von Richthofen' in English.

[51] Mrs. Daisy May White (1882?-1969), BAC heiress, then Mrs. Major E.J. Hudson.

[52] See 'Altitude Aircraft Developments'in Sect. 4.3.2 and—see Irving, The rise and fall, p. 64: On 1 July 1937 Lord Trenchard,—see Wikipedia, 'Hugh Trenchard, 1st Viscount Trenchard' in English,—the founder of the RAF, visited Berlin. He asked if Germany would ever use poison gas and—acc. to Irving—Milch gave a solemn undertaking that Germany would not initiate such warfare.

[53] Up to late in the war when even the first operational turbojet aircraft *Messerschmitt Me 262* had to be built also in this version.

[54] The first German aircraft adopted for that role was the single-engined *Heinkel He 70*, soon replaced by the twin-engined *Dornier Do 17* and the *Junkers Ju 88*. Most successful was the bomber version of

faster than any foreign fighter taking part. In general, however, the *Schnellbomber* lacked the payload and range of heavy bombers which put them at considerable disadvantage for strategic bombing.

After Ernst Udet had been sent to the RAF air show display at Hendon, 26–29 July 1937, there was again one confirmed counter-visit to Fedden's two already mentioned trips to Germany during 1937. A high-ranking German delegation onboard Milch's personal He 111 V-16 prototype (D-ASAR, with elliptical wing, but still without the glazy *'greenhouse'*nose, typical for later versions and two newly-developed 910 hp DB 600 C engines) flew to RAF Mildenhall on 17 October 1937, where a RAF band intoned (Hitler's favourite) *'Badenweiler'*and *'Alte Kameraden* (Old comrades)'marches. There was a first welcoming reception by then Air Minister Philip Cunliffe-Lister (Lord Swinton) on the following day,[55] before the official round trip up to 25 October 1937 touched the unique *'shadow factories'* in the Midlands—producing motor-cars and engines in peacetime, but ready for rapid conversion to aircraft production under war demands;[56] in addition, there was also a stop at Rolls-Royce Derby.

6.1.3   Berlin, 11–15 October 1938 Or: *Circus Lilienthal* At The Zoo Palace

The foregoing narrative, interwoven with indicated elements of espionage and counter-spionage, culminated quite naturally in an international aeronautical convention in mid-October 1938 at Berlin, to which the Lilienthal Society for Aeronautical Research[57] had invited. Though the applications for that conference had been issued certainly well in advance, the atmosphere was still highly influenced by the foregoing *Munich Conference* and the signed *'Peace-for-our-time' Agreement*, concluded on 30 September 1938. Generally, it is said, Europe celebrated the *'outbreak of peace'* because it prevented the war threatened by Hitler on last minute, by allowing Nazi Germany's annexation of the *Sudetenland*, a region of western Czechoslovakia inhabited by more than three million people, mainly German speakers. Hitler announced it was his last territorial claim in Europe, and the choice seemed to be between war and appeasement.

---

the British *de Havilland Mosquito,* which retained a speed advantage over its enemies for much of the war and—very late—the jet-engined Arado Ar 234 Blitz, sometimes dubbed ‚Schnellstbomber (the fastest bomber)'.

[55] A link to a corresponding photograph can be found by means of the internet search key words <Cunliffe-Lister Udet Milch>.

[56] Visited were the Austin Motor Works at Longbridge, Birmingham as well as the Standard Motor Co. and Humber Ltd at Coventry. See Irving, The Rise and Fall, p. 67, and Footnote 14 (there).

[57] The *'Circus Lilienthal'*in the header has been taken deliberately in view of the extraordinarily high *'spy concentration'*at this conference of the Lilienthal Society, in appreciation of John Le Carré's 1974 spy novel *'Tinker tailor soldier spy'*, where the London office of the British Secret Intelligence Service is located at *'Cambridge Circus'*, and the corresponding location of the event at the—see Wikipedia, 'Ufa-Palast am Zoo' in English.

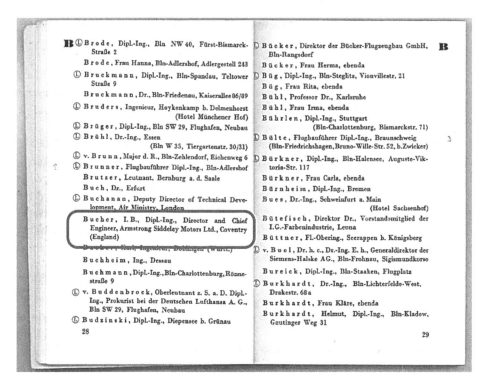

Fig. 6.5 Int'l. Convention of the Lilienthal Society for Aeronautical Research, Berlin 12–14 Oct. 1938, pp. 28/29 of the 182 p. *'Booklet of registered conference attendees'* © Lilienthal Museum Anklam

The aeronautical conference took place in the centre of Berlin's New West, in the *Ufa-Palace am Zoo*, and had nearly 3000 participants,[58] Fig. 6.8, of which some 150 came from abroad (IT 32, GB 26, US 18, JP 14, PL 11, FR 8). Certainly the youngest of the British contingent was Armstrong-Siddeley's John B. Bucher (26), whose entry in the 182 page *'Booklet of conference participants'* has been highlighted in Fig. 6.5.[59] His visible professional promotion to *'Director and Chief Engineer'* after just two years at ASM may be surprising, but can perhaps be explained by his special mission, which might have required some seniority, at least on the business card. In the early 1990s Richard Hodgson could not interview Jack Bucher any more directly, but found hints about him in *Public*

---

[58] The official max. cinema capacity since 1926 were 2200 seats. One can imagine that part of the conference sessions was swapped to neighbouring institutions, but it is definitively known that the conference was shifted to DVL Adlershof on Sat 15 Oct.1938, combined with visits of that aeronautical research institution and a *'farewell party'*.

[59] The encircled "L" in front of some names stands for members of the Lilienthal Society.

Record Office[60] files which had been released after 50 years, though the original classification period had been set to even 70 years. In addition, he could speak to Bucher's near relatives.

The information combined created the impression that Bucher—especially after the Vorkauf steam engine license negotiations were already closed in 1937—had been sent to attend this Lilienthal Convention not only on behalf of ASM. As indicated before, there is the possibility that Heppner was seriously interviewed by the British secret service about his turbojet contacts and development observations before he had left Germany.

If there grew at the same time already strategic business plans at the Air Ministry to plant the asset *'Heppner'* mid-term at ASM, it could have been a nearly self-suggesting idea to set Bucher with very suiting preconditions (mechanical engineer, excellent German language skills, ASM background) on a verification mission to Berlin about the collected Heppner information.[61] Section 12.2 has a detailed excerpt of in total 58 names out of a listed number of ~2700 registered participants, which all show up in this book in various interrelations, sufficient contact possibilities for Bucher just on the convention venue. Moreover, ideally information might have been received directly from Hellmut Weinrich on the spot, Heppner's presumed engineering partner before 1935, who lived then at Berlin, and a *'casual street contact'* could have certainly be arranged. Hodgson observed in the ASM/PRO context, files with a blocking classification period of 100 years (!),[62] not unlikely that there is still some information about the Heppner case to be expected. The official Convention programme of the Lilienthal Society foresaw two occasions for award ceremonies, on the evening of Tue 11 October 1938 at *'Neues Palais* (New Palace)'*, Potsdam and during the convention opening ceremony at *'Zoo Palace'* on Wed 12 October 1938 morning; both will be reviewed in detail in the following.

Besides the ceremonial highlights there were numerous smaller honours distributed among the conference guests, as there is e.g. Major John Buchanan,[63] visible in Fig. 6.5 in a booklet entry just above *'Bucher'*, who in 1933 was instrumental in issuing the Air Ministry's first monoplane fighter specification, which led to *'Hurricane'* and *'Spitfire'*. Major Buchanan together with Major George P. Bulman,[64] who led the Air Ministry's

---

[60] The *Public Record Office PRO* in the City of London, was the guardian of the national archives of the United Kingdom from 1838 until 2003, when it was merged with the Historical Manuscripts Commission to form *The National Archives*, based in Kew.

[61] Most interesting must have been information on the status of the counter-rotation BMW 109-002 project, Fig. 5.11, at that time under intense development at BMW Berlin-Spandau.

[62] Telephone interview with R. Hodgson on 28 March 2020.

[63] In the post-war period 1950–1951 John Buchanan served also as RAeS President.

[64] Bulman has been already addressed in the foregoing Sect. 6.1.1 and at some length at Chap. 5, crossing Heppner's path after *'November 1939'*. Bulman apparently did not attend the conference; nothing is known about his and Buchanan *'merits'* for the German case, but it is possible that these were in support of Fedden's still—with respect to contents—secret two 1937 *'sales missions'* (beyond the German aero industry inspections) together with Captain Ken Bartlett and Baron William de Ropp. The initiative on the German side for honouring two influential representatives of the Air

DED Department of Engine Development between 1928 and 1944, became on that occasion *'corresponding members of the Lilienthal Society'*.

The illustrated page of the *'Booklet'*, Fig. 6.5, has also in the left column as third entry fom the top Dipl.-Ing. (Bruno) Bruckmann, then head of the air-cooled piston-engine development at Bramo Brandenburgische Motorenwerke Berlin-Spandau, which was merged in 1939 with the BMW aero engine development activities to BMW Flugmotorenbau GmbH, Munich. Bruno Bruckmann, a long-time acquaintance of Roy Fedden, gave him during the conference also important advance information about the German aero gas turbine development plans.

While Jack Bucher in John Le Carrée's spy listing belonged rather to the category of *'tinkers'*, the English conference delegation had also a real, coming *'master spy'* in their rows: William F. Forbes-Sempill, 19th Lord Sempill (1893–1965),[65] who attended the conference mostly as former RAeS President 1927–1930 in support of the then actual RAeS President Roy Fedden, but also as known, extreme-right wing pro-Nazi politician.[66]

The handling of Sempill's spy case over time developed quite strangely and it was not until the release of intelligence records by the Public Record Office in 1998 and 2002 that his activities as a spy during the war and in the 1920s respectively became common knowledge. In the early phase of the 1920s, Sempill began giving military secrets to the Japanese and although his activities were uncovered by British Intelligence, Sempill was not prosecuted for spying and allowed to continue in public life. On the outbreak of war in 1939, Sempill was given a position in the Air Department at the Admiralty, the government office for marine affairs. This gave him access to both sensitive and secret information about the latest British aircraft, which was obviously forwarded to Mitsubishi's London office. Sempill was also probably passing on detailed information about the British government. In August 1941, Winston Churchill and President Franklin D. Roosevelt held a meeting in Newfoundland aboard HMS*Prince of Wales* to discuss the military threat posed by the Japanese; soon after, communications between the Japanese Embassy in London and Tokyo were deciphered by the *Bletchley Park* code breakers. The decrypted messages were transcripts of the conference notes. Finally, on 13 December 1941, six days after the attack on Pearl Harbor, Sempill's office was raided. A search revealed various secret documents that he should have handed back over three weeks earlier. Two days later Sempill was discovered making phone calls to the Japanese Embassy. Despite the evidence of treason in wartime, no arrest or prosecution was ordered; instead, Sempill agreed to retire from public office.

Ministry came certainly from the Milch/Udet camp. In 1939, a few months after the—from Nazi perspective—successful conference, Goering appointed Milch as *'Honorary President'* and Udet as *'President of the Lilienthal Society'*.

[65] See Wikipedia, 'William Forbes-Sempill, 19th Lord Sempill'in English.

[66] Sempill's active role as then RAeS President in bringing Ludwig Prandtl to England in 1927 for delivering the prestigious *'Wilbur Wright Memorial Lecture'*has already been described at the end of Sect. 4.2.1. After his first term as RAeS President 1938–1940, Roy Fedden was elected a second time 1944–1945.

Fig. 6.6 Dr. Hugo Eckener and his Zeppelin airship D-LZ127 (l + m), recipient of the *'RAeS Gold Medal 1936'* on 11 Oct. 1938 (m), Roy Fedden, RAeS President and recipient of the *'Lilienthal-Ring'*, 12 Oct. 1938 (r)

Besides these impressive decade-long spy affairs, Sempill was a record-breaking aviation pioneer and regular guest at the *Wasserkuppe* glider community from early on. On 2 April 1936 Col. Sempill flew a *'B.A.C. Superdrone'*, an ultra-light single-seater motorglider-type aircraft of 178 kg empty weight, fitted with a 2 cyl., 750 cm^3, 19 hp Douglas motorcycle engine, manufactured by British Aircraft Co., Maidstone, Kent—in the meantime owned by the German-British glider champion Robert Kronfeld—from Croydon-Airport to Berlin-Tempelhof in 11 h flying time, corresponding to an average flight speed of 88 km/h, on 64 l of petrol or 6.5 l/100 km. At Tempelhof, Lord Sempill was greeted by Konstantin Freiherr von Neurath (1873–1956), German ambassador to Britain 1930–1932 and thereafter Reichsminister for Foreign Affairs 1932–1938. The return flight on 4 April took him nine hours via Rotterdam to Canterbury; this was a new long-haul-flight record along straight line for aircraft below 200 kg empty weight.[67]

On 1 November 1928, German pilot Dr. Hugo Eckener (1868–1954), the most successful airship commander in history, landed the LZ-127 *'Graf Zeppelin'* airship in Friedrichshafen, Germany after a 71-h flight, Fig. 6.6.[68] The 6384.50 km flight, which

---

[67] 'Flugsport', illustrated magazine, Vol. 28, No. 8, 1936, p. 176.

[68] The figure shows the *'British Gold Medal for Aeronautics 1936'*, 50 mm diam. silver-gilt, designed by British medallist Allan G. Wyon sc. (1882–1962), with the circumscription *'Awarded for Advancement of Aeronautical Science'*, half-portrait of Sir George Cayley, Father of Aeronautics 1773–1857 and First model airplane, 1804. At present (2022), RAeS has 'dumped' Eckener from the List of Gold Medalists.

set world records for both *'duration'* and *'distance'* flown in an airship, began in Lakehurst, NJ, USA on 29 October of the same year. Ratified by *FAI*, the two records set by Eckener's transatlantic flight still stand today—making them the longest held records in FAI history. Thereafter in August 1929 the *'Graf Zeppelin'* launched from Lakehurst to a tour round the world which was accomplished after a flying time of 12 days, 12 h and 13 min—with stops 21+ days—covering 33,234 km, the fastest circumnavigation of the globe at the time. Highly safety oriented, Eckener nevertheless followed as former journalist of the *'Frankfurter Zeitung'* his instinct for spectacular effects: after crossing the Pacific, he delayed over the coast at San Francisco's *'Golden Gate'* so as to come in near sunset for aesthetic effect. And, on 26 April 1930 *'Graf Zeppelin'* flew low over the FA Cup Final at Wembley Stadium,[69] dipping in salute to King George V, then briefly moored alongside the larger *'R100'* at Cardington Airfield, Bedfordshire—the English loved it. Eckener was responsible for many innovative aviation developments, notably the trans-Atlantic passenger services offered by the airships *'Graf Zeppelin'*[70] and *'Hindenburg'*, for which the RAeS awarded him its Gold Medal in 1936,—but the actual award ceremony was put on hold.[71]

An anti-Nazi who was invited to campaign as a moderate in the 1932 German presidential elections against Hitler, he bowed out when Paul von Hindenburg (1847–1934) decided to run for a second term. Consequently, Eckener was blacklisted by the Nazi-regime after 1933 and eventually sidelined. After various quarrels with Joseph Goebbels, Reich Minister of Propaganda, Eckener was *persona non grata*, feeling temporarily—as he stated in a 1948 *Spiegel* opinion page[72]—threatened to be put in concentration camp himself, he took his last chance in March 1938 to deliver on Goebbels's request a short, radioed loyalty address, by praising Hitler's *genius* after the Austrian *'Anschluss* (connection to the Reich)'. Obviously this cleared also the way in time for a 2-year-belated award ceremony to deliver the *'British Gold Medal for Aeronautics'* to Hugo Eckener on Tue 11 October 1938, during a *'Herrenabend* (Gentlemen's evening reception)' at the *'New Palace'*[73] Potsdam before opening the Lilienthal Convention on the next morning.

---

[69] Arsenal beating Huddersfield Town 2:0; the passing occurred at the start of the second half approximately 600 m above the 92,499 stunned spectators, apparently *'low'* in comparison to *Graf Zeppelin's* length of 236.6 m,—see Wikipedia, 'FA Cup Final' in English.

[70] See Wikipedia, 'LZ 127 *Graf Zeppelin*'.

[71] The hidden policy developments of the Reich Ministry of Propaganda against Hugo Eckener can be neatly reconstructed from the issued press directives, e.g. on 2 May 1936 it was prohibited to take over what the *Times, p. 14* had reported the same day: *The British Gold Medal for Aeronautics has been awarded by the Amulree Committee on the recommendation of the Council of the Royal Aeronautical Society, to Dr. Hugo Eckener for his technical achievements in lighter-than-air craft.* ... See Toepser, NS-Presseanweisungen.

[72] See 'Eine SPIEGEL Seite für Hugo Eckener', 18 Dec.1948, in German https://www.spiegel.de/spiegel/print/d-44420964.html

[73] The *'Neue Palais'* is a palace situated on the western side of the *Park Sanssouci* in Potsdam, some 30 km in the south-west off the conference location at the city centre in Berlin-West. The building was begun in 1763 under King *Friedrich II—the Great* and was completed in 1769; it is considered to

On that Wednesday the *Dundee Evening Telegraph*,[74] copied from the *Times*, reported on the event of the foregoing evening under the header *'British Honour for Zeppelin Expert'*:

Sir Nevile Henderson, the British Ambassador in Berlin, presented the Gold Medal of the Royal Aeronautical Society to Dr Hugo Eckener of airship fame.

The presentation was made at a dinner given by Field-Marshal Goering in the new Palace at Potsdam for the German and foreign delegates to the annual congress of the Lilienthal Society for Aeronautical Research, which is being held in Berlin.

Speaking in German, the British Ambassador mentioned that this was the first time that the Gold Medal of the Royal Aeronautical Society had been awarded outside the British Empire. The award, he added, was to be taken as marking British appreciation of the high place Dr Eckener had won for himself in the sphere of aeronautical research.

The British delegates to the Congress include Lord Sempill and Colonel Etherton.

The United States is represented by, among others, Colonel Lindbergh, who, accompanied by Mrs Lindbergh, arrived at the Templehof Aerodrome piloting his own machine.

The delegates hope to have opportunities of visiting German aircraft factories and experimental establishments.

In view of this official newspaper clip, it becomes clear that memories to the events on 11/12 October 1938 in Fedden's official biography must have been somewhat *scrambled*[75]—neither

- *received he the Lilienthal Ring from Hitler* on that evening (done by Udet the next morning during the conference opening ceremony, Figs. 6.8 and 6.9), nor
- did he *present the RAeS Gold Medal to Zeppelin commander Dr Eckener* (done by Ambassador Henderson in German—with RAeS President Fedden standing aside).

However, Gunston's book has two remarkable scenes in this context,

---

be the last great Prussian Baroque palace.The award ceremony took place '*im kleinen Kreis* (in a small round)' with 400 invited participants, mostly from abroad, after dinner in the *Upper Gallery* and the *Marble Hall*. General Milch delivered the welcoming address in the name of General Field Marshal Goering, which Lord Sempill replied for the British guests and RAeS. Before, there was a concert arranged in Frederick the Great's Theatre in Rococo style, see picture in Wikipedia, 'New Palace (Potsdam)' in English, Wilhelm Kempff and the Berlin Philharmonic played Beethoven and Mozart. At that time, Kempff and his family lived in the Orangery of Park Sanssouci. See Wikipedia, 'Wilhelm Kempff' in English. The Luftwaffe top *clique* preferred the vast New Palais, because of their apartments for private usage, especially advantageous after somewhat longer *soirées*.

[74] Newspaper clip, courtesy to John Bailey, 1973 NUJ president, 21 Oct. 2019.

[75] See Gunston, Fedden, p. 210.

- Fedden meets *at that dinner Group Captain* (John L.) *Va(t)chell, Henderson's air attaché in Berlin,... to whom he was related,*[76]—in view of the following an important hint to the attending British Embassy personnel, and *his encounter with*
- *Bruno Bruckmann, Technical Director of the great BMW engine firm, whom he had known for years. With disarming frankness the German engineer told Fedden* "We have decided we do not need to acquire a licence for your Hercules motor. Not only do we hope to rival it with our own engines but we are developing a new type of engine for aircraft, a gas turbine".[77]

Of that remarkable evening exists a rare photo document, Fig. 6.7, which in the official press release and consequently, the Bundesarchiv (Federal archive) files identified correctly from left-to-right: Ernst Udet—Erhard Milch—Ernst Heinkel, but left the attentatively looking person to the right—unidentified.

After some dead-end searches, it became clear that the British Embassy Berlin was that evening not only represented by the Ambassador Nevile Henderson (1882–1942)[78] and the Air Attaché Group Captain John L. Vachell (1892–1947),[79] but also by the Military Attaché Colonel Noel Mason-MacFarlane (1889–1953)[80]—and not to forget the whole accompanying spectre from *'tinker'* J.B. Bucher to *'master spy'* Lord Sempill.[81] Mason-MacFarlane spoke fluently French and German when growing up under the influence of a German governess hired by his father. He was educated at Rugby School and attended the Royal Military Academy, Woolwich. Until 1909 his surname was Mason, when he

---

[76] See Wikipedia, 'Ada Vachell' in English.

[77] A credible statement with respect to the timing, see Sect. 6.2, and in view of a similar after dinner confession of Hermann Goering to Charles Lindbergh in the American Embassy one week later, on Tue 18 Oct. 1938: Goering *'said the new Junkers 88 bomber (which no one we know has seen) is far ahead of anything else built...(it) did 500 kilometers per hour and that it was not "a magazine figure",* and *'He said they expected to have a plane which would make 800 kilometers per hour in the near future (at critical altitude)'*. Before, during cocktail small talk Goering handed out to Lindbergh a little red box *'by order of the Fuehrer'* which he later found out was the *German Eagle Order (with Star),* endowed as an Order of Merit to foreigners by Hitler in May 1937. Henry Ford had received it already to his 75th birthday on 30 July 1938. See Lindbergh, The Wartime Journals, p. 102/103, and—see Wikipedia, 'Verdienstorden vom Deutschen Adler' in German.

[78] See Wikipedia, 'Nevile Henderson'in English, who must have received during these days in Oct. 1938 his cancer diagnosis, which however did not initiate his removal from this decisive post.

[79] The internet has several biographical entries about <Air Commodore J L Vachell>.

[80] See Wikipedia, 'Noel Mason-MacFarlane' and—see Rankin, Defending the Rock, both in English. The positive identification was finally successful on the basis of several other photographs on the internet, showing characteristically a slight scar at the lower lip centre, presumably a leftover from *Mason-Mac's* severe car accident in 1933. In addition, it was a special pleasure for the Author that the present, now 'Defence Attaché' since Feb. 2015 at the British Berlin Embassy, still at Wilhelmstrasse 70, Brigadier Rob Rider agreed on 20 Sept. 2019 *'The resemblance seems striking. And a good fit.'*

[81] The *'Who is who'* of that evening, the *Booklet of conference participants* can be seen lying on the table, and Ernst Heinkel is browsing his personal copy.

Fig. 6.7 Lilienthal Conv. evening reception, New Palace Potsdam, 11 Oct. 1938. L-t-r: Col.-Gen. Ernst Udet, Gen. Erhard Milch, Dr.-Ing. Ernst Heinkel, Col. Noel Mason-MacFarlane © Bundesarchiv/Wikimedia Commons

hyphened it to Mason-MacFarlane out of pride in his Scottish heritage as MacFarlane was the maiden name of his Scots mother. Interested readers may find their way from Mason-MacFarlane's Wikipedia article to a bunch of related anecdotes, for which typically a *'Mirror Online' article from 22* Februar 2016 stands: *'It summarized a Mason-MacFarlane proposal to use a sniper to shoot the German leader from his flat in the centre of Berlin, put to the British government in March 1939, six months before German troops stormed into Poland, under the headline:* "Britain's secret plan to shoot Hitler could have prevented war—but the government decided it was "unsportsmanlike"'. The actual detection of this historic gem goes back to the German news magazine *'Der Spiegel'* which reported on 4 August 1969 under the header *'Widerwillen gegen Morde* (Distaste against murder)"[82] about an unpublished article draft which then General Mason-MacFarlane after his service as Governor of Gibraltar during WW II wrote in 1952—and which was found in his collected papers, since June 1969 at the Imperial War Museum London.

---

[82] See https://www.spiegel.de/spiegel/print/d-45740913.html.

Fig. 6.8 Int'l. Lilienthal Convention Berlin, opening ceremony, 12 October 1938: Ufa-Palast am Zoo (l), on stage: Lilienthal-Ring ceremony with E. Udet, RLM and R. Fedden, Bristol Motors (r) © Austrian National Library (r)

The main session of the International Convention of the Lilienthal Society was opened at the Berlin *Ufa-Palace am Zoo* in the morning of Wed 12 October 1938, Fig. 6.8.[83] Nearly 3000 attendants listened to the short welcoming address of General Milch, followed by a lengthy speech, devotional to *'Fuehrer'* and Goering, by the President of the Lilienthal Society Adolf Baeumker (1891–1976), just newly appointed Ministerialdirigent in the Ministry of the Reichswehr and generally responsible for the coordination and direction of the aeronautical research resources in line with the Nazi regime policies. Thereafter, the head of the RLM Technical Office Col.-General Ernst Udet opened the Lilienthal award ceremony.

Lilienthal Memorial Medals, which were handed out after 1937 only for the second time, received:

- Dr.-Ing. E.h. Claudius Dornier for the improved, structural design of metal aircraft,
- Prof. Dipl.-Ing. E.h. Heinrich Focke for the creation of the first flight-worthy helicopter,
- Prof. Dr. phil. Albert Betz for his fundamental work on wing and propeller theory,
- Dr.-Ing. Heinrich Ebert for the successful development of the variable-pitch propeller,
- Albert Patin[84] for his advanced contributions to telemetry and remote control (*Patin compass*),
- Adolf Beck for fundamental improvements of Al-Mg alloys for aero applications.

---

[83] A detailed description of the events on 11/12 Oct. 1938 appeared in print,—see Deutsche Luftwacht, Luftwissen, Hauptversammlung 1938, in German.

[84] Albert Patin was a French engineer, living and staying in Germany during WW II. Though the Lilienthal Memorial Medals (*'Denkmünzen'*) were per definition reserved to excelling German nationals, nothing is known about a conflict. A situation, somewhat similar to the colported anecdote

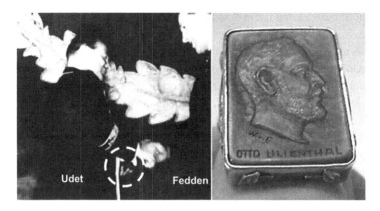

Fig. 6.9 Close-up view of *'Lilienthal Ring'* hand-out from Udet to Fedden (l), and *'Otto Lilienthal Ring'*, onyx cameo, 18x25 mm (r), designed and made W.v.E. Wilhelm von Eiff © Austrian Nat'l Library (l) and Lilienthal Museum (r)

As ceremonial highlight, Udet presented the first time issued Lilienthal-Ring, Fig. 6.8 (r), to

- *Herr Fedden, director and chief designer of the Bristol Works, who earned during the past two decades extraordinary merits in flight engine design and construction. It is due to his creativity, that the air-cooled star motor* (radial engine) *succeeded again and again in new design solutions. The German aviation keeps friendly relations with Herr Fedden since years. By happenstance he became just recently President of the Royal Aeronautical Society, with which we have many targets in common.*[85]
- and thereafter, the Lilienthal-Award (5000 RM) of the Ministry for Science and National Education was presented to studing. Reimar Horten (23), Bonn for—together with his brother Walter (25)—impressive model flight contributions to the *flying wing* concept.[86]

Fedden himself had remembered the handing out of the Lilienthal-Ring, according to his biographer Bill Gunston *'at the impressive ceremonial dinner at the Neue Palais at Potsdam, at which he received the Ring from Hitler and presented the RAeS Gold Medal to Zeppelin commander Dr Eckener, there were other non-Germans present and they appeared to be completely overshadowed. At the centre table there was a gorgeous riot of colour from the Nazi uniforms. The only guests in evening dress were Sir Arthur (sic!) Henderson, British Ambassador; Col. Charles Lindbergh, representing the USA; and Fedden.*[87] But as the close-up view of Fig. 6.9 clearly shows, with the Ring-containing box encircled in Udet's left hand, it was Udet and not Hitler, at the Zoo Palace and not at

---

about Goering's comment, hearing rumours about Milch's Jewish father after 1933: *'I decide, who is Jewish . . .'.*

[85] Translated from—see Deutsche Luftwacht, Hauptversammlung 1938.

[86] See Wikipedia, 'Horten brothers' in English.

[87] Quotation from—see Gunston, Fedden, p. 210, published in 1999.

the New Palace, in Berlin and not in Potsdam, on Wed 12 October morning and not on Tue 11 October evening.

This allows also a closer look on this unique Ring and its own history, Fig. 6.9. It was endowed by Goering in 1936 as the highest international, scientific decoration of the newly formed Lilienthal Society on the occasion of merging and renaming German aeronautical societies, which actually dated back to 1912, second after the Royal Aeronautical Society of 1866.[88] Though otherwise planned, this Ring was never awarded before, and—after the tribute to Roy Fedden, his engineering achievements and indirectly through his presidency also to the RAeS in 1938—not thereafter. The centrepiece of the Ring carries a *carnelian*[89] cameo, embellished with an engraved relief portrait of the German aviation pioneer Otto Lilienthal (1848–1896). The 18x25 mm stone is in a 525/14 k gold mounting, weighing alltogether 12 g. The stone face has along the bottom rim the designating name in stone engraving *'intaglio'* technique and above, the long time unknown abbreviation of the artist's name W. v. E., who has been identified only recently as the German glas artist Wilhelm von Eiff (1890–1943).[90]

A. H. Roy Fedden (1885–1973) was born in the Bristol area to relatively wealthy parents; he attended the *Clifton College* in a suburb of Bristol and started thereafter an apprenticeship as engineer which he finished in 1906. Many elite engineers were trained by apprenticeship into the interwar years. Fedden's 3-year premium apprenticeship cost his family £250, in addition to the attended college evening classes.[91] He quickly became established as a car designer, and during the 1st World War as an aero-engine designer. After some intermediate professional steps where he mostly gained his practical engineering experience, he established the engine building department of the Bristol Aeroplane Co. in 1920, where he was chief engineer until 1942. According to David Edgerton, he became a very wealthy man in the interwar years, perhaps even Britain's highest paid engineer. Initiating a famous range of piston engines, including the *Jupiter, Taurus, Pegasus, Hercules* and later the *Centaurus*, he was especially notable for his unique development of the sleeve-valve engine.[92]

---

[88] As outlined in more detail in the Attachment-Sect. 12.2.

[89] Also known as *red/reddish chalcedony*.

[90] See Wikipedia, 'Wilhelm von Eiff' in German. Von Eiff's work is closely related to the tableware manufacturer WMF Württembergische Metallwarenfabrik, Geislingen near Stuttgart, a surprising bridge to the axial turbojet developments in Germany, Sect. 8.1. At the end of WW II, WMF adapted their establshed deep-drawing technology for cooking pots for the manufacture of air-cooled hollow turbine blades, e.g. for the BMW 109-003 A-1 axial turbojet engine.

[91] See Edgerton, England and the Aeroplane.

[92] Distinct from the usual poppet valve, it was first introduced to pre-World War II luxury cars due to quietness of running and the very high mileages without servicing. A sleeve valve takes the form of machined sleeves, fitting between the piston and the cylinder wall of an internal combustion engine, where it rotates and/ or slides. *'Ports'* in the side of the sleeves come into alignment with the cylinder's inlet and exhaust ports at the appropriate stages in the engine's cycle. Fedden took a simplified car system and applied it after thorough development to the famous high-power British aero piston engines of the 1930/1940s.

Accordingly on 12 October 1938 afternoon, Fedden gave at the plenum of the Lilienthal Society convention an invited lecture *'The development of the mono-sleeve valve for aero engines'*, which appeared thereafter in a unique bi-lingual print.[93] He held a variety of governmental and international posts until 1960, and was—as mentioned—two times President of the Royal Aeronautical Society; Fedden was knighted in 1942. He was childless, so after his death in 1973 his personal belongings were apparently spread and the Ring showed up in 2016 in Pennsylvania, where Dr. Bernd Lukasch, long-time head of the excellent Otto-Lilienthal-Museum at Anklam managed to acquire it for the Museum exhibition.[94] Anklam, Otto Lilienthal's native town, lies some 180 km north of Berlin and 30 km south of the Peenemuende rocket test site.

Before World War II, Roy Fedden and many other experts throughout the world shared the assumption that better aircraft engines would result from small improvements of components of the existing internal combustion engine. Because the aircraft engine was an adaptation of the automobile engine, radical innovations were expected to appear first in the automobile engine. Roy Fedden wrote in a 1933 SAE article:[95] *'I do not anticipate any radical changes in the type of four-cycle internal-combustion engine as used today. When the present form of gasoline engine is superseded by a radically different power unit, it seems logical that this development will most probably be accepted first in the automotive field before it is introduced into aircraft'.* As known, Fedden's prediction was wide of the mark, for it was precisely the independence from the automotive background of traditional power plant expertise that enabled the various, at that time just forming engineering teams in Germany and England to seek a new, revolutionary internal combustion engine uniquely suited for flight.

6.1.4   Another Booklet and a Cartoon

Just 20 months after an insisting plea for a better understanding of peoples (Milch) at the Lilienthal Convention at Berlin, and its accompanying 182 page *'Booklet of registered conference attendees'*, containing approximately 2700 names, another similarly designed *'Black Book'* or officially *'Special Wanted List Great Britain'*[96] was prepared by SS-RSHA (Reich Sicherheits-Haupt-Amt, Reich Main Security Office) Berlin, this time comprising 279 pages and 2820 names for not so human targets. The secret list named prominent British residents with their presumed addresses, which were to be arrested after a successful

---

[93] See Fedden, The development of the mono-sleeve valve.

[94] Link to the multi-lingual website of the Otto-Lilienthal-Museum Anklam: http://www.lilienthal-museum.de/olma/home.htm

[95] See Fedden, Next Decade's Aero Engines.

[96] See Wikipedia, 'The Black Book' in English and—see Wikipedia, 'Sonderfahndungsliste' in German.

German invasion of Britain, code-named *'Operation Sea Lion'*—amongst these also eleven names (as originally listed, including the corresponding Black Book page number) which have been mentioned in the foregoing text:

- C47a Christie, [M. Graham(e)], British intelligence service agent, London, p. 32,
- C49 Churchill, Winston S., prime minister, Chartwell Manor, Kent, p. 32,
- E6 Eden, R. Anthony, war minister, London, p. 51,
- H101 Henderson, Nevil[l]e M., former British ambassador at Berlin, p. 84,
- K15 Kantorowicz, Hermann, 1877, Prof., England, p. 105,
- K17 Kantorowicz, Otto, Dr., 1906, London, p. 105, [Heppner's 1932 pat. partner],
- M88 Mason-MacFarlane, F. Noel, colonel, aviation attaché, Engl., p. 131, also F13
- M177 Morton, [Desmond], Brit. major, presumably England, p. 134,
- T55 Trenchard, ~~Barnett~~ [Hugh, 1st Viscount Trenchard], RAF marshal, p. 204,
- V1 Vachell, J.[ohn] L., colonel, aviation attaché, England, p. 213,
- V6 Vansittart, Robert, British intelligence service leader, London, p. 213.

To the few, in the meantime detected *follies* of the *'Wanted for arrest'* list belongs:

- F114 Freud, Sigmund, Dr., Jew, 6.5.[18]56 Freiburg (Mähren), London, p. 62, who had died already on 23 Sept. 1939 in London.

Actually the *'Special Wanted List G.B.'* had been planned to be issued as a supplement to a generally introducing *'Informationsheft G.B.* (information brochure Great Britain)', both of which to be distributed in a print run of 20,000. Whoever had collected the required information over time, it went to details like an assessment of the British school system, in this context combined with a warning to German officers not to promise *'Eton'* to their children prematurely, since the prestigious school was known to be booked out up to 1949.

In context of the apparently fading memories of Fedden and his biographer Gunston, as described in the foregoing, there is an interesting statement of one of Britain's leading historians David Edgerton, who tried under the header *'The Politics of Aeronauts'*[97] a kind of political classification of a few, here familiar British engineers:

'Whittle's class background was reflected in his politics. He was a socialist, like a number of other government-employed aeronautical engineers of the 1930s, including A.A. Griffith, humbly-born and Liverpool University educated, and R.M. McKinnon Wood, educated at Cambridge. But amongst aeronautical engineers as a whole this appears to have been unusual: Barnes Wallis[98] *and Roy Fedden, whose cases are best documented, were clearly on the right, for example. Other senior men in the industry, were, as one*

---

[97] See Edgerton, England and the Aeroplane

[98] Sir Barnes N. Wallis (1887–1979), English scientist, engineer and inventor, best known for the bouncing *'Dambuster'* bombs and the *'Tall Boy'* for deep-penetration bunker bombing.

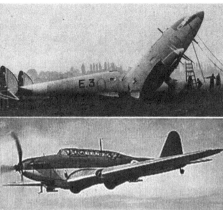

Fig. 6.10 The Aeroplane team ~1935: Charles G. Grey, Constance Babington Smith, Felix Kelly (l), broken DH.91 Albatross prototype 1937/8 (r, top), Fairey Battle bomber plane (r, bt.) © D.Simanaitis (l), Waynos (r, top), Wiki Commons (r, bt.)

would expect, also right-wing figures and a number ..., were on the far right. Of course, it is difficult to generalize about the politics of an industry, but the aircraft industry was in one important way exceptional. Its leading trade paper, The Aeroplane, *which was also read by many enthusiasts, was quite openly pro-Nazi, pro-fascist and anti-Semitic.* [99]

This said, it is nevertheless worthwhile to have a closer look to the team behind this influential aero journal, Fig. 6.10, and to a remarkable *Aeroplane cartoon* which appeared in a nearly historic moment, in the last issue before the outbreak of WW II, Fig. 6.11.

Charles Grey GREY (1875–1953), Fig. 6.10 (l), was the founding editor of the British weekly *The Aeroplane* in 1911, holding this post till November 1939,[100] and in addition that of the second editor of *Jane's All the World's Aircraft*, 1916.1940. He wrote his leading weekly aviation article for 28 years without missing a single issue—as a man of decided opinions. Unfortunately for him these included strong support for the fascist dictators of Italy and Germany, which played no small part in his leaving the magazine.[101]

---

[99] The *Aeroplane* chief editor Charles G. GREY attended also this Lilienthal Convention 1938, see Sect. 12.2, and was officially and explicitly welcomed, as *'old friend'*.

[100] Always an admirer of Germany, his position as editor was no longer bearable when he predicted in exactly this historic issue of 30 Aug. 1939, that Britain would never again go to war with Germany. He was always a critic of RAF and RAE Farnborough in its early days, regarding it a timewasting institution. When the Queen was going to use RAF communication aeroplanes for travelling to official engagements, he could only conclude by stating *'God Save The Queen'*. After a steady downturn, the final issue of *The Aeroplane* was published in October 1968. The Author found the Aeroplane cartoon actually in a blog by Erik Lund about—see Giffard, Making Jet Engines, which will be discussed in more detail on her peculiar view on the Me 262 and the German turbojet engines in Sect. 8.1, http://benchgrass.blogspot.com/2017/08/technical-appendix-for-june-1947.html

[101] See Wikipedia, 'C. G. Grey' in English.

Fig. 6.11 '*OUR PUZZLE—Every picture tells a story! Ah! But what?*' The Aeroplane, 30 Aug. 1939, p. 293 by *fix*—Felix Kelly

Aside of C.G. Grey in the shown Aeroplane team photo is Constance Babington Smith (1912–2000), a journalist and writer,[102] who is probably best known for her wartime work *in* imagery intelligence. Her knowledge of aircraft took her into the *WAAF Women's Auxiliary Air Force*, where in December 1940 she had been posted to the Photographic Reconnaissance Unit (PRU) at Medmenham, Buckinghamshire. In June 1943 she met there Frank Whittle and showed him then recently taken high-altitude reconaissace pictures of the Peenemuende Luftwaffe test site on the Baltic shore, where parallel scorch marks on the runway grass were first indications for the English side of the twin-engined Me 262 jet

---

[102]CBS—originally a trained milliner, worked for *Vogue*, London, before venturing into journalism at Aeroplane magazine; combining both talents, she created and wore at a RAeS aerodrome garden party in 1938 a daring pillbox hat with a vertically banked model *Spitfire* on top—and meticulously tailored elliptical wings.

fighter aircraft.[103] The abbreviation *-fix-* in the lower right-hand corner of Fig. 6.11 leads to the cartoonist Felix R. Kelly (1914–1994), also visible on Fig. 6.10 (l), a New-Zealand born graphic designer, painter, stage designer, interior designer and illustrator, who lived the majority of his life in the United Kingdom. At the time of the *Aeroplane* cartoon he was just 25, so that it is self-suggesting that the depicted subjects follow direct input from the Aeroplane editor C.G. Grey. The cartoon arrangement allows to differentiate six scenes/ stories for which a possible explanation will be outlined in the following, starting in the upper left-hand corner, and then for the visible three horizontal lines from left to right, and then top-down:[104]

(I) *Air Ministry on 'Cloud Nine'*[105] Somewhat detached from reality, the ministerial 'holy angels' apparently play a game of bridge, with the chap on the right taking a winning trick, while a snobbish, ignorant clerk *pulls the ladder up after him*, got his selfish will.

(II) *Latest I.C. Engines* Internal Combustion Engines are sold to the public media, represented by *bowler man* C.G.Grey (?), Fig. 6.11, as *piston engines*—with labels *advanced ignition, radial, inversed* (hanging valves) and *in-line*—in front of the fence, while behind, industry gets acquainted with so far secret *jet propulsion*— opening stacked surprise packages, labelled *turbine* and *diesel*.[106]

(III) *Shadow Factory* These were built by the British government in time adjacent to automotive plants to go into military aircraft production when war came. Not too secret then, as recalled by corresponding visits of German military delegations, there

---

[103] The Me 262 first flight, powered by two Jumo 004 turbojet engines, was on 18 July 1942 by Messerschmitt chief pilot Fritz Wendel from the Leipheim air field, some 60 km west of the Messerschmitt main plant at Augsburg-Haunstetten. There is a 3 min *Youtube* video in German, https://www.youtube.com/watch?v=p5JYn3yW8qM where the scorching by surplus fuel after landing of the then still tail-supported aircraft is clearly visible (after 1:40 min). Over time Peenemuende proved to become an area of successes for CBS's photo interpretation unit: At the time of the Me 262 traces also the first aerial photographs of the launching pad of the V-2 guided ballistic were taken there and, on 28 Nov. 1943, a RAF Mosquito brought back a picture in which the CBS group was able to discern a kind of stunted aircraft on a launching ramp, thus identifying a V-1 flying bomb for the first time; —see Wikipedia, 'Constance Babington Smith' in English.

[104] The Author owes helpful suggestions in the context of this interpretation to the English insider know-how of Andrew Nahum and John Bailey, 23 May 2020.

[105] In English—a state of bliss, could be etymologically based on the International Cloud Atlas (1896) with ten cloud definitions, of which the ninth, *Cumulonibus*, rises highest to 10 km, whereas the German equivalent 'Wolke 7 (cloud 7)'has its origin in Aristoteles's theory of celestial spheres, e.g. *'in seventh heaven'*, 350 BC.

[106] *Diesel* is mentioned certainly as pro-GT-argument in view of its cheap availability, while at the same time preparations for the operational usage of 100 octane fuel started for fighter piston engines, which was actually introduced on the British side on the eve of the Battle of Britain, while the *Luftwaffe*, with small exceptions, stayed with 85 octane fuel, resulting in a considerable speed/ performance deficit.

were apparent problems to make the puzzle work, while a business man in flashy pin-striped suit wants impatiently access, watching with contempt to the undecided, foggy Air Ministry, in view of a profitable business ahead.

(IV) *Chicken coming home to roost*[107] The English proverb goes back to Chaucer's *'Parson's Tale'* (~ 1390), meaning, one day one has to face the consequences of one's past evil or foolish deeds. The Luftwaffe and RAF chicken aircraft roost on top of full arsenal nests, wondering somewhat surprised how the sudden confrontation could have happened.

(V) *Foreign trade in appeasement times* Prime Minister Chamberlain was not averse to military spending per se, but the challenge was to pursue rearmament in an affordable and politically acceptable way. Officials feared that lucrative defense contracts would draw British industry's attention away from overseas markets, worsening Britain's balance of trade. And, doubly so, given that rearmament itself would require an increase in imported raw material.[108] Consequently, while the British trade tube is only dripping, at the same time the wells of the Americans and Germans, eying each other suspiciously, are fully spouting; especially the latter pushed sales of military goods against desperately required valuta.[109]

(VI) *The Bombing Turnaround* The sketch at the lower right-hand corner of the cartoon appears to be about the developing bomber philosophy, indicated by the bomb pile to the right. Again, a wall marks a secret decision. The unarmed *'Schnellbomber* (fast bomber)*'*[110] gets a bloody nose by crashing at the closed door, because behind the brick wall already the idea of a heavy, turret-armoured bomber takes shape. The pitiful view of the broken Albatross DH.91 prototype, Fig. 6.10 (r, top), might have been a stimulus for Felix Kelly to pick this motive,[111] while he decided to depict the

---

[107] There is no directly equivalent proverb in German: *'Das dicke Ende kommt erst noch* (Wait for the negative consequences)*'* has a related meaning.

[108] See Trubowitz, When states appease, p. 24.

[109] Footnote 22 of this Section referred already to the French desire to buy at the end of 1938 Daimler-Benz aero engines—under the brokerage of Charles Lindbergh.

[110] See Wikipedia, 'Schnellbomber' in English.

[111] De Haviland's *DH.91 Albatross* was a four-engine British transport aircraft in the 1930s, which flew for the first time in May 1937,—see Wikipedia 'de Havilland Albatross' in English. The second prototype broke in two during overload tests, https://ww2aircraft.net/forum/threads/problematic-aircraft.24070/, but was repaired with minor reinforcement and was thereafter operated by Imperial Airways. The aircraft was notable for the ply-balsa-ply sandwich construction of its fuselage, later used in the most successful *'Schnellbomber'* of WW II, the *DH.98 Mosquito,* challenged only to the very end of the war by the Me 262 turbojet aircraft, as illustrated in an impressive 7 min RAF Museum Cosford video https://www.youtube.com/watch?v=0m9XfJ91fWk

then winning concept by the single-engined two-seat fighter bomber *Fairey Battle* with the characteristic long canopy, Fig. 6.10 (r, bottom), in which much hope was originally placed.[112]

This extraordinary cartoon might not be complete without a remaining, unresolved aerodynamic riddle. By comparing the crumpled, presumed '*Albatross*' in Kelly's illustration with the real configuration, there is a clearly deviating tail arrangement to the actual twin fin empennage of the DH.91, which may fall in the category of artistic freedom or intentional camouflage. However, stunning are the pronounced swept wings of the sketched aircraft, clearly different to the nearly straight wing leading edge of the original. A fact first highlighted by Andrew Nahum: '*Wing sweep as a way of ameliorating transonic flow phenomena was first announced internationally by Prof. Adolf Busemann,*[113] TH Dresden (1901–1986) *at the 5th Volta conference in Rome, in his presentation* 'Aerodynamischer Auftrieb bei Ueberschallgeschwindigkeit (Aerodynamic lift at supersonic speed)' *on 2 October 1935—the last time that the international aerodynamics community met before the war. But it was not implemented in Britain until post-WW2. When British and American investigators and aerodynamicists discovered German swept wing work in 1945/46 at Focke-Wulf, LFA Voelkenrode, and AVA Goettingen it came as a complete revelation. . . . Does the cartoon mean that -fix- was more* au fait *with the latest aerodynamics than the aircraft professionals?*'.[114]

6.2   Early Axial Turbojet Developments in Germany

Looking at the chronology of beginning turbojet engine developments in Germany and England during 1930–1935, the synchronicity of several comparable events in a relatively short timeframe of several years is most striking. Theoretical technical historians may speculate that a certain level of technological maturity determined the further evolution. In Germany there was a clear shift in military aeronautical paradigms[115] in the beginning

---

[112] See Wikipedia, 'Fairey Battle' in English.

[113] See Wikipedia, 'Adolf Busemann' in English.

[114] Although the proceedings of the Volta Congress were published in 1936 and von Kármán himself had chaired the session with Busemann's presentation, this groundbreaking idea went by unnoticed outside Germany. Consequently, von Kármán was taken by complete surprise when he visited Busemann's LFA Institute for Gas Dynamics at Voelkenrode on 9 May 1945 (the day after the German capitulation) as head of the US Intelligence '*Operation LUSTY*' (LUftwaffe Secret TechnologY), Figure 10.1. Corresponding British 'Missions' have been described by—see Nahum, I believe . . . Quotation from Nahum communication with the Author, 22 May 2020.

[115] The term '*paradigm shift*' was first introduced by the US scientific historian and philosopher Thomas S. Kuhn (1922–1996) in his 1962 book '*The Structure of Scientific Revolutions*' which was also adopted by Edward W. Constant II in his 1980 book '*The Origins of the Turbojet Revolution*'. He

1930s, leaving the concept of high-altitude flight of the late 1920s to the future dominance of flight speed. This strategic move was considerably pushed by the charismatic, independent mind of Major Wolfram Freiherr (Baron) von Richthofen in his post as head of the RLM aircraft development department, who has been already presented in Fig. 6.4.[116] Richthofen, the Dr.-Ing. engineer, thought far beyond the immediate requirements of the Luftwaffe. He believed the time was not far off when air forces would be equipped with rockets and high-altitude *'interceptor'* rocket planes, to dash up to 12+ km within unheard of three minutes to attack high-altitude bomber streams. Since 1936 in general and after 1938 at the newly founded test area Peenemuende-West, he provided Luftwaffe support for Wernher von Braun's rocket research. This resulted also in the first proof-of-concept demonstration flight of a liquid-rocket powered He 176 aircraft on 15 June 1939,[117] starting from this Luftwaffe test site—and just 2½ months later in the first take-off from the Heinkel airfield at Rostock-Marienehe of a turbojet powered He 178 in the early morning hours of 27 August 1939.[118] Both historic firsts were achieved by RLM's first jet test pilot Erich Warsitz (1906–1983) at the controls,[119] thus executing a private venture of the Heinkel aircraft company in accordance with the fanatically speed-minded director Ernst Heinkel (1888–1958).[120] For the same reason Heinkel had also hired the young physicist Hans von Ohain (1911–1998) for a *'flying work start'* on 14 April 1936 in pursuit of his proposed all-radial aero gas turbine project, then after a record development time of 3 years and 4 months for the first time successfully tested in flight as *HeS 3B* turbojet engine of 490 kp take-off thrust.[121]

Similarly, the jet impulse engine that became the engine for the V-1 cruise missile began with a contract issued by Richthofen, again first tested at Peenemuende, where it was also first detected by means of aerial reconnaissance evaluation by Constance Babington Smith, as outlined in the foregoing Sect. 6.1.4. Finally, the rocket plane developments initiated by Richthofen in 1935 led to the first serial-produced Messerschmitt *Me 163 Komet*[122] rocket

---

points out that the inventors of the first turbojet engines in the early 1930s were not responding to any current failure of the conventional piston engine, which had by no means reached its development limits. At the time of a *'presumptive anomaly'*, they rather foresee by scientific analysis the future flattening of technological progress, Fig. 2.3, and finally the break-down of the established system, so that a new and radically different aircraft propulsion system could and should be built in time.

[116] See Footnote 49 in the foregoing;—see also Neufeld, Rocket Aircraft and the *'Turbojet Revolution'* which combines already the dual impact of 'speed' in the header and—see Corum, The Other Richthofen.

[117] This confirmed date is acc. to—see Koos, Heinkel, Raketen- und Strahlflugzeuge, p. 41; Wikipedia dates the first rocket-powered flight later to 20 June 1939, though without further proof. See Wikipedia, 'Heinkel He 176' in English and Koos, Heinkel He 176.

[118] See Wikipedia, 'Heinkel He 178' in English.

[119] See Wikipedia, 'Erich Warsitz' in English.

[120] See Wikipedia, 'Ernst Heinkel' in English.

[121] See Wikipedia, 'Heinkel HeS 3' in English.

[122] See Wikipedia, 'Messerschmitt Me 163 Komet' in English.

fighters, designed by Alexander Lippisch, pulling up in a 70 deg. climb angle and operating above 15 km, which broke for the first time the *'magic 1,000 km/h'* speed mark on 2 October 1941 again at Peenemuende-West, piloted by the 1937 glider world champion Heini Dittmar (1912–1960).

For the period 1935–1939 under consideration, the assessment of the emerging turbojet developments will deal first with Germany in this Sect. 6.2, followed by a view on Great Britain in Sect. 6.3 and on the United States and other countries in the final Sect. 6.4. Though separate country views are worthwhile to grasp the individual national circumstances and dynamics, it is evident that these national developments became more and more espionage interrelated in the late 1930s and the rising threat of a military conflict between the parties. As a unique document, Fig. 6.12[123] illustrates a *'comparative timetable of jet engine developments'* with data taken from a German report of 2 November 1944, which Hugh Dryden[124] detected as scientific deputy to the Scientific Advisory Group of the US Army Air Forces, better known as the Allied *von Kármán Mission* to Germany, immediately after the end of WW II at Brunswick on 9 May 1945, Fig. 10.1. The print was translated unchanged, and provides thus not only a first survey of the parallel turbojet developments in Germany, Great Britain and the United States, but also a revealing insight to the quality of German intelligence up to the last months of the war.

While the entries for the German side appear to be correct throughout,[125] there are a few deviations to the actual Allied developments. The indicated *'1ˢᵗ* (British) *Jet Engine Running'* concurs with the actual date of Whittle's first test run of his *Power Jets WU engine* on 12 April 1937. Thereafter, the *'Contract for Jet Fighter by R.A.F.'* could stand for the Air Ministry's issuing of a specification to Gloster for an aircraft to test one of Frank Whittle's turbojet designs in flight. The Gloster E.28/39 designation originates from the aircraft having been developed in conformance with the 28th *'experimental'* specification issued by the Air Ministry in September 1939, i.e. actually four months after the corresponding German intelligence information. Comparably, *'1ˢᵗ Flight of «Squirt»'* stands for the first flight of the Gloster E.28/37 on 15 May 1941, nicknamed around the Gloster plant also as *«Squirt»*—and thus reflecting a table mark three months ahead of reality.

Similarly for the 'United States' column, the first German tabulation entry *'Information brought over by General Arnold'* is precisely in-line with Arnold's reported three weeks stay in Great Britain during April 1941. Thereafter, apparently nothing is as yet known about a *'1ˢᵗ Airplane* (delivery) *to U.S.A.'*, but the following *'1ˢᵗ Engine to U.S.A.'* is again

---

[123] See Tsien, Technical Intelligence Supplement, p. 76.

[124] Hugh L. Dryden (1898–1965), then NACA advisor to the US Air Force, became Director of NACA Aeronautical Research for the period 1946–1958.

[125] Somewhat sensitive for the neck-and-neck race based on Whittle vs. von Ohain assessments is von Ohain's first engine run date in reference to Whittle's firm 12 April 1937. Ohain's biographer—see Koos, Heinkel—has two dates, that with hydrogen combustion in March/April 1937, p. 50, and shown here, the first fuel combustion test of the flight-worthy engine HeS 3B end of August 1938, p. 53.

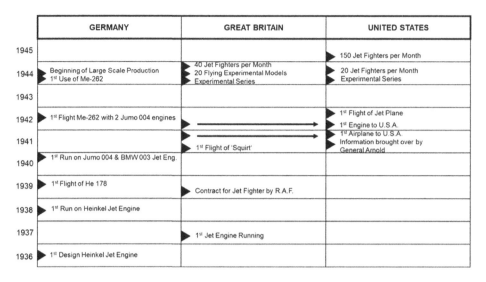

Fig. 6.12 Comparative Timetable of Jet Engine Development, *Data taken from a German report, dated 2 November 1944*

close to the real transfer of Whittle's W.1X engine in October/November 1941, as well as the indicated *'1ˢᵗ Flight of Jet Plane'* corresponds nicely with the actual first flight of the Bell P-59 Airacomet on 1/2 October 1942, powered now by the WU-copied GE I-A as first US-made turbojet engine.

Contrary to Edward W. Constant's assessment of 1980, who saw the *'turbojet revolution'* primarily on the shoulders of Frank Whittle and Hans-Joachim Pabst von Ohain, this noble circle certainly has to be expanded—as will be outlined in the following two sections—by the names of Herbert Wagner, Helmut Schelp and Hayne Constant. And, these technical leaders revolutionised aircraft propulsion, not because of a knowledge of superior aerodynamics, but through their insight that the combination of compressor and turbine was uniquely suited as a power plant for flight. Although the potential problem of compressibility as propeller tip speeds neared the speed of sound should have had an influence, there is little evidence that this relatively unstudied phenomenon in the 1930s was actually a factor in their decision to look for a radically new type of aircraft propulsion.[126]

Rather, it appears that they were drawn to the turbojet because of its simplicity and potential as a lightweight, high performance power plant, specially adapted to aircraft. It was their independence from the automotive-aircraft engineering tradition that enabled

---

[126]Extensive experimental investigations specifically directed towards an empirical understanding of propeller tip compressibility were undertaken in the United States by Hugh L. Dryden and others at the NACA and in Great Britain by Hermann Glauert and others at RAE during the 1920s and early 1930s.

them to think outside the paths of accepted practice. When they began their work, the aircraft reciprocating engine had become a mind-boggling example of mechanical complexity. To attain higher speeds, designers added cylinders to increase power. To compensate for the decreased density of the air at higher altitudes that reduced power, they added superchargers. With added cylinders and superchargers, cooling became a problem, solved by the addition of yet another component, the intercooler. In contrast to the difficult to maintain array of components and accessories, the turbojet offered the possibility of a lightweight, efficient engine of extraordinary simplicity. The engine, as conceived by this new generation of turbojet engineers, potentially could perform better at the higher altitudes that caused problems for the reciprocating engine, and its efficiency was likely to increase with speed, Fig. 2.3.[127]

Technically, the steam turbine is especially relevant for the *'turbojet revolution'*, if for no other reason than that it is in fact also a gas turbine, just as the turbojet is. As has been outlined in detail in the introductory Sect. 4.1 on the basis of the preparatory work at Brown Boveri, the accumulation of data, know-how and technology during the foregoing steam turbine revolution up to the 1920s laid also the ground for successful turbojet developments in the 1930s.[128] Typical, work done on blade design, gas duct flows, shaft and bearing loads, rotor balancing, casing temperature distributions, lubrication systems, control governors, material and manufacturing techniques, test procedures and instrumentation, and countless other facets of design and production could be immediately applied to gas turbine development. With the exception of *aerodynamics*, the steam turbine heritage was perhaps the most important element in the unique historical environment in which the *'turbojet revolution'* was to occur.[129] As a general, credible explanation for Germany's strive for revolutionary propulsion systems, Robert L. Perry pointed out that both RLM and industry realised as of the early 1930s their lagging behind in advanced piston-engine technology which certainly supported the rapid acceptance of the work on turbojet engines.[130]

Besides the fundamental influence of Ernst Heinkel, Fig. 6.7, on the preparation of the turbojet era in Germany due to his general emphasis on aircraft *speed*[131] and by making his company in due course host to already two of the most promising German turbojet development projects, it is mostly the structural impact of the Reich Air Ministry RLM Technical Office, represented by the *'Special Engines (Sondertriebwerke)'* department LC

---

[127] Based and adapted according to—see NASA, Engines and Innovation.

[128] For more details—see Eckardt, Gas Turbine Powerhouse, p. 77: *The Path to the First Power Generation Gas Turbine, 1927–1945.*

[129] See Constant, E.W., The Origins, p. 82.

[130] See Perry, Innovation, pp. 15–18 and 22–27.

[131] Heinkel himself describes the moment when he was speed *'intoxicated'*—attending the 1927 Schneider Trophy competition at Venice: *'I saw of course that the floats must disappear . . . I had to build fast land planes with a moderate landing speed, the least possible wind resistance and the smoothest possible surfaces.'* See Heinkel, Stormy Life, p. 111.

8, since 1938 under Hans A. Mauch and his deputy, after Mauch's leave in 1939 predominant Helmut Schelp, who triggered and coordinated early development activities to a success. Besides the highlighted, speed-driven influence of Heinkel and Schelp, it is worthwhile to remind here also at Max Koenig and his far-sighted, prophetic gas turbine lecture at Newcastle in 1925, Chap. 3, who was led by relatively simple weight consideration to the conclusion of a fundamental change towards revolutionary, rotating turbomachinery concepts and a near-term downturn of the piston age in aeronautical propulsion.

The early turbojet engine projects and their responsible designers are listed in the following:

- *He S30*, highlighted by an advanced five-stage axial compressor—goes back to Herbert Wagner's private project start in 1934, turned to become a Junkers project since 1936 when Wagner joined that company, mutating into a Heinkel/RLM project under Max A. Mueller after 1939, from where it reached still a first run on 19 April 1942, but the project was stopped after Mueller left Heinkel in June 1942;
- *HeS 3B*, an all radial design, rooting in Hans Joachim Pabst von Ohain's private project start in 1934, which he took to Heinkel in 1936, where it received RLM attention and funding since December 1937, to power the first turbojet aircraft He 178 with first flight from Rostock-Marienehe on 27 August 1939, a success which Hans von Ohain could not repeat in full with his follow-on engine projects HeS 8 and HeS 011 up to the end of WW II;
- *Jumo 004*, an all-axial turbojet design around an eight-stage AVA axial compressor from the responsible engineer Anselm Franz, was a Junkers/RLM initiated project since 1939 and became as of February 1944 the first series-produced turbojet engine with 6000+ units, mainly for Messerschmitt Me 262, Arado Ar 234 and Horten Ho 229 aircraft applications. Variants and copies of the engine were produced in Eastern Europe and the USSR for several years after WW II;
- *BMW 003*, an all-axial turbojet design, originally equipped with a seven-stage AVA compressor. Hermann Oestrich was design team leader of this BMW/RLM project in parallel to Jumo 004 since 1939; about 750 BMW 003 engines were built in Germany, but very few were installed in aircraft like the *Heinkel He 162* and the *Arado Ar 234C*. The engine also formed the basis for turbojet development in Japan during the war, and in the Soviet Union following the war. BMW 003C and D versions with improved 7- and 10-stage axial compressors from Brown Boveri Mannheim were planned at war's end which were used as basis for the successful post-war French Snecma ATAR, Sect. 9.3.
- *DB 007*, the first turbofan engine with counter-rotating axial turbomachinery was designed by Karl Leist, as already outlined in the context of Figs. 5.12 and 5.13, and a Daimler Benz/RLM project since 1939. After a few test runs in May 1943 the project was put on halt by RLM in May 1944.

Deduced from the foregoing, the period 1934–1938 is characterised by two, at first unrelated activities, those of Herbert Wagner and of Hans Joachim Pabst von Ohain, who must have started within weeks in parallel, while the remainder of activities after

1938 fell in RLM responsibility and supervision. Consequently for the description of details, there follow five Sections on—Herbert Wagner and his axial engine designs (Sect. 6.2.1),—Hans von Ohain, going radial at Heinkel (Sect. 6.2.2),—the RLM turbojet programme and its management (Sect. 6.2.3),—the BMW 003 turbojet activities (Sect. 6.2.4) and—the start of the Junkers Jumo 004 turbojet development (Sect. 6.2.5).

6.2.1 Herbert Wagner and His Axial Engine Designs

Presumably in the reference year 1934, Herbert Wagner (34) and Hans von Ohain (23) ventilated similar thoughts about a new high-speed aircraft propulsion system within weeks, some 300 km apart at their university institutes at Berlin-Charlottenburg and Göttingen. Both had an invention forerunner in the English Frank Whittle (27), who had started some 5 years earlier—and later all three had to admit that basic thoughts of their inventions—either in the direction of an all-axial turbomachinery design or towards the revolutionary turbojet propulsion principle—had been filed unnoticed already in a French patent of 1921 by Maxime Guillaume (46).

Therefore, it may appear somewhat arbitrary to start the review of German efforts in this area with Herbert Wagner, Fig. 6.13 (l), but from the beginning he decided not only to use the, later-on prevailing, all-axial turbomachine concept, instinctively he chose here also the right/advantageous *'degree of reaction'* design principle. Independently, the established Von Ohain vs. Whittle historic perspective will anyhow always favour the Heinkel camp for achieving the first turbojet flight and thus the ultimate demonstration of the reaction gas turbine principle. Nevertheless, the events at the Baltic coast were for the rest of the emerging German turbojet community rather an interesting side effect; and—not to underestimate—Wagner's reputation as an excellent and acknowledged aircraft designer paired by his humorous, optimistic 'can do' attitude, played apparently an essential role to get the *'RLM project band wagon'* rolling during the decisive weeks in autumn 1938, as outlined in Sect. 6.2.3.

Born in Graz (Austria) in 1900, Herbert A. Wagner became a cadet of the Austrian Imperial and Royal Naval Academy[132] at Rijeka/Fiume[133] between 1914–1917, reaching the end of WW I as ensign of the Austrian Navy in the Adriatic Sea, where he survived the

---

[132] Potentially, Wagner still met Peter Salcher (1848–1928) at Rijeka, who taught at the Naval Academy till Nov.1909 and stayed thereafter. At the request of Ernst Mach (1838–1916), who had not managed to experimentally prove his shockwave theory, Salcher in 1886 at Rijeka conceived the experiment and the world's first ultrafast photographic recording of a shock wave phenomenon accompanying a grenade flight through air,—see Wikipedia 'Peter Salcher' in English and Chap. 2, <1887>.

[133] *Rijeka*, Italian *Fiume*, became famous after WW I by the Italian war hero and writer Gabriele D'Annunzio (1863–1938). On 9 Aug. 1918 D'Annunzio commanded a nine plane fighter squadron in a 700-mile round trip to drop propaganda leaflets on Vienna. After the 1918 Peace Conference he led

Fig. 6.13 German turbojet engine pioneers I: Herbert Wagner (1900–1982, l), Hans Joachim Pabst von Ohain (1911–1998, m), Hans A. Mauch (1906–1984, r)

sinking of a torpedoed battleship.[134] After the political collapse of Austria-Hungary in 1918, Wagner studied mechanical engineering at the TH Graz, before he went to the TH Berlin-Charlottenburg to study naval engineering, receiving his Dipl.-Ing. degree in 1922. He remained at that institute as teaching assistant to Professor Karl Krainer, his future father-in-law, who kept the chair for steam turbines and propellers. Wagner stayed up to his Dr.-Ing. doctorate in 1924 with a thesis *'On the Arising of Lift on Aerofoils'*. From 1924 to 1927 Wagner was a designer for the Rohrbach Metal Aeroplane Company, which pioneered the development of all-metal flying boats. There he undertook a theoretical investigation of stress-carrying thin metal sheets, which led him in 1925 to the *'Theory of the diagonal-tension field beam (Schubfeldtheorie)'* to describe the behaviour of stressed-

---

troops to reclaim the once Italian port of Fiume, which he briefly ruled, calling himself *Il Duce*, with a retinue of black-shirted followers—before Mussolini.

[134]This was the 22,000 t Austrian battleship HMS *'Szent István* (Holy Stephen)'with a crew of 1000, of which 89 sailors drowned after an Italian torpedo boat attack on 10 June 1918 near the Croatian island of Premuda, 60 sm south of Fiume,—see Wikipedia 'SMS *Szent István'* in English. The battleship's sinking was one of only two on the high seas to ever be filmed: https://www.youtube.com/watch?v=5pSiCjfhUUw

On 12 July 1960 Wagner received a honorary doctorate from TU Berlin which he commented: *'I absolved my entry exam in naval engineering by striding around a capsizing battleship. A neighbouring ship shot a nice motion picture and I saw this clip often in movies or at more serious occasions. Then I could realise that my 'striding' was not very measured.'*—See Knothe, Herbert Wagner, p. 11.

Besides his profiling in turbojet and aircraft design, Wagner remained attached to the torpedo/ glide bomb idea when he developed the *Henschel Hs 293* anti-ship radio-controlled glide bomb, operational since 1943—with a liquid-propellant rocket *HWK 109-507* slung underneath, a forerunner of Wagner's post-war cruise missile developments in the United States. See Wikipedia 'Henschel Hs 293' in English.

skin structures, the analytical foundation for a revolutionary change in aircraft structures during the 1930s.

From 1927 to 1930 Wagner was a Professor of Aeronautics at the TH Danzig, before returning from 1930 to 1938 to the TH Berlin-Charlottenburg. Significantly, during the 1920s and 1930s, the *DVL 'Deutsche Versuchsanstalt für Luftfahrt (German Research Institut for Aviation)'* at Berlin-Adlershof and the *TH Berlin-Charlottenburg* were virtual sister institutions, with some professors and students doing work in both. Consequently, Wagner led the Flight Technical Institute (B) in the faculty for mechanical engineering in parallel to Wilhelm Hoff (1883–1945), then head of DVL and of the THB Flight Technical Institute (A). In addition, since 1934 Wagner became a member of the *'Military Engineering Faculty'*, a subsidiary institution of the *HWA Heereswaffenamt* (Army Ordnance Office) at the TH Berlin-Charlottenburg.[135] The THB university calendar 1933/1934 has Wagner as *'Senator'* and head of the 'Flugtechnisches Institut', Franklinstrasse 29, Fig. 6.14 (l).[136] Wagner took a leave of absence from TH Berlin-Charlottenburg as of September 1935[137] to work for Junkers Aircraft Dessau—and it appears that by advance notice, his coming responsibility for the airframe development, especially the *Junkers 'Entwicklungs-Flugzeug* (development aircraft)' *EF 61* high-altitude aircraft developments influenced already his considerations also towards high-altitude propulsion systems, at least a year earlier when still at Berlin.[138] From early on, he understood that the gas turbine

---

[135] See Wikipedia, 'Herbert A. Wagner' in English; he is said *'to have been one of the most brilliant (and versatile) engineers in German aviation'*,—see Schlaifer, The Development, p. 379. Besides his salient contributions to aeronautical technology, he was involved in early studies and applications of *FEM Finite Element Methods*,—see Wagner, Bauelemente, prepared at Henschel Berlin the first computer usage in Germany by Konrad Zuse and—as mentioned before already in Sect. 4.3.2, proposed the building of a super-cyclotron as essential pre-condition for German nuclear weapons, —see Beisel, Bomben-Stimmung, in German.

[136] There were extension plans to erect here at Franklinstrasse 27/29 as part of *WTF Wehrtechnische Fakultät* (Military engineering faculty) under the direction of General Karl Becker also the Institute for Gas Chemistry, which however, were not realised after the outbreak of war,—see Schmalz, Kampfstoff-Forschung, p. 110.

[137] Effective 1 April 1936 Wagner was accompanied by a small core team of institute employees around chief engineer Dipl.-Ing. Max Adolf Mueller, who was installed at *MWF 'Magdeburger Werkzeugmaschinenfabrik* (Machine Tool Factory)', a subsidiary plant of *Jumo Junkers Motorenwerke*, Dessau, just two weeks ahead of the beginning of Hans von Ohain's activities at *EHF Ernst Heinkel Flugzeugwerke,* Rostock-Marienehe,—see Friedrich, Vom Wagner/ Mueller RTO. Axial compressor designer Dipl.-Ing. Rudolf Friedrich (1909–1998), who had joined Junkers Dessau after his studies in mechanical engineering at TH Hannover in 1933, was then also transferred to Magdeburg for Mueller's support.

[138] Wagner's work has already been addressed in the context of the Junkers high-altitude aircraft developments in Sect. 4.3.2 and in view of the *'degree of reaction'* design principle in Sect. 4.3.3, sub-section 'Degree of Reaction'.

Fig. 6.14 Impressions of H. Wagner's R&D sites: TH Bln.-Charlottenburg, Flight Technical Institute, Franklinstr. 27/29, 1934 (l), Magdeburg Junkers administration building, Wasserkunststr.10/11, 1937/8 (r)

would be ideally suited for high-altitude propulsion—however, he was solely focused to the turboprop principle then. He studied the complex question of the optimum ratio of propeller to engine exhaust jet energy, which he preliminarily settled to a 2:1 ratio. Wagner prepared the next step in his professional career quite thoroughly: he founded an '*OHG Offene Handels-Gesellschaft* (general commercial partnership)', consisting of a draftsman and himself, mainly—as he confided to Hans von Ohain[139]—to keep his aero gas turbine design activities completely separate from his THB Institute. This applied also especially to the—in hindsight—important question in axial compressor design of the best choice of the '*degree of reaction*' R: energy transfer and pressure rise solely in the rotating blading corresponding to R 1 or steadily splitted between rotors and stator equal to R 0.5. Wagner's familiarity with reaction steam turbine design from his former involvement with the Austrian Imperial and Royal Navy and subsequent studies at TH Graz and TH Berlin-Charlottenburg, where he became assistant at the chair for steam turbines, appear to be sufficient proof why he followed a track completely independent from the AVA design.

In 1980, when he was explicitly questioned about this fundamental difference to the otherwise German standard axial compressor design philosophy of AVA's Walter Encke, he stated in his short memoirs under the header '*My turbojet engine works*':

My lecturing (at TH Berlin) *comprised also aircraft propulsion systems and <u>the concept of the stationary gas turbine</u> was known to me. In 1934 I applied this to aircraft propulsion, where this engine type achieves a higher efficiency because the stagnation pressure*[140] *reduces the necessary compressor work and the exhaust gas reaction produces a forward thrust. I executed then without any support from my Flight Technical Institute a detailed drawing of*

---

[139]Von Ohain, Turbostrahltriebwerke, p. 26.

[140]Though right, the reference to the beneficial inlet stagnation pressure has a somewhat detrimental effect to the further progress of the coming Junkers turbojet development activities at Magdeburg under the lead of Max Adolf Mueller, as outlined in the following.

such an engine. It was meant to drive a propeller, ...The basic patent claim was directed towards the mentioned (propeller/nozzle) *thrust ratio of 2:1.*[141]

Given Wagner's background as Austrian steam turbine engineer, the hint towards '*the concept of the stationary gas turbine*' (underlined above) points with high likelihood to the Swiss *Brown Boveri* and their all-axial *Velox* boiler gas turbine concept, up for sale since 1932. The photographs of Fig. 6.15 are taken from a Brown Boveri publication,[142] from which the competent reader or correspondingly, a plant visitor in front of a disassembled Velox turboset could easily get the basic R 0.5 design message, as illustrated in the 'symmetric' blading scheme at the bottom.[143] This is a first strong indication of a Brown Boveri design influence on German turbojet activities as early as 1934, which will culminate 8–10 years later with several advanced compressor design contracts from RLM to Brown Boveri Mannheim, resulting from an Allied post-war perspective in the best/most efficient German axial compressor design.

Most interestingly, Wagner's idea of an all-axial engine concept was traced back by Volker Koos[144] to a 1927 presentation of Martin Schrenk, DVL Berlin-Adlershof, which has been addressed already before in Sect. 4.3.2;[145] thereof, Fig. 6.16 (m) illustrates Schrenk's proposal of a turbo-charged altitude piston engine—with a single stage radial blower and—unconnected—a single stage axial exhaust gas turbine.[146] In addition,

---

[141] Wagner's turboprop patent GB495,469/ DEX495,469 with priority 8 Feb.1936 (and also his later turbojet patent DE724,091 with priority 14 Aug.1938) made no reference to the axial compressor degree of reaction, but his test team at Junkers Magdeburg investigated solely R 0.5 axial compressors according to Wagner's design input. Consistently, the literature has overlooked the trifold function of the shiftable gearbox in Wagner's turboprop patent; (a) to reduce cranking power during starting, (b) to keep propeller speed constant, when the turboset is turning faster at altitude and (c) to uncouple propeller and turbojet engine at altitude, so that the latter propels the aircraft solely in the 'rocket'/reaction mode.

[142] See BBM, Rueckblick, p. 35 and Abb. 93 with the caption '*Turbocharger of a Velox steam generation plant. The special axial blower* (pressure curve) *characteristics have been applied beneficially to the Velox steam generator*', see also Fig. 4.49.

[143] For the different, optical impression of a R 1 compressor blading,—see Figs. 4.45 and 6.15.

[144] See Koos, Heinkel Raketen- und Strahlflugzeuge, p. 86. The Schrenk engine was not the only and straightforward stimulus to Wagner's all-axial turbojet configuration. Max Adolf Mueller's work at Junkers/Heinkel comprised a sundry of other schemes, amongst these as a kind of fallback position to Schrenk the ML Motor-Luftstrahl (jet) piston-driven ducted fan configuration—in Heinkel jargon known as '*Marie Louise*', the MLS i.e. ML with *Staustrahltriebwerk* (subsonic afterburning) and the MTL as ducted fan in combination with piston engine and turbine drive.—See Kay, German Jet Engine.

[145] See Schrenk, Probleme.

[146] Schrenk's described configuration was realised mostly as advanced two-stroke motorcycle engine under the term '*Split-single engine/Doppelkolbenmotor*', patented by the Swiss racing engine designer Arnold Zoller (1882–1934) as US2,014,678 with priority 24 Oct. 1931. For animation,— see Wikipedia, 'Doppelkolbenmotor'in German, https://de.wikipedia.org/wiki/Doppelkolbenmotor

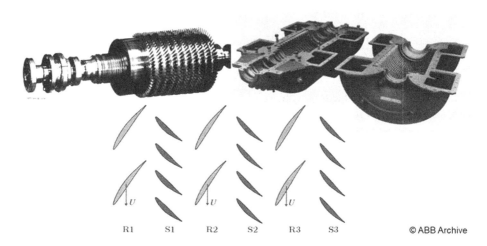

© ABB Archive

Fig. 6.15 Herbert Wagner: *'I knew the stationary gas turbine concept'*, Brown Boveri GT compressor with R 0.5 blading scheme, 1933

Schrenk suggested to exploit the exhaust gas energy/thrust by directing the exhaust piping accordingly against the flight direction. In line with this interpretation Wagner's innovative step towards an aero gas turbine, Fig. 6.16 (r), appears relatively straight-forward to replace Schrenk's unsteady piston-combustion arrangement by a continuously operated combustion chamber. However, following these thoughts, it is highly intriguing to compare Schrenk's proposal of 1927 with an even earlier idea of 1905, which the Swiss Alfred Buechi (1879–1959), often considered as the inventor of turbocharger technology in general, had been filed in his first patented turbocharger scheme, Fig. 6.16 (l). This, presumably earliest related ancestor on the axial turbojet engine pedigree combines on one shaft an eight-stage axial-flow compressor with a four-stage axial-flow turbine.[147]

The further axial turbojet development is closely related to Hugo Junkers and his aircraft and engine companies, especially those at Magdeburg and Dessau.[148] In 1913 the *'Junkers Motorenbau* (motor construction*) GmbH'* had been founded at Magdeburg, was relocated by shareholder decision in 1919 intermediately to Dessau—and back to Magdeburg in 1923. In the early 1930s the Junkers group underwent financial troubles. When the Nazis came into power in 1933 they requested Junkers and his businesses to aid in the German re-armament. When Hugo Junkers declined, the Nazis instrumented by Goering's deputy Erhard Milch initiated his complete expropriation, denied him further access to his works and implemented Heinrich Koppenberg (1880–1960) on 24 November 1933 as chairman

---

[147] See Buechi, 'Ueber Verbrennungskraftmaschinen' and 'Geschichtliches', Fig. 2; for more details—see Eckardt, Gas Turbine Powerhouse, p. 54.

[148] See Wikipedia, 'Junkers', 'Magdeburg' and 'Junkers Jumo 004' in English.

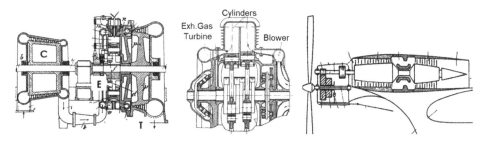

Fig. 6.16 Axial Turbojet Engine Pedigree: Buechi turbocharger patent 1905 with C—axial compressor, E—piston engine, T—turbine (l), Schrenk altitude piston engine proposal 1927 (m), Wagner axial turboprop patent 1935 (r)

instead, a trained fitter, up to 1906 mechanical designer at *Krupp* and in the meantime self-made manager within the *Flick Group*. In 1934 Hugo Junkers was placed under house arrest, and in due course died at his home at Gauting near Munich on 3rd February 1935, his 76th birthday, during negotiations to give up the remaining stock and interests in Junkers.

In mid-1936 the so-far independent Junkers aircraft and engine divisions were merged on Koppenberg's initiative to the JFM Junkers Flugzeug- und Motorenwerke AG. Already in April 1935 the new Junkers had acquired the MWF Magdeburger Werkzeug-maschinenfabrik (Machine tool fabrication), where the famous series of Junkers in-line flight piston engines was produced—the Jumo 210, 211 and 213. In this context it can be assumed that the pick of Herbert Wagner with responsibility for all further Junkers airframe developments—and as a counter-balance to Junkers long-time head of engine developments Otto Mader (1880–1944)[149]—was initiated[150] by Goering's deputy, State Secretary E. Milch and only formally executed by H. Koppenberg. Beginning 1934/1935 Magdeburg, together with Dessau, was in the focus of a gigantic production capacity increase of both Junkers airframes and engines under the direction of Koppenberg, far beyond any imaginable civil demand. As an example Fig. 6.14 (r) illustrates the new,

---

[149] See Wikipedia, 'Otto Mader' in German, and—see Google <Calum Douglas German Aviation Personalities> in English.

[150] Wagner describes his first contact with Junkers in 1934, when the Amtsgericht (district court) Berlin asked for his expertise (as one of the rare specialists who had not yet worked for Junkers), when Handley Page sued Junkers for violating a *slotted wing* patent of Wagner's fellow-Austrian Gustav Lachmann,—see Chap. 5 for Lachmann's internment. Based on Wagner's written expertise, the court suggested a compromise settlement; Koppenberg accepted somewhat disgruntled, but contacted Wagner thereafter.

within one year erected administration building at Magdeburg's Wasserkunststrasse on the ground of the former machine tools factory.[151]

On Wed 1 April 1936, Professor Herbert Wagner initiated a group of three young engineers from his THB institute[152] to their future turbojet development tasks in the directorate room of MWF Magdeburger Werkzeugmaschinenfabrik, which officially had been acquired at the same day to the Junkers companies group. He presented a large general arrangement drawing[153] of his patented turboprop engine to outline the concept of this ~1200 shp engine, Figs. 6.16 (r) and 6.17 (r),—with in flow direction a planetary gearbox, a 14-stage axial compressor, an annular combustor and a two-stage axial turbine, expanding to ambient for the intended turboprop simulation. By early 1937, however, systematic studies done jointly by Wagner and Mueller showed that without propeller a very light engine would result and that an acceptable efficiency could be attained with gas temperatures >600 °C that were easily low enough for then available blade materials. Accordingly Wagner, again alone, filed a turbojet patent which was granted as patent DE724,091, (Fig. 6.25 started from that patent original drawing), and a priority date 14 August 1938—with the single claim that beneficially the jet exhaust nozzle area should be in the same order as the turbine exit area. Wagner continuously supported the turbojet developments of his engineering group at Magdeburg up to his leave in late 1938 from his aircraft development directorate office at Junkers Dessau.

All this work was kept completely secret until about the middle of 1938, and was thought entirely(?) financed during this period from the funds of Wagner's new employer, the Junkers airframe division. 'Entirely', however is doubtful, after Mueller mentioned in a 1952 interview,[154] that in 1934 Wagner's THB Institute had about 30–40 employees—on RLM payroll, and since those working at Magdeburg obviously did this just on a kind of lease basis, which alleviated also their uncomplicated return after their Junkers mission ended in mid-1939, at least part of the personnel cost came from a public budget. For the

---

[151] Responsible for the architecture of this steel frame construction was Werner Issel (1884–1974), from 1906 when he started his career under Peter Behrens and the project of the renowned—see Wikipedia, 'AEG Turbine Factory', Berlin-Moabit up to 1966 responsible for many industrial and power plants. In the 1930s Issel also became the preferred architect for many Junkers projects at Dessau.

Magdeburg's position within the 11 Junkers manufacturing sites is reflected by few, rather rarely passed on turnover data of in total 217 million Reichsmark in the second quarter of 1943: Magdeburg 41%, Dessau 28%, Koethen 17%—in comparison to 1.9 billion Reichsmark new orders.

[152] The group comprised Max Adolf Mueller as group leader, Wilhelm Peppler and Rudolf Friedrich; all working at Magdeburg still as employees of the THB Institute, up to Wagner's leave of his university chair in Nov.1938 which in due course also ended soon the group's activities at Junkers. Wagner's co-author of the 'Bauelemente' book Dr.-Ing. G. Kimm led the Institute between 1938 and 1942.

[153] The corresponding patent priority for his invention of an all-axial turboprop engine GB495,469/DEX495,469 was achieved—just in time—on 8 Feb.1936.

[154] Heinkel archive file # FA001/0025 at Deutsches Museum Munich.

rest Heinrich Koppenberg revealed a surprising solution after the war.[155] After inevitable, internal confrontations between the airframe and engine sides after the dismissal of Hugo Junkers, it was clear that Otto Mader and his engine resources would not stand up for expenditures on behalf of the new head of the airframe division. The proceedings for the allocation of a financial budget for Mueller's engineering group were comparably as *'secret'* as their work in a remote test barrack with integrated tiny office space on the Magdeburg factory premises. The development budget was provided by the commercial director August Muehlen[156] of JFM Junkers Flugzeug- and Motorenwerke AG, the holding structure above airframe and engine divisions which Koppenberg had just introduced in mid-1936.

A first effort by Mueller to get official support via DVL failed in 1937—with the explanation that neither necessity nor feasibility of such propulsion concept looked encouraging. When Wagner realised about the middle of 1938 that Hans Mauch of the RLM was talking favourably of aero gas turbine developments, he revealed the project to the Air Ministry in the hope, if not subsidies to get at least some priority in material supplies. The response was the same as that made to Heinkel's similar request, that development of an aircraft engine could be carried out satisfactorily only by a qualified aircraft engine manufacturer. Mueller remembered the financial basis as extraordinarily *'thin'* up to summer 1938—without official Junkers project approval. Under the additional pressure of a steel shortage, it was decided to go to the ultimate, and *'present'* the new revolutionary engine ideas and data to Reich Marshal Hermann Goering during the Nuremberg NSDAP party convention, 5–12 September 1938. Nothing is known about immediate consequences of this desperate step, but it is imaginable that the beginning RLM activities towards a coordinated turbojet development programme have been considerably accelerated.

The project became known as *'RT0 Rueckstoss-Turbine Null/Nullserie* (Reaction Turbine Zero)', Fig. 6.17; first RT0 test runs were carried out in spring 1939, at a time when Wagner had already left. However, a decisive peculiarity overshadowed the whole test effort:

Though there was already a clear preference for the turbojet concept, the actually used RT0 test set-up reflected the early turboprop design with clumsy inlet gearbox casing, 14-stage axial compressor and the two-stage reaction turbine—expanding to ambient conditions and thus without significant exhaust jet.

Most details about the axial compressor development have been reported by Rudolf Friedrich (1909–1998), who possibly met his 8-years-older, future superior Max Adolf Mueller (1901–1962) during the final stages of their Dipl.-Ing. mechanical engineering

---

[155] See Koppenberg, Das Wagner-Triebwerk.

[156] August Muehlen was—with short-time interruptions—commercial director on the JFA/ JFM Board between 1931 and 1945.

Fig. 6.17 Junkers RT0 engine rig arrangement at Magdeburg 1939: e-powered *'ram'* blower (l), gearbox casing, 14-stage axial compressor, annular combustor, two-stage axial turbine (r)

studies at TH Hannover in 1933:[157] *'We investigated a number of Goettingen airfoils with various inlet angles and blade numbers on a model stage,* Fig. 6.17 *(l), followed by a 3 (4–5?) stage test rig. A complete stage characteristic was determined even beyond the maximum mass flow by means of a suction blower, which allowed to calculate the envisioned compressor map for various speed lines and thus the stepwise design of the 14-stage axial compressor. . . . Besides . . . the* 'AVA Goettingen Results' *and Bauersfeld's seminal paper . . . the doctoral thesis of Curt Keller at Ackeret's ETHZ Institute was of special importance to us. . . . We realised clearly the limitations*

- *in blade loading in diffusing flow with pressure rise, corresponding to the C_A-values of single aerofoils,*
- *in the profile inlet velocity near the speed of sound,*
- *in the interaction of individual stage characteristics and the narrow working range.*

Of special importance was the manufacturing of twisted blade and vane profiles, for which we used a circular copy milling process by means of a 6-times enlarged model profile on a "Deckel Pantograph". [158]

---

[157] Friedrich's further professional stations were—1933 Junkers Dessau, 1936 Junkers Magdeburg, 1939 Heinkel Rostock, 1941 Turbine factory Brueckner & Kanis Dresden, 1948–1964 *'father'* of the Siemens AG power generation gas turbine, Muelheim/Ruhr. Dr.-Ing. in 1949 at TH Hannover, 1964–1976 Professor of Thermal fluid machinery at TH Karlsruhe, in parallel to Hermann Reuter (1911–1981) Professor for machine design and construction; in 1944 Reuter and Friedrich took part in a tank engine/gas turbine competition on behalf of the Weapon SS, which will be addressed in Sect. 8.3.4. In 1935, Max Adolf Mueller became chief engineer at Wagner's THB institute in search of a doctorate, which he never achieved—apparently also due to significant gaps in his foregoing engineering education; nevertheless, his engineering and political (!) influence was so widespread over the years, that he will be presented in a personal portrait in parallel to Helmut Schelp (1912–1994) in Chap. 7.

[158] See Friedrich, Vom Wagner/Mueller RT0; Prandtl, Ergebnisse; Bauersfeld, Die Grundlagen, and Keller, Axialgeblaese. *'Friedrich Deckel'* is the name of a Munich precision machine tool manufacturer since 1903.

Much less is known about the applied design principles for the RT0's annular combustor and the two-stage axial turbine. Operation with propane gas had been planned, but for the follow-on design of the Heinkel HeS 30 turbojet, Fig. 6.19, came the retreat back to 10 single *'cannular'* burners. Finally, the 2-stage axial turbine provides the impression as if Herbert Wagner had tried to transfer a crude steam turbine design into the jet age. A first row of 24 high-turning turbine nozzle vane channels from thick sheet metal is followed by an excessively turning first rotor with 80+ blades, etc., Fig. 6.18. This information is based on a low-quality photo-documentation *'Magdeburger PTL-Gerät (Flugtechnisches Institut der TH Berlin) 1938'* which reveals—completely unexpected—the application of an action/impulse or constant pressure turbine design principle, in contrast to the usual, present-day reaction/over-pressure turbine design concept for gas turbines. Nothing is known about the details of the later *HeS 30* turbine redesign in 1942, but one can assume that preventing excessive turbine flow losses alleviated the then finally successful compressor/turbine matching.

In spring 1939 the first *'Junkers turbojet engine'*, actually a heavy turboprop installation without propeller, was ready for static tests, but did not run continuously and self-sustained! Nothing is said about a special starting aid, except the described, apparently insufficient pressurised air injection.[159] As can be speculated, there was either a compressor/turbine capacity matching problem or, more likely, the combustion efficiency in Mueller's direct responsibility was still inadequate. The deficiency was not solved then—and in hindsight can be brought also in connection with the electro-powered *'ram'* blower of Fig. 6.17 (l). Still inspired by Herbert Wagner's assessment, that the axial compressor work would be alleviated at high flight speed by a stagnation pressure/ram effect of approximately 0.25 bar, this presumably commercially bought axial blower belonged to the RT0 installation from early on, at first as a kind of pneumatic starter assistance, but then under demonstration pressure it helped to run up the RT0 to about 50% speed (and excessive fuel consumption), but never self-sustained and not always unnoticed by impressed, nevertheless puzzled visitors.[160]

---

[159] See Gersdorff, Flugmotoren, p. 299, but other than described, this *'pressurised air start'* was not successful, and other, later introduced techniques, unavoidable for the starting of a 14-stage axial compressor, like bleed valves, VGV variable (compressor) guide vanes are not mentioned. Obviously, Mueller realised these deficits soon, and consequently his follow-on *HeS 30* set-up at Rostock had—first of all—an electric starter motor—see Gersdorff, Flugmotoren, p. 278, and additionally variable geometry elements like a row of variable turbine inlet vanes and an adjustable exhaust nozzle to lower the pressure resistance during starting, Fig. 6.19.

[160] For further reactions and negative consequences—mostly for the reputation of Max Adolf Mueller,—see Chap. 7.

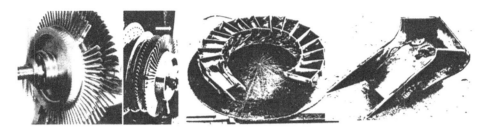

Fig. 6.18 Junkers RT0 engine two-stage action turbine, Magdeburg 1939: two-stage rotor (l), entry stator with 24 nozzle vanes (r)

The engine project was transferred to Heinkel Rostock[161] after a fundamental design restart as pure turbojet under the new designation *HeS 30*—now with a sufficiently strong electrical starter and with single and two-stage turbines, up to the point when Max Adolf Mueller left the Company in early June 1942. Salient highlight of this engine was Rudolf Friedrich's courageous step to reduce the compressor configuration down to five highly-loaded axial stages. After the restart at Rostock the HeS 30 compressor mechanical design in Friedrich's group was the responsibility of Willi Kastert, who had joined the team from BBC Mannheim, an early prospect of the company's contributions to advanced turbojet designs on the German side.[162] A new, possibly reaction turbine design approach with compressor-adapted, higher flow capacity by Hans Stabernack at Heinkel-Hirth, Stuttgart-Zuffenhausen matured slowly, so that in November 1942—already after the official RLM programme stop for the HeS 30 in July 1942—for the first time and only shortly up to approximately 850 kp thrust could be achieved, considerably off the originally targeted 1125 kp,[163] but in the order of the Jumo 004 (900 kp) and BMW 003 (800 kp) competitor engines.

Although the HeS 30 was abandoned at a fairly early stage in its development, it was the most promising of all Heinkel engines and with some development persistence, it might have been developed even to a real engineering showcase. Its thrust to weight ratio was far better than e.g. the Jumo 004 B, and in terms of specific fuel consumption and frontal area

---

[161] The otherwise well-informed Edward Constant, taking in parts information from Schlaifer, and Robert Schlaifer himself are wrong in placing the building *'of the first Junkers turbojet ... with five-stage axial compressor'* (HeS 30) still in 1938 at Junkers Magdeburg,—see Constant, E.W., The Origins, p. 204 and—see Schlaifer, The Development, p. 380/381. Unsuccessful test trials took place at Magdeburg in spring 1939 solely with the RT0 test engine set-up with 14-stage compressor.

[162] See Friedrich, Erinnerungen; W. Kastert left Heinkel together with R. Friedrich in May/ June 1941 to join H. Weinrich at the company Brueckner-Kanis, Dresden for work on high-power 5000 and 10,000 hp marine gas turbines.

[163] This actual thrust value is reported by—see Friedrich, Vom Wagner/Mueller RT0 and by—see Kay, German Jet Engine, p. 33, who provided also the He S30 specification, while—see Koos, Heinkel Raketen- und Strahlflugzeuge, p. 94 has only 650 kp for the achieved maximum thrust.

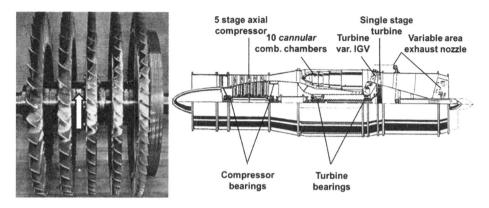

Fig. 6.19 Wagner/Mueller Heinkel *HeS 30* turbojet, 1939–1942, five-stage axial compressor PR 3 (l), cross section (r), F 860 kp, n 10,500 rpm, L 2.72 m, D 0.62 m, W 390 kg © V. Koos

per unit thrust, it is claimed by Antony L. Kay *'that it was not equalled anywhere until 1947.'*[164]

Therefore, it is worthwhile to have a closer look to this advanced HeS 30 turbojet concept, Fig. 6.19. The PR 3 compressor was of the five-stage axial type, driven by a single-stage axial turbine;[165] by assuming—according to Friedrich—a somewhat beneficial centrifuging effect on the hub boundary layer, an increased degree of reaction of R 0.55 appeared permissible. The best compressor efficiency was determined to 87%. Ten individual combustion chambers were provided, at the exit of which the novel feature of variable turbine inlet guide vanes was introduced. Secondary air was admitted to each combustion chamber by allowing it to pass under five overlapping, progressively larger, tubular sections.[166] The exhaust nozzle was of the variable-area type, a feature to become common on German axial turbojets—to provide as far as possible optimum operating conditions.

Even if it looks as if the Magdeburg RT0 turbojet engine could have been a forerunner of the Jumo 004 B, the latter was mostly an independent development at Junkers Dessau under the directive of Otto Mader, with Anselm Franz as project leader at a time, when Mueller and most of the Magdeburg team had left Junkers already. Therefore, a lot of

---

[164]The non-plus-ultra would have been then the HeS 30A, a turbofan bypass version of 1050 kp thrust with 890 mm diam., of which no drawing survived;—see Gersdorff, Flugmotoren, p. 282.

[165]A draft letter from Heinkel to Udet, dated 17 July 1940, contains information from Mueller, by which the turbine variable IGV Inlet Guide Vanes allow to vary the aerodynamic loading of the—and now explicitly—constant pressure (action) turbine. This design feature is claimed to contribute also to an *'economic throttled flight'*. Part of *'Thorwald file'*, see Footnote 171.

[166]This design solution follows BBC patent US2,268,464 for Claude Seippel with Swiss priority starting 29 Sep. 1939,—see Eckardt, Gas Turbine Powerhouse, p. 209.

potentially useful know-how was lost and unconciliatory up to 1989,[167] both teams then still represented by Anselm Franz and Rudolf Friedrich could not achieve a compromising common view on their design merits. In the meantime it is obvious that a combination of both design approaches would have (potentially) saved time, but definitively would have been close to optimum.

In late 1938 Herbert Wagner left Junkers, following a disagreement within the board of directors, but certainly also affected in his reputation after the *EF 61* development failure and settled at Henschel Flugzeugwerke in Berlin.[168] In May 1939 Wagner was replaced at Junkers by the returning Heinrich Hertel (1901–1982), who had worked for Junkers already up to 1932—and who had to leave EHF works at Rostock after a hefty dispute with Ernst Heinkel, just in time.[169]

After a short visit of Otto Mader at Magdeburg,[170] of which it is not known if he also saw a failed RT0 start test, and the Junkers internally communicated RLM request to stop the turbojet development activities at Magdeburg, Max Adolf Mueller contacted Ernst Heinkel, whom he claimed to have met years before at Bristol(!). Heinkel asked Hans von Ohain to participate in this historic meeting which von Ohain remembered to have happened at Rostock approximately in February 1939.[171] Nearly certainly, it can be taken for granted that Mueller stressed in this meeting the superior compressor design

---

[167] Reference is made to the DGLR two-day-conference *'50 years of turbojet flight'* on 26/27 October 1989 at the Deutsches Museum Munich. See DGLR, 50 Jahre Turbostrahlflug.

[168] After the war, Wagner was the first of many German scientists brought to America as part of *Operation Paperclip*, arriving at Frederick, Maryland already on 18 May 1945, where he started working for the US Navy, until he formed his own technical consulting company. In 1957 he returned to Germany to take up a position as Professor of Technical Mechanics and Space Technology at the RWTH Aachen.

[169] Of course Hertel's position at Heinkel was not left open for long, on 1 June 1939 Robert Lusser (1899–1969) joined Heinkel after a foregoing six-year-stay at Messerschmitt, where he developed amongst other the successful *Me 109*. At Heinkel he designed the first twin jet fighter aircraft, the *He 280*, which lost the competition with the Messerschmitt *Me 262*. Still in 1941, Lusser moved on to participate in the development of the *Fieseler Fi 103*, otherwise known as *V-1 flying bomb*. There he met in the final stages of series production preparations Hans A. Mauch, in 1938/1939 at RLM still responsible together with Helmut Schelp in the emerging turbojet engine development programme;— see Sect. 8.1. In hindsight, an astonishingly breathless, not *folkish* but selfish—and certainly not effective—personnel carousel with an early appeal of present day top stars in arts and sports.

[170] See Kay, German Jet Engine, p. 58. Though colported by Mueller, it could have been that meeting in which Mader said: *'Look Mueller, when I was still a young man like you, I also wanted to build a gas turbine; if you occupy yourself longer* (with that matter), *you will agree ... ';*—see Heinkel archive file # *FA001/0025* at Deutsches Museum Munich.

[171] Von Ohain, Turbostrahltriebwerke, p. 23. What von Ohain not knew, however, was the *kind* of Mueller's contact. After rumours spread in autumn 1939, Heinkel had perhaps *'headhunted'* Mueller to weaken Junkers position, Ernst Heinkel himself clarified in a letter to Udet, dated 20 Oct. 1939, that in early May 1939 Mueller had replied simply by letter to a coded Heinkel personnel search ad in a newspaper. This letter belongs to the *'Thorwald file'* which E. Heinkel transferred to his biographer

approach in comparison to Heinkel's competition, in addition to the other advanced features of his turbojet engine concept. Negatively, he claimed to have objected to continue this work at Dessau under Otto Mader, whose thinking he assessed as too conservative and solely piston engine oriented;[172] however, he presumably was already informed that Anselm Franz, head ot the Junkers turbocharger development, had been picked as the new project leader at Junkers Dessau.

The further development is known; Heinkel hired the highly qualified team of 14–16 engineers and scientists under Mueller's lead to come to Rostock in August 1939 and to carry out the assumed, relatively small amount of corrections necessary—for a successful introduction of a turbojet with—as Ernst Heinkel hoped—for the first time advanced turbojet aircraft and engine from one source and as soon as possible.[173] End of 1939 the new advanced engine project received the official RLM designation <109–006>, better known in Heinkel terminology as *HeS 30*.[174]

6.2.2   Hans von Ohain, Going Radial at Heinkel

Striving for technical and economic dominance was certainly a salient feature in the ambitious character of Ernst Heinkel (1888–1958) after his principal decision to start his

---

J. Thorwald (actually H. Bongartz) on 27 July 1953, and of which the author gratefully received copies from Lutz Budrass, Ruhr-Uni Bochum in Sept. 2018.

[172] This widely accepted argument in literature appears to follow solely Mueller's interpretation. On the basis of Schelp/ Mauch quotations in the context of Fig. 7.14, it is more likely that both persuaded Otto Mader to pick Anselm Franz to lead the future Junkers turbojet engine development group.

[173] As known, this hope did not materialise, but it is stunning to observe that Herbert Wagner ventilated exactly the same thought of a single company source for advanced aircraft and engine, before he came to his new employer Junkers in 1934 and at that time, still altitude−/turboprop-oriented, while Heinkel five years later and following the general German military strategy change, had already the high speed turbojet aircraft in mind.

[174] See Wikipedia, 'Heinkel HeS 30' in English. The location designation *'Rostock-Schmarl'* of this Wikipedia article corresponds in principle with the established *'Rostock-Marienehe'*, where the first turbojet flight took place and also the development facilities of Ohain's group were located, on the west bank of Unterwarnow, the estuary of River Warnow, some 4 km north of the Rostock city centre on the way to Warnemuende; effectively, Schmarl is some 1.5 km north of Marienehe railway station.

In 1942 heavy Allied air raids destroyed Heinkel's aircraft factory; remnants were brought to Soviet Union after the war, and GDR erected here the harbour of their ocean-going fishing fleet, which employed with fish processing industry a work force of 8000. Between 1967 and 1975 a preferred fishing ground was *Georges Bank*, 200 km east of Cape Cod, where the GDR trawlers caused considerable conflict with Maine lobstermen; in 2005, basis of a *Discovery Channel* documentary/reality show *'Lobstermen: Jeopardy at Sea'*.

own aircraft business in 1922.[175] This was not only in view of the national competition—the Junkers Flugzeug- und Motorenwerke, but also in comparison to England and their leading aviation concerns—Bristol Aeroplane Company and Bristol Engine Corporation (Sir R. Fedden, born 1885) and Sir Thomas Sopwith's (born 1888) Hawker Siddeley/ Armstrong Siddeley Motors in mind, the big deal of aircraft manufacturing, equipped with engines from own production.

Though Junkers's piston engine experience could not be caught up, but the emerging ideas of advanced aircraft propulsion concepts of the 1920s and 1930s found Heinkel's full attention, ready for fully committed investment wherever an opportunity opened up. The first example in this context is the aviation steam engine which went to test at Heinkel Warnemuende in 1928. Volker Koos in his book has no details about the designer of this 3-cyl. radial steam engine, Fig. 6.20,[176] but according to an internet search, the actual design origin should be (or close to) the *Great Lakes Aircraft Corporation*, in a certain similarity to the turbojet engine development start, as already described for Armstrong Siddeley Motors in 1936.[177]

Figure 6.20 (m) of the Heinkel test set-up is in principle agreement with a functional sketch on Fig. 6.20 (r), which illustrates the 3-cyl. radial steam engine in the hub of a three-bladed propeller, a combination which in 1932 was described as the *'simplest "irreducible" power plant. ... The only perfect power plant using reciprocating motion without reciprocating weight.'*[178] Apparently, the enthusiastic author had somewhat underestimated (a) the difficulties of feeding steam to the rotating system, as similarly (b) the effect of steam expansion was ignored in the sketch of Fig. 6.20(r), which is incompatible with the shown steam inlet and exhaust duct areas.

After the failed steam propeller approach, Heinkel continued his strive for innovative high-speed aircraft propulsion by getting involved in the *HWA* (Army Ordnance Office) activities of applying rocket propulsion, which led to the first flight of Heinkel's He 176 rocket-powered aircraft on 15 June 1939[179] at Peenemuende-West, 2½ months

---

[175] See Wikipedia, 'Ernst Heinkel'in English.

[176] Figure 6.20, left and middle, courtesy to Volker Koos, 27 July 2020 and—see Koos, Raketen- und Strahlflugzeuge, p. 43. Koos has there also the information that this engine-propeller combination has been tested in a Heinkel HD 22 biplane, at times when the HD 40 *'newspaper delivery aircraft'* was also under consideration to be produced in the USA;—see Wikipedia, 'Heinkel HD 22' and 'Heinkel HD 40' in English.

[177] See Sect. 6.1.1, Fig. 6.1 (r).

[178] See Hilburn, Steam. To the positive side-effects of this invention belonged certainly, that propeller anti-icing measures became obsolete.

[179] Acc. to—see Koos, Heinkel Raketen- und Strahlflugzeuge, p. 41, the literature first flight date of 20 June 1939 could not be verified. A He 176 presentation flight took also place on 3rd July 1939 at the Luftwaffe test centre Rechlin, with Hitler and Goering attending, where also the He 178 was shown as ground display.

Fig. 6.20 3-cyl. radial steam engine/propeller combination of Great Lakes Aircraft Corporation: Heinkel test set up, 1928 (l, m) and published functional principle, 1932 (r) © V. Koos (l, m)

ahead of the first turbojet flight, again with Erich Warsitz at the controls.[180] However, the HWA secrecy curtain around that project and the prospects for its low industrial usage limited Heinkel's enthusiasm soon; in this respect, Professor Pohl's letter of recommendation for Hans von Ohain and his radically new air-breathing propulsion concept, issued on 3rd March 1936, reached Heinkel just in time.

Like Frank Whittle, Hans Joachim Pabst von Ohain (1911–1998), Fig. 6.13 (m), started with the idea that flight required a power plant specially adapted to motion through the air—and with passenger comfort in mind. His enthusiasm stemmed from the insight that an engine that burned continuously was *'inherently more powerful, smoother, lighter and more compatible with the aero-vehicle'* than the clumsy four-stroke cycle of a piston engine. However, he soon realised that in order to get an efficient engine he needed to separate the two phases of compression and expansion, and arrived at a turbojet configuration very similar to that of Whittle.[181] But even before von Ohain's engine powered the first flight of a He 178 turbojet plane on 27 August 1939 from the Heinkel Airfield in Rostock-Marienehe, the *independent followers* Schelp, Wagner and H. Constant had already corrected the initial development shortcut to the aerodynamically more challenging, but lasting all-axial turbojet engine configuration.

Hans von Ohain's early professional life appears to be characterised by a certain ambivalence between phases of highly concentrated dedication to a given task and certain *'stand-by'* phases, when he apparently waited that somewhat confused situations sorted

---

[180] Warsitz had *'flown'* the He 176 aircraft with 5 m wingspan already between 9 and 13 July 1938 in the large AVA wind tunnel at Goettingen, with an elliptical nozzle of 7.0×4.7 m, speed limited to 32 m/s to prevent a higher vibration level, just to simulate take off and landing.

[181] Potential patent considerations in this context will be discussed at length in the Attachments, Sect. 12.1.

out, mostly not on his initiative; perhaps an early example of a successful work-life-balance? His decision to join the Heinkel works at Rostock in April 1936 is described as a decision in favour of Heinkel's preference for speed, and his interest in innovative, uncommon engineering solutions. On the other hand, for von Ohain it was also a return to happy student times at Rostock University in the summer semester 1932—with lots of sailing and leisure along the Baltic coast.[182] This expectation at least did not materialise under Ernst Heinkel's demanding *24/7 time regime* after his working head start together with his Goettingen car mechanic Max Hahn in April 1936.

'Dedication and coincidence' appear also to be ruling principles in Hans von Ohain's personal environment, Fig. 6.21. It is obvious that von Ohain's success was indebted to the comprehensive support by the Heinkel engineering team and facilities, but also to the infatiguable enthusiasm and encouragement by Ernst Heinkel himself. As a young man Ernst Heinkel became interested in aviation, and was impressively driven in the aircraft direction after he experienced in a crowd of 50,000 the Zeppelin *'catastrophy of Echterdingen'*.[183] In October 1909 he attended at Frankfurt/M. the *ILA*, the first *'International Airship Exhibition Zeppelin'* in Germany, then the largest international event of this kind, for 100 opening days after 10 July 1909. Just one year later Heinkel started building his first own aircraft, based on a set of construction plans from Henri Farman and a 50 hp flight engine *'on loan'* from local Daimler-Benz. After Ernst Heinkel flew the first time on 12 May 1911 as passenger on board of a Rumpler *'Dove'*, piloted by his life-long friend Hellmuth Hirth (1886–1938), he undertook his first self-piloted flight on the *'Wasen'* (meadow, airfield) at Stuttgart-Bad Cannstadt on 9 July 1911, Fig. 6.21 (l). Only ten days later, after several successful flights, he crashed his biplane from an altitude of 40 m, incuring severe injuries. After a recovery period of five weeks in hospital, the highly indebted Heinkel was forced to cancel his further studies and to commence his dedicated working career as aircraft designer at Berlin.

Similarly, the important, decade-long relationship between Hans von Ohain and his *'assistant-mechanic'* Max Hahn (1904–1961) grew somewhat coincidental—still at Goettingen. In 1930, during von Ohain's first year at Goettingen University, his 4-year-elder step-brother Herbert transferred the ownership of his automobile to him, most unusual

---

[182] Hans von Ohain spent his sixth semester between May and July 1932 at Rostock University, http://matrikel.uni-rostock.de/id/200030553

[183] After an emergency landing on open ground at Stuttgart-Echterdingen in the morning of 5 Aug. 1908, 136 m long LZ4 (Zeppelin airship, model 4) was torn from its moorings by a gust of wind in the afternoon and in due course completely destroyed, when the hydrogen gasbags were damaged and immediately ignited. The cause of ignition was later ascribed to a static charge being produced when the rubberised cotton of the gasbags was torn. The disaster produced an extraordinary wave of nationalistic support for Zeppelin's work. Unsolicited donations from the public poured in: enough had been received within 24 h to rebuild the airship, and the eventually donated total was over six million Marks, at last providing Zeppelin with a sound financial base for his further experiments.

Fig. 6.21 Hans von Ohain's professional career: Dedication and coincidence, E. Heinkel 1st flight 1911 (l), Von Ohain's Opel 4/16 PS 1931 (r)

for a student at that time. It was a two-seater *Opel 4/16 PS*,[184] Fig. 6.21 (r), which required regular maintenance. Hans von Ohain picked for this job the auto repair shop of Bartels & Becker[185] and the chief mechanic Max Hahn became soon an important discussion partner to him—not just on his car but also on his emerging ideas for a new aero gas turbine concept. Hans von Ohain's professional career as 25 year-young physicist from his entry at Ernst Heinkel Flugzeugwerke at Rostock on Tue 14 April 1936[186] to the first turbojet-powered flight just 1230 days later on Sun 27 August 1939 is assumed to be known;[187] Fig. 6.22 highlights just the decisive milestones on this path. Already on 15 April 1936 started wind tunnel tests at Heinkel Marienehe with the Goettingen *'Garage Model'* which Hahn had built on von Ohain's instructions and which they had brought with them. Target was to simulate the positive effect of a flight stagnation pressure at engine intake, very similar to how the Magdeburg RT0 tests had been influenced, as described in the foregoing Section. The circumferential speed of the model rotor was limited to 100 m/s or approximately 3000 rpm only, not to endanger the wind tunnel installation by Hahn's rather soft back-to-back sheet metal rotor. The results were again insufficiently low component

---

[184]The official name *Opel 4/16 PS* referred to the fact that the car's engine size gave it a 4 tax horsepower *(Steuer-PS)* rating, while the second number stands for the actual horsepower, here for a production period 1927/28; claimed top speed was 70 km/h. This was the first German car to be assembled on a Ford inspired production line and corresponding volume production, which was achieved with approx. 120,000 sold units. The two-seater *Opel 4 PS* when launched in 1924 bore an uncanny resemblance to the little Torpedo-bodied *Citroën 5 CV* which had been launched in 1921 in a name-associating citron-yellow. Correspondingly, the German Opel appeared in a light grass-green, which led to the car's nickname *Laubfrosch* (tree frog) and the phrase *'Dasselbe in Gruen* (The same thing in green)'.

[185]The (now disappeared) *garage* was located then on Reinhaeuser Landstrasse 22A (which still exists unchanged), some 700 m away from Prof. Pohl's first Physikalisches Institut, then still at Bunsenstrasse (today DLR Goettingen).

[186]The Easter holidays were on Sun 12 and Mon 13 April 1936.

[187]See Constant, E.W., The Origins, p. 194 f., Conner, Hans von Ohain; Kay, German Jet Engine, p. 18 f.; Meher-Homji, Pioneering; von Ohain, Turbostrahltriebwerke.

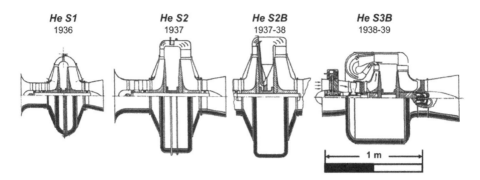

Fig. 6.22 Heinkel/Hans von Ohain turbojet engine development 1936–1939 © Deutsches Museum Munich

efficiencies, so that mid-1936 a re-launch of the activities in two, timely parallel directions *He S1* and *He S2* was decided, Fig. 6.22:

- **He S1** intended a much sturdier mechanical design, stiffened by ring flanges produced from the neighboured *'Neptun'* shipyard—with the overall diameter reduced to 0.7 m. The design followed still von Ohain's filed patent idea of a *'Nernst turbine'* (rotating U-tube channel). Compressor and turbine rotor were now axially separated for smoother flow—with the sheet metal blades mounted on separate wall disks. However, the flow channel through-flow area from compressor exit to the radial turbine was considerably increased,—thus already leaving the Nernst principle. The test results as documented on 19 December 1936[188] remained unsatisfactory, especially since the functional operation test required still constant auxiliary drive support.

- **He S2** with hydrogen combustion[189] demonstrated for the first time self-sustained operation in the (presumed) time-frame mid-January to April 1937. This success contributed considerable to the further project motivation of Ernst Heinkel and the whole von Ohain team, which in due course became independent in the Heinkel organisation since June 1937.

  The H_2-combustion went smoothly and led to an equal temperature distribution at turbine entry, if the through-flow velocity was sufficiently reduced. This required the installation of a vaned compressor diffuser, which was supplemented by a row of turbine inlet guide vanes; consequently, the diameter of the He S2 test facility grew to 0.97 m. Both measures further left the basic Nernst idea of von Ohain's filed patent application

---

[188] Dates follow those provided in—see Koos, Raketen- und Strahlflugzeuge, who generated a commendable, reliable time structure of the historic activities from comprehensive Heinkel Archive evaluations at Deutsches Museum Munich.

[189] Both von Ohain and Whittle suffered the constraint of having no adequate test facilities for components and thus had to develop their engines as a whole. Von Ohain's choice of hydrogen as an experimental fuel largely accounted for the nearly 2-year development lead he had built up by late 1939.

and of preventing power losses by unnecessary energy transformation from velocity to pressure, and vice versa.

- **He S2B** was a further development milestone on the way to kerosene combustion; the illustration of Fig. 6.22 shows an experimental variant (of four) with a fuel injection blower between compressor and turbine disk. These configurations with a maximum diameter of 1.06 m showed first succesful combustion test results after mid-September 1937.
- **He S3B** became the final configuration, with 500 kp take-off thrust[190] in early March 1939 ready for the successful first flight on 27 August 1939 on board of the He 178 aircraft, piloted by Erich Warsitz[191]—after Max Hahn had introduced two significant inventions, a) a evaporation tubes, practically deduced from the operation principle of a soldering torch, which—correctly positioned—allowed the first time, failure-free operation. Both ideas, as illustrated in Fig. 6.23, were patented for Max Hahn, the reverse annular combustor with priority 27 May 1938 and the fuel evaporation tubes on 5 February 1939.[192]

The work of developing the *He S1 Heinkel Strahltriebwerk* (jet engine) *#1,* as the demonstration engine was known, was carried out on great secrecy in an isolated, newly erected shed on the Heinkel Marienehe airfield, which was ready for occupation in mid-June 1936. Because von Ohain, at that time, had no engineering training he was assigned Dipl.-Ing. Wilhelm Gundermann (1904–1997) to work with him. Gundermann had studied mechanical/turbomachinery engineering at the TH Berlin-Charlottenburg under Professor Hermann Foettinger (1877–1945), who presumably at the same time had also Fritz Heppner (1904–1982) amongst his students. Gundermann's team at Heinkel comprised six to eight draughtsmen and stress analysts. Calculations were made using, as a starting point, the characteristics of centrifugal pumps, the *Francis* radial-inflow water turbine and also steam turbine design methods. Meanwhile, Max Hahn soon had six to eight fitters and mechanics in the workshop. As seen, Hahn also played a decisive role in the design work, especially for the combustion system.[193]

---

[190] Acc. to Constant, E.W., The Origins, p. 200, at a total pressure ratio of PR 2.8 and 13,000 rpm speed, for 360 kg weight and a specific fuel consumption of 0.163 kg/(kN h).

[191] For Warsitz's personal account—see https://www.youtube.com/watch?v=TlUk-WWCPa0

[192] The surprising fact that Hans von Ohain's principle invention of the all-radial turbojet engine concept—with only limited similarity to the actual flight hardware—became an unpublished *Geheimpatent* (secret patent) *317/38* with priority of 10 Nov. 1935, while Max Hahn's realised invention of the same subject was internationally filed and published, requires explanations which—to the best of the author's knowledge—will be discussed in the Attachments, Sect. 12.1.

[193] However, the statement—see Kay, German Jet Engine, p. 19—that Hahn *'had been very active previously in this field at Göttingen and had filed several patents'* is at least questionable. Besides the two patents mentioned in the foregoing, no further patents can be found for Max Hahn at that time, and only one during his post-war employment at Heinkel Stuttgart-Zuffenhausen about a *'Mixing procedure for plastic glues'*, DE1,023,540, priority 17 April 1957.

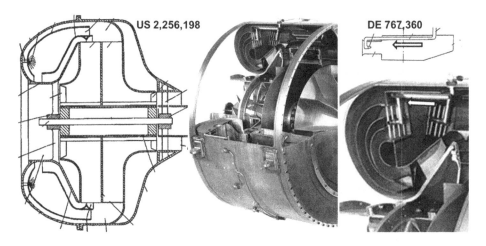

Fig. 6.23 HeS 3B turbojet engine by Hans von Ohain/Max Hahn 1939, M. Hahn patents: Reverse annular combustor US2,256,198 (l,m), fuel evaporation tubes DE767,360 (r) © Deutsches Museum Munich

The personal relationship between Hans von Ohain and Max Hahn was certainly determined from mutual respect, modesty and appreciation for their ideally complementing, theoretical and practical talents; but there could have been more. . .

Max Hahn descended from a family with long metal worker's tradition. His father Carl Hahn (1872–1950) was turner, and had been still *'on the tramp'* before he settled at Goettingen in a *Reichsbahn* (state railway) repair shop. His paternal grandfather Johann August Hahn, born 1841 in Silesia, moved already to Goettingen in 1879, where he also was employed at the railway station. Johann August's wife and Max Hahn's grandmother was Wilhelmine Amalie Hahn, a born Katz, who—though a Protestant at the time of her marriage in ~1869/70—was born Jewish. Max Hahn, in the sense of the *Nazi 'Nuremberg race law'* of 15 September 1935, was a *'quarter-Jew'*[194]—a fact, presumably without any restrictive consequences at the time of von Ohain's decision to seek employment at Heinkel also for Max Hahn, later and during the war this employment at the liberal Heinkel Company might have provided some additional personal security to the Hahn family.[195]

By chance, looking to the ancestors of the English turbojet inventor Sir Frank Whittle (1907–1996) yields also some surprising observations in this context. He was born as the eldest son of Moses Whittle and Sara Alice Garlick, and though he underwent in his professional life several drawbacks and discriminations, none of these was antisemiticly founded. Actually, his parents belonged to the *(John) 'Wesleyan Chapel'* which around

---

[194] In 1939 there lived in Germany (without Austria) still 330,000 Jews, 64,000 'semi-Jews' and 48,000 'quarter-Jews',—see Wikipedia, 'Jüdischer Mischling' in German. In 1939 the registration card of Max Hahn's father Carl was *'earmarked'* at the Goettingen registry office accordingly.

[195] Information by courtesy of Mrs. Dr. S. Dahmen, Stadtarchiv Goettingen, 28 Nov. 2016.

1800 had separated as *Methodists* from the *Church of England*, and administered the tradition of biblical naming. In this respect also *Frank* was originally foreseen by the Whittles to become a *Moses,* until the Garlick branch of the family intervened for a more modern naming.[196]

The intense development progress towards the all-radial concept at Heinkel after the start of the von Ohain/Hahn team in April 1936, shifts activities for the usage of axial compressor components somewhat in the background, though both Ernst Heinkel and Hans von Ohain must have realised that *'axial'* was the way to go, driven by aerodynamics and latest, when the RLM preferences of Hans Mauch and the newly hired Helmut Schelp spread after August 1938. Only the straightforward simplicity of the von Ohain concept, the stunning project head-start and the relentlessly kept *'Heinkel tempo'* nourished the hope for a merit in being *'first in air',* even with the *'wrong concept',* with corresponding attention, recognition and orders.

Heinkel's axial activities started seriously in June 1938 with the appearance of the HeS 8 as a more powerful follow-on concept to the HeS 3B. It was the first Heinkel turbojet to receive official financing and was accordingly given the RLM number of 109–001. The development aimed at producing again an all-radial engine with about 700 kp thrust. A reduction in diameter was accomplished by the introduction of a new layout, whereby the combustion chamber was allocated between radial compressor and turbine wheel. Thus, the previous reverse-flow combustion chamber was abandoned for a straight-through design, and the thrust-to-weight ratio grew from F/W 1.25 to 1.84, Fig. 6.24.

Although full information is lacking, it is believed that the HeS 9 engine was designed as part of a family concept around the HeS 8. The first power increase was achieved in the HeS 8 V15[197] with one additional axial compressor stage after the main centrifugal compressor stage.[198] The HeS 9 represented the next step in power growth, still with a radial-inflow turbine similar to the He S8, but with a new compressor set-up of smaller diameter, consisting of an axial inducer, a *diagonal-flow* compressor, followed by a two-stage axial compressor. This arrangement fits the HeS 9 into the flow path development sequence leading to the HeS 011 engine (which will be further discussed in Sect. 8.1), but the only facts known for certain are that ten HeS 9 engines were ordered by the RLM, in

---

[196]Information by courtesy of Ian Whittle, son of Sir Frank, 29 Nov. 2016: *'The Whittles were a working-class family. My antecedents had worked as 'weavers' in the cotton industry for many generations. Grandfather Moses left school when he was 11 and worked as a «Grease Monkey» on the machinery in the cotton mill—lubricating the mechanism whilst it was dangerously in motion. They chose particularly small boys for this work so they could crawl in amongst the working parts ... very dangerous. He learnt the rudiments of mechanical engineering as he grew up, becoming a skilful engineer and becoming a shop-floor foreman before he prospered enough to buy his own small engineering business. He was a very inventive man and so were two of his sons—my father and my Uncle Albert.'*

[197]'V15' stands for the 15th Versuchsaufbau or experimental build.

[198]The HeS 8 engine family concept is shown as Fig. 2.21 in—see Kay, German Jet Engine, p. 41.

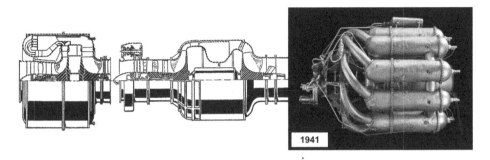

Fig. 6.24 General arrangement of HeS 3B, 450 kp thrust, D_{max} 0.93 m (l) HeS 8, 600 kp thrust, D_{max} 0.775 m (m) © V. Koos and Whittle W.1X, 560 kp thrust, D_{min} 1.12 m (r) © Smithsonian

addition to three HeS 10 bypass engines.[199] The HeS 9 was foreseen to be applied for the *He 180* aircraft, for which an RLM order was received already in June 1939 and which materialised in the *He 280,* with first flight on 30 March 1941, powered by two HeS 8A engines. Hans von Ohain reached out to AVA Goettingen and its axial compressor expert Walter Encke in mid-1938,[200] just within weeks before the new RLM axial turbojet engine strategy with common, type-of-standardised seven and eight-stage AVA compressors for the official Junkers and BMW engine projects was presented by Helmut Schelp at Goettingen. At first especially Walter Encke believed to be able to serve both interested

---

[199]Principle data for the geared Heinkel HeS 10 turbofan are provided in—see Koos, Heinkel Raketen- und Strahlflugzeuge, p. 70, in addition to a von Ohain patent of an all-axial geared turbofan DE767,258, priority 12 Sept. 1939, with the additional extras of a row of fan VIGVs (variable inlet guide vanes), and/or a variable thrust nozzle. An afterburner in the bypass duct was optional. Koos mentions also an AVA report by W. Encke, dated 13 Sept. 1939 on basic HeS 10 fan tests and analysis. Gersdorff refers to the HeS 10 take-off thrust of 740 kp, 1.64 m casing diameter and 500 kg weight; on top, there existed a turbofan version of Mueller's all-axial engine as HeS 30A with 1050 kp thrust and 0,890 m outer diameter;—see Gersdorff, Flugmotoren, p. 282. While von Ohain highlighted in the patent text mostly the mechanical advantages of driving the geared fan/ compressor combination with only one turbine, Heinkel stressed in a letter to Udet, dated 18 July 1940, that the turbofan promised an expected thrust increase and a reduction of specific fuel consumption of 30 percent; part of 'Thorwald File',— see Footnote 171.

[200]There appears to be a slight discrepancy between von Ohain's own memory, expressed in a letter to Ed. Constant on 31 July 1970, that *'he was fully aware of Betz and Encke's work on axial compressors at the AVA'* and a comment of Volker Koos that Heinrich Helmbold (1899–1973), responsible for the He 178 aerodynamic design after W. Guenter's deadly car accident on 21 Sept. 1937, had introduced von Ohain to the axial compressor design principles and had also recommended to contact W. Encke at AVA Goettingen. Hans von Ohain had met Helmbold at his first Heinkel working place, the wind tunnel building, where Helmbold had built a new high-speed wind tunnel with two-stage counter-rotation (CR) axial compressor. Helmbold came from AVA so that the ideas for this design might have emerged in parallel to Encke's own CR activities, as addressed already in Sect. 4.3.3, sub-section 'Degree of Reaction'. See Constant, E.W., The Origins, p. 196 and—see Koos, Heinkel Raketen- und Strahlflugzeuge, p. 68.

groups; correspondingly, a set of AVA-produced He S9 axial wheels was still delivered to Heinkel on 19 January 1939. However thereafter, the AVA axial compressor activities were more and more directed towards the main clients Junkers and BMW, in line with the official policy of the RLM Technical Engine Office (Eisenlohr, Mauch, Schelp), that the new turbojet engines required the manufacturing expertise of established engine manufacturers. Initially Ernst Heinkel and Hans von Ohain hoped to (over-) compensate the emerging axial compressor design and competence gap by hiring Max Adolf Mueller and his team in February 1939. After a short time of euphoria, they all had to admit that the variety of engine projects (HeS 6, HeS 8, HeS 9, HeS 10, HeS 30 and HeS 30A, not to mention Mueller's rather exotic engine concepts HeS 40, HeS 50 and HeS 60)[201] considerably overstretched the development capacity of the small Heinkel engine team. In a last exertion of forces, it was again Hans von Ohain, who fulfilled Udet's requirement of a successful He 280 flight as precondition for Heinkel's acquisition of Hirth Motoren GmbH at Stuttgart-Zuffenhausen in April 1941. On paper this finally brought the status of an established engine manufacturer for Heinkel, but in the meantime the Company had lost the original development advantage of an early start. Udet's suicide on 17 November 1941 shook Heinkel's protected position and thereafter, Ernst Heinkel's autocratic and erratic management style had left the Company as an economic *'basketcase* (Sanierungsfall)'.

6.2.3   The RLM Turbojet Programme and Its Management

To develop the turbojet engine required not only a certain level of technical development and ability, but also considerable resources to pay for skilled teams and special equipment. Germany's aviation industry suffered after the First World War, but by 1926 when aviation restrictions of the Treaty of Versailles had been largely removed, the industry was again picking up and by 1933, when Hitler's Nazi party came to power and brought with it general expansion, the industry was among the best in the world, including the aero-engine companies of Argus, BMW, Bramo, Daimler-Benz, Hirth, Junkers and Walter. For engine development work, government support was available, but few at that time—except the just addressed industrial examples of Heinkel and Junkers—had any interest in jet engines. Few probably realised that the turbojet engine would need many years of intensive R&D work before it became usable, but only the government could initiate correspondingly long-term programmes in anticipation of future high-performance aircraft.

The fact that Germany had this extensive turbojet programme at all was due to the early, energetic efforts of just two men within the RLM—Hans Adolph Mauch (1906–1984) and Helmut Schelp (1912–1994); the reason, why both belong to the salient category of *'turbojet engine pioneers'*, where Hans Mauch has already been depicted in Fig. 6.13 (r)—and Helmut Schelp will get his deserved personal portrait in the following Chap. 7,

---

[201] See Kay, German Jet Engine, p. 18 f.

Fig. 7.9 (m), together with the ingenious, though repetitively somewhat overshadowed character of Max Adolf Mueller (1901–1962), Fig. 7.15.

Hans Mauch was born at Stuttgart-Bad Cannstadt and commenced his studies at TH Stuttgart in 1924 in mechanical and electrical engineering, where he also belonged to the *'Sonderbund'*, a *pietistic* students's fraternity of 1859. He finished his studies at TH Berlin-Charlottenburg in 1929 in the upper 2% of his class. After a failed start with a doctorate thesis at the Institute of Professor Georg Schlesinger,[202] he joined the Siemens daughter company E. Zwietusch, before in 1935 he took a job in the RLM Ministry of Aviation.

According to Schlaifer,[203] German government financing of any form of jet propulsion began in 1931. There were several forerunners of these *'pulsojets'*, which use resonant combustion and harness the expanding combustion products to form a pulsating exhaust jet that produces thrust intermittently. Munich engineer Paul Schmidt (1898–1976) pioneered the more efficient *'valved pulsejets'*—in German simply called *'Schmidtrohr'* (tube) and during the war known as *'V-1 flying bomb'*—based on modification of the intake valves (or flaps), earning him early government support. The first grants came still via Adolf Baeumker[204] from the *RVM Reich Verkehrsministerium* (Traffic Ministry) and beginning 1935 continued intensified by the RLM Research Division until the end of the war.[205] After

---

[202] Professor Georg Schlesinger (1874–1949), since 1904 first chairholder and scientific founder of the *'Institute for machine tools and factory organisation'* at the TH Bln.-Charlottenburg, led since 1916 the *'Prostheses test establishment'* of VDI, the German Association of Engineers, and developed at that time together with the famous Prof. Ferdinand Sauerbruch (1875–1951) arm and leg artificial replacements. Interestingly, another prostheses development route existed between Sauerbruch and Prof. Aurel Stodola since 1917 in Switzerland, also by investigating the *'iron fist'* of the German robber-knight *Goetz von Berlichingen* (~1480–1562). This collaboration marked one of the first documented examples of a surgeon and engineer merging efforts; Sauerbruch said *'Henceforth, surgeon, physiologist, and technician (prosthetist/engineer) will have to work together'*. On 2nd May 1919 Sauerbruch and Walter D. Boveri (1865–1924), co-owner of Brown, Boveri & Cie. founded at Singen, Germany the *DERSA Deutsche Ersatzgliedergesellschaft Sauerbruch GmbH.* (German prostheses society Sauerbruch). In 1930, Mauch skipped the doctorate at Schlesinger's Institute, after he learned about a similar, already finished work at another university and progressed at the Siemens daughter company Zwietusch AG, Berlin (in the vicinity of Prof. Herbert Wagner's THB Institute at Franklinstrasse) to develop high-speed, pneumatic tube mail distribution systems, for which he acquired some 25 patents in the period 1930–1935. From 1935 to 1939 lasted Mauch's decisive phase for organising the German turbojet developments at RLM, described above, before he left RLM officially to form a private consulting company, which may be only partially true. What is known about Mauch's professional career between 1940 and 1945 and thereafter will be described in short in Sect. 8.1.

[203] Schlaifer, The Development, p. 381 f.

[204] See foregoing Sect. 4.3.2 in Baeumker's initial involvement in altitude aircraft developments as of spring 1926.

[205] Though 1935 coincides with Mauch's entry in the Aviation Ministry and though Research (LC1) and Development (LC8) were closely related in the terms of the 1937/38 valid organsation as Divisions of Udet's Technical Office, Mauch claimed—after the war—that it was not until the middle of 1939 that he heard of the Schmidt pulsejet;—see Schlaifer, The Development, p. 382. Strange—

one of many RLM reorganisations a small section *'Sondertriebwerke* (Special engines)' was established within the RLM power plant group. On 1 April 1938, when Dipl.-Ing. Uvo Pauls (1902–1989) left Berlin to take over the newly erected Luftwaffe *Entwicklungsstelle* or in short *E-Stelle* (development establishment) at Peenmuende-West, a position which he kept up to 1st September 1942,[206] he was replaced in the *'Special Engines'* responsibility by the energetic, by practical training highly qualified young engineer Hans A. Mauch. To his first propulsion system task belonged the Walther rocket engines for JATO Jet-Assisted Take-Off which Helmut Walther (1900–1980) had started in 1937, again as a private initiative.[207] At the same time Mauch in his new function learned again rather by accident of Heinkel's privately financed turbojet development activities since April 1936.[208] The *'Heinkel Chronology'* at the Archive of the Deutsches Museum (V. Koos) has a date in May 1938[209] for a *'project presentation for RLM (Mauch)'* which demonstrated already *'liquid-fuel testing and the HeS 3B flight test configuration'*. By learning the—then surprising—increasing propulsion efficiency with rising aircraft speed, Mauch—then still LC8[210]—discussed the subject of aircraft speed with members of the neighbouring Airframe Department LC7. He soon was under the impression that flight speeds were rapidly

---

like Mauch's sudden disappearance from RLM/ government service a few months later,—and his equally sudden reappearance in 1944 as one of the highest decorated civil servants in the Third Reich;—see Sect. 8.1.

[206] See Neufeld, Rocket Aircraft, p. 228. At that time Pauls became *'Sonderbeauftragter* (special commissioner)' for Logistics of the *Fieseler Fi 103* (V-1) air-breathing cruise missile which brought him again in close contact with Hans A. Mauch, then up to 1944 responsible for the final V-1 series production preparation;—see Sect. 8.1.

[207] Luftwaffe's official interest in rockets as main power plants for *'interceptors'* arose during the *Heinkel He 176* programme and two years later with the *Messerschmitt Me 163 Komet* which was flown in 1940.

[208] Different to the *'saga'* of the *uninformed* RLM, Heinkel revealed the turbojet project to the RLM latest still in Dec. 1937 to receive priorities for materials in the tightly regulated Nazi economy. In addition, the DVL was approached in view of possible performance improvements (higher turbine entry temperature and aerofoil air cooling, eventually by the DVL Institute of Prof. FAF Schmidt). This resulted in a first contact visit on 1st Feb. 1938 of Karl Leist (1901–1960), then responsible for the DVL turbocharger development and of the centrifugal compressor specialist Werner van der Nuell (1906–1993), who was also head of the DVL Institute for Fluid Machinery.

[209] On 27 August 1938, exactly one year ahead of the first flight, there was a second presentation at Heinkel with *'He 178 aircraft mock-up ready'*; in addition, to what had been presented to Mauch already in May, now presumably for General-Engineer Wolfram Eisenlohr (1893–1991), Director of the RLM Engines and Accessories Dept. (1 Jan. 1938–31 Jan. 1944), and accompanied by Mauch and Schelp(?). Acc. to—see Schlaifer, The Development, p. 383, at that time Wolfram Eisenlohr had been informed of the turbojet development at Heinkel and probably also of the pulsejet and ramjet projects of the Research Division, but he seems to have considered them as long-range projects—without no immediate importance.

[210] A new, so-called *'General-Luftzeugmeister organisation'* (1939–1941) with GL/C as Technisches Amt after Ernst Udet's corresponding appointment became effective on 1st Feb. 1939—with GL/C2 Airframes and GL/C3 Engines.

approaching a turning point when the efficiency of jet propulsion would be high enough to be attractive. Mauch announced then that the RLM was seriously interested in jet propulsion, and in due course he was approached by Herbert Wagner in spring 1938 to discuss the engine developments which Max Adolf Mueller and his group were carrying out at Junkers Magdeburg and, if possible, to lay the groundwork for future government support for these activities.

In summer 1938[211] Mauch, himself just 32, came by chance within the Air Ministry in contact with Helmut Schelp, a 26-year-young engineer, who was in the final stages of his *'Flugbaumeister'* education and training programme at DVL, and whose career, studies and conclusions thereof will be outlined in detail in the following Chap. 7. Within DVL there was little enthusiasm for turbine engine research, so Schelp had been sent already in August 1937 as a temporary employee to the RLM Research Division, where intermediately he was put in charge of the then funded two programmes—the *Walther rocket engine* and the *Schmidt tube*. At their meeting one year later, Mauch was as a result of what he had seen and heard already in the meantime very much interested in the general jet propulsion principle. This interest grew when the well-prepared Schelp laid out before him the whole range of propulsion possibilities with their theoretical foundations. Mauch immediately invited Schelp to leave the Research Division and come as his deputy in the Development Division and the so-far isolated Schelp, convinced that he could accomplish little where he was, agreed.

Presumably at the same time, Schelp learned of the highly promising axial compressor activities at AVA Goettingen, as an essential precondition for the realisation of his turbojet development plans without much extra-ado. He knew that a useful turbojet would require compressors with efficiencies over 70% at pressure ratios of at least 3:1. Since German superchargers achieved in 1938 only 65–70% at PR 2, the AVA announcement of well over 70% efficiency for their axial compressors was of very considerable importance that a useful gas turbine could be built at this time. With all likelihood still in August or early September 1938 Schelp went for the first time to Goettingen to inform Professor Albert Betz and his compressor expert Walter Encke about their role in the governmental planning. This was apparently just two months after Hans von Ohain's first visit at Goettingen on 2 July 1938—to secure axial design support for the HeS 3B impeller stage. The inherent conflict must have been felt already early on and the RLM position was not to compromise.

---

[211] Both standard references—see Schlaifer, The Development, p. 383 and—see Constant, The Origins, p. 205 put this first meeting between Mauch and Schelp to Aug. 1938. However, a document from Schelp's personnel file, dated 10 Aug. 1938, mentions an order of 15 July 1938—with all likelihood from Mauch (Gen.-Eng. Eisenlohr, superior to Mauch, noted on that document somewhat reproachful that he had not been involved in the transfer),—which put Schelp in agreement with DVL to lead group LC 8 VIIa *'Development of jet engines'* in Mauch's 'Special developments' division LC 8, so that with some probability the first Mauch/Schelp meeting was already in early July 1938.

Impressions of the somewhat shaky turbojet development activities at Heinkel and Junkers Magdeburg confirmed Mauch's and Schelp's view that the time was ripe for serious turbojet engine development, but they were also convinced that only experienced aero-engine manufacturers should and could develop/manufacture a practical and reliable engine. Therefore, they refused to support the existing Junkers and Heinkel projects (wholeheartedly)[212] and in the fall of 1938 Mauch and Schelp started a kind of *'road show'* by visiting all four major German aero-engine manufacturers to interest them in a government-funded turbojet engine development programme—Junkers Motors Dessau, BMW Munich, their near-term subsidiary Bramo Berlin-Spandau, and finally Daimler Benz Stuttgart.

As a rather rare original document of these negotiations survived in the Jumo archive material a *'poster'* which contains a summary of the technical development targets in compact form, Figs. 6.25 (original) and Fig. 6.26 (translation).[213] Minutes of meetings indicate a sequence of meetings amongst Helmut Schelp, Professor Otto Mader, the coming man Dr. Anselm Franz and the still self-reliantly attending Max Adolf Mueller in the OMW Otto-Mader-Works office building at Dessau, Fig. 6.28 (r), beginning in late autumn 1938 and continuing for several months till spring 1939. The principle, simplified sketch of a turbojet engine on that *'poster'* was obviously still available in Anselm Franz's group of engineers from Herbert Wagner's just recent turbojet patent application DE724,091—with a priority on 14 August 1938, i.e. just only a few weeks before the first Schelp meeting at Junkers. One can imagine that basic notes about *A Zielsetzung* (development target) and *B Arbeitsweise* (working principle) were added in the first meeting, written by hand (on a blackboard?) and then for the follow-on meetings[214] with black ink and by means of a writing template transferred to the permanent poster.

To understand the pedagogic effect of this *'script'*, one may imagine young, enthusiastically lecturing RLM *freshman* Helmut Schelp in front of the sceptical though dutybound, engineering institution Otto Mader—*'Follow me, line by line …'*:

- *Turbinenluftstrahltriebwerk*, addressing the German abbreviation TL for *'turbojet engine'*, in German rather 'Turbine air jet propulsion unit', and then assertively confirming (*'ohne Verwendung einer Luftschraube'*—without propeller),

---

[212] Deviant from Schlaifer's and Constant's accounts, there are indications that at least Mauch sought contact to Ernst Heinkel, e.g. on 23 Jan. 1939 during a visit at Rostock, to discuss a kind of financial settlement for Heinkel's advance provisions.

[213] See Mueller, Junkers Flugtriebwerke, p. 221.

[214] Theoretically, one could even imagine that Schelp took copies of that poster as a kind of standardised *target setting* also to the other follow-on company meetings.

Turbinenluftstrahltriebwerk

(ohne Verwendung einer Luftschraube)

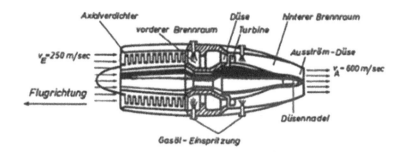

A Zielsetzung bei Einbau in 2 mot. Jäger:

bei einer Fluggeschwindigkeit von 950 km/h erreicht das
T.L.-Triebwerk eine Leistung von 550 kg Schub, dies
entspricht einer Motorleistung von 2000 PS.
   Brennstoffverbrauch: etwa 800 gr/Psh
   Leistungsgewicht: etwa 0,17 kg/PS

B Arbeitsweise:

1   Durch Axialgebläse wird Luft angesaugt.von 1 ata auf
    3,5 ata verdichtet und in die vordere Brennkammer
    ausgeschoben.
2   In dieser erfolgt Einspritzung und Verbrennung von
    Gasöl.
3   Aus der Brennkammer strömen heisse Gase durch
    die Turbine in die hintere Brennkammer.
4   Turbine treibt Axialverdichter an.
                    (Gastemperatur in der Turbine 600°C)
5   Nochmalige Einspritzung von Gasöl in die hintere
    Brennkammer.
6   Die Gase, die daher auf 1000°C erwärmt werden,
    treten dann durch die Ausströmdüse mit v=600 m/sec
    in die Atmosphäre.
7   Die Erhöhung der Luftgeschwindigkeit von 250 m/sec
    auf 600 m/sec ergibt als Reaktion einen Schub von 550 kg.

Fig. 6.25 RLM—Junkers turbojet *'script'*, late autumn 1938 German original poster © R. Mueller

- the following figure, presumably known to Mader from Wagner's recent turbojet patent
  is modified after combustor exit by adding the single turbine stage, the second *'gasoil
  injection'*, the *'aft combustor* (afterburner)*'* including *'nozzle plug'* and explains the

Turbojet Engine
(without propeller)

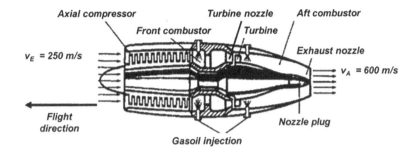

A Target for 2 engine fighter :

at a flight velocity of 950 km/h the turbojet has
550 kg thrust, corresp. to 2,000 hp engine power.
Fuel consumption: ~ 800 g/ hp h
Power weight: ~ 0.17 kg/ hp [~ 2.67 hp/ lb]

B Mode of operation :

1 Axial blower sucks air in, compresses it from 1 at
to 3.5 at and pushes it into the front combustor.

2 Here – injection and combustion of gasoil.

3 Hot gases flow from front combustor through the
turbine to the aft combustor.

4 Turbine drives axial compressor.
(Turbine gas temperature 600° C)

5 Another gasoil injection into aft combustor.

6 1,000° C hot gases are ejected with v=600 m/s
through the exhaust nozzle to atmosphere

7 Increase of air velocity from 250 m/s up to 600 m/s
results in a reactive thrust of 550 kg.

Fig. 6.26 RLM—Junkers turbojet 'script', late autumn 1938 English translation of original poster
© R. Mueller

sequence of major turbojet components, kindly assisted by the helpful hint on 'flight
direction',

- the header 'A Target for 2 engine fighter' collects the essential assumptions of a (high)
flight speed v of 950 km/h and a turbojet thrust F of 550 kp, from which the equivalent
engine power P = F x v ~ 2000 hp could be deduced. Schelp's performance calculations

provided then the given values of *'fuel consumption'* and *'power weight'* which Mader presumably put automatically in relation to his best piston engine, the *Junkers Jumo 211*,[215] with considerably lower sfc ~ 210 g/(hp h) and correspondingly higher power weight ~ 0.53 kg/hp,

- the wording under *'B Mode of operation'* appears to be self-explanatory, except that *'4 Turbine drives axial compressor'* was apparently still worth to be explicitly mentioned,

   '5 Another gasoil injection into aft combustor' describes possibly for the first time an axial turbojet with afterburner arrangement[216] and

   '7 the resulting reactive thrust' is calculated with the formula $F = m \, \Delta v$ and an engine mass flow at altitude of m ~15 kg/s.

During summer 1938, simultaneously to Mauch and Schelp and their setting out to create an official jet engine development programme, Hans Martin Antz (1909–1981) joined the RLM Airframe Development Group LC 2. Like Schelp, with whom he was closely acquainted, he participated in the *'Flugbaumeister'* programme. After a 2-year-stay at MIT, Cambridge, MA, and at von Kármán's GALCIT[217] in Pasadena, CA between mid-1933 and mid-1935, followed a second trip to Pasadena to accomplish his master examen in December 1936.[218]

Immediately thereafter, he achieved his 'Flugbaumeister' examen with DVL in January 1937, and commenced on 1 February 1937 as *Referent* at the *Luftwaffe Rechlin Test Centre* and came thereafter to RLM, similar to the foregoing professional development move of Schelp.[219] At first Antz encountered a certain amount of opposition in the Ministry, but he

---

[215] See Wikipedia, 'Junkers Jumo 211' in English.

[216] See Golley, Whittle, p. 258: Whittle's patent GB 471,368 *'Improvements relating to the propulsion of aircraft'* with priority date 4 March 1936 mentions solely generally—and not for an axial arrangement—that his invention *'may incorporate* known *features to improve the efficiency or power, such as heat exchangers, or additional fuel combustion . . .'*, but a search for earlier afterburner schemes was not successful.

[217] See Wikipedia in English, 'Massachusetts Institute of Technology' and 'Guggenheim Aeronautical Laboratory' at the California Institute of Technology (GALCIT). Antz's master thesis is part of— see White, Tests. In addition, he co-translated O. Schrenk's *'Experiments with suction-type wings'* which appeared as NACA TM 773 in Aug. 1935. Antz's data from his personnel file BA-MA PERS 6/137433.

[218] Courtesy to Richard Fuchs and his thorough genealogical search, Antz was detected as passenger on board of *Norddeutscher Lloyd's SS 'Bremen'*, departing Bremen to New York on 13 Nov. 1936. The 51,656 BRT *SS 'Bremen* (IV, of 1929)' won the *'Blue Riband of the Atlantic'* on its maiden voyage in July 1929 with on average 27.83 kn, thus requiring for the 3200 sm Atlantic crossing between *Bishop Rock Lighthouse (Scillies)* and *Ambrose Light, NY*: 4 days, 18 h and 50 min, powered by 4× *Parsons* steam turbines of 135,000 hp.

[219] For details of the *'Flugbaumeister'* 'programme—see Chap. 7 for Schelp's personal course. Antz achievd the C2 pilot's licence and finished his military training equally as Reserve officer candidate.

succeeded fairly quickly. In the first round in the early fall of 1938, he contacted Robert Lusser, then still chief of development for Messerschmitt Augsburg, to get him acquainted with the newly projected kind of engine and to make preliminary investigations. Schelp and Antz had discussed engine installation problems early on and came to the conclusion, largely on the basis of anticipated difficulties with the long tailpipe of the single-engine *He 178*, that an aeroplane with two wing-mounted engines would be superior. This automatically led them to rule out the centrifugal, a decision immediately underlined by an efficiency comparison.[220]

On 14 October 1938 Antz issued a secret 10 page-memorandum for his department head LC 7/III (2 copies for Aircraft Dept.), entitled *'Gegenwärtiger Stand und künftige Entwicklungsarbeit auf dem Gebiete des Schnellfluges mit Strahltriebwerk'*[221]—with the first sentences: *'A further rise in performance of fast-flying-aircraft with combustion engine and propeller is limited on the base of todays' knowledge, on the one hand side by engine size and weight and on the other hand by decreasing propeller efficiency at high Mach number. This barrier is assumed to exist close to 800–850 km/h. Additional performance increases i.e. in the vicinity and beyond the speed of sound are possible by using 'special jet engines'. This report is focussed on aircraft with this special propulsion system. . . . and refers to a corresponding LC 8* (Engine Dept., author Schelp) *report.'*

After reviewing the status of aircraft development with rocket propulsion at Peenemuende, Antz concludes at first Schelp's calculation example of Fig. 6.26 under the header *'A Target for 2 engine fighter'* by comparing a single-seater 1 h fighter mission with *'standard* (piston) *engine'* and with *'special* (turbojet) *engine'*:

1 h fighter mission	Standard (piston) engine	Special (turbojet) engine
Power, resp. thrust	2000 hp	600 kp
Engine weight	1075 kg	350 kg
Specific fuel consumption	0.25 kg/(hp h)	0.5 g/(kp s)
Mission fuel	750 kg	1080 kg
Mission weight (engine + fuel)	1825 kg	1430 kg

Schelp's assumed specific fuel consumption of 80 g/(PS h), Fig. 6.26, corresponds to 0.81 g/(kp s); both values are considerably overestimated in comparison to the actually achieved 0.38 g/(kp s) for the Jumo 004 B engine

In summary, Antz stated: *'The engine developments initiated at LC8 let expect an increase of jet velocities in or beyond the sonic range. Regretfully, it must be said that the airframe side is not yet ready to exploit such engine advantages.'* and *'To re-emphasise,*

---

[220] See Schlaifer, The Development, p. 391, and p. 391, FN 25; in his early analysis of 1949, Schlaifer had assumed that the British double-sided centrifugal had a considerably smaller frontal area than von Ohain's single-sided configurations, but a view to Fig. 6.24 tells a different story.

[221] *'State and future development of high-speed flight by means of turbojet engines'.*

this is not utopia or a probability of a distant future, but according to recent insight a realistic near-term possibility of high-speed flight.'

Therefore, he suggests,

- urgent completion of all high-speed wind tunnel projects at the research institutions,
- urgency *'S'* for research activities towards a family of high-speed wing profiles,
- expansion of single-seater high-speed aircraft development capacity beyond Ernst Heinkel Flugzeugwerke—apparently Messerschmitt AG is best suited due to preparatory activities,
- development of a stagnation pressure-dependent steering transmission for all control surfaces, so that at best the rudder forces would be kept constant for all flight speeds.

After he had demonstrated a significant *mission weight* advantáge of (at least) 22% in favour of the new propulsion system, late in 1938, Antz issued a contract to Messerschmitt with Schelp's figures of *thrust-specific weight and dimensions* as a kind of specification, but without prescribing the actual aircraft configuration,—just the basic demand for 1 h flight endurance and 850 km/h flight speed was set. In January 1939 the RLM issued the first *'Preliminary Technical Specification'* for this type of aircrafts.[222] Woldemar Voigt (1907–1980), whom Antz presumably knew already from common study years at TH Darmstadt, was the responsible project engineer at Messerschmitt. He investigated a number of possible arrangements, with one and two engines. Voigt's plans were first drawn up in April 1939, and the original design was very different from the aircraft that eventually entered service, with wing root-mounted engines, rather than podded BMW 003s with 600 kp thrust each, when submitted in June 1939 as part of *Project P.1065* offer for 900 km/h top speed. The aircraft was renamed in March 1940 to *Messerschmitt Me 262 Schwalbe* (swallow), when the first three test aircraft were ordered. Because the engines were slow to arrive, Messerschmitt moved the engines from the wing roots to underwing pods, allowing them to be changed more readily if needed; this would turn out to be important, both for availability and maintenance. Since both the BMW 003 and the Jumo 004 jet engines proved heavier than anticipated, the wing leading edge was swept slightly from originally 7.0 deg. to 18.5 deg., to accommodate a change in the center of gravity.[223]

After the Me 262 was first flown with a conventional engine and propeller in 1940; the first flight with Jumo 004 A jet power ocurred for Me 262-V3 on 18 July 1942 at the Messerschmitt *Leipheim airfield*, piloted by test pilot Fritz Wendel; this was almost 9 months

---

[222] Accordingly, Heinkel received a preliminary order for its He 280 fighter aircraft, in competition to what should become the Me 262, on 17 October 1938;—see Schabel, Die Illusion, p. 54.

[223] In hindsight this resulting Me 262 backsweep was also termed as *'sweep by accident'*, while the first actual and intentional swept-back design in-line with Busemann's aerodynamic ruling was the Messerschmitt P.1101 after mid-1944 with ground-adjustable sweep of 35°–45°.

ahead of the British *Gloster Meteor*'s first flight on 5 March 1943, powered by two substituted *de Havilland Halford H.1* engines, owing to problems with the intended *Power Jets W.2* engines. The first Meteor became airborne at *RAF Cranwell*, piloted by Michael Daunt.[224]

Eventually—still one year away from the successful first turbojet demonstration flight on 27 August 1939—the jet engine concept started to become more widely known within the RLM, and Schelp and Mauch started to contact the engine manufacturers and to push for the immediate development of a flightworthy axial engine model. By spring of 1939 three of the major engine companies—BMW, Bramo and the Junkers engine division Jumo—were willing to begin the developmet of turbojet engines. Surprisingly, Mauch and Schelp then learned that still another gas turbine project was in progress—a Hellmut Weinrich of Chemnitz was developing a gas turbine for the Navy. By investigation—certainly supported by Weinrich's former fellow student at Chemnitz Helmut Schelp and as outlined in short aready in Chap. 5—it appeared that Weinrich, already in 1936 after Heppner's leave to England, had sent to the RLM plans for a turboprop with counter-rotating compressor and turbine. However, no attention had been paid to the proposal then, up to the point of complete ignorance in the Air Ministry.

In this situation, Mauch and Schelp developed a two-objective strategy:

- at first, all types of engines were to be adequately studied, before it became really clear which would be best for service, and
- second, the *'basso continuo'* came up again, that all the developments should be carried out by experienced manufacturers of aircraft engines.

The industrial-political suggestions were radical:

- *Jumo*, Junkers engine division was to integrate the Junkers airframe division's axial turbojet, turboprop, and piston-driven ducted fan,
- *Daimler-Benz*, so far blank, was to take over the Heinkel centrifugal turbojets,
- *BMW Spandau*, the former Bramo plant, was to develop Weinrich's counter-rotating turbojet and
- *Argus Motoren Gesellschaft*, a small—in 1938 *'arianised'*—company at Berlin-Reinickendorf and now to 100% owned by the mighty Junkers chairman Heinrich Koppenberg, was to take over the Schmidt pulsejet development; a decision of far-reaching consequences, not the least to the *'mastermind'* Hans Mauch personally.

---

[224] See Wikipedia, 'Messerschmitt Me 262' and 'Gloster Meteor', both in English. In addition in the late summer of 1939, the He 280 jet fighter development programme was begun also with RLM support. Details of these developments will be outlined in Sect. 8.1.

At first this programme was opposed by General-Engineer Dipl.-Ing. Wolfram Eisenlohr (1893–1991), director of LC8,[225] now known as GL/C3 after yet another re-organisation, who felt that a longer-term project was needed to develop such a new concept. Mauch gave in to this conflict and left RLM with the beginning of WW II to form a consulting firm,[226] and Schelp took over the development programme. Eventually matters came to a head[227] when Ernst Udet (1896–1941), director of the RLM Technisches Amt, still influenced by his personal acquaintance Hans Mauch[228] and also a long-time personal friend of Ernst Heinkel, over-ruled Eisenlohr, allowing engine developments to continue, including the shaky Heinkel turbojet efforts.[229] By 1941 the engines appeared to be maturing quickly and even Eisenlohr was convinced the project was worthwhile, becoming a strong supporter.

A convincing proof of Mauch's successful lobbying activities in favour of the new turbojet propulsion systems is reflected in the development of the budget numbers[230] in the responsibility of the RLM aero engine department:

Accounting year	[Million RM]	1940	1941
LC 3IIA	Liquid-cooled piston engines	30	30
LC 3IIB	Air-cooled piston-engines	25	25
LC 3/7	*Special engines (turbo- and pulsojets)*	*10*	*15*

---

[225] Eisenlohr kept this post officially between 1st Jan. 1938 and 31 Jan. 1944. In 1942/43 he and other high-ranking RLM personnel were court-martialed after Udet's suicide for *'failed management'* of their jobs—mostly in view of the defeat in the Battle of Britain, but released after the accusations were judged unfounded. Eisenlohr, who voluntarily refused to apply for retirement was granted leave on 1st Feb. 1944. From 1st May 1944 to 30 June 1945 he held the post of Technical Director for propeller design at VDM Propeller Works, Frankfurt/ M.

[226] Mauch's activities after leaving RLM are described in Sect. 7.4.

[227] The sequence of events is quite transparent: On 20 Oct. 1939 Ernst Heinkel wrote a letter to Ernst Udet, asking for his continued *'special patronage'* and blaming unnamed, young RLM technical officers of *'nearly hostile highhandedness'*. As early as 25 Oct. 1939 W. Eisenlohr and a RLM delegation including H. Schelp and E. Waldmann of Schelp's team (see Chap. 7)—but without Mauch, who had been replaced by the also participating, bland Ernst Beck—had to show up at Rostock to correct the negative impression of the influential industrialist, and also to negotiate Heinkel's compensation for upfront development payments; the negotiations continued on 2 Nov. 1939 to discuss the financing of the rocket propulsion programme with W. Kuenzel and M.A. Mueller's broadly structured future engine developments. Heinkel appeared to be the *'winner'*, but two years later and without Udet's protection fiercely started a new game.

[228] See Schlaifer, The Development, p. 394.

[229] According to—see Schlaifer, The Development, p. 394, especially FN 29, *'Udet personally gave Mauch authority to proceed with the plan'* (in favour of the new turbojet propulsion) and *'It is worth remarking that the receptive attitude toward turbojets of Udet, a tactical officer with no engineering background whatever, corresponds exactly to the attitude of General H.H. Arnold in the United States, who was extremely anxious to promote the development of turbines in that country as soon as he learned in early 1941 that they were being developed in England.'*

[230] See Schabel, Die Illusion, p. 42, cost survey, dated 4 Feb. 1939.

Fig. 6.27 German turbojet
engine pioneers II: Hermann
Oestrich, BMW (1903–1973, ~
1940, l), Anselm Franz, Jumo
(1900–1994, ~ 1955, r)

The following Sections deal with the two main Companies on which the RLM turbojet
engine planning mainly relied in the coming years, BMW/Bramo and Junkers
Flugmotoren,—and especially with their picked project leaders Hermann Oestrich
(1903–1973) at Bramo Brandenburgische Motorenwerke, Berlin-Spandau and Anselm
Franz (1900–1994) at Junkers Flugmotoren at Dessau, Fig. 6.27. Besides the mentioned,
political support by Ernst Udet, presumably the biggest asset for the Companies to
participate in the RLM turbojet programme was the offer to receive a ready-made axial
compressor design by AVA Goettingen. Though the details of the first contact between
RLM's Helmut Schelp and the AVA representatives are not known, the timing of Schelp's
start in the LC 8 Development department puts this to August/September 1938.

In principle, the importance of Prandtl's Institutes at Goettingen grew considerably after
1933. Only weeks after Hitler was appointed Chancellor, the institutes received 2.5 million
Reichsmark to expand their facilities and construct a large wind tunnel. The AVA staff
grew from 80 employees in 1933 to more than 530 by 1938 and reached more than 700 by
1940. In April 1937, the AVA was separated from the Kaiser Wilhelm Institute for Fluid
Dynamics and refounded as a registered corporation. One of the effects of this was that
Prandtl and with him the Kaiser-Wilhelm-Society was relieved of the burden of the greatly
increased administrative work in connection with the growing staff. Vice versa, the
administrative load on Albert Betz as AVA Director grew considerably, and his already
scarce support to Encke's (uninspired) turbomachinery activities diminished completely.
Up to Helmut Schelp's sudden contact in August/September 1938, nothing was done at
AVA to improve their fundamental turbomachinery technology level, since Encke had left
this area with a few counter-rotating compressor efforts, which he had to give up due to
mechanical difficulties latest in 1936.[231] Most notably, nothing was done to scrutinise their
axial compressor design standard with degree of reaction R 1. After carefully re-reading
Albert Betz's Lilienthal paper *'Axiallader* (Axial Superchargers)' of 28 Feb. 1938, one
could conclude that he had some idea of what might be asked from his institution in the

---

[231] See Sect. 4.3, Sub-Section 'Degree of Reaction'.

near future—but rather vaguely. His review goes back to 1928/30 compressor data, he speculates undecidedly about possibilities to shorten the axial length, he cconsiders the exhaust gas supercharger up to 15 km flight altitude and a total pressure ratio PR 8 with intercooler(!), he addresses the counter-rotation axial compressor set-up, he realises the vicinity of the critical speed of sound and the accordingly rising flow losses, but the simple idea to reduce the relative Mach numbers by changing the rotor/stator loading by switching from R 1 to R 0.5 does not touch his mind,—? Somehow Betz and Encke appear to be on their guard, offering their services, but not knowing what for. The author's mood might be reflected in the last paragraph of his paper:

> I tried to provide a short survey about the axial supercharger in its present form and about possibilities for improvement. I believe that the axial supercharger, even today with its inherent drawbacks, is applicable for many tasks—and can provide decisive advantages especially for the flight at very high altitudes.

Vision—sounds different. In 1938, the AVA was reorganised once again and divided into eight institutes, of which interest here in the moment just Professor Dr. Albert Betz's *'Institute for Theoretical Aerodynamics'*, Professor Dipl.-Ing. Reinhold Seiferth's *'Institute for Wind Tunnels'* and Dipl.-Ing. Walter Encke's *'Institute for Turbomachines'*.[232] In addition, in mid-1938, Albert Betz's attention was—again—distracted by a patent case, though direct financial compensation could no longer have been the immediate drive, since Betz was appointed civil servant of the Reich since 1 August 1937. At first he applied for a patent on the invention of the *swept wing* alone, in full awarenes of Busemann's foregoing presentation at the 5th Volta Congress, Rome on 1 October 1935, and international publication within the conference proceedings. After a lengthy struggle with the Reich Patent Office, the problem was solved and a *'secret Reich patent'* #732/42 was issued with priority 9 September 1939—for Albert Betz and Adolf Busemann.

Walter Encke also tried to sharpen his design tools by patenting a *'Multi-stage axial compressor for high circumferential speed'*, DE707,013 with priority 27 May 1936, which apparently found its application in the Jumo 004/BMW 003 compressor annulus, Fig. 8.1. Again it is illustrated that the AVA axial compressor design philosophy relies on one to two single (basic) aerofoils, which are repetitively (a) modelled by geometrical similarity laws, (b) the through-flow area are reduced stage-by-stage in line with the corresponding gas conditions and (c) the stage diameter is steadily increasing in flow direction, so that the circumferential Mach number (circumferential speed/local speed of sound) remains close to a found optimum; these three rules are then stretched to a multi-stage arrangement—always with unchanged *degree of reaction R 1.0*.

---

[232] See Maas, Scientific Research.

Already in 1930/1931 commenced from talks between HWA Heereswaffenamt (Army ordnance office) and aero industries what later became the *RLM Typenliste* (type list), a standardised system for aircraft and engine designations. The category < aircraft > was marked by <8 >, example 8–109 for Messerschmitt Me 109, sailplanes became category <108>; analoguous piston-engines were category <9> and consequently, turbojet and rocket engines <109>. In 1939 the Air Ministry gave each turbojet development a government identification number and stated the responsible project leaders in charge:

The 1939 RLM Engine Programme

RLM Designation	Contractor's Designation	Project Leader
109-001	Heinkel HeS 8	Hans von Ohain
109-002	BMW P.3304 (counter-rot.)	H. Oestrich/H. Weinrich
109-003	BMW 003	Hermann Oestrich
109-004	Jumo 004	Anselm Franz
109-006	Heinkel HeS 30	Max A. Mueller
109-007	DB 007	Karl Leist

The universal picture of the RLM Technical Office would not be complete without mentioning its role in planning, financing and erection of living space for the rapidly growing workforce in aero industries. Beginning in 1935 this became a central task for RLM department LC III, which kept the relevant industrial companies completely out, instead of there emerged public utility housing enterprises which carried the formal responsibility. The financing of the building construction works was shared among RLM, housing enterprises, the aero company, the community and the individual acquirer or settler. The programme capital was secured by a Reich debt guarantee and the RLM financed on the average 30% of the construction cost out of its budget. Between 1/1935 and 12/1940 the RLM financed 34,150 housings at the main sites of the aero industries. In 1935 the largest RLM-financed housing estates were the *'Junkers settlement'* with more than 500 houses and at Rostock a three-storey building block of 588 flats for Heinkel employees.[233] Leading Junkers employees should put their attention to their works, and not to private affairs like housing.[234]

---

[233] See Budrass, Flugzeugindustrie, p. 449.

[234] This attitude was also true when in Oct. 1946 Junkers engine designer Ferdinand Brandner (1903–1986) and his engineering team (600 with family members) were brought from Dessau to Uprawlentscheski near Kuibyshev, to continue their engine developments for the Soviet Union,—see Sect. 10.2.1. Their flats were ready for occupation with their name plates at the doors, erected by German prisoners-of-war in the familiar German housing society style.

6.2.4 Initiating the *BMW 003* Turbojet

In the fall of 1938 Mauch and Schelp visited all four of the major engine companies in an attempt to interest them in the new field of propulsion. The whole range of possibilities was sketched out and each was invited to make a detailed investigation of the most promising engine type from its perspective. Though the reaction of the engine companies to the RLM proposals was far from enthusiastic, it was not completely negative.

In general the chief difficulties foreseen were with the turbocomponent efficiencies and with turbine blade materials. The efficiency of radial supercharger compressors was like that of their turbines not over 70%, and experienced engineers believed the necessity of a quantum improvement of 10% if the gas turbine was to have any real possibilty of success. At this time the turbosupercharger for gasoline engines was far from an established success in Germany.[235] Werner van der Nuell (1906–1993), head of the Institute for Fluid Machinery at DVL, Berlin-Adlershof had been studying turbosuperchargers since about 1936,[236] but only BMW was continuously developing them. Problems with turbine blades were even more serious in Germany than elsewhere, owing to the shortage of such strategic materials as cobalt and nickel, so the development of air-cooled, hollow blades was the only way-out.

Helmuth Sachse (1900–1971),[237] then head of development in BMW Flugmotorenbau (Aircraft engines) GmbH, Munich, a subsidiary of BMW Bayerische Motoren-Werke AG, was certainly positive to the RLM activities, since he had left the Air Ministry as predecessor and long-time colleague of General-Engineer Wolfram Eisenlohr only end of 1937. He received Mauch and Schelp a little less sceptical than Daimler-Benz and Junkers, and accepted a study contract willingly. In parts this was perhaps due to the foregoing groundwork: BMW had the mentioned experience with the development of turbosuperchargers for gasoline engines, and had developed also already a turbine wheel with hollow air-cooled blades which had run at inlet temperatures of 900 °C.[238] The work was put in charge of their research head, Kurt Loehner (1900–1978).[239] The decision was

---

[235] For the early turbocharger history in Germany,—see Eckardt, Gas Turbine Powerhouse, pp. 53–65.

[236] Similar to—see Sherbondy, Textbook, some 20 years earlier,—see Sect. 4.1.2—he measured the pressure heads and collected data of 275 different turbosuperchargers in 1937. In 1948, Van der Nuell came to Garrett AiResearch Co., Los Angeles and developed it together with chief engineer Helmut Schelp to the leading turbocharger manufacturer worldwide.

[237] For details of Sachse's career,—see Sect. 4.3.3, header 'Government Aviation Organisation'.

[238] See Schlaifer, The Development, p. 388. The outset of turbine cooling technology, essential patents together with an air-cooled turbine test, comparable to that of BMW, at Brown Boveri, Baden, CH, in 1938 up to a TET 909 °C are described in detail in—see Eckardt, Gas Turbine Powerhouse, pp. 397–407.

[239] K. Loehner came to BMW engine research in 1932 and was then involved in design improvements of their radial engines, manufactured under a Pratt & Whitney license. Since 1937 he had contributed to the 2000 hp BMW 801 double-row radial engine of which 21,000 units have been produced (Do217, Fw190). End of 1942 Loehner received the Chair of Piston Engines at TH Braunschweig,

made that valuable experience could be gained by the actual construction and testing of an experimental turbojet engine, utilising, where possible, previous experience that BMW had with their exhaust-driven turbosuperchargers for piston engines. This work was from 1937, largely in the hands of the then still Dipl.-Ing. Alfred H. Mueller-Berner (1907–1979),[240] working for BMW as an independent consultant. The elements of Loehner's straightforward, experimental turbojet engine were a single-stage axial turbine with hollow, air-cooled blades, driving a two-stage centrifugal compressor in line with his experience and a combustion system with the implied high working temperature of 900 °C. While this engine, which Kay[241] believed to have received the BMW-internal project designation P.3303, was being designed in more detail, experiments proceeded with the turbine wheel in which the first difficulties arose. According to Schlaifer,[242] *'Loehner ultimately reached the conclusion that in order to have better efficiency a service engine should have an axial compressor and a two-stage turbine'*. The observed difficulties experienced with his first experimental turbine convinced him that the combustion temperature should be reduced to 700–750 °C. These considerations led to drawings from the same Loehner/Mueller BMW design office, Munich, for a turbojet engine with the project designation F.9225,[243] with seven-stage axial compressor, annular combustor and two-stage turbine.

After the outbreak of war, new priorities were set within BMW: The whole turbojet developments were concentrated at the new subsidiary company BMW/Bramo at Berlin-Spandau, Fig. 6.28 (l),[244] while the BMW activities at Munich were focused soley to piston engines. From early on the Bramo engine company was the easiest of all to persuade by the

---

which he kept for 30 years. In 1960 his research addressed as one of the first the automotive pollutant emissions.

[240] A. Mueller(-Berner) will reappear in Sect. 8.3.4 in the context of tank gas turbine developments. After the war there are several corresponding patents in his name for British Leyland Motor Corp. and thereafter again for Daimler-Benz, Stuttgart, now with a Dr.-Ing. degree. See Kay, German Jet Engine, p. 97 refers in this context to A. Mueller's patent DE682,744 with priority 22 Jan.1938; however, this patent does not, deviating from Kay's assessment, protect the design of hollow air-cooled blades. Therefore, it appears that there is besides the mentioned BBC patents no other patent coverage for the turbine blade cooling arrangements as used in Jumo 004 and BMW 003 turbojet engines.

[241] See Kay, German Jet Engine, p. 97.

[242] See Schlaifer, The Development, p. 389.

[243] See Von Gersdorff, Flugmotoren und Strahltriebwerke, p. 284.

[244] On the left, Brandenburgische Motorenwerke GmbH (Bramo) in Berlin-Spandau, previously Siemens Apparate- und Maschinenbau GmbH, and BMW merged the development of air-cooled aero-engines. In summer 1939, BMW took over Bramo and integrated the Spandau plant as BMW Flugmotorenbau Brandenburg GmbH—or in short—BMW Flugmotoren, Entwicklungswerk-(development works) Spandau into BMW AG; today BMW Motorradwerk (motor cycles), Am Juliusturm, Bln.-Spandau. On the right, building #200, general administration building of Junkers Flugzeugwerke, today Kühnauer Str., former Hermann-Göring-Str., as planned in Bauhaus-style still by Hugo Junkers, realised by Werner Issel's architecture office, Berlin—up to summer 1935, see also Fig. 6.14 (r); OMW, building #403, is approx. 1.3 km in the south.

Fig. 6.28 The German turbojet engine development centres 1939–1945: BMW/Bramo Brandenburgische Motorenwerke Berlin-Spandau (l), Junkers Flugzeugwerke Dessau (r)

RLM emissaries. This was not, however, because Bruno W. Bruckmann (1902–1997), the engineering manager, and Hermann Oestrich (1903–1973), the head of research, were any less sceptical than the responsible engineers of the other companies, but because Bramo was in imminent danger of losing all its piston-engine business by a withdrawal of government support.

Oestrich studied at the Technische Hochschule Hannover and in Berlin. After completing his studies, he first went to the DVL Deutsche Versuchsanstalt für Luftfahrt in 1926, where he remained as assistant to Wunibald Kamm (1893–1966), between 1926–1930 head of the DVL engine department,[245] until he moved to Bramo Brandenburgische Motorenwerke in 1935. There he became chief engineer after earning a doctorate at the TH Berlin-Charlottenburg in 1937. As part of the new development of jet engines, Oestrich began research in this field by investigating the ML Motor-Luft/jet principle; still in 1938 a 900 hp Bramo 323 'Fafnir' 9 cyl. Radial engine was combined with a ducted fan. In 1939, he was appointed head of the development of jet engines in the BMW plant in Berlin-Spandau. Bramo had begun its studies with ducted fans[246] driven by conventional gasoline engines, internally known as project P.3301. At that time Oestrich believed that the efficiency of a gas turbine would be too low to be practical.

Before the end of 1938 and after the first RLM contacts on turbojet propulsion, however, the programme was broadened to include theoretical studies of gas turbines and actual

---

[245] This early cooperation laid ground to a later intensive relationship up to 1945 between Hermann Oestrich and his BMW turbojet developments, and Kamm's '*FKFS Forschungsinstitut für Kraftfahrwesen und Fahrzeugmotoren Stuttgart* (Research institute for automotive engineering and vehicle engines Stuttgart)' in S.-Untertuerkheim, in the latter case represented by Bruno Eckert, Fritz Weinig and as guest Hermann Reuter from BBC Mannheim; all these will show up again in Sect. 8.3 and Chap. 9.

[246] So called '*ML Motor-Luft Geraet* (motor-air device)' which belonged to the RLM-sponsored research package at BMW/Bramo, as similarly besides the turbojet engine concept investigated in much broader scope by Max A. Mueller at Junkers/Heinkel.

experiments with corresponding combustion systems. After Hellmut Weinrich's concept of a counter-rotation turbojet engine had been integrated on intervention of Hans Mauch as project P.3304, or in official terminology BMW 109-002, as described already in the context of Fig. 5.11, intensive work was being performed on the 109-002 by 1940; the available resources were notwithstanding limited. As said, there were only about twenty designers and engineers available for this complicated development at Spandau. Individual stages of the counter-rotating compressor were built and tested. Forged aluminium alloy compressor blades came from WMF Wuerttembergische Metallwaren-Fabrik Geislingen, and an annular combustion chamber was agreed on in principle, though not developed in detail.[247] These studies led with some delay to two fundamental conclusions, (1) that a counter-rotating engine would be the best, seen from the weight and size perspective, being as light as the centrifugal and even smaller in diameter than the axial, as impressively illustrated in Fig. 8.8. But (2) that this type of engine would certainly be slow and difficult to develop. Consequently, the company should first develop the simpler single-shaft configuration, which originally had been considered the second most advantageous form, just to gain the required experience, but now became more and more the preferred project concept P.3302 of the coming BMW 109-003. As experience was gained, the realisation dawned that the technical difficulties in developing even a straightforward axial compressor turbojet were demanding enough.

The actual design of this axial engine, laid down by Bramo between December 1938 and April 1939, was considerably affected by the imminent merger of Bramo with BMW. As soon as Bruckmann had staved off the Bramo shutdown in the summer of 1938, he had begun promoting a merger of his company with BMW, and although BMW did actually not finalise the Bramo acquisition until the summer of 1939, an arrangement for pooling their development capacities had been reached already still in 1938. The most important immediate effect was that the design of the 109-003 axial turbojet was modified to use the single-stage air-cooled turbine wheel of Loehner's project P.3303, being developed in Munich.

Six letters from BMW Flugmotorenbau to AVA Goettingen in the timeframe 21 March—2 November 1939 allow a more detailed evaluation of the activities of the BMW/Bramo team during this period:[248]

---

[247] See Kay, German Jet Engine, p. 97 and—see Bruckmann, Unexpected, p. 78, who describes the *'rotating combustor'* of the P.3304 as a *'nightmare'* and *'after a short period of experimenting with this design, mechanical and other difficulties seemed insurmountable.'* By early 1942 the 109–002 project was abandoned on Bruckmann's suggestion to the Air Ministry in order that all resources could be concentrated on the significantly simpler BMW 003, see Sect. 8.1.2.

[248] Courtesy to Helmut Schubert, 2018, who copied this part of correspondence from BMW Group Archive, Munich.

- It is revealing that Oestrich's favourite personal project, the *'ML Geraet* (Project P.3301)' is still followed in a detailed preparatory test specification with some urgency to R. Seiferth, AVA Wind Tunnels, up to the very end of 1939,[249]

- On 31 May 1939 is agreed, that BMW pays 10,000 RM (Reichsmark) for the AVA compressor know-how in general, in addition to 10,000 RM for manufacturing and testing of three/four model wheels with 150 mm diam. and blade numbers 12, 18 and 24, to be tested on Encke's small test rig, Fig. 4.43, bottom. The order is repeated on 19 September 1939.

- On 24 July 1939 for the first time some urgency is put on the timely delivery of (six) compressor rotor wheels of the 600 mm diam. *'TL Turbo-Luft Geraet* (turbo-air device)' from AVA to BMW—with reference to a RLM *'urgency cat. I'*. Finally, on 12 September 1939 Oestrich's deputy Frerichs, Sect. 12.2, accepts the delivery of *'the 6stage blower'* for November 1939—now with reference to *'utmost urgency from RLM'*.

- On 5 October 1939 Oestrich confirms the reception of *AVA rotor drawings E 78* etc., apparently the same axial blower profiles which Encke had introduced as *'5stage axial blower E 78'* for paid cooperation with Von Ohain during a visit at Heinkel on 13 September 1939.

After Loehner's P.3303 centrifugal turbojet project was stopped at Munich, it was decided to incorporate solely its air-cooled turbine wheel into the 109-003 engine design at Spandau. With this hollow-bladed turbine wheel, it was assumed that the combustion temperature could be raised to the demonstrated though high figure of 900 °C—and reduce the compressor pressure ratio accordingly. By this time the RLM had ordered ten experimental engines 109-003 (V1 to V10).

The then still six-stage axial compressor, designed by Walter Encke, AVA, had the typical $R \sim 1.0$ AVA design with an aerofoil pattern very similar to the eight-stage Jumo 004 compressor, as depicted in Fig. 8.1. The compressor gave a relatively low PR 2.77 at 9000 rpm. As seen in Fig. 6.29 for the later seven-stage configuration after BMW redesign,[250] the rotor blades had the AVA-typical, nearly repetitive high-speed profiles, and test stand results indicated up to 80% compressor efficiency.[251] An annular combustion chamber was chosen for the 109-003 which resulted in an engine diameter of only 0.670 m, in line with the RLM demand for lowest possible frontal area. However very soon, experiments with the turbine wheel led to the conclusion that the selected TET 900 °C was too high, so that it was lowered to 750 °C still before the end of 1939. When

---

[249] See Constant, The Origins, p. 212—with the information that *'tests at the AVA proved the ducted fan impractical'*.

[250] Picture, courtesy to Olaf Bichel, 15 Aug. 2020, 23.32 h, taken at Deutsches Museum Munich, Flugwerft Schleissheim, where this seven-stage compressor version of BMW 003A is on exhibition (with just six stages visible).

[251] See Kay, German Jet Engine, p. 100.

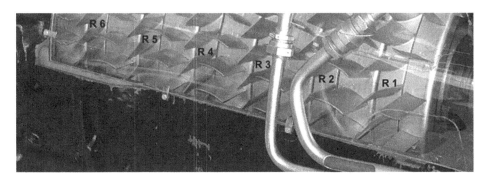

Fig. 6.29 W. Encke, AVA Goettingen: six stages of seven-stage axial compressor of BMW
109-003A, ~ 1944 © Olaf Bichel

BMW 109–003 V1 first ran in August 1940, the result was highly disappointing, and only
150 kp thrust was achieved at 8000 rpm on the test bed.

With this compressor version, the specified pressure ratio could not be realised, only
after adding a further (R ~ 1.0) stage, the design pressure ratio PR ~ 2.9 could be obtained.
But, the engine did not yet produce the desired thrust. In this situation, based on the gained
experience and in view of increasing aircraft thrust demands, a new compressor design with
a 30% higher mass flow was undertaken—now with seven stages from the beginning. The
design was carried out at Berlin-Spandau under the special responsibility of Hermann
Hagen's newly founded computational group.[252] At this point the information transmission
of key technical historians deviates: While three (von Gersdorff, Kay, Hirschel/Hoheisel)
maintain correctly that the AVA design was given up and a degree of reaction R 0.7 was
introduced instead, three others, not less renowned (Schlaifer, Constant and most importantly, Hagen) confirm the fact of a redesign, but know nothing of the important R—
adaptation.

Though the general BMW R 0.7 redesign has been confirmed by various BMW
003 photo-sources, its effect must have been somewhat unsatisfactory. Otherwise, the
RLM would have had no need to launch alternative compressor design work (*Hermso*
programmes I–III) on the basis of R 0.5 by BBC Mannheim as early as April 1941, as
addressed in Chap. 9.

This re-designed BMW 003 ran for the first time at Berlin-Spandau on 20 February
1941, four months after the Jumo 004 on 11 October 1940. The new BMW seven-stage
axial compressor design with now thicker NACA rotor blades yielded:

---

[252] With some likelihood two FKFS design reports No. 395 on seven-stage, and No. 399 on six-stage
axial turbochargers, contributed to this work,—see Eckert and Weinig, Berechnung, and—see Eckert
and Kobel, Berechnung.

- mass flow rate ṁ 18 kg/s
- total pressure ratio PR 3.14
- rotational speed n 9500 rpm
- circumferential velocity u 265 m/s
- isentropic ad. efficiency η 78.5% (80% max.)
- engine weight 570 kg
- degree of reaction R 0.7

The result is impressively good, given the fact that the compressor design took place without pretests and a hardly validated blade cascade theory.

6.2.5   Start of the *Junkers Jumo 004* Turbojet Development

Besides the Wagner/Mueller activities at Junkers Magdeburg, Sect. 6.2.1, the Junkers Dessau involvement with the—for the turbojet perspective—important axial compressor know-how can again at best reconstructed from three discoveries in the Junkers Archive in the Deutsches Museum Munich:

- On 26 October 1936 a letter to Professor Betz, AVA Goettingen, signed by Otto Mader with respect to the *'Blower order J 2644'*, summarises test results of the first month of measurements at Junkers with a four- to five-stage axial supercharger of AVA design, 0.125–0.150 m diam. and 0.5 kg/s mass flow in 6 km flight altitude. The achieved total pressure ratio is with PR 1.82 slightly below the expected 1.9; the efficiency was measured, but not repeated in the letter. In this context back to December 1935 both parties discussed also the implications of the *'BBC patent'*, Sect. 4.3.3, Sub-Section *'In the Patent Arena'*; for AVA's Walter Encke presumably a reminder to stay on the safe, so far unchallenged ground of R 1.0 designs.
- After two years of silence Anselm Franz, in the meantime responsible for turbocharger development at Junkers Dessau meets Walter Encke at the Lilienthal Conference at Berlin, Sect. 6.1.3. In a follow-on letter, dated 25 October 1938, Franz outlines the concept of a four-stage axial blower of outer diameter D 0.179 m, diameter ratio hub/tip d/D 0.5 and 26,700 rpm; target PR 1.938 in 7 km altitude. He asks for an offer about the manufacturing of the four wheels from AVA, under the assumption that the sheet metal vanes can be manufactured again at Junkers, based on AVA drawings.
- Finally, a nearly historic meeting took place on 4 April 1939 at Junkers Dessau, Fig. 6.28 (r), with the Dipl.-Ings. Mauch and Schelp from RLM, Junkers Dessau represented by Prof. Mader, Franz and Schorlemmer, and Junkers Magdeburg by Mueller, *KoBü Konstruktions-Büro* (design office) *II*. The following—in hindsight surprising—decisions were made:
  - The started turbojet engine activities (Wagner/Mueller) towards 550 kp thrust should continue with *'Priority 1'*,

- Mader to cooperate with potential turbojet aircraft company (Messerschmitt),
- Wagner's turboprop concept has only *'Priority 2'*,
- Mueller's ML concept on basis of the 24 cyl. Jumo 222[253] with 2000/3000 hp keeps also *'Priority 1'*,
- Turbojet engine activities (at KoBü II) should keep aspects of boundary layer suction on aircraft wings in mind,
- RLM will compensate JFM for all past and future KoBü II Magdeburg activities,
- RLM and JFM support the transfer of KoBü II incl. test facilities to Dessau within 6–12 months.

All this being discussed in the presence of Professor Mader's favourite Anselm Franz, his potential, just 1-year-older, though in Mueller's assessment largely inexperienced, successsor as project leader. One can assume that Mueller's search for professional alternatives started immediately, which resulted in answering Heinkel's *'Wanted'* ad in May 1939. And—later reproaches towards Mueller of following a technically too broadly designed research programme[254] have to be re-considered—at least in mid-1939 he knew the RLM and Heinkel on his side.

Anselm FRANZ (1900–1994), Fig. 6.27 (r), born Austrian, is definitively a German turbo-jet engine pioneer, not so much by his actual mechanical design contributions, but for his persistance in realising the first Jumo 004 turbojet engine series production—against all difficulties. He studied mechanical engineering at the TH Graz up to the master's degree (Dipl.-Ing.) in 1924. From 1924–1928 he was assistant to the renowned Professor Karl Federhofer (1885–1960), who had the chair for general mechanics, hydro- and aeromechanics at TH Graz since 1923. For the years 1928–1929 Franz moved on to ETH Zurich to continue his studies in designing hydraulic torque converters under Professor Fritz Dubs (1880–1963), who had been installed there as Professor for Water Turbines and Water Power Facilities in 1927.[255] At ETH Zurich Franz also attended lectures of Robert Gsell (1889–1946), then freshly installed Assitant Professor for Aeronautics, whose early

---

[253] See Wikipedia, 'Junkers Jumo 222' in English.

[254] Besides the HeS 30 turbojet, resulting from the RT0 project, Mueller oversaw in his 33 months with EHF the constant-volume HeS 40 turbojet engine—see Eckardt, Gas Turbine Powerhouse, p. 72 *'The Holzwarth Gas Turbine'*, the HeS 50 ducted-fan engines (ML) and the HeS 60, a turbo-compound ML concept, all these with derivatives; for details,—see Kay, German Jet Engine, pp. 30–40.

[255] Between 1907 and 1927 Dubs became chief engineer at the Escher-Wyss AG, Dept. *'Water turbines'* as immediate forerunner in this position of Jakob Ackeret. He founded the famous Escher-Wyss Zurich *'Test establishment for turbine research'*—successfully in use up to 1963. There, Dubs developed the *Francis Turbine* further to the so-called *'Dubs's X-Wheel'*, an early form of the *Kaplan Turbine*. One can assume that Franz actually failed with his original intention to achieve a *'doctorate'*, both at his stations in Graz and Zurich. However, given the standard length of 8–10 years of such an endeavour at that time, especially in Switzerland, the turbojet engineering community may praise in hindsight *'Il forza del destino* (The power of fate)'. Somewhat belatedly, Franz received his *'Dr.-*

Junkers F 13 flight altitude record achievements have already been described in the context of Fig. 4.30. Consequently, for the period 1929–1935 Franz went as development engineer for hydraulic torque converters to Maffei-Schwartzkopff Werke, Berlin-Wildau,[256] some 25 km south of Berlin city centre. Originally a joint locomotive plant, the locomotive business—for which the hydraulc torque converters were mainly applied—deteriorated after the economy crisis of 1930, and the company went bankrupt. Beginning 1933 the whole area was increasingly occupied with production for military goods, in 1936 AEG took over and since 1938 the plant became a complete subsidiary of Junkers Dessau. Aircraft fuselages were transported by train to the 12 km distant Henschel Flugzeug-Werke, Berlin-Schönefeld[257] for final assembly. It appears that Anselm Franz could here already steadily grow relations to AEG and Junkers, before he moved in 1936 actually to Junkers Flugmotoren Dessau, with responsibility for turbocharger development. In 1939 he was put in charge of the Jumo 004 turbojet engine design.

The further development is well known. Even before Mueller had reached a definitive agreement with Heinkel, Franz came to Magdeburg about May 1939 to inspect the RT0 turbojet,[258] which Mader in his foregoing visit had indicated willingness to sponsor. Franz, however, after a few tests came to the conclusion that this engine, which still could not get fully started, was not worth further development.[259] It is not clear—even on the basis of post-war interviews—if Franz ever grasped the significant influence of the axial compressor degree of reaction on engine performance. Given the multi-compromised RT0 design, his decision to start from a *'clean sheet of paper'* was partially right, but it was a blunt mistake to skip Wagner/Mueller's R 0.5 axial compressor design—apparently born out of mutually over-shooting aggressiveness.

Another example of Franz's misjudgement becomes visible, when he blames Mueller for committing absolutely unrealistic completion deadlines towards the RLM, in a *'Magdeburg transfer documentation'* on 5 August 1939.[260] However, the *'official deadline comparison'* (which Franz could not know in full) of turbojet engines with 600 kp thrust for P.1065/Me262 and He 280 applications reveals that the—in hindsight consistently

---

Ing.' degree in aeronautical engineering in 1940, then rather casually from TH Berlin-Charlottenburg, Military Engineering Faculty.

[256] Today the area is occupied by the campus of TH Wildau (FH).

[257] Today *BER*, location of the new airport Berlin-Schoenefeld, opened on 31 Oct. 2020.

[258] See Constant, The origins, p. 211, the statement *'Franz examined the Wagner-Müller engine, and decided that its design was far too advanced ...'* would have made only sense in view of Mueller's later HeS 30 design, the realisation of which at Heinkel Rostock was delayed, however, till 1941; at Junkers Magdeburg in 1939, Franz could only investigate the RT0 test engine.

[259] Franz is indirectly quoted in an Allied C.I.O.S. interrogation on 16 May 1945 in his home at presumably 'Ballenstedt (misspelled Allanstedt)', Harz, of having said, this (RT0) being *'a half-hearted effort and made no significant progress'*.

[260] See Schabel, Die Illusion, p. 43; project/deadline tabulation, p. 42—dated 16 Nov. 1939.

unrealistic—time-pressure till first sample deliveries had been clearly built up from the RLM representatives:

Company	Turbojet engine	Delivery	Remarks
BMW	109–003 [BMW 003/Oestrich,1× TU]	1 July 1940	4 test objects
	109–002 [P.3304/Weinrich, CR]	1 Oct. 1940	8 test engines
	109–003 [F.9225/ex-Loehner, 2× TU]	1 Dec. 1940	7 test engines
Junkers	109–004 [Jumo 004, Franz]	1 July 1940	8 test engines
Heinkel	109–001 [HeS 8, von Ohain]	1 April 1940	0-Series (60 units)

Of course, the *'degree of reaction disaster'* cannot be blamed to the unexperienced, nevertheless responsible project leader Franz alone, but falls more on the side of the AVA *'compressor experts'*, who failed to correct this dubios design concept during a more than 10-year preparatory phase.[261] Mauch at first strongly resisted the idea of abandoning the results of all Mueller work, but after Heinkel took it over, a contract was given to the—in the meantime united—*'JFM Junkers Flugzeug- und Motorenwerke, Motorenbau Stammwerk* (main works) *Dessau'* in the summer of 1939 for a new axial-flow turbojet, ultimately known as the *'Junkers Jumo 004'*,[262] Fig. 8.1.

Again, also on the expert side there was no open discussion of the issue up to 1948; instead, there were weak arguments in favour of a R 1.0 design collected, e.g. the benefit that the unloaded, just turning vanes could be manufactured by cheap sheet metal. Fact is, that worldwide nobody else followed this AVA design path and Bill Bailey's original observations can be re-animated by digesting the repetitive compressor vane/blade/vane design pattern of Fig. 8.1: *'The first captured 004 was sent to Pyestock sometime in 1944. The author* (Bailey) *recalls waiting, in great anticipation, to see the stripped compressor. Here was the rival German axial technology at last. What was seen, left a feeling of deflation. Instead of a new approach to a high efficiency aircraft gas turbine compressor, all that appeared was blading designed for about 0/100% reaction, with a large part of the stator made from shaped steel plate. It was very disappointing . . .'*[263]

The only sign of *'corrective action'* is the RLM decision in 1941 to activate Brown Boveri Mannheim's aero turbomachinery design group of Hermann Reuter for an advanced R 0.5 alternative axial compressor in 1942, which came too late for the Jumo 004 programme, but was developed in full in 1944 as 7- and 10-stage replacement for

---

[261] An early, somewhat convincingly explanation of this *'riddle'* provided—see B.I.O.S. Overall Report No.12, p. 17, in 1949, where the AVA design philosophy was deduced back to its origins in single-stage blower design, as illustrated in Fig. 4.38; if true, it would nevertheless certify to the *'experts'* at Prandtl's respectable Institution a remarkable resistance and inflexibility against fundamental aerodynamic insight.

[262] See Schlaifer, The Development, p. 396.

[263] See Bailey, The Early Development, p. 56.

Fig. 6.30 *Adastral House*, Home of the Air Ministry, Aldwych and Kingsway, London, ~ 1950

the advanced BMW 003 C/D turbojet versions (*Hermso programmes I–III*), as will be further outlined in Chaps. 7 and 9, and Sect. 8.3.

6.3   Early Turbojet Developments in Great Britain

6.3.1   Through Adversity to The Stars

'Although the atmosphere in German jet engine circles was rich with rancour and suspicion, the jet effort was never so poised on a knife's edge as it was for Power Jets and Whittle.'[264]

This situation was not the least the result of a long-time hesitant and undecided posture of the Air Ministry. But the situation changed, slowly but decisively, in the present period under consideration, so that the RAF motto *'Per Ardua ad Astra'—Through adversity to the stars* increasingly rightly decorated *'Adastral House'*, the Air Ministry building[265] at London's Kingsway, Fig. 6.30.

---

[264] See Jones, The Jet Pioneers, p. 57.

[265] At the beginning of the *Air Ministry* stood the adversity and lack of co-ordination of the Army's Royal Flying Corps and the Navy's Royal Naval Air Service by 1916. A growing number of German air raids against Great Britain led to public disquiet and increasing demands for something to be done. A new air service was recommended to be formed that would be on a level with the *Army* and *Royal Navy.* The *Royal Air Force RAF* was to receive direction from the new *Air Ministry,* which was formed in due course on 2 Jan. 1918 with Lord Rothermere as first Air Minister. The Air Ministry initially met in the *Hotel Cecil* on the*Strand*; later, in 1919, it moved to Adastral House on Kingsway.

Decisions on aeronautical research and development issues were prepared here by discussions in and studies for the *ARC Aeronautical Research Committee*—under this name since 1919—and especially the ESC Engine Sub-Committee. On the main committee a majority of the members were *Fellows of the Royal Society FRS*, on the Engine Sub-Committee, the proportion was lower, but was never less than 40%.[266] During the deteriorating international situation of the 1930s, ARC was given fresh impetus with the appointment of Sir Henry Th. Tizard, Fig. 6.31 (m). One of the committee's most important decisions was to speed and implement the development of a national system of air defence based on *radar*.

His foresight and his trust in the new technology became perhaps the decisive factors in the *Battle of Britain* 1940. In terms of overall influence, Henry Tizard[267] was the undisputed lead, able as a *'gentlemanly professional'* to interact with politicians and civil servants as an equal, to negotiate state committee business with ease. Trained as a chemist, he had joined the ARC after the First World War, becoming its chairman in 1933, after he had chaired the Engine Sub-Committee already since 1925. After taking up the rectorship of Imperial College in 1929, he remained committed to his committee work; described as a *'stimulating chairman'* and a *'master of the searching question,'* he was able to cut to the heart of difficult technical issues.

As already addressed in the foregoing,[268] Frank Whittle, Fig. 6.31 (l), applied for and was granted his first patent in 1930. He applied for assistance to the Air Ministry and several private firms, but was turned down by all of them. Lacking the necessary financial support required to perform any actual test work, Whittle continued to do paper work from 1930 to 1936, when his company Power Jets, Ltd, was formed on 27 January 1936.[269]

The first engine designed by Whittle after this restart, the *WU* (Whittle Unit), was intended as a bench demonstration engine, not as a flight prototype, Fig. 6.32.[270] It was to produce a rather arbitrarily chosen figure of sea-level static thrust. The compressor wheel was of the double-sided centrifugal type with 0.483 m (19″) diameter, because it was easier

---

From the roof, Sir Arthur T. *'Bomber'* Harris (1892–1984) observed the *Blitz* on the Docks of London, 8 km in the east on 7 Sept.1940, allegedly with the old testamentary words *'They are sowing the wind ...'*

[266] See Whitfield, Metropolitan Vickers, p. 36 f.

[267] See also details of Tizard's life in Chap. 2, *'April 1930'* and *'16 March 1937';* as outlined already in the corresponding footnote for the latter date, Tizard had studied in 1913 at Humboldt University Berlin with the later Nobel Prize winner Walther Nernst, and Whittle's *'virtual'* competitor Hans von Ohain had started his patenting efforts in 1935 with the related concept of a *'reactor-Nernst-tube'*—a rotating U-tube flow path, consisting of an out-flowing centrifugal compressor, the combustor in the horizontal bend and a centripetal turbine in the in-flowing branch—which will be further outlined in Sect. 12.1.1 'Hans von Ohain's Secret Turbojet Patents'.

[268] See Sect. 4.2.3 'Griffith and Whittle—a Patent Conflict?'.

[269] See Wikipedia, 'Power Jets' in English.

[270] Courtesy of Ian Whittle, 14 June 2018.

Fig. 6.31 British turbojet engine pioneers I: Sir Frank Whittle (1907–1996), l © Coventry Ctrl. Library, Sir Henry Th. Tizard (1885–1959), m © NPG 81706, London, Sir David R. Pye (1886–1960), r © NPG 156952, London

to develop and lighter than the axial type, double sided in order to reduce the diameter[271] and thus the drag of the engine with notwithstanding adequate air-flow volume. No diffuser vanes were used, reliance being placed in the *vortex space* diffuser in accordance with conclusions drawn from Whittle's centrifugal supercharger investigations in 1931. This first WU configuration had a single combustion chamber, feeding into a single-stage axial turbine through a nozzleless scroll turbine inlet. The turbine was only 0.406 m (16″) in diameter, and from it Whittle had to get the equivalent of just over 3000 hp to run the compressor—more than the net power then produced by any piston engine.[272]

Whittle expected that he could achieve extraordinary performance from his components, particularly the compressor, from which he expected a total pressure ratio PR of 4:1 at an efficiency of 80%, when the best that had been achieved to date was a PR ~ 2.5 with 70% efficiency.[273] For the only part of the design which Whittle considered critical in view of demands far beyond existing experiences, the combustion system, he sought external industrial support. The order for the actual construction of the first engine, apart from the combustion system, was given to the former GE-daughter company, British Thomson-Houston (BT-H), in June 1936, and the engine was ready for test in April 1937.

Testing of the first engine began on 12 April 1937, and continued until 23 August, at which time it became apparent that the combustion chamber presented a major problem and the performance of the turbomachinery components was far below expectations.[274] The mere fact that such a bravely designed, unorthodox engine concept actually ran on its first attempt was, however, extremely encouraging to all concerned.

---

[271] A diameter comparison of Whittle's W1X engine (1941) vs. Hans von Ohain's engines at Heinkel has been provided in Fig. 6.24.

[272] See Constant, The Origins, p.188.

[273] See Schlaifer, The Development, p. 348.

[274] See Kay, Turbojet, Vol. 1, p. 22.

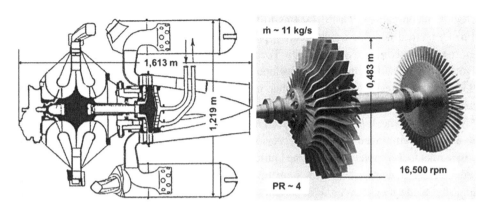

Fig. 6.32 Whittle *WU* first experimental engine, Rugby 12 April 1937, cross-section with water-cooled turbine disc (l), dual-flow compressor rotor (r), design data: F ~ 630 kp, target efficiencies $\eta_C \sim 0.8$, $\eta_T \sim 0.7$ © I. Whittle

Whittle, meanwhile, was completing his tenure at Cambridge. After his final examen in spring 1936, he managed to wrangle from the Air Ministry an additional graduate year at Cambridge, and thereafter the Air Ministry assigned him back to Power Jets. Thus, though the Air Ministry *'did refuse any direct support for the turbojet project, it did provide indirectly the most precious of all resources, Frank Whittle.'*[275]

In March 1936, Whittle had met Henry Tizard at a Cambridge Air Squadron Dinner—and thereafter several times later that year. The chairman of the Engine Sub-Committee of the Aeronautical Research Council was probably the most influential scientist to the Air Ministry. Tizard had supported Griffith's work up to 1930, and in late 1936, largely at Tizard's urging, the Engine Sub-Committee reopened the whole issue of aero gas turbines. One result was the authorisation for the RAE to pick up the gas turbine investigation abandoned by Griffith in 1930. The ESC also asked Griffith to write an evaluation of the Whittle engine, which he did with some delay, submitting it in February 1937. As could be expected, his evaluation was positively icy. After criticising Whittle for the overly optimistic centrifugal compressor efficiency of 80%, he proceeded to compare the expected performance of the Whittle engine with that of a conventional propeller/piston engine combination and concluded: *'In its present form, the proposed jet propulsion system cannot compete with a conventional power plant in any case where economical flight is demanded. . . . It is of value only for special purposes, such as the attainment of high speed at high altitude for a short time, in cases where take-off requirements are not stringent.'*[276]

---

[275] See Constant, The Origins, p. 190.

[276] Quoted after Constant, The Origins, p. 190.

Quite unconsciously, Griffith seems to have adopted then the just emerging new Air Ministry tactical doctrine as engineering principle. Ironically, Griffith was presenting his report to a sub-committee chaired by Tizard, who since 1934 had also been intimately involved in England's ultra-secret development of radar, which just then, in the spring of 1937, was proving its value in bomber interception. The question still was, by what technical means?[277]

To complete the *WU history*, a redesigned engine was started on 16 April 1938; the rebuilt engine employed a vaned compressor diffuser, a redesigned combustor, and a new turbine inlet to a completely redesigned turbine of free-vortex design. These turbine blades were constructed with a *'twist'* to compensate for differential radial velocity and pressure along the blade height. Whittle apparently had independently arrived at the free-vortex blade design before 1936, as had Griffith at the RAE in 1929. Griffith's analysis of the new blade design concept was, however, contained in his still secret 1929 report.[278] BT-H had built the turbine of the first *WU* engine according to steam turbine practice, which determined also the Brown-Boveri axial turbomachinery design for quite some time.[279] Whittle had to argue quite vigorously to convince BT-H that the new blades offered better efficiency. Besides the already outlined negative effect on German axial compressors by AVA's strict R 1, degree of reaction design preference, the higher turbine efficiency which derived from the superior blade design of the British turbojets accounted certainly also for the lower fuel consumption of the British engines over their German counterparts, which used axial turbines of conventional steam turbine origin throughout the war.[280] Tests of the rebuilt Whittle engine continued until June 1938, when the engine was severely damaged by a turbine blade failure.

---

[277] This question has been addressed in a comprehensive special study—see Nahum, Two-Stoke or Turbine, about the Rolls-Royce *'Crecy'sprinter* engine and corresponding turbojet alternatives. The *Crecy*, an experimental 26 l, 1400 hp, two-stroke, 90 deg. V12 liquid-cooled piston engine will be further addressed in the following Sect. 6.3.2. Tizard himself was a proponent of a high-powered *sprint engine* for fighter aircraft and had foreseen the need for such a powerplant as early as 1935 with the threat of German air power looming. It has been suggested that Tizard influenced his personal friend Sir Harry Ricardo (1885–1974) to develop what eventually became the—see Wikipedia, 'Rolls-Royce Crecy'.

[278] See Griffith, The Present Position.

[279] 'Radial equilibrium' designs—with the radial acting forces on the fluid in equilibrium—were verified relatively early by BBC,—see Eckardt, Gas Turbine Powerhouse, p. 116. However, these tests showed that the method for calculating the resulting radial blade twist only had a small influence on the compressor efficiency and did not strictly need to be fulfilled. This considerably facilitated the task of the designer and allowed for simpler and cheaper blade forms.

[280] *'Free vortex'*designs found early application in the US and UK,—see Hawthorne, The Early History. He outlines in a Footnote on p. 59 of that book: *'Free vortex or constant circulation blading for steam turbines was introduced to Westinghouse—and such apparently also to US turbojet designs—by the Prandtl scholar Oskar Tietjens (1893–1971) in the late twenties, Griffith had used it on his model rotor and Constant on the axial supercharger'*.

Finally, a third model of Whittle's *WU* experimental engine was placed on the test stand in October 1938. With the third reconstruction, Whittle's engine took on the basic form it would keep throughout the war. The new version used ten reverse-flow can-type combustion chambers, but the same compressor impeller, slightly modified, together with basically the same diffuser. Testing of the third version began in October 1938 and continued until February 1941, when it too was finally destroyed by a turbine failure. Decisively however, on 30 June 1939, David R. Pye, Fig. 6.31 (r), Director of Scientific Research at the Air Ministry, visited Power Jets' new premises at Ladywood and witnessed a run-up of the engine to 16,000 rpm. Pye, who had been rather skeptical before, was now convinced that Whittle *'had the basis of an aero-engine'*.

Two weeks later, the Air Ministry placed a contract with Power Jets, Ltd, for a flight engine known as the *W1*, of which the testing did not begin until 1941, and Gloster Aircraft received a contract for an experimental airframe, designated after the specification *E.28/39*. At the same time—and after Griffith had left RAE to Rolls-Royce—the RAE focus was redirected from turboprop toward the development of a turbojet. Consequently since 1936, two independent groups in England began work on design, construction and testing for aero gas turbines: One group at Power Jets, Ltd, led by Whittle, worked on a turbojet engine using a centrifugal-flow compressor, and a second group at RAE Farnborough, now under the sole direction of Hayne Constant, worked on the construction and testing of axial-flow turbocompressors and corresponding engine applications, as will be retraced in Sect. 6.3.3.

Summarising, Whittle's role as Britain's turbojet engine pioneer #1, Fig. 6.31 (l), can be unequivocally confirmed, not only due to his stand-alone initiative for his own project, but by his charismatic emanation on the long-time hesitant scene and his pushing and demanding example. The Air Ministry did not resurrect the work until they possibly became uneasy when work began on a turbojet project outside of their influence. Apparently, it was only when Whittle's Power Jets, Ltd, was formed in 1936, that ARC-chairman Sir Henry Tizard decided that the RAE should restart work on aero gas turbines. Hayne Constant, who had been at RAE for some years, had left to become a lecturer at Imperial College—and it required Tizard, rector of this College, to persuade him to return as an assistant to Griffith. At that time, Henry Tizard clearly was the essential catalyst, diplomatic mediator and encouraging inspirator for the turbojet project, so that his classification as turbojet engine pioneer should be also acceptable, Fig. 6.31 (m). Opinion about the justification of the third pick in this context, Sir David R. Pye, Fig. 6.31 (r), could be disputed and would certainly so by Frank Whittle, given the many real or imagined discriminations which he had to endeavour from the Air Ministry and especially from the Directorate of Scientific Research, in the responsibility of Pye since 1927 as deputy director and from 1937 to 1943 under his full lead. In the meantime, Pye's account derived from official papers contradicts somewhat Whittle's lament, as recently stated by the English expert-historian Andrew Nahum: *'Pye did his utmost to secure funding for Whittle using imaginative routes in spite of difficulties*

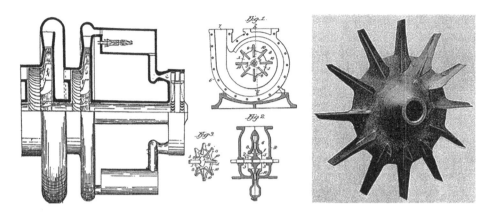

Fig. 6.33 F. Whittle compressor concept Oct. 1929 (l), A. Rateau *Blowing Machine*, Pat. US827,750, 7 August 1906 (m), Brown-Boveri-Rateau dual-flow turboblower impeller, 1907 (r) © A. Kay (l)

imposed by official Air Ministry procedures.[281] Pye's special role will be highlighted in Sect. 6.3.3, where there is justified assumption that the reinsuring support from Swiss Brown Boveri to RAE and Metrovick developments was resulting from his initiative.

Before leaving Whittle's turbojet development process up to the outbreak of WW II, which has been investigated and described in depth in a vast technical-historic literature,[282] it is worthwhile to have a closer look to the enigmatic design feature of the dual-flow centrifugal compressor impeller which determined the Whittle's engine designs throughout, beginning with the first experimental concepts—without any indication of its origin.[283] As stated in the context of Fig. 4.9, the attempt of the *Elling camp* to construct a relationship between both designs on the basis of the strikingly similar dual-flow impeller configurations has been already rejected by Frank Whittle in 1968—without providing any alternative explanation. It is known that Whittle's turbojet concept as presented to Griffith and the Air Ministry in October 1929 consisted of a two-stage centrifugal compressor, as illustrated in Fig. 6.33 (l). His supercharger studies in 1928 and his proposals for a free-wheeling, self-propelled supercharger might have helped him on his way to the double-flow, back-to-back solution, possibly emphasised as a means to reduce weight and, most importantly, the total length of the shaft. However, it appears, that the actual inspiration for Whittle's double-sided arrangement of 1935/36 had not been found yet.[284]

---

[281] Personal communication with Andrew Nahum, 17 Sep. 2020.

[282] See Kay, Turbojet, Vol. 1—technically by far the most reliable source; for rather narrative accounts, e.g.—see Golley, Whittle and—see Jones, The Jet Pioneers.

[283] Whittle got the dual-flow centrifugal impeller patented in GB456,976 *'Improvements relating to centrifugal compressors'* with priority 16 May 1935.

[284] Communication with Ian Whittle, 14 June 2018.

After a comprehensive patent search, there is significant likelihood that the French turbomachinery designer and inventor Professor Auguste C.E. Rateau (1863–1930) received patent coverage with a priority as early as 7 August 1906 for a corresponding concept, as depicted in Fig. 6.33 (m), also as basis for a subsequent, licensed hardware realisation by Brown-Boveri in 1907, Fig. 6.33 (r).[285]

6.3.2   The Emerging *Swiss Connection*

After the first, still unsuccessful industrial gas turbine activities at the beginning of the twentieth century, these engines showed too inefficient to be useful in ordinary applications. However, very soon after a development restart after World War I, component efficiencies and available materials improved considerably, so that special, close relatives of the gas turbine could be of use in cases where the heat which drove them was a by-product which would otherwise completely lost. The Swiss firm of *BBC Brown Boveri & Cie.* was preeminent in the development of such machines throughout the period from about 1905 to 1945—from early failures to partial successes to the first 4 MW power generation gas turbine.[286]

The first steam-generating, *Velox boiler* with all-axial turboset, ordered in 1932 from the *Société Métallurgique de Normandie* to burn blast furnace gas for the *Mondeville ironworks* near Caen, France, was already equipped with two axial compressors of similar sizes, one for air and one for blast furnace '*waste gas*'—with a row of inlet guide vanes to set the optimum air/gas ratio, adjustable before starting.[287] The 11-stage axial turbo-blowers supplied the air/gas mixture to a combustion chamber, including closed steam generation tubing,[288] and the combustion products drove a five-stage reaction turbine, which in turn drove the compressors according to the gas turbine principle. The compressor

---

[285] See Eckardt, Gas Turbine Powerhouse, p. 47. The illustrated dual-flow turboblower impeller was manufactured for the Siemens-Martin steelworks '*Rote Erde*' (Red Soil), Aachen, Germany. The 1.3 m diam. impeller from a *Krupp* steel casting weighed 245 kg, and generated—under the power of a 750 hp steam turbine drive—at 2600 rpm a mass flow of 14 kg/s and a total pressure ratio of 1.2. Given Rateau's foregoing patent coverage, the patent claim #8 for a dual-flow centrifugal impeller in Whittle's patent for a '*Powerplant for aircraft*', CH188,758 with priority 16–18 May 1935 should not have been granted.

[286] See Eckardt, Gas Turbine Powerhouse, *3 Gas Turbine Forerunners*, p. 43 f. and *4 The Path to the First Power Generation Gas Turbine, 1927–1945*, p. 77 f.—and here especially Sect. 4.3 *The Early BBC Gas Turbines*, p. 158 f.

[287] The first mechanism to allow for vane angle variation *during* operation of turbomachinery was implemented in a BBC axial fan for mine ventilation before 1940,—see Eckardt, Gas Turbine Powerhouse, p. 108, there also with reference to later claimed, corresponding '*firsts*' for turbojet applications (M.A. Mueller, GE J79, etc.), see also Sect. 12.1.2.

[288] The steam generating process was excessively accelerated by combustor exhaust gas velocities of up to 200 m/s, which drove the heat transfer rates up to 10–20 times higher than that of conventional

design data of PR 3.3, V 3.5 m^3/s at 12,000 rpm and 83% efficiency were quite impressive for the first commercial axial compressor product worldwide. The Velox boiler concept was promoted by its operational simplicity, inherent reliability and lack of general mainte- nance demands, leading to impressive sales figures of approximately 75 units in the first 10 years of delivery.

Although the first Velox turbines required some e-motor backup to turn the—then mostly still centrifugal—compressors, by the mid-1930s the turbine not only ran the then axial compressors, but delivered a certain excess of power, the e-motor became an electricity generator. Beginning in 1936 Brown Boveri built turbo-blowers for use in the *Houdry cracking process*.[289] The primary purpose of these machines, as of the blast- furnace blowers, is to supply compressed air. In many chemical processes industrial supercharging was impossible due to the high cost of compressing the air, but if gases from the process could be expanded in a turbine with enough power to drive the compres- sor, then supercharging would become economical. The Houdry process installations had no combustion chamber of their own, but used sequentially several *CCC Catalytic Crack- ing Cases* (10 m high, 3 m diam.) during the catalyst regeneration cycle, which simply meant burning off the oil tar deposits from the catalyst surfaces. Like the Velox boilers the Houdry turbo-blowers had already a net power output to generate electricity. On 31 March 1937 Houdry unit *'Eleven Four'* came on line at the Marcus Hook oil refinery at the Sun Oil Co., Philadelphia, USA—with brand new BBC turbomachinery equipment, see also Figs. 6.53 and 6.54. The sheer size of the required Houdry turbo-set was a challenge for BBC, with a volume flow of nearly 20 m^3/s at a pressure ratio PR 4.2. The compressor unit with 20 stages was a considerable step forward in comparison to the previously built Mondeville unit in 1932. The excellent performance of the delivered turbomachinery meant the 4.4 MW compressor driving power was more than compensated by the 5.3 MW turbine output, and a gearbox-coupled generator on the turboshaft produced electricity accordingly. Between 1936 to 1943, 36 such Houdry sets were built, six of these in Baden, while the remainder was licensed by BBC for Allis-Chalmers's US production, lying the basis for that Company also to become a frontrunner in US turbojet developments, Sect. 6.4.

By the end of the 1930s the efficiencies of BBC-developed turbo-components were remarkably high, each being in the neighbourhood of 85%. It was the substitution of axial for centrifugal compressors which brought a GT net power output for the first time; nevertheless, the overall efficiency e.g. of Brown Boveri's first power generation gas turbine at the Swiss Neuchâtel (Neuenburg) was as low as 17.4% only—chiefly because the initially used temperatures were very low. For the 4 MW unit rotating at 3000 rpm, the seven-stage turbine provided 15.4 MW with an inlet temperature of 550 °C, of which the

---

steam generating boilers, with the additional benefit of a very compact design and considerably shortened heat-up times, e.g. 6 min in comparison of 6 h for conventional navy boilers.

[289] See Eckardt, Gas Turbine Powerhouse, p. 169 f., named after the French engineer and inventor Eugène J. Houdry (1892–1962).

23-stage compressor absorbed 11.4 MW at an air inlet temperature of 20 °C. However, shortly thereafter considerable progress in material technology brought turbine inlet temperatures of 750 °C in reach, which meant that with a turbomachinery design standard as verified for Neuchâtel, the overall plant efficiency would rise up to 26% accordingly. While Brown Boveri had clearly paved the path in demonstrating the general feasibility of the gas turbine during the 1930s, the detailed job of substantially increasing the aerodynamic stage loading remained the main challenge for turbojet engine designers to be accomplished at the beginning of the 1940s.

Another side product of the early gas turbine work was the turbosupercharger, whose early history in the final stages of WW I has already been described in Sect. 4.1.2. In the context of rearmament, aircraft engine supercharging awoke new interest from the mid-1930s onward; BBC was always interested in new markets as the economic crisis came to an end.[290] Aircraft engine supercharging remained only an episode though, for the turbojet engine ousted the large aircraft piston engine within a decade. It was nevertheless an interesting prologue to the coming turbojet drama—here with Brown Boveri and Rolls-Royce on stage. As internally documented, discussions took place between BBC's French daughter company C.E.M. Compagnie Électro-Méchanique, Paris and the French engine manufacturer Hispano-Suiza on aircraft turbocharging in 1934, and the initial contacts with Rolls-Royce with regard to turbocharging a '$12V \times 137.2 \times 152.4/2,040$ rpm (cyl. $\times$ bore [mm] $\times$ stroke [mm]/rpm)' engine of what became the 1500 hp 'Rolls-Royce Merlin' must also have taken place during this period. In the following years BBC's test department TFVL produced numerous test reports for an aircraft turbocharger, in BBC terminology VTF 280 (Verbrennung/combustion—Turbomachinery—Flug/flight, and the number indicating the compressor wheel diameter in mm), its Rolls-Royce applications mostly camouflaged by the term 'Kestrel type'. By mid-1943 the 27 l Merlin was supplemented in service by the 37 l, more powerful 1700–2400 hp RR Griffon,[291] though within nearly the same dimensions, which consequently determined the BBC/RR cooperation in the late 1930s, incorporating several design improvements and ultimately superseding the Merlin. To cool the VTF 280 turbine, approximately 45% of the compressor exhaust air was fed to two turbine cooling segments, while 55% went directly into the cylinders as combustion air. The compressor thus had to provide considerably more air than the engine itself needed. In addition a direct turbine blade showerhead film-cooling was tested for these RR turbocharger applications, Fig. 6.37, a remarkable 'first' for turbojet developments in general, some 25 years before this breakthrough technology was actually introduced by the established turbojet engine manufacturers.

As illustrated in Fig. 6.34b[292] the VTF 280 radial compressor wheel of 1935 is 'shrouded' with radial blades between coverplates. The incidence adaptation at entry was

[290] See Jenny, The BBC Turbocharger, printed in 1993 in both English and German on 300 pages, a marvellously presented, bibliophilic rarity.

[291] See Wikipedia, 'Rolls-Royce Griffon' in English.

[292] See Jenny, The BBC Turbocharger, p. 71, Fig. 4.20.

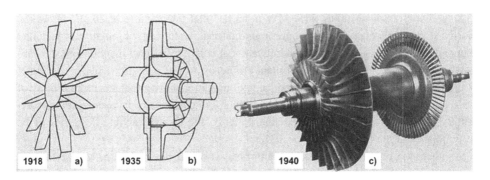

Fig. 6.34 The evolution of centrifugal compressors for aero turbochargers: (**a**) Moss (GE) impeller, (**b**) BBC VTF280 shrouded impeller with axial inducer, (**c**) Hispano Suiza integral ax/centrifugal impeller

brought about by means of an axial inducer, aerodynamically a significant progress in comparison to Moss's impeller design of 1918, Fig. 6.34a. Test experience at BBC in 1937 revealed that Hispano's coming open/unshrouded rotors, Fig. 6.34c, were lighter and attained a higher PR 2 with better efficiency.

Though the actual start of the Rolls-Royce turbo-supercharger collaboration with Brown Boveri cannot be reconstructed exactly (~ 1934), there survived RR original drawings for that period, Figs. 6.35 and 6.36, which illustrate a *'Brown Boveri Turbo Compressor'*, actually the *VTF 280* as described in the foregoing. The RR drawing *Advanced Projects Design APD 416* dates from July 1939 and represents a re-traced BBC *V.92797*, originally issued in December 1936. With all likelihood the reason for this unusual re-tracing was a simple communication problem between power industry experiences on the Swiss side, which knew the *'imperial system'* usage from their British competitors, not realizing that at least Rolls-Royce was already clearly on *metrics* then; apparently an imperishable misunderstanding for the time, reappearing in Sect. 6.3.4. The cross-section of Fig. 6.35 (l) shows the compressor intake on the right and the turbine exhaust on the left. The turbocharger shaft runs in two ball bearings: one on the cold compressor entry section and one on the hot turbine exhaust side. This heat-exposed bearing is supported by four air-cooled bearing struts, of which the symmetrically arranged four 90 deg. tubes are clearly visible in the cross-sections which guide a considerable share of the compressor air to the air-cooled turbine bearing struts.

Though direct insight into the RR/BBC turbocharger collaboration is scarce, there is an expert observer report from independent third side. As addressed already in Sect. 4.1.2, Sanford Moss (1872–1946)—following the work of A.C. Rateau and E.H. Sherbondy— developed for General Electric the exhaust-driven turbocharger (1921), which led to the broad use of turbo-charged piston engine aircraft in the Second World War. In the second half of 1935 General Electric's chief scientist S. Moss (63) toured Europe for turbocharger

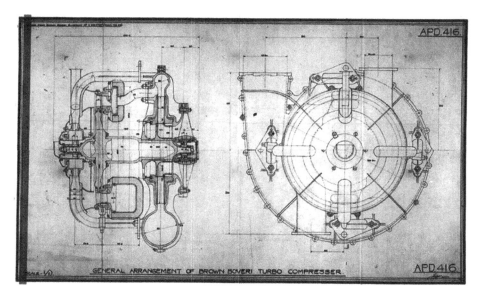

Fig. 6.35 Rolls-Royce General Arrangement APD 416 of *'Brown Boveri Turbo Compressor'* VTF 280 dated 29 July 1939, traced from BBC drawing No. V.692797, dated 7 Dec. 1936 © RRHT Derby

'news' and wrote an intriguing eight-page travel report;[293] besides Rolls-Royce and RAE, he visited the GE daughter companies BT-H[294] and AEG, Berlin, but he got not the requested *permission* for visiting Brown Boveri at Switzerland.

Before outlining his own trip, Moss refers to three, immediately preceding visits from England at GE, which indicates somewhat the spirit of renewed interest in advanced engine technology on the English side, as mentioned before:

- *'In 1935, Mr. James E. Ellor, now in charge of superchargers in the Aviation Engineering Department of the Rolls-Royce Company, visited the United States and collected all the information he could about turbo-superchargers. He visited our supercharger factory at Lynn but at the express request of the Army Air Corps was shown nothing about turbo-superchargers. ... He made overtures to purchase by the Rolls-Royce Company of a turbo-supercharger and was told by our International General Electric*

---

[293] A copy of that report entitled *'Turbo-superchargers in Europe'* had been detected by Rolls-Royce Heritage Trust (RRHT) member Dan Whitney at the *National Air and Space Museum*, Washington DC (Doc. ref. NASM B4-84000 Wright field file D52.42/15), published in *'Archive'* of the RRHT Derby & Hucknall branch, No. 81, Vol. 27, Issue 2, Aug. 2009, and was sent in this form to the author in 2010, courtesy to Dave Piggott, RRHT Derby. Moss had issued the report on 7 Jan. 1936 from Thompson Research Laboratory, GE Lynn, Mass.

[294] Rumours that Moss might have seen then already first hardware parts for the Whittle WU engine are apparently unfounded, given an actual order date of mid-1936, at least six to eight months after Moss's European journey.

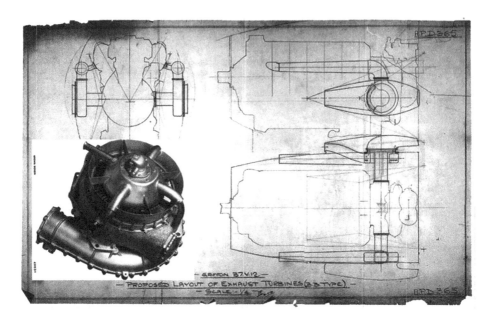

Fig. 6.36 Rolls-Royce Griffon 37.V.12 Installation: *'Proposed Layout of Exhaust Turbines (B.B. Type)'*, APD 365, dated 29 April 1938—with inset of Brown Boveri VTF 280/225 turbo-supercharger

Company that this would not be possible on account of Prohibition of the United States Army Air Corps.

- *During 1935 Mr. R.D. Pye, Fig. 6.31 (r), one of the Chief Engineers of the British Air Ministry visited the United States. He made some observations of exhaust superchargers, but so far as we know in not as much detail as Mr. Ellor. Mr. Pye afterward complained to one of the engine companies that he had not been shown a great deal at our River Works Plant.*
- *Major Bulman, Chief of the Engine Divison of the British Air Ministry visited Wright Field in 1935 and saw General Electric turbo-superchargers on Consolidated P-30 planes.'*

The cocky and overconfident Moss continues:

- *'Alfred Bu(e)chi, now collaborating with the Brown Boveri Company in Switzerland, devised a divided manifold scheme ... now in use in many exhaust superchargers for diesel engines made by the Brown Boveri Company for motor ships and rail cars. I tried to visit the Brown Boveri Company but could not obtain permission. However, I was informed that the Brown Boveri Co. have never made exhaust superchargers for aviation engines. The design for diesel engines, of course, differs greatly from that necessary for aviation engines in that the temperatures are very much lower and there is*

no restriction regarding weight. The use of the divided manifold also eliminates the
necessity for best efficiencies.'

As noticed before, at the time of Moss's statements, Rolls-Royce and Brown Boveri had
started their secret and innovative turbo-supercharger collaboration at least one year before;
following the Moss report, none of these activities leaked out at Derby and the leading
engineers at Swiss Brown Boveri had already lifted the *'draw-bridge'* in time: *Moss—*
Sherbondy—Seippel, there were certainly still open accounts accumulated over the past
15 years, perhaps not all were aware of them.

One of Moss's visits was to Rolls-Royce, Derby:

- *'... There was also exhibited to the writer a great deal of data which Mr. Ellor had*
  collected on exhaust superchargers while in the United States. The extent of this was
  rather surprising in view of the supposed restriction on such information. ... The
  request was again made that the General Electric Company furnished the Rolls-
  Royce Company with an exhaust supercharger. This however, was made more definite
  than had previously been the case and a specified design was asked for quite different
  from anything which had been made in the United States. This was the use of two
  separate superchargers, one on each side of a twelve cylinder liquid cooled engine with
  a divided manifold. ... The object of this was to secure all of the advantages of the Buchi
  divided manifold system and also to avoid any cross-over pipes ... The proposed
  English scheme offered a much simpler way of using divided manifolds than anything
  which has been discussed in America. ... All of these points led the writer to think that
  the English scheme might have some advantages to compensate for the increased weight
  of the two superchargers, and certainly was worthy of investigation. All of the details of
  the scheme were quite different from anything which had been proposed in America, so
  that a supercharger designed according to it would be wholly different from anything
  furnished by the General Electric Company to the United States Army Air Corps.'

Glad to observe that Mr. Moss's trip to *'merry old England'* apparently still became a
success in the end. After his competent introduction it is not difficult to assess the contents
of the following Fig. 6.36 which shows as designated the installation of a Brown Boveri
VTF 280 bi-turbo arrangement on a 12 cyl. Rolls-Royce Griffon engine.[295]

The BBC exhaust gas turbocharger *VTF 280*, was investigated at Baden and also at
Derby. The two-stage mechanical radial supercharger of the *Merlin*, of which the second
stage was switched on only after the aircraft was airborne, was thought to limit the high-
altitude performance of the *Spitfire* and *Hurricane* fighter aircraft near the 8000 m mark,
while the free-wheeling VTF 280, later first time operational with turbine air-cooling,

---

[295] The drawing belongs to a series of three with BBC reference in the RRHT Archive, comprising in
addition to the BBC also a RAE turbocharger for *Griffon*, courtesy to Dave Piggott, RRHT
Derby, 2009.

Fig. 6.37,[296] opened up the range to flight altitudes of 10,000 m and beyond. The 100 hp VTF 280, which weighed only 40 kg, produced with the single stage centrifugal compressor PR 2.8. The 45% cooling air for the VTF 280 turbine bearing cooling, as described before, was considerably brought down to 25% by the introduction of the mentioned, new 'shower-head' film cooling concept for the axial turbine blades, so that in due course also the radial compressor impeller diameter shrinked to 225 mm and the total unit weight of a *VTF 225* was only 32 kg.

BBC's Hans Herger[297] provided a humorous episode of these joint test activities at Rolls-Royce, where the red-hot turbine backend close to 900 °C could be observed/photographed and speed-measured by means of a stroboscopic technique ('Strobotak'). Rumours about the new turbocharger spread and drew sometimes high-ranking visitors of the Air Ministry. Herger claimed, he saw during one of these axial observations a beginning crack in the—visually standing—turbine blade which within seconds, before he could throttle back the rig, became a chunk of red-hot blade material, disappearing horizontally into the horse-hair-padded shoulder of a bystander's heavy military coat. The turbocharger ran smoothly on; it had elastic and damped bearings and was designed to run supercritically. Herger presumably could smell the event in the wake of the luckily unharmed, but certainly impressed, leaving visitors.[298]

The results of the VTF 225 prototype testing with 'showerhead', air-cooled turbine blades were taken at Brown Boveri's test department, Baden, CH, in 1938/9—together with the first time use of thermocolours for determining surface temperatures. Figure 6.37 shows the correspondingly evaluated surface temperatures at a turbine test point with 15,700 rpm speed, a gas temperature of 909 °C at turbine entry and a corresponding cooling air temperature of 150 °C; the first time applied 'showerhead' film-cooling principle is illustrated schematically with the blade cross-section inset at the root of the turbine disk: the cooling air 'film' is ejected through a row of cooling holes, radially distributed over the turbine blade leading edge:[299]

---

[296] See BBC, Flugzeug-Aufladegruppe VTF 225.

[297] Hans Herger (1911–1991) had joined the BBC turbine test centre TFVL in 1936 and got the RR liaison task immediately thereafter. In 1941 he became responsible for turbocharger engineering in the Gas Turbine Dept. under Cl. Seippel. In 1969 he was the first head in charge of Brown Boveri's new verticalised T Division (Thermal Machines, including turbochargers).

[298] See Jenny, The BBC turbocharger, p. 71.

[299] The film-cooling variants, investigated for the first time at Brown Boveri in the mid-1930s, are described in—see Eckardt, Gas Turbine Powerhouse, p.402 f. The special VTF225 'showerhead' cooling configuration followed a patent idea from G. Darrieus, US2,951,014 with priority 29 Jan. 1934. A protective fluid/layer is discharged through one or more orifices/slits, located in the well-rounded leading edge of a turbine airfoil and acc. to the patent 'spills smoothly and nearly equally on either side of this edge. Both faces of the blade ... are covered by a protective laminar and continuous layer of cooler or non-corrosive fluid which separates them from the hot of corroding working fluid'.

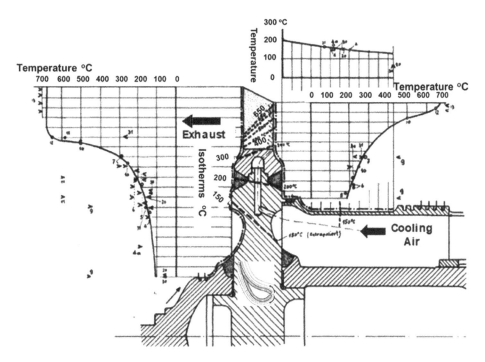

Fig. 6.37 BBC air-cooled turbocharger turbine wheel VTF 225 for Rolls-Royce, 1938: n_{max} 31,000 rpm, surface temperature distribution @ 15,700 rpm, turbine entry temperature 909 °C

- to the right (upstream) of the turbine rotor, the temperatures of the front rotor blade view and of the outer disk section are shown with a variation between 670 °C at tip and 190 °C at the inner measuring diameter. In addition, the wall temperature distribution of the outer surface of the cooling air entry duct is shown, axially rising in the flow direction from approximately 140 °C to 190 °C at the blade disk front face,
- for the cooled turbine blade and the outer disk section, isotherms are drawn, which characterise a turbine blade leading edge variation from 290 °C at the hub to 655 °C at the tip, while the corresponding temperatures at the blade trailing edge vary between 390 °C at the inner duct and 680 °C at the tip,
- finally, on the exhaust side, the rear face temperatures of the blade row and of the outer disk section are depicted, with 2/3 of the blades's rear outboard section being constant at 680 °C, until the temperature curves decrease down to 110 °C at the inner-most measured temperature of the turbine disk.

The general drawback of Darrieus's cooling configuration becomes clearly visible; a pronounced film-cooling effect over the radius can only be observed for the whole blade front zone, while the trailing edge situation confirms the known concept weakness, where a cooling effect is only visible for approximately one quarter of the inner trailing edge duct height.

The highlight of an intensifying, though always secret turbocharger cooperation between Rolls-Royce and Brown Boveri lay still ahead. At best it can be reconstructed by a combination of very reliable, though often unspecific information on the English side—the minutes of regular ESC meetings—with the Swiss technical input, normally without hints to the corresponding applications. Jakob Whitfield's 2012 thesis[300] is a specifically rich source in this context. As addressed before, the operational environment of the late 1930s meant that high-powered engines were needed for interceptor aircraft. The gas turbine was but one of the potential powerplants that might be useful for this goal, at first Ricardo's unconventional two-stroke 'Crecy' engine concept was in the focus, as the Air Ministry's desire to develop a 'sprint engine for Home defence'.[301]

The corresponding Rolls-Royce Crecy was a two-stroke 90-degree V12 liquid cooled aero engine of 26.1 l capacity, featuring sleeve valves and direct petrol injection. Single-cylinder development began in 1937 under project engineer Harry Wood using a test unit designed by Sir Harry Ricardo (1885–1974). The Crecy was originally conceived as a compression ignition (diesel) engine and Rolls-Royce had previously converted a Kestrel engine to run on diesel. By the time they started development of the Crecy itself, in conjunction with the Ricardo company, the decision had been taken by the Air Ministry to revert to a more conventional spark-ignition layout, although still retaining fuel injection. The engine was named after the Battle of Crécy (26 August 1346); Henry Tizard was a proponent of the engine as Chairman of the Aeronautical Research Council. The power of the engine (1500–2500 hp) being interesting in its own right, but also the exhaust thrust at high speed and altitude making it a useful stop gap between engines such as the Rolls-Royce Merlin and anticipated jet engines. The engine research was terminated in December 1945, Crecy no. 10 achieved 2500 hp on 21 December 1944.[302]

The most crucial auxiliary engine technology discussed by the ESC was the engine supercharger or compressor, which compressed the intake air. Deviating from the original altitude-dependent operation principle of aero turbochargers at the end and after WW I, the situation changed in the 1930s, when high-power engines began to be designed to use superchargers to boost the engine's power at all heights, including ground level. By the mid-1930s, the best current superchargers gave a compression ratio of about 2, but it was clear that for the high-powered engines then being conceived this would have to be raised. The RAE's experts suggested that maximum blower compression ratios would be 3 for an exhaust driven compressor, and about 2.5 for one driven off the engine. In the following and intensively in the 99th ESC meeting on 31 March 1936 the 'supercharger', specifically

---

[300] See Whitfield, Metropolitan Vickers, Chapter 'Origins of the gas turbine', p. 48 f. Surprisingly, though he is very close, Whitfield omits mentioning the 'Rolls-Royce Crecy'. The following ESC references are owed to Whitfield's meticulous search and documentation.

[301] 98th ESC meeting, 25 Feb. 1936.

[302] Excerpted from, for further details—see https://www.gracesguide.co.uk/_Engines:_Crecy

in the form of an axial compressor, came to the ESC attention.[303] At that meeting, David Pye gave an overview of the latest work being done under the auspices of the Air Ministry. He informed the committee that the *Bristol Engine Company* was working on a two-stage radial supercharger, and that the RAE had also been doing research on two-stage blowers. In addition, rather in a side-remark, he mentioned that Rolls-Royce had asked the Swiss electrical engineering company *Brown Boveri & Cie.* to design them *'a multi-stage axial blower for one of their newest high powered engines (1500 H.P.)'* and after informing that Brown Boveri's supercharger for Rolls-Royce was planned to provide the necessary design pressure ratio of 2.5, apparently Pye added, that Brown Boveri were *'confident that it would have a good efficiency'*. All this in immediate context and nevertheless, for the minutes sufficiently vague, could only mean an application for the *'desired sprint engine'*.

Whitfield's following comment *'In the mid-1930s, Brown Boveri was internationally the company with the greatest experience of axial turbomachinery; like Metrovick, the company manufactured steam turbines, but they also had extensive experience of building axial fans and blowers. Working with the eminent aerodynamicist Professor Jakob Ackeret of the Swiss Federal Institute of Technology, Brown Boveri had built the compressors for the Institute's high-speed wind tunnel.'* ... though in principle justified, requires some clarification. The close connection of Ackeret and Brown Boveri, if the latter being mentioned at all, appears to be a peculiarity of English technical historians. In this case, the origin of this perception may even date back to the early contacts between RAE and Brown Boveri, as dealt with in the following Sect. 6.3.3. Technically—with respect to an insinuated, collaborative exclusivity in applying aerodynamic science and turbomachinery design—the described cooperation was non-existent. In the early 1930s, Ackeret's Institute for Aerodynamics at ETH Zurich installed two wind tunnels, one large subsonic wind tunnel, for which the design of the two identical drive compressors came from Ackeret's assistant Curt Keller, while the then unique, first closed-loop supersonic wind tunnel was ordered from Brown Boveri, including the drive compressor for this application, which was already described in Fig. 4.26.[304]

Whitfield continues,[305] still with reference to the 99th ESC meeting: *'Returning to the Rolls-Royce blower* (from Brown Boveri), *Tizard asked whether the Brown Boveri design's pressure ratio could be increased to 4, whilst still retaining a reasonable efficiency. Griffith responded that if he had to aim for that ratio with a low risk of failure he would choose a two-stage centrifugal blower, but for the highest possible efficiencies he would go for the*

---

[303] 99th ESC meeting, 31 March 1936. In addition, to the Brown Boveri axial compressor, there was also informed that the Company was working on an exhaust-driven blower; as already discussed in Fig. 6.35 f.

[304] See Eckardt, Gas Turbine Powerhouse, *'The First Supersonic Wind Tunnel at ETH Zurich'*, p. 126 f.

[305] See Whitfield, Metropolitan Vickers, p. 53.

axial type, albeit at a heightened risk of failure. In response, Swan[306] *suggested* "that if the Committee desired to pursue the axial flow type it might be a good plan to call in a firm such as Metropolitan-Vickers as consultants"; *Tizard suggested that Metrovick should be asked to manufacture the compressor, but should be* "allowed to use the RAE as consultants."' So the conclusion might be drawn that Metropolitan-Vickers was brought to the emerging axial turbojet group as apparently best available equivalent to Brown Boveri and its expertise.

Half a year later, in the 102nd ESC meeting on 29 September 1936, quoting Whitfield for the last time, *'Pye stated that the position in regard to superchargers had been reviewed and arrangements made for Dr. Griffith and Mr. Constant to visit the works of the Metropolitan Vickers Company and discuss with them any of their supercharger proposals. Pye also gave further details for the* proposed *Brown Boveri—Rolls-Royce axial compressor, as well as for the other firms that were developing blowers. One of these was the British Thomson-Houston Company; Pye noted that* "no recent progress had been made although that firm was building a compressor for a client at Cambridge."' The client was, as is known from the foregoing, *Power Jets*, and the compressor was for Frank Whittle's first jet engine.

Two other points of Pye's statement of 29 September 1936 are worth to put attention to (if the reported dates are correct):

- apparently, as seen below, the announced RAE visit at MV was considerably delayed up to 3 June 1937, and
- the *'proposed'* in Pye's foregoing statement allows the interpretation that the actual order from Rolls-Royce was issued thereafter, presumably still at the end of 1936.

Obviously, the test preparations, including the shipping of an especially manufactured light-metal casing and of a light-weight aero gearbox with gear ratio 8.4 from Rolls-Royce to BBC Baden, required some time, which explains that the reference test programme was issued on 12 January 1938, so that a delivery to Rolls-Royce might have been planned originally still in 1939. The Rolls-Royce order confirmed the ESC-mentioned offer drawing for a 7-stage axial turbo charger of 0.24 m rotor inlet diameter and 0.305 m axial length. For the first time, BBC left the classic reference to Prandtl's profile catalogue for this challenging design and used high Mach number cascade data instead—from NACA.

Rotor blade height was steadily decreasing from ~28.6 mm at inlet to ~20.6 mm at exit, with a cylindrical hub. The design tip clearance was 0.3 mm. Due to the extreme overall weight target of 40 kg, the design speed had to be increased from intended 24,000 to 25,500 rpm, a fatal move. Figure 6.38 (l) shows the test set-up at BBC Baden, with the

---

[306] In the mid-1930 the head of RAE's engine department was Andrew Swan, who was often joined at the ESC meetings by his subordinates A.A. Griffith, who was to become head of department after Swan left the RAE, and by Hayne Constant.

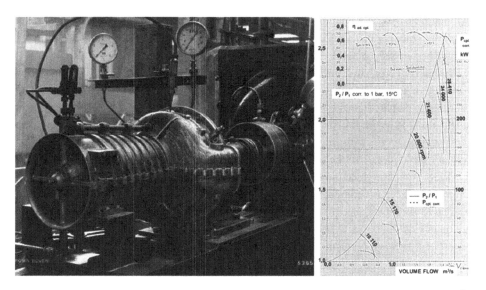

Fig. 6.38 Brown Boveri seven-stage axial aero blower for Rolls-Royce 1939: PR 2.4, V 1.8 m³/s, n 25,410 rpm, $\eta_{ad,\ max}$ 0.75, 280 kW/200 hp drive power, W 40 kg © ABB Archive

cylindrical Rolls-Royce gearbox in the rear, in front of the driving DC motor. The originally measured compressor map, Fig. 6.38 (r), received for better readability new descriptive labels: total pressure ratio p_2/p_1 stretches vertically from 1.0 to 2.5—with PR 2.4 at the design point, the horizontal Volume Flow scale cover the range 0 to 1.9 m³/s. Besides the solid pressure speed lines up to n 25,410 rpm, there are also dashed lines for the measured coupling power P_{cpl}, drawn as right ordinate with values up to 280 kW.

The measured adiabatic efficiencies at the motor coupling $\eta_{ad,\ cpl}$ reach their maxima at 0.72 and 20,000 rpm, which have to be corrected by a coupling efficiency in the order of 0.96 … 0.975 so that a maximum blower efficiency of 0.75 resulted. The BBC test report[307] mentions as prime test target to investigate the compressibility effect on efficiency, pressure ratio and volume flow in the vicinity of the speed of sound, and in fact, especially the efficiency curves deteriorate considerably beyond the best point at 20,000 rpm. Due to the lightweight casing, the testing procedure required to limit the exhaust pressure p_2 < 1.5 bar and T_2 < 120 °C. Consequently, to simulate the required entry conditions of p_1 0.575 bar and T_1 −15 °C the tests were carried out with entry throttling and by installing a small expansion turbine upfront. After the test programme was accomplished, the test rig was destroyed during a planned further speed rise up to 27,000 rpm. Detailed failure analysis in 1941 revealed that one second row rotor blade was broken in the root section due to mechanical overload, removing all blading—except for the front and end rows—and penetrating even the especially manufactured casing, so

---

[307] See BBC, Axialgebläse für Rolls-Royce.

that finally this PR 2.4 supercharger could not be delivered.[308] In the meantime the axial compressor race for PR > 4 had gained priority, in which Brown Boveri again participated, as described in Sect. 6.3.4.

Especially the exhaust turbocharger tests at and for Rolls-Royce finally yielded good results, so a cooperation was thought in reach. While BBC looked for work and had accordingly prepared in the meantime a workshop for aircraft engine turbochargers in Baden, Rolls-Royce insisted the manufacturing to take place in England and finally the plans were given up with England's involvement in hot war activities in 1940. Interestingly, a double-picture of the self-illuminated Brown Boveri turbocharger test rig at Baden CH, 1938, was reproduced in G. Geoffrey Smith's popular *'Gas Turbine and Jet Propulsion for Aircraft'*, 3rd edition, April 1944 (!), at a time when the turbocharger cooperation between Rolls-Royce and Brown Boveri had been put on hold since at least four years. The caption of Fig. 48, p. 66, in that book reads: *'A Brown Boveri exhaust turbo blower set on the test bed, running continuously at 30,000 rpm with a gas temperature of 1000 °C.'*[309]

6.3.3 Royal Aircraft Establishment, Metrovick and Brown Boveri

The indicated, second period of aero gas turbine developments, focusing on axial compressors, began in July 1936, when RAE obtained authority to build a small scale model compressor, actually a six-inch-diameter (152 mm) version of eight stages of the inner portion of the Griffith engine, Fig. 8.28, named 'Anne', potentially also applicable for a piston engine supercharger. It was originally built with adjustable blow-off at each stage and on the delivery diffuser, to avoid blade stalling during starting and for boundary layer control. Work on this unit proceeded slowly, it was not completed until early in 1938 and

---

[308] Axial turbochargers were not uncommon for the high-power class of RR engines. Another small axial was commissioned by RR/Air Ministry from RAE/ Metrovick around the same time. No engine name was mentioned, but D. Piggott from the RR Heritage Fund suspected, that this charger might have been also for the *Crecy*, a two-stroke V12, 26 l 'sprinter' (H.T. Tizard) engine for a *Supermarine Spitfire* application, and with ~2500 hp somewhere between the upper Merlin class and a gas turbine. This unit was with PR 3.7 and a mass flow of 2.5 kg/s considerably bigger though than the BBC axial supercharger.

Though the timing of the patent dispute between AVA Goettngen and BBC, as described in Sect. 4.3.3 *In the patent arena*, decided by the Reich Patent Office pro-BBC on 28 Feb.1936, and the delivery of exactly such a product from BBC to England may insinuate a connection between these two events, especially since AVA was forced in the process to provide drawings of their five-stage axial compressor of 0.15 m diameter and 0.5 m length, Figs. 4.44 and 4.45, for evaluation to BBC, there is no proof for such a link close to espionage. And, of course, this BBC compressor design like all her others differentiated from AVA by the used degree of reaction.

[309] A picture of the VTF 201 turbocharger, renamed to VTF 280, operating under load on the test bench at Baden, CH, in 1938, has been also reproduced in—see Eckardt, Gas Turbine Powerhouse, p. 404.

stripped its blading 30 s after starting, because of a faulty oil seal and resulting rapid overheating. It was rebuilt without the blow-off arrangements and with larger clearances between the blade rows. It ran again in October 1938, and gave some first insight into the many of the problems of multi-stage compressors, until it was destroyed during a German air raid by Ju 88 bombers of the Luftwaffe's KG 54 on 13 August 1940, otherwise known as *'Adler Tag* (Eagle Day)', when the Luftwaffe's all-out air assault on Britain began.

According to Constant,[310] who was responsible for the design, *Anne's* blades had 50% reaction at the inner radius—with a twist to give free vortex flow, apparently from the beginning, though the same feature reads by Kay, *'From then on* (after Anne's re-build and a possible, intermediate BBC influence), *RAE compressors used 50% reaction blading* ...'.[311] This uncertainty aside, quite promising results were achieved: The tested speed range went up to 24,000 rpm, with an estimated design point at PR 2.45 and ṁ 1.03 kg/s. Best adiabatic efficiency was achieved for 16.000 rpm with η_{ad} 0.806, but similar to the BBC axial supercharger before—with a relatively steep decrease to η_{ad} 0.76 at design speed.[312] Independent of the actual origin of this design concept, the fact that both sides followed finally the same R 0.5 principle was certainly encouraging—for both sides; and, the inherent substantial performance advantage was kept secret on both sides, till the Howell 1945 IMechE paper.[313] Claude Seippel waited even more than 20 years before he handed out his corresponding test results with systematic R-variation for publication by John Horlock,[314] again one of these appreciated Englishmen.

Eight months after his return to Farnborough, Constant was given the task of writing a feasibility report on the gas turbine.[315] As Frank Armstrong put it,[316] *'Constant, though regarded by some as rather aloof, even a somewhat arrogant figure, had a sense of humour which he was prepared to use in writing as well as verbally. In his report, which concluded strongly that there was now a prospect of constructing a turbine engine well competitive with reciprocating engines, he also wrote* "The magnitude of the advantage it has to offer, associated with the repeated failures to achieve a practical design, have given the impression that the Internal Combustion Turbine is merely a convenient medium on which to work off the surplus energy of imaginative inventors"'.

At this point there was some debate as to how to proceed. The team, which included besides Alan Griffith, as head of the RAE Engine Department and Hayne Constant, responsible for RAE's Turbocharger Division under Griffith, Fig. 6.39, since 1938 also two, later highly

---

[310] See Constant, The Early History, p. 411.

[311] See Kay, Turbojet, Vol. 1, p. 14.

[312] See Roxbee Cox, British Aircraft Gas Turbines, p. 413.

[313] See Howell, Design of Axial Compressors, Fig. 95.

[314] See Horlock, Axial Flow Compressors, Figs. 4.11 and 4.12 in that book.

[315] See Constant, The Internal Combustion Turbine. In this RAE Note E 3546, dated March 1937, Constant coined the term *'ICT (Internal Combustion Turbine)'* which very soon replaced also the name of his *'Supercharger Section'* to *'ICT Section'* at RAE,—see Bailey, The Early Development, p. 11.

[316] See Armstrong, Farnborough.

Fig. 6.39 British turbojet engine pioneers II & *visitors* at BBC, *1937/38:* Alan A. Griffith (1893–1963) l, © NPG 8348 London, Sir William S. Farren (1892–1970) m, © The Hooke family, Hayne Constant (1904–1968) r © RAeS

renowned axial compressor designers, A. Raymond Howell (1917–1988)[317] and A. Denis S. Carter, studied a number of approaches to buiding a complete engine. Finally, three of these schemes (A–B–C) were picked for a first contact with Metrovick on 3 June 1937:[318]

- A two-stage centrifugal compressor driven by a high-pressure turbine, with a low-pressure turbine to drive an airscrew,
- A low-pressure axial compressor driven by a low-pressure turbine, with a high-pressure axial compressor driven by a high-pressure turbine, later known as the *'Dispersed Compound'* turbine engine, and
- A contraflow turbo compressor with a separate parallel gas flow to power the turbine.

Besides the lack of industrial experience, RAE was also not equipped for large-scale manufacture, therefore it was arranged that especially detail design and manufacture should be carried out at Metropolitan-Vickers Electrical Company (in short, MV or Metrovick), Old Trafford, Manchester in the future cooperation. Metrovick and Brown Boveri were obviously similarly structured companies. Both were globally acting firms with a strong power generation and especially steam turbine background, always open to related new technology areas on the basis of a science-minded engineering team. This was not sheer coincidence, but had also in parts a consequence of common roots.

---

[317] Alun (a Welsh boy's name) Raymond Howell was one of the leading compressor scientists at RAE Farnborough. He played a major role in the evolution of successful axial compressor design methods which would be used by the first post-war UK gas turbine engines. In 1946, he was appointed Head of Aerodynamics Dept. of the RAE successor organisation NGTE National Gas Turbine Establishment which he led for more than 20 years,—see Dunham, Howell.

[318] See Jones, The Jet Pioneers, p. 75. Acc. to—see Whifield, Metropolitan Vickers, p. 59, the detailed design discussions of an MV team at RAE Farnborough started only on 13 Dec. 1937.

Fig. 6.40 Metrovick—from Electric Machinery to Turbojet Engines: Philip A. Lange (1856–1937) l, Karl Baumann (1884–1971) m, David M. Smith (1900–1983) r

Loose ties between these companies commenced to develop already before WW I. Out of the Brown Boveri engineering executive team at Baden—Adolf Meyer (1880–1965), Fig. 6.47 (r), - Claude Seippel (1900–1986), Fig. 3.5,—Walter G. Noack (1881–1945),[319] the latter, inventor of the Velox boiler and important contributor to the Brown Boveri gas turbine status, had the longest personal experience with that Manchester engineering community. After finishing his engineering studies in 1906, graduating with a diploma examined by Professor Stodola, he went to England, where he received his first job as an *'expert'* for MAN[320] gas engines at Lilleshall Co., a MAN licensee. Before returning to Switzerland in 1909 to begin 36 years of uninterrupted and successful work for BBC in Baden, he was offered a position at British Westinghouse Electrical and Manufacturing Co. in Manchester, to build up the gas engine department there. This connection was revived nearly 30 years later in the emerging field of aero gas turbine propulsion. Noack decided to leave England then, following news from his mountain-climber friend and Meyer's deputy Paul Faber, that somebody was needed to seriously explore the possibilities for a gas turbine. And, of course, for the weekends also a few challenging high-Alpine tours were still waiting,—a *'rope party'* pattern which will show up in this context again.

British Westinghouse was led at that time by the US-delegated German Director Philip A. Lange (1856, Rostock—1937, London), Fig. 6.40 (l). Lange came to the USA in 1882 via Siemens Woolwich and joined Westinghouse, East Pittsburgh, PA in 1886 as an electrician, later becoming superintendent of the detail department. In 1906, he was sent to British Westinghouse, which became Metropolitan-Vickers. In 1909, after the hiring of

---

[319] See Eckardt, Gas Turbine Powerhouse, P. 88 f.

[320] Maschinenfabrik Augsburg-Nuernberg,—see Wikipedia, ‚MAN SE‘in English. Since 1893, Rudolf Diesel puzzled for 4 years with future MAN engineers in a laboratory in Augsburg, until his first diesel engine was completed and fully functional.

Walter Noack had failed, Lange[321] travelled to Zurich to talk to Stodola and hired Karl Baumann in due course, Fig. 6.40 (m), another of Stodola's famous scholars. After a short-time assistant to Professor Stodola in 1906/7, Baumann had at that time already left ETH Zurich and was also employed by MAN, so that a foregoing communication with his colleague Noack is probable. Baumann was Metrovick's chief mechanical engineer since 1912, and Technical Director from 1927 till his retirement in 1949.[322] Baumann appointed David M. Smith, Fig. 6.40 (r), to lead the design, development and manufacturing team in the establishing collaboration with RAE. Then in the early 1940s, Smith played a decisive role in the development of the *Metrovick F.2*, the first British turbojet engine based on an all-axial turbomachinery design.[323]

Hayne Constant had written the report for the seminal ESC meeting on 16 March 1937, which successfully gained thereafter not only Air Ministry's approval for RAE to proceed with gas turbine research for *'propeller power'*, but released also somewhat the financial restrictions on Power Jets's activities. Nevertheless, the RAE programme for the period 1937–1939 bears still strong hallmarks of being directed restrictively by Alan Griffith:

- The small ICT Section must have spent many valuable man-hours on Griffith's contraflow counter-rotating gas turbine and, since,
- Griffith in his earlier reports had concluded that an aircraft gas turbine would require, to overcome starting problems, a low- and a high-pressure compressor, separately driven, each with PR 2.0–2.5 for an overall target PR 4.0–5.0, thus, all experimental compressors designed and run in this period were of this low pressure ratio.

For the time-wise assessment of the further developments, it is worthwhile to take two travels[324] from RAE key personnel to Brown Boveri, Baden, CH into account, with the travellers pictured, united on Fig. 6.39:

- In 1937, the influential head of RAE's Engine Department Alan A. Griffith went—either as a result of discussions within the Engine Subcommittee of the ARC or, more likely,

---

[321] *'Papa Lange'*, though seen as the honoured *'Father of Metropolitan-Vickers'*, had to leave the Company in 1917 when war tension between Great Britain and Germany could not be left out of company dealings any longer.

[322] Throughout his career, Baumann was at the forefront of power station steam-cycle development, pioneering the use of increased turbine inlet pressures and temperatures and, in 1916, introducing multi-stage regenerative feed-water heating and the *'Baumann split-exhaust steam turbine'*,—see Stodola, Dampf- und Gasturbinen, p. 570 f. *'Die Baumann-Turbine der Metropolitan-Vickers El. Co. Ltd'*, his 105 MW set for *Battersea A* power station, installed in 1933, was for many years, the largest single-axis steam turbine unit in Europe,—see Grace's Guide, 'Karl Baumann' in English.

[323] See Sect. 8.2.2 for continuation.

[324] Regretfully, no independent confirmation of these trips appears to be possible from BBC's side; Griffith's trip in 1937 has been mentioned—see Kay, Turbojet, Vol. 1, p. 14 and—see Constant, The Origins, p. 214. The information about the Constant/Farren travel in 1938 results from Constant's personal statement, issued on 10 Sept. 1942,—see Constant, (Internal) RAE Report.

by direct order out of the Directorate for Scientific Research[325]—alone to Baden, followed

- in May 1938 by Hayne Constant, accompanied by (Sir) William S. Farren, then (1937–1939), Deputy Director of Scientific Research to (Sir) David R. Pye at the Air Ministry.[326]

Unnoticed from technical historiography so far, another English visitor was—somewhat undercover—at Brown Boveri in summer 1937. Robert *'Bob'* Feilden (1917–2004) joined Power Jets, Rugby in 1940 and managed there Whittle's engine test programme.[327]

With respect to the timing of Griffith's trip to Baden in 1937, there are two framing dates—the 16 March 1937 of the all-initiating and -accelerating ESC meeting and RAE's first Metrovick contact on 3 June 1937, where—compared to the foregoing undecidedness—rather radical and courageous decisions were made. The further collaboration with Metrovick was focused to the *'Dispersed Compound'* engine concept only, Fig. 6.41, while the prepared alternatives with double-centrifugal and Griffith's counter-rotating contraflow arrangements were put aside first, before they were dropped completely from the RAE work programme with Griffith's leave to Rolls-Royce in June 1939.[328] There is a strong probability that these developments had been influenced by Griffith's *'Damascus experience'* at Brown Boveri in the meantime. The multi-stage axial compressor with a stage pressure ratio PR $_{Stage} > 1.09$ was now apparently state-of-the art at Brown Boveri, as illustrated by the Company's delivery records of Fig. 4.24, and the then standard Velox turbomachinery groups could be seen on the BBC shop floor in a kind of series production, for which Fig. 6.44 is a typical example. After this visit it was clear that

---

[325] See Schlaifer, The Development, p. 349.

[326] After an early stay at the Royal Aircraft Factory 1915–1918, he returned to Farnborough as RAE Director 1941–1945,—see Wikipedia, 'William Farren (engineer)'in English. As scholar of Perse School Cambridge 1905–1910, considered one of the leading schools, specialising in languages by using the *'direct method'* even for the classics, Farren commented: *'Possibly my greatest pleasure was in French and German, which I learned to speak fluently and well, and to read widely in their literature'*;—see Thomson, William Scott Farren. Apparently, his command of the German language was not only helpful in 1938 at BBC, but also in his earlier mission to Prandtl in 1928, Sect. 4.2.2, and after WW II, when he was appointed to lead *'Operation Surgeon'*, to exploit German aeronautics and deny German technical skills to the Soviet Union. A list of 1500 German scientists and technicians was created, with the goal of forcibly removing them from Germany *('whether they like it or not')*. Actually, of the scientists relocated from 1946 to 1947, 100 chose to work for the UK,—see Wikipedia, 'Operation Surgeon'.

[327] See Wikipedia, 'Bob Feilden' in English. During the long vacation at Cambridge, Feilden went to work for two months during summer 1937 at the Brown Boveri Co. at Baden, CH. He went on his motor bike, finding it necessary to fit two new piston rings *en route* and still catching the ferry on time. Brown Boveri's early gas turbine activities certainly whetted Feilden's interest for this new kind of prime mover, but without noting the intensive high-level information exchange at the same time between Baden and Farnborough.

[328] See Kay, Turbojet, Vol. 1, p. 19.

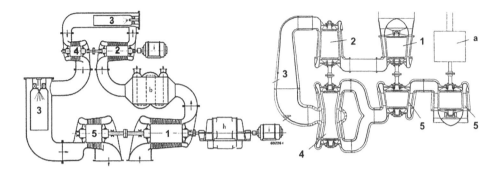

Fig. 6.41 Dispersed Compound Engine Concepts: Brown Boveri two-stage 10 MW power generation plant Bucharest 1943–1946 (l), RAE two-stage turboprop concept 1938 (r) 1 lp compressor, 2 hp compressor, 3 combustor, 4 hp turbine, 5 lp turbine, a gearbox, b intercooler, h generator, i starter motor

Tizard's ESC target of PR 4 was definitively achievable as a multi-stage axial compressor—with a considerably better efficiency outlook than the double-centrifugal alternative—and without any of the feared starting problems.

Given Griffith's known, somewhat cranky character, it can be assumed that the trip to Baden needed a *protocolary* preparation. Again, there is no proof, but the fact that Pye's deputy was on board in 1938, insinuates a certain role for *'Master Pye'* himself in 1937, who had clearly demonstrated his determination to push things ahead in the ESC meeting on 16 March 1937. Pye's biographer Sir Owen Saunders emphasised besides Pye's numerous professional merits, his astonishing, lasting reputation as enthusiastic mountain climber:[329] *'He appears to have started serious climbing at the age of twenty-five while at Oxford. ... From then on he scarcely missed a season in the Alps. ... He made his last Alpine expedition in 1947, but continued to walk in Austria or Switzerland until 1956. ...*

---

[329] See Saunders, David Randall Pye. To Pye's lasting climber achievements belong the first time *'Cioch direct'*, 150 m long, difficulty S 4b and thereafter the severe *'Crack of Doom'*, today a HS 4b (Hard-Severe) tour in 1918 on Isle of Skye, in north-west Scotland. He did several extreme tours together with his personal friend George Mallory (1886–1924),—see Wikipedia, 'George Mallory' in English, who took part in the first three British expeditions to *Mount Everest*, in the early 1920s. During the 1924 expedition, Mallory and his climbing partner, Andrew 'Sandy' Irvine, disappeared on the north-east ridge of the Everest. The pair were last seen when they were about 245 m from the summit. Mallory's ultimate fate was unknown for 75 years, until his body was discovered on 1 May 1999 by a search expedition; whether Mallory and Irvine had reached the summit before they died remains a subject of speculation and continuing research. In 1927, David Pye published *'George Leigh Mallory, A Memoir'*, a still available bestseller, regarded as some of the finest mountaineering literature. Mallory, seen from Pye's engineering perspective, quote: *'There is no doubt that all his life he enjoyed taking risks, or perhaps it would be fairer to say doing things with a small margin of safety'.*

*He usually climbed without guides and often as leader of the party, most of his expeditions
being in the Pennines, the* (Swiss) *Oberland and around Mont Blanc.'*

Saunders mentions Pye's *'biographical notes'*, which perhaps would provide a hint, if
he combined his usual Alpine trip in spring 1937 to Switzerland with a stint to Baden, not
excluded to a direct mountain comrade in BBC's upper management—with the enthusiastic Alpinist Adolf Meyer as prime candidate[330] or the already mentioned Paul Faber.
Griffith certainly met A. Meyer, Director of the BBC Thermal Department, and Claude
Seippel, the *'father'* of the established BBC axial compressor design philosophy. It can be
taken for granted that Meyer and Seippel realised the military-strategic importance of the
aero gas turbine topic from the beginning, and provided comprehensive and wholehearted
support within their power.[331] The principle feasibility of axial compressors with the
required pressure ratio range and efficiency had been demonstrated at Brown Boveri, the
task of a significant weight reduction together with the verification of aero-operational
flexibility lay still ahead of RAE.

In view of Hayne Constant's visit at Baden in May 1938, just one year after Alan
Griffith's reconnaissance inspection, it is worthwhile to check how he himself described
this decisive phase in hindsight. There are only two reports with the corresponding
information. The only generally published paper *'The Early History of the Axial Type of
Gas Turbine Engine'*[332] appeared in 1945 and described the time after the catastrophic test
of *Anne* in spring 1938—and its restart in October 1938: *'After this accident, immediate
steps were taken to redesign the compressor. During the interval since the first design had
crystallized, certain rumours, information, and changes in outlook had occurred. We had
heard of the success of the Brown Boveri axial compressor, which could be started and
operated satisfactorily without air bleeds. ... In the light of the Swiss information,*
(we decided,) *to eliminate all air bleeds in the new design. Later experience proved this
to be a retrograde step ...'* Apparently, neither this sounds as if he had been personally on
the spot in May 1938, nor does this apply for an alternate version in the internal RAE
report, dated 10 September 1942: *'In the interval between the original design and the
reconstruction* (of Anne) *we had received from Ackeret* (sic!) *in Switzerland two pieces of*

---

[330] Courtesy to Dr. York-Peter Meyer, AM's grandson, the author had the opportunity to review
A. Meyer's comprehensive mountaineering photo documentation on 11 Nov. 2016, but regretfully
without any hint to D. Pye.

[331] Especially on Seippel's side, influenced already by his father—see Wikipedia, 'Paul Seippel'in
German, (1858–1926), salient Swiss advocate for the *Voelkerbund* (League of Nations) idea after
WW I, biographer and personal friend to pacifist and literature Nobel Prize winner 1915 Romain
Rolland (1866–1944), *antifascism* belonged in the 1930s nearly automatically to the family-DNA.
Claude Seippel's father died in 1926, when he acquired the axial compressor design basics at
Sherbondy's engineering office at New York City. There and in the same year his son Olivier
(1926–2012) was born; the naming *'Olivier'* occurred after Rolland's self in his Nobel Prize novel
'Jean Christophe'.

[332] See Constant, The Early History, p. 412.

information. The first was that Brown Boveri were running an axial compressor which started without stalling with no blow-off, the second was that to ensure reasonable efficiency the axial gap between blade rows must not be less than 1/3 the blade chord.'[333]

Very strange, this type of vague and indirect, hear-say communication, based on *'certain rumours'*, at a time when RAE top management sought for and received considerable support from Brown Boveri. Was it somewhat inopportune in 1945, to admit, that there had been phases in the past when Brown Boveri's achievements were encouraging, and the Company's readiness for open discussions presumably highly appreciated?

The seriousness and urgency of the formal RAE/Air Ministry approach to Brown Boveri in May 1938 illustrate Constant's own, revealing words: *'I also paid a visit with Mr. Farren in May 1938 to Brown Boveri at Baden in Switzerland to see what we could learn from them. Owing to this firm being unable to guarantee not to pass their experience on to Germany, the Ministry felt itself unable to give them a contract to develop a turbine for aircraft use. They decided instead to get Brown Boveri to construct a power station type of unit, similar to one already supplied to America, so that we could study the design and get running experience. By the time this was running in RAE in 1940 our own development had outstripped Brown Boveri's, so that the knowledge we gained was not great.'*[334]

Again, Constant's wording is jaw-dropping, to end one sentence with *'to see what we could learn from them'*, and then to continue by asking for a de facto, exclusive license of Brown Boveri's aero-propulsion know-how.[335] The acquisition deal for a 2000 hp *'power station type of unit'*, obviously belonging to BBC's Velox-type turbomachinery, will be further outlined in the following Section. The reference to *'America'* in the foregoing, must refer to the six *Houdry* turbosets delivered out of BBC Baden between 1936–1938, while the remainder US-production of up to 36 units in total had been licensed to Allis-Chalmers. A thorough comparison, how far RAE had actually *'outstripped'* Brown Boveri, Baden and other developments on the Continent will be shown in Sect. 8.2, for the *'British turbojet engine developments'* during the War years 1940–1945.

Interestingly, there exists another statement on the RAE–BBC collaboration at a time when the corresponding need could be felt still more imminent, and the public naming of Brown Boveri apparently had a positive connotation for the RAE representatives. On 23 September 1938 as part of an US-UK information exchange programme on piston engine technology, especially the reliability of sparkplugs and bearings, Alan A. Griffith

---

[333] As said before, the reference to Ackeret as a source of advanced multi-stage axial compressor design insight for potential turbojet applications, either from his Institute or about BBC, appears to be improbable, if not intentionally introduced to mislead. In an address to the *Deutsche Akademie der Luftfahrtforschung* at Berlin on 4 Dec. 1942, Ackeret had still stated: *'The pure gasturbine will have to prove itself as land and ship engine first, before an aircraft application can be considered,'*—see Ackeret, Auf dem Weg zur Gasturbine.

[334] See Constant, Influence of the R.A.E.

[335] For a patented, *'light-weight, high-efficiency'* turboprop engine design by Brown Boveri's Cl. Seippel, US2,326,072, priority 28 June 1939, see Sect. 8.3.2 and Fig. 8.40.

visited the American government engine research facility at Wright Field, OH. The minutes of that meeting have the following side comment: '*He* (Griffith) *stated that the British Air Ministry is interested in the exhaust turbine as a prime mover and they are doing work with the Brown-Boveri Company on this type of powerplant. Dr. Griffith stated that there were 140 men at Farnborough working on aircraft engines.*' Short and precise ...[336]

The origin of the '*Dispersed Compound Engine*' concept, next in the row of RAE development steps, raises also some questions. English standard technical historiography deduces the concept of a lp/hp turbomachinery split to Griffith's long-time fear of trying to develop too high a pressure ratio in a single compressor, not least because of difficult starting. The idea was therefore mooted of a compound engine having two mechanically independent compressors, one low-pressure followed by a high-pressure one, each driven by its own turbines. Constant had been investigating the effect of dual turbo-compressors, and suspected apparently that independent compressors would have better average compressor efficiency under varying load and running conditions. As these loads were typical of those which would be encountered in an aircraft plant, he supported the choice of a multi-compressor scheme.[337] These ideas presaged the modern twin-spool, bypass turbojets, however, the mechanical complexity of the concentric shafts was not in RAE's reach at that time, so that even the idea of at least a parallel arrangement of low- and high-pressure shafts was abandoned early on, instead—as outlined by Constant—'*we decided to avoid the mechanical difficulties of the concentric arrangement by dispersing our units. . . . In this (new) layout, the aerodynamic requirements were sacrificed in order to give a simple mechanical arrangement.*[338]

Alternatively, a view to Fig. 6.41 could shorten these lengthy concept-finding considerations, which show a stunning similarity between a Brown Boveri powerplant arrangement, actually realised in 1946, and the discussed RAE scheme of 1938. As known from the power generation gas turbine history,[339] the GT concept became a rather '*low-hanging* (development) *fruit*', finally sold and realised as the 4 MW Neuchâtel plant in 1939/1940 only, when during Velox boiler development up to 1935, the turbine power capacity surpassed the compressor driving demands for the first time, so that extra-electricity power generation came into reach. Consequently, the principle concept of the pure power generation gas turbine will have emerged on Brown Boveri's drawing boards already as early as in the mid-1930s, and in due course, also the next step in plant size from 4 to 10 MW, which then was shifted to a post-war realisation at Bucharest-Filaret.[340]

The flow of ideas can be reconstructed in this case by means of a small design feature, Fig. 6.42, which Brown Boveri kept secret for more than a decade. In reality, the GT

---

[336] See Douglas, The secret horsepower race, p. 67.

[337] DM Smith 'Report on Conference, 1 Feb. 1938',—see Whitfield, Metropolitan Vickers, p. 74.

[338] See Constant, The Early History, pp. 415–416.

[339] See Eckardt, Gas Turbine Powerhouse, p. 158 f. 'The Early BBC Gas Turbines'.

[340] See Eckardt, Gas Turbine Powerhouse, p. 191 f., 'First BBC Power Generation Gas Turbines'.

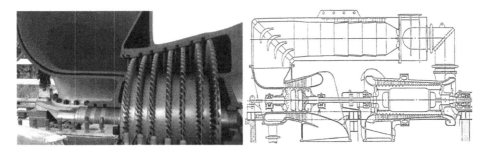

Fig. 6.42 BBC air-cooled double-wall structure in Neuchâtel 4 MW GT 1939: axial turbine entry (l), general arrangement (r)

hot-gas sections were mostly double-walled with a cooling air gas flow in between, as seen for the 550 °C turbine air entry collector of the Neuchâtel plant on the left, and also on the General Arrangement of that plant—with a strictly limited distribution of that information outside of the Company. In fact, also the BBC lay-out of the 10 MW Filaret plant, Fig. 6.41 (l) and others, omits any hint in this detail design direction, whereas RAE's drawing of their *'dispersed engine'* copy on Fig. 6.41 (r), clearly shows this feature for the hot-gas sections 3–4–5. The same observation of a kind of BBC *'master design'*—and a correspondingly adapted RAE version can be drawn from the *'4 bearings'* turbine—compressor, back-to-back configuration, typical for all Brown Boveri GT General Arrangements for the 1935–1970 period, Fig. 6.43 (top) and the RAE *B.10* turbine-compressor solution, Fig. 6.43 (bottom). The only significant deviation are the *'double-cone'* end pieces of the turbine rotor, which RAE introduced to relieve thermal expansion stresses, where BBC relied to their already established thermally-balanced, welded rotor design practice, heavy though proven as reliable, since the first realisation in a steam turbine in 1929.[341]

Besides the indicated conceptual influences of a large scale, the English/Swiss collaboration produced also plenty of detailed information—and misunderstandings. Constant describes one *'Swiss information'* during the *Anne* reconstruction interval, that *'to get maximum efficiency it was desirable to operate with an axial clearance* (spacing) *between the blade rows of not less than one-third chord.'* Since however, *Anne's* axial rotor length was fix, he came to the—predictably negative—conclusion to reduce the average blade chord length from 14 to 11 mm.[342] Another *'rule of thumb'* which Constant had brought back from Baden was to observe the lift coefficients in the aerofoil hub sections, differently for rotating blades (with beneficial, boundary layer centrifuging) and stationary vanes. As this was a three-dimensional effect, it would not show up in RAE's aerofoil cascade tests.

---

[341] See Eckardt, Gas Turbine Powerhouse, p. 321 f., 'The Welded Rotor Design and Manufacturing'.

[342] See Constant, The Early History, p. 412.

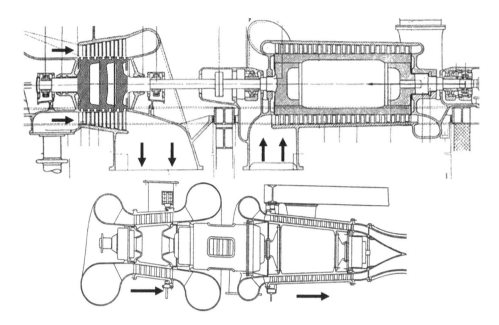

Fig. 6.43 Turbine (l)—Compressor (r) *'Four Bearings Arrangements'*: Brown Boveri Velox/gas turbine turbo-sets 1935–1970 (top), RAE/MV *B.10 'Betty'* turbo-set, August 1938 (bottom)

From the very beginning this kind of external information was also immediately shared between RAE and Metrovick.[343]

Summarising, one can state the expectation, that the RAE axial compressor design will surpass Brown Boveri's fundamental industrial design approach—without doubt and basically as the consequence of thorough cascade wind tunnel research up to 1945. But, for the present phase under consideration, 1935–1939, it appears to show RAE in relation to Brown Boveri still in a benevolent dependence, largely as recipient of ideas, somewhat happy about a certain guidance how to plan and proceed the next steps. Hayne Constant may have had reasons to remember that period after the War differently.

6.3.4   The Power Gas Turbine Re-Insurance

In retrospective it is sometimes difficult to cut through historically built-up *'strata'* and to imagine e.g. a sufficiently realistic picture of RAE's ICT Department before WW II; here

---

[343] See Whitfield, Metropolitan Vickers, p. 73, with reference to Minutes taken by D.M. Smith, MV, containing BBC information as early as 1 Feb. 1938.

the rather rare accounts of *'eyewitnesses'* are helpful, as in this Section by the former RAE employees Bill Bailey and A. D. Baxter.[344]

Before joining the RAF in mid-1940, Bailey had eight years of academic training in mechanical engineering and was a graduate of Durham University. During a 4-year apprenticeship at Richardson Westgarth—Brown Boveri, Hartlepool, he worked in the drawing office and assembly shop on BB steam turbines, multistage centrifugal compressors and *'also on the assembly and testing of a large axial flow gas turbine as part of a Velox boiler for the Royal Navy.'*[345] On its foundation, the ICT Section had two graduate engineers: Constant, as Head of Section, and Baxter. In late 1940, Bailey increased the group of engineering graduates to five in total, a somewhat meagre team for the major development ahead, as Constant had described the Section's effort during his introduction: The development of the axial turbojet engine and primarily the axial compressor. Shortly, he referred to *Power Jets'* competing centrifugal engine concept, giving the impression that the latter was not a serious contender because of its low compressor efficiency. *'I was an easy convert, knowing that Brown Boveri had reached the same conclusion; I was partnered with Ray 'Taffy' Howell'.* (Bailey)

Howell explained the method of designing axial flow compressors, similar to steam turbine design with the added problem of diffusing flow in the blade rows. The urgent requirement was to establish a design method for the nine-stage axial compressor for the F.2 engine.[346] To obtain 4:1 pressure ratio with an efficiency of 85% in a single unit required an in-depth knowledge of compressor cascade performance and this was not than available.

By 1940, to accomplish this complex research programme, the ICT Section had—after its founding in 1936—only two pieces of test equipment: A cascade wind tunnel and a single stage axial compressor. *Anne*, the multi-stage experimental compressor that Constant had designed, using 1928 cascade results, had been destroyed in the meantime. Consequently, the decision to order from Brown Boveri a 2000 hp gas turbine unit filled a perceivable gap for the time being, Fig. 6.44—after exceeding expectations towards an exclusive axial compressor design license were not achievable during the Constant/Farren visit in Baden in May 1938.

Bailey and Howell shared the tremendous task of cascade data generation, the former did the testing, the latter the data evaluation. To reduce the number of variables, they fixed the blade profile—leading and trailing edge radii, thickness/chord ratio and aspect (blade height/chord) ratio—so that only two blade variables pitch/chord ratio and camber needed testing with varying angles of incidence and stagger. Since the used RAE cascade wind

---

[344] See Bailey, The Early Development, p. 21 f. and—see Baxter, Professional Aero Engineer, p.183 f.

[345] A confirmation, long-time overdue; in 2012, when—see Eckardt, Gas Turbine Powerhouse, p. 164, was finished, the existence of a separate Velox test boiler for the Royal Navy was still uncertain.

[346] To be addressed in Sect. 8.2.2.

Fig. 6.44 The Deal: Brown Boveri 2000 hp gas turbine at RAE Farnborough, 21-stage axial compressor PR 4.2, five-stage axial turbine, picture past Nov. 1941 shows the remaining 20 C-stages © ABB Archive

tunnel of 1928[347] did not have a large enough air supply to test at critical Mach numbers, this range was left open.The operational history of the 2000 hp gas turbine from Brown Boveri in England could be resolved, thanks to the tenacious search of the late John Dunham in the RAE archives still in November 2000, and by additional information from Frank Armstrong in June 2009. Here an excerpt of these former NGTE/RAE officials's feedbacks:

'According to notes written in 1976 by Raymond Howell, former head of RAE, the unit was regarded by Brown Boveri as commercially very confidential, therefore no details (except for installation) were given to RAE, and no information about it was given to Whittle or Metropolitan Vickers. The contract for this unit was placed on 8 November 1938 with Richardson Westgarth and acceptance tests were run successfully at Baden from 11 to 16 August 1939. First run at RAE was on 3–6 May 1940, but due to problems with a brake dynamometer (no BBC part), *real running under load did not take place until 1 August 1941. By this time, however, the RAE research team's own experimental compressors were providing considerably higher pressure ratios per stage, and much higher gas temperatures were being used in the combustion and turbine components. Operation of the BBC unit was therefore limited to the gaining of some practical running experience.*

---

[347] See Sect. 4.2.2. Acc. to—see Bailey, The Early Development, p. 13, it was not until 1944 that a high-speed wind tunnel was available for studying compressibility effects in compressor cascades. In his view, *'This lack of appreciation of an important development of the late 1920s era, the cascade tunnel, was a major setback to the development of the axial compressor in this early critical period.'* From German perspective the scientific underpinning of turbomachinery design know-how at the responsible AVA Goettingen appeared even more deplorable, where even the essential clarification of the degree of reaction impact on axial compressor performance was omitted, an investigation which Seippel had initiated at Brown Boveri as early as 1933.

After about 70 h of running, a first stage compressor blade failed in Nov. 1941, and as a result the first stage was removed entirely. In late Sept. 1942 the turbine casing became too distorted to run further without considerable modification, and in 1943 the partly disman-tled unit was sold back to the suppliers,' i.e. Richardson Westgarth.[348]

The present timing underlines that the gas turbine sales deal had been the second best outcome after the failed license negotiations at Baden in May 1938 by Constant and Farren. The relatively short time of just one year between order intake at Richardson Westgarth, Hartlepool and the gas turbine delivery at Farnborough still in 1939 suggests that the unit belonged to the BBC Velox series production types of that time. The timing also illustrates that the UK order was executed at Baden with highest priority. The famous certification run of the first Neuchâtel power generation gas turbine by Professor Stodola took place on 7 July 1939, Fig. 6.47(l), immediately followed by the acceptance tests for the UK GT unit between 11 and 16 August 1939. And, while the Neuchâtel engine was thereafter at display—with daily full-speed demonstration runs—at the 1939 Swiss National Exhibition *'Landi'* Zurich, the UK GT was prepared for shipment by rail through France, presumably to one of the channel ports, Le Havre or Rotterdam.[349]

After arrival at Farnborough still in October 1939, the unit rested—based on the foregoing information—unused for 22 months, which indicates a serious interruption of the whole development progress in parallel to the outbreak of the War. Even 60 years later Dunham's wording about the *'limited use'* of the BBC GT unit appears to indicate a certain change in attitude towards an object which two years earlier was still desperately required. This impression is underlined by Dunham's earlier report from September 2000:

I found an internal RAE Engine Dept. report dated Sept. 1939, by W.L. Taylor and A.D. Baxter, which is a formal report on the acceptance trials carried out in Baden on 11–16 Aug

---

[348] The fate of the BBC UK GT doomed already before delivery, when Constant was quoted in the ESC Minutes, *'... that the Brown Boveri axial flow compressor which has been ordered by the Air Ministry is designed for a pressure ratio of 4.2 to 1 with 21 stages and he believes that a higher pressure ratio will be obtainable by overspeeding.'* At that time the delivery to England was planned for June 1939; Fig. 6.44 shows only 20 compressor stages, so that the picture has to be taken acc. to this account after Nov. 1941.

[349] Visiting the *'Landi 1939'* between Sat 6 May and Sun 29 Oct. 1939 was considered a kind of national pilgrimage, expessed by the stunning fact, that 10.5 million tickets were sold, measured against a total Swiss population of close to 4.2 million at that time. There exists a silent movie which shows the first BBC gas turbine amongst a crowd of exhibition visitors, women with prams nearby, a BBC head-foreman in uniform, stepping forward, taking order from hatted management to start the engine (without generator), salutes—and buzzes it up to 3000 rpm, with an intermediate pan shot to the instrumentation board ... The GT presentation at the national exhibition delayed the actual installation of the Neuchâtel emergency power plant up to spring 1940. An impression of a Velox steam generator shipment—comparable to the UK GT unit—via Baden Railway Station in Oct. 1939 is shown,—see Eckardt, Gas Turbine Powerhouse, p. 164, Fig. 4.58 of that book.

1939 to satisfy the contract acceptance conditions and witnessed by two RAE staff (presumably the authors).[350] The running time was 8 hours, and the trials included operational tests (starting and stopping) and performance tests, and were followed by a strip inspection. The report was favourable. Starting needed great care, but the running was free from vibration and noise, and the performance guarantee of 15% efficiency was met. The mechanical condition after the test was generally good.

Unfortunately, I found no report of the tests in Farnborough, which Constant's note ... says took place. In another progress report on the RAE gas turbine programme by Constant, dated August 1940, he quotes the compressor and turbine efficiencies of the BB unit, and speculates that they may be low because the blading was not of the free vortex type. But the efficiency figures he gave were quite different from those given in the Taylor/Baxter acceptance report, so it does seem that the unit was tested at RAE.[351]

Compared to Constant's rather grim assessment, Baxter's memories to his short trip to Switzerland in August 1939, just weeks before the outbreak of war hostilities reads friendlier—though also 'over-shadowed':

'The last leg of our journey was the comparatively short trip (from Basle) *to Baden but even this ended with another misfortune for me. My suitcase, which had seen better days, was creaking in all its joints and the long journey with the numerous lifts, and other none too gentle treatment which it had to suffer, proved too much. Just as we descended from the train at our destination, the handle broke and the case swung sideways and downwards. It first hit my knee so that I fell off the step on to the platform and it then crashed alongside me. This sensational descent was hardly noticed by my companions who were ahead and already looking for the Brown Boveri car which was to take us to the hotel. It was, however, observed by other passengers and by the station porters. They quickly rushed to my assistance, lifted me from my undignified sprawl on the ground, dusted me down, and revived me with a powerful schnapps, which seemed to appear from nowhere. Fortunately the hinges and locks on the case had survived and I was able to gather the case under my arm and proceed, slightly dazed, and with expressions of thanks in my limited French and German.'*

And, ...

'My first duty on arrival was to sit down with the chief test engineer, Friedrich Stu(e)rmer, and examine the readings and test results of the previous day's running. These were all presented in the engineering units generally used in Britain, especially in steam machine practice. This

---

[350] Apparently John Dunham was then not aware of A.D. Baxter's book *'Professional Aero Engineer...'*, first published in 1988, and which contains on p. 181 f. a detailed, colourful report of his mission to Baden, together with RAE's Head of Engine Dept., W.L. *'Jock'* Taylor and by the Chief Engineer of Richardson Westgarth, F.R. *'Foxy'* Dean, later assistant to W.S. Farren at RAE. A few sections from Baxter's book will be quoted in the following.

[351] Remarkable, that Dunham then still kept the component efficiencies secret, but they must have been close to those of the 4 MW Neuchâtel gas turbine with 23-stage compressor and seven-stage turbine, measured under the supervision of Aurel Stodola on the same test stand, just four weeks earlier: overall thermal efficiency 17.38%, compressor efficiency ad. 84.9%, turbine efficiency ad. 88.4%,—see Eckardt, Gas Turbine Powerhouse, p. 185.

puzzled me because all the readings had been taken with Continental meters marked in degrees Centigrade, kilogram weights, and so on. In the British aeronautical world it was usual to work in the Centigrade system of units, and I found that it was necessary for me to convert the Swiss figures into this system to appreciate their significance. When the second day's test results appeared in the same form, I enquired why this was done. Sturmer then told me that it was intended for my benefit; he believed that I would only understand the old fashioned clumsy British system of measurements and so he had spent hours in the evening converting all the day's readings and calculations from Centigrade to Fahrenheit! When he realised that I had to convert them all back again, we got on much faster.'

Dunham's statement that RAE's compressor development progress had finally outperformed Brown Boveri's PR 4.2 design with approximately 85% adiabatic efficiency is certainly true, but will be checked more closely later[352]; at the end of 1939, when the Brown Boveri GT unit arrived at Farnborough the actually achieved total pressure ratio of the RAE *Betty*[353] *B.10* compressor with nine stages marked PR $\sim$ 2.0 only, and also the follow-up 17-stage *Doris D.11* compressor design—all planned for turboprop configurations—was roughly still on the level of the BB unit. It was only the switch to pure turbojet concepts, first time realised for test compressor *Freda*—on the way to the first successful axial turbojet engine *Metrovick F.2* in 1943, Fig. 8.26, when the approach of Brown Boveri, Baden, was significantly bypassed. At this time, however, the centre of gravity of turbojet-oriented developments had also been shifted on Brown Boveri's side from Baden, CH, to Mannheim, Germany, Fig. 8.25, as will be addressed in detail in Chap. 9—with reference to the chief designer Hermann Reuter.

6.3.5 London Calling . . .

The end of 1938 nearing, the threat of war became more and more real—and the ambiguous memories of the International Lilienthal Conference, Berlin still in mid-October of that year faded rapidly. Against all odds, a few fought the apparently inevitable—and the longest aero ties between Germany and Britain in existence, established during the 1920s and early 1930s among the Wasserkuppe glider community, appeared to last longest: Alexander Lippisch (1894–1976) and Beverley Shenstone (1906–1979) had, as shown on Fig. 4.40, worked jointly to develop the *'Delta I'* tailless aircraft during winter 1930/1931 at the

---

[352] See Sect. 8.2.2, Fig. 8.25.

[353] As seen in the context of Fig. 8.25, six or more of the RAE compressor rigs had girls' names; for *Betty* the association with A.A. Griffith's first daughter, born in Sept. 1926, might be obvious; tragically she died in 1946 when an undergraduate at King's College,—see Rubbra, Alan Arnold Griffith, p. 118.

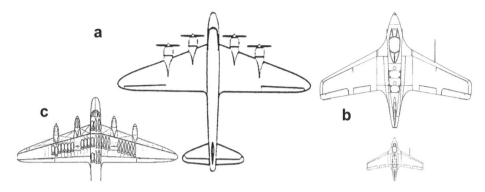

Fig. 6.45 *'Delta'* wing heritage: (**a**) Supermarine *Type 316,* 4-mot. bomber 1936–1940, (**b**) Messerschmitt *Me 163A 'rocket interceptor'* 1941 (+ in *Type 316* scale), (**c**) *Type 316* tanks/bombs arrangement

'Wasserkuppe' glider centre. They had met again at DFS Darmstadt in 1938, before both attended the Lilienthal Conference.[354] During this visit Lippisch had presented openly his most recent scientific achievements, a series of high-speed flow visualisation movies which he had recently taken in the new DFS smoke-generator wind tunnel.[355]

At this time Shenstone was employed as a senior scientific officer for the director of civil aviation in the Air Ministry, but before he had been—up to Reginald Mitchell's premature death, aged 42 on 11 June 1938—responsible chief aerodynamicist on the *Supermarine B.12/36* proposal of a British prototype four-engine heavy bomber design that was destroyed by a *Luftwaffe* air raid before completion on 26 September 1940 on Supermarine's Woolston works. The Mitchell/Shenstone design, the *Type 316*, Fig. 6.45a, c,[356] was a single-spar, mid-wing aircraft; the leading edge was swept back but the trailing edge was straight. Bombs were carried in both the wings and the fuselage, while fuel tanks strengthened the wing leading edge by a torsion resisting 'D' shaped box—familiar to glider designers.[357] In January 1939 Lippisch's design bureau left DFS and was integrated as *'Department L'* to the Messerschmitt works, Augsburg—on the way to develop the Messerschmitt *Me 163 'rocket interceptor'*, Fig. 6.45b, with flight testing

---

[354] See Sect. 12.2, 'List of Selected Participants'—with Shenstone's correct Air Ministry affiliation.

[355] See Cole, Secrets of the Spitfire, who has also that Captain Joseph *'Mutt'* Summers (1904–1954),—see Wikipedia, 'Joseph Summers' in English, chief test pilot of *Vickers-Armstrong* and *Supermarine,* accompanied Shenstone on that trip to Germany, but no further details are known in this context. There, Cole reports another unusual German-English encounter, regretfully without a date: *'Claudius Dornier .. flew to Mitchell's Supermarine works at Woolston, Southampton, and moored his flying boat on the river Itchen outside the factory. There may have been a* "Mr Mitchell I presume" *moment when Dornier walked up the slipway to meet the world's greatest, Schneider Trophy winning, float plane designer.'*

[356] Courtesy to John Shelton, 12 October 2020.

[357] See Wikipedia, 'Supermarine B.12/36' in English and—see Shelton, R.J. Mitchell.

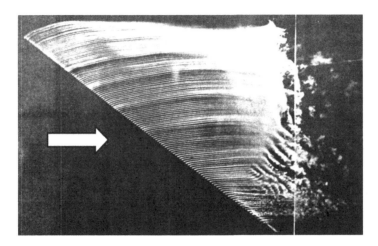

Fig. 6.46 Lippisch movie presentation at RAeS London, 15 Dec. 1938: Lam./turb. airflow transition on swept/delta wing, DFS low-turb. smoke wind tunnel, Zeiss high-speed camera with 3000 frs./s

during August/September 1941 and subsequent speed record setting of 1004.5 km/h, at Peenemuende-West, piloted by *'Heini'* Dittmar on 2 October 1941.[358]

As illustrated, Alexander M. Lippisch was a unique pioneer of aircraft aerodynamics who made important contributions to the understanding of tailless aircraft, delta wings and the ground effect. He developed and made all concepts with delta wing to function practically in supersonic's delta wing fighter aircraft as well as in hang gliders. His most famous design is the Messerschmitt Me 163 rocket-powered interceptor; people, he worked with, continued the development of the delta wing, especially the supersonic flight concepts over the twentieth century.[359] One of the earliest (subsonic) applications of the delta-concept abroad was certainly the Supermarine Type 316.

Before the *Orcus of War* opened, ending finally the dream of an international aeronautical scientific community, Shenstone invited Lippisch to come to London, who accepted to provide a presentation at the Royal Aeronautical Society on Thu 15 December 1938, Fig. 6.46,[360] in which he demonstrated his recent work in wind tunnel development, smoke flow, scale effects and wing design. Lippisch was unique in developing and

---

[358] See Wikipedia, 'Messerschmitt Me 163 Komet'in English.

[359] Examples are supersonic transport aircraft (Tupolev Tu-144, British-French Concorde), winged re-entry vehicles (US Space Shuttle Orbiter), hypersonic airbreathers, modern fighter aircraft (e.g. Eurofighter Typhoon, Dassault Rafale, Saab JAS 39 Gripen, most current US and Russian fighter aircraft with hybrid wings) and Unmanned Aircraft Systems (UAS) with very high agility and at the same time low Radar and IR signature; for an excellent survey,—see Hirschel, Separated and Vortical Flow.

[360] See Lippisch, Erinnerungen, p. 191.

applying the smoke wind tunnel,[361] and in 1938, he was the first in his making of super-slow motion film footage of wind tunnel smoke flows, using the then brand-new, spring-wound Zeiss-Ikon '*Movikon 8*' motion-picture camera for documentation, then operated for Lippisch exclusively by the Zeiss' professional specialist '*Herr Koenig*'. Figure 6.46 shows a single frame of the London lecture, taken out of a high-speed recording sequence of 3000 individual frames per second, mainly illustrating the length of stable laminar, parallel streamlines from the smoke injection spots along the (thin) delta wing leading edge over the delta-wing surface, before they are getting '*hazy*' and turbulent, and finally, the flow shows signs of break-off separation in the lower, right-hand corner. A unique feature of Lippisch's work was also the development of periodic smoke injection to produce '*time lines*' in the test section. All this gave him unrivalled '*real*' airflow analysis '*at a glimpse*' and led him to state, that in his view, mathematical assumptions would always tend towards an approximated solution. Physical findings mathematically expressed as visible outcomes of wind tunnel testing were Lippisch's great leap forward. He himself described the advantage of this approach: '*The value of the smoke tunnel consists chiefly in demonstrating the effects of a given body on the flow and to show the true course of flow in the case of special devices. Without carrying out tedious force measurements we are, therefore, in a position to develop aerodynamically good designs or to observe a special flow phenomenon which enable us, for instance, to determine the true causes of stalling effects.*' There was, evidently, no serious consideration given then to obtaining quantitative data from such experiments.

It is remarkable to observe that this integral approach of flow visualisation was also followed and even trumped by the next generation of German all-wing aircraft designers, the then 23- and 25 years old Horten brothers, Reimar (1915–1994) and Walter (1913–1998). No amount of small-scale testing in a wind tunnel could in the 1930s have revealed the localised effects of design features. This was why the Hortens were adamant as had been Lippisch on actual flight testing in real air that avoided the '*scale effect*' of wind tunnel work, and preferred instead e.g. to monitor tuft-testing on an all-wing model in flight, from a following photographic chase aircraft.[362]

Beverley Shenstone had been instrumental in getting Lippisch to the RAeS[363] and acted, on the basis of his good technical German, also as his interpreter. At the end of Lippisch's London lecture, much cordiality was expressed, and the desire for scientists to work together for peace rather than war; the audience rose for standing ovations. From the RAeS-documented discussions afterwards, there is known that British aviator and tailless aircraft designer Geoffrey T.R. Hill (1895–1955) attended, then Professor of Engineering

---

[361] See Mueller, On the Historical Development of Apparatus and Techniques for Smoke Visualization. Interestingly in 1932, also William S. Farren used a modified smoke tunnel, originally developed at the NPL National Physical Laboratory, for his lecturing at Cambridge University.

[362] See Wikipedia, 'Brüder Horten' in German,—see Cole, Secret Wings, p. 82 and https://www.youtube.com/watch?v=QD3JWVSkDuU

[363] Shenstone became Fellow of RAeS in 1942, and was RAeS President 1962–1963.

Science at London University, as well as Professor Sir B. Melvill Jones (1887–1975), known for his 1929 seminal RAeS-paper *'The Streamline Airplane'* and Dr. Herbert C.H. Townend (1896–1943), in 1929 inventor of the *'Townend Ring'*, a narrow-chord cowling ring round the cylinders of a radial engine to reduce drag and improve cooling. Besides many British aircraft, also the mid-engine of the *Junkers Ju 52/3 m* carried this characteristic feature.

While the collaboration between RAE and Brown Boveri might have been initiated by Sir David R. Pye in spring 1937, there is no doubt that the following invitation to Brown Boveri's General Manager Adolf Meyer (1880–1965) for lecturing at London's traditional IMechE Institution of Mechanichal Engineers[364] came from Meyer's Swiss countryman, Metrovick's chief mechanical engineer Karl Baumann (1884–1971), Fig. 6.40 (m). They were both university assistant colleagues at Zurich in the 1904–1906 timeframe, Baumann with Stodola, and Meyer, who had studied electrical engineering, thereafter at the chair for Machine Design and Mechanical Engineering at the Polytechnikum Zurich with Professor Albert E. Meyer-Schweizer.[365] One can assume the early acquaintance of Baumann and Meyer was instrumental, even in view of a principle competition between both Companies, to establish with the benevolent involvement of future, 1952 IMechE President Sir David R. Pye, Fig. 6.31 (r), direct contacts between the RAE axial compressor designers and BBC Baden from 1937/8 onwards. It also prepared the platform for Meyer's seminal lecture on Fri 24 February 1939 at the Institution's headquarters, purpose-built in 1899 on 1 Birdcage Walk in central London, Fig. 6.47(m). The described, long-time acquaintance of Swiss engineering with Great Britain as motherland of engineering determined also Meyer's IMechE lecture. It is at best illustrated by the outfit of the 80 year-old Professor Stodola, while commanding the certification test run of Brown Boveri's first 4 MW power-generation gas turbine on Fri 7 Juli 1939 at Baden, Fig. 6.47, l—with frock coat and

---

[364] IMechE maintained strong ties to the Continent back to its foundation at the Queen's Hotel, Birmingham in 1847, where the German—see Wikipedia, 'Charles Beyer' in English, born as Carl Friedrich Beyer at Plauen, Vogtland in 1813, celebrated German-British locomotive designer and builder, belonged to the group of founders, besides the normally named, renowned *'Father of Railways'* George Stephenson (1781–1848). Brown Boveri and its founders maintained continuous connections to IMechE: Charles Brown sen. joined IMechE in 1879, his son BBC co-founder Charles E.L. Brown was elected a Foreign Member of the Institution in 1892 and became a regular Member in 1911, while Adolf Meyer received the IMechE *'George Stephenson Research Prize'* for *'best achievement of the year'* in 1939 and 1942. In 1954, Baumann became IMechE 'Honorary Fellow', an honour received together with Prof. Dr. Aurel Stodola 1911, Sir Charles Algeron Parsons 1925, Orville Wright 1942, Air Commodore Frank Whittle 1944 and Lord Hives of Duffield 1953.

[365] The *Polytechikum* became *Eidgenoessische Technische Hochschule* (Federal Technical University) in 1911. Meyer was awarded a Honorary Doctorate in 1935 from Stevens Institute of Technology in Hoboken, NJ, USA and in 1941 from ETH Zurich, while Baumann received his ETH Honorary Doctorate in 1951, presumably again with a strong recommendation from Brown Boveri's side.

Fig. 6.47 *'Auld acquaintance':* Prof. A. Stodola (80, l) with bowler hat, first GT test Baden, 7 July 1939, and A. Meyer (59, r), first BBC GT lecture, 24 February 1939 at IMechE, London (m) © IMechE, ABB Archive

traditional bowler hat of the senior engineer, a kind of salute to the founding fathers of the industry, typified by Charles Brown sen. and Sir Charles A. Parsons.[366]

Meyer's presentation was entitled *'The Combustion Gas Turbine: Its History, Development and Prospects'*[367] and addressed broadly gas turbine usage for standby and peak load power generation, supercharging for chemical processes (*Houdry process*), the gas-turbine electric locomotive, ship propulsion, wind tunnels, blast furnace plants, gas turbine and steam plant combined (!) and has a *'glimpse into the future'* which contained recuperator and sequential combustion GT modifications,—but there is nothing about flight applications.

Meyer provided no absolute turbo-component efficiency numbers in the main section of the paper, but answered corresponding questions in the discussion openly, with typical *Houdry turbo-set* data of PR 4 and $\eta_{ad, C}$ 0.845 and $\eta_{ad, T}$ 0.86 of a Mr. H.O. Farmer and Engineer-Commander W.G. Cowland, who could be identified as *ESC-member 1939,*[368] so that there could be no doubt about the expectable performance of the UK GT on order from Brown Boveri. David M. Smith, Metropolitan-Vickers asked about the relative importance of compressor and turbine efficiencies, and received Meyer's confirmation that achievable turbine efficiencies should surpass those of compressors by about two points. In addition,

---

[366] See Eckardt, Gas Turbine Powerhouse, p. 183 f.

[367] See Meyer, The Combustion Gas Turbine. The author owes the valuable discussion section of that IMechE Proc. paper version, courtesy to Ronald Hunt, 31 Jan. 2012.

[368] The ARC ESC Engine Sub-Committee members in 1939 were—Sir H.T. Tizard (Chairman),—Prof. L. Bairstow, Mr. E.S.L. Beale, Major G.P. Bulman (Air M.), H. Constant (RAE), Eng.-Comm. W.G. Cowland (R.Navy, subm.prop.), Prof. A.Ch. Egerton, W.S. Farren (Air M.), Major F.M. Green, D.R. Pye (Air M.), H.R. Ricardo, W.L. Tweedie (Air M.).

Smith addressed the recent publication in *'Engineering'* of the Hungarian engineer György Jendrassik on an all-axial 100 hp gas turbine with heat exchanger.[369]

A highlight of Adolf Meyer's lecture was undoubtedly the prepared discussion contribution of Frank Whittle at the very end. John Golley knows that Whittle had attended together with Mogens L. Bramson and Leslie Cheshire.[370] Besides Whittle and friends, one can speculate with some probability that A.A. Griffith, H. Constant, W.S Farren, H.T. Tizard, H. Roxbee-Cox and A.T. Bowden might have attended that historic meeting as well, but no attendance list prevailed at IMechE. Otto Zweifel (1911–2004), inventor of the *Zweifel number* for optimum turbine blade count,[371] participated for Brown Boveri, who stayed in England at that time anyhow.

Before Whittle entered the discussion, Commander Cowland had already set the tone, after congratulating Meyer *'on his ability to interest engineers in the possibilities of the gas turbine as an alternative to existing prime movers for almost all purposes where large power is required. It has to be noted, however, that he did not suggest it for aircraft!'* From where the discussion write-up continued: *'Squadron Leader F. Whittle, R.A.F. (Rugby) remarked that so far little had been said about the gas turbine in relation to aircraft, which he regarded as a most hopeful field for it, for, as the gas turbine was taken up into the air the reduced atmospheric temperature made possibly very much higher efficiencies than could be obtained on the ground.'* Giving an example, he made his point: *'With a turbine efficiency of 85 per cent and compressor efficiency of 70 per cent* (sic!—apparently a centrifugal design), *a maximum temperature of 1100 deg. C abs., and an adiabatic temperature rise in the compressor of 200 deg. C the overall efficiency would increase by 46 per cent, and the power per pound of air per second by 66 per cent on rising from sea level to 35,000 feet.'*[372]

Meyer commented David Smith's discussion note with respect to a strive for higher turbine temperatures: *'Air cooling might be ... found ... interesting, and a cooling system of this kind, where the boundary layer of gas around the blade was continuously replaced by fresh and comparatively cold air supplied to the hollow blade and emitted through a slit along the leading edge promised to give good results'*—echoing Brown Boveri's blade cooling test for Rolls-Royce in 1938, as illustrated in Fig. 6.37. He denied the question for a heat exchanger installation, since the small number of working hours for the (Neuchâtel) emergency powerplant would not justify such an investment. However, *'An experimental gas turbine set, burning pulverised fuel* (coal), *which had been run for several months at*

---

[369] See the following Sect. 6.4.4 and—see Engineering 1939, Vol. 147, p. 186.

[370] See Golley, Whittle, p. 122. Mogens L. Bramson and especially his positive—see Meher-Homji, Bramson Report, 1935, laid the first solid financial ground to the Whittle turbojet project; Leslie Cheshire was a senior engineer of Whittle's team, on lease from BTH.

[371] See Eckardt, Gas Turbine Powerhouse, p. 117 f.

[372] While Meyer's paper with attached 9 discussion contributions has been reproduced in the Proceedings of IMechE already in June 1939, Whittle's discussion alone got the special attention of being reproduced still in—see Smith, Gas Turbines, 3rd ed., April 1944, p. 60 f.

Messrs. Brown, Boveri's testing plant, had given such encouraging results that the firm decided to build a bigger experimental set of about 2,000 h.p.' After this concept of a 1.6 MW coal-dust-fired gas turbine had been investigated by Walter Noack up to 1943 at Baden, this keen idea re-emerged on the German side as a joint BBC/AVA *'project of dispair'* during 1944 for a means of long-distance *'America bomber'* propulsion, discussed in Sect. 8.3.5.

While Meyer responded to all other discussion notes openly and with precise terms, he apparently struggled with the vision of an aero gas turbine at that time: *'Squadron Leader Whittle's remarks on the possibilities of the gas turbine as a prime mover for aircraft seemed rather optimistic even to himself—and he was not very bashful—but Squadron Leader Whittle no doubt knew much more about this field of application, so that one might hope that his forecast would soon come true.'* As Golley commented this *'Clash of generations'*, nearly 15 years after Max Koenig's courageous appearance at NECIES Newcastle,[373] and the then BBC prophecy almost coming true, but now with reversed *'Battle formation'*: *'Whittle was, of course, careful to avoid any reference to its application as a jet propulsive device, or to give any indication of the work he was doing.'* And regretfully, Meyer—in the meantime 59—lacked the intuition that the time, when his heavy machinery adapted to wings, was just round the corner. Perhaps the general outlook of the future ahead was too dark—in the spirit of the last lines of Baxter's book:[374] *'Because I started this story with a broadcast message from the Prime Minister, I should like to end it with part of one made a few months later by the King, who quoted:*

> I said to the man who stood at the gate of the year: Give me a light that I may tread safely into the unknown,—and he replied: Go out into the darkness and put your hand into the hand of God. That shall be to you better than light and safer than a known way.[375]

---

[373] See Chap. 3, Excursion I: Max König (1893–1975) and Claude Seippel (1900–1986).

[374] See Baxter, Professional Aero Engineer, p. 188.

[375] See Wikipedia, 'The Gate of the Year'in English. The poem by Minnie Louise Haskins (1875–1957), was written in 1908 and privately published in 1912. It caught the public attention and the popular imagination when King George VI quoted it in his 1939 Christmas broadcast to the British Empire, then thought to be brought to his attention by his wife, Queen Elizabeth, the Queen Consort. However, in the book *'The Servant Queen and the King She Serves'*, published for Queen Elizabeth II's 90th birthday, and with a foreword by the late Queen, says that it was the young Princess Elizabeth herself, aged 13, who handed the poem to her father.

6.4   Early Turbojet Developments in the USA and Other Countries

6.4.1   Pioneering US Turbojets: A Difficult Delivery

Before and during the Second World War the USA lagged behind Europe in the turbojet development by some five years—and not for technological reasons. Even when the principles for the turbojet were appreciated, their usefulness was in parts wrongly interpreted or the development was held back by lack of official backing or in due course, a lack of technical interchange between the companies and institutions that were working on the aero gas turbine projects. A complicated story which—in principle comparable to the highly emphasised, in-depth historical analysis of the early works of Whittle and von Ohain—has been addressed in specialised literature,[376] with facets actually beyond the scope of this book, which focuses rather on interrelations within the various European developments and the closing of gaps in the established historic narrative. For a first insight in the accumulated reasoning it is sufficient to quote Schlaifer's summary in this context:

> Development of the turbojet began almost simultaneously in Britain and Germany in 1935–1936. All its elements were old, and even the precise combination of elements was patented over a decade before development began.[377] In both Britain and Germany the actual beginning of development was due to the enthusiasm of engineers who were not employed by the aircraft engine industry, and in both cases the projects were backed by capital which came from neither the aircraft engine industry nor the government. In both countries this private work first awoke the interest and obtained the support of government, and it was still later that the engine industry began to take a hand. In the end, however, development of turbojets in these two countries to full readiness for service was accomplished only by experienced builders of conventional engines.
>
>   The United States was five years behind Britain and Germany in undertaking the development of turbojets, but it would seem that no particular significance can be attached to this fact: the actual beginning of work in those two countries depended on chains of accidents which simply did not happen to occur in the United States. The development could have technically have begun in all three countries years before it did. The fact that the United States was still behind in turbojet development in 1945 was due in parts to its late start, and the record of General Electric's development of turbojets starting from British data in 1941 is nearly as good as the British record itself. In part, however, it was due to ill-advised policy of the services, particularly the Navy, which prevented any interchange of information among the interested firms.[378]

---

[376] See e.g. Schlaifer, The Development, especially Chaps. XII–XVII, pp. 321–508;—see Constant, The Origins, especially Chap. 6 'National Patterns in the Pursuit and Utilization of Scientific Knowledge', pp. 151–177; for overall information,—see Kay, Turbojet, Vol.2, Part IV, United States of America, and—see St. Peter, The History, and for production issues,—see Giffard, Making Jet Engines, Chap. 2, p.134 f.

[377] Approximately true for the Whittle turbojet development, the corresponding timing for the von Ohain and Wagner cases was considerably faster.

[378] See Schlaifer, The Development, p. 493.

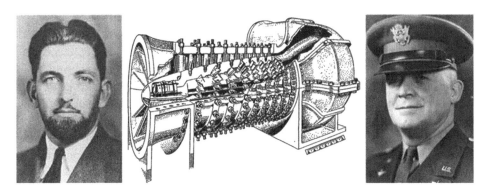

Fig. 6.48   United States turbojet engine pioneers I: Eastman Jacobs (1902–1987) l, NACA Langley eight-stage axial compressor (m), General Henry H. Arnold (1886–1950) r

The confusion on the US side in the early 1940s, which is indicated above by terms as *'the usefulness of turbojet principles* (was) *in parts wrongly interpreted'* and *'ill-advised policy of the services'* camouflages a fundamental conflict about the best engine concept—on NACA's side, a motorjet, suggested by their leading scientist Eastman Jacobs,[379] Fig. 6.48 (l), and therefore also called *'Jake's Jeep'*, while very few, informed insiders tried to propagate under extreme secrecy the Whittle engine, the transfer of which had managed General Henry H. Arnold,[380] Fig. 6.48 (r), after a visit in England in April 1941. Arnold personally inspected the Whittle engine several weeks before its first flight and arranged to have General Electric's Supercharger Division at West Lynn, MA take on the further American development of the engine prototype. In addition, Arnold selected Bell Aircraft of Buffalo, NY, to work concurrently on an airframe for a fighter, what became Bell's *P-59 Airacomet* twin-engined fighter aircraft. The first flight happened in record time on 1 October 1942, powered by GE's copy of the *Power Jets W.1X* engine of 550 kp thrust, later produced as power-upgraded *GE J31*. This date lay well before either the production Whittle engines or the Gloster Meteor itself had flown.[381] While the P-59 was not a great success, the type did give the USAAF experience with the operation of jet aircraft, in preparation for the more advanced types that would shortly become available.

Extreme project secrecy caused some kind of factual disinformation amongst the US stake-holders which started already a year before, when Sir Henry Tizard, in his new role as science advisor to the British Ministry of Aircraft Production, visited the United States in September 1940. Although it is well known that Tizard informed his American allies of British technical advances in radar and opened discussions concerning British cooperation in the development of atomic energy, he also brought with him the first news of the British developments in the new field of jet propulsion. Tizard met with both the NACA Chairman

---

[379] See Wikipedia, 'Eastman Jacobs'in English.

[380] See Wikipedia, 'Henry H. Arnold'in English.

[381] See Constant, The Origins, p. 222.

(since 1938) Vannevar Bush and the NACA Director of Aeronautical Research George W. Lewis. Apparently, Tizard revealed very little except the seriousness of British efforts in jet propulsion. As Bush later recalled: *'The interesting parts of the subject, namely the explicit way in which the investigation was being carried out, were apparently not known to Tizard, and at least he did not give me any indication that he knew such details.'*[382] Especially, this meant, that it remained unclear if the British generous technology offer was based on an axial or a centrifugal compressor concept.

In this situation, Eastman Jacobs completely on his own began research in earnest, unaware of any other projects due to the government *'compartmentalising'* them, meaning that research was conducted independently with each party being unaware of what the others were doing. Jacobs returned to what he had learned about a ducted-fan engine design in 1935, when he took part as lecturer at the Volta Congress in Rome.[383] The engine concept had been first conceived in about 1931 by the Italian Secondo Campini (1904–1980), a hybrid scheme of a conventional, ducted piston engine in combination— on demand—with a fan-bolstered afterburner stream. The Campini engine had a two-stage centrifugal compressor, which Jacobs replaced by a two-stage axial fan arrangement. After the maiden flight of the *Caproni Campini N.1*[384] in August 1940, and some experimental flight speed tests up to 515 km/h,—on 30 November 1941, the second prototype was flown from Milan to Rome's Guidonia Airport, in a highly publicised event that included a fly-past over Rome and a reception with Italian Prime Minister Benito Mussolini.

Similarly, Jacobs designed an engine and airframe that came to be known as the *'Jeep'* in accordance with the motorised vehicle of the US Army for which the term meant something that could basically go anywhere and be adopted to just about any purpose. For his aircraft Jacobs utilised a 850 hp *Pratt & Whitney PR-1535 Twin Wasp jr.* piston engine, driving a gas turbine of his own design, consisting of a two-stage axial fan with additional duct afterburner, heat exchangers, etc.

A side effect of this—in the end—vain effort, put Jacobs nevertheless in the position of an US axial turbojet pioneer. As early as 1938, NACA recognised—later also in search of an efficient *'Jeep'* compressor component—that high efficiency axial compressors could be designed by the application of aerodynamic principles. And a look in the corresponding literature list reveals, that three Swiss publications of that time by Keller, Meyer and

---

[382] See Parker Dawson, Engines and Innovation, p. 46.

[383] E.N. Jacobs' presentation was entitled: *'Methods Employed in America for the Experimental Investigation of Aerodynamic Phenomena at High Speeds';* he presented wind-tunnel tests at high subsonic speeds, noted extremely large increases in drag beyond certain Mach numbers and showed for the first time Schlieren pictures of a shock wave propagating over an aerofoil surface.

[384] See Wikipedia, 'Caproni Campini N.1' in English.

Seippel delivered decisive stimulations.[385] Eastman Jacobs and Eugene Wasielewski (1912–1972) began an investigation at Langley Field, VA, for the purpose of determining the performance of an axial-flow compressor based on the current information gained from extensive research on airfoils. Basically, Keller's theory was applied to the design and construction of an eight-stage compressor and preliminary tests were conducted in 1941/2. The compressor, Fig. 6.48 (m),[386] achieved a total pressure ratio PR 3.42, a volume flow of 3.65 m^3/s and an impressive adiabatic efficiency η_{ad} 0.87.[387]

In March 1941, Chairman of the NACA Vannevar Bush asked the then 82-year-old William F. Durand (1859–1958) to head a committee to study and develop jet propulsion for aircraft.[388] From industry, the committee was composed of members from General Electric, Westinghouse and Allis-Chalmers, and agreed early that the three companies would work separately developing jet engines to promote diversity in design. But while Durand and Bush in line with Jacobs unanimously supported the motorjet concept, General Henry H. Arnold stubbornly defended the finally successful turbojet introduction, which justifies his listing as US turbojet pioneer, Fig. 6.48 (r); in his own words:

> I was told in England in April [1941] that in ten years there wouldn't be any more poppet valves or, as a matter of fact, any type of gas [piston] engine as we now have in pursuit planes, and another five years would see the end of that type of engine in all types of aircraft. ... I do not believe that we are ready at this time to start a development program tending towards the production of the jet propulsion engine on the same scale as we now have for the conventional type of gas [piston] engine. I am of the opinion, however, that it will be much easier to reach the 4000 to 5000 horsepower with the jet propulsion and gas turbine than it will be with the conventional type of engine. Everything points in that direction. The turbine has everything to its advantage.[389]

After the first flight of the turbojet-powered *Bell 'Airacomet'* in October 1942, it lasted still up to mid-1943 before NACA also stopped Jacobs *'Jeep'* activity at Langley, VA. The end of his long-time project in sight, Jacobs journeyed to Great Britain, where he also saw the

---

[385] See Curt Keller, 'The Theory and Performance of Axial-Flow Fans: Adapted for the Use of Fan Designers by Lionel S. Marks with the assistance of John R. Weske', 1937; Weske, a student of Lionel S. Marks at Harvard University, had worked with his professor to produce the first axial-flow compressor based on the isolated airfoil design theory, an approach that later came into standard use by American designers. See Adolf Meyer, The Combustion Gas Turbine, 1939 and—see Claude Seippel, The Development of the Brown Boveri Axial Compressor, 1940.

[386] No reliable geometrical data could be found, except for the constant rotor hub diam. of 14 inches.

[387] See Sinnette, Performance of the NACA Eight-Stage Axial-Flow Compressor.

[388] See Munzinger, Duesentriebwerke, p. 13, outlines in this context that NACA acted then on demand of General H.H. Arnold, who had received already in 1940 US secret service information about turbojet engine development activities at Junkers and BMW, and due to their all-axial design had requested research along these lines, which automatically brought design work of Brown Boveri and its US licensee Allis-Chalmers into focus.

[389] See Parker Dawson, Engines and Innovation, p. 51

aerial photographs from Peenemuende—with the parallel scorch marks on the ground of the Me 262 jet exhausts' tracks. Jacobs also found out a good deal more than what he already knew about the Whittle engine and its application. With this new insight into the state of foreign turbojet technology came a new argument from frustrated Jacobs[390] in protest of the military's decision to kill his experimental airplane idea. In subsequent reports to the NACA about his visits, he observed that mistakes were repeated: *'the mistake of applying the new power plants to more or less conventional airplanes rather than giving careful consideration to essentially new extreme-performance types',* made possible through the use of the new turbojet engines. He feared that otherwise an imbalance between power plant and airframe might occur, as to make both propulsion and aerodynamics ineffective.[391] A problem, which on the German side, Helmut Schelp (26) and Hans Antz (29) had identified jointly as early as in October 1938,—and solved by a coordinated turbojet (Jumo 004, BMW 003) and advanced airframe (Me 262, He 280) development programme.

After the war and Jacobs' early retirement, the *'Jeep'* airframe/engine concept regained considerable importance for the further developments of US axial turbojet engines with afterburner.

6.4.2 BBC: At the Cradle of US Turbojets

The transfer of Frank Whittle's turbojet technology from the UK to the USA in 1941 on the initiative of US General Henry H. *'Hap'* Arnold, Chief of the Air Corps, is well known in line with the foregoing. Not so well known, however, are earlier, similar, independent US initiatives, and BBC's involvement with them. BBC's leading role in the area of the industrial gas turbine became acknowledged in the US at the end of the 1930s. In 1949, Schlaifer[392] opens his chapter on *'Aircraft Gas Turbines in the United States'* with the statement: *'... it was in Switzerland, specifically in the firm of Brown-Boveri et Cie, that the Velox boiler was developed at the beginning of the decade and the Houdry-process turbine in the middle of the decade. The first practical industrial gas turbines in the United States were built under Brown-Boveri license: these were Houdry-process turbines built by the Allis-Chalmers Manufacturing Company beginning in 1938.'*[393]

---

[390] In a form of—at that time—unusual *work-life balance* in mind, Jacobs (42) retired already in 1944—and started in 1958 a restaurant on his property at Malibu, CA, became known throughout the 1960s as *'Jake's Diner'*.

[391] See: history.nasa.gov/SP-4305/ch8.htm

[392] See Schlaifer, Development of Aircraft Engines, p. 443.

[393] Before BBC licensed Allis-Chalmers (AC) to produce Houdry axial turbosets beginning 1938, a first license out of BBC's Swiss home cantone Aargau for a *'turning plow'* had established the contacts to AC as early as 1910–1914 and contributed to the emerging line of AC's agricultural machinery. Turning ploughs had been developed in Switzerland at the beginning of the twentieth

Before, there were six *Houdry turbosets* delivered directly out of Baden, CH, since the turning of 1936/1937.[394] Allis-Chalmers Manufacturing Co. (AC) of West Allis, Wisconsin, USA is an American company known for its past as a manufacturer with diverse interests. The company eventually divested its manufacturing businesses and today is based in Houston, Texas, as Allis-Chalmers Energy. In the 1920s through the 1960s, AC Power House and Industrial equipment was very much competitive with industry giants like General Electric and Westinghouse. The company would also play a major part as a manufacturer in the World War II building pumps for uranium separation as part of the *Manhattan Project* and building—like BBC Mannheim in Germany—electric motors for submarines; AC also built triple expansion marine steam engines for *Liberty Ships*.

Slowly, by the end of the 1930s, it was beginning to be thought that the low limit on turbine inlet temperature was the last obstacle to be overcome, to the production of gas turbines with competitive efficiencies to steam or diesel engines. A good deal of work was done on internal turbine cooling, both with water and by air,[395] and alloy manufacturers were claiming much improved high-temperature strength for a number of new materials. In 1940, a committee set up by the American *National Academy of Sciences* reported that turbine operation at 800+ °C might soon be possible, even without extra-cooling. With compressor and turbine components of the Neuchâtel quality, an increase in turbine inlet temperature from 650 °C to 820 °C would raise the overall efficiency from Brown Boveri's measured 17.4% to 26%, or about up to 2/3 of the efficiency of the much heavier and more expensive diesel engine.

The Schlaifer quotation continues: *'One reason for the pre-eminence of Brown- Boveri in the gas-turbine field was undoubtedly that Brown-Boveri had led in the development of axial compressors, which finally gave efficiencies high enough (of the order of 85%) to make possible the construction of turbines with positive efficiency despite the use of the very low blade temperatures required for economical life. It was only at the very end of the decade that various American engineers began to write and speak of the possibility that gas turbines might be used as industrial prime movers.*

In 1939, as a result of the inspection of foreign gas turbines by a Navy officer, the Bureau of Ships requested the National Academy of Sciences to investigate the possibility of the use of gas turbines for ship propulsion. A special committee was set up in 1940 and reported in January 1941 that the promise of gas turbines for ship propulsion was very good indeed and that development should be begun at once. The report added, however, that gas turbines were completely out of the question for aircraft, since they would weigh at least 13 pounds per horsepower, against little over one pound for conventional aircraft engines. The error was due largely to the fact that no member of the committee had any

---

century; after finishing a furrow, the ploughman could change direction for the next furrow immediately by turning the plough-share.

[394] See Eckardt, Gas Turbine Powerhouse, p. 169 f.

[395] See Eckardt, Gas Turbine Powerhouse, p. 397 f.

knowledge of aircraft engine practice, either as regards lightweight construction or as regards the reduced durability which is acceptable in that field.'

It appears that *'The Committee on Gas Turbines'*, appointed by the (American) *National Academy of Sciences*, issued its apodictic judgement, different from Schlaifer above, already as early as 10 June 1940, notwithstanding in a very clear wording:

> *In its present state, and even considering the improvements possible when adopting the higher temperatures proposed for the immediate future, the gas turbine engine could hardly be considered a feasible application to airplanes mainly because of the difficulty in complying with the stringent weight requirements imposed by aeronautics. The present Internal Combustion Engine equipment used in airplanes weighs about 1.1 pounds per horsepower, and to approach such a figure with a gas turbine seems beyond the realm of possibility with existing materials.*

The *Committee*, apparently under the leadership of Professor Lionel S. Marks of Harvard, included renowned scientists like Theodore von Kármán and Robert A. Millikan of the California Institute of Technology, as well as Charles F. Kettering, head of GM Research. This negative classification of the gas turbine as quite inappropriate for aeronautical applications reflects and resembles the historically ill-reputed assessments of Dr. William J. Stern at the Air Ministry Lab and RAE 20 years earlier, and somewhat later by Edgar Buckingham of the NBS National Bureau of Standards for the US-side,[396] but the technological progress in the meantime was completely ignored; technology was already available to make aircraft turbines much lighter. Presumably Adolf Meyer, though honourable member of the US National Academy of Arts and Sciences from 1950 onwards, did not belong directly to this illustrious circle. However, he certainly might have influenced the decision in that unfortunate direction by means of the Allis-Chalmers vote, as will be further outlined in the following. Consequently, the stark, overall positive assessment of Brown Boveri's influence on the introduction of the aero gas turbine might be put somewhat into perspective. This is underlined by Fig. 6.49, which shows a 20-stage axial compressor from Allis-Chalmers, presumably the first US product of that kind in serial production, still in the typical heavy mechanical-engineering design.[397]

In line with the committee decision, Allis-Chalmers had been given a *Navy* contract for a marine gas turbine and not long thereafter, Westinghouse and GE took Navy contracts for

---

[396] See Chap. 2, <Sep. 1920>.

[397] See Lowell, Gas Turbine Design. This original BBC mechanical design is characterised by the general usage of roller bearings, specifically for the rough Houdry refinery operation and deviating from normal power generation design, based on journal bearings. Three rows of coupled ball bearings were installed for axial thrust compensation, see feature (X) at the high-pressure end (shown enlarged in centre space), which will reappear in German turbojet engine designs, Sect. 8.1.1 and Fig. 8.2 for Jumo 004.

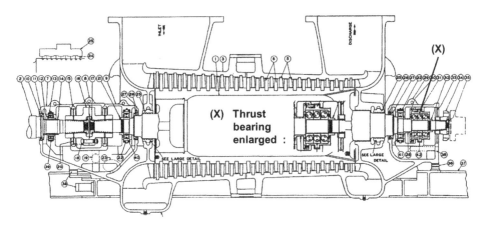

Fig. 6.49 Allis-Chalmers 20-stage axial compressor, PR 4.2, V 18.9 m³/s, rotor diameter $D_{1,Tip}$ 0.82 m, for Houdry gas turbine in BBC license production, ~1939,—with (x) thrust bearing enlarged

marine gas turbines as well. In addition, Schlaifer[398] continues *'Two if not all three of the companies were actively studying the possibility of gas turbines for locomotives since Brown-Boveri had built one in 1939.'*

After World War II, Allis-Chalmers also herited Brown Boveri's role as No. 1 provider of wind tunnel blowers in the United States; with some likelihood it can be assumed that AC adapted the BBC compressor facility after the *Kochel 1×1 m hypersonic windtunnel*[399] had been transferred from Bavaria to AEDC Tullahoma, TN, in 1946. Figure 6.50 (l) was reproduced in the 1954 *Brown Boveri Review* with the accompanying text *'Maximum output 510,000 m³/h with a pressure ratio of 3.2:1. The air flow is axially, the discharge connection is located horizontally on the right-hand side'*, but without any further clarification about the wind tunnel location. The quoted pressure ratio comes close to the estimation of the Kochel/Tullahoma lp front row compressors, and the compressor volume flow correlates perfectly with an interview between von Kármán and Seippel of June 1945.[400]

Theodore von Kármán himself became a potential Brown Boveri wind tunnel customer from early on. In the period 30 September until 6 October 1935, he attended the *5th Volta Congress* at Rome, dedicated to *'High Velocities in Aviation'*. The Volta conference was concluded with an excursion of all participants to Guidonia, approximately 30 km north-east of the downtown conference site at Villa Farnesina on the road to Tivoli, where the huge Italian high-speed test centre *'Citta dell' Aria'* (City of Airstream) took shape, with the highlight of the then largest 0.4 × 0.4 m, close circuit Mach 2 supersonic wind tunnel—

[398] See Schlaifer, Development of Aircraft Engines, p. 459 f. For the BBC locomotive gas turbine, — see Eckardt, Gas Turbine Powerhouse, p. 196 f.

[399] See Eckardt, Gas Turbine Powerhouse, p. 137 f.

[400] See Eckardt, Gas Turbine Powerhouse, p. 146 f.

Fig. 6.50 Post-war large wind tunnel developments 1946–1954: BBC casting of axial compr. casing, upper half (l), NASA Langley 7-st. axial compressor, 4 × 4 ft SS wind tunnel (r) © ABB, NASA Archives

with 13-stage axial compressor and 2.1 MW DC motor from Brown Boveri. The facility was to a large extent a 1:1 copy of Ackeret's Zurich facility, except for the drive power, which was more than doubled and expanded the researchable Re number range considerably. Th. von Kármán was so impressed by the Guidonia wind tunnel, that he contacted Brown Boveri immediately after the conference to discuss the acquisition of a copy of this wind tunnel for his Institute for Fluid Dynamics at the *California Institute of Technology*. Negotiations for Brown Boveri led their New York representative *Paul Sidler*, and after the whole deal was considerably delayed, he wrote to von Kármán on 2 May 1940, ascertaining Brown Boveri's unchanged interest: '*. . . So far the work at Baden has kept up on practically a peace-time schedule. They have large stocks of raw materials and have meanwhile organized the purchase of additional steel and other metal products in this country. The deliveries which Baden has quoted in recent times are hardly different from those that they offered during peace time and I am fully convinced that they could build the wind tunnel power plant with considerably shorter delivery than [. . .] other companies in this country.*'[401]

The wind tunnel project was further delayed after the War, then the 4 × 4 ft Supersonic Pressure Tunnel at NASA Langley, Va., was a hurried necessity and designing began in February 1945 with completion scheduled by the end of 1945. NACA launched the project of building a Mach 2 tunnel within 10 months. Contractors were hired for the mechanical design and actual fabrication of the tunnel, comprising with all likelihood also Allis-Chalmers. Near the end of 1945, the objective was near completion but a 2-year strike halted production of the compressor and operation did not begin until 1948. Many

---

[401] See Nickelsen, Theodore von Kármán, p. 170.

important military and space vehicle tests were conducted in the 4×4 ft. Supersonic Pressure Tunnel. Century Series fighters like the F-102 and F-105, the B-58 supersonic bomber, and the X-2 research aircraft were all tested here. It was such an asset to NACA that it underwent modifications in 1950 for new drive motors which increased the horse-power from 6000 to 45,000 (continuous) and 60,000 (intermittent), generating correspondingly by means of the gigantic seven-stage axial-flow compressor at PR 2 and 1300 rpm, a maximum volume flow of 1,461,149 m³/h, Fig. 6.50 (r).

As described in the foregoing section, in the first half of 1940 the strict opposition against GT aero applications changed and encouraged by Durand's sub-committee, all three turbine builders were asked to proceed with studies of whatever type of aero propulsion engine seemed most promising.[402] In the summer of 1941, Glenn Warren and Alex Stevenson, engineers from GE Schenectady, were invited by Durand to discuss changing GE's research priority from locomotive to aircraft gas turbines. GE subsequently decided to delay their research effort on locomotives, finished up their work on naval gas turbines, and initiated a small group to begin looking at aircraft gas turbine engines. The three company proposals received in July 1941 were all based on axial concepts, and again Schlaifer has the explanation: *'This was certainly due at least in part to the fact that the NACA had begun development of this type of compressor in 1938 and was showing considerably higher efficiencies than could be obtained with centrifugal compressors. In addition, Allis-Chalmers had had experience since 1938 with axial compressors in the Houdry turbines which it built on Brown-Boveri license, and its experience was certainly reported to the* (Durand) *committee.'* Westinghouse had decided for a pure turbojet, Allis-Chalmers proposed a ducted turbofan and GE a turbo-prop, internally first known as the TG-100,[403] and later under the official designation as the T-31.

Edward Constant has interesting details about additional European influences on this first US turboprop, mainly in reference to an interview of one of its designers, Glenn B. Warren (1898–1979), Fig. 6.52 (l):[404] *'The TG-100, for which an Army contract was granted on 8 Dec. 1941, had a 14-stage axial compressor (based essentially on published Brown Boveri designs), nine 'can' combustion chambers, and a single-stage axial turbine.'*[405] The GE design team at Schenectady, NY, finished the design layout of the TG-100 by 23 December 1941, which laid the foundation for GE's pedigree of successful

---

[402] Membership on Durand's special committee went to the usual consulting sampling of government and academic experts, as well as to three representatives of the commercial firms engaged in turbine development: one from Allis-Chalmers, one from Westinghouse, and one from General Electric, a fact which did not go unnoticed, and consequently, was criticised.

[403] See Constant, The Origins, p. 222, and—see Wikipedia, 'General Electric T31'.

[404] See Wikipedia, 'Glenn B. Warren' in English; in his life-long GE carreer, Warren made it Vice President and General Manager of the Turbine Division, and in the year 1959–60 he served as President of the American Society of Mechanical Engineers.

[405] Interview with Glenn B. Warren by Ilan Kusiatin, Harvard Business School, in Schenectady, NY, 18 Sept. 1975.

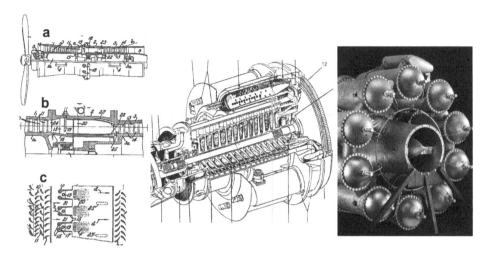

Fig. 6.51 General Electric 1200–2200 shp turboprop *TG-100*, 1941 (m), BBC turboprop patent 1939 (l), Whittle W1X *'can'* combustors, 1937 (r) © Kay (m), Smithsonian (r)

axial-flow turbojet designs. The baseline of the TG-100, Fig. 6.51 (m), reflected GE's studies of Navy PT boat gas turbines during 1938–1940—with the dominant single-stage, large diameter power (impulse) turbine to handle gas temperatures of up to 1000+ °C, to the right. The 14-stage axial compressor had a constant tip diameter of 0.421 m and an overall length of 0.635 m. The compression ratio at a typical design speed of 13,000 rpm achieved PR 5...6.15 at ṁ 10.8 kg/s, the measured isentropic compressor efficiency was below 80%.[406]

For the indicated Brown Boveri influences there are two possible, both from Claude Seippel:[407] In May 1940 appeared his seminal paper about *'The development of the Brown Boveri axial compressor'* in the international *Brown Boveri Review*, which was certainly available at GE Schenectady shortly afterwards, and somewhat surprisingly, he filed a patent US2,326,072 about a turboprop *'Gas Turbine Plant'*, with priority as early as 28 June 1939, which already foreshadowed Brown Boveri's coming ambitions in the area of aero gas turbines, and which will be discussed in Sect. 10.3. Figure 6.51 (l) has the essential patent drawings, of which *detail a)* illustrates a rather conventional axial set-up, with which the z-folded, reverse combustor flowpath of the TG-100 has not much similarity. However, *details b)* and *c)* offer a number of helpful suggestions for a GE design adaptation. Besides the attractive use of Whittle's single-can combustor pattern, Fig. 6.51 (r), which simplified the development considerably, the patented can-type combustor arrangement shows the used, characteristic double-wall structure—for cooling ventilation, and typical for future turbojet designs—in both cases, a central *'doghouse'* bearing

---

[406] See Wallner, Preliminary results and—see Smith, Axial Compressor.

[407] See Chap. 3 and—see Eckardt, Gas Turbine Powerhouse, p. 92 f.

chamber. The BBC rear compressor stages have their standard degree of reaction R 0.5 blading, which presumably was also typical for the NACA compressor master-design. The TG-100 turbine nozzle had a Prandtl-Busemann type of convergent-divergent supersonic flow channel, which allowed for a large gas temperature drop upfront of the rotating turbine blading and at the same time provided very high exhaust gas velocities.

Based on the TG-100 experiences, the real cornerstone of GE's axial compressor technology after 1943 became the 1.300–1.800 kp thrust, *TG-180 (J 35)* axial turbojet at Schenectady, its 11 compressor stages with constant tip diameter delivered a PR 4.0 at $\dot{m}$ 34 kg/s, with a polytropic efficiency in the upper 80% regime.[408] With a first run on 21 April 1944, this development becomes already a noticeable success for General Electric: Up to 1955 in total 13.000 *J 35* were built—mostly for the Republic *F-84 'Thunderjet'* and the Northrop *F-89*. Since however, GE's production capacity was limited then, they just built 140 units themselves, while the rest were placed at Allison, and even Chevrolet.[409]

While Allis-Chalmers dropped out of the beginning aero engine development race already at the end of 1943, GE and Westinghouse continued the building of aircraft GT engines, for which standard technical literature may be used as reference.[410]

Here, the short outline of US turbojet developments shall be closed in view of the *Northrop Aircraft Inc.*, not so much through the—together with the *Lockheed Aircraft Corp.*—remarkable involvement of airframe manufacturers in the new engine business, echoing comparable endeavours by Heinkel and Junkers on the German side, but mainly for three reasons:

- the achieved, technologically astonishing, high standard of the Northrop turboprop engines,
- the intended and openly reported, intensive discussion of European influences in this context, and finally,
- the personality of the Czech-born chief designer Vladimir Pavlecka, somewhat in anticipation of the planned illustration at the end of this Chapter on the contributions of *'Other countries'* on the turbojet development.

The *Northrop Aircraft Corporation* was founded in March 1939 by John (Jack) K. Northrop, after he had left the *Douglas Aircraft Corporation* of which his company had been a subsidiary. One of the original employees of this company was the Czech engineer, Vladimir H. Pavlecka, Fig. 6.52 (r), who—according to Schlaifer[411]—*'had acquired considerable enthusiasm for the gas turbine during his work in the aircraft*

---

[408] See Smith, Axial Compressor.

[409] Lichtfuss, 75 Jahre Turbostrahlflug.

[410] See St. Peter, The history of aircraft gas turbine engine development in the United States, and—see Kay, Turbojet, Vol. 2.

[411] See Schlaifer, Development of Aircraft Engines, p. 446.

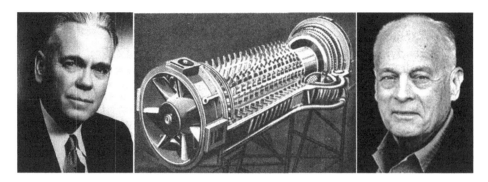

Fig. 6.52 US turbojet engine pioneers II: Glenn B. Warren (1898–1979) l, Northrop-Hendy XT37 turboprop, 10,000 shp @ S.L., 1944 (m), Vladimir Pavlecka (1901–1980) r

industry abroad'. This statement appears to be questionable in view of Pavlecka's own resume,[412] which puts him to Douglas, Sta. Monica, Ca. (1934–1939), Metalclad Airship Corp., Detroit, MI (1927–1933) and General Motors Research Corp., Detroit, MI (1926)—after his immigration to the USA in ~1925. With all likelihood, Pavlecka received his training in turbomachinery engineering still during his studies at TH Prague (1920–1925) with the renowned Prof. Jan Zvoníček (1865–1926), specialised in steam turbine and compressor design, who had invented in 1907 a counter-rotation radial turbine (Pat. US876,422), similar to the Ljungstroem turbine configuration as depicted in Fig. 5.9 (l).[413]

The technology transfer between Brown Boveri and Allis-Chalmers from stationary to aircraft gas turbines was more or less a natural outcome of the foregoing turbomachinery licensing agreement between the partners for the Houdry refinery equipment. At the same time, BBC or better its General Manager Adolf Meyer were part of another US turbojet development string. Again Schlaifer[414] has the basic facts: *'The earliest complete designs and serious proposals for the aircraft gas turbines in the United States, ..., came not from builders of engines of any sort, but from builders of airplanes. This fact is of considerable significance: the two earliest gas turbine projects in Germany likewise came from builders of airplanes.* [Heinkel & Junkers] *The first of these designs originated in Northrop Aircraft, Inc.* [the other came from Lockheed].' Pavlecka considered the gas turbine a superior replacement for the reciprocating engine in driving a propeller, an observation underlined by Kay's cryptic remark: *'His inspiration had come from seeing industrial gas turbines at Neuchâtel in Switzerland'.*[415] Pavlecka convinced the head of the company, *'Jack'*

---

[412] https://vladimirhpavlecka.com/

[413] Apparently, Pavlecka originated from Bohemia in the Austrian imperial dual monarchy, which would imply with some probability a good German-speaking background.

[414] See Schlaifer, Development of Aircraft Engines, p. 446 f.

[415] See Kay, Turbojet, Vol. 2, p. 75. It is unclear, if Pavlecka had visited Brown Boveri and/or the 'Landesausstellung' Zurich still in 1939 or the construction site of the first 4 MW power generation gas turbine at Neuchâtel, CH, in spring 1940.

Northrop, that such an engine should be developed chiefly on the grounds that it was simpler, somewhat lighter and freer of vibration. Northrop/Pavlecka hoped to attain a specific fuel consumption of only approx. 30% higher than the cruising consumption of an average conventional piston engine at that time, mainly by the use of an unheard of, very high pressure ratio of PR 10.5 and by the development of axial turbocomponents with efficiencies of 85%. Again the only encouragement to strive for such high efficiency levels was the demonstrated business success of the existing BBC designs, which nearly automatically must have attracted a rapprochement of both camps, but very soon Pavlecka recognised that there were extra problems hidden in the aircraft engines, compared to the much heavier and robust industrial gas turbine, e.g. small clearances and the resulting demand for accurate machining.

The Californian Lee Payne wrote for internet publication a short, nevertheless very informative story[416] entitled *'The Great Jet Engine Race ... and how we lost'*, which in a kind of story-board confronts the parallel lifes of Whittle, Pabst von Ohain and Pavlecka, who obviously contributed direct personal information in his late years—and which contains also the following historic *'nugget'*.

As a private venture the Northrop/Pavlecka initiative started obviously too early to get whole-hearted official support by the Navy, and also a recommended cooperation exercise with MIT Massachusetts Institute of Technology and P&W Pratt & Whitney ended in misunderstandings and frustration. Later Pavlecka recalled *'almost universal skepticism about the jet engine'*. But Pavlecka was not discouraged: *'Never,'* he says, quoted by Payne, *'I knew the history of the gas turbine from Armengaud in France to Lysholm in Switzerland.*[417] *Dr. Adolph Meyer, the chief engineer at Brown-Boveri had been a guest in my home, though he didn't believe the gas turbine could ever be made light enough to fly. I knew the history. The experts at MIT and Pratt & Whitney didn't and this meant they would miss out on the beginning of this new industry. I knew I was right.'*

This, in this context, surprising reference to Adolf Meyer triggers imagination; we see them in intensive discussions in German in Pavlecka's home overlooking the Long Beach shore. A. Meyer's strange skepticism towards GT aero applications has already been addressed in the foregoing Sect. 6.3.5, where Meyer's non-commenting of aero gas turbines in his otherwise comprehensive IMechE lecture at London on 24 February 1939 had obviously provoked already the attending Frank Whittle. Now, Pavlecka provides this little piece of supplementary and revealing information.

On the other hand, one can assume that Meyer had received before his trip to the US West Coast brand new test information about BBC's seven-stage axial turbocharger, Fig. 6.38, to clearly encourage Pavlecka about the feasibility of his daring GT engine

---

[416]https://www.airforcemag.com/article/0182engine/

[417]Alf Lysholm (1893–1973), chief engineer at Ljungstroms Angturbin AB, Sweden (not Switzerland) realised in 1935 a twin-screw supercharger which had been originally invented nearly half a century before by Heinrich Krigar in Hannover, Germany. For the Armengaud-Lemâle gas turbine,—see Eckardt, Gas Turbine Powerhouse, p. 70 f.

design, now designated *'Turbodyne'* or *XT-37*. Besides the mentioned PR 10.5 for the 18-stage compressor version, Fig. 6.52 (m), there exists a second XT-37 dataset for a 14-stage axial compressor configuration with a constant blade tip diameter D_{Tip} 0.787 m, generating PR 7.5 with a design mass flow of ṁ 46.3 kg/s, and a *'demonstrated efficiency'* of 87%. These data, representing a stage pressure ratio of 1.155, can be reviewed later in Fig. 8.25 in comparison to the best achieved stage pressure ratios of that time in England and Germany, illustrating a nearly equivalent technology level already then.

Pavlecka resigned in late 1943 and moved on to the more prospectfully looking *Lockheed* engine camp, where the *L-1000 turbojet* was under development. The name of the *'Turbodyne'* engine was later used for the Northrop subsidiary *Turbodyne Corp.* which succeeded as company to the Northrop-Hendy partnership in late 1949. Northrop-developed 'Turbodyne' engine, a turboprop gas turbine to deliver 10'000 hp at sea level, Fig. 6.52 (m), was intended to drive huge counter-rotating propellers on Northrop's *B-35 Flying Wing*. This power plant XT-37, which eventually had received Air Force support, was also considered for the Boeing B-52 bomber. But the shift of the *B-52* design in 1949 from turboprop to turbojet (introducing *Pratt&Whitney's J57* engine with the characteristic double-pod mounting, Fig. 11.2) and the discontinuance of the B-35 programme brought an end to these orders and in due course also to Northrop's engine activities. In 1950 Northrop disbanded Turbodyne Corp. and the Air Force ordered TC to turn over all patents, name and technical data to GE Schenectady, the winner takes it all.

6.4.3   The Aeronautical Impact of *Houdry* 100 Octane Fuel

BBC's single most important axial turbomachinery contribution to the Allied air warfare in WW II was presumably the punctual delivery of the new axial turbomachinery to complete the revolutionary *Houdry* refinery process.[418] In a 2005 article, the US author Tim Palucka[419] wrote about the impact of the Houdry process on the *Battle of Britain (BoB)*: *'That process would make a crucial difference in mid-1940, when the Royal Air Force started filling its Spitfires and Hurricanes with 100-octane gasoline imported from the United States instead of the 87 octane it had formerly used. Luftwaffe pilots couldn't believe they were facing the same planes they had fought successfully over France a few months before. The planes were the same, but the fuel wasn't. In his 1943 book The Amazing Petroleum Industry, V. A. Kalichevsky of the Socony-Vacuum Oil Company explained what high-octane gasoline meant to Britain:* "It is an established fact that a difference of only 13 points in octane number made possible the defeat of the Luftwaffe by the R.A.F. in the fall of 1940. This difference, slight as it seems, is sufficient to give a plane the vital *'edge'* in altitude, rate of climb and maneuverability that spells the difference

---

[418] See Eckardt, Gas Turbine Powerhouse, p. 169 f.

[419] See Palucka, The wizard of octane.

between defeat and victory.'" Typically, improvements of 20–30% in these fighter aircraft performance parameters are quoted in the literature as a benefit of these extra 13 octane points.[420]

In the first third of the twentieth century, high-octane gasoline paved the way to high compression-ratio engines, higher engine performance, and greater fuel economy. However, by far the most dramatic benefit of the earliest Houdry units was in the production of 100 octane aviation gasoline, just before the outbreak of World War II. The Houdry plants provided a better gasoline for blending with scarce high-octane components, as well as by-products that could be converted by other processes to make more high-octane fractions. The increased performance meant that Allied planes were better than Axis planes by a factor of 15–30% in engine power for take-off and climbing, 25% in payload, 10% in maximum speed, and 12% in operational altitude.[421] In the first six months of 1940, at the time of the Battle of Britain, 1.1 million barrels per month of 100-octane aviation gasoline was shipped to the Allies. Houdry plants produced 90% of this catalytically cracked gasoline during the first two years of the war.[422]

The original Houdry process embodied several innovative chemical and engineering concepts that have had far-reaching consequences. For example, the improvement of the octane rating with catalytic processes showed that the chemical composition of fuels was limiting engine performance. Further, aluminosilicate catalysts were shown to be efficient in improving the octane rating because they generated more highly branched isoparaffins and aromatic hydrocarbons, which are responsible for high octane ratings. From an economic standpoint, the catalysts could be regenerated after a short usage time, thus

---

[420] See Wikipedia, 'Rolls-Royce Merlin' in English, with Sect. 1.2.4.4 on *'Improved fuels'*, which slighty deviates from Palucka's assessment by stating that *'.. in the first half of 1940 the RAF transferred all Hurricane and Spitfire squadrons to 100 octane fuel'* from the USA, West Indies, Persia and domestically from the UK. Merlins II and III, originally producing 1000 hp on 87 octane aviation fuel boosted their power output to 1310 hp. Somewhat in contradiction, a more *not-invented-here*, pro-British view credits the Merlin II+ superiority after mid-1940 primarily towards the introduction of the constant-speed, variable-pitch propeller. In addition, the present pro-US position might be balanced by a critical UK analysis of the supply situation,—see Gavin, The Narrow Margin. Also for the German side,—see Wikipedia, 'Aircraft of the Battle of Britain' in English, reports in a Sect. 1.1.7 *'100 octane aviation fuel'* about a corresponding introduction of superior 100 octane fuel C-2 for the German side (and of suitably adapted engines like the *DB 601N*) at approximately the same time, however, with the conclusion that the German re-powering programme was somewhat slow, meaning in July 1940 only three Bf 109E/F fighter squadrons (36 aircraft!) were equipped with the new engines. Strategic analysis of BoB is rather complex, and the availability of 100 octane fuel in sufficient quantities is just one aspect, although important. For a more comprehensive insight into the German wartime fuel set-up,—see Jantzen, Betriebsstoffe, and—see Birkenfeld, Der synthetische Treibstoff, p. 74, summarising: *'The intentionally slow expansion of the Isooctane capacity in 1935–1938 essentially contributed to the fact that the Luftwaffe up to 1945 never received a quality fuel, corresponding to that of the Anglo-Saxon countries.'*

[421] The significant contribution of 100 octane fuel on Allied air superiority has been also underlined by—see McFarland, To command the sky; this reference, courtesy to Lutz Budrass, 1 May 2018.

[422] See ACS, The Houdry Process, p. 3.

Fig. 6.53 *Sun Oil, Marcus Hook* Refinery, Philadelphia, USA: (**a**) *Catalytic Cracking Case* (CCC), (**b**) *Houdry* turbine house (l) 1938, CCC replacement (m), typical catalyst pellets (r) 1950

returning the catalyst to full activity without having to add additional material. Figure 6.53 (l)[423] shows a rare 1938 outside view of the Houdry catalytic cracking unit at Sun Oil's Marcus Hook Refinery with a—one of three Catalytic Cracking Cases, typically 10 m high and 3 m in diameter, and b—the Houdry turbine house in the rear, while Fig. 6.53 (m, l) provide an impression of catalyst replacement activities.[424]

The timely axial turbomachinery development at Brown Boveri and the considerable extra risk taken when contacted by Sun's engineering in 1936 led to the unique convergence of two independent branches of mechanical and chemical engineering. The BBC contribution has not been explicitly noticed by the community of chemists yet, which rightfully praises the Houdry process and gave the Marcus Hook plant National Historic Chemical Landmark status in April 1996.[425] From start, the plant was capable of cracking 15,000 barrels of residuum feedstock left over from a thermal cracking unit per day,[426] yielding 48% of 81 octane petrol, i.e. twice as much as the previous thermal cracking output. Thick residuum from the fractionating tower was pumped into a still case where it was heated to 470 °C and vapourised. This vapour rose at a low pressure through the active catalyst cases where the long-chain hydrocarbons were cracked into smaller molecules. Another fractionation separated the resulting mixture into petrol and fractions with higher boiling points that could be put through the system again. In 1939 Sun Oil already had ten Houdry plants in operation, based on further turbomachinery deliveries from Baden and the first, licence-produced equipment from Allis-Chalmers. During World War II, Sun Oil

---

[423] See Eckardt, Gas Turbine Powerhouse, p. 176

[424] For details of the catalyst regeneration cycle,—see Eckardt, Gas Turbine Powerhouse, p. 172 f.

[425] See ACS, The Houdry Process.

[426] Rising up to a maximum of 175,000 barrels per day near the end of the century under Sunoco Inc. In 2004 the plant began producing the official fuel for NASCAR, followed by INDYCAR, the National Hot Rod Association etc., and it remains in Marcus Hook as a tenant to an Energy Transfer subsidiary, which now owns the facility.

Fig. 6.54 First BBC gas turbine in *Houdry* turbine house at Marcus Hook Refinery, Philadelphia, USA 1936 (l), World War II billboard at the Marcus Hook Refinery (r)

Co. employees at Marcus Hook processed more jet fuel for the Allies than any other refinery. Nine times during 1942 and 1943, tankers of the Sun Oil fleet were struck by German U-boat attacks, and four were sunk. Those encounters cost the lives of 141 Sun seamen, Fig. 6.54 (r). The attacks, while devastating, did not prevent the Sun fleet from shipping more than 41 million barrels of high-octane aero gasoline over 2.3 million miles of ocean during the war.[427]

The original fixed-bed Houdry Process units have been outmoded by engineering advances that transformed the fixed-bed to more economical fluidized-bed systems and introduced the use of crystalline aluminosilicate catalysts to provide higher yields of gasoline. Yet it is remarkable that, 70 years after Houdry's discovery of the catalytic properties of activated clay to convert petroleum fractions to gasoline, the same fundamental, mechanical and chemical principles that made the process a success are still the primary basis for manufacturing gasoline worldwide.

6.4.4   G. Jendrassik and the First Turboprop Engine: From Hungary[428]

A brilliant Hungarian engineer, György/George Jendrassik (1898–1954), Fig. 6.55 (l), was yet another who started work on gas turbines in the early 1930s,[429] especially remarkable,

---

[427] See Energy Transfer, Marcus Hook.

[428] Besides the US-Czeck V. Pavlecka in Sect. 6.4.3, the Hungarian G. Jendrassik will be the only other aero gas turbine protagonist considered from '*other countries*', especially due to his close interrelation with developments in Switzerland, Germany and England. For a comprehensive survey of the 'other country' aero engine developments, – see especially Kay, Turbojet, Vol. 2. Actually, as documented—see Eckardt, Gas Turbine Powerhouse, p. 2, the author's preoccupation as engineer-historian started on 5 April 2003 at ETH Zurich on the occasion of the late Prof. Georg Gyarmathy's 70th birthday, an enthusiastic Hungarian fellow-countryman to G. Jendrassik and his early, often neglected contributions to aero GT propulsion.

[429] See Wikipedia, 'György Jendrassik' in English.

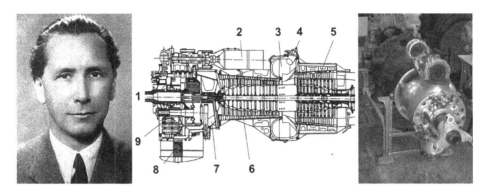

Fig. 6.55 György Jendrassik (1898–1954) l, first turboprop engine Cs-1, 1000 hp, 1940, (m), Cs-1 on display at Museum of Technology, Budapest, r © A. Kay (m)

since from these small-scale activities resulted the world's first turboprop engine. Jendrassik graduated as Dipl.-Ing. in 1922 at the Joseph Technical University in Budapest, after studying also at TH Berlin-Charlottenburg, and joined the *'Danubius Machine, Wagon and Ship Works, est. 1844'*, later in short the *Ganz Works*[430] in that city, making amongst other diesel engines for rail traction. He brought about and patented substantial improvements in that area, and since 1927 the production of 550 locomotives with Ganz-Jendrassik engines were licensed world-wide. His increasing interest in the gas turbine led him to leave Ganz, to found his own company in 1936, and to start testing his first 100 hp experimental gas turbine already in 1937. *Engineering* (London) reported on 17 February 1939 about *'The Jendrassik Combustion Turbine'*, just one week ahead of A. Meyer's IMechE London lecture *'The Combustion Gas Turbine'*, Sect. 6.3.5, apparently a nice, competitive PR coup of the multi-talented Hungarian. Jendrassik had described his considerations away from Holzwarth's intermittently-operated, constant-volume system[431] to a constant-pressure gas turbine, which *Engineering* welcomed consequently as *'a new form of combustion turbine'*, apparently unaware of the early attempts with the constant-pressure gas turbine principle in the 1899–1906 time-frame (e.g. Stolze, Armengaud-Lemâle) and, more recently, of Brown Boveri's *Velox* and *Houdry* gas turbines.

Jendrassik's measured test data at a design speed of 16,400 rpm show outstanding performance for an initial programme—with a turbine inlet temperature into his seven-stage turbine of only 475 °C, a leaking steel-plate heat exchanger, and a poor, separately mounted combustor, a thermal efficiency of 21.2% was claimed. The efficiency of his

---

[430] See Wikipedia, 'Ganz Works'in English. On the Ganz *Danubius Shipyard* at Fiume also the 22,000 t Austro-Hungarian battleship 'SMS Szent-István' was built, the capsizing of which with the young sea cadet Herbert Wagner on board, has been described in Sect. 6.2.1.

[431] See Eckardt, Gas Turbine Powerhouse, p. 72 f.

10-stage axial compressor with PR 2.2 was η_{pol} $0.865.^{432}$ It is assumed that Jendrassik's turbo-components had—in this respect comparable to Brown Boveri—its roots in steam turbine practice.

In July 1939 George Jendressik turned to the possibility of an aeronautical gas turbine. He went on to design and build a 1000 shp turboprop engine *Cs-1*, Fig. 6.55 (m, r), of impressively advanced design—with 1—propeller shaft, 2—starter motor, 3—annular, reverse-flow combustion chamber, 4—six fuel injectors, 5—11-stage axial turbine, 6—15-stage axial compressor, 7—annular air intake, 8—oil cooler, 9—turboprop reduction gear, and a fixed area exhaust nozzle (not shown). The initial rating of the 1000 shp engine was configured for 13,500 rpm and a simple rotor mounting on just two bearings. A speciality of Jendrassik's design was air cooling of the turbine discs and the patented feature of turbine blades with extended roots to reduce heat transfer to the disc. The annular air intake surrounded a reduction gear for propeller drive takeoff, thus reducing the gearbox heat rejection.

The Cs-1 stirred interest in the Hungarian aircraft industry with its potential to power a modern generation of high-performance aircraft, and correspondingly construction was begun of a twin-engined fighter-bomber, the *Varga RMI-1 X/H*. The first bench run took place in 1940, becoming the world's first turboprop engine to run. Work on the engine stopped in 1941, when the Hungarian Air Force selected the Messerschmitt Me 210 for the heavy fighter role, and the engine factory converted over to the Daimler-Benz DB 605 to power it; the prototype RMI-1 was later fitted with these engines in 1944.

Jendrassik left Hungary after WW II to re-establish Ganz's interrupted international business contacts, still officially employed up to 1948 as Ganz's General Director. Eventually he came to Britain, then already prosecuted by the Communist regime back in Hungary. His newly founded *Jendrassik Developments Ltd* worked shortly on pressure exchangers for Power Jets in 1949, before he became a director in Metrovick's power department, up to his early death in 1954. In this context, one can speculate that he had lifelong ties to Switzerland—and its leading representatives in power engineering, Karl Baumann, Metrovick and Adolf Meyer, BBC; as with the Norwegian Elling, Sect. 4.1.2, Jendrassik is often not given the credit that is due to engineers from smaller countries.

---

[432] For an early evaluation of Jendrassik's 100 hp gas turbine with basic performance data,—see H. Constant, Gas Turbines, p. 18.

Excursion III: Helmut Schelp (1912–1994) and Max Adolf Mueller (1901–1962)

Helmut Schelp and Max Adolf Mueller influenced the German turbojet engine developments as individual personalities and as members of their working administrations in a different, though in any case outstanding manner. Both were mechanical engineers, Schelp mostly as conceptual organiser and ministerial administrator, while Mueller moved rather arbitrarily and unsettled through a bunch of industrial project opportunities before and during the war.

Helmut Schelp was—though not by official title (which often prevented his youthful age), but effectively—the visionary director of advanced engine development at the RLM's T-Amt (technical division) leading up to and during WW II. He used his office to fund a widespread programme in turbojet engine development, which led to many of the engine concepts still used today. In particular, he was instrumental in favouring the use of axial-flow compressors from the very beginning—and especially before the first gas turbine powered flight in August 1939—over the simpler but bulkier centrifugal compressors. Unlike in England where the turbojet engine had no single champion within the Air Ministry, and their efforts were long delayed as a result, Schelp can be directly credited with the advancement and refinement of the jet engine in Germany over the decisive first years.

Consequently, Schelp's influence on the German turbojet engine programme in general is of salient importance; therefore with some emphasis the following three sections are dedicated—in Sect. 7.1, to his early life as teenager up to the age of 23,—in Sect. 7.2, to his professional development as one of the rare, highly specialised *'Flugbaumeister'* (master aircraft/engine builders/ designers) accomplished in 1939, and—in Sect. 7.3, his following years as *'Referent'* (Technical officer/director) in the *RLM Technisches Amt* (technical division) and *'Flieger-Stabsingenieur'* (flyer staff engineer). All this, being largely based

D. Eckardt, *Jet Web*,
https://doi.org/10.1007/978-3-658-38531-6_7

on his luckily preserved, 142 page personnel file from *Bundes-Militaerarchiv Freiburg* (federal military archive).[1]

To the contrary, Mueller's life up to 1945 could be reconstructed in Sect. 7.4 only on the basis of sporadic and distributed information. Somewhat even better qualified than Schelp as a versatile mechanical engineer, he is apparently always at or close to the turbojet engine development hot spots between 1935 to 1945, but—contrary to the smart Schelp—his rush personality in combination with what looks like a deep, inflexible Nazi conviction prevents him to succeed very often in team efforts which required a more steady and sustained engagement.

7.1   Helmut Schelp's Youth Up To 1935

Helmut Schelp was born on 11 June 1912 at Goerlitz, Saxony, then located in *'Middle Germany'*, some 320 km west of the German-Polish border. The house, where he grew up and lived in Goerlitz, since April 1918 as half-orphan,[2] together with his mother at *Biesnitzer Strasse 13*, exists still today, now only just 1 km west of the Oder-Neisse river line of the present German-Polish border.

Between Easter 1918—in the month of his father's death—and Easter 1927, he attended the *RRG Reform-Realgymnasium Goerlitz*,[3] where he finished with a secondary school level I certificate, as an average pupil with some preference for mathematics and a noted, specific excellence in *'linear drawing'*. Following a classic German engineering education, and as a precondition for his following mechanical engineering studies at Chemnitz, Saxony, he commenced on 1 April 1927 an apprenticeship for 2½ years, just 450 m away from his living place, in the local *Maschinenfabrik* (machine factory) *Raupach*, which produced machinery for the local, Lausitzer (Lusatian) ceramics industry.

---

[1] Personal-Akten Helmut Schelp, Flieger-Stabsingenieur, Bundesarchiv PERS 6/162977, 142 p., ordered 15 July 2017. Schelp's personnel file covers actually the documented period from 30 March 1927 (leaving certificate of Reform-Realgymnasium Goerlitz) to 26 Oct. 1940 (certificate of marriage) only. The remainder up to May 1945 might have been removed for unknown reason either from the German side or thereafter, from Allied authorities. However, for the period up to 10 March 1945 exist various minutes of meetings with H. Schelp's attendance, which will be used in this and in Sect. 8.1.

[2] Schelp's father, the merchant Albert Schelp (1879–1918) died in April 1918 at Armentières, an area of heavy fighting during the *Battle of the Lys/4th Flandernschlacht*, 7–29 April 1918. In March 1918 within fifteen hours the Germans shelled 20,000 grenades with mustard gas on Armentières, so that *'liquid mustard gas flowed like heavy rainfall in the street gutters'*. British troops had to be evacuated and the Germans could not enter Armentières for 2 weeks due to the heavy contamination.

[3] The RRG was founded in 1913 as a professional education-oriented institute; today it is the *BSZ Berufliches Schulzentrum* (professional school centre) at Carl-von-Ossietzky Strasse, Goerlitz with some 1500 students.

Between October 1929 and March 1933, Helmut Schelp studied mechanical engineering at the *Staatliche Akademie fuer Technik* at Chemnitz.[4] This academy played a major role in helping Chemnitz to become one of the industrial centres at the beginning of the twentieth century. At this time Chemnitz (Auto Union), the neighbouring Zschopau (DKW), 15 km south-east of Chemnitz, and Zwickau (Audi,[5] Horch), 50 km in the south-west, were the heartland of an emerging motorbike/automotive industry. It had the highest number of patent registrations in the whole of Germany and hence also world-wide. By the late 1920s the *Zschopauer Motorenwerke J.S. Rasmussen AG* and its brand *DKW Dampf-Kraft-Wagen* (steam-powered cars) was with 65,000 bikes annually the world's largest motorcycle manufacturer.[6] Helmut Schelp finished the Technical Academy with an overall *'Sehr Gut'* (very good), underlined by a special certificate for *'excellent achievements'*. This qualified him on the initiative of the Director of the Technical Academy Chemnitz, Professor Dr. Heinrich Paul Schimpke (1880–1970)[7] for a follow-on scholarship from the prestigious *'Studienstiftung des Deutschen Volkes* (German National Academic Foundation)'; at the same time the exam represented also the general qualification for university entrance, which Schelp used in a further kind of *'university hopping'* to achieve in a shortened process of just three semesters the 'Vordiplom' (intermediate diploma examination) in mechanical engineering at TH Dresden between April 1934 and May 1935, followed by the *'Master of Science'*, equivalent to the Dipl.-Ing. degree, from Stevens Institute of Technology, Hoboken NJ, as result of his US stay between September 1935 and June 1936, which will be outlined in detail in this Section.

It appears that Sun 7 June 1931 could have been a *'moment of personal, technical and political awakening'* for the 19-year-old Schelp, when Adolf Hitler spoke to a roaring audience of 16,000 at the *'Suedkampfbahn'* (sports arena), Chemnitz. In due course on 1 September 1931, Schelp became a NSDAP party member and member of the students organisation NSD-Studentenbund,[8] which pushed its membership at Chemnitz as a result of this event from 30 to 100 students. Schelp's NSDAP membership #636'353 resulted shortly thereafter in his affiliation to the *SA 'stormtroopers'* with pilot training (A certificate) between August 1932 and August 1933. Fig. 7.1 (l) shows him still as low-level *SA Scharführer* (squad leader) with the standard Nazi party badge, before he was

---

[4] See Wikipedia, 'Chemnitz University of Technology' in English.

[5] For the complex early history of the company, see Wikipedia 'Audi'in English.

[6] The motorcycle and car industry at Zschopau and Zwickau has already been addressed in the context of Fritz Heppner's lengthy studies at TH Dresden, as a possible interruption in the mid-1920s, see Chap. 5. While Heppner was eight years older than Schelp, and thus presumably out of contact distance, there is some likelihood that Schelp already met Helmuth Weinrich as a student at Chemnitz, some ten years before their professional paths crossed again in Berlin, at Bramo, Brandenburg and Daimler Benz, Stuttgart.

[7] See Wikipedia, 'Paul Schimpke'in German.

[8] See Wikipedia, 'National Socialist German Students League' in English.

Fig. 7.1 Helmut Schelp (1912–1994): *SA/DLV*-Flyer ~1933 (l), Dresden Air Show—in front of Udet's *'Rhönbussard'*, 7 July, 1935 (m), Poster *'Deutschlandflug 1935'*, ~ He 50 (r) © M. Schelp (l, m)

transferred from *SA* to *DLV Deutscher Luftsport Verband* (German aviation assoc.), Chemnitz. Thereafter, he started for the next five years a considerably overloaded life, as an enthusiast in the world of flying, which with all likelihood required his strong political Nazi engagement in parallel, and his activities as a part-time engineering student at TH Dresden. *Speed* characterised his life; before flying became his main interest, it was on tuned/racing motorcycles—of course, then practised without protecting helmet.[9]

As a kind of stimulus package after economic depression, in 1928 the national government exempted motorcycles below 200 ccm from tax and the demand for a driver's license. DKW's answer was its most popular model, the 2-stroke, single cylinder *'Luxus 200'* with 4 hp and 198 ccm displacement. Widely known under the nickname *'blood blister'* for its distinctive bright red petrol tank, more than 37,000 units of this commuter bike were produced between 1929 and 1933. Schelp's bike had a maximum speed of 70 km/h on paper, but soon he knew how to squeeze out some 20% extra. The only precondition for getting that experience was that the driver was older than 18,—and somehow had managed to get the required 750 RM purchasing price.

With some probability, Schelp tested his newly acquired bike during a 500 km one-way summer trip in July/August 1931 from Chemnitz to Beuthen/Bytom, still on the German side of the newly drawn German-Polish border in 1920. Officially, he worked and earned some money as fitter at the machine shop of Schikora & Gerdes at Korf, Upper Silesia, while he visited and lived then with his grandfather from his mother's side, Paul Jakisch, *1863, a professional miller, who lived at Beuthen nearby.

The practical training for continuing his mechanical engineering studies up to 'Vordiplom' (intermediate diploma examination) at TH Dresden (THD) between May 1934 and June 1935, he completed with apparent enthusiasm at the DKW works Zschopau,

---

[9]Information, courtesy to Tom Schelp, 25 Sep. 2015.

first between March 1933 and April 1934 as *'designer assistant'*, then in a second stint between August and October 1934—as he proudly stated—as fully accepted *'detail designer'* in DKW's motorcycle racing design team. Helmut Schelp received his Vordiplom certificate on 28 June 1935 with the overall grade *'Gut'* (good), but he admitted in a written CV of the time that his examen preparations were limited to a few night hours, while during the days he was mostly occupied with his other tasks as deputy leader of the THD student body, member of the students's leader council—and in his way, closely associated under the new regime and different to the academic glider movement[10] of the 1920s—his new full-time hobby of flying engine-powered aircraft.

This personal wish was in full agreement with the official policy after 1933, where *'Flugsport'* (flight sports/aviation) was a preferred means to kindle the technical enthusiasm among young people. In Saxony, the *'1st N.S.-Grossflugtag'* (national-socialistic air show) took place at Dresden-Heller on Sun 25 June 1933, immediately followed by similar Sunday events at Chemnitz on 2 July and 17 September 1933. Figure 7.2 illustrates the 1933 highlight at Dresden, the visit of the Junkers G.38 giant aircraft.[11] The location is the same as that of Fig. 5.4 (m) at the heather airfield of Dresden 'On the Heller', where Professor Trefftz and the FVD model glider group with Fritz Heppner and Emil Pohorille had met in ~1927. In the following Section Erich Trefftz will reappear as fatherly friend to Helmut Schelp.

A rare snapshot of that time shows Helmut Schelp, Fig. 7.1 (m), on Sun 7 July 1935 at the *3rd Dresden Air Show* in front of Ernst Udet and his *'Rhönbussard'* (Rhoen buzzard, reg. D-Udet) glider[12]—in waiting of the airplane tow start. Udet prepared there for a glider aerobatic flight in expectation of the 1936 Olympics at Berlin, where he flew at Berlin-Staaken a similar demonstration programme close to ground, for *'sailplane gliding'* to become a future potential Olympic discipline.

Helmut Schelp's emotional highlight of his crowded schedule in the years 1934 and 1935 was certainly his participation in the *'Deutschlandflug'* (Flight over Germany), a traditional annual flying event since 1911. The target setting for the pre-war *Deutschlandfluege* 1934—1938 (not in the Olympic year 1936) changed considerably, so that no longer the individual aeronautical qualification was tested, but the group performance of three to nine aircrafts.

---

[10] In his 1965 interview—see Ermenc, Interviews, p. 101—Schelp put the start of his glider activities to the year 1929, which would have been still at Goerlitz. However, it is more likely that his first contact to the Akaflieg (academic glider group) scene was rather in 1930 in Chemnitz, from which he quickly proceeded to the—in his view, more attractive—military-oriented, professional offer of the *SA storm-troopers* towards motor-aircraft piloting.

[11] See Wikipedia, 'Junkers G.38' in English, and Junkers *Youtube* movie of first flight (3:25 min) https://www.youtube.com/watch?v=oK_EQwzCabA

[12] The *'Rhönbussard'* was a Hans Jacobs design, Fig. 4.40, with a strong, aerobatic-suitable structure,—see Wikipedia, 'Schleicher Rhönbussard' in English.

Fig. 7.2 First '*N.S. Grossflugtag*' Dresden-Heller, 25 June 1933: Junkers G.38 (D-2500) on display, wingspan 44 m, 34 pax 2-deck

The *Deutschlandflug 1934* took place between 21 and 24 June 1934 over a total distance of 4701 km with daily start and return from and to Berlin-Tempelhof airport. Helmut Schelp participated as '*Orter*' (navigator) on board a Focke-Wulf, two-seat biplane *Fw44 Stieglitz*,[13] piloted by Dr. Göttmann,[14] which was part of the team *G9*. In that year the flight programme of the first day, Thu 21 June 1934, stretched after an early start at 3.15 h in the morning from Berlin–Stettin/Szczecin—through the '*Polish corridor*' to Danzig/Gdansk and East Prussia via Swinemuende/Świnoujście—Greifswald back to Berlin.

Between Swinemuende and Greifswald was an additional turning point mapped as '*Nsp. Usedom*' which with some probability can be interpreted as '*Nordspitze Usedom*',[15] since 1936 site of the HVA Heeresversuchsanstalt (Army research centre) Peenemuende.[16] The date of this overflight during the *Deutschlandflug 1934* is roughly 1.5 years in advance of the official detection of the place by Werner von Braun for his V-2 rocket development programme,[17] and might be pure coincidence then. However, it is not unlikely that among the many flight crews were also decision makers in this context, who at least later remembered the remote location from personal flight experience.[18]

---

[13] See Wikipedia, 'Focke-Wulf Fw 44 Stieglitz' in English.

[14] In 1938 RLM-Fliegerstabsingenieur (flight staff engineer) Dr. Göttmann is traceable as department leader E13 'Ground equipment' at the Rechlin test centre.

[15] The author owes this hint to Ph. Aumann, HTM Hist.-Techn. Museum Peenemuende, 6 Jan. 2020.

[16] See Wikipedia, 'Peenemünde' in English.

[17] See Neufeld, The Rocket, p.49. According to this colourful account the hint to the suitable location at the northern corner of Usedom came actually from von Braun's mother during his Christmas holidays 1935. Like the German aviation pioneer Otto Lilienthal, she was born at Anklam, some 30 km in the south of Peenemuende, and remembered that his grandfather had preferred the area for duck-hunting.

[18] The author was unable to find a list of participants for the *Deutschlandflug 1934*; however, already the *Deutschlandflug 1935* names numerous high-ranking RLM personnel, as e.g. then Colonel W. Wimmer (1889–1973), head of the RLM Technical Office before E.Udet, as pilot for the RLM flight group *F3* together with Colonel H. Felmy, as navigator, then head of all Luftwaffe pilot training centres. In addition, RLM was presented by Ritter von Greim (in 1945, General Field Marshall and successor to H. Goering), General Engineer R. Lucht and General J. Kammhuber, then RLM group leader.

The second day covered from Berlin the north-western sector Bielefeld—Hamburg—Flensburg—Berlin. The third day went via Schelp's hometown Goerlitz to Upper Silesia (Breslau/Wroclaw), where Göttmann and Schelp were interviewed for an article in the *'Nordschlesische Tageszeitung* (North-Silesian Daily Newspaper)', Glogau, reporting that some of the participating aircraft could nearly not carry all the gifts which had been prepared by enthusiastic crowds at the various landing points. On the final day, Sun 24 June 1934, the course—designed by *'the Minister of Aviation General Hermann Goering'* went southward from Berlin via Bayreuth to Ainring near Berchtesgaden, where after a tank stop the formation of the remaining 70 aircraft flew over the marker *'Obersalzberg'*, where *'the Fuehrer took off the air parade—standing at the window'*,[19] before the fleet returned via Munich and Bamberg back to Berlin.

The *Deutschlandflug 1935* took place on six flight days between Tue 28 May and Sun 2 June 1935, illustrated in Fig. 7.1 (r) by an announcement poster, showing a group of diving Heinkel He 50 biplanes.[20] The total attendance had nearly doubled in comparison to the end result in the foregoing year: there were 30 groups with 161 aircraft. Schelp participated again as *'Orter'*, now for the group *'B7 Dresden'* of five, then newly developed two-seater Klemm L25d VIIR with 80 hp Hirth HM60R engine.[21] Fig. 7.3 (l) shows him on the first day, which was flown under the motto *'Flight to Silesia'*, at the airfield of his hometown Goerlitz, where all aircraft had to make an obligatory landing. The second day went from Guben to Koenigsberg/Kaliningrad (East Prussia), followed on the third day by a leg from there to Bremen, continued on the fourth day from Bremen to Freiburg in the most south-western corner of the *'Reich'*, and finished on the fifth day, from there via Munich to Erfurt. The sixth day was organised as a common group flight of all the remaining 140+ aircraft under the lead of the RLM State Secretary and then General Lieutenant Erhard Milch from Erfurt to Berlin-Tempelhof.

The photo of Fig. 7.3 (r) was taken there after the competition, showing Helmut Schelp together with his pilot Hermann Steckhan,[22] an experienced Dresden-born test pilot for

---

[19] See Mueller-Romminger, Geflogene Vergangenheit; this source is based on kind input from Hedwig Sensen, 7 June 2016. The electrically retractable *'panorama window'* of 8×4 m was one of the earliest of Hitler's numerous architectural follies; since positioned above the garage where the Mercedes car park had to be warmed up, more or less continuously, it could nearly never be used opened for a long time. For the *Deutschlandflug 1934* itinerary,—see Huebner, Der Deutschlandflug 1934; the author Dipl.-Ing. Walter Huebner was head of the *Deutschlandflug* organisational team.

[20] See Wikipedia, 'Heinkel He 50' in English.

[21] See Wikipedia, 'Klemm Kl 25' in English.

[22] His flight book has between 17 April 1934 and 27 March 1945 entries for the aircraft types Ar 234, Me 262, He 162, Fw 200, Ju 290, Ju 292, Ju 352. The *'Volksjäger'* Heinkel He 162 and the tragic death of the test pilot Gotthold Peter during the second flight at Vienna-Schwechat on 10 Dec. 1944, just 73 days after project start, has been described—see Wikipedia, 'Heinkel He 162' in English. In parallel, as outlined in a detailed master's thesis in German,—see Huemer, The 'Volksjäger', p. 77. Hermann Steckhan, chief test pilot at Junkers Bernburg, flew the Mittelwerke-manufactured He 162 approx. 20 times successfully, Fig. 8.11, until he crashed with the first aircraft of that series,

Fig. 7.3 *Deutschlandflug 1935*, flyer group *'B7—Dresden'*, fourth position: H. Schelp, navigator on *Klemm L25d* at Goerlitz (l), H. Schelp and Hermann Steckhan, pilot, at Berlin (r) © M. Schelp

Junkers Dessau and Bernburg[23] in the 1934–1945 time frame. The commonly achieved fourth place in the overall ranking for the *'Dresden B7'* group was a considerable success; winner was *'Danzig B6'*, led by the Rechlin chief test pilot Otto Cuno,[24] ahead of *'Stuttgart G3'* with the largest number of nine Klemm L25d VIIR for one team. Besides the personal recognition, Helmut Schelp drew certainly the biggest and lasting advantage from the *Deutschlandfluege* by creating a considerable network of contacts and acquaintances.[25] Emil Waldmann, then a student in mechanical engineering at TH Stuttgart, member of *Akaflieg* (academic flyer group) and as such navigator of one of the Stuttgart team aircraft became after 1938 one of the closest members in Schelp's small RLM staff.[26]

---

severely injured on 27 March 1945; his cause of accident was similar to that of Peter's deadly crash—a failure of the wood structure in the high speed range of 700+ km/h. It is remarkable that the early acquaintance of Hermann Steckhan with Helmut Schelp interconnects over a period of less than ten years the turbojet engine development from first definitions to nearly the very last application during WW II.

[23] In March 1945 at Junkers Bernburg, some 30 km in the west of Dessau, a He 162 assembly line started operation, but already between 12 and 14 April 1945 US troops overran the facility and ended aircraft production. On 21 July 1945 the control of the plant was handed over to the *Red Army,* which started immediate dismantlement, finished in 1950.

[24] Cuno's participation in the Heinkel *'Blitz'* He 70G transfer to England, some ten months later, has already been described in Sect. 6.1.2.

[25] To the *'Dresden B7'* belonged also Major a.D. (ausser Dienst/ out of service) Walter Stahr, after his time as head of the secret German flying school at Lipezk, Russia, 1925–1929 (Sect. 4.3.1), then occupied with *'special tasks'* at RLM—and who might have been instrumental to establish the contact of Schelp to R. Otto Fuchs and the RLM/ DVL *'Flugbaumeister'* programme,—see the following Sect. 7.2.

[26] See Sect. 7.3 for an overview of the known working members and their wartime activities. Another pilot of the *'Stuttgart G3'*, later a well-known aerodynamicist at BMW Flugmotorenbau Munich up to 1945, and thereafter for 20 years at General Electric, Evendale, OH was Dipl.-Ing. Peter G. Kappus.

7.2   The Making of a *'Flugbaumeister'*

Helmut Schelp earned, if not as *the first*, certainly as one of the first ones, the newly created title of a *'Flugbaumeister'* (flight master builder). *'The Making of . . .'* in the header is not meant to vilify his impressive track record thereto, but to indicate a kind of well-meaning system support which apparently accompanied this effort.

Shortly after the government change in Germany at the end of January 1933, and the immediately following decision to expand massively the—still secret—*Luftwaffe*, emerged the question of how to recruit and train the accordingly required number of pilots and other personnel. A group of available experts was set-up at DVL Deutsche Versuchsanstalt für Luftfahrt (German test establishment for aviation) at Berlin-Adlershof, who had developed a corresponding education and training strategy in the foregoing years in the context of personnel recruitment for the German flight test facility at Lipezk, Russia, as outlined in Sect. 4.3.1. The programme was designed and coordinated by a task force under R. Otto Fuchs (1897–1987),[27] Fig. 7.9 (l), a colourful and extraordinary versatile character, who over decades was an influential aviation pioneer and manager of flight education and training in Germany. Since 1916 together with his brother a fighter pilot,[28] Fuchs studied after 1919—in that row—philosophy, literature and agriculture before in 1924, he decided to go for aviation technology at TH Darmstadt, in combination with the nearly inevitable glider activities on *Wasserkuppe*.[29] Between 1927 and 1930, he joined Walter Stahr (1882–1948), who since June 1925 had built up the secret *Reichswehr* (army) flying school at Lipezk, Russia. Thereafter, Stahr in 1929 and Fuchs in 1930 returned, Stahr to build up, again in secrecy, the *Rechlin* flight test centre,[30] while Fuchs finished his engineering studies. After 1934 both met again at DVL Berlin-Adlershof in the ominous *'Department M'* (military).[31]

---

[27] See Wikipedia, 'Otto Fuchs (Luftfahrtpionier)' in German.

[28] Marking their aircrafts by *'foxes'* (German: Fuchs)*;* information courtesy to V. Koos, 2018.

[29] Saying goes that he suggested—after his return from Russia—as one of the first the use of motor-planes for towed glider starts, as a precondition for uncoupling starting ground and the availability of strong thermic currents. In this context he also demonstrated first-time glider thermic flights over Berlin;—see Zacher, Otto Fuchs, p. 978.

[30] See Wikipedia, 'Rechlin-Lärz Airfield' in English. Rechlin is located in a thinly populated area some 130 km in the north of Berlin, to which nevertheless the RLM established at times a Junkers Ju 52 shuttle service from Berlin-Tempelhof. The first known personal organisation of the newly founded RLM, 1934, has *Flight Commodore Stahr* as commander of all test centres and as commander of Rechlin Airfield. The RLM *Technical Office T* was then led by *Dipl.-Ing. Reidenbach*, and *Group T2 Triebwerk* (engines) by *Dipl.-Ing. Eisenlohr*. In 1937, Stahr's successor was *Major Carl-August von Schoenebeck* (1898–1989), a dedicated flyer and fighter pilot of WW I, who learned hang gliding at the age of 77,—see Wikipedia 'Carl-August von Schoenebeck' in English; Reidenbach was now head of Rechlin, and Eisenlohr led the same Group, now renamed as E2.

[31] Documentation about the early DVL history is very fragmentary, most files are still presumed in Russia. There is one written testimony,—see Wefeld, 75 Jahre Akaflieg Berlin, pp. 57/58, that Fuchs

The existence of this department and its support function for the secret *Lipezk flying school*, as such a violation of the Treaty of Versailles,[32] was revealed for the first time in an article *'Windiges aus der deutschen Luftfahrt* (Windy news from German aviation)' of the weekly magazine *'Weltbühne'* on 12 March 1929, which led on 23 November 1931 to the conviction of the accused *Weltbühne* editor Carl von Ossietsky (1889–1938)[33] and the investigating pacifist journalist-engineer Walter Kreiser (1898–1958) for *'crimes against ... the law on disclosure of military secrets'* to 18 months prison.[34] Von Ossietsky accepted, before he was pardoned prematurely after seven months in jail in December 1932, while Kreiser fled to France immediately after the sentence. In the following weeks Carl von Ossietsky continued to be a constant warning voice against militarism and Nazism, before he was arrested in February 1933 and thereafter put to various concentration camps, where he was severely mistreated, while deprived of food. On the initiative of Willy Brandt (1913–1992), then in Norway and later Chancellor of the Federal Republic of (West) Germany from 1969 to 1974, von Ossietsky received the 1935 Nobel Peace Prize (which Brandt himself received in 1971), which the Nazi government denied by refusing him to travel to Oslo. On 4 May 1938 Carl von Ossietsky died in a Berlin hospital, still in police custody of tuberculosis, and from the after-effects of the abuse he suffered in the concentration camps.[35]

---

led the DVL Dept. M in 1943. In addition,—see Zacher, Otto Fuchs, p. 977, has the information that Fuchs founded in 1934 and was Director of a DVL Dept. *'Flugwerk* (airframe)'* with additional responsibility for setting up aero-education and -training programmes for young technicians and engineers up to war's end; in 1944 he became also a DVL board member. After the war he was highly influential to rescue and restart the idea of a centralised West-German aeronautical research establishment (now DLR); however, the early chapter of *Lipezk*, the *DVL Dept. M* and implications with the *Weltbuehne-Prozess* still remain to be investigated.

[32] Though it can be taken for granted that Reichswehr plans for practising poison gas bombing under the camouflaging term *'Schaedlingsbekaempfung* (pest control)' at or near Lipezk, as outlined in Hauptmann Student's travel report of 10 Sep. 1926, Sect. 4.3.1, remained just plans, it is more than likely that both the Lipezk school head W. Stahr, as well as in his partial succession O. Fuchs, had been informed accordingly; the latter published in 1933 his WW I memoirs in a 251 p. autobiography, entitled *'Wir Flieger* (We flyers)'*, where his Russian post-war chapter at Lipezk was explicitly removed and remaining names alienated.

[33] See Wikipedia, 'Carl von Ossietsky' in English. Von Ossietsky was married to Maud Lichfield-Woods, born to a British colonial officer and granddaughter of an Indian princess in Hyderabad. With some probability Kreiser knew at least Fuchs personally from the glider camps at *Wasserkuppe* in the mid-1920s. During the process, the counsel for the defendants pointed out that the information they had published was true and more to the point that the budgeting for *DVL Dept. M* had actually been cited in reports by the Reichstag's (parliament's) budgeting commission. In addition, they named 19 international witnesses, amongst these Charles Grey Grey, Fig. 6.10 (l), founding editor of the British weekly *The Aeroplane* and the second editor of *Jane's All the World's Aircraft* to testify, that the claimed secrets were already known abroad, which the court discarded.

[34] For details,—see Wikipedia, 'Weltbühne-Prozess' in English.

[35] The memorial grave of Carl and Maud von Ossietsky is on the Berlin-Pankow, *Cemetery IV* at Herthaplatz, Berlin-Niederschoenhausen.

After 1 August 1933 when the 'Abteilung für Ingenieurnachwuchs (Dept. for Junior Engineers)'[36] was established, Fuchs and his small team at DVL initiated successfully numerous programmes to intensify the recruitment of young aeronautical engineers and pilots with background from the student's flying activities at technical universities and engineering academies, but his most ambitious plan was the 'Flugbaumeister'. According to Schlaifer,[37] 'The course had been established to fill the need for engineers who would be neither highly specialized theorists nor practical handbook engineers, but men broadly trained in all aspects of aviation and aeronautical engineering, and who it was expected would soon make their way into responsible positions in industry or government. The course involved theoretical work both in courses and in independent research, practical experience in industry, and as training as a pilot.'

Though reliable facts about the 'Flugbaumeister' programme are still missing, Schelp's CV illustrates five decisive preconditions in this context, in his case for the picked specialisation in *power plants*:

(I) Dipl.-Ing. (Master degree) in Mechanical Engineering at technical university level,

(II) piloting licences, originally apparently up to Class C (multi-engine aircraft beyond 5 t Take-Off Weight, TOW), but early confined as in Schelp's case to B1 level, 1–4 seat aircraft up to 2.5 t TOW),[38]

(III) a *Luftwaffe* military career, at least as in Schelp's case, to *ROA Reserve-Offiziers-Anwaerter* (reserve officer candidate),

(IV) two semester university studies in a foreign language,

(V) participation in a compact lecturing course of 30 months duration with individual specialisation in aerodynamics, power plants or instrumentation/electronics and successful final examination.[39]

To start with the last point V, Schelp began his personal 'Flugbaumeister' programme at DVL Berlin-Adlershof on 1 November 1936—and finished it, passing successfully the final examination on 15 April 1939. Before that official start, he had already accomplished

---

[36] See Hirschel, Aeronautical Research in Germany, p. 75 f. Fuchs's DVL department cooperated also directly with the research department LC1 of the RLM Technical Office, then led by—see Wikipedia, 'Adolf Baeumker' in English, which explains also Helmut Schelp's smooth, first professional move out of the DVL 'Flugbaumeister' training directly to this RLM department in August 1937.

[37] See Schlaifer, The Development, p. 383.

[38] See Zegenhagen, Schneidige deutsche Mädel, Anhang 8.1 Luftfahrerscheine, p. 451, courtesy to Hedwig Sensen, 5 Jan. 2021: Up to 1945 German motor pilot licence classification comprised—Class A1 (1-2 seat aircraft up to 500 kg TOW),—Class A2 (1-3 seat ac. up to 1 t TOW),—Class B1,—Class B2 (1-8 seat ac. w. 2.5–5 t TOW), and—Class C.

[39] It goes without saying, that the candidate had to pass also the first—of several following—thorough *GeStaPo* (secret state police) background screenings which attested Dipl.-Ing. H. Schelp of being *'einwandfrei* (impeccable)' on 7 Jan. 1937.

the listed pre-conditions I–IV. In this context, a few significant deviations between Schelp's own recollections[40] and documented dates in his personnel file are noteworthy. Schelp: *'I had heard of this new* ("Flugbaumeister") *training program before I came to the States* (in Sep. 1935)' and *'I got my Bachelor of Science degree in Germany and then went after the Master degree in Dresden, Germany. Just before I got the Master degree in Germany, I had a chance to come over to this country* (USA)....'. Also Schlaifer in his assessment,[41] *'After theoretical training in engineering both in Germany and the United States, Schelp had been one of a small group chosen in 1936 to follow a new advanced course in aeronautical engineering at the German Research Institute for aeronautics (DVL) in Berlin'*, mixes up the official November 1936 starting date and the foregoing preparations which imply that Schelp's *'Flugbaumeister'*-oriented career planning must have started at least 2½ years earlier. After a thorough review of Schelp's CV, it is more likely that he was informed about plans for the emerging *'Flugbaumeister'* training programme possibly already as early as during the *Deutschlandflug 1934*, 21–24 June 1934, where the participants returned daily back to the flyer camp at Berlin-Tempelhof—and also the newly installed DVL pilot recruiting programme developer Otto Fuchs must have appeared. At that time Schelp had already started his three semester *Vordiplom* (Bachelor degree) studies at TH Dresden in the time frame May 1934 to June 1935, where he passed the examination with a final grade 'Good' on 28 June 1935. As early as May 1934, Schelp should have met also Walter Stahr in the Dresden flyer community, so there is the additional possibility that Stahr introduced and recommended Schelp to Otto Fuchs as a hopeful candidate.

Since nothing has been found in relation to the *'Flugbaumeister'* programme, it is difficult to estimate its overall effect—beyond its limited number of 50–100 participants. In any case it was a short-living programme up to 1939/1940 only, when increasing conflicts about the military authority (*'Befehlsgewalt'*) brought it to an end and generated the professional profile of a *'Luftwaffe engineer officer'* instead. After the military basic training in this case there were plans for a flight training up to Category C2/'Blind flight 2', followed by six semesters of mechanical engineering studies at TH Berlin-Charlottenburg and later at the *Technische Akademie der Luftwaffe* in Berlin-Gatow. In reality followed after the flight training during wartime immediately the *'Frontbewaehrung* (front line probation)' with the effect that at the turning of 1943/1944 approximately 50% of the 1940 programme applicants were dead or missed in combat—and the survivors then received no longer a permission for such a *'study escape'*.[42]

---

[40] See Ermenc, Interviews; the interview collection with *'German contributors to aviation* history' appeared in 1990 only, though the tape-recorded interview with H. Schelp on pp. 97–126 took place already at AiResearch Manufacturing Co., Phoenix AZ on 22 April 1965.

[41] See Schlaifer, The Development, p. 383.

[42] Courtesy to personal memories of Prof. Dr.-Ing. Gert Winterfeld (1924–2021), DLR Cologne, 12 March 2018. For Luftwaffe personnel recruiting and training—see Stilla, Die Luftwaffe, especially about pilot training pp. 207–226.

One might speculate, if Schelp's further career on the way to the *'Flugbaumeister'* was at least partially influenced by his extraordinary political engagement far above-average. Latest at TH Dresden and thereafter his CV shows such a breadth of nearly superhuman, parallel activities in flying and practical engineering training, university studies and politicising, that the evidence of a significant self-confidence implies the backing of a kind of inherent system might. And, the sequence of narrowly clocked events leaves no room for any randomness. As part of the *Vordiplom* qualification at TH Dresden, he had to show 28 months of practical mechanical engineering training, of which he absolved the largest part between March 1933 to April 1934, and then again from August to October 1934 at DKW Zschopau, Saxony. Especially, these last 3 months as responsible designer in the DKW motor-cycle racing department were in his own view important and represented in hindsight again an invaluable precondition for his stunning success at the Stevens Institute of Technology, 1½ years later. Another example of the 23-year-old Schelp's incredible multi-tasking capability is his successful attendance—always in parallel to his *Vordiplom* preparations end of June 1935—of a *'Flight theory teacher'* short course in March/April 1935, the acquired knowledge of which he immediately implemented up to mid-August 1935 in several training courses with an officially registered expertise of his scholars *'far above the past average'*.

Finally on 26 August 1935, he received, facilitated by *DAAD Deutscher Akademischer Austausch-Dienst* (German academic exchange service), his foreign exchange scholarship from the *IIE Institute of International Education*, New York[43] for a 10-months stay at *Stevens Institute of Technology,* Hoboken NJ,[44] one of the oldest technological universities in the United States, and the first college in America solely dedicated to mechanical engineering.

Helmut Schelp belonged to a group of 17 exchange students, who were checked in at Ellis Island, NY, on Thu 5 Sep 1935, 7.35 pm as incoming passengers on board of the *HAPAG Hamburg-Amerikanische Packetfahrt-Actien-Gesellschaft* (Hamburg-America) liner *'S.S. Albert Ballin'*, Fig. 7.4 (l).[45] The ship had been named after the Jewish shipping

---

[43] See Wikipedia, 'Institute of International Education' in English.

[44] See Wikipedia, 'Stevens Institute of Technology' in English—with the Latin motto *'Per aspera ad astra'* (Through adversity to the stars), similar to that of the Royal Air Force, Fig. 6.30; as outlined in this article, the Stevens Institute of Technology opened in 1870 and initially was dedicated to mechanical engineering. The founder John Cox Stevens was the first commodore of the New York Yacht Club; he and his brother built the yacht *America* and were aboard its 1851 regatta victory in England, later recognised as the first winner of the *America's Cup*.

[45] See Wikipedia, 'SS *Albert Ballin*' in English. The departure from the Hapag Cuxhaven Terminal had been on Fri 30 Aug.1935, 11.35 h, and then via Southampton and Cherbourg in 6 days to New York. In spring 1934, the 22,000 BRT Hapag steamer received a new bow design which prolonged—as visible in Fig. 7.4—the overall length by 15 m to 206.5 m in total; in the bow section was also the third class (students) compartment. Information about Schelp's Ellis Island arrival, courtesy to Richard Fuchs, 14 Aug. 2020, who detected also the Atlantic crossing of Hans M. Antz (1909–1981), Schelp's *'Flugbaumeister'* colleague with responsibility to Me 262, on board of 52,000

Fig. 7.4 H. Schelp at New York/Hoboken NJ, Sep. 1935–June 1936: SS *'Albert Ballin (Hansa)'* 5 Sep. 1935 (l), Lower Manhattan—as seen from Stevens Inst., 1938 (m), typical students' hall (r)

magnate A. Ballin (1857–1918), who had made Hapag for a time to the world's largest shipping company. Ballin's name was also closely tied to the history of German aviation, as described already in short in Chap. 5: the bronze eagle on top of the *'Flyer Monument'* on *Wasserkuppe* originally decorated Ballin's garden at Hamburg and was a gift of Ballin's widow together with a substantial financial donation, handed over to Otto Fuchs in 1922; Fuchs was then secretary of the *Ring Deutscher Flieger e.V.* (Ring of German Flyers) which erected the fiercely westward-looking monument—and also the nationalistic-pathetic monument inscription came from Fuchs.[46] Shortly after the last Atlantic crossing as SS *'Albert Ballin'* in September 1935, on 1 October 1935 the ship was renamed to Hapag SS *'Hansa'* on the insistence of Goebbels' Ministry of Propaganda to eliminate the memory to Albert Ballin; nothing is known about Otto Fuchs's and Helmut Schelp's reaction then.

As illustrated in Fig. 7.4 (m), which shows New York's Lower Manhattan skyline as of 1938, the Stevens Institute of Technology is located on the west-bank of the—here— 1.2 km wide Hudson River, just 3.5 km in the west of Empire State Building, in the background centre; correspondingly, Fig. 7.4 (r) shows a typical students' hall of that time on the Stevens Institute campus.

From Sun 13 Oct. 1935 dates a unique survey letter[47] which Schelp wrote from *'Stevens Castle'* to his *'Lieber Herr Professor! (Dear Prof. Trefftz)'*, describing first the stormy Atlantic crossing with several days *Beaufort 10–11* and 90% of all passengers seasick, followed by his first technical New York impressions, on the one hand deeply impressed by the *'riesige Felsengebirge* (huge rocky mountains)' of the *Empire State Building* and the

---

BRT Norddeutscher Lloyd *'S.S. Bremen'* between 8 Nov. (Bremerhaven) and 13 Nov.1936, 10.40 am (Ellis Island) as transit passenger on his way to GALCIT Pasadena, Ca.

[46] The Flyer Monument literature refers in this context also to a *Reichswehr-Oberleutnant* (first lieutenant) *Ottfried* Fuchs, but the agreeing life data (1897–1987) indicate to the same R. Otto Fuchs.

[47] Letter, courtesy to Lutz Müller, 18 April 2020, from TUD-Universitaetsarchiv <Trefftz Files>.

1000 m span of the *George Washington Bridge*,[48]—but also somewhat dismissively mentioning the old-fashioned steam ferryboats with standing beam-engines (*'Balancier'*) and telephones *'to be seen in German museums'*. He appears to be somewhat disappointed in comparison to his foregoing TH Dresden experiences by the *kindergarden* type lecturing with regular in-class tests, but the homework requirements are higher. He picked four courses in *Industrial Engineering*, *Economics of Engineering*, *Advanced Thermodynamics* and *Internal Combustion Engines (ICE) in Research*, the latter two apparently presented by Professor Eugene Hector Fezandié (1897–1957) to Schelp's explicit liking. Fezandié was also his master's thesis supervisor on an ICE research subject: *'The anti-knock effect of alcohol additions to gasoline'*, as alternative to the then spreading use of TEL TetraEthylLead. Different alcohols had to be investigated in different concentrations on their effect on the fuel octane rating, on exhaust temperature and engine power. With some likelihood one can assume that Schelp had at least already heard of the topic during his stay at the DKW racing team at Zschopau, one year earlier, if he had not even such motorbike tuning practice experiences himself. As he describes in his letter to Trefftz, he alone had access to a newly acquired *CFR Cooperative Fuel Research Engine*[49] for these investigations. After his weekly progress reports to Professor Fezandié developed positively, Fezandié suggested to write an SAE Society of Automotive Engineers paper[50] on this subject for the East Coast chapter annual prize competition, for which Schelp presented his paper at the annual SAE meeting already in January 1936, and finally won the first prize, awarded in April 1936.[51] After reporting his steadily improving command of the

---

[48] Designed by—see Wikipedia, 'Othmar Ammann'in English, (1879–1965), a Swiss-American civil engineer, whose bridge designs include also the Verrazzano-Narrows Bridge and assisting support to the Golden Gate Bridge; see also Chap. 9 on the model bridge across the small *Argen* creek near Kressbronn, and in the vicinity of Lindau-Rickenbach, birthplace of the French ATAR turbojet engine.

[49] The *CFR Engine* is also listed as #50 in the *ASME Technical Landmark* programme for the year of its first introduction—1928:

   https://www.asme.org/about-asme/engineering-history/landmarks/50-cooperative-fuel-research-engine

[50] See Schelp, Alcohol-Blends. Though Schelp believed that this topic was partially new, one can assume that the Stevens library was also well vested with related literature as e.g.—see Tizard, The character of various fuels.

[51] This episode is also in—see Ermenc, Interviews, p. 107 f., but wrongly with <*Fernandez*> as the supervisor's name. The Stevens Inst. of Technology established in 1958, one year after his death, an *'Eugene Fezandie Award'* for this distinguished and popular Professor of Mechanical Engineering. Prof. Fezandié's Paris-born father J. Hector Fezandié (1856–1943), a successful author of children's books and—as *Edgar Morette* of detective stories—had used/coined in an 1894 graduation speech at *Stevens Institute of Technology* (sic) the later popular proverb (which he might have brought along from France): *'With great power goes great responsibility'*. This proverb was considerably popularised since 1960 by the *Spider Man* comic books and films, and most recently requoted by President B. Obama in a press conference at Perth, Au, on 23 Dec. 2010 in reference thereto,—see Wikipedia, 'With great power comes great responsibility'.

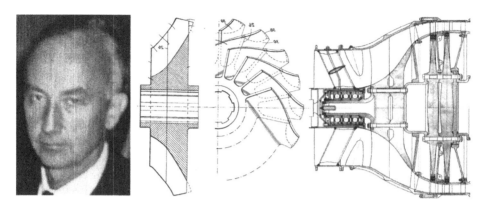

Fig. 7.5 Rudolph Birmann (l), his diagonal impeller pat. US1,959,703 in 1932 (m), Heinkel HeS 011, 1200 kp *'Class II turbojet'* with Ohain's/Schelp's first diagonal/axial compressor design, July 1942 (r) © V. Koos, r

English language, he comes to the essential, daring message: *'I feel really relieved that I left the formal student body and other obligations behind at Dresden—and can now work freely again. It was high time that I left, it was neither one thing nor another.'*

Unnoticed at first, Schelp carried along a *souvenir* from his US stay, which came to fruition only six years later.[52] Possibly in context with his SAE activities, he must have come in contact with the Swiss-born engineer Dipl.-Ing. Rudolph Birmann (1899–1968), Fig. 7.5 (l),[53] who worked at the time of Schelp's US visit and continuously between 1923 and 1940 for De Laval Steam Turbine Co., Trenton, NJ, some 100 km in the south-west of the Stevens Institute at Hoboken, NJ. Birmann had studied mechanical engineering at ETH Zurich between 1918 and 1922. For his diploma thesis he had submitted design and calculations for a free-piston, compound gas turbine—similar to the later Pratt & Whitney PT-1 turboprop test engine of 1941,[54] but with a radial turbine wheel of the centripetal type, *'believing that it would show a better efficiency than the ordinary axial-flow turbine, be capable of handling larger pressure ratios, and operate at higher speed for a given flow, thus exposing less blade surface to heating by high-temperature gases and simplifying the cooling of the wheel, which was necessary with the materials then available. In addition, the centripetal turbine was free from all problems of attaching the blades to the disk, since*

---

[52] And, consequently, belongs in full to Sect. 8.1 *'German turbojet developments'*, 1940–1945.

[53] The unique German axial thrust compensation concept for early turbojet engine designs with three rows of ball bearings, including this first HeS 011 design from July 1942,—see Koos, Heinkel Raketen- und Strahlflugzeuge, p.71, will be discussed in Sect. 8.1 for the Jumo 004 application.

[54] For PT-1, P&W's first turbine engine animation:
   https://aviationtrivia.blogspot.com/2015/02/the-pt1-pratt-whitneys-first-turbine.html

the blade and disk could be machined out of a solid forging.[55] Based on patent US1'959'703 with priority 26 January 1932, Birmann transferred these ideas also to the De Laval turbocharger concept—a one-sided cantilevered design with typically a diagonal compressor wheel on the cold bearing side, and an overhanging centripetal turbine.[56] Since Hans von Ohain's first turbojet engine Heinkel HeS 3B relied on the centripetal turbine concept as well, one might speculate about possible influences directly from Birmann's 1922 ETH thesis. In addition, as illustrated in Fig. 7.5 (r), Birmann's idea of a diagonal compressor wheel reappeared after 1942 in the Heinkel HeS 011 engine concept of Germany's first *'Class II turbojet'*, according to Schelp's (coming) classification system. At that time it might have been not too difficult for Schelp to suggest the diagonal compressor concept to von Ohain, who remembered Birmann's name perhaps still from his first engine design.

Finally, Helmut Schelp received his Master of Science certificate, equivalent to a German Dipl.-Ing. degree, from the Stevens Institute of Technology, signed by the President of the College Harvey N. Davis (1881–1952)[57] on the *'thirteenth day of June in the Year of the Lord one thousand nine hundred and thirty-six'*, which in RLM transcription was humourlessly shortened to *'13.6.1936'*. In one of his conveyed CV's, dated 15 October 1936, Schelp describes this achievement: *'In der Mindestzeit von 2 Semestern gelang es mir, trotz Sprachschwierigkeiten, zum "Master of Science" zu promovieren* (In the demanded minimum time of two semesters, I managed, language problems notwithstanding, to achieve the Master of Science degree)', in which *promovieren* in German is exclusively used for the earning of a doctorate, and a hint to an—depending on interpretation—unworthy or desperate spectacle, which soon would unfold.

Presumably in contact with Otto Fuchs, Schelp was informed that his *'Flugbaumeister'* training at DVL Berlin-Adlershof was planned to start on 1 November 1936 only, so that there was still time for a slow return to Germany and further action thereafter.

In the mentioned CV he wrote modestly: *'For getting back to Germany, I signed in as cook's mate on board of the Hapag steamer 'Reliance' to return with some detour back to Hamburg.'* In fact, this was the official cruise *'Hapag-Nordlandfahrt 1936 to the Northern Wonderlands and Russia'* on board of the 187.4 m long, 19,980 BRT S.S. 'Reliance',

---

[55] See Schlaifer, The Development, p. 454 f., where Schlaifer up to p. 457 deals at length also with Birmann's role 1937–1944 as Vice President and Chief Engineer of TEC Turbo Engineering Corp., Trenton, in a bunch of turbo-supercharger and gas turbine programmes under US Navy contracts.

[56] Birmann was in fact a master inventor with approx. 72 filed turbomachinery patents, mostly in the area of exhaust gas turbochargers in view of similar BBC and Buechi concepts with radial compressor and turbine wheels, with which he claimed for a De Laval application of having achieved a total pressure ratio PR ~ 4.

[57] See Wikipedia, 'Harvey N. Davis'in English, the third President of the Stevens Institute of Technology from 1928 to 1951, and the 57th President of ASME American Society of Mechanical Engineers in the year 1938–1939.

Fig. 7.6   H. Schelp's *rolling home* from New York to Hamburg, 26 June–30 July 1936: Cook's mate on Hapag SS *'Reliance'* (l), *Fourth of July* 1936 festivities on board (r)

Fig. 7.6 (l),[58] for which exists still the special *Fourth of July 1936* programme in the Reliance Winter Garden, Fig. 7.6 (r).[59]

After all preconditions for a successful start in the *'Flugbaumeister'* programme had been fulfilled, it appears that Schelp had changed his mind or became shaky in his commitment thereto. A possibility could have been that the course now was him too closely linked to the regime's targets, and he looked rather for a prolonged stay in the university environment.

On 23 August 1936 he wrote from his Goerlitz home address to the *Reich Ministry of Science, Education and Popular Education*, Berlin, describing his recent achievements up to the *Master of Science* degree at the *Stevens Institute of Technology*, and asking consequently for permission to continue with a doctorate in engineering. This letter was forwarded from the Ministry and answered on 16 December 1936 by the Rector of *TH Berlin-Charlottenburg*, that a doctorate in this case could be only allowed after two additional study semesters. This was apparently a relatively recent development, and

---

[58] See Wikipedia, 'Reliance (Schiff, 1920)' in German. The cruise left New York on 26 June, to continue via Reykjavik, Iceland, Spitsbergen, Hammerfest/ North Cape, the Lofoten Islands, Trondheim, Bergen, Oslo, Tallinn, Leningrad, Helsinki, Stockholm, Visby, Copenhagen and to arrive at Hamburg after 34 days on 30 July 1936; the minimum fare was then 495 USD, including the transportation back to New York. Part of the passengers might have continued to the 1936 Summer Olympics, Berlin, 1–16 August 1936.

[59] The *fourth of July* programme started at 3.00 pm under the honorary chairmanship of Captain H. Kieff, Commander (Kommodore) and the official event chairman Mr. J. W. Kellogg—with the audience singing for the opening *'America, the beautiful'*, followed by *'Stars and Stripes forever'* from the Reliance Orchestra, . . ., an address by Rev. George C. Moor, *'Songs you like to sing'* and the *'Star-Spangled Banner'* as the audience's closing song. Captain Hans Kieff was *'newly named staff captain of the Hapag Lloyd liner Reliance, a job which is eluding him like a will o' the wisp'*, as the *New York Times* wrote on 10 Jan.1936, before Kieff directed the *Reliance* for a 4-month-tour around the globe. John W. Kellogg (*1883) was the first son amongst eight children of John H. Kellogg (1852–1943), best known today for the invention of the breakfast cereal cornflakes, originally intended to be an *anaphrodisiac*, made by JHK's brother Will K. Kellogg.

after Schelp in a return letter, dated 10 February, 1937, pointed out, that he had in good faith already received a (verbal?) agreement for a doctorate without foregoing extra-studies directly from the Ministry, also the TH Berlin-Charlottenburg accepted him accordingly in a letter from 17 February, 1937.

In the meantime however, the official *'Flugbaumeister'* programme had started on 1 Nov. 1936, so that Schelp as employee of DVL Berlin-Adlershof had to give up his *Dr.-Ing.* promotion plan. The length of the Flugbaumeister programme was limited to 30 months up to the final examination in April 1939. The first stations are known— November 1936 to February 1937 flight engine trainee at Daimler-Benz AG, Work 4, Berlin-Marienfelde, March to May 1937, military flyer training at Zeithain, 40 km in the north-west of Dresden, which he finished as *Private of the Reserve* and *Reserve Officer Candidate*. According to a personnel questionnaire, dated 28 June 1938, he put himself as *Assistant Referent* to the RLM Research Department between August 1937 and August 1938, and thereafter as *Referent* and *Flugbaufuehrer* (one level under the *Flugbaumeister*) to the RLM Development Department under Hans Mauch. For the trainee stay at Daimler-Benz exists—triggered by a RLM request from 8 November 1937—a somewhat strange collective attestation with solely six uninspired lines about Helmut Schelp, which confirmed bluntly that he had shown up (without providing dates) with the prognosis that he might become *'qualified for a job in the machine shop or in flight operation'*.

Another event that has to be taken into account to structure Helmut Schelp's life between November 1936 and his official working start in the RLM development department LC 7/8 after Hans Mauch took the responsibility for this department on 1 April 1938, is the *Fourth international flight meeting* at Duebendorf, Switzerland between 23 July and 1 August 1937. This event and the exhibited *Messerschmitt Me 109* speed performance is important, since it roughly fixes also the starting date of Schelp's first study task, in which he simply was asked to investigate *'what would be required for aircraft* (propulsion), *if you wanted to double their speed. This was the only specification I was given'.*[60] *'Duebendorf'* therefore separates a *'preparatory phase'* upfront of that study of approximately six months from November 1936, first up to the end of February 1937, once in a while interrupted by a parallel *'visit'* at Daimler Benz Marienfelde not to endanger his obligatory industrial internship there, and then intensely continued in the months June and July 1937 after his military flyer training, and—after *'Duebendorf'*—the actual study phase of also some six months, roughly between September 1937 and February 1938. Main purpose of the *'preparatory phase'* was a thorough personal training course in his picked prime specialisation subject *'power plants'* for *'Flugbaumeister'* qualification at DVL Berlin-Adlershof and the responsible *DVL-Institute for Engines, Working Methods and Thermodynamics,* then under the lead of Dr.-Ing. habil. Fritz A.F. Schmidt (1900–1982).[61]

---

[60] See Ermenc, Interviews, p. 102.

[61] See Wikipedia, 'Fritz Schmidt (Ingenieur)' in German.

Fig. 7.7 DVL Berlin-Adlershof extension programme 1933–1935, I: Large subsonic wind tunnel (l), 2 MW axial blower of 8.5 m diam. with adjust. blades (m), free-spin wind tunnel, 20 m height (r)

The *DVL Deutsche Versuchsanstalt fuer Luftfahrt* (German Test Establishment for Aviation) had been founded in 1912, and was located from the very beginning at the *Johannisthal Airfield* in Berlin-Adlershof.[62] In 1933 there were approximately 500 employees at this site, a number which grew up to 1940 to 2100, thus rather only weakly reflecting the overall aviation personnel increase in research and industry in Germany in that period from 4000 to 300,000.[63] After 1933, immediately greater financial resources were available for a considerable DVL extension plan with test facilities of impressive size even today.[64] On the northern campus of the original DVL site a new closed-loop, subsonic wind tunnel, Fig. 7.7 (l, m) was erected after 1933, covering an area of 45×58 m; the exchangeable elliptical measuring section had 5×7 m, and 6x8 m respectively, the maximum nozzle velocity was 55 m/s. In addition, a high-speed wind tunnel was commissioned in 1938 with 13 MW drive power. The test section for tests close to Ma ~ 0.95 measured there 2.7×2.7 m. In addition, a free-spin wind tunnel, Fig. 7.7 (r), was erected for spin testing of aircraft models up to 1.5 m span, within an upward directed, vertical air stream of 4 m diameter and up to 40 m/s wind speed. The whole tower could be pressurised up to 3 bar abs. for keeping the aerodynamic similarity conditions.[65]

---

[62] Approximately 8 km in the north of the newly opened, BER Berlin-Schoenefeld airport.

[63] See Hirschel, Aeronautical Research in Germany, p. 72.

[64] The development of the Science and Technology Park Berlin-Adlershof is maintained by *GBSL Soc. for the Preservation of Historic Sites of German Aviation History,*—see www. luftfahrtstaetten.de

[65] The models were inserted with a rotatable telescope mechanism and then electro-magnetically released, the model rudder positioning was radio-controlled. Several high-speed cameras recorded spin initiation, the balanced, continuously adjustable vertical drift and the spin recovery; the movie light installation had 170 kW. Results of these investigation were summarised in Feb. 1935 in an internal, classified report,—see Richter, Das Trudeln, of which a copy in Russian was intercepted by *GeStaPo* (secret police); another source indicates an information 'leak' to England as well. The author Willi Richter survived the war in the infamous *'Zuchthaus* (jail) *Brandenburg'* solely due to

Fig. 7.8 DVL Berlin-Adlershof extension programme 1933–1935, II: Universal piston engine altitude test stand, air- or water-cooled, max. 1500 hp, max. 5 m propeller diam., today (l) and ~ 1938 (r)

Figure 7.8 provides an impression of Helmut Schelp's working environment at Professor *FAF* Schmidt's Institute for Engines in the first half of 1937. Most impressive—still today—is the noise-damped engine test facility which was newly erected between 1933 and 1935, also for altitude simulation up to 24,000 m. Both, air- or water-cooled engines up to 1500 hp in combination with propellers of up to 5 m diameter could be tested on a pendulum-frame set up. The engine test facility was designed for mass flows up to 1.2 kg/s, later to be increased to 3 kg/s. The altitude control allowed for inlet temperature variations between -30 °C to $+55$ °C. The idea to install an expansion turbine for combustion air reduction had been suggested by the DVL pioneer of altitude flight Asmus Hansen, who also contributed to the development of numerous other altitude engine test facilities. Main contractor for this DVL demonstration project was BBC Brown, Boveri & Cie., Mannheim, which thereby preserved the right for license installations. Air-cooled flight engines were tested by means of an additional 0.5 MW axial blower which generated a cooling air slip stream of up to 70 m/s. The facility's air intake and exhaust towers and the—no longer preserved—water cooling towers, visible in Fig. 7.8 (r), were 15 m tall.[66]

Fritz Anton Franz Schmidt, during his lifetime only known as *FAF*, was described as efficient and authoritarian type of German professor.[67] Since 1925, he was assistant to the

---

his continuing, indispensable work on *'aircraft spinning'*. After the war he became part of the *Baade Ba 152* turbojet aircraft project at Dresden—see Sect. 10.2.1, and was a Professor for Flight Mechanics at TU Dresden,—see Altenburg, Ingenieur.

[66] For a comprehensive description—see Hirschel, Aeronautical Research, pp. 217/218. Out of these developments, the unique *'Herbitus'* turbojet engine altitude test facility at BMW Munich-Milbertshofen (today AEDC Tullahoma, TN) will be described in Sect. 8.3.3.

[67] See Jenny, The BBC Turbocharger, p. 69. Jenny had met Schmidt after the war at RWTH Aachen: *'Nobody spoke without permission and in the evening at about 7.00 the assistants cautiously inquired whether they were still needed, or if they could go home. Everything was done briskly and correctly. Many of Prof. FAF Schmidt's assistants who later had successful careers were grateful to him for*

Fig. 7.9 Making a DVL *'Flugbaumeister'*: R. Otto Fuchs (1897–1987) l, Helmut Schelp (1912–1994) m, Heinrich Kühl (1906–2000) r

renowned Professor Wilhelm Nusselt[68] at TH Munich, Institute for Thermodynamics, where he received his Dr.-Ing. degree in 1927. In 1929 he came to Berlin as director of the power station of the machine and locomotive manufacturer *A. Borsig GmbH*; in 1930 he habilitated at TH Berlin in thermodynamics, before in 1933 he was installed as director of the newly founded DVL Institute. Looking for qualified and reliable scientific support during the extension phase of his institute, he came back to Nusselt's TUM Institute, where he hired as the DVL Institute's *Oberingenieur* (chief engineer) Dr.-Ing. Heinrich Kuehl, who had received there his Dr.-Ing. in 1935.[69] As thesis supervisor and a kind of personal tutor during Schelp's *'Flugbaumeister'* training he had indirectly also considerable influence on the further turbojet developments. Correspondingly, Fig. 7.9 shows the *Making of a 'Flugbaumeister'* at DVL, with Helmut Schelp (25)[70] at the centre, framed by his mentor R. Otto Fuchs to the left, and his scientific advisor Heinrich Kuehl[71] to his right.

---

what they had learned under his hard institutional regime, and his book (with several editions between 1939 to 2013) *was extremely useful.'*

[68] See Wikipedia, 'Wilhelm Nußelt'in German. *Nu*, the Nußelt similarity number describes the convective heat transfer between a solid surface and an attached fluid flow.

[69] See Kuehl, Die Dissoziation. This thesis is remarkable due to its 30 p. brevity, which reflects—though the kind and polite Kuehl had nothing in common with his master's (Schmidt) authoritarian style—the generally determined, no-nonsense style of the Institute's spirit. The thesis deals with the reaction chemistry of combustion processes, an indication of Kuehl's extraordinarily broad theoretical background. He compares his study results—with good agreement with those of Tizard and Pye 1924—see Tizard, The Character, a hint that Schelp's knowledge of relevant gas turbine and turbojet literature was also with all likelihood collected at that DVL Institute.

[70] The only available picture of Schelp as RLM employee, from a personnel questionnaire, dated 28 June 1938, part of Schelp's BA-MA file.

[71] See Kuehl, Die Brennkammer, with a portrait taken during his Turboméca stay after the war.

After Kuehl's arrival he soon must have been integrated in the Institute's comprehensive publication activities, which certainly supported Schmidt's official professorship in 1940. Already in 1939 appeared the first edition of Schmidt's '*Verbrennungsmotoren* (Combustion engines)' with 328 pages, which grew over time in several editions up to 649 pages in 2013.

A speciality of the book are numerous calculation examples[72]—though not officially published—they were certainly available for Schelp's training for the necessary comparative performance calculations of the assessed engine concepts in his study effort. The Institute's research comprised also topics[73] close to Schelp's own, foregoing small-scale research at the Stevens Institute—and might have alleviated his familiarisation.

An immediate indication how close Schelp's '*Flugbaumeister*' study—which in its short conciseness was certainly defined by Schmidt and/or Kuehl—was associated to the Institute's ongoing research activities is a six page paper[74] for the *DVL Jahrbuch 1937* (annual book) on the '*Applicability of various piston-engine processes for altitude and long-distance flight*' by H. Kuehl and FAF Schmidt, where they compare the performance of four- and two-stroke diesel engines as well as four-stroke Otto (ignition) engine, all with exhaust gas turbocharger—and alternatively, with mechanically coupled turbocharger. The—as stated—very approximative results for the six investigated engine concepts are summarised in a final '*Fig. 10*', here only described as chart, which shows for an invariable flight speed of 400 km/h at altitudes of 4, 12 and 16 km the plots of a '*Leistungsgewicht* (power weight)' on the ordinate over 0–20 h flight hours. The calculated results are for every altitude case a bunch of nearly equally inclined, three to six straight lines. The unique *Leistungsgewicht* is defined as the '*required weight for engine and fuel for one hp of power, in function of the flight endurance, and for the assumed constant flight speed of 400 km/h.*' The conclusive summary for this piston-engine study states first, that the sought flight range correlates with the flight time, so that this unit is used for graphic abscissa. At low altitude (4 km) and short flight times, the Otto engine with both investigated turbocharger versions is superior. Longer distances demonstrate the advantage of the two-stroke, turbocharged diesel engine. At 12 km altitude the turbocharged Otto engine and for longer distances, the four-stroke diesel engine with exhaust gas turbocharger have the largest ranges. Finally at 16 km altitude, the exhaust gas turbocharged Otto engine is generally superior, while the two-stroke diesel engine should be excluded from operation at these altitudes.

---

[72] See Schmidt, Verbrennungskraftmaschinen. In the foreword (page v), Schmidt mentions explicitly the contributions of Drs. H. Kuehl and M. Scheuermeyer for the numerical examples; visible in the Table of Contents of the 2013 edition are—Otto flight engine with turbocharger p. 464 f., gas turbine engine p. 480 f., rocket engine thermodynamics p. 485, and in addition, Schelp study topics like— combined ML piston-jet engines p. 415 and—ram jets p. 421.

[73] See Schmidt, Zündverzug.

[74] See Kuehl, Die Eignung, p. 437.

After Schelp had developed some familiarity with the basic tools for aircraft-engine performance analysis in the first half of 1937, the time neared that the *Flugbaumeister* candidate had to be given his first study assignment in his selected special field of *'flight engines'*,[75] which was—as Schelp remembered in 1965—*'what would be required for aircraft if you wanted to double their speed'*.[76] Constant[77] mentions this in the context of the Zurich-Duebendorf fourth International Flight Meeting, Fri 23 July to Sun 1 August 1937 (Swiss national holiday), for which Germany had announced in total six of their new *Messerschmitt Me109* single-seater fighter aircraft, and especially Ernst Udet had ventilated hopes for a new official speed record on his bright-red painted Me 109 with the new 950 hp DB 601 injection engine. Unofficially, he had flown the 210 km transfer distance between Messerschmitt's Augsburg airfield and Duebendorf in 23 min, corresponding to 540 km/h on average—and thus indicating already a considerable improvement to the Me 109 initial performance in April 1936 of 470 km/h, powered by the 695 hp *RR Kestrel* engine. Disappointing for Udet, his new engine set-up caused problems during the first leg of the speed competition on the opening Sunday in front of 75,000 spectators, so that he dropped out and the final ranking saw another Me 109 with 640 hp Jumo 210 G engine, piloted by Carl Francke, Rechlin, as first with on average 409.6 km/h only, in front of Charles Gardner on the British racer aircraft Percival Mew Gull, powered by a 205 hp Gipsy engine, still achieving 350.7 km/h. The expected boost in Me 109 flight speed was finally achieved on 11 November 1937, when Dr.-Ing. Hermann Wurster (1907–1985) flew his modified Me 109 with a tuned, 1660 hp DB 601 injection engine along the straight railway track Augsburg–Buchloe to a top speed of 610.95 km/h, and thus breaking the foregoing speed record of Howard Hughes of 567.1 km/h of 13 September 1935. In these cases, the *Messerschmitt Me 109* (Bf 109) generally followed its design principle to install the highest engine power in the relatively smallest aircraft fuselage.

Consequently, therefore the question arose within the DVL Engine Institute of what, in view of the new, aerodynamically improved airframes, the limits of aircraft performance really were; it was the starting point of Schelp's DVL study. In a first step, he deduced from the plane wing that the limit with all likelihood was set by compressibility and the correspondingly observed aerodynamic drag rise at sea level, at a Mach number Ma ~ 0,82,—or just slightly above 1000 km/h, according to *'existing wind tunnel data'*.[78] A number, which in view of the—in the meantime established—subsonic

---

[75] In addition, the *'Flugbaumeister'* exam requested a shorter treatise outside of the selected special field, for which Schelp addressed a ramjet design with interpretation of test results, see Fig. 7.11.

[76] See Ermenc, Interviews, p. 102.

[77] See Constant, The Origins, p. 205.

[78] The second, speed limiting factor which Schelp had to take into account, was the propeller efficiency degradation at high subsonic Mach numbers. See Wikipedia, 'Helmut Schelp' in English, where one of his (unnamed) professors at DVL is claimed for a propeller efficiency demonstration of only η 0.71 at Mach ~0.82,—see also Madelung, Beitrag (DVL 1928).

commercial jet traffic at that level was an excellent choice. Nearly automatically came the next question, how to bridge the appearing gap between the speed at which the propeller became inadmissibly inefficient and the compressibility limitation to airframe speed. Schelp, presumably as a result of his DVL environment/library, was familiar with some early work on stationary gas turbines, especially that of René Armengaud and Charles Lemâle, and apparently also with the reaction propulsion patent of the French inventor Maxime Guillaume.[79] In addition, he was aware of publications *'concerning big stationary gas turbines from companies like Brown-Boveri and Escher-Wyss'* and he had heard of the Campini first flight, as he remembered afterwards to have happened at the turning of 1938/1939.[80] Schelp realised clearly that for flight at high (subsonic) Mach numbers, two preconditions were absolutely necessary: In view of Fig. 2.3, a power plant with a significantly higher power-to-weight ratio P_{max}/W than current engines to overcome increased high speed drag, and some means of dispensing with the propeller. As a result of his comprehensive study effort, latest in the autumn of 1937 he concluded that the propulsion system most likely to offer the desired characteristics would follow some reaction principle, not requiring a reciprocating engine. Schelp carried out besides the conventional propeller/piston engine combination as reference a systematic analysis[81] of nearly all thinkable reaction propulsion engine types, mainly for the four fundamental turbomachinery concepts of Fig. 7.10,[82] i.e. with his original German designations:

---

[79] See Chap. 2, <1906> and especially <24 Feb. 1910>, where—acc. to W. Noack, BBC—Marcel Armengaud even prophetically appears to have recommended the design switch from centrifugal to axial compressors by BBC; for Guillaume, see <1921>. Justifiably so,—see Constant, The Origins, p. 205, Schelp has been praised for this rather rare qualification among the turbojet pioneers. Of course, it goes without saying that he was not misled by the heavy weight aspects of these designs.

[80] See Ermenc, Interviews, p. 105, for <Campini>,—see also Chap. 2 <1908> in the context of R. Lorin, and the actual dates of the Campini flights in 1940/1941,—and p. 108, for the BBC and Escher-Wyss reference. However, Ermenc in the Schelp interview uncovers also astonishing knowledge gaps about Berlin's local technical heritage, where he (and presumably also DVL) neither has heard of the work of *Franz Stolze* (1872, 1904),—see Chap. 2 <1899>, nor of the *Nernst tube* (1908,—see Sect. 12.1.1, with reference to H. von Ohain's secret patents) and also not of the Huettner aviation steam power plant;—see Sect. 6.1.1. Nothing is known, if he read Max Koenig's 'prophecy' in his 70 page Newcastle gas turbine lecture, which appeared in print in 1925,—see Chap. 3, some 12 years before Schelp's breakthrough.

[81] The secret original study of 1937 has apparently not survived, but 23 pages of—see Schelp, Luftstrahltriebwerke, from a DAL Deutsche Akademie fuer Luftfahrtforschung lecture series of 31 Jan. 1941, published as part of an equally secret 248 p. book, corresponds largely with the original study, but makes of course reference to intermediate results, missing still in the 1937/1938 original. The author owes his copy #44 of this rather rare document to the generosity of the late Prof. W. Albring, TU Dresden, 2 July 1993.

[82] Original drawings, reproduced in the same scale from—see Schelp, Luftstrahltriebwerke, p. 22–26.

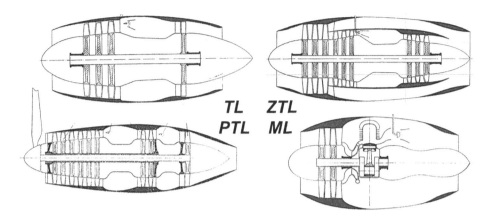

Fig. 7.10 H. Schelp DVL Study 1937: Investigated, all-axial turbo-engine configurations

- *TL Turbo-Luftstrahltriebwerk*, the pure turbojet engine,[83]
- *ZTL Zwei-Kreis-Turbinen-Luftstrahltriebwerk*, the turbofan engine,[84]
- *PTL Propeller-Turbinen-Luftstrahltriebwerk*, the turboprop engine,[85]
- *ML Motor-Luftstrahltriebwerk*, a ducted fan, powered by a piston engine.[86]

In addition to these described four turbo engines, which exclusively use axial turbo components[87] to exploit their considerably lower pressure drag due to the smaller frontal area, Schelp investigated also Lorin's simple ram-jet configuration of 1913,—with the

---

[83] The study/ paper engine was considered as being sufficient for a fighter single-engine propulsion, however, in Schelp's words *'with a certain drawback that development engines* (of 1941) *with a compressor PR ~ 4 producing 600 kp thrust, at 900 km/h weighed approx. 400–500 kg and had a specific fuel consumption sfc of 0.5–0.55 g/ (kp s), all at sea level SL.'*

[84] For the ZTL, Schelp set the Low spool PR ~1.19 and the HPR ~8; he expected for 600 kp thrust at 900 km/h a weight of 700 kg and a sfc of 0.4 g/ (kp s), all at SL.

[85] In Schelp's PTL concept is the second combustor in flow direction fuel-controlled and the rear working turbine has a variable speed to increase the power capability for *take-off* and *combat*, and thus limit the overall weight demand. The SL flight speed is reduced to 720–800 km/h to limit again the plant weight; functionally, the propeller corresponds to the ZTL LP compressor.

[86] For the ML, the axial LP fan is powered by an efficient piston engine, followed by a centrifugal compressor, unique for Schelp's study,—and an additional afterburner. Here, he expects the best sfc, but the weight will be close to that of the conventional propeller/piston engine. Consequently, the ML will find its application only in speed ranges of decreasing propeller efficiency.

[87] Nothing is known about the actual timing of this emerging concept, Schelp himself noted—see Ermenec, Interviews, p.108: *'I became convinced later on in the game that the future of the turbojet would be in axial flow machinery, …'*, a straightforward insight in view of a demanded power increase, for which the frontal diameter of a centrifugal compressor has to grow, while the axial compressor arrangement keeps the diameter in principle and grows in axial length only with higher stage count.

known disadvantages of high fuel consumption, low performance in the investigated subsonic regime and the necessity for extra take-off power. After a discussion of some basic advantages of the constant-volume-cycle for gas turbines in comparison to the here generally used constant-pressure-cycle, Schelp opens a niche application for the *Schmidt Rohr/*(later V-1) pulse jet—especially due to its superior fuel consumption of 0.7–0.9 g/ (kp s) in comparison to the ram-jet.

Finally, Schelp produces a statement, which later with the availability of the turbojet-powered *Messerschmitt Me 262,* and additionally heated up by incompetent interference of Adolf Hitler gained some importance after 1944, about the reasonability of the turbojet-powered *'Schnellbomber'* (fast bomber).[88] In agreement with his assessment of the long-range-aircraft in general, he states in 1941 *'Long distances are at best flown with fuel-efficient PTL engines, at relatively low speed and—where possible—with some extra combat boost.'* Somewhat cautiously towards his largely conservative audience, Schelp formulates his main study message: *'While in the past the piston/ propeller engine was unchallenged in the complete application range from fighter to long-range aircraft, this new insight allows to specify for every engine concept solely based on the decisive* "aircraft total weight balance" *its optimum operation range'.* And adds soothingly as last sentence: *'A last warning, I do not want to push exaggerated hopes, the piston/propeller engine of today will not be replaced completely. As shown, this concept is a natural part of the investigated row of engine configurations, and will always find its place in the range of lower flight speed and for extreme long-range flight.'*[89] As indicated before, the *'Flugbaumeister'* examination demanded two written treatises, a longer test paper (*'Klausurarbeit'*) in the candidate's *'special field'*—for Schelp, *'Possibilities and limitations for performance improvements of flight engines'*,[90] and a shorter, so-called *'Baumeisterarbeit'*, outside of the selected *'special field'*, for which Schelp had to address *'Design and calculation of a ramjet—with usage and interpretation of test results'*.[91]

After Schelp had prepared his engine concept comparison for his test paper by deducing the necessary data, in 1937 mostly on the basis of relative plausibility calculations (and only in very few cases based on measurements),[92] he went back to the described, comparative *Kuehl scheme*[93] of *Leistungsgewicht* (power weight) *vs. flight time* and just adapted the *Leistungsgewicht* definition accordingly, to the *'required weight for engine and fuel for one kp of thrust, in function of the flight endurance, and for the assumed constant flight*

[88] See Schelp, Luftstrahltriebwerke, p. 31.

[89] See Schelp, Luftstrahltriebwerke, p. 37.

[90] In German: 'Möglichkeiten und Grenzen der Leistungssteigerung der Flugmotore'.

[91] In German: *'Berechnungsunterlagen eines Luftstauantriebes'*.

[92] Especially the axial compressor performance had to be estimated, since the contact to the AVA Goettingen expert group happened in summer 1938 only, approximately just 9 months after Schelp's *'Flugbaumeister'* study accomplishment.

[93] See Kuehl, Die Eignung, Fig. 10, p. 437.

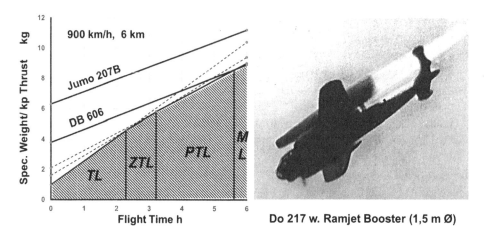

Fig. 7.11 H. Schelp *'Flugbaumeister'* Study Results: DVL 1937 (l): *Total aircraft weight/kp thrust vs. flight time*; DFL 1942 (r): E. Saenger's ramjet net thrust verification

*speed of—*now selected*—900 km/h'*. There are two, very similar charts in Schelp's study results for 6 and 0 km flight altitude, of which the first one for the combination *900 km/h, 6 km* has been redrawn in Fig. 7.11 (l).[94] The shown five straight result lines cross the ordinate at the power weight/kp thrust, increasingly at ~1 kg for *TL*, ~ 1.6 kg for *ZTL* and ~ 2.1 kg for *PTL(ML)*. In addition, there is the in 1941 most recent, best fuel-efficient piston engine Daimler-Benz DB 606[95] with ~3.8 kg and the Junkers Jumo 207B[96] of 1936, with ~6.3 kg. The relative inclination of these lines indicates the specific fuel consumption in g/(kp s), which marks the general superiority of the piston engines under this aspect. The result is shown as a slightly *'kinked'* optimum envelope curve, which represents the then state-of-the-art and which covers and sorts the various turbo-engine concepts from the pure turbojet TL at left, for short flight times up to 2.2 h, followed by ZTL, PTL and finally closed on the right rim for the *motorjet* ML, and flight times beyond 5.6 h. Obviously, the TL turbojet has the most pronounced benefits in comparison to the piston/propeller engines, and especially so for the short endurance fighter applications. Since he had concluded that a much lighter power plant was needed for high speed, and since he believed that propellers would be very inefficient at these speeds, the logical solution was the turbine-driven engine in its simplest form of jet propulsion. Thus by mid-1937, Schelp,—except for some initial DVL support—single-handed, had deduced the essential

---

[94] Redrawn from—see Schelp, Luftstrahltriebwerke, Fig. 6, p. 28.

[95] The Daimler-Benz DB 606 was a 2700 hp double-engine of two gear-coupled, speed-reduced DB 601 F/G engines, mostly used for the ill-fated He 177 dive bomber application,—see Wikipedia, 'Heinkel He 177 Greif' in English.

[96] The mid-1930 high-altitude version of the ~600 hp, Junkers Jumo 205 two-stroke flight diesel engine with two inline centrifugal superchargers and a precooler.

turbojet conclusion from aerodynamic first principles, independent of contemporary industrial teams as Whittle in England and at Heinkel and Junkers in Germany, and in the following move was about to inherit the correspondingly influential RLM technology control and decision centre.

In a separate paragraph[97] he describes the fundamental differences with respect to the altitude behaviour of all kinds of turbojet engines in comparison to the conventional piston engine. He knows that the power output of the turbojet configurations decreases for rising flight altitudes mainly proportional to air density, which in parts is compensated by the decreasing temperature with increasing altitude. This effect allows to generate more power by increasing the fuel burn up to the fixed upper material temperature limit: *'This behaviour is completely different from the piston engine with stoichiometric combustion throughout, so that the advantage of a temperature decrease over altitude cannot be exploited, while the turbojet engines are always operated with considerable excess air for material reasons.'*

Interestingly, Schelp's second *'Flugbaumeister'* treatise about *'ramjet design'* which he presumably accomplished in parallel to his RLM activities in the second half of 1938, caused some unexpected *'trouble'*.[98] Based on theoretical considerations, he had stated that a ramjet (as isolated propulsor body) would have a net propulsive thrust which could propel an aircraft. Afterwards he admitted that this had been a *'very questionable conclusion'*, which came to the attention of Eugen Saenger (1905–1964), a highly qualified Austrian ramjet and rocket specialist,[99] who since 1936 was at DVL Adlershof, and at the time of Schelp's ramjet paper in transition to build on RLM invitation his own rocket research institute at Fassberg/Trauen in the Luneburg Heath, which would become part of the newly founded DFL Deutsche Forschungsanstalt fuer Luftfahrt (German research establishment for aviation) near Brunswick. A scientific controversy emerged, which culminated in a Saenger paper to document the apparent flaws in Schelp's rationale that there could not be any thrust left after taking the ramjet propulsor drag into account. This argumentation stirred up considerable interest, however with the positive side effect, that Saenger obtained funds to build a demonstration ramjet and thus settle the argument—finally in 1942. A large ramjet with 1.5 m diam. was built at Fassberg and mounted on top of the fuselage of a Dornier Do 217 E *'flying pencil'* bomber[100] with dive-bombing capability, Fig. 7.11 (r). Schelp described this as *'hairy operation'* which, however, fulfilled its purpose: Saenger discovered that the ramjet produced a net thrust, retracted his criticisms and conceded that young Schelp was right.

---

[97] See Schelp, Luftstrahltriebwerke, p. 29.

[98] Schelp uses the word *'trouble'* in his 1965 interview,—see Ermenc, p. 107, for an affair which due to its positive outcome might have increased his professional standing in RLM office.

[99] See Wikipedia, 'Eugen Sänger' in English.

[100] See Wikipedia, 'Dornier Do 217' in English.

The document of Helmut Schelp's successful *'Flugbaumeister'* examination, which he passed with the best possible 'Sehr gut (very good)', carries the date of 22 May 1939[101]:

- The ramjet paper was judged by Prof. A. Busemann and Dr. K. Leist,
- the engine performance assessment reviewed also Prof. A. Busemann alone,
- *'Staatskunde'* (Civics) and *'Air Law'* evaluated the RLM attorneys MinR Dr. jur. R. Schleicher and MinR Dr. A. Wegerdt,
- orally examined were 'Gasdynamics' by Prof. A. Busemann, 'Flight engine thermodynamics' by Prof. FAF Schmidt, 'Flight and engine controls' by Dipl.-Ing. F. Hoppe, 'Civics' and 'Air Law' by Drs. Wegerdt and Schleicher, and finally 'Foreign language' by Dipl.-Ing. W. Wendland.

The last contribution to Schelp's *'Flugbaumeister'* personnel file came from his early-on mentor R. Otto Fuchs as a favourable judgement: *'... a perfect character with confident and skilful demeanour. His friendliness and agreeableness, his vivacious appearance in combination with his well-founded proficiency will ascertain him life-long success. His natural comradeship secures his popularity with superiors and colleagues. Schelp has average abilities as flyer; but, intellectually controlled and with regular practise he will be able to perform also in this area. Well-trained in sports and powerful ... Well-qualified for a leading position in aviation.'*

The further development, after Schelp's study had produced such a favourable result for the pure turbojet, is nowhere clearly outlined, but requires not much imagination. Schelp, definitively convinced that he *'had something'*, immediately must have looked for an opportunity to demonstrate and realise the new engine concept, this required first of all 'funding' and—even more important—a system familiarity with the actual industrial players, which he, the RLM *'rookie'* (25), could not have. Predictably, his first pro-turbojet offensive still at the DVL Institute must have been rebuffed; at that time within DVL there was apparently little enthusiasm for research in turbine engines. The smart aleck Schelp was sent from DVL in August 1937 still as temporary employee[102] to the RLM Research Division LC1, where he was put in charge of the two ongoing developments of this sort in that Division, the Walther ramjet and the Schmidt pulsejet, but he was unable to obtain support for his favourite air gas turbine research. As will be seen, there was an ongoing DVL reservation towards the new turbojet concept up to

---

[101] Confirmed by Schelp's *'Flugbaumeister'* certificate of appointment, dated 15 June 1939. From that date he was officially employed at RLM as *'Referent* (technical officer)'. Presumably during 1940 he gained in addition the title of a *'Flieger-Stabsingenieur* (Flyer staff engineer)' which corresponded to the military ranks of *'Captain/ Major'*,—see Budrass, Flugzeugindustrie, p. 504.

[102] Schelp describes his LC1 position as *'Hilfsreferent'* (technical officer assistant). In April 1939 only, officially still with the *'Flugbaumeister'* degree, he entered the government service as *'Referent'*/technical officer at RLM LC8.

1943,[103] which however had no delaying impact on the initial industrial turbojet development programmes, mainly launched at *Junkers Jumo* and *BMW/Bramo* by Mauch and
Schelp during 1939.[104] A significant exception to the general DVL conservatism against
the new engine concept demonstrated Schelp's *'Flugbaumeister'* supervisor Heinrich
Kühl, who dedicated his further research activities fully to the aero gas turbine.[105]
Officially, Kuehl and Schelp could be jointly traced up to October 1938 as participants
of the International Convention of the Lilienthal Society, Berlin, as described in detail
in the foregoing Sect. 6.1.3—and in the corresponding *List of Selected Participants*,
Sect. 12.2.

7.3   Further Theoretical Work and Programme Structuring

For keeping continuity the following addresses in short Schelp's later known publications
on turbojet basics up to mid-1943, before Sect. 7.4 returns to the RLM activities in 1939
and thereafter.

On 31 January 1941 the German Academy for Aviation Research[106] invited to a
Working Session *'Jet Engines'* in which 12 presentations provided a broad survey on the

---

[103] At best seen in the seven discussion contributions of Schelp's 1943 publication,—see Schelp,
Hochleistungstriebwerke, p. 27 (p. 17 in English version). Here, in July 1943, Schelp's former DVL
colleague W. von der Nuell states rather uninspired about selecting the right engine concept: *'A few of
my colleagues have been working for about the past three months, on the comparatively same notion
concerning diverse types of power plants. It turns out to be difficult to illustrate the attained findings,
or rather to exemplify the comparing analysis in such a manner that satisfies the engine and airframe
manufacturers.'*

[104] See Sects. 6.2.3, 6.2.4 and 6.2.5.

[105] In the early 1940s, H. Kuehl re-oriented fundamentally towards GT engine steering and control, —
see Kuehl, Grundlagen der Regelung, at best known for a control mode on the basis of *'Kühlschen
Geraden'* (lines of constant T_4/T_2 in the gas generator compressor map),—see Bauerfeind, Steuerung
und Regelung, p. 21, and after the war at the French Turboméca (TM)—see Wikipedia, 'Safran
Helicopter Engines' in English, as responsible designer and combustion specialist together with
G. Oberlaender, especially for the *'Artouste'*, TM's first really successful helicopter engine. After his
return to Germany in 1952 and a few years in the Daimler Benz gas turbine development, in 1959—
up to 1973, then Prof. Heinrich Kuehl became the first director of the newly founded DVL/DLR
Institute for Propulsion Technology at Cologne-Porz, where the author had the privilege of his kindly
understated scientific guidance.

[106] The *Deutsche Akademie der Luftfahrtforschung (DAL) was* founded on 24 July 1936—under the
presidency of H. Goering, with E. Milch as vice-president and A. Baeumker as academy secretary.
The German aero industry was represented up to 1942 by W. Messerschmitt as vice-president,
thereafter by K. Tank. L. Prandtl used his reputation to establish the required international contacts
which listed for 1938 the following *'corresponding members'*:—for Great Britain: L. Bairstow,
B.M. Jones, D.R. Pye, E.F. Relf, R.V. Southwell, G.I. Taylor, Sir H.T. Tizard;—for Italy:
G. Caproni, G.A. Crocco, A. Eula, G. Gabrielli, M. Panetti, E. Pistolesi, C. Rosatelli;—for Sweden:

status of this newly emerged area of aircraft propulsion. The welcoming speech held General-Engineer Wolfram Eisenlohr, the responsible authority for all aircraft engines within the RLM, followed by a keynote speech by Helmut Schelp, which in short has already been documented in the foregoing:

- *W. Eisenlohr Engine design for fast flight and long distances*
- *H. Schelp Turbojet engines for high-speed aircraft*
- *O. Lutz Evaluation basics for turbojet engines*
- *A. Busemann Stationary and periodic thrust*
- *R. Lusser Turbojet engines in aircraft design*
- *H. von Ohain Radial turbojet engines*
- *M.A. Mueller Axial turbojet engines and the motorjet*
- *H. Oestrich TL engines at BMW Spandau*
- *K. Loehner TL engines at BMW Munich*
- *A. Franz TL engines at Junkers*
- *K. Leist ZTL bypass* (turbofan) *engines*
- *F. Gosslau The pulse jet engine*

Schelp's presented *'turbojet classification system and terminology'*, Figs. 7.10 and 7.11, was accepted then, and remained so since then. In January 1941 Schelp summarised the situation as follows: *'The development of TL engines for fighter aircraft is of first priority at present, since here the difficulties of the conventional piston engine are most conspicuously. The development of these engines at Ernst Heinkel Flugzeugwerke GmbH, BMW Flugmotorenwerke Brandenburg GmbH und Junkers Flugzeug- und Motorenwerke AG is today so advanced, that we can expect the pre-production series to start still during this year.'*

On 4 November 1942 largely the same participants as in January 1941 met again in a DAL Working Session *'Engine planning'*,[107] where again H. Schelp held the keynote speech: *'In agreement with the airframe manufacturers, the TL engines were originally designed for 600 kp thrust and a SL flight speed of 250 m/s* (corresponding to the *Me 262* specification with 850 km/h and 1 h flight endurance). *Actually, the Jumo engine will start the series production with 640 kp, though there are sufficient reserves for an upgrade to*

H.K.A. von Euler-Chelpin;—for Swizerland: J. Ackeret;—for the USA: J.S. Ames, L.J. Briggs, W.F. Durand, J.C. Hunsaker, G.W. Lewis and C.B. Millikan.

[107] According to a 84 p. secret document, distributed in 60 copies, GL/C-E3 I, GL/C-Nr. 4885/42 the meeting took place at RLM Berlin, Room 4280 on 4 Nov. 1942, 10 h, with an *'Introduction'* by Lt.-Col. Georg von Pasewaldt, Luftwaffe General Staff, followed by a *'Summary Dev. Planning'* by Gen.-Ing. W. Eisenlohr and finally, *'The Engine Dev. Planning'* by Flyer Staff Eng. H. Schelp. Remarkably, Pasewaldt used internally the term *'total war'* more than three months before Goebbels coined the term in public in his—see Wikipedia, *'Sportpalast* speech' in English. Continuing *'We will not win the war by quantity, but our focus has to be quality as well.. I cannot see any qualms for striving to this target.'*

1,000 kp. The design of the BMW engine was somewhat more demanding from the beginning—with smaller dimensions and weight. This will limit the achievable thrust to 800 kp, though this implies already long-range features like the air-cooled turbine. Notwithstanding to a certain optimism at programme start, we can state today that the actual results surpassed the expectations especially with respect to the achievable efficiencies.' He then continued *'There are plans to introduce an additional fighter power class with 1200 to 1600 kp thrust and 250 m/s SL flight speed'*, which resulted in the official *'RLM turbojet development programme 1942'*:[108]

Engine Class	~ Thrust [kp]	Compressor PR	Turbines (TL/PTL)
I	600—1000	3.5	1/–
II	1200—1600	5.0	2/3
III	2000—2400	6.0	2/3
IV	3000—4000	7.0	3/5

Schelp used the 1942 conference also to reveal the critical Mach number Ma 0.82 to a wider audience: *'Consequently, there is the question, by which of the available engine types for aircraft propulsion this limit can be achieved'*. No longer shy, Schelp drew the conclusion: *'The engine power survey indicates clearly that the future PTL engines will replace the piston engines beyond 5,000 hp'*, and, *'It is a prime finding of the first development phase that it will be sufficient in the future to cover the engine variants TL, ZTL and PTL with just one design effort.'* Finally, *'After project investigations of Dipl.-Ing. Ritz, AVA Goettingen, it appears to be feasible to achieve for a 5000 hp gas turbine with heat exchanger a fuel consumption of 150 g/ (hp h) at 10 km altitude; these development activities are shared between BBC Mannheim and AVA.'*[109]

On the basis of this engine thrust classification, Schelp told Heinkel to stop working on the Class I Heinkel HeS 8 and Heinkel HeS 30 engine designs, and concentrate only on the Class II Heinkel HeS 011. At the time, in 1942, this decision made sense considering that two other Class I engines appeared to be ready to enter production. The eventual 3-year delay before the BMW 003 or Jumo 004 entered service may have meant the HeS 30 would have beaten them to service, and in the end the HeS 011 would never leave the prototype phase, as outlined in detail in Sect. 8.1.

Again on 2 July 1943 on the occasion of the *fifth DAL Scientific Session*, Schelp presented a comprehensive paper *'High performance turbojet engines'*—in view of

---

[108] Basically the turbojet thrust class differentiation was not new, it followed a similar structure which Helmuth Sachse had already introduced for aircraft piston engines in the late 1920s at the HWA Av. Dept. (see Sect. 4.3.3): Class I 800–1100 hp, Class II 1200–1600 hp, Class III 1700–2400 hp.

[109] See Sect. 8.3.5 'Regenerated Aero Gas Turbine Developments'. Schelp certainly knew that the realisation of a flight-worthy heat exchanger was no easy task, but the quoted sfc target must have stunned his audience, e.g. in comparison to the sfc ~ 170 g/(hp h) of the proven Jumo 207A opposed piston diesel engine, which powered the Ju86P high-altitude reconnaissance aircraft.

intermediate successes and correspondingly grown self-consciousness now for an enlarged scope: Ma < 0,82, flight altitude <20 km and flight range < 15,000 km.

The piston engine receives only a short side-comment: '*Past work showed that this concept has reached its limit; consequently, propulsion research and development has to be directed more and more towards turbomachinery . . . In the future the engine designer shall not optimise the* "shaft power" *but the* "effective thrust S_e", *i.e. the difference between the* "nominal engine thrust S_n" *and the* "engine nacelle drag W_T".

With $S_n = \dot{m}$ (c - v), i.e. '*engine mass flow $\dot{m}$*', '*turbojet engine exhaust velocity c*' and '*flight speed v*' follows the non-dimensional '*specific thrust coefficient* $c_e = S_e/(q\ A)$'—with '*flight stagnation pressure q*' and '*engine max. cross section A*'. In addition, the '*specific fuel consumption b_e*' was determined; Fig. 7.12[110] shows redrawn from Schelp's originals over the flight speed the c_e -curves on the left, and the corresponding b_e distributions to the right.

In the data analysis Schelp emphasises the still preliminary character of the results, often based on simplified assumptions instead of thorough measurements (especially for the engine nacelle drag), nevertheless with general validity. For the intended survey he compares as high-power reference, the then advanced 24 cyl., 3000 hp, 55.5 l Junkers *Jumo 222 C/D* piston engine with TK 9 turbocharger[111] with

- a Jumo 004C turbojet of 1000 kp TO thrust,[112]
- a BMW 018 turbojet of the new Class IV with 3500 kp TO thrust,[113]
- a BMW 028 turboprop of 8000 eshp at 800 km/h at 8 km altitude.[114]

---

[110] See Schelp, Hochleistungstriebwerke, Figs. 3 and 4. This paper is available in German and English, the latter version is hampered by a translation error; the German term '*Kritische Geschwindigkeit*' (critical speed, Ma ~ 0.82) was incorrectly translated to '*supersonic*'.

[111] The *Jumo 222* was a German high-power multiple-bank in-line aircraft piston engine from Junkers Motorenwerke, designed under the management of Ferdinand Brandner (1903–1986),—see Wikipedia, 'Ferdinand Brandner' in English. The *Jumo 222* was a massive and very costly failure; 289 examples were built in total, none of which saw active service,—see Wikipedia, 'Junkers Jumo 222' in English.

[112] Thouh Schelp describes the Jumo 004C as '*today existing*', he considers it rather as an intermediate solution towards the next generation turbojets HeS 011, Jumo 012 and BMW 018. After Schelp realised the substantial drawbacks of the AVA R 1 compressor design as early as 1941, redesigns of a Jumo 004C and BMW 003C engine on the basis of an advanced seven-stage axial compressor with changed degree of reaction R 0.5, PR 3.5, $\eta_{ad} > 0.85$ from BBC Mannheim were launched, which A. Franz considered as too late for Junkers, so that progress up to war's end was reported only from the BMW/ BBC cooperation,—see Chap. 9, as part of the H. Reuter portrait. Finally, the ultimate beneficiary of the broader BMW turbojet development became the French Snecma, where the BMW team continued after the war.

[113] According to Schelp's development plan, Heinkel-Hirth worked on the Class II turbojet HeS 011, Junkers on the Class III turbojet Jumo 012 and BMW was chosen for a larger still turbojet in Class IV.

[114] Initiated in 1940 by the BMW project department, the BMW 028 turboprop had the backing of the *Technisches Amt* (RLM Technical Office) by early 1941, thereafter, however, the design was never finalised.

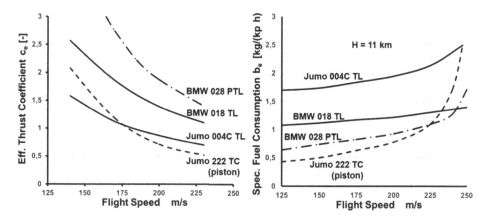

Fig. 7.12 H. Schelp, Typical Performance Study Results 1941–1943: *'Thrust coefficient and fuel consumption for specific powerplants'*

Figure 7.12 (l) illustrates with $c_e = f(v)$ the effective thrust variation with flight speed and the local gradient of the curves indicates the relative acceleration capabilities. For horizontal c_e—curves the engine thrust would grow equivalently to the engine drag. The c_e value of the piston engine is at $v \sim 235$ m/s already close to zero, i.e. the engine power is nearly completely compensated by the engine/propeller drag, which clearly highlights the deficiency of the piston engine in this speed range. The superiority of the turbojet engine concept is most striking, if the advanced *BMW 018* is compared with the *Jumo 222* piston engine.

Figure 7.12 (r) *'shows the specific fuel consumption $b_e = f(v)$ mainly to correct the erroneous belief that TL engines have a considerably higher fuel consumption than piston engines. (For an altitude of 11 km) the relevant fuel consumption b_e is at 230 m/s for the BMW 018 equal to that of the piston engine Jumo 222, and for higher speeds even lower, though the Jumo 222 curve is based on the standard value of 220 g/(hp h). Again this chart should prevent to follow long-time traded creeds about the engine characteristics too easily.'*

In a 1945 interview Schelp remembered the DVL-internal, pre-1937 speed limit for propeller-driven aircraft at 480 mph/772 km/h, in Fig. 7.12 corresponding to 215 m/s, which is close to Fritz Wendel's speed world record of 26 April 1939 with a severely tuned Me 209 aircraft of 755 km/h.[115] Antz's speed specification for a twin-engine fighter aircraft in 1939 was at 236 m/s (850 km/h)—for 1 h endurance.

---

[115] See Constant, The Origins, p. 204. For the present speed record status of piston-engine powered aircraft,—see Wikipedia, 'Fastest propeller-driven aircraft' in English.

In summarising his 1943 DAL conference paper, Schelp stated[116]: *'It is necessary to lay-out a clear path for the future developments; three 4-year time-periods are foreseen for accomplishing the postulated requirements:*

1. 1939–1942 *The first development phase provided the necessary fundamentals and allowed the turbojet to be brought to some maturity without having it initially produce extreme power outputs.*
2. 1943–1946 *The second development period will aim at the TL turbojet with increased thrust level, the ZTL turbofan and the PTL turboprop engine, and the pulse jet. In addition, preliminary research activities will address the GTW gas turbine with heat exchanger and the PML propeller ML motorjet.*
3. 1947–1950 *The third period should bring GTW development solutions for long-range applications which ends the major developments for the time being … It is difficult to predict the future, especially an appointed programme cannot be given for supersonic flight within this third main phase, since the necessary requirements are not yet defined; however, with respect to the powerplants, there is no problem.'*

7.4   Mauch and Schelp: At and Off RLM After 1939

As said, in August 1937 Schelp joined the RLM Technical Department's section LC1, their short-lived pure-research arm. Neither LC1 nor DVL shared his enthusiasm for the jet engine, so without any backing nearly one year passed without any real turbojet progress. But when the RLM was re-organised in 1938, he found himself in the LC8 division for *'special propulsion engines'* development. Here he found an ally in the assertive Hans Mauch, in charge of rocket and pulsejet development within LC8 since 15 April 1938.[117] Mauch had seen a demonstration of Hans von Ohain's engine at the Heinkel works shortly thereafter.[118] Mauch was adamant that engine companies work on such projects, however, and refused official funding for Heinkel's developments as they were taking place at an

---

[116] See Schelp, Hochleistungstriebwerke, p. 24 f.

[117] Mauch followed Uvo Pauls, who was then put full time in charge of planning and later operation of Peenemuende-West, the Luftwaffe section of this test facility, mostly used during the V-1 flying bomb development,—see Wikipedia, 'Peenemuende Airfield' in English. Mauch started his RLM career already in 1935. Though there are apparently no details known about his intermediate occupation, he presumably collaborated with Pauls on the Peenemuende-West planning from early on. Schelp claimed complete ignorance about Mauch's background and confirmed explicitly that Mauch had joined RLM *after* him,—see Ermenc, Interviews, p. 109, which in his view might have increased his importance as turbojet expert for Mauch.

[118] Mauch's visit was presumably on Wed 4 May 1938, based on a note of E. Heinkel to H. von Ohain, dated 29 April 1938, in which a *'Herr from the Ministry'* was announced, *'nothing special, the 'Aggregate'* (engine set-up) *should turn,* (you) *could use also hydrogen'.*

airframe company. After Mauch and Schelp had met rather accidently at RLM in early July 1938 and Schelp had explained the whole theoretical rationale of the turbojet to him, they quickly took the initiative. After Schelp had had a positive, his engine concepts reassuring visit at AVA Goettingen[119] in August 1938, they started contacting the larger engine companies, notably BMW/Bramo, Jumo and Daimler-Benz. None of whom proved to be terribly interested at first, mostly because they were in the midst of bringing new piston designs into production. After repetitive contacts during which especially Schelp gained in stature at the side of Mauch, by tactically not *'selling'* his own study results, but rather by encouraging own investigations within the Companies along these lines: *'I felt that they must convince themselves'*, but *'I also gave them, as a background, the survey I had made of all possible power plants in the future.'*[120] In addition, there were obviously more or less subtle methods applied—first to get the industry interested, and shortly thereafter, to get their whole-hearted *'buy-in'* to the new engine development plans. Even though Schelp with his all-axial turbomachinery preference was not too enthusiastic about von Ohain's radial engine concept, there was certainly the positive example highlighted to the other hesitant companies, how convinced and dedicated Heinkel proceeded with this new propulsion concept. And in 1939, when the successful first flight of the He 178 was accomplished, these developments were certainly also used as a lever which demonstrated the feasibility of the suggested ideas and their practicality, and got the industrial development finally moving.[121]

More subtle, if not perfidious were speculative considerations, which Bill Bailey ventilated on the English side apparently for the first time in a 2004 paper to the Royal Aeronautical Society about *'The Early Development of Aircraft Jet Engine'*.[122] The starting point of Bailey's speculation is the report about A.A. Griffith's trip to Brown Boveri, Baden CH in summer 1937,[123] where presumably for the first time cautious discussions about the possibility of an exclusive aero license of Brown Boveri's axial compressor technology began, which were intensified in May 1938 during the follow-on visit of Hayne Constant and Sir William S. Farren, resulting in *'the deal'*, Fig. 6.44. But while Bailey's

---

[119] Schelp had analysed that a turbojet compressor should achieve at least PR 3 at efficiencies better than 70%. In 1938, German supercharger compressors achieved 65% on average at PR 2; Goettingen's promise of axial compressor efficiencies well over 70% was therefore of considerable importance. Actually, for Schelp it was a *'.. difficult situation. In the DVL all turbomachinery developments related to* (centrifugal) *turbo-superchargers... DVL was not very amenable to research and development on axial compressors at all. So I had to get the support of Encke, Betz and Prandtl at Go(e)ttingen. Sides were drawn up and there was quite a battle before we came finally to a decision and could say, "yes, our future is with axial flow compressors and turbines." '*,—see Ermenc, Interviews, p. 111.

[120] See Ermenc, Interviews, p. 114.

[121] See Ermenc, Interviews, p. 110.

[122] See Bailey, The Early Development, p. 17.

[123] Bailey refers to Ed. Constant's book in general, actually—see Constant, The Origins, p. 214.

speculation strays somewhat away, when he presumes that Brown Boveri sales department might in due course have taken the opportunity to offer a 2000 hp demonstration gas turbine also to *'Berlin'*, he overlooks the simple fact that information about intensified contacts from RAE to Brown Boveri to get access to axial compressor technology might have represented a very valuable lever in internal RLM—industry discussions at that time. The Swiss Brown Boveri certainly kept these sensitive contacts to Great Britain *'secret'* in the tense state of affairs in the pre-war era, simply by limiting the contacts to the upper management level (Ad. Meyer, Cl. Seippel) and with some certainty and against Bailey's presumption, even the German branch of BBC Mannheim was not informed. However, somewhat tragically belonged to the very *'inner management circle'* at Baden, CH, also the German Dipl. Ing. and Dr. h.c. ETH Walter G. Noack (1881–1945),[124] deserved inventor of the *Velox boiler* with all-axial turbo set, who had back to 1919 excellent connections to the Reichswehr Berlin-Charlottenburg and -Adlershof,[125] and in this context with all likelihood also to (the later General-Engineer and Schelp's superior) Dipl.-Ing. Wolfram Eisenlohr, and in case, they would have found a way of confidential communication; certainly, an element of surprise in this (imagined) indirect British contribution to an acceleration of the German turbojet affairs in 1939.

Perhaps not in 1939, but looking ahead to e.g. 1943 and imagining a scenario, where the Swiss-British relations might have developed positively, while a German technology shortfall could have been traced back to some industrial ignorance towards an RLM offer a few years earlier, could have resulted in horrible consequences. Section 8.1 will address court-martial cases against General-Engineer Eisenlohr and even down to the engineering level of Hans von Ohain.

Eventually the jet engine concept started to become more widely known within the RLM, and Helmut Schelp together with his new *'Flugbaumeister'* partner Hans Antz on the airframe side started to push for the immediate development of a flightworthy model. Schelp was supported by a small RLM expert team to which belonged Emil Waldmann (1911–1975),[126] a flyer acquaintance from his *'Deutschlandflug 1935'* past, Sect. 7.1, and

---

[124] See Eckardt, Gas Turbine Powerhouse, Sect. 4.1.2 'Prominent Engineers', pp. 84–96.

[125] See Wikipedia, 'Idflieg' in English, the abbreviation for 'Inspection der Fliegertruppen',—and Noack's affiliation to the *'Flugzeugmeisterei'*, Sect. 4.1.2 'Turbocharger development history', Figs. 4.14 and 4.15.

[126] See Kay, German Jet Engine, p. 17, and S. 150, where in Aug. 1944 Waldmann is cited with the proposal of a new on-board starter concept for the DB 021, a turboprop version of the Heinkel-Hirth HeS 011. Presumably in 1943 Waldmann's home at Berlin was destroyed by bombing and he found occupation at Junkers Dessau. In Oct. 1946, E. Waldmann and his family was deported for several years to work in the design office of N.D. Kuznetsov near Samara on the River Volga, SU. Information, 5 Nov. 2018, courtesy to Hans Waldmann, son of E. Waldmann and the author's former colleague at MTU Aero Engines, Munich.

Walter Brisken,[127] who continuously scanned the administered *'turbojet engine cata-logue'*, e.g. for the possibilities to develop turboprop engines from the larger Class II turbojet engines with extra turbine stage/s added for the airscrew drive, and also, in later years, to assess the benefit of adding a regenerative heat exchanger to a PTL engine cycle. These studies for a long-range *'America bomber'* with speeds up to 600 km/h and flight altitudes at 10,000 m will be reviewed in the context of the mentioned, joint AVA/Brown Boveri activities in Sect. 8.3.5. Although the industry took up the development of such engines, the 16-year time span of Schelp's carefully worked out development schedule extended beyond the end of the war and could not be covered.

On 4 August 1938, W. Eisenlohr and H. Mauch jointly signed a work description for H. Schelp at his new position as *'Referent Strahlantriebe* (Jet propulsion) *LC 8 VIIa'*, in which he received full and independent responsibility for *'reaction and turbojet engines including accessories'*. Though agreed in principle already for the turning of the year 1938/ 1939, his full RLM employment had to wait up to the date of his successful *'Flugbaumeister'* exam on 15 June 1939. His salary at that time was with all extras after taxes 559,—RM/month, a considerable jump from the foregoing 200,—RM during his *'Flugbaumeister'* training, but in comparison considerably less than the 1200,—RM which H. von Ohain received at Heinkel after 1 Oct. 1940 and, even more impressive (if true), the first salary offer by E. Heinkel to M.A. Mueller of 1000,—RM during their first contact, in Mueller's memory presumably still in 1938 at their meeting in Bristol.[128]

In spring 1939, Hans Mauch and in his wake Helmut Schelp were convinced that the time was ripe for jet propulsion and had formulated a development plan accordingly, which addressed not only the engine engineering field in its full breadth, but cared also for the right manufacturing basis from the very beginning. This implied automatically consider-able strategic decisions at Junkers, between the engine and aircraft division, the acquisition of Bramo by BMW and foresaw, most radically, also the participation of Daimler Benz in resolving Heinkel's engine manufacturing deficiency. Within RLM there was strong opposition to this plan from Mauch's superior General-Engineer Wolfram Eisenlohr, the head of power plant development. He certainly had been informed about von Ohain's activities at Heinkel, and probably also of the pulsejet and ramjet projects of his Research Division, but he seems to have considered them at that time as long-range projects which could be of use only far in the future, if at all, and had attached no immediate importance to them.[129] Consequently, Eisenlohr had been inclined to attach little importance to jet propulsion at the time Mauch came into the power plant group, in April 1938. Not only

---

[127] On 10 March 1945 'Fl. Haupting. Brisken' led the presumably last documented BMW 003 devel-opment meeting at Staßfurt, with 'Dir. Dr. Oestrich and Fl. Stabsing. Schelp' attending, Sect. 8.1. After the war, Walter R. Brisken came like H. Schelp as part of *'Operation Overcast/Paperclip'* first to Dayton, OH and from there to General Electric, Evendale/ Cincinnati, OH.

[128] Interview with Max A. Mueller, ca. 1952—Heinkel Archive, Deutsches Museum FA 001/ 0025 and FA 001/ 0314..

[129] See Schlaifer, The Development, p. 383.

technical but personal differences had been growing between the two men, the more so when Mauch started turning as described, the big strategic' wheel after a few months. Mauch, however, was personally acquainted with Ernst Udet, since 1936 head of the Technical Office within RLM and thus in charge of all R&D activities, and apparently it was Udet personally, who gave Mauch his backing to proceed. In this context Schlaifer placed an interesting side-comment: *'Udet tended in general to favour radical developments … the receptive attitude toward turbojets of Udet, a tactical officer with no engineering background whatsoever, corresponds exactly to the attitude of General H.H. Arnold in the United States, who was extremely anxious to promote the development of turbines in that country as soon as he learned early in 1941 that they were being developed in England.*'[130]

Hans Mauch left the RLM Air Ministry to establish his own consulting engineering office in Berlin presumably in March/April 1939,[131] which he headed until the end of World War II. Schelp commented this incisive event rather laconically: *'His main reason for leaving was that he did not work very well with Eisenlohr who was over us. As a matter of fact he became quite ambitious and wanted Eisenlohr's job. He wound up as winner of second place left the Air Ministry for industry'* (sic).[132] Mauch was replaced as RLM Department Head by Ernst Beck, who came from industry with some experience in solid propellant rocketry, but otherwise no specific qualification for the job. Officially Schelp considered himself as too young for such a job promotion, and *'Beck was available'*. But of course, he was more than ready to take over and must have been deeply frustrated by the decision for Beck.

Schelp's reaction reached his superiors on 10 August 1939 as a *'Letter of resignation'* from RLM, effective 31 December 1939, which he cancelled in written form already on 7 September 1939 with reference to the intermediately occurred state of war.[133] No further details are known about Schelp's preliminary hiring at Junkers Dessau presumably in early 1939. With all likelihood, Schelp saw himself sufficiently qualified even as a potential successor to Herbert Wagner as chief engineer, who had left the Junkers Aircraft Division

---

[130] See Schlaifer, The Development, p. 394.

[131] See Murphy, Hans Adolph Mauch, p. 261. As part of Schelp's personnel file, there is as last entry from Hans Mauch a *'Buergschaftserklaerung* (suretyship declaration)', dated 17 March 1939, for Schelp's *'political trustworthiness'*, now at *Ref. LC3, 7a*, up to the corresponding, then still outstanding *GeStaPo* declaration.

[132] See Ermenc, Interviews, p. 115; it appears to be revealing that Schelp again is uninformed about Mauch's whereabouts, which of course could be also a result of Mauch's secret manoeuvring.

[133] The original document has been removed from Schelp's personnel file, and can be reconstructed only indirectly. Schelp's reasoning was, that by this cancellation he might contribute at best to the *'final success'* if he continued to apply his experiences as *Luftwaffe* engineer. Closer to the truth of an enforced change of mind might be Eisenlohr's hand-written comment on that document: *'Z.Zt. sind wohl Abwanderungen zurückzustellen. Schelp .. ist einverstanden. (At present migration moves - to industry - are apparently to be put on hold. Schelp .. agrees.).'*

in late 1938, and was finally replaced by Heinrich Hertel in May 1935, Sect. 6.2.1. After this development Schelp restricted his ambitions perhaps to the Junkers Engine Division as a potential successor to Professor Otto Mader, before he finally resigned and had to give up his plans to leave the Air Ministry.

The further intricate development can be reconstructed only indirectly. In a letter, dated 20 October 1939 from Ernst Heinkel to *'General Udet'*,[134] he complains about Schelp and Schelp's RLM team, who in his view obviously tried to obstruct the engine developments at his Company. In return, Heinkel had stopped the information flow about the engine developments back to the Air Ministry, the more so, since he had learnt after Max Adolf Mueller came to Heinkel from Junkers Magdeburg during summer 1939, that *'in March 1939 Professor Mader had told Mueller that Schelp planned to take over the turbojet developments at Junkers, if he received the RLM permission to leave'*.[135] Internally, this was for the time being certainly another decisive piece in Heinkel's argumentation to restrict the information flow to Schelp as future employee of his competitor.

In this context it is worthwhile to summarise in short what is known about the somewhat mysterious whereabouts of Hans A. Mauch after his official leave from RLM, up to the end of WW II and thereafter. According to his biographer,[136] *'he developed testing equipment and aviation* (transmission equipment) *and automotive engine accessories; he also acted as a consultant to the German Air Ministry, where he was later placed in charge of the terminal development of the V-1 buzz bomb, the first guided missile. He knew, and sometimes differed with, many of the top German engineers of the period, including Wernher von Braun.'* The latter remark points to Peenemuende-West again, where the RLM after considerable delays finally launched the series development of the Fieseler Fi-103 cruise missile programme, commonly known as *V-1*, under the camouflaging designation *FZG 76* (Flak-Ziel-Gerät 76) on 19 June 1942.[137] Thereafter, the project became a salient example of development speed, a comprehensive production management network—and of extreme secrecy.[138] The first catapult launch happened already

---

[134] The letter belongs to a series of seven documents, which Ernst Heinkel sent to his *'ghost writer'*—see Wikipedia, 'Jürgen Thorwald' in English, on 27 July 1953 for use in his biography—see Heinkel, Stormy Life. Courtesy to Lutz Budrass, 16 Sep. 2018, who had detected these separated documents in 2017 in the Heinkel Archive files at the Deutsches Museum Munich, FA 001/0035.

[135] *'Secret'* teleprinter communication from Lusser to Heinkel, dated 20 Oct. 1939, in the file attached to the Heinkel letter to Udet.

[136] See Murphy, Hans Adolph Mauch, p. 261.

[137] See Wikipedia, 'V-1 flying bomb'in English, and a 1944, 28 min *Youtube* video https://archive.org/details/FZG76 in German, a 'Top Secret' training film of E.d.L. (*Erprobungsstelle*/ test centre, *der Luftwaffe Karlshagen*/ Peenemuende) which describes in detail the technique and handling of this completely new weapon system.

[138] The V-1 operations 1943–1945 were directed by Col. Max Wachtel (1897–1982), fifth Flak Div., who for reasons of counterintelligence against British agents carried several identities as e.g. *Martin Wolf* and *Michael Wagner*,—see Aumann, Vernichtender Fortschritt.

six months later on 24 December 1942 at Peenemuende-West, and in September 1943, 15 months after development start and actually before this was accomplished, the series production started at the *Volkswagen* plant Fallersleben.[139] The production programme coordination put the RLM in the hands of Dipl.-Ing. Hans Mauch and his *'Arbeitsstab* (task force) *FZG 76'*. Four assembly plants and fifty further supply companies were organised in four *'Industrieringe* (industrial groups)' which were coordinated by the influential Dipl.-Ing. Karl Frydag.[140] Mauch's task force had 70 members and met regularly once per month in the large RLM conference room. The shooting against London from 64 starting catapults, which had been erected in Northern France between Calais and Rouen began one week after the Allied landing in the Normandy in the night of 12/13 June 1944—with a failure. The first successful attack was in the night of 15 June with 244 started Fi-103. On 20 July 1944—in German history generally known for the failed attempt of Col. Claus Count von Stauffenberg (1907–1944) to assassinate Adolf Hitler at the military headquarter *Wolf's Lair*[141] and remove the Nazi Party from power[142]—Hans Adolph Mauch received the *'Knight Cross to the Merit Cross with Swords'*[143] from the German Air Ministry, then the highest non-political decoration for civilians.

The standard V-1 had an operational range of 250 km, flying in an altitude between 600 and 900 m, and achieving there a maximum speed of 640 km/h. The average manufacturing costs for the produced 33,000 Fi-103 units were in the order of 5000 RM/piece (for the 400 kp thrust *Argus-Schmidt pulsejet As 014* alone only 770 RM) and thus considerably less than the 140,000 RM for W. von Braun's V-2 rocket. Due to the high fuel consumption there were plans to replace the pulsejet by an expendable turbojet, for which Porsche developed in 1944 the *TP-300*, official designation *109-005*

---

[139] See Benecke, History of German guided missiles, and—see Benecke, Flugkörper, p. 86 f. Today Fallersleben belongs to the township of Wolfsburg, Lower Saxony; it is the native town of the German poet—see Wikipedia, 'August Heinrich Hoffmann von Fallersleben' in English, who wrote in 1841 on the island of Heligoland, then in British possession, the *'Lied der Deutschen* (Song of the Germans)', its third stanza now being the national anthem of Germany.

[140] Since 1941 K. Frydag, Technical Director of Henschel Aircraft Works Berlin, was member of Goering's (failed) *'Industrierat'* to duplicate aircraft production. Acc. to—see Mommsen, Das Volkswagenwerk, p. 694, Frydag implemented Mauch in May 1941, then an Industrierat-associated engineer, as *'assault patrol leader'* within Volkswagen to accelerate the V-1 production.

[141] Located east of the small East Prussian town of Rastenburg, in present-day Poland,—see Wikipedia, 'Wolf's Lair' in English.

[142] See Wikipedia, 'Claus von Stauffenberg' in English.

[143] In German *'Ritterkreuz des Kriegsverdienstkreuzes mit Schwertern, RK KVK mS';* Mauch received the decoration as *'Engineer of the V-1 project'* together with—see Wikipedia, 'Walter H.J. Riedel' in German, and thus belongs to an illustrious circle of roughly 90 persons including Walter Dornberger, Karl Frydag, Max Wachtel, Ernst Brandenburg, Ludwig Prandtl, Hellmuth Walter, Ernst Leeb (*'with swords'*) and Willy Messerschmitt, Alexander Lippisch, Ferdinand Porsche and Theodor Morell (*'without swords'*),—see Wikipedia, 'Träger des Ritterkreuzes des Kriegsverdienstkreuzes (1939)' in German.

with 500 kp thrust; it was planned to use a nine stage axial compressor from the BBC tank gas turbine project GT 102, Sect. 8.3.4; the RLM had ordered Max Adolf Mueller to support the Porsche development team.

Thereafter Hans A. Mauch left two traces in the German *'War diary chief TLR from 18 December up to war's end'*[144] on 31 January and 5/11 February 1945 in minutes of meetings of the *'Rüstungsstab bei HDL Saur'*[145] on the subject of *'ruthless priority work for 8-262'* (Me 262), with reference to priority installation of *EZ42*,[146] the *'gyro gunsight'* visor system for permanently installed weapons, where power of procuration is asked by the *'Herr Reichsmarschall* (Goering) *to Dipl.Ing. Mauch'*—again a kind of high priority, last-minute technical task force activity.

Mauch's post-war biographer makes believe that his *'third major consulting activity was original research and development in the fields of aviation medicine and prosthetic devices. In this bioengineering sector, Hans Mauch cooperated closely with Ulrich K. Henschke, a radiologist with a Ph.D. in physics. Head of the Aeromedical Institute in Munich, Dr. Henschke worked on a variety of physiological problems.'* There are even credible technicalities that *'They worked on a mass-production artificial leg that could be quickly adapted to an individually fitted socket, as well as various ways to stabilize the knee for above-knee leg replacements.'* However, somewhat suspiciously this text omits that Dr. med. Dr. phil. Ulrich K. Henschke (1914–1980) and his planned *'Institute for aviation physiology'* had been prospectively allocated to the *LFM Luftfahrtforschungsanstalt Muenchen* (Aeronautical Research Establishment)[147]—under construction at Munich-Ottobrunn (the later MBB/EADS/Airbus site) since 1940.[148]

As described already in the foregoing Sect. 6.2.3, Hans Mauch must have come in touch with the subject of *'protheses and artificial limbs'* already at Georg Schlesinger's Institute at TH Berlin-Charlottenburg in 1929 and thereafter, but—it could have been also an ideal camouflage for his V-1 and other military activities after his leave from RLM and later, immediately after the war, in view of the fact that V-1 production and deployment was

---

[144] In German 'Kriegstagebuch Chef TLR, Technische Luftrüstung (Technical Air Armament) at OKL Oberkommando der Luftwaffe (Luftwaffe High Command)', led from Aug. 1944 to April 1945 by Luftwaffe Major General—see Wikipedia, 'Ullrich Diesing' in German.

[145] See Wikipedia, '*Rüstungsstab*'in English, and—see Wikipedia, 'Karl Saur' in English; *HDL* stands for *HauptDienstLeiter*,—see Wikipedia, '*Dienstleiter* (NSDAP)' in English.

[146] *EZ 42—Einheits-Zielvorrichtung 1942* (standard visor) which increased the fighter hit distance from 400 m up to 1500 m, manufactured by *Steinheil*, Munich; the introduction was considerably accelerated, when a *'gyro gunsight'* had been detected in a captured *P-47 Thunderbolt* in 1942.

[147] Since the erection of LFM never materialised in full, the course catalogue of the University Munich in the summer semester 1944 has still as *Dozent*/lecturer *'Henschke, Ulrich, Dr. med. habil., Dr. phil. (12. 3. 42) for medical physics, conservator of the radiology dept. of the first university gynaecological hospital, Maistr. 11'*, but—see Borck, Das künstliche Auge, knows that Henschke's *'Med. Forschungsinstitut* (Med. research institute)' was transferred to Garmisch-Partenkirchen in August 1944.

[148] DLR online archive catalogue *'LFM 1936-1945'*, correspondence A. Baeumker.

Fig. 7.13 Post-war impressions: Charles Lindbergh, 1945 (l), Ulrich K. Henschke, Sepp Zott and Hans Mauch, 1946, l-t-r (m), Helmut Schelp, ~1980 (r) © T. Schelp (r)

closely SS-related. In the beginning Mauch's frequent travels from Berlin to Munich might have been officially directed to Henschke, while he maintained and intensified contact to the pulsejet inventor Paul Schmidt; even the open conflict with his superior General-Engineer Wolfram Eisenlohr could have been somewhat made up to mislead observers of the scenery. When the war ended, the US Army Air Force brought many aviation engineers, including Mauch and Henschke, Fig. 7.13 (m) to Heidelberg.[149] The Henschke group and in its wake H. Mauch had apparently changed sides on war's end at or near Garmisch-Partenkirchen, approximately some 90 km south of Munich. After 1943 the group studied advanced methods for steering aircraft, designed a new instrument panel and proposed a system of aircraft controls eliminating the use of the feet. In this context there must have been ventilated also far-reaching ideas for steering impulses by means of electro-mechanical supported telepathy. It becomes evident that the artificial limb was rather a medical by-product first, which over time gained considerable USAAF attention, especially when the idea of telepathic control was applied here as well.[150] While at Heidelberg, the two wrote a major chapter on human factors and cybernetics entitled *'How Man Controls'* in a two-volume work on *German Aviation Medicine—World War II*.[151] The work created

---

[149] There is a 64 p. report of the *'USAAF Aero Medical Center in Germany, 1945–1947'*, with the depicted photo of Henschke, Zott (with unknown function) and Mauch at Heidelberg,—see http://resource.nlm.nih.gov/14130150R

[150] For the present status,—see Wikipedia, 'Neuroprosthetics' in English. Actually, also Ch. Lindbergh, Fig. 7.13 (l), had visited Henschke's small institute still at Garmisch-Partenkirchen on Thu 7 June 1945, apparently with mutual sympathy, and explaining to Lindbergh also his strive for eliminating the use of feet for steering/ aiming: 'It is best to use only one group of muscles', as Lindbergh could confirm in a test on the spot; the difference was several 100% in favour of Henschke's new technique. See Lindbergh, The Wartime Journals, p. 977 f.

[151] See Henschke, How man controls. The book *'German Aviation Medicine'* was edited by—see Wikipedia, 'Hubertus Strughold' in English, the chief of aeromedical research for the *Luftwaffe*, who

some attention, so that in due course Mauch and Henschke were brought to Dayton, OH, to work at the USAAF Aeromedical Laboratory, where both were involved almost entirely in advanced prosthesis research.[152] In 1957, Mauch formed his own medical consulting firm, which was incorporated in 1959 as '*Mauch Labs, Dayton, Oh.*'

As a kind of humorous reminiscence to his former life, Hans Mauch filed with priority 4 May 1955 a patent US2,940,213 for a '*Jet-propelled balloon*', with five toy variants of the described kind, '*having an inflation opening* (and) *therein, a preformed blade-like element ... whereby on inflation of the balloon and release thereof, a propulsive jet air stream will be metered ... effecting a spinning stabilized rocket-like ascent thereof to substantial heights.*' In 1973, he was elected to the US National Academy of Engineering and served on its General Engineering Peer Group from 1976 to 1978.

Ulrich K. Henschke, after developing together with Hans Mauch the customised artificial leg incorporating the '*Mauch S-N-S Swing and Stance*' hydraulic automatic movement, became in the following years a pioneer in radiation cancer treatment equipment, for which he received a number of national and international awards. Both physician and physicist, he had been since 1955 Director of the Department for Radiation Therapy at the Memorial Sloan Kettering Cancer Center, New York, and since 1970 Professor of Radiotherapy at Howard University Medical School, Washington D.C. On 29 June 1980, three days before his official retirement date and a planned permanent move to Africa, this '*modern Albert Schweitzer*' died as a result of an airplane crash in Ngorongoro Crater, Tanzania.[153]

Back to Helmut Schelp,—still in 1939 and with his known *drive* he had issued '*The 1939 RLM Engine Programme*', Sect. 6.2.3, which will be further dealt with in Sect. 8.1. The development activities started in any case within the responsibility of the Companies, however, in the early phase with secondary priority, due to the expected short length of the war. In addition, after Mauch's departure, Schelp lacked increasingly attention and backing from his immediate RLM superiors. The official jet propulsion programme had gained much support from Ernst Udet, and it appears that Helmut Schelp personally benefitted of a late Udet decision in the fall of 1941, when he finally replaced Beck as head of the RLM section which directed all jet and (aircraft) rocket engines.[154] But, following Udet's suicide under great pressures on 17 November 1941, decisions on development and production

---

also was brought to the United States as part of *Operation Paperclip* in 1947, and later became known as '*Father of Space Medicine*' for pioneering the study of the physical and psychological effects of manned spaceflight. Following his death in 1986, Strughold's activities in Germany during World War II came under greater scrutiny and allegations surrounding his involvement in Nazi-era human experimentation greatly diminished his reputation. Another critical case in this respect amongst the German doctors at the USAAF Aero Medical Center Heidelberg is—see Wikipedia, 'Siegfried Ruff' in English, and—see Wikipedia, 'Siegfried Ruff (Mediziner)' in German.

[152] See Wikipedia, 'Hans Mauch' in English.

[153] Information, courtesy to Dr. med. Maria Kissler, former Head of Radiotherapy at St. Josef Hospital-Clinic, Ruhr-University Bochum on 5 April 2021.

[154] See Schlaifer, The Development, p. 402.; nothing in Schelp's personnel file after 1940.

were more and more taken by Erhard Milch. Milch was not initially in favour of jet aircraft and preferred to see the large production figures attainable for conventional aircraft. On top after Udet's suicide, war court processes were ordered against some leading engineers in 1942/1943, accused of failure of fulfilment of their job; General Engineer Dipl.-Ing. Wolfram Eisenlohr (1893–1991), who kept the post of Director of the Department for Aircraft Engines and Accessories in the RLM Technical Office officially between 1 January 1938 and 31 January 1944 was one of them.[155]

A certain turning point with respect to development urgency was reached, when Milch assembled his department heads on 25 May 1943, to decide on the mass production of the Junkers Jumo 004 B turbojet engine. His question on the state of development was answered by the head of aero-engine development, Eisenlohr, by stating that it was at that time *'technically more mature than any aero-engine'*.[156] Initially, Milch was not in favour of jet aircraft and preferred to see the large production figures attainable for conventional aircraft. It was not until November 1943 that another proponent of jet aircraft, the competent Oberst Ing. Siegfried Knemeyer (1909–1979) was made responsible for airframe and engine development at the RLM Technical Office, from 1941 onward GL/C-E2 and -E3, later known as TLR/FL-E. As a pilot with own operational experience and simultaneously technically qualified, he insisted that a speed advantage of at least 150 km/h over enemy aircraft was essential in view of Germany's critical air defence position; therefore, Knemeyer proposed the abandonment of most conventional aircraft in order to concentrate on Messerschmitt Me 262 jet fighters and Arado Ar 234 bombers—with Jumo 004 and BMW 003 turbojet engines.[157]

Due to the already described gaps in Schelp's personnel file up to 1945, there are then considerably less documented events associated with his name. In October 1940 he married his later wife Gertraud Ladwig at Dresden-Cossebaude, which required a complicated month-long preparation to obtain a *GeStaPo* marriage licence. On 18 July 1941 the RLM

---

[155] During the process enquiries ordered resulted in the recognition that the accusations were unfounded; General Judge Kraell managed that Reichsmarshall Goering ceased the process after interrogating Eisenlohr concerning the 20 charges. Thereafter, General Engineer Eisenlohr, who refused to voluntarily apply for retirement, was granted leave on 1 February 1944 and was transferred into retirement on 30 April 1944 ahead of schedule. From 1 May 1944 to 30 June 1945 he was Technical Director of the Department of Propeller Design and Manufacturing at *VDM Vereinigte Deutsche Metallwerke* (Combined German Metal Works), Frankfurt/M. It appears that Eisenlohr was pushed aside as Schelp's superior by General Engineer Dipl.-Ing. Franz Mahnke (1900–1975), now as RLM Dep. Director (13 Oct. 1941–31 July 1942) and later Office Group Chief GL/C-B3 (Engines, 1 Aug. 1942–3 Nov. 1944), responsible for engines and accessories procurement, but also for related workforce questions.

[156] RLM Amtschefbesprechung (Office chief meeting), 25 May 1943, Bundesarchiv-Militärarchiv RL 3/20, fol. 438, *'Es ist in der Serie weitaus reifer als jeder Motor'*.

[157] See Kay, German Jet Engine, p.17. In due course, 1942 actions executed by Schelp stopped the turbojet engine developments of Heinkel's *HeS 30* and *HeS 8* (with the new Class II *HeS 011* instead), Weinrich's counter-rotating *BMW-002* and the Daimler-Benz turbofan *DB 007*.

Technical Office issued a specification for a bomber to be powered by a new turboprop power unit, planned by Schelp's department. It consisted of two gas turbines which acted as gas producers to feed a third power turbine driving a variable-pitch airscrew. The design power of Schelp's *'bomber engine'* was 7000 hp at 800 km/h in 8 km altitude, with a target fuel consumption of 270 g/(hp h). The planned power split was to load 60% on the propeller, while 40% should have been forward thrust from the exhaust jet.[158] Though the dual turboprop unit originally proposed never materialised, work went ahead on the single *HeS 011*, or *109-011* turbojet engine, which Schelp became keen to have developed into the first Class II unit in its own right, and for which he got personally uncommonly deep involved in the detailed design process in October 1941,—by adding as a kind of *'souvenir'* from his US stay five years earlier, Birmann's diagonal compressor, Fig. 7.5. This was apparently also a kind of compensation for Schelp's termination of the HeS 30 and HeS 8 developments at Heinkel—and laid ground to a special relationship, if not friendship with the just one year elder Hans von Ohain.

Presumably already in April 1941, remarkably following a suggestion from the RLM Technical Office's airframe section GL/C-E2, the industrial firm of BBC Brown Boveri & Cie., Mannheim, were given the task of designing and developing a new axial compressor with the degree of reaction corrected to R 0.5 for the advanced BMW 003 C/D versions, for which Schelp's department issued in the coming years the development contracts *'Hermso I–III'* to BBC, named after BMW's chief engineer Hermann Oestrich.[159] The object was to achieve greater efficiency, air mass flow and pressure ratio while still being interchangeable with Encke's original compressor from AVA, which was achieved.[160] In the minutes of the—already addressed in the foregoing—meeting on 25 May 1943, Schelp's position has been documented for the plan to introduce gas oil/diesel fuel for *safety* reasons for the new generation of turbojet engines, to which E. Milch concurred by ordering the switch of two of the coal hydrogenation plants reserved to the *Luftwaffe* over to diesel fuel.[161]

Throughout the development of turbojet engines in Germany there was a serious shortage of qualified personnel, which also affected Schelp's activities. The allocation of manpower among the various types of engines being developed was within the power of General Engineer Wolfram Eisenlohr; although in opposition to the new engines at first, he

---

[158] See Koos, Heinkel Raketen- und Strahlflugzeuge, p. 71.

[159] In Sect. 6.2.5 had been outlined that a corresponding application to the Junkers Jumo 004 came too late. Antony Kay,—see Kay, German Jet Engine, p. 212, indicates a preference for the R 1 axial compressor design and *'mistrust of reaction compressors'* by H. Schelp himself, which could explain this initiative from outside the engine department by another of Schelp's *Flugbaumeister* colleagues, here Walter Friebel. Acc. to Kay, *'Later, this mistrust was belied by the promising progress of the Heinkel 109-006 (HeS 30) turbojet and the success of the work under Hermann Reuter of the Brown Boveri company in developing new compressors for the BMW 109-003 C turbojet.'*

[160] The HeS 011 development will be discussed further in Sect. 8.1, while the BBC axial compressor design activities for the BMW 003 C/D versions will be part of Chap. 9.

[161] See Budrass, Review, p. 184,—and Sect. 8.1.

had by 1941 become convinced and did all that he reasonably could to give it adequate resources. In 1941 Schelp was successful in getting a relatively large addition of manpower to his field when he could arrange the purchase of Hirth Motors, Stuttgart-Zuffenhausen in favour of Heinkel.[162] By 1942 the gas turbine side was faced with a still more difficult problem in the attitude of the Ministry's top officials—like Milch. Because Milch and others were completely unaffected by technical arguments from engineers, advocates of gas turbines resorted to trying to influence the upper ranks through the opinions of non-technical flying officers with a name. In Schelp's own words: *'From the military viewpoint naturally it was very desirable to obtain an extremely high performance airplane which was much better than, for instance, the Messerschmitt 109. We convinced Galland, the Commanding General of our fighter forces by coaxing him to fly one of the prototypes himself; he was very enthusiastic afterward.'*[163] And, with respect to the strategic decision pro- Me 262: *'A turbojet engine does not need the highly refined fuel which you had to have for the reciprocating engines. And further, more jet fuel can be made from a barrel of crude oil than gasoline for a piston engine. This was a vital factor.'*[164]

The turbojets nearing production in view of Germany's deteriorating war fortune meant that Schelp from Eisenlohr's GL/C-E3 (E—Entwicklung/development), now associated to the RLM Technical Office engine procurement department GL/C-B3 (B—Beschaffung) under General Engineer Franz Mahnke, also got involved with the horrible aspects of *'forced labour'* at the Sachsenhausen-Oranienburg concentration camp,[165] as described by Daniel Uziel. Initially, Milch had prepared the ground by writing to the responsible SS officer at Oranienburg: *'An adequate supply of manpower for the Luftwaffe industry is of crucial importance for the successful outcome of the present war'*, before Schelp appeared on the scene: *'In mid-August 1943, Staff Engineer Helmut(h) Schelp, of department GL/C—B3 of the Technical Office, dealing with experimental engines, and engineer Schaller, who had worked at the Oranienburg plant in 1942 and was now a hangar director at the Rechlin Flight Test Center, the central military flight-testing organization,*

---

[162] See Schlaifer, The Development, p. 399.

[163] The historic flight of Adolf Galland (1912–1996), in the age of H.Schelp, took place on 22 May 1943; at that time he was the youngest general in the German Army, and had 94 air victories.

[164] See Ermenc, Interviews, p. 117. With respect to the Me 262 testing of General Lt. Adolf Galland, acc. to—see Irving, The rise and fall, this influential test flight happened on Milch's initiative. It was only on 22 May 1943, that Galland had tested the Me 262, then still with tail wheel. Thereafter, he called Milch, coining the famous *'.. that aircraft flies like pushed by angels!'* In a one-page summary report, which Milch read to his staff on 25 May 1943, Galland wrote *'The aircraft is a big design hit, which represents an unimaginable operational advantage (except during take-off and landing), if our adversaries stay with the piston engine concept'*.

[165] See Wikipedia, 'Sachsenhausen concentration camp' in English.

met Hans Jüttner, head of the Leadership Main Office (SS-FHA) *of the SS.*[166] *They briefed him of the highly confidential development of jet propulsion and of the forthcoming production of jet engines, mainly by BMW. Based on Schaller's first-hand experience with concentration camp inmates while working in Oranienburg, they suggested using this manpower reservoir also in the production of the new engines. Schaller's role in this affair is revealing and exemplifies how some important initiatives in the Third Reich were based on personal contacts and early political affiliations. As Jüttner later reported to Himmler about this meeting, he mentioned that present in the meeting was one of his staff officers, Standartenführer Hoffmann, who knew Schaller from the good old 'time of struggle' (Kampfzeit). This acquaintance was used to support Jüttner's advice to Himmler to approve the enterprise by implying that Schaller was an* 'old fighter' (Alter Kämpfer)[167] *of the Nazi Party and therefore could be trusted. Jüttner obviously appreciated the importance of the matter and ordered his staff to look into it.*'[168]

Interestingly, the family traditions of both Helmut Schelp and Hans Antz have it, that both flew the Messerschmitt Me 262.[169] The aforementioned *'War diary'* (Kriegstagebuch Chef TLR) has the confirming, corresponding entry on 20 January 1945, that due to an imminent lack of *'fly-in'* personnel *'Luftwaffe and industry will suggest suitable candidates'*. A further indication on the desperate (personnel) situation; piloting briefings were generally given on the ground by servicemen, who themselves had often no own flight experience.

Apparently the *'War diary'* has also the last documented traces of Helmut Schelp's war-time activities, now as *'GL (group leader) in <Chef TLR/FL-E, F 2>'. 'January 1945'* starts with a surprisingly positive statement: *'The whole area of high-performance and performance-enhanced aircrafts shows considerable improvements.'* But the tone changes abruptly on 31 January: *'The surprisingly fast Russian thrust towards Berlin caused alert level 1; lack of accommodation facilities creates barracking problems. Chief TLR orders to destroy immediately all unused files'*, and specifically the entry for 8–14 January 1945 with respect to *'Subject W2: Black powder and explosives'* reads alarmingly: *'Again test problems with solid propellant castings for booster rockets at Karlshagen/Peenemuende; production release from FL. E/F2 urgently required, since otherwise the demand for powder-assist rockets cannot be fulfilled.'*

---

[166] See Wikipedia, 'Hans Jüttner'in English. In 1961, Jüttner testified for the prosecution in the trial of Holocaust architect Adolf Eichmann.

[167] See Wikipdia, *'Alter Kämpfer'*in English.

[168] See Uziel, Arming the Luftwaffe, p.183.

[169] Information, courtesy to Tom Schelp, 25 Sep. 2015, and a 500 year family pedigree of the Antz family at Pfeddersheim/ Mainz in German:

   https://www.worms.de/de/kultur/stadtgeschichte/wussten-sie-es/liste_familien_firmen/Antz.php

   Data for Hans Antz shown as *XIII.8*; however, the statement there, that Antz piloted also the Me 262 first flight is wrong.

In early February 1945 approximately 25% of the RLM Technical Office were no longer operational after heavy bombing raids, and part of the organisation was switched to the Flight School Doeberitz, some 10 km in the west, where in 1910 the still Royal-Prussian Air Force had commenced its activities. In the following weeks Schelp like many other dissolving military units found his way from Berlin, 600 km to Munich in the south—and to the expected American occupation zone. US troops of the 42nd and 45th Divisions who had liberated the Dachau concentration camp in the afternoon of Sun 29 April, were fighting in Munich the next morning and by nightfall had, along with XV Corps's other three divisions, captured the city that was the capital of Bavaria and the birthplace of Nazism. Not an industrial centre, this association nevertheless made Munich a target for air attacks, and in the end 80% of the city was damaged or destroyed.[170] Three weeks later H. Schelp's location can be determined already again in Charles Lindbergh's *'Wartime Journals'* on Tue 22 May 1945—in a small, one-storey house at Munich-Solln—*'with an American "half-track" with loaded machine guns in front'*; a military precaution, because a number of Russians had moved in next door to Schelp, and there was concern that the Soviet government might have tried to get him into their territory because of his scientific and technical experience. Actually, Lindbergh found Schelp unlucky then, since *'his wife and 6-month-old child* (boy) *were still in Dresden: "No one thought the Russians would come that far west."'*[171]

In due course, Schelp alone was brought to USAAF Wright Patterson Air Force Base, Dayton OH, for the first time in February 1946 presumably on initiative of Lindbergh's team (Lt. Robinson) as one of the *'Operation Paperclip'*[172] engineers, only to return to Germany one year later, and then transferring his complete young family out of Russian-occupied East Germany,[173] arriving in the United States as documented by border control records on 10 April 1947, on board of *SS Zebulon B. Vance.*[174]

From 1946 to 1951 he worked at Wright Field's Laboratory on the development of small gas turbines. In 1951 Schelp joined the Garrett Corporation, CA,[175] as a staff engineer, before he was sent to Arizona in 1954. Since 1958, he served as chief engineer at Garrett's AiResearch Manufacturing Company in Phoenix, AZ, until 1967, as he led the company

---

[170] See Ziemke, The U.S. Army, p. 253.

[171] See Lindbergh, The Wartime Journals, p. 957.

[172] See Wikipedia, 'Operation Paperclip' in English.

[173] In an US Army truck,—the then 2½ year old son *'smuggled'* in a basket under the bench. Information, courtesy to H. Schelp's grandson Tom Schelp, 25 Sep. 2015.

[174] During WW II one of 2710 US-built—see Wikipedia, 'Liberty ship' in English, named after *Zebulon Vance*, the two time Governor of North Carolina, lawyer and Confederate Army Officer. It was modified and renamed USAT *Zebulon B. Vance*, carrying military dependents between the United States and Europe until 10 Dec. 1948.

[175] See Wikipedia, 'Garrett AiResearch' in English.

through development of its first turboprop and turbofan engines.[176] Schelp then became technical assistant to the corporate vice president of engineering until his retirement in 1977, Fig. 7.13 (r). He served as consultant until 1986. It is said of Helmut Schelp (1912–1994) that without him and a few others, the Garrett divisions of Allied-Signals Aerospace Company, now part of Honeywell Aerospace,[177] would not be in turbine jet business; a statement that historically could be applied with some generosity to that of a turbojet engine industry in total. Looking back to his German turbojet experiences he contemplated:

> During the late 30s and 40s I was young and brash enough, and certainly didn't have the experience to know what difficulties to expect. If I were to do it all over again today I would be scared stiff![178]

Before switching to a personal portrait of Max Adolf Mueller (1901–1962), it is worthwhile to highlight his relationship to Helmut Schelp, and under a broader scope with the inclusion of Ernst Heinkel, Hans von Ohain, Hans Mauch and even Rudolf Friedrich, on the basis of a rare file note from January 1941.[179] A first *foul-play* against Schelp has been described already before, when Mueller leaked prematurely, certainly confidential information received from Otto Mader at Junkers in spring 1939 to Heinkel, that Schelp was about to change sides from RLM to Junkers; an information, which considerably must have tangled and hampered Schelp's image of a neutral turbojet programme manager, if not his further career within the (military) ranks of the Air Ministry.

The note has two parts, the first quotes Schelp saying to von Ohain during a visit at (Heinkel) Marienehe (1940?), literally: *'The TL apparatus at Junkers never ran* (really). *Herr von Ohain, you know, I will not forget how Müller has fobbed us then.'*[180] In the second part, Hans Mauch is cited, saying to von Ohain, in the presence of R. Friedrich, over a glass at the *'Alte Post'*, Stuttgart, when the (1940?) talk touched the personality of Mueller: *'You know, what he did then at Magdeburg to the* (Technical) *Office, was a*

---

[176]The first small gas turbine Schelp developed for Garrett, the *85 series* auxiliary power unit, has become the most produced small gas turbine in history, with more than 30,000 engines delivered since 1954.

[177]See Wikipedia, 'Honeywell Aerospace' in English, generating with 40,000 employees approximately $ 10 billion in annual revenues from a 50/50 mix of commercial and defence contracts.

[178]See Ermenc, Interviews, p. 125.

[179]Internal note from Hans von Ohain to Ernst Heinkel, Marienehe, 13 Jan. 1941, in Heinkel files at Deutsches Museum, Munich.

[180]Schelp quotation in German: *'Die Sachen bei Junkers sind ja nie gelaufen, dieses TL-Gerät (Turbinen-Luftstrahl). Wissen Sie Herr von Ohain, das nehme ich dem Müller doch schwer krumm und kann es ihm nicht so leicht vergessen, weshalb er uns damals so beschwindelt hat.'*

Fig. 7.14 *Alte Post*—Stuttgart, 1938, opposite Alte Stiftskirche (l), Restaurant 'Stiftsstube' 1938 (m), Stiftskirche 1945 (r)

very dirty affair. He was wanting us to believe that the device was running, but in reality it was externally driven. He provided numbers which turned out as absolute humbug.[181]

For an atmospheric impression of that meeting environment, Fig. 7.14 shows the traditional beer and wine guesthouse '*Alte Post* (Old mail station, left, from the outside) at Stuttgart's city centre, opposite the '*Alte Stiftskirche* (Old collegiate church)', the foundations of which date back to the tenth century.[182] A wall mural in the upper part of the façade illustrated a postillion on a stagecoach, drawn by four horses, and in the gable above hang a traditional '*brew star*' in the form of a hexagram, as a medieval brewers' guild emblem, indicating here a house brewery.[183] The middle picture of the wood-panelled '*Stiftsstube* (Collegiate chamber)' behind bulls's eye window panes transmits the then still cosy and relaxing atmosphere, before the complete destruction in 1945, right.

7.5   The Talented, But Unsteady Max Adolf Mueller (1901–1962)

Max Adolf Muller was born on 10 August 1901 at Metz, Lorraine, near the tripoint along the junction of France, Germany and Luxembourg; after the Franco-Prussian War 1871 up to the end of World War I, the city became part of the German Empire. Mueller was forced

---

[181] Mauch quotation in German: '*Wissen Sie, was er damals in Magdeburg* (1938) *mit dem Amt gemacht hat war eine sehr unsaubere Angelegenheit. Da hat er uns weismachen wollen, dass das Ding läuft und in Wirklichkeit war es fremd angetrieben. Er hat uns Zahlen angegeben, die sich absolut als plumpen Schwindel herausstellten.*'

[182] See Wikipedia, 'Stiftskirche, Stuttgart'in English.

[183] See Wikipedia, 'Brauerstern' in German.

to interrupt his school education during wartime, and was finally expelled from Lorraine in 1921.[184] Consequently, *'loss'* overshadowed Mueller's young life, as it had that of young Schelp by the death of his father in 1918, and similarly, that of young Fritz Heppner, who had to leave his West-Prussian home at Posen/Poznań together with his parents after 1919.

Up to 1926 Mueller was forced to make his living as engineer-fitter and engine assembler, while he attended evening courses to qualify for subsequent mechanical engineering studies in 1927/1928 at the *Höhere Technische Staatslehranstalt* (Technical college), Cologne, also known as *FH Fachhochschule*. After he had received the qualification for continuing technical university studies on the *TH Technische Hochschule* level, followed another working year as mechanical designer and structural engineer in steel-framed building construction. It was only with considerable delay that in 1929, he finally could commence studies of Physics at TH Aachen. For the year 1933, Mueller can be traced as *Hilfsassistent* (assistant aide), Fig. 7.15 (l), at the Institute for Experimental Physics, since 1927 led by Professor Hermann Starke. With the coming into power of the Nazi-regime in early 1933, Mueller belonged to a group of students, who aggressively requested radical *'purification'* of the professorship.[185] Instrumental for a staged-up intrigue by Starke's potential successor in March 1933, Mueller accused the *Physical Institute* of being a *'communist trash and dirt centre of Soviet friends'*, and Starke directly of corruption. In due course, Gerhard Harig (1902–1966),[186] then one of Starke's main assistants, who before 1933 openly had revealed his political preferences to the left, was shortly arrested—and managed in October 1933 to flee to Leningrad, where he started research on nuclear fission at the renowned *Ioffe Physical-Technical Institute of the Russian Academy of Sciences*.[187]

Professor Starke, however, remained steadfast,[188] and discharged Mueller, who—additionally threatened by university relegation—decided to leave TH Aachen in summer 1933, to continue without success first at TH Brunswick and TH Hanover,[189] before he found a

---

[184] See Koos, Heinkel Raketen- und Strahlflugzeuge, p. 109, where Mueller's CV is provided which will be used in the following—with corrections.

[185] See Kalkmann, Die Technische Hochschule Aachen, p. 232 f.

[186] See Wikipedia, 'Gerhard Harig' in German, and—see Bernhardt, Zu Biographie. In 1938, Harig was accused as a German spy, and possibly as part of a NKWD/Gestapo deal sent back to Germany, where he was imprisoned at the Buchenwald concentration camp, 1938–1945; his following career in East-German *GDR* led up to a position of State Secretary for Higher Education, 1951–1957.

[187] See Wikipdia, 'Ioffe Institute' in English.

[188] After a defamation process against Mueller had been turned down in August 1933, Starke even became a member of the Nazi party NSDAP on 1 May 1933 to initiate a party process, which was also stopped after Mueller's leave.

[189] Mueller certainly struggled—like Schelp, in the end in vain—to accomplish his doctoral thesis. As H. Wagner revealed in a 1970 letter to—see Constant, The Turbojet, p. 203 and there Footnote 45: *'Max Adolf Mueller . . . had fulfilled all requirements for his doctorate except mathematics'*; with all likelihood a hint, that Mueller had difficulties to provide the (informally, ~20%) obligatory theoretical section of his planned thesis, presumably still a result of his patchwork education in the 1920s—and

Fig. 7.15 Max Adolf Mueller (1901–1962): At FH Cologne/TH Aachen, ~1928/1930 (l), at Junkers, Magdeburg ~1936 (m), at Heinkel-Hirth, Stuttgart ~1942 (r) © V. Koos

new employment as assistant of Professor Herbert Wagner, and as his deputy at the Flight Technical Institute of the TH Berlin-Charlottenburg. Mueller's activities at that Institute and in due course at Junkers Magdeburg, Fig. 7.15 (m), up to mid-1939 have already been described in detail in Sect. 6.2.1, while his engine developments at Heinkel in Rostock up to his leave from Heinkel-Hirth, Stuttgart-Zuffenhausen, in June 1942 will be covered in Sect. 8.1.

From early on at Junkers 1937–1939, the Wagner/ Mueller team followed a number of advanced engine concepts, of which over time the turbojet (*RT0/HeS 30*) took the development lead—and in its wake Mueller's ducted fan engines '*ML/MTL*', which reached already mock-up status then. One of the first realised, large Junkers aircraft under the new development responsibility of Herbert Wagner was the *Ju 90,*[190] a 40 seat, four-engine airliner and transport which descended directly from the Junkers *Ju 89*, a contender in the long-range, strategic *'Ural bomber'* programme. This concept was abandoned by the RLM, Reich Aviation Ministry in April 1937 in favour of smaller, faster bombers. During the war the *Luftwaffe* used the 18 built Ju 90 units for military transport purposes. In the late 1930s as a Ju 90 development spin-off emerged the project of a *Ju EF100*[191] six-engine, 100 passenger trans-Atlantic airliner with a range of 5000 km for post-WW II use. The aircraft was to be of all-metal construction, with a pressurised fuselage having a wide-body 'double-circle' cross-section. As standard, six 24 cyl., 2500 hp *Jumo 223* diesel engines were to be installed, but Mueller prepared for this application at flight speeds of 720 km/h and mission flight times of 10–11 h a *ML motorjet* configuration with a superior fuel consumption, better than the basic diesel motorisation. In line with Schelp's principle

---

against all odds, a possible psychological explanation of his highly stressed relationship to Dr. von Ohain, with a similar professional background in Physics.

[190] See Wikipedia, 'Junkers Ju 90' in English.

[191] *EF Entwicklungs-Flugzeug*, development aircraft.in English.

assessment, Fig. 7.11, it was thought that these units, with their airflow rate somewhere between that of the conventional airscrew and the turbojet, would prove more advantageous than the piston-driven airscrew. Heinkel's first ML unit, internally known as '*Marie Louise*', was to use a diesel engine, but due to its insufficient power-to-weight ratio, it became necessary to develop a high-speed two-stroke piston engine, specially suited to the ducted-fan units. The corresponding research was put in the hands of Ernst Heinkel's long-time friend of study times, Professor Wunibald Kamm (1893–1966) of the *FKFS Forschungsinstitut fuer Kraftfahrwesen und Fahrzeugmotoren* (Research institute of automotive engineering and vehicle engines) at TH Stuttgart. On 31 January 1941, Mueller stated in a DAL book: '*It is to be expected that trial runs of the first ML full power unit will take place shortly. Already the results attained with the single-cylinder engine make the fostered hopes appear fully realisable.* '[192]

The FKFS diesel configuration *HeS 50d* was superseded by a new petrol engine design, the *HeS 50z*, Fig. 7.16, which was simpler, lighter, smaller, and was expected to have a far more even thrust curve with more thrust at the high end of the speed range[193]; on top it had a higher speed and higher-temperature exhaust gases to heat the ducted air. This 16-cylinder engine had a rated power of 800 hp, at take-off 1000 hp. A three-stage axial fan with a large-diameter hub was provided at the duct intake which led back between the 4×4, X -shaped engine cylinder banks. A separate ram duct provided the air flow for a 0.17 m diam. five-stage axial supercharger, visible in Fig. 7.16 (r), at the centre of the engine back-end. For the ducted-fan unit as a whole, the ducted fan pressure ratio was PR 1.5 . . . 1.9, whilst the piston engine was operated at a compression ratio PR 3 . . . 6. Basic data for the HeS 50z were:

- thrust 550 kp at 0 km/h, 400 kp at 800 km/h,
- fan speed 6000 rpm,
- weight 370 kg,
- diameter 0.62 m,
- length 1.47 m.

The first HeS 50z V-1 prototype was expected to be ready-to-test in April 1942, but due to Mueller's lay-off in early June 1942 the project came to a halt.[194] The capture of a German engineer, Franz Warmbrünn, on 1 January 1944 on the road between Vitebsk and Orsha provided the Soviet Union immediate insight to the German jet work, since Warmbrünn had been with Heinkel, and he also knew of Messerschmitt's jet aircraft activities.[195]

---

[192] See Mueller, TL-Triebwerke, p. 129.

[193] Illustration in—see Koos. Heinkel Raketen- und Strahlflugzeuge. p. 97.

[194] See Koos, Heinkel Raketen- und Strahlflugzeuge, p. 99.

[195] See Kay, Turbojet, Vol. 2, p. 17. Vitebsk (Wizebsk, Belarus) lies nearly 500 km west of Moscow, and 1100 km east of Berlin. The mentioned road from Vitebsk stretches 80 km to the south to Orsha.

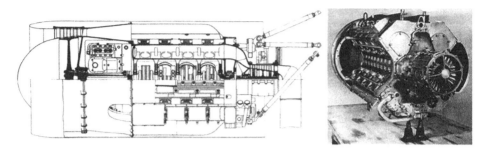

Fig. 7.16 Heinkel *HeS 50z* turbo-compound *motorjet*, ducted fan with supercharged FKFS 16 cyl. two-stroke petrol engine, 550 kp TO thrust (l), prototype hardware, April 1942 (r) © FKFS

According to Kay, the result were orders on the Soviet side to proceed with the development and construction of a turbojet of 1250 kp static thrust. However, the timing of Warmbrünn's capture could also have initiated or restored Russian interest in a jet augmentation system for application to airscrew-driven aircraft, known as *Kholshchyevniko's Accelerator*. This mixed motorjet propulsion unit consisted of a 1400 hp Klimov VK-107 piston engine driving a conventional airscrew, but with a rearward extension shaft from the engine to a geared axial compressor,—and thus somewhat comparable to Mueller's ML concept.[196] Consequently, in spring 1944 experimental fighter aircraft to test this motorjet/boost compressor combination were ordered from the *Sukhoi* and *MiG* design bureaus. *N1*, the first prototype of *MiG's I-250*, as it was designated, first flew on 3 March 1945, piloted by A.P. Dyeyev.[197] On the third flight of the armed I-250 N1, the jet drive was turned on and an increase in speed of about 100 km/h was achieved—from 525 km/h to 625 km/h at sea level and—claimed— 725 km/h to 825 km/h at 7 km altitude.

Besides the turbojet engine *HeS 30* and the motorjet *HeS 50* in the discussed configurations, Mueller had brought along from Junkers for further planned engine development activities at Heinkel:

- *HeS 40* an explosion gas turbine concept with intermittent constant-volume combustion, following ideas of Hans Holzwarth (1877–1953) in Germany for early stationary, power

---

Kay's spelling of <Warnbrünn> is somewhat uncommon, more popular in Bavaria/ Silesia would be <Warmbrünn>, the presently picked version, or perhaps <Warmbrunn>; however, also search in Russian archives provided no further clarification.

[196] See Kay, Turbojet, Vol. 2, p. 15/16; a principle sketch has here a multi-stage axial compressor, but a detail engine drawing in—see Goepfert, Mikojan, p. 50, shows only a single axial booster stage. Thanks for that latter source go to Hedwig Sensen, 21 April 2021.

[197] OKB-155 (Mikoyan) test pilot A.P. Dyeyev was killed on 19 May 1945, when the aircraft broke up under high G loads (Kay), other sources refer to noise fatigue of the tailplane fin.

generation gas turbines,[198] foreseen as a potential HeS 30 variant, also as a turbofan (ZTL),

- *HeS 60* a combination turbo-motorjet *MTL*, which combined a gas turbine with a three-stage axial fan and a driving diagonal turbine, and a 32 cyl. two-stroke piston engine, as a direct derivative of the HeS 50z.

Given the early turbojet engine development status and the strained wartime situation, Mueller's complete development set-up was apparently much too complex and overloaded, which caused repetitive interferences and corrective re-focussing actions from Schelp's side.[199] In addition, his development time estimates appeared to be highly unrealistic, as his successor at Junkers Anselm Franz noted after Mueller's leave to Heinkel: *'In view of the actual technical achievements of Mueller's group, the promised delivery deadline April 1940 for the Messerschmitt P.1065* (the later Me 262) *is incomprehensible'.*[200] Besides a mixture of short- and long-term projects, it appears that Mueller's programme comprised mainly backup solutions and alternatives to circumnavigate or replace eventually the demand for an effective highly loaded axial compressor; possibly, a reflection of the joint investigation which Wagner and Mueller had carried out at Junkers Magdeburg in 1937.

Mueller's just quoted DAL paper provides also a good opportunity to study his technical understanding and thinking. Though in 1940 his advanced turbojet concept *HeS 30* resembles *'a stroke of genius'*, includes the best possible axial compressor concept[201] and he and his team were enjoying Heinkel's wholehearted support,[202] he remained hesitant and undecided amongst the various engine options. He discusses R 0.5 compressor design as a possibility for weight reduction, but ignores the corresponding performance potential and at the same time, he is afraid of the *'excessive backflow losses'*—without indicating where this information might come from. It is doubtful, if he understood what a simple velocity triangle analysis would have revealed as the inherent performance benefit of a *'symmetric'* R 0.5 blading, a still relatively moderate Mach number level, and the otherwise steep Mach number rise towards Ma 1. Finally, he fixes his crooked perspective: *'At present, the most urgent question of axial compressor design is how to increase the pressure rise per stage by simple means—without diminishing the stage efficiency at the*

---

[198] See Eckardt, Gas Turbine Powerhouse, Ch. 3.3 The Holzwarth Gas Turbine, p. 72 f.

[199] However, in this respect Mueller had the full backing from Ernst Heinkel and his development philosophy.

[200] See Koos, Heinkel Raketen- und Strahltriebwerke, p. 89 and—see Schabel, Die Illusion, p. 43, who sees the root cause of the early timing difficulties in Schelp's unrealistically short time settings.

[201] As said before, designed by R. Friedrich (1909–1998), but the decision towards a degree of reaction R 0.5 came certainly from H. Wagner and his steam turbine experience.

[202] In 1940 Mueller's team comprised correspondingly already some 50 persons, while von Ohain managed his tasks still with 6–8 engineers and technicians,—see Koos, Heinkel Raketen- und Strahlflugzeuge, p. 90.

same time. Several solutions are available, which correspond largely with measures to increase the wing lift coefficients, but there is no clear trend yet. Given today's low compressor efficiencies, the explosion gas turbine represents without doubt a not to be underestimated competition to the constant pressure gas turbine at high compression levels. Consequently, the turbofan and turboprop engines with explosion gas turbine will be of some importance.[203]

Mueller's already mentioned CV tabulation[204] has for April 1939 his return from Junkers Magdeburg to the TH Berlin Institute, and continues in September 1939: *'Head of jet engine development at Ernst Heinkel Flugzeugwerke Rostock'*, an often repeated point of controversy, which in view of Hans von Ohain's then already established position and further career at Heinkel, cannot be correct. Actually, Heinkel established only on 13 March 1941 a re-organisation with separate responsibilities of Hans von Ohain for *SoE Sonderentwicklung (Special development)* I, and Max A. Mueller for *SoE II*.[205]

On Schelp's short-term initiative[206] and after the successful first flight of the *He 280* fighter aircraft with von Ohain's *HeS 8* engine (and not as originally planned, Mueller's *HeS 30*) on 30 March 1941,[207] Heinkel was able to acquire the *Hirth Motoren GmbH*[208] in Stuttgart-Zuffenhausen after notarisation on 30 April 1941, which put the new *Heinkel-Hirth Motoren GmbH* now officially as *flight engine manufacturer* in a position to continue the successfully started turbojet engine developments at Rostock-Marienehe with own production capacity, correspondingly experienced personnel and machine tool equipment some 650 km in the south at Stuttgart-Zuffenhausen.

On 13 June 1941, H. Schelp (*RLM-LC3*) came to Marienehe to discuss *HeS 8* difficulties and the *HeS 30* status; on the following day he suggested Director Harald Wolff[209] of

---

[203] See Mueller, TL-Triebwerke, p. 126.

[204] See Koos, Heinkel Raketen- und Strahlflugzeuge, p. 109.

[205] See Koos, Heinkel Raketen- und Strahlflugzeuge, p. 105.

[206] On 18 July 1940 Ernst Heinkel had asked *Generalluftzeugmeister (Luftwaffe Director-General of Equipment)* Ernst Udet officially to get the acquisition permission, which was agreed on the basis of demonstrating the He 280 airworthiness in advance. Consequently, Udet's suicide on 17 Nov. 1941 was also a considerable blow against Heinkel's influence and standing.
The take-over was for Ernst Heinkel also very emotional, remembering to his Suebian home country, *'where my family can be found in the church registers of Kirchheim-Teck and Remstal as craftsmen and winegrowers for more than 300 years'*. Helmut Hirth was his close friend since 1903, and he reminded to his aircraft crash on *Cannstädter Wasen*, Fig. 6.21.

[207] A special demonstration flight was arranged for Udet and RLM representatives on 5 April 1941.

[208] See Wikipedia, 'Hirth' in English.

[209] Actually H. Wolff started his new task at Heinkel-Hirth. Stuttgart as early as August 1941,—see Koos, Raketen- und Strahlflugzeuge, p. 106, but apparently it was only in March 1943, that he was fully implemented there on order of Field Marshal E. Milch as *'RLM commissioner/ controller'* of Heinkel-Hirth turbojet developments, officially only temporarily up to the *'completion of the first HeS 011 series engines'*, but in fact disempowering Ernst Heinkel,—see Kay, German Jet Engine, p. 40. The founding of EHAG Ernst Heinkel AG on 1 April 1943 saw him only in a representative function

BMW-Bramo, Berlin-Spandau as new superior principal for, and moderator amongst Hirth's Technical Director Schif, as well as Mueller and von Ohain after their SoE units had been relocated at Zuffenhausen. There were even discussions that von Ohain may stay somewhat longer at Rostock to finish the HeS 8 development, and start preparations for a possible professorship at Goettingen in 2- to 3-years-time, which Ernst Heinkel was willing to support. The southward relocation of Mueller's SoE II group was planned for 25 August 1941—to work there primarily on the HeS 30 engine, while the ML projects were put on hold. This freed capacity stayed at Rostock to accelerate the series-readiness of the HeS 8 engine.

The relocation influenced also Mueller's management position at Heinkel-Hirth, which was renegotiated together with the Heinkel General Manager Dr. Josef 'Jupp' Koehler,[210] who sent the related minutes on 27 August 1941 from Heinkel Works, Jenbach[211] to Ernst Heinkel at EHAG headquarters, Rostock. Koehler installed Mueller in agreement with Director Wolff as head of *SoE II* unit at (still) *Hirth Motoren G.m.b.H.*—and listed Mueller's following wishes (in that order):

- he wants to eat regularly in the *Casino* (agreed),
- fuel privilege mark-up of his car—*'bewinkelt'* (agreed),
- procuration—*'Handlungsvollmacht'* (agreed),
- promotion to *'Director'* after series-readiness of his engine (with hand-written comment on the archived document from Ernst Heinkel: *'Only if he fulfills the necessary preconditions with respect to character and leadership!'*),
- asks for contract modification, since the relocation causes delays which endanger his expected bonuses (kept open up to Koehler's return).

In April 1941, after considerable delay still the first test run of Mueller's HeS 30 was accomplished at Rostock, but then—the relocation ahead—not very much happened in the second part of 1941. Immediately after the described meeting with Koehler, Mueller must have fallen severely sick with *bilateral pneumonia* and had to stay in hospital for 12 weeks.

---

as head of the supervisory board, a trauma and humiliation which followed E. Heinkel beyond war's end.

[210] Koehler belonged to Heinkel's eldest and most trusted acquaintances, whom he met already 1920–1922 at Caspar Works, Travemuende; he was primarily responsible for the upper personnel management.

[211] Jenbach lies 36 km east of Innsbruck. In 1938 after Austria's 'Anschluss', Fritz Reitlinger (1877–1938), owner of the *'Jenbacher Berg- und Hüttenwerke* (Mine and metallurgical plant)' died under unclear circumstances. In due course the works were 'arianised' and came to the Ernst Heinkel AG. Wartime production comprised the Walther rocket engine 'HWK RII 211' for the Messerschmitt Me 163 and the potassium manganate vaporiser drive unit for the V-2 rocket main fuel turbo-pump. Between 2003 and 2018 *Innio Jenbacher* belonged to *General Electric Distributed Power*. Fritz Reitlinger was a cousin of British art historian—see Wikipedia, 'Gerald Reitlinger' in English.

Finally, end of January 1942 the promised component tests at Zuffenhausen started with a separate measuring campaign on the five-stage HeS 30 axial compressor; the combustor test bench was delayed to a possible start in February 1942.

In February 1942 a severe personal dispute broke out between Ernst Heinkel and Max A. Mueller. Heinkel apparently became nervous, after nearly three years he expected substantial progress from Mueller, the more so since competitor Junkers had passed successfully with the *Jumo 004* already in December 1941 the 10 h endurance test. On the other hand Mueller had again managed to raise broad opposition from the local Hirth organisation, in a complaint letter to Heinkel he laid down work on 20 February 1942 in view of the ineffective organisation. Heinkel in return demanded that Mueller had to integrate himself in the existing organisation. He had enough of Mueller's frustrating empty promises and excuses, when he should have simply worked under pressure, culminating in: *'Of all my 50,000 employees, you are the most trouble!'*[212]

Though the first HeS 30 full test could be accomplished at the Hirth plant in Zuffenhausen on 19 April 1942, a renewed quarrel with Heinkel in May initiated finally Mueller's resignation in early June 1942, Fig. 7.15 (r).[213]

The Heinkel Archive keeps a 1952 interview[214] of Max A. Mueller, then 51, which can be exploited cautiously for a kind of indirect psycho study of his somewhat volatile and showy character, with which with some likelihood he over-compensated an inferiority complex, presumably based on his patchy engineering education. All following statements come directly from Mueller:

- Interestingly, Mueller starts with a historic review, making reference to the Romanian inventor and aerodynamics pioneer Henri M. Coandă (1886–1972),[215] who built an experimental aircraft *Coandă-1910*[216] in France, which was powered by what Coandă called a *'turbo-propulseur'*, consisting of a conventional piston engine driving a multi-bladed centrifugal blower, which exhausted into a duct. In the mid-1950s Coandă claimed this as the world's first jet aircraft. Mueller knew that this aircraft was captured by German troops in northern France in 1917, and from resulting descriptions originated his interest in the *ML concept*,[217]

---

[212] See Conner, Hans von Ohain, p. 105.

[213] Mueller's axial compressor designer R. Friedrich took the opportunity to leave Heinkel for Dresden and marine gas turbine developments at the Brueckner, Kanis & Co. turbine factory.

[214] Heinkel Archive at Deutsches Museum Munich, FA 001/ 0025.and FA 001/ 0314.

[215] See Wikipedia, 'Henri Coandă' in English.

[216] See Wikipedia, 'Coandă-1910' in English. During this aircraft crashing, Coandă detected the—see Wikipedia 'Coandă effect' in English, and—see also a 1:40 min YouTube demonstration video of a *Dyson Airwrap* ® hairstyler https://www.youtube.com/watch?v=X1kQFJcKEaM.

[217] Mueller refers also to a first 1913 paper on the air-cooled two-stroke piston engine for flight applications by—see Baumann, Motorensysteme,—and see Wikipedia, 'Alexander Baumann (aeronautical engineer)' in English, a German aircraft designer, credited with being the first full professor

- (Junkers General Manager) Koppenberg had hired Wagner, and it was on his request that the first *'ML compound engine'* evaluations were carried out:
  - Which compressor concept up to which flow capacity?
  - Can the piston engine serve as combustion chamber?
  - (GT) turbine operation with constant-volume or constant-pressure combustion?
- before the move to Heinkel, he ordered on the basis of Junkers preferential purchasing conditions a prestigious *Opel Admiral* luxury car,
- beginning spring 1939 there were first activities for Heinkel at the TH Berlin flight technology institute, but after August 1939 more and more of the personnel was drawn into military service, so that Heinkel's *(sic)* personnel administration suggested a move of his team from Berlin to Rostock-Marienehe to return to stable working conditions. The relocation took finally place in October 1939 and 120(!) employees[218] followed this call. The topic *ML* and corresponding machinery was then immediately transferred from Heinkel further to FKFS Stuttgart,
- his superior material know-how shifts the endurance of von Ohain's *HeS 3B* engine from originally *'several minutes to hours'*, and—to relax the somewhat concerned He 178 test pilot Erich Warsitz before the first flight on 27 August 1939, Mueller demonstrates a seven min *'bull ride'* on the engine test rig,
- the RLM (Schelp) knew the Wagner/ Mueller all axial turbojet engine design (RT0/HeS 30) since autumn 1938, consequently the similar, since 1940 appearing Jumo 004 and BMW 003 configurations are direct derivatives thereof,
- he (Mueller) suggested to replace the CR counter-rotation *BMW TL 109-002* concept (Weinrich) by the much simpler axial compressor with 50% reaction blading of the *BMW TL 109-003*. Though difficulties with *'that impossible CR concept'* were known since 1938, it was stopped only in 1943/1944 after some seven Million Reichsmark had been wasted (*'verpulvert'*). In 1943, Schelp confessed to him, that prior, he had not understood his arguments, but *'now, it dawns on me! (jetzt, ist mir eine Stall-Laterne aufgegangen!')'*,
- Heinkel had promised to build the desperately required expansion of turbomachinery test facilities at Zuffenhausen; there were plans to transfer a 14 MW, 12,000 rpm compressor test facility from Bramo/Siemens, complete with Siemens DC pendulum machine. When Mueller returned from hospital, H. Wolff had been implemented by RLM as head of Hirth Zuffenhausen, and one of his first decisions was to give away the Siemens test equipment to the Goettingen University in anticipation of von Ohain's

---

of aeronautical engineering in aviation history as the *Chair of Airship Aviation, Flight Technology, and Motor Vehicles* at the Royal Institute of Technology, Stuttgart in 1911; during WW I, Baumann designed the *Zeppelin-Staaken R.VI 'giant aircraft'*, Fig. 4.14, then the largest, quantity-produced bomber. In 1930, Baumann's successor at TH Stuttgart became Professor Wunibald Kamm, who founded then *FKFS*.

[218] See Koos, Heinkel Raketen- und Strahltriebwerke, p. 89, where realistically the transfer rate has been estimated to 15–18.

coming professorship there; other comparable *'gifts'* went to Dresden, Munich and Stuttgart,[219]

- a half year after leaving Heinkel, he claims involvement in the erection of a 60–100 MW high altitude test facility, which—given the extraordinary power demand—points to the huge LFM wind tunnel facility at Oetztal,[220] In addition, he claimed coordination responsibility on behalf of RLM for all turbojet production activities, before the SS-FHA[221] leadership main office asked him to coordinate the development of a tank-propulsion gas turbine[222] (according to Mueller, a *'Mueller idea'*),

- Mueller ends the interview by the lapidary statement *'On 1st October 1942 the HeS 30 achieved its design data with 500 kp continuous thrust, and 750–800 kp (max. 920 kp) take-off thrust. 'Gez.[223] M.A. Mueller, 20. XI. 1952'.*[224]

In addition to the stations above, Mueller's official CV[225] mentions him after his leave from Heinkel associated as Technical Consultant to *Friedrich Goetze AG* at Burscheid, some 25 km north-east of Cologne, an established manufacturer for metal seals and piston rings. Apparently, the company was also a secure haven for his engineering activities after the war, when he tried to make a living as inventor. In this respect Mueller was with some 163 patents in his name quite successful. The inherent pattern becomes already visible by comparing the 42 inventions in his name before war's end to the 121 thereafter, many of the later ideas had been applied for the first time already during the war, when there was simply

---

[219] Ernst Heinkel's archive copy of Mueller's interview carries here a hand-written comment: *'In addition, cancelled and returned some three million Reichsmarks which had been already agreed from RLM for planned enlargements and test fields at Zuffenhausen.'*

[220] Ernst Heinkel added here in the 1950s *'now installed abroad'*, which applies to the Oetztal facility which was re-erected as the ONERA Modane S1MA continuous-flow, atmospheric Mach 0.05 to Mach 1.0 wind tunnel,

https://www.onera.fr/en/windtunnel/s1ma-continuous-flow-wind-tunnel-atmospheric-mach-005-mach-1

which comprised in the original planning also a copy of the 'Herbitus' high-altitude engine test facility, which was used up to 1946 at BMW Munich-Milbertshofen, and thereafter transferred to AEDC Tullahoma, TN, where it is still in use—now extended up to Mach 3+,—as *'J-1'* Engine Test Cell,—see Eckardt, The 1×1 m hypersonic wind tunnel.

[221] See Wikipedia, 'SS *Führungshauptamt*' in English.

[222] See Sect. 8.3 *'BBC aero gas turbine activities and derivatives'*. Design activities took place at Porsche, Hoechst-Rheinau on the southern, still Austrian shore of Lake Constance, approx. 20 km north of Feldkirch, and at the *'KTL Kraftfahrtechnische Lehranstalt (Automotive academy)'* of the Waffen-SS at St. Aegyd near St. Poelten, Austria. In this context Mueller participated also in corresponding derivative developments of the Porsche TL 109-005 (internally Porsche TP 300), a loss engine with 400+ kp take-off thrust for long range Fi-103 (V-1) applications,—see Sect. 8.1.

[223] Gez. = abbreviation for 'gezeichnet (signed)'.

[224] The Heinkel HeS 30 development in the early 1940s will be addressed in Sect. 8.3.4.

[225] See Koos, Heinkel Raketen- und Strahlflugzeuge, p. 109.

no time for the necessary paperwork. What originally was a heavy tank as invention object, became afterwards a (*Heinkel*) farm tractor, and also the ownership questions of intellectual property were often difficult to assess. This approach to *'fortune'* was not unusual for the immediate post-war era, but others like BMW's former chief engineer Hermann Oestrich were in this respect much more successful than Max A. Mueller.[226] For the period 1945–1948 Mueller ran his own engineering office, in which he carried out a number of industrial compressor design contracts. In 1949 and a few years thereafter, as reflected by the 1952 interview, there were renewed activities together with the Ernst Heinkel AG, Stuttgart-Zuffenhausen, the only plant still owned by the Heinkel family after the war, while six large works including the parent company at Rostock-Marienehe and further 27 subsidiary plants spread all over Europe were lost by the end of WW II. After the restart Heinkel's financial situation remained precarious after he was unsuccessful in getting 300 million Marks from the new Federal Republic of Germany,—for open bills which he claimed still unsettled by the Third Reich at war's end.

One possibility to compensate foregoing losses by renewed military engagements in the beginning *'Cold War'* period was strictly prohibited in Germany by Allied rule; however, Heinkel managed to collect his specialists in concealing camouflage offices in Switzerland and Liechtenstein. The first offer came from Tito's Yugoslavia to develop a new fighter aircraft for the new *'Yugo Airforce'*, superior to their outdated MiGs. According to his printed CV, Mueller participated in a corresponding Heinkel turbojet engine project, and presumably prolongated this engagement, when Heinkel was contracted—officially out of Vaduz, Liechtenstein—to a considerably larger fighter aircraft/engine project with Nasser's Egypt.

Though there existed again a small camouflage design office at Vaduz, the larger part of the developments was carried out by some 170 aircraft and engine designers at Feldkirch on Austrian territory, some 15 km in the north of Vaduz,—and at Heinkel Zuffenhausen, where a total of 30 supervising Egyptian air force personnel, including a *'vice air marshal'*, were stationed.[227] The project was abruptly stopped, when Nasser nationalised the Sues Canal on 26 July 1956, Egypt switched sides and military equipment was delivered from Russia thereafter.

---

[226] As outlined in Chap. 9, Oestrich is supposed to have earned up to 14 million Euros in royalties from German intellectual property, which he sold to the French state after the war. Mueller's patenting activities will be reviewed in Sect. 12.1.2, his corresponding revenues remained presumably modest.

[227] See Google <Heinkel AG, Umwege über Vaduz> in German: Der Spiegel 40/1958, 30 Sep. 1958.

The War Years 1940–1945

8

8.1 German Turbojet Engine Developments and Related Activities

This chapter deals with the German turbojet engine developments during the war period 1940–1945, and separately, in continuation of Sect. 6.2.5 for Junkers Flugzeug- und Motorenwerke AG, Dessau and the JFM team of chief designer Dr. Anselm Franz, in Sect. 8.1.1,—and in continuation of Sect. 6.2.4 for BMW Bayerische Motoren Werke AG, Berlin-Spandau and the activities of Dr. Hermann Oestrich and his development team, in Sect. 8.1.2. Already in 1941, the aero engine group of BBC Brown Boveri & Cie., Mannheim, led by chief designer Hermann Reuter was contracted by RLM to support BMW by alternative, exchangeable axial compressor designs; related information can be found in Chap. 9 as a highlight of Hermann Reuter's engineering work. Finally, wartime work at Heinkel-Hirth at Rostock-Marienehe and Stuttgart-Zuffenhausen of chief designer Dr. Hans-Joachim Pabst von Ohain and his relatively small team, supplemented by Max Adolf Mueller and his larger group up to his leave in early June 1942, are collected in Sect. 8.1.3.

8.1.1 The *Junkers Jumo 109-004* Development and Production Programme

The only German turbojet engine to achieve real mass production was the Junkers Jumo 109-004, designed and built by the Junkers Engine Division. In July 1939 Otto Mader had finally accepted a RLM development contract for a turbojet programme. The Junkers Jumo 004 design activities at *'KoBü 1'* (Konstruktions-Büro/design office), Dessau commenced in October 1939, led by chief engineer Fritz Boettger,[1] who had been also responsible for

---

[1] See Irving, The Rise and Fall, p. 289 f.

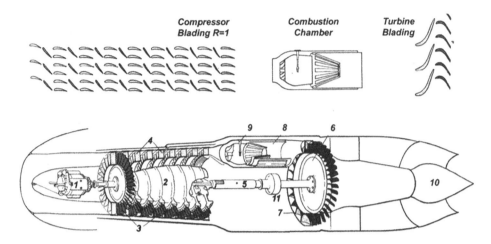

Fig. 8.1 Junkers Jumo 004 B turbojet components, September 1944: 1—Riedel starter, 2—compressor, 3—compressor blades, 4—compressor vanes, 5—hollow shaft, 6—turbine, 7—turbine vanes, 8—combustion chamber, 9—fuel injection, 10—nozzle plug, 11—turbine front bearing © Deutsches Museum Munich

the successful 12 cyl., liquid-cooled Jumo 210 piston engine, in production up to 1938. The first engine test run of the *Junkers Jumo 004 A (TL)* turbojet engine followed in record time already on 11 October 1940. Based on the AVA design by Walter Encke, the Jumo 004 eight-stage axial compressor with R 1.0, Fig. 8.1 (top, for Jumo 004 B, status Sep. 1944), had the following data:[2]

• mass flow rate ṁ	21.2 kg/s
• total pressure ratio PR	3.14
• rotational speed n	8700 rpm
• circumferential velocity $u_{1,\,Tip}$	248 m/s
• isentropic ad. efficiency η_C	0.78

supplemented by the following engine data:[3]

• combustor efficiency (burnout)	0.95
• turbine entry temperature TET	755 °C
• turbine isentropic ad. efficiency η_T	0.795
• thrust	910 kp
• specific fuel consumption sfc	1.4 kg/(kgh)
• engine weight	745 kg
• engine diameter	0.76 m
• engine length	3.86 m

---

[2] See Hirschel, Aeronautical Research, Chapter 'Axial Flow Compressors', p. 231 f.

[3] See Franz, Der Weg.

Though the aerofoil pattern in the *General Arrangement*, Fig. 8.1, is drawn—except for the number of inlet and exit vanes—absolutely regular and repetitive, it was only nearly so, but close. The rotor blade count was identical for R1—R2: 27 and for R3—R8: 38, constant and unchanged for all these rear rows; the tip stagger of the rotor blades remained fairly constant over the first six stages, but was increased over the last two stages. Besides the number of inlet guide vanes IGV 32, the vane count V1—V8 varied between 56 and 71 only. All this relatively close to Betz's and Encke's original design intent of simply adding stages on the basis of only two to three tested master stages, added up to the desired total pressure ratio.[4]

The aerodynamic design of the compressor was provided by W. Encke, AVA Goettingen, and was such—as had been discussed in foregoing degree of reaction R 1 considerations—that all the stage pressure rise occurred across the rotor blades. Apart from the inlet guide vanes and the last row of stator blades which acted as flow straighteners, the stators were arranged as impulse blading, set at approximately zero stagger angle, just to redirect the air into the next row of rotor blades.[5] The tip diameter of the first compressor stage was 0.544 m, that of the last stage 0.582 m; as Heinz Hoheisel observed: *'The external diameter of the compressor stages increases from front to rear and in this way the circumferential velocity grows also. Because of the rise of the speed of sound with $\sqrt{T}$, where T is the static temperature in °K, one obtains thereby a rise of the stage work from front to rear at a nearly constant Mach number level.'*[6] And—this is precisely the mechanism of Encke's patent DE 707,013 with priority 27 May 1936, as described in the Sect. 6.2.3. Overall, this design was not the best for efficiency but permitted of a simple sheet metal construction for the stator vanes, and did not call for fine axial clearances between the stator shrouds and the disc rims.[7]

---

[4]In 1949, J.W. Adderley, author of—see B.I.O.S, German Gas Turbine Developments, p. 17, reasoned then already correctly: *'Encke's background was originally in the field of single-stage fan design and it was therefore almost inevitable that he should extend this work to axial compressors leading to a design in which the pressure rise took place almost entirely in the moving blades.'* With all likelihood, Encke carried out already the single stage Siemens-Betz blower tests in 1928, Fig. 4.38.

[5]Modern axial compressor represent a combination of these design principles on the basis of degree of reaction: Since the flow normally enters the compressor axially and leaves axially, there are advantages to having the front and back stages close to axial; correspondingly the reaction starts high (R ~ 1), goes down to around R ~ 0.5 in the mid stages and rises again towards the last stage.

[6]See Hirschel, Aeronautical Research in Germany, p. 231.

[7]In Fig. 4.48 an efficiency advantage of 7–12 percentage points was shown for a R 0.5 compressor design in comparison to R 1.0, deduced from BBC measurements, 1933,—and a theoretical assessment of R. Howell, 1945. These numbers are correct on the basis of the then limited aerodynamic understanding. However, as outlined by compressor expert, em. Prof. Nick Cumpsty, Imperial College London, in a discussion with the author on 15 July 2021, *'As for the difference in efficiency of seven points, it is not at all clear how much of this is due to reaction. There are so many things which could be different, notably loading and flow coefficient, but also blade solidity and thickness.'*

The aerofoils for the first jet engines were also manufactured in the AVA model workshops at Goettingen—by a high-quality copy milling technique. In general, characteristic for Franz's in every respect cautious, though in the end successful approach is his statement:

> In consideration of the novelty of the object, in order to guarantee as much as possible the greatest chance of success, it was deliberately avoided to aim always at the possible maximum
> . . .

The single-stage turbine of Fig. 8.1 was—in line with the standard steam turbine design practice of the time—of the impulse-type (degree of reaction $R < 0.2$) with 35 nozzle vanes and 61 turbine blades. It was designed largely by Franz's fellow-Austrian Professor Dr.-Ing. Ernest A. Kraft (1880–1962), then head of the AEG Turbine Factory, Berlin-Moabit, whose acquaintance he possibly could have made already at Schwarzkopff, Wildau a few years before.[8] In this context, Schlaifer in his post-war analysis[9] pointed out: *'The aerodynamic design of the turbine wheel, . . ., was probably the poorest feature in all the German engines and was certainly by far the most important factor in their high fuel consumption. The efficiency of the turbine of the 004 was only 79% or 80%, 'total to total',* while the Welland* [Whittle's W2B] *turbine had an efficiency of 87% on this basis'*, and added in a footnote, that already the *'Whittle Unit WU'* of 1938 had a corresponding turbine efficiency of 84%. While the actual turbine loading in terms of u^2 is not far apart between W2B and Jumo 004,[10] normally Frank Whittle stressed here his invention and usage of the *'free vortex blading'* with radial equilibrium. Certainly untwisted blades and vanes would not be very good, but Antony Kay[11] knows about the similar BMW 003 turbine design, which was influenced by BBC Mannheim: *'In order to ease production, the profiles of both inlet nozzles (. . .) and turbine blades were kept uniform along the whole radial length and had no twist. BMW had made comparative tests of both twisted and untwisted blades. They found that twisted* (nozzles and) *blades gave a higher peak efficiency by giving optimum angles of attack throughout their length, but this efficiency fell off much faster once above or below optimum conditions than was the case with untwisted blades. The untwisted blades were, therefore, a compromise in efficiency while being most suitable for mass production.'* A view also on *'radial equilibrium'* designs, with which Brown Boveri in

---

[8] See Wikipedia, 'Ernest Anton Kraft'in German, and—see AEG, Wir stellen vor.

[9] See Schlaifer, The Development, p. 432.

[10] In general, an axial engine turns more slowly than a centrifugal, while turbine efficiency falls off as speed decreases. Correspondingly, also the turbine of the British axial engine F.2 had an efficiency slightly below that of the turbines of the British centrifugal engines. Another factor was that the 004 was designed to have a certain residual pressure after the turbine, since this was thought to facilitate planned afterburning.

[11] See Kay, German Jet Engine, pp.112–113.

Switzerland concurred from early on.[12] In addition to the foregoing arguments, it had to be taken into account that the German reference turbojet engines Jumo 004 B and BMW 003 had air-cooled turbine vanes and blades, Fig. 8.4, with an estimated impact on the adiabatic efficiency of the turbine stage by 1–2%.[13]

For the combustion system, six individual can-type combustion chambers were chosen, largely on the grounds of the more rapid development that single-can tests promised.[14]

For first engine testing, it was decided to design and test a scaled-down engine model so that the compressor absorbed only 400 hp at speeds of ~30,000 rpm and to allow the use of already existing test facilities. The set-up was completed in late 1939, but the small scale gave inadequate combustion and the engine suffered from vibrations. Thereafter, independent tests were then made on the compressor alone, but this burst then at high speed.[15] After this failure, it was decided that studies would be as well, if performed, on a full-size engine, and work on this, the Jumo 109-004A, was begun still in December 1939.

The compressor construction of the Jumo 004 was obviously designed for the utmost ease of production. The rotor blades were mounted on eight discs held together by a tension bolt through the centre and spaced apart and spigotted together at about their mid-radius. Each row of stator vanes was mounted in inner and outer shroud rings—in two semi-circular halves. The first two stator rows were light metal stampings and were brazed into the shroud rings. The remaining vane rows were sheet metal pressings, mounted in the shroud rings either by bent-over tabs, a long-time standard in German sheet metal toys, or by brazing. The assembled half rings were then mounted in the longitudinally split compressor casing.[16]

Both the Jumo 004 and BMW 003 rotors were supported in four bearing compartments and the compressor and turbine shafts were connected by an internally splined sleeve which permitted relative longitudinal movement between the two shafts. This however precluded some *axial thrust balancing* between both components, which in turn necessitated the use of thrust bearings for both compressor and turbine. Especially the R 1 compressor design gave rise to high end thrust and complex front bearing arrangements had to be adopted to

---

[12] See Eckardt, Gas Turbine Powerhouse, p. 116, and there especially Footnote 74.

[13] Further investigation of the Jumo 004B turbine efficiency deficit revealed in—see Decher, Die Entwicklung, p. 43, that thermal casing deformation required a turbine rotor tip clearance increase from nominal 2 up to 6.5 mm!

[14] For a general construction survey of the various Jumo 004 engine versions—see Gersdorff, Flugmotoren in German and—see Kay, German Jet Engine, pp.57–95; by far the most detailed information has still—see Power Jets, The Junkers Jumo 004.

[15] See Kay, German Jet Engine, p. 60.

[16] See B.I.O.S, German Gas Turbine Developments, p. 24. The usage of sheet metal for compressor vanes was also to circumvent ever existing forging bottlenecks. BMW had experimented with reinforced plastic material for the first two compressor stages, but finally selected sheet metal for all rows, easily handled by women working in the BMW automobile shop,—see Bruckmann, Unexpected, p. 78.

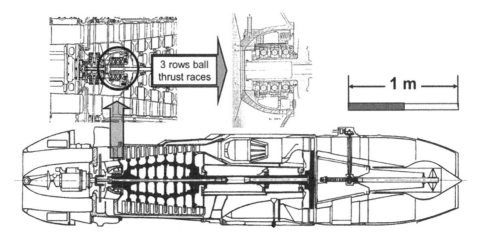

Fig. 8.2 Junkers Jumo 004 B turbojet engine,—with axial compressor front thrust ball bearing arrangement, enlarged © H. Schubert

meet this, Fig. 8.2.[17] The resulting end thrust was calculated to 2270 kp at maximum speed at sea level,[18] which demanded for a complex three-row ball bearing arrangement, the inner races of which were mounted on a sleeve which fitted over and was keyed to the compressor shaft. Each other race of the bearings was mounted in a separate sleeve, and the three sleeves nesting one within the other. The assembly could then be preloaded, a hole drilled through the sleeves and a holding pin inserted and the end of the sleeves on which the thrust was taken could be machined flush. In this way it was ensured that thrust was divided evenly between the three bearings. The three bearings of this thrust compensation package were identical—with 65 mm inner diameter, 120 mm outer diameter and 23 mm wide, with 15 balls of 15.87 mm diameter.[19] As for other design practices it is worthwhile in the spirit of an interconnected *'Jet Web'* of design ideas, to look for potential foregoing applications of this remarkable concept. And in fact, exactly the same thrust bearing arrangement had already been highlighted in Fig. 6.49 for an Allis-Chalmers 20-stage axial compressor, a 1939 Brown Boveri license product for a Houdry gas turbine.[20] In general, industrial turbomachines are built with journal bearings, but the rough refinery operations required roller bearings with which all Houdry compressor designs since 1936 had been equipped.

---

[17] See Gersdorff, Flugmotoren, pp. 299–300.

[18] See B.I.O.S, German Gas Turbine Developments, p. 24.

[19] See Power Jets, The Junkers Jumo 004, p. 342. The three-race front bearing arrangement was also used for the BMW 003, Fig. 8.9, and for the first Heinkel HeS 011 design, Fig. 7.5 (r), before it was replaced by a single, heavy cyl. roller bearing for the production standard.

[20] See Eckardt, Gas Turbine Powerhouse, p.169 f.—there, with the 20-stage axial compressor as Fig. 4.64.

However, it is worth to mention that this detail never showed up or was described in any Brown Boveri publication; so Lowell's SAE paper[21] of 1940 was apparently a first time publication, repeated after the war on several occasions e.g. by Kruschik in the first (1952) and the second (1960) edition[22] of his gas turbine book. The missing communication link to the Junkers and BMW turbojet engine designers during the early Jumo 004 and BMW 003 definition phase (~ 1940) could have been either out of BBC Mannheim, or, with higher likelihood out of the German ball bearing centre from *SKF* and *F&S (Fichtel and Sachs), Schweinfurt,*[23] which presumably collaborated with the Swiss Brown Boveri already in the mid-1930s to develop a feasible solution of axial thrust compensation for the new Houdry axial compressors.

Obviously, the Allis-Chalmers publication had no influence on German design decisions. Nevertheless, the company's La Porte, IN, plant appeared for their anti-aircraft gun production on a list of 21 potential German bombing targets in North America, including GM's Allison Division in Indianapolis, IN and Colt Manuf., Hamilton Std. and Pratt & Whitney Aircraft, all at East Hartford, CT.[24] The author of that target list, Colonel-Engineer Dipl.-Ing. Dietrich Schwencke,[25] Fig. 8.3 (l), had an early career at the Rechlin test centre between 1929 and 1937 with responsibility as department head (on the level of W. Eisenlohr) for '*Ausrüstung* (equipment)', until he was called to the RLM, after Udet's death reporting directly to Field Marshal Erhard Milch (since 4 Sept. 1940), the Luftwaffe's air armaments chief.

Obviously in parallel to his RLM occupation, the Colonel operated a test and evaluation centre at Rechlin, where captured and downed enemy aircraft were brought to be either dissected for closer examination, or repaired and test-flown by German pilots. Schwencke himself once flew a captured American B-17, when that huge four-engine bomber was generating highest interest among the Luftwaffe leadership.[26] Eventually, he employed a group of 200 selected Russian prisoners-of-war at Rechlin for the task of dismantling captured airplanes. Schwencke had once served as assistant air attaché in London and, surprisingly, was also son-in-law to Erhard Milch.[27] In 1942, the Allied superiority in aircraft production (fighters and bombers) became evident for the first time in reliable

---

[21] See Lowell, Gas Turbine Design.

[22] See Kruschik, Die Gasturbine.

[23] Also known as *VKF, Vereinigte Kugellagerfabriken* (United ball bearing fabrication) *AG, Schweinfurt.*

[24] See Wikipedia, 'Amerika Bomber' in English and—see Forsyth, Messerschmitt.

[25] Schwencke's name is often misspelled as 'Schwenke',—see e.g. Irving, Rise and Fall. Figure 8.3 (m) has also Schwencke's signature, with which he was identified as author of a secret RLM GL-C (Milch) report, special print in 50 copies, issued on 15 March 1944, '*Strahltriebwerke des Auslands* (Foreign turbojet engines)', used to interpret corresponding '*Signals across the Channel*', Figs. 8.23 and 8.24.

[26] See Duffy, Target: America.

[27] No official source, but reliable and probable information, courtesy to Lutz Budrass, 13 Aug. 2021.

Fig. 8.3 Strategic cleverness against material superiority: Colonel-Engineer Dietrich Schwencke (l), his signature (m), *Cryolite* mine Ivigtut, Greenland, summer 1940 (r)

numbers—the German annual output of 12,000 units was confronted to 43,500 from the Western allies.[28] This situation triggered on the German side a frantic search to increase the own production capacity, but also to radically innovative approaches against this threatening, war-decisive unbalance. Countermeasures ranged from Schwencke's unconventional ideas up to serious considerations of the new revolutionary turbojet propulsion.

In dealing with the United States, Schwencke noted that American aluminium works and aero-engine plants, propeller factories and armament plants could be attacked only by the four-engine Messerschmitt Me 264, fitted with the newly developed, 3800 hp Daimler-Benz DB 613 engines.[29] Typical for Schwencke's strategic thinking is the focus on *Cryolite*, necessary for the energy-efficient aluminium production. Target #15 on Schwencke's US bombing target list of 21 in total, was the Cryolite Refinery at Pittsburgh, PA, but the real bottleneck of the US aluminium production appeared to be the single source location of raw cryolite in larger quantities at the south-west end of Greenland, the tiny mining community of *Ivigtut*, Fig. 8.3 (r).[30]

---

[28] Split, 60% from the USA and 40% from UK. In 1944, the German output increased to 34,100 aircraft, but was outperformed by 96,600 units from Allied production, with 77% from the USA alone. Source: Statista Res. Dept. 1998.

[29] Since March 1941 there was USAAF presence on Greenland to establish a shorter, though weather-sensitive aircraft route to England, backed by an agreement with Denmark against attack by a *'non-American nation'*. Schwencke's project failed to come to fruition later; the Allied bombing was so intense near the end of the war, it disrupted the German supply chain, which anyway was running low on supplies, particularly fuel and kept what little was left for defence, and—see Wikipedia, 'Daimler-Benz DB 603' in English; the DB 613 was a twin DB 603 package.

[30] The cryolite mine at Ivigtut is depleted since 1987; now synthetic sodium aluminide fluoride is produced from the common mineral fluorite, a technique which the German Reich possessed exclusively in the 1940s.

Cryolite[31] is an uncommon mineral, which was historically used as an ore of aluminium and later in the electrolytic processing of the aluminium-rich oxide ore *Bauxite*. The difficulty of separating aluminium from oxygen in the oxide ores was overcome by the use of cryolite as a flux to dissolve the oxide mineral(s). For smelting pure aluminium the industrially dominant *Hall-Héroult process*[32] involves dissolving aluminium oxide (alumina)—obtained most often from bauxite, aluminium's chief ore, through the *Bayer process*[33]—in molten cryolite, and electrolysing the molten salt bath, typically in a purpose-built cell.

Schwencke was also highly influential in the context of the German Me 262 series production decision on 25 May 1943, which will be discussed as part of the following continuation of the Jumo 004 production preparation.

Without a radical design change of the axial compressors, away from the AVA preference for R 1 blading, Schelp's planned thrust enhancement for Class II+ turbojet engines would have been limited to a maximum pressure ratio PR < 5. Consequently, the decision for a development start of alternative, replacement compressors with R 0.5 (and correspondingly smaller axial thrust) from Brown Boveri Mannheim in April 1941 was just in time.[34] In this context the development history of the first Class II turbojet, Heinkel HeS 011 [35] illustrates Schelp's dilemma quite clearly. While the first HeS 011 drawing of July 1942 contained still the three rows ball bearing arrangement, Fig. 7.5 (r), later engine versions beginning with the 109-011 V6 test version, developed at Heinkel-Hirth in Stuttgart-Zuffenhausen in early 1944, and the corresponding first production engine HeS 109-011 A-0 had either a heavier, two rows ball and cylindrical roller bearing combination

---

[31] Chemically, *Cryolite* stands for—Na_3AlF_6—sodium hexafluoraluminate. Prior to the Hall-Héroult process the cost to produce aluminium was very high, higher than for gold or platinum. Production costs came down over time, but when Al was selected for the lightning rod atop the Washington Monument in Washington D.C., it was still more expensive than silver.

[32] See Wikipedia, 'Hall-Héroult process' in English. The process was invented independently in 1886 by the American chemist Charles M. Hall and the Frenchman Paul Héroult—both 22 years old. In 1888, Hall opened the first large-scale aluminium production plant at Pittsburgh, the roots of Alcoa Corp.

[33] See Wikipedia, 'Bayer process'and 'Carl Josef Bayer', both in English. Bayer, a born Austrian, invented the process of extracting alumina from bauxite in 1888, essential to the economical Al production up to this day. Bayer's solution, along with the Hall-Héroult process, caused the price of Al to drop about 80% in 1890, from what it had been in 1854. This, in turn, made it possible for pioneers like Hugo Junkers to utilise Al—and AlMg—alloys to make metal airplanes in large quantities.

[34] As outlined in Sect. 8.3.1, the aero engine design department TLUK/Ve at BBC Mannheim, led by Hermann Reuter, was established in April 1941.

[35] See Sect. 8.1.3 'The Heinkel HeS 011 Development Programme'.

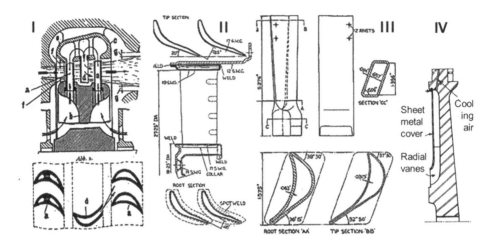

Fig. 8.4 Junkers Jumo 004 B advanced turbine design: I BBC cooling patent 1920, II Jumo 004B turbine nozzle assembly, III Jumo 004B-4 hollow turbine blade, and IV *TOBI*

for the -V6, or what appears to have been a lately introduced, robust single-row ball bearing for the A-0.[36]

Different to the rather mediocre turbine aerodynamics, which have been discussed for the Jumo 004, the German designs were far ahead of other countries in the field of *turbine cooling*. Because of the poor materials available, cooled turbine vanes and blades were essential if reasonable peak temperatures were to be used, and all turbojet engines in production or development at the end of the war had hollow air-cooled aerofoils. In addition, these had the advantages of lightness, cheapness and ease of manufacture, and correspondingly, also the turbine discs could be made lighter, Fig. 8.4.

The selected low degree of reaction for the turbine blading, $R \sim 0,2$, enabled cooling air to be drawn up over the blade roots. Although solid turbine blades were first used for the Jumo turbojet versions 004 B-1 and B-2 for quick results, air-cooled, hollow vanes and blades were put under development from the beginning. Especially, since it was realised that for a given limiting stress in the turbine wheel, a higher rotational speed and thrust was possible with the lighter, hollow blades at higher temperatures and thus lower fuel consumption. In addition, all versions of the engine used air-cooled, hollow turbine inlet nozzles.

The main need to develop hollow, air-cooled turbine blading arose, of course, because of the German shortage of heat-resisting materials. Compared with solid turbine aerofoils, hollow nozzles and blades needed less material and turned out to be easier and cheaper to

---

[36] See Kay, German Jet Engine, pp. 43–47. A single-row, cylindrical roller bearing was foreseen for the Class III Junkers Jumo 109-012 A with 2780 kp take-off thrust, and a single-row ball bearing for the Class IV BMW 109-018 with, 3500 kp static thrust,—see Kay, German Jet Engine, p. 90 and p. 132.

manufacture, while their air-cooling permitted either a higher working temperature with increased efficiency and thrust, or a constant temperature increased engine life correspondingly, which in the average was fixed to 25–35 h. According to a post-war assessment, the Jumo 004 cooling air amounted to about 4% of the engine air throughput, permitting this engine to operate at a turbine inlet temperature of approximately 760 °C.[37]

The early beginnings of turbine air-cooling date back to turbocharger developments of Christian Lorenzen at Berlin-Neukölln at the end of WW I,[38] which in 1926 were continued at DVL Berlin-Adlershof. But while Lorenzen's turbocharger, and the related gas turbine concept did not succeed up to 1938, mostly because of internal leakages and corresponding flow losses at the coupling of the radial compressor with the radially outward, hollow turbine wheel, one feature of Lorenzen's work was beneficial, as Constant[39] states correctly: *'his work on air-cooled blades proved invaluable to the Germans* (turbojet engine developments) *during the Second World War.'* However, supposedly on the surface only, since the decisive patents[40] as illustrated by Fig. 8.4 I, and also applicable for Lorenzen's patented 'Gas turbine', came from Brown Boveri since 1920. The stated influence of these early gas turbine cooling activities on the German first turbojet configurations is clearly deducible. As an example, Fig. 8.4 II, illustrates the turbine vane and (III) the turbine blade of the Junkers Jumo 004 B turbojet engine. Clearly, there is a principal agreement between the patented air-cooled vane configuration—I d)—of 1920 and the realised engine turbine vane of 1942, as well as the simple, straight air-cooled blade of the B-4 engine version from the end of 1944, which has a high similarity to the corresponding BBC patent—and the later Lorenzen turbocharger/gas turbine. Finally, Fig. 8.4 IV, highlights a Jumo 004 B-4 turbine feature which has been described for the first time by a C.I.O.S. [41] team which found one Jumo 004 engine in a Junkers piston-engine repair shop at Strassbourg (Matford) still in December 1944: *'Air cooling of blades, through holes drilled in the stubs on turbine disc, is aided by a built in centrifugal compressor, arranged as shown ...* '[42] This feature found its way to modern turbojet design under the designation *TOBI Tangential On Board Injector,* i.e. turbine cooling air

---

[37] See Hafer, Gas turbine progress report, p. 130.

[38] See Eckardt, Gas Turbine Powerhouse, p. 56 f.

[39] See Constant, Turbojet Revolution, p. 148. Similarly,—see Wilson, Turbomachinery, p. 39, stated: 'Christian Lorensen (sic) in Berlin began experiments on axial-flow turbines and hollow air-cooled blades, which were made by Brown Boveri, in 1929. His work led directly to German turbojet cooled blade designs.'

[40] The German patent DE346,599 on *'Gas turbine cooling method and device'* had a priority of 21 July 1920, and was deduced from BBC's original Swiss patent CH92,250—with equal priority. Eighteen months later Lorenzen's *'Gas turbine'* was patented as CH101,035 with priority 12 January 1922. A 1947 statement of Ad. Meyer insinuates, that Ch. Lorenzen was accordingly a BBC licensee in this context,—see Eckardt, Gas Turbine Powerhouse, p. 400.

[41] C.I.O.S. Combined Intelligence Objectives Sub-Committee.

[42] See C.I.O.S., Description of Junkers .004, p. 29.

is delivered by a tangential on-board injector that turns the air so that the loss coming on board is minimal. Normally, the air is turned and pressurised in such a way that the incoming velocity is close to that of the turbine blade itself.

The various manufacturing processes for the hollow turbine aerofoils of the turbojet engines Jumo 004 B, BMW 003 and Heinkel HeS 011 have already been described in the literature,[43] and shall not be repeated here. Junkers first tried making hollow blades by folding and welding of *Tinidur*[44] sheet metal, which had the following background. In 1939, when development work on the Jumo 004 started, a high-temperature Krupp steel known as P-193 was available. This material, which contained Ni, Cr, and Ti, could be given good high-temperature strength by means of solution treating and precipitation hardening. Krupp developed an improved version of P-193 known as *Tinidur*. It was of the same type as *Nimonic 80*, which was used in British gas turbines from 1942, but contained over 50% iron (in Nimonic 80 replaced by Ni) and this caused a rapid drop in creep strength at ~580 °C (~ 680 °C for Nimonic 80). While Krupp knew that Tinidur could be improved by increasing the Ni content from 30 to 60%, there was a recognition that Ni would not be available. The Ni content was therefore left at 30%. Similarly, work on cobalt-based alloys was also shelved due to a shortage of cobalt.

Junkers first trial for series manufacturing failed, when Tinidur proved unsuitable for welding. Therefore, in February 1943, Junkers sought on recommendation of the turbine designing AEG the assistance of the William Prym firm[45] of Stolberg near Aachen, which already in February 1942 had started manufacturing hollow turbine blades for the Junkers Jumo 207 piston-engine turbo-supercharger. After Prym had produced the first 70 hollow blades for the Jumo 109-004 on 24 April 1943, it became obvious that a satisfactory job could be done, and a decisive kick-off meeting was held at the Stolberg works on 11 May 1943: Not only were there RLM officials present, but—following Schelp's strive for information sharing—also personnel from Junkers, Dessau; AEG, Berlin; Krupp, Essen; Brown Boveri, Mannheim; DVL and BMW, Berlin; Heinkel-Hirth and Daimler-Benz, Stuttgart. For 1943, 140,000 blades were scheduled, and by mid-1944, 225,000 blades—

---

[43] See Hirschel, Aeronautical Research in Germany, with H. Schubert's comprehensive section on *'Turbine—The Hollow Metal Blade as Solution for Material Shortage'*, pp. 244–251, and—see Kay, German Jet Engine, pp. 75–78 (Jumo), p. 105 and pp. 112–114 (BMW) and p. 49 (Heinkel).

[44] *Tinidur* chemical composition: 15 Cr, 30 Ni, 2 Ti, 0.8 Si, 0.7 Mn, 0.15 C, balance Fe. Acc. to—see Schlaifer, The Development, p. 426, the decision in favour of *Tinidur* as preferred turbine material was taken by Heinrich Adenstedt as early as summer 1939 (!), who made corresponding assessments still in M.A. Mueller's group at Junkers Magdeburg, and continued with materials responsibility in A. Franz's Jumo 004 development team at Junkers Dessau.

[45] Founded in 1530 at Aachen by the goldsmith William Prym (1490–1561), the company is Germany's oldest industrial family enterprise. In 1642 during a religious civil war, the protestant Prym family had to leave their guild in catholic Aachen, and was forced to resettle with others at Stolberg, 10 km in the east of Aachen. The company became world-wide known for *'Prym's press button'*, basically invented in 1885 and in 1903 improved by Hans Prym (1875–1965), by insertion of the 'crown spring'.

with a daily target of 500. However, this was never reached, and the production for August 1944 for example, came close to 50% of that target only.

Whilst Prym were developing their Tinidur deep-drawing process, the Sächsische Metallwarenfabrik (Saxonian Metal Works) August Wellner Söhne at Aue/Saxony were working on a process for producing hollow turbine blades from *Cromadur*,[46] another heat-resisting alloy developed by Krupp as substitute for Tinidur, using manganese instead of the scarce nickel[47]—and with good welding capability. Wellner produced the hollow blades by folding and welding down the trailing edge. Though planned as a competitive approach, in the end neither Prym nor Wellner could achieve the requisite production rate, so that both Tinidur and Cromadur hollow blades were kept in production for the Junkers Jumo 004 B-4, Fig. 8.4 III.[48]

At Junkers Dessau, in the renamed flight engine division since 1939, *Otto-Mader-Works (OMW)*, Anselm Franz and his development team created an unique system of '*L reports*'[49] which documented the essence of the turbojet development progress between 11 September 1939 (L/001) and 28 March 1945 (L/361). Accordingly, the Jumo 109-004 development milestones can at best be structured by the various versions over time:

- The *Jumo 109-004 A* with eight-stage axial compressor, six separate combustion chambers and a single-stage turbine with solid Tinidur blades was intended to be rapidly developed in a pre-production engine. This prototype was first test run on 11 October 1940, without an exhaust nozzle (L/009);[50] it reached 430 kp of thrust at full speed of 9000 rpm by the end of January 1941 (L/017 and L/022).
- During the first half of 1941 Franz's design team struggled with compressor stator vibration problems which were finally overcome by changing the stator material from aluminium alloy to steel. The design thrust of 600 kp was demonstrated for 5 min during a test on 6 August 1941 (L/042), on 24 December 1941 the first 10 h endurance run was successfully accomplished, and on 23 January 1942 the 1000 kp thrust mark was reached (L/070). The Jumo 004 A was first flight-tested—after 1941 attempts had failed to fly the Me 262 V1 prototype with two Heinkel HeS 8 and thereafter, with two BMW 003 turbojets—on 15 March 1942 under a piston-engined Me 210 (L/078), and the first

---

[46] Cromadur chemical composition: 18 Mn, 12 Cr, 0.65 V, 0.5 Si, 0.2 Ni, < 0.12 C, balance Fe.

[47] The actual scarcity of Nickel supplies has been sometimes disputed, especially in view of the small amount required for turbojet engine production, but the volatility between 1942 and 1944 (Finland) was high, and in fact the advantages of blade fabrication from sheet metal might have outweighed the corresponding nickel savings,—see Kay, German Jet Engine, p. 16.

[48] See Kay, German Jet Engine, p. 75 f.

[49] Available as nearly complete collection at Deutsches Museum Munich,—see Mueller, Junkers Flugtriebwerke, p. 223.

[50] In the following this so called '*T1 Gerät*' (device) was presented in 'idling operation' immediately to Prof. Mader, mainly to change (successfully) his reserves against the new turbojet propulsion, to Helmut Schelp and some members of the Junkers board (L/010).

all-jet flight of this fighter aircraft was made on 18 July 1942 by the Me 262 V3, lasting 12 min with top speeds up to 600 km/h, and piloted by Messerschmitt's chief pilot Fritz Wendel (1915–1975) from the Leipheim airstrip[51] (L/108). First flight of Me 262 V2 followed on 1 October 1942 (L/116); this month saw also the first successful 50 h demonstration run (L/120 and L/122).

At this time an increase in Me 262 take-off weight and a higher demand in flight performance put the focus on further engine thrust augmentation. The correspondingly designed *Jumo 012* turbojet engine[52] with 2780 kp TO thrust was already in Schelp's Class II category with typically three-times higher mass flow and thrust in comparison to the Class I engines under development. As a rapidly to realise measure for thrust increase of the existing Jumo 004, additional fuel injection was tested between 24 September and 8 October 1942.[53] Since a first test with six additional fuel injection nozzles 60 mm downstream of the turbine rotor failed—with partial combustion after the exhaust nozzle, in a second trial the boost fuel was injected *upstream* of the turbine. Starting from *'full load'* at 9000 rpm the gas temperature rose from 650 °C up to 1100 °C, and correspondingly the thrust grew by—at that time significant 22%—from 900 to 1100 kp (L/115 and L/124). However, after this principle demonstration of *'afterburning'*, it was decided not to introduce this extra-complexity into the series configuration, mostly due to a lack of development time for the difficult fine-tuning problem of the two engines in parallel operation.[54]

A special—often underestimated feast within Jumo 004 design is the development of an *'indirect speed governor'* by A. Franz and H. Moellmann (L/092, dated 6 June 1942), which demanded decisive attention after mid-1942 for nearly one year, and which can find here naturally only a cursory mention. This so called *'all speed governor'*, in Junkers-internal terminology an *'Isodrom'* governor and today at best known as PI (proportional-integral) control unit, implied that the throttle did not regulate directly the fuel flow, but rather in relation with shaft speed. After another definition, it was a *'flyweight speed governor with hydraulic augmentation and soft/damped feedback'*, which indicates that the parameter modification started moderately, accelerating and strengthening in the second action phase, while the end phase of the control shift is

---

[51] Leipheim lies on the Munich-Stuttgart Autobahn, some 60 km in the west of the then Messerschmitt Works at Augsburg-Haunstetten. In a second flight on the same day the top speed rose without difficulty to 720 km/h.

[52] See Kay, German Jet Engine, p. 89.

[53] L/018, dated 30 Dec. 1940, contains presumably a first T,s-representation of a complete Joule gas turbine cycle for the Jumo 004 turbojet engine with afterburning.

[54] There were plans to implement 20+ percent thrust by afterburning to a Jumo 004 E type in July 1945, on the basis of a, 1000 kp Jumo 004 C, which in itself was a re-designed B-4 model with detail refinements,—see Kay, German Jet Engine, pp. 87–88.

again damped and softened.[55] Though far from perfect, the Junkers governors were definitely superior to the Allied wartime systems, where fuel flow was directly regulated by the pilot, often resulting in deadly flame outs or bursting in flames.[56] The only problematic turbine speeds for German engines were those below 6000 rpm/idle, where the governor was not active, and those speeds were only for engine start up and taxiing. The first governor tests under flight conditions were on 21 May 1943 (L/162). By RLM order the Junkers turbojet governor became also standard for the following BMW 003 and Heinkel HeS 011 developments.[57] The governor saw its high-time only after the war, when it spread with the Jumo 004 in the Soviet Union, and with the BMW 003 as basis for the Snecma ATAR engine family in France up to 1975. In the SU the Junkers governor replaced also the RR Nene/Lucas governor *'with rigid/inflexible feedback'* in the re-engineered versions of Klimov RD-45 and Vk-1 from MiG-17 F up to MiG-21.

- On 22 May 1943[58] followed General A. Galland's famous test flight—*'It was as angels were pushing'*—on Me 262 V4, which already on 25 May 1943 triggered the decision of Generalluftzeugmeister Erhard Milch to launch the Jumo 004 B mass production immediately. As illustrated in Fig. 8.5, the flight speed advantage of 230 km/h of that Me 262 V4 prototype with the new Jumo 004 A turbojet engine in comparison to the standard fighter Me 109G with 1500 hp Daimler Benz DB 605A piston engine was too inviting.[59] The 004 B series production in large scale was considerably facilitated by an early form of *modular design*, which allowed to produce individual engine sections in specialised works, to be finally integrated in central assembly shops.
- The layout of the *Jumo 109-004 B-0* production model was chosen in December 1941, and its detail design completed by October 1942. Chief alterations mainly in the interest of màss production were a modified compressor construction with separate rotor discs, replacement of castings with sheet metal, wherever possible, and substitution of more

---

[55] See Kay, German Jet Engine, pp. 72–74; a better explanation of the Junkers '*Isodrom*' governor provides—see Mueller, Junkers Flugtriebwerke, pp. 231–234.

[56] Slow acceleration was a problem with all early jet engines because compared with a propeller driven plane a pure jet engine produces little thrust at take-off.

[57] An example of RLM-enforced cooperation amongst the turbojet engine manufacturers,—see Schlaifer, The Development, p. 436, who highlights the significant cooperation differences in comparison to the English industry. And typical for the 'Junkers secrecy', it was only in March 1942 that W. Encke was allowed to witness—one time—a full-scale test of a compressor based on his work at Dessau.

[58] See Wikipedia 'Adolf Galland' in German; other sources have the 23 May 1943, and—see Wikipedia, 'Messerschmitt Me 262' in English,—and Green, Warplanes, p. 622, put Galland's flight already to 22 April 1943.

[59] See Budrass, Review; p. 184; circumstances and decisions of that RLM meeting on 25 May 1943 have already been described in Chap. 7.

than half the weight of *strategic material* in comparison to the 004 A (but solid turbine blades were still installed).[60]

- Compressor vibrations were still a problem, and a suggestion from Encke to use thin *'Goettingen 684'* blading throughout was turned down; it was found better to use thin, wide-chord blades for the first two stages, and comparatively thicker, narrow-chord blades for the last six stages. The first *Jumo 109-004 B-1* production engines were delivered in early June 1943, powering the Me 262 V6 in its maiden flight in October 1943. For the first time this aircraft had a fully retractable nosewheel undercarriage.

- In mid-1943 the design of the *Jumo 109-004 B-2* represented another trial by Junkers, assisted by AVA, to cure the ongoing compressor vibration problems, while at the same time seeking an improved altitude performance. Despite the effort, in the end the compressor suffered its own form of blade vibrations, and the attempt was given up. The existence of the *Jumo 109-004 B-3* appears to be uncertain;[61] it could have been a forerunner version in the process of introducing hollow turbine blades.

- The fundamental feature of air-cooled, hollow turbine blades was finally introduced to series production, when by the end of 1944, the *Jumo 109-004 B-4* version replaced the foregoing B-1 standard. Besides a small change of the cooling air tapping after the last compressor stage, this engine was capable of a higher working temperature and thrust, but the major benefit was certainly their considerably *faster and cheaper production rate* in comparison to the earlier engine version with solid turbine blades.

Confusingly, it was exactly this aspect of *'faster and cheaper production'* which—investigated by technical-historian writer Hermione Giffard[62]—has put the whole technological turbojet achievements on the German side in a strange, doubtful light; wrongly so, as can hopefully be convincingly argued and debunked by the following comments.

Besides her well-researched 2016 book *'Making Jet Engines'*, H. Giffard produced two supplementary articles which contain apparently already in the title her firm convictions in this context: German turbojets are *'Engines of desperation'*, while *'Britain's aero-engine manufacturers* (produced) *a series of innovative, world-beating axial jet engines by the end of the Second World War.'*[63] No recognition of the Germans' courageous, early and fundamental move from piston to jet engines, no indication of leaving the dead-end of centrifugal turbojet engine developments wholeheartedly to an all-axial design approach

---

[60] L/055, dated 23 Oct. 1941, mentions a saving of 60 kg Nickel per engine, and an overall dry weight reduction of 100 kg.

[61] See Kay, German Jet Engine, p. 75.

[62] See Giffard, Making Jet Engines (2016),—Engines of Desperation (2013) and,—Engine of innovation (2019). Already the 2013 publication was extensively quoted in the German news magazine 'Der Spiegel', 7/2014, under the title 'Wundertüte am Himmel' (instead to a wonder weapon, 'Wundertüte' refers to a grab bag), and the believing journalist saw Giffard *'smashing the legend of the Me 262'*, in her view as *'shoddy, unproven and dangerously faulty'*.

[63] See Giffard, Engine of innovation, abstract.

already in 1942, no praise of the first time integral design of an aerodynamically advanced aircraft fuselage with (later) swept wings and the revolutionary axial turbojet in a low-drag nacelle, no mentioning of the innovative, lasting effect of introducing turbine air-cooling in aero gas turbine engines after 1943, admittedly caused by a lack of material resources, but does this hamper the degree of innovation? On the other hand and at the same time (1943), nothing on the British 'despair' to find an alternative path to Whittle's Power Jets approach, when Griffith's hilarious, contra-flow ducted-fan turbojet[64] received still considerable attention, before *'the scheme was abandoned in 1944 owing to insuperable problems, but it formed the foundation for later very successful engines.'*[65] No revelations about aerodynamic backwardness due to ignorance on the importance of wind tunnel work and the correspondingly slow adaptations of the *Gloster Meteor's* high speed buffering of the short *RR Welland* nacelles.

Giffard has made up her mind: *'Far from wonder weapons, Germany's jet engines, first deployed in July 1944, were engines of desperation, unreliable but well fitted to the conditions of production in the National Socialist war economy. Despite the chaos of late war weapons production in Germany, Albert Speer's Armament's Ministry oversaw the manufacture of large numbers of jet engines, exploiting in particular the Mittelwerk weapons factory. The jet did not allow Germany to regain air superiority, but jet engines optimized for production enabled the regime to produce aero-engines as efficiently as possible with its remaining resources. Great Britain also deployed jet aircraft in mid-1944, but its government de-emphasized short-term engine production in favour of a broad development programme. Comparison of the different paths followed by the two nations to jet engines highlights how the design and production of jet engines in the Third Reich reflected a compromise typical of the regime's last years: Manufacturing masses of inferior weapons, whose virtue lay in the fact that they were easy to build.'*[66]

The argument above of *'manufacturing masses of inferior weapons'* might have been countered by the undeniable *'delta in speed'* of the shown 230 km/h in comparison to all other piston-engine powered aircraft, Fig. 8.5. Other cases of Giffard misinterpreting German data make speechless, nevertheless, are apparently not causing her to reconsider cautiously some of her rash judgements. At the end of the foregoing Sect. 7.3, a 12-year turbojet development programme had been outlined, which Schelp had presented at a DAL academy meeting in July 1943.[67] Giffard makes this a 16-year programme and deduces obviously from the presented time scope with the disparaging remark that *'a relatively low-level office for special propulsion systems (Sondertriebwerke)'* in the RLM, run since 1941 (actually since the end of 1939 after Mauch's leave) by Helmut Schelp, *'who viewed jet engine development as a long-term activity that would come to fruition only after the*

---

[64] See next Sect. 8.2 'British Turbojet Engine Developments'.

[65] See Kay, Turbojet, Vol. 1, p. 19.

[66] See Giffard, Engines of Desperation, abstract.

[67] See Schelp, Hochleistungstriebwerke.

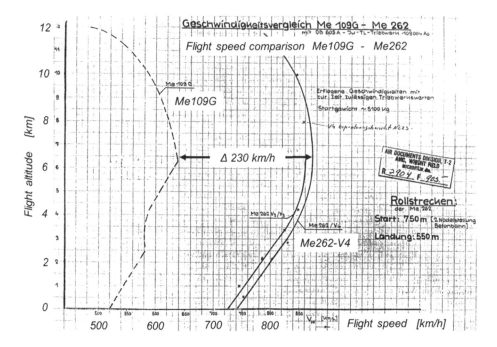

Fig. 8.5 Messerschmitt flight test results, ~ May 1943: Me 109G with DB 605 A vs. Me 262 V4 with Jumo 004 A-0

war. He fitted his office's contracts into a 16-year national plan for German jet engine development, in which Germany's jet engines gradually achieved higher powers and longer ranges.'

Reading Schelp's ambitious, multi-task loaded CV in Chap. 7 might already be sufficient to put Giffard's attribute of a *'leisurely development'* (42) [68] fashion ad absurdum. Schelp's own words corrected this fundamentally wrong impression when he described the first turbojet development phase 1939–1942 in his DAL speech above: *'The first period had to create the necessary basics and to develop the TL engine up to series readiness, without going to the extremes … This first period is finished, and brought revolutionary results not only for the whole propulsion area, and not alone for the Luftwaffe.'* [69]

On at least two occasions Giffard's argumentation comes close to arousing the suspicion of making up evidence to proof her claims, both in the context of manufacturing hollow, air-cooled turbine aerofoils from folded sheet metal. She describes (58) the design

---

[68] Page numbers in parenthesis of quotations from—see Giffard, Making Jet Engines.

[69] Schelp in German on 2 July 1943: 'Der erste Abschnitt galt der Schaffung der nötigen Grundlagen und der Entwicklung von TL-Triebwerken bis zur Serienreife, ohne von vornherein Extremwerte anzustreben. … Der erste Abschnitt ist abgeschlossen und hat Ergebnisse gebracht, die revolutionierend auf das gesamte Triebwerksgebiet, nicht allein für die Luftwaffe, sein werden.'

transition from Jumo 004 A to B correctly, that '... *Complex forged parts were replaced by ones folded out of sheet metal that were easier and faster to make, although these parts increased the frequency of fatal, catastrophic engine failures'*, but provides immediately hereafter a supposedly confirming reference of these negative statements in Constant's book,[70] pages 208–211, who however writes rather contradictorily: *'The 004 was a remarkable technical achievement in the minimal use of strategic materials. The first experimental 004 engines had solid turbine buckets, but later engines employed hollow, internally air-cooled buckets that permitted the use of manganese alloys that could be fabricated simply by folding and welding sheet stock.'* Nowhere, any mentioning of resulting *'fatal, catastrophic ... failures'*. And three pages later, Giffard states (61): *'The operational cost of these engines of desperation was high. Routine flights were often fatal. The fraction of German jet engines that made it off the ground and into service had extremely short service lives, and many exploded in use'*, again concluding this sentence with a supposedly confirming reference, here to Pavelec's book,[71] page 91, which however describes the facts neutrally, *'... a full third of Me 262s were lost to ground accidents due to their unstable landing gear, and another third were lost in transport or from engine failure, there were a full third delivered to combat squadrons for aerial operations'*, however, nowhere, any reference to Giffard's specific *'many (engines) exploded in use'*.

To put the R&D capacities on both sides in perspective, Bailey produces another indirect assessment to judge the turbojet research and development intensity for the year 1943 between Germany and Great Britain.[72] In Sect. 6.1.4, there is a June 1943 meeting described, in which Frank Whittle learnt by visiting the *PRU* Photographic Reconnaissance Unit at Medmenham, Buckinghamshire first hints (pictured parallel scorch marks) to a twin-engined turbojet at the Luftwaffe test ground at Peenemuende-West. This materialised later-on to the first, indirect appearance of the Messerschmitt Me 262. Apparently, Whittle filed thereafter an internal *'concern'* on German turbojet activities, which found its way up to the desk of the then Minister for Aircraft Production, Sir Stafford Cripps, where it was put *'on ice'*, intentionally under the motto: *'We must avoid exciting the uninitiated about this sort of thing.'* On 6 October 1943, Prime Minister Winston Churchill—informed by his son-in-law Duncan Sandys, and concerned by the threatening loss in air superiority—intervened personally with an unequivocal command: *'Recent evidence shows that the Germans are working hard on jet propelled aircraft and accentuates the need for the utmost pressure to be put on their development here.'* Under this *'utmost pressure'* the reconciling bureaucracy proposed for a corresponding development task force *'a maximum of 100 men to be detached for 6 months'*, at a time, when the Messerschmitt turbojet development team was in the order of 1400 already.

---

[70] See Constant, Turbojet Revolution.

[71] See Pavelec, The Jet Race, p. 91.

[72] See Bailey, The Early Development, p. 57.

In building up her black and white argumentation, on the one hand the Germans hastily *'manufacturing masses of inferior weapons'* up to war's end, on the other a British government which *'de-emphasized short-term* (and implied 'unreliable') *engine production'* during the war, Giffard must have somehow overlooked the further consequences of this cautious British government's turbojet production policy, exemplified by the *Gloster Meteor*, proudly classified as *'The World's First Operational Jet Aircraft'.*[73]

Of the nearly 4000 *Meteors* built, a total of 890 were lost in RAF service (145 of these crashes occurring in 1953 alone), resulting in the deaths of 450 pilots. Contributory factors in the number of crashes were the poor brakes, failure of the landing gear,[74] the high fuel consumption and consequent short flight endurance (less than one hour) causing pilots to run out of fuel, and difficult handling with one engine out due to the widely set engines. The casualty rate was exacerbated by the lack of ejection seats in early series *Meteors*; the much higher speed that the aircraft was capable of meant that to bail out pilots might have to overcome high *g forces* and fast-moving airflow past the cockpit; there was also a greater likelihood of the pilot striking the horizontal tailplane.[75] So in summary, it appears to be difficult to confirm Giffard's argument of severe unreliability under the given circumstances for the German Me 262 in comparison to the British Meteor.

Without repeating here all well-founded points of Lutz Budrass's *'Review'* of Giffard's book,[76] one point of general importance is worth a more detailed discussion. First Budrass contradicts with some emphasis the qualification of the German turbojet engines as—*'a low quality product'*, in Giffard's terminology (61) *'made worse by shoddy construction work'.*[77] Thereafter, Budrass stresses the point that the re-design effort from Jumo 004 A to the B version, amongst other to save *'strategic materials'* (see above), was originally under the decisive precondition, that *'Goods made of "Heimstoffe* (home materials)*"* had to be as good, if not better than those made of "imported raw materials".*[78] ... That enterprise would probably have failed if the "Heimstoff" debate within the reach of the German Air Ministry had not taken another turn at the height of the war. An investigation of the service time of German aircraft in 1942 confirmed that their average lifespan was considerably shorter than those 100 hours then still demanded for aero-engine time*

---

[73] See Kay, Turbojet, Volume 1, p. 61,—and Wikipedia, 'Gloster Meteor' in English.

[74] Similarly, the most frequent non-enemy affected cause of Me 262 crashes.

[75] See Wikipedia, 'Gloster Meteor' in English.

[76] See Budrass, Review, p. 181.

[77] A qualification, additionally accentuated in—see Giffard, Response on (Budrass's) review, p. 190: *'National Socialist Germany designed ... the jet engine, to be produced under concentration camp conditions, which did go against everything industry believed to be good engineering practice. ... That these engines did not live up to the engineering standards of production by a peace-time aero-engine firm is unsurprising.'*

[78] Budrass refers to—see Maier, Forschung, p. 366 and—see Flachowsky, Von der Notgemeinschaft.

between overhauls.[79] *. . .That realization gave rise to an overall review of the tight quality standards in German aircraft production which had until then remained on a level comparable to civil aviation. The results of that movement which spread over the whole of the aircraft industry placed it on a completely new footing. Apart from the fact that lowering quality standards in production made the employment of a higher proportion of unskilled and semi-skilled slave workers possible, it also facilitated the use of raw materials which had been considered unfeasible before, namely all kinds of secondary aluminium from aircraft scrap. All aircraft and all aero-engines were subject to that redefinition of quality standards which did not affect performance but adapted the lifespan of German aircraft to the realities of war.*[80] *That, however, was most important for turbojet development. Despite numerous efforts German metallurgists did not find 'Heimstoff' alloys for the turbine blades and the turbine rotor which were as heat resistant as the British Nimonic alloy which contained some 50 per cent of nickel. For that purpose, Tinidur sheet, containing 30 per cent of nickel, and later Cromadur sheet, containing no nickel*[81] *at all, lasted for just 25 hours. But that deficiency was controllable by applying tight overhaul intervals. The Jumo 004 Bs of the Me 262 had to be replaced after 25 hours of service—and due to the efficient series production Luftwaffe units had more than enough spare Jumo 004 B to follow that order.'*[82]

Interestingly, the *quality issue of German workmanship* has also been addressed by Hitler in one of his last documented speeches from his resort at Obersalzberg, Berchtesgaden, in early July 1944 to the collected leaders of German war industry, as quoted in parts by Brendan Simms[83]: *'It addressed the problem of production, and was given in the shadow not only of the Allied bombing but also of the "Morgenthau Plan". Speer seems to have seen, or got wind of, an early version of the plan of US Treasury Secretary Henry Morgenthau to deindustrialize Germany after the defeat.* "Should the war be lost," *Speer wrote in his notes for Hitler, the result would be* "merciless extirpation of German industry, to eliminate competition in world markets." *The Führer's speech also . . . warned the industrialists that the only choice they would have then* (after defeat) *would be whether to commit suicide or to allow themselves to be* "strung up" *or be sent to work in Siberia . . .' . . . 'Hitler spelled out what this meant for German industry.* "This is not just a war of soldiers," *he explained,* "but especially of engineers". *By this Hitler meant not only inventors, but also the machine tool engineers needed for mass production. He expressed some admiration for the quality and quantity of Soviet output, particularly with regard to the legendary T-34 tank, but his principal concern was the Anglo-Americans. Their*

---

[79] In 1942 a Messerschmitt Bf 109 on the Eastern front was in service on average for just 65 h before it was lost or damaged beyond repair, a Focke-Wulf Fw 190 on the Western front for just 25 h.

[80] See Budrass, Flugzeugindustrie, pp. 818–829, and—see Budrass, Ideology and Business.

[81] The chemical composition of Cromadur has 0.2% of Nickel, see Footnote 46.

[82] See Mueller, Junkers, pp. 235–236 and 244–245.

[83] See Simms, Hitler, p. 514 f., who requotes from—see Von Kotze, Es spricht der Führer.

Fig. 8.6 Junkers turbojet engine series production, ~ November 1944: Jumo 004 prod. line at *'Zittwerke'*, Zittau (l), forced labour in repaired Messerschmitt aircraft plant (r) © US Nat'l. Archives (l), D. Uziel (l, r)

technological advances in the field of detection had basically neutralized the U-boat arm. They had also achieved a "mass production of aircraft", *which threatened to* "crush" *the Reich. He now called upon German industry to support the mass production of new weapons,'* ... *'For this reason, Hitler wanted a move away from* "quality work"—*what he also called* "German workmanship"—*towards* "mass production" *for war purposes.'* And, "'The engineers of the Americans", *he claimed,* "are for the most part of German origin," *especially of* "Swabian-Alemannic blood". *Astonishingly, Hitler sought to counter this threat by insisting that he had German engineers as well.* "Superiority of the enemy," *Hitler scoffed,* "there you can see how little they are superior.'"

Giffard's foregoing accentuation of the *Mittelwerke* underground production site (with horrible working conditions) might have resulted from her intention to prove a *'concentration camp design standard'*, however, the actual Jumo 004 series production start commenced simultaneously before August 1944 in the Junkers plants, Dessau with 96 units, Koethen (25) and Muldenstein (189), while Mittelwerke (Nordhausen) followed in October 1944 with 10 units only (and rising then to a maximum of 815 units in March 1945).[84] In total the Jumo 004 B production was spread over seven sites, of which the location *'Zittwerke'* at Zittau was opened as replacement for the bombed sites at Junkers, Dessau and Magdeburg on order of Reich Minister of Aviation H. Goering on 20 September 1944, Fig. 8.6.

The Junkers workforce at Zittau has been estimated to 2500; for security reasons and to prevent from sabotage and espionage there was only a limited work scope for slave workers

---

[84] See Kay, German Jet Engine, p. 78.

and concentration camp inmates.[85] *Zittwerke* delivered the first 12 Jumo 004 B engines in November 1944 (unmistakably heard by the parallel operation of eight engine test rigs), increasing up to 120 units in March 1945. Fig. 8.6 (l) shows the Jumo 004 B series production at *Zittwerke* in November 1944: The rails, on which the engine trolleys travelled along the production line, are visible on the floor to the right; correspondingly, Fig. 8.6, (r) shows slave workers in an unknown Junkers plant.[86]

Beginning in January 1945 the completely assembled and tested turbojet engines from Zittau were shipped by rail to the Messerschmitt *'forest plant'* at Obertraubling near Regensburg, for integration in the Me 262 fuselages.[87]

A salient role in the complete Jumo 004 production network played the precision task of compressor blade (and vane) mass manufacturing—in view of in the end 6010 produced turbojet engines which meant on the basis of 282 blades (and 533 vanes) per engine an impressive total of 1.7 million milled blades (and 3.2 million vanes from sheet metal). At the forefront was the Junkers Magdeburg plant, where R. Walter Moebius (1901–1976) led the special machinery development[88] and realised there in 1944 also the first fully-automatised production *takt streets*.[89] The basic copy milling technique for blade manufacturing has been described in its early form by R. Friedrich in context of the RT0 project,[90]—and belonged certainly since that time to the Magdeburg manufacturing capacity. Another, and presumably even earlier nucleus for the copy milling process could have been AVA's Goettingen workshops, led by the talented chief manufacturing engineer Karl

---

[85] In summary, eight Junkers production sites employed 864 concentration camp inmates in 1944,—see Wikipedia, 'Zittwerke' in German, of which also the following related information has been taken.

[86] Both pictures from—see Uziel, Arming the Luftwaffe, who has also on p. 232 a direct quotation in this context: *'Another woman survivor described the conditions in Junkers aero-engine factory in Zittau as "paradise", when compared to Auschwitz.'*

[87] See Wikipedia, 'Obertraubling'in English. Author—see Uziel, Arming the Luftwaffe, p. 202, knows about the last few operational months, that at Messerschmitt Obertraubling Airfield *'aircraft were moved around primarily with horses and oxen because there was no gas for the tow tractors'*, and beginning mid-January 1943, Messerschmitt's Regensburg plant organised privately *'factory protection flights'*, with two fighters at constant readiness and four more aircraft in reserve during daytime, normally tasked to operate within a 30 km radius around the factory.

[88] See Vierhaus, DBZ Vol. 7, p. 134. Moebius's organisation was, in German: *Junkers KoBü Sondermaschinenbau, MZM Motorenbau-Zweigwerk Magdeburg*.

[89] Documented is a 25 m long, fully automatised *takt street* for the manufacturing of Jumo 213 cylinder heads—with automatic quality control, designed by Friedrich Wahl (1907–1973). In addition, there were several such units for the mass production of Jumo 004 parts (without further details),—see Mueller, Junkers Flugtriebwerke, p. 240. Wahl, after he had been responsible for the Jumo 004 production quality at Nordwerke AG, the 40,000 m^2 Junkers section in the north of the underground 'Mittelwerke' site at Kohnstein in the Harz Mountains (mostly used for V-2 production), escaped after the war from being deported with the rest of the Jumo team to Kuibyshev near Samara, SU by staying in the Harz region as forest worker between 1946 and 1948.

[90] See Friedrich, Vom Wagner/Mueller RT0 Versuchsgeraet. For the RT0 project,—see Fig. 6.17.

Grothey. He had also the responsibility for designing and developing the requisite special milling machines for the blading of the numerous AVA axial compressors since the early 1930s, which were later transferred during the war to the AVA's dispersal at Reyershausen, approximately 12 km in the north-east of the AVA site in the surface building of a disused salt mine. Starting from a master blade, up to 20 copies could be milled at a time either on—later removed—rotor discs or inside of material rings. One can estimate that Junkers must have controlled and operated some 20 of these blade copy milling centres, and potentially Grothey's Reyershausen installation was either used for process development and/or a kind of training centre.[91]

Milch's decision for Me 262 series production on 25 May 1943[92] was extraordinarily short-termed;[93] still in January 1943 there was no general staff demand for the turbojet fighter. In spring 1943, Messerschmitt prepared the series production for the new Me 209 [94] as planned mid-term successor (January 1945) for the established Me 109G. It was only on 14 May 1943 that Milch received first disappointing test reports for the piston-powered Me 209, which was 50 km/h slower than the Focke-Wulf Fw 190D, and offered actually no improvements with respect to climb speed and general handling characteristics, compared to the current Me 109G.[95] Suddenly the next fighter generation appeared to be undefined,

---

[91] On 27 June 1944, 1.30 am, the small mining village of approx. 800 inhabitants was attacked by a British air-raid, and an oversize bomb exploded in village centre, approx. 500 m from the AVA compound, killing 19 inhabitants. In—see Eckardt, Gas Turbine Powerhouse, pp. 132–134, where especially the V-3 super gun threat to London had been emphasised, like in the community was speculated since then about possible reasons; the relocation of the described AVA copy milling equipment to Reyershausen puts now a new target into focus, in addition to AVA high-speed wind tunnel investigations by Prof. O. Walchner. However,—see Middlebrook, The Bomber Command, still claims an attack of 35 'Mosquitos' on Goettingen railway workshops in that night, though there were no bombs on Goettingen then, except the disastrous one, 12 km off on Reyershausen. As of mid-Nov. 2022 (manuscript deadline) this 'riddle' has been fully clarified, waiting for a future account.

[92] Milch's corresponding diary entry reads: 'Massage, GL conf. with department heads. Lunch with Speer. Afternoon telephone Göring: "Drop the Me 209, put Me 262 in its place. I propose an anti-invasion [air] corps"'—see Irving, The Rise and Fall, FN 19, p. 438 f., who comments: 'The latter corps was to consist of airborne troops, special fighter and fast bomber units fitted with special weapons for combating an Allied invasion attempt wherever it might be made.' Applied for the invasion theatre with expectable Allied air superiority, Hitler's ill-fated 'Schnellbomber (fast bomber)' mission demand for the Me 262 stands in a new context. However, on D-day, 6 June 1944, no Me 262 were available operational, and it was only at the end of that month that 26 Me 262 s with engines rolled off the production lines,—see Pavelec, The Jet Race, p. 94.

[93] Two months later, Milch himself called his decision 'very brave and courageous (sehr tapfer und mutig)'; to put this in perspective, one can quote W. Messerschmitt on 2 June 1943: 'For a long time, perhaps never, a turbojet aircraft (262) will replace the normal (piston-powered) fighter aircraft';—see Schabel, Die Illusion, p. 36.

[94] See Wikipedia, 'Messerschmitt Me 209 (1943)' in English.

[95] On 1 May 2018, Johannes Schubert, former head of the MBB Space Division, provided a comment, that in mid-war also for German standard fighter aircraft, the performance became increasingly uncompetitive due to the basic 85–87 octane fuel—and the shortage of anti-knock-TEL tetraethyl

the radical jump to the Me 262 became a possibility, to which finally, Galland, Schelp and Eisenlohr contributed considerably with their positive comments; not so much noted, however, is the British share in this ... Irving[96] has an episode only three days after the historic decision day, when Colonel-Engineer Dietrich Schwencke had disturbing news for Milch—a reliable English prisoner-of-war had observed just recently at Farnborough a British turbojet aircraft in *'very fast flight'*.

With some likelihood the observed *'very fast flight'* of a turbojet-powered aircraft at RAE Farnborough was during three weeks after mid-April 1943, so that the Gloster Meteor which had its maiden flight on 5 March 1943 at RAF Cranwell only, is out of this competition. The focus goes to the two built prototypes of the Gloster E.28/39,[97] of which the first took off on 15 May 1941. Gloster's chief test pilot, Flight Lt. Gerry Sayer flew the aircraft, jet-powered by Whittle's W1 for the first time from RAF Cranwell, 225 km north of Farnborough, in a flight lasting 17 min. In this first series of test flights, a maximum true speed of 563 km/h was attained, in level flight at 25,000 ft and 17,000 turbine rpm. It was the fourth jet to fly, after the Heinkel He 178 (27 August 1939), one year later the Italian Caproni Campini N.1 motorjet (27 August 1940), and the Heinkel He 280 (30 March 1941). Having proved that it was indeed possible to build a jet powered aircraft, the two E.28 s were then used as test beds for further development. A dedicated airfield was created at Edge Hill, 110 km north-west of Farnborough, where prototype #1 (W4041) was used to test new versions of the W.1 engine, as well as testing the W.2/500/3 engine in 1942. It was also used to familiarise new pilots with jet aircraft and made its last flight on 20 February 1945. It is now on display in the Science Museum in South Kensington.

The second E.28, (W4046) had a much shorter life. It did not make its first flight until 1 March 1943, nearly two years after the W4041, powered by a Rover W.2B/110 engine. On 17 April 1943 the W4046 made the first cross country jet flight in England, from Edge Hill to Hatfield, 36 km north of the *cabinet war rooms* in London-Westminster, where on Mon 19 April it was demonstrated to Winston Churchill, before returning to Edge Hill on 20 April. From 3 May until 30 July 1943 W4046, now operated by RAE Farnborough and newly equipped with a 680 kp thrust W.2B, achieved a maximum speed of 750 km/h. A speed, when observed by the unknown British PoW in early May 1943, would have

---

lead, which was then produced still under US licence in two *I.G. Farben* plants in Brandenburg. He considered the strategic importance of TEL for German warfare even higher than the heavily attacked ball bearing production at Schweinfurt; the Brandenburg plant(s) with an annual TEL capacity of 3600 t remained unscathed from Allied bombing. Special thanks for intermediation to Ernst H. Hirschel. In the meantime, a 2020 publication—see Douglas, The secret horsepower race, p. 65., located the two secret TEL plants at Gapel, 60 km west of Berlin, as the first site erected still with (US/ Esso?) licence, and an IG Farben 'copy plant' at Frose, 40 km south-east of Magdeburg.

[96] See Irving, The Rise and Fall, p. 245.

[97] See Wikipedia, 'Gloster E.28/39' in English, and http://www.historyofwar.org/articles/weapons_gloster_e28_39.html

certainly created the impression of a *'very fast flight'*. The second prototype of the Gloster E.28/39 made a total of 111 flights, and thus more flights in five months of its existence than the W4041 did in 4 years. On 30 July 1943, while on a high-altitude test flight, the second E.28 prototype was destroyed in a crash, resulting from an aileron failure.

The interrogation of the British prisoner-of-war mid-May 1943 and its unquestionable, accelerating impact on the German decision on Me 262 series production—with the immediate and explicit appointment of Colonel Edgar Petersen (1904–1986),[98] former head of all test centres, now as solely responsible coordinator of the jet engine production—highlights the *'Jet Web'* type character of intelligence-based development information among the various players.

Comparably, the three weeks visit of then Major General Henry H. Arnold (1886–1950)[99] in England in April 1941, where he saw the Whittle engine for the first time, was also prepared and charged by information which he had received immediately before about German turbojet activities. Arnold was accompanied on this trip by Major Elwood R. Quesada (1904–1993),[100] who at that time was chief of the Foreign Language Section, USAAF Intelligence Division—and it is more than likely that the information covered not only the first flight of the Heinkel He 178 in August 1939, but also the highly visible, twin-engined Heinkel He 280 turbojet fighter development flights at Rostock-Marienehe in spring 1941, Fig. 8.12. Before the first flight of the He 280 with Hans von Ohain's delayed HeS 8 turbojet engines on 30 March 1941, there commenced unpowered, towed drag tests of the He 280 V-1 fuselage with aerodynamic (dummy engine) nacelles, behind a Heinkel He 111 tow-plane, first on 22 September 1940 at the test centre Rechlin-Roggentin, but already on 2 October 1940 continued at Marienehe. A He 280/HeS 8 development status report, dated 5 February 1941, lists not less than 16 towed flights of the He 280 V-1 and 32 gliding flights with tip speeds up to 700 km/h.[101] The aircraft was definitively *'ready for the engines'* then, but certainly the attention of US intelligence had also been sufficiently alerted.

One outcome of Arnold's visit in England was the setting up of a programme for training British pilots in the US, which subsequently became known as the *'Arnold Scheme'*, which ran from June 1941 to March 1943, Fig. 8.19.[102] Among the most significant results of Arnold's first trip to England since World War I was his being

---

[98] See Wikipedia, 'Edgar Petersen' in English.

[99] See Wikipedia, 'Henry H. Arnold' in English and—see Houston, American Air Power (Arnold diaries). The trip was between 9 April to 1 May 1941, of these 23 days nine were spent for travelling Washington—Bristol, and back, most impressively Arnold's Atlantic crossing on board a long-range flying boat—see Wikipedia, 'Boeing 324 Clipper' in English, NY-Bermuda-Azores-Lisbon, eighteen passengers and a crew of 11.

[100] See Wikipedia, 'Elwood Richard Quesada' in English.

[101] See Koos, Heinkel Raketen- und Strahlflugzeuge, p. 130.

[102] See Houston, American Air Power, p. 140. At the request of the AAF, Pan American Airways opened a school to train navigators, the first class commencing instruction in Coral Gables, FL, in

given the plans and specifications of the Whittle jet engine. That Arnold did not record any information neither about this generous gesture nor about received turbojet intelligence from Germany in his diary is not surprising in view of the secret status of that subject. In any case Arnold's contact with the turbojet issue definitively accelerated the acceptance and attention to the new propulsion concept considerably. By Arnold's account in his memoirs, he immediately brought in Larry Bell of Bell Aircraft to produce an aircraft and D. Roy Shoults of General Electric to provide an engine, sharing with them the materials he had been given by the British. With utmost secrecy, in October 1941 a small dedicated GE team—called the *'hush-hush boys'*—at Lynn, Massachusetts, began the intensive development of the first US jet engine. GE initially referred to their engine as the Type I. [103] Already on 18 April 1942, 28 weeks after official work began, GE's engineers successfully ran the first Type I engine. With their vast turbocharger experience, General Electric turned their expertise to improving the Type I. A modified version, the Type I-A, incorporating partitions in the blower casing to separate the air flow into each of the individual combustion chambers at the suggestion of Whittle, who consulted at Lynn. GE I-A testing began on 18 May 1942, and developed a thrust of 570 kp, at an overall pressure ratio of 3:1. On 1 October 1942, a Bell XP-59A aircraft, powered by two GE I-A turbojet engines, made its first flight at the *Muroc Army Air Field* in California. Ultimately, General Electric found they could produce a thrust of 730 kp from a package the same size and weight as the I-A, which they called the I-16; later, when the Bell P-59[104] went into production for a first batch of 50, the aircraft was fitted with J31s, which was the USAAF designation for the I-16. Under Arnold's command, the US air arm grew from 22,000 officers and enlisted personnel with 3900 aircraft to nearly 2.5 million men and 75,000 planes.

In reference to Arnold, Whittle and Shoults—and in continuation of US turbojet engine developments as described already in Sect. 6.1.4, NACA historians detected a special 'Jet Web'-type facet:[105]

'At the April 22 [1941] *meeting, members of the* [Durand] *Special Committee also heard summaries of British reports on the development of axial-flow compressors, probably those of A. A. Griffith and Hayne Constant of the Royal Aircraft Establishment, who had been working for many years, with limited success, on axial compressors. Their goal was a gas turbine engine to drive the aircraft's propellers. What the Special Committee could not know was that Whittle had chosen not to use the more complicated axial configuration for his turbojet engine. By the May 8 meeting, Arnold had returned to the United States, and the Special Committee expected to be briefed on the latest developments. Instead, because of the British imposition of a "most secret" classification on the project, the committee was*

August 1940. After the AAF opened its own school, RAF students were admitted to the Pan American course in 1941.

[103] See Wikipedia, 'General Electric I-A' in English.

[104] See Wikipedia, 'Bell P-59 Airacomet'in English.

[105] See https://history.nasa.gov/SP-4306/ch3.htm

merely asked to suggest the names of two engineers to be sent to England to "make contact" with British developments in jet propulsion. Durand, betraying his mistaken belief that the new British propulsion system used the axial compressor, suggested D.R. Shoults of General Electric, "an expert in matters relating to axial-turbo compressors, which type of equipment forms the core of the British development." *At the same meeting, Lewis, sharing the same prejudice in favor of the axial-flow compressor, referred to the eight-stage compressor developed by Eastman Jacobs and Eugene Wasielewski intended primarily as a supercharger. He revealed that General Electric* "was interested in developing this compressor to its full capacity since the Committee's tests had been limited to low speeds and the use of only six of the eight stages which had been provided." *At this point, all the signs indicated that an axial compressor would be a significant component of any jet propulsion scheme, a presumption shaped by the influence of Jacobs and the knowledge of the publications of the British aerodynamicists, Griffith and Constant. Future engineering practice would vindicate this decision, since the axial compressor did eventually prevail over the centrifugal. For short-term wartime needs, however, Jacobs underestimated the axial-compressor's recalcitrant problems. The NACA would pay dearly in terms of lowered prestige for its early commitment to axial compressor development and its failure to recognize the definite advantage of the compressor-turbine combination embodied in Whittle's turbojet.'*

After the foregoing, three monthly dates have been isolated from which information about substantial turbojet development activities must have reached the other enemy side— and correspondingly, considerably pushed own endeavours:

- Latest in *March 1941,* German flight test activities—now identified as the Heinkel fighter prototype He 280 V-1—without and with engines at the Heinkel works along the Baltic coast must have become known to US intelligence sources,
- on 28 *May 1943,* Colonel Dietrich Schwencke reported to Field Marshal Erhard Milch reliable information of a captured RAF member, who had observed just weeks before at the RAE test centre Farnborough a jet aircraft in *'very fast flight',* now with all likelihood identified as the second prototype of the first English turbojet demonstration aircraft Gloster E.28/39,
- and finally, as described already in Sect. 6.1.4, and nearly at the same time (*May 1943*), when the Farnborough news spread in Germany, Constance Babington Smith had evaluated high-altitude reconnaissance pictures of the Luftwaffe test site Peenemuende-West, which showed the characteristic parallel torch marks of one of the first, still tail-supported Messerschmitt Me 262 prototypes.[106] During June 1943, she informed a visiting Frank Whittle at her PRU Medmenhem, but nothing is known about

---

[106]The tricycle, front wheel supported landing gear was only introduced by a design change of the Messerschmitt Me 262 prototypes V-5 (fixed) and V-6 (retractable), after Oct. 1943.

any impact of this *'glimpse over the enemy's fence'* on the English turbojet engine development schedule.

Returning to the Jumo 004 B production schedule, it entered pilot production in mid-1943, and—with monthly increasing rates—was ordered into full-scale production near the end of that year. Under rapidly deteriorating circumstances it is quite surprising that aviation factories were able to produce anything at all during the early months of 1945; in fact, at least until March 1945 output was surprisingly high. The aviation industry produced a total of 3185 aircraft in January 1945 and 2479 in February; total documented output for 1945 was 7539 aircraft, among them 4935 conventional fighters, and 947 turbojet and rocket fighters.

Between February 1944 and March 1945 about 6010 Jumo 004 engines were built, in 1945 alone 2388,—mostly as B-1 and B-4 versions, of which documented 4752 units have been delivered before the end of the war.[107] Operationally, however, just some 200 Messerschmitt Me 262 aircraft[108] and a few Arado Ar 234[109] came to the front at the same time, Fig. 8.7,—in view of the Allied air superiority an insufficient number. Even on the most active day for the Me 262, 10 April 1945, only 55 Me 262 were available against an attack by over 2000 Allied aircraft.[110] Nevertheless, it is technically worthwhile to mention, that two more Jumo 004—powered turbojet aircraft types were brought to first flights:

- on 16 August 1944, the first of two prototypes of the heavy long-range turbojet bomber *Junkers Ju 287*[111] with forward-swept wings, and powered by four Jumo 004 B-1 engines, took off from Brandis-Waldpolenz Airfield,[112] with a 2.2 km long, 30 m wide concrete runway, 60 km in the south-east of the—with just 1 km—insufficiently short Junkers Dessau runway, and

---

[107] See Gersdorff, Flugmotoren, p. 301.

[108] See Wikipedia, 'Messerschmitt Me 262' in English.

[109] See Wikipedia, 'Arado Ar 234' in English. After the Me 262, this turbojet aircraft became the second in series production. Overall from mid-1944 until the end of the war a total of 210 aircraft were built. The first Ar 234 prototype, a design of WW I flight ace Walter Blume (1896–1964)—see Wikipedia, 'Walter Blume (aircraft designer)' in English, made its first flight, powered by two Jumo 004B, on 15 June 1943 at Rheine Airfield near Muenster/Westphalia. The four-engine Ar 234 V8 with BMW 003A turbojet engines, shown in Fig. 8.7 (r) with 'twinned' engine nacelles, flew first on 4 Feb. 1944, also first take-off with four turbojet engines.

[110] See Pavelec, The Jet Race, p. 105.

[111] See Wikipedia, 'Junkers Ju 287' in English.

[112] See Wikipedia, 'Flugplatz Brandis-Waldpolenz' in German; this military airport, up to 1989 used by the SU Airforce, was located 20 km in the east of Leipzig.

Fig. 8.7 German turbojet aircrafts, ~ 1944: Messerschmitt Me 262 fighter aircraft with 2x Jumo 004B (l), Arado Ar 234 C bomber/reconnaissance aircraft with 4× BMW 003A (r)

- on 2 February 1945, the *Horten Ho 229*[113] all-wing aircraft, also known as Horten Ho IX and Gotha Go 229, started from the Oranienburg Airfield,[114] under the power of two Jumo 004B turbojet engines.

Historically, on 26 July 1944 the first aerial victory by a jet fighter was claimed by Luftwaffe Lt. Alfred *'Bubi'* Schreiber—or nearly so. Schreiber intercepted and attacked an unarmed DH 98 *Mosquito* reconnaissance aircraft of No. 540 Squadron RAF at high altitude over Munich with his Me 262 A-1a, *'White 4'*. In several high-speed encounters Schreiber fired on the British plane, lost it in the clouds, assumed the plane destroyed, and submitted an aerial victory claim after landing. Actually, the front hatch had come off the Mosquito, hitting its wing and tail, but instead of returning to England, the British crew F/L A.E. Wall and F/O A.S. Lobban limped the damaged plane to Fermo Airfield near Ancona, It., where it was lost in a crash landing.[115]

In total, about 1400 Me 262 planes were produced, but—as said before—only a maximum of 200 were operational at any one time. Me 262 pilots claimed a total of 542 Allied aircraft shot down,[116] with the Allies destroying about one hundred Me 262 s in the air, mostly during low-speed flight during take-off and landing.

As World War II drew to its end and mid-Germany was squeezed between the vices of Allied armies from East and West, there was at best personnel evacuated, before the plant was overrun. In some cases, factory managers lost their nerves and fled long before the arrival of the Allied ground troops, leaving their workers behind. Dr. Anselm Franz, the

---

[113] See Wikipedia, 'Horten Ho 229' in English.

[114] See Wikipedia, 'Flugplatz Oranienburg' in German; this military airport, jointly used by Luftwaffe and Heinkel's Oranienburg Works, had a 2.2 km long, 52 m wide concrete runway, and was located approx. 25 km in the north-west of the Berlin city centre.

[115] Schreiber (1923–1944) was credited with a further four aerial victories before being killed on 26 Nov. 1944 in a crash landing at Lechfeld Airfield, making him the first *'jet ace'* in history.

[116] See Wikipedia, 'Messerschmitt Me 262' in English.

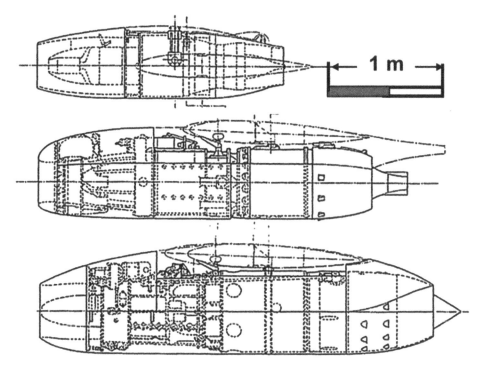

Fig. 8.8 Size comparison of German Class I turbojet engines: 1941 BMW 109-002 (top), 1944 BMW 109-003A (mid), 1943 Jumo 109-004 B-0 (bottom) © A.L. Kay

Austrian director of the Junkers aero-engine division and chief developer of the Jumo 004 jet engine, fled from the main factory at Dessau to a private home some 100 km further west, two weeks prior to the arrival of US troops on 29 April 1945, and nearly three months before the actually threatening occupation by Soviet Russian forces began on 1 July 1945, *'leaving behind workers extremely bitter over this action'.*[117]

8.1.2   The *BMW 109-003* Development and Production Programme

The initiation and development of the BMW 003 turbojet programme up to the early 1940s has already been described in Sect. 6.2.4. A size comparison on the same thrust basis illustrates in Fig. 8.8[118] the technologically more demanding role of the BMW 003 (mid) in relation to the already addressed Jumo 004 (bottom). Stunning is Weinrich's compact, counter-rotation concept of the BMW 002 (top), which was given up at Spandau in early

---

[117] See Uziel, Arming the Luftwaffe, p. 204.

[118] Excerpt from—see Kay, Turbojet. Vol.1, p. 176.

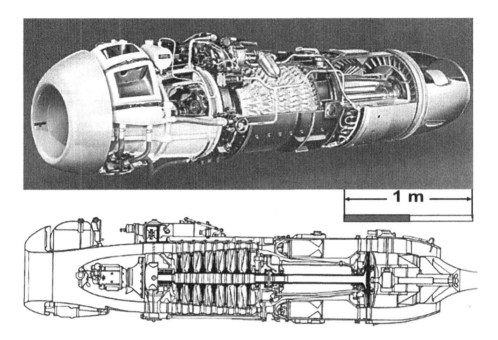

Fig. 8.9 BMW 109-003 A axial turbojet engine—for Arado Ar 234 and Heinkel He 162 aircraft
© H. Schubert

1942, while, nearly at the same time the idea of counter-rotating axial compressors was re-animated by the German engineer Fritz Heppner at Armstrong-Siddeley, Coventry, as will be outlined in the following Sect. 8.2.

A detailed description of the BMW 003 design has been provided by A. Kay,[119] Fig. 8.9. The essential layout of the 109-003 A-1 main production model followed the general lines of the original prototypes, thus confirming the thorough upfront analysis of chief designer Hermann Oestrich and his team. The compressor casing was cast in Electron alloy and, different to the Junkers design, not split longitudinally. The axial compressor had seven stages and used, as discussed before at the end of Sect. 6.2.4, with all likelihood a small amount of reaction. The change of the degree of reaction to $R < 1.0$ reduced first of all the axial thrust of the compressor rotor, because part of the reaction forces were transferred to the casing via the stator vanes. However, more important is the fact that in view of critical Mach number restrictions—as shown below—for the same relative entry velocity the circumferential tip speed $u_{1,\,Tip}$ of the BMW 003 compressor increased to 273 m/s, what was 248 m/s for the Jumo 004. This results, according to Euler's equation of motion immediately to a higher pressure rise per stage. Thus, the R-reduction of the

---

[119] See Kay, German Jet Engine, p. 96 f.

003 compressor from V11 engine onwards was the fundamental measure to achieve with only seven stages instead of eight, as for the Jumo 004, a total pressure ratio of 3.1.

As shown in Fig. 6.29, the blades had almost symmetrical profiles and a constant 40 mm chordal length from root to tip. Blade thickness varied between 5% of chord length at the tip and 12% at the base. Each rotor blade was secured to its disk groove by a single pin through a tongue in its root. The rotor was supported by three ball races at the front and a single roller race at the rear, very similar to the Junkers solution for axial thrust compensation—and its drawbacks, Fig. 8.2. Eight rows of stator vanes were secured between inner and outer shroud rings. The hot air tapped off from the last compressor stage was used for the long-time established BMW turbine cooling. Most remarkable, no surging problems were experienced with this compressor.

Further downstream, the engine was largely of a sheet metal construction. The annular combustion chamber, the main difference to the Jumo 004 six can combustor arrangement, had forty inner and forty outer hollow *'fingers'* to feed in secondary air after the sixteen burner cones. These injected J2 fuel downstream and imparted some swirl to better fuel atomisation. These hollow fingers were made from heat-resisting steel to cope with temperatures up to 800 °C, but the rest of the combustion chamber was constructed, again comparable to the Jumo 004, from aluminised mild steel. These combustion chambers were reported of having a life of about 200 h, considerably more than the just 50 h for the Jumo 004 combustor can solution, though without convincing explanation.

The single stage turbine wheel, originally designed on the basis of Brown Boveri's steam turbine know-how, was bolted to a flange on the hollow rotor shaft, supported at each end in a single roller race. Again, very similar to Jumo 004 and driven by the selected thrust compensation method, a splined coupling connected the turbine shaft to the rear of the compressor rotor. The turbine life was expected in the order of 50 h, with a 10 h inspection interval. The turbine inlet vanes and rotor blades were made from heat-resisting steel,[120] and had a hollow construction to permit air cooling. The hollow rotor blades had an insert to reduce the internal air space, so that higher through flow velocities improved the heat transfer—and also providing friction damping, when the blades flexed. The blades were attached to the turbine wheel by means of pins and wedges. It was normally after 50 h that the blades hardened and cracked, the life was used up, but a turbine wheel could be exchanged within 2 h only.

In general, the RLM demand to select for the BMW 003 the more advanced design solutions was achieved in many cases. A comparison of Figs. 8.2 and 8.9 illustrates e.g. the lighter and more 'slender' rotor design of the BMW 003, which contributed considerably to the weight advantage of 570 kg (003) vs. 745 kg (004). The Ministry specifically required that the man-hours for production were not to exceed 500 and the quantity of nickel was

---

[120] BMW developed a turbine blade rolling/ welding process, similar to the one described before from Wellner & Söhne for the Jumo 004 turbine, from *Boehler FBD* sheet stock (17 Cr, 15 Ni, 2 Mo, 1.15 Ta-Nb, 0.9 Mn, 1.0 Si, 0.1 C, balance Fe)—with quantity production by WMF Wuerttembergische Metallwaren-Fabrik, Geislingen an der Steige, 50 km east of Stuttgart.

again as for Jumo 004 to be reduced considerably. The large-scale substitution of sheet metal for castings led to a weight reduction of more than 150 kg.

Given the challenging circumstances during wartime, it must be acknowledged that the BMW team tried to support the development of all turbojet engine components by systematic research work; for each engine component there was at least one component test rig available.[121] And, testing of a compressor on the test bed at GHH Gutehoffnungshuette Oberhausen showed that the efficiency could still be improved by small adaptations of blade angles.

As mentioned in Sect. 8.1.1, the RLM had ordered the Junkers all-speed governor, which regulated fuel flow and engine speed, to become general standard. In September 1944 Captain (Hauptmann) Josef Bisping of BMW's flight test section flew an Ar 234 prototype with BMW 003 A-1 engines, Fig. 8.7 (r), and installed all-speed governor up to 13,000 km altitude. Flame outs would occur if less than full power was used at such height and restarting was only possible at or below 3500 m.

The 109-003 A-1 turbojets were designed for underwing installation (Ar 234), while the 109-003 E-1 allowed for above-fuselage installation (He 162). In a listing similar to that for the Jumo 004 in foregoing Sect. 8.1.1, the characteristic seven-stage axial compressor data for the BMW 003 A-1 were:

• mass flow rate $\dot{m}$	19.3 kg/s
• total pressure ratio PR	3.1
• rotational speed n	9500 rpm
• circumferential velocity $u_{1,\,Tip}$	273 m/s
• isentropic ad. efficiency η_C	0.78–0.83 (opt.)

supplemented by the following engine data:[122]

• combustor efficiency (burnout)	0.97
• turbine entry temperature TET	800 °C
• turbine isentropic ad. efficiency η_T	0.75–0.8
• thrust	800 kp
• specific fuel consumption sfc	1.33 kg/(kgh)
• engine weight	570 kg
• engine diameter	0.69 m
• engine length	3.565 m

After an early Me 262 V1 flight attempt on 25 November 1941 with BMW 003 turbojets—additionally equipped with a 1200 hp 'safety' Jumo 210G piston engine and airscrew in the nose,—had to be stopped after both turbojets failed at take-off rpm, and

---

[121] See Hagen, Zur Geschichte des Strahltriebwerkes.

[122] See Kay, German Jet Engine, p. 106 and 116.

a similar *'déjà vu'* on 25 March 1942, did much to sound the death knell of BMW's chances of providing engines to power future Me 262 s. On that day, Messerschmitt's chief test pilot Fritz Wendel cautiously prepared the Me 262 V1 for take-off, again with nose-mounted piston-engine and the two improved BMW 003 turbojets of 550 kp TO thrust running: *'The aircraft managed to take off, however, although greatly underpowered, but at a height of 50 m, the turbine blades of the port engine failed and the starboard engine immediately followed suit. With great skill, Wendel brought the stricken aircraft in to land after the shortest possible circuit on the piston engine alone.'*[123]

Thereafter, BMW 003 turbojets were solely applied to the four-engine Arado Ar 234 bomber, Figs. 8.7 (r) and 8.9, and the single-engine Heinkel He 162 'Volksjaeger', Fig. 8.11. As outlined before, the RLM decided to put jet-powered fighters into mass-production toward the end of 1943. At this time the Jumo 004 both produced 900 kp thrust against the 800 kp of the BMW 003, and was a more fully developed and reliable engine. However, the 003 was believed capable of ultimately giving the same thrust as the 004, and promised in addition, a smaller frontal area and—what was much more important—it was much easier to maintain, while being both lighter and more economical to produce. Consequently, the Ministry froze the development of the 004 and ordered it in immediate production, while Oestrich's team[124] was instructed to hasten the 003 development. The number of workers in the BMW turbine shop increased from roughly 700 in 1942 to nearly 1000 in 1945; the total number of trained engineers and designer scarcely increased, but they could now be more and more diverted to turbojets from the development of reciprocating engines, and increased here accordingly from 100 in 1942 to about 400 at the end of the war. Production of service engines began in the second quarter of 1944, but production levels never reached those of the Jumo 004.

The Luftwaffe *Kampfgeschwader 76*[125] began conversion to the Arado Ar 234 in June 1944. By 1 December 1944 it had 51 of these machines on strength, nearly one-quarter of the entire number of Ar 234 production aircraft to ever be built. The third group of Kampfgeschwader 76, i.e. *III./KG 76,* operated over France and the Low Countries until the end of the war. It flew some of the first jet bomber missions in history on 24 December 1944 against rail targets in Belgium. Troop concentrations were attacked around Liège and Bastogne on 26 and 31 December respectively, in support of German forces during the *'Battle of the Bulge/Ardennes'.*[126] Consequently, the parallel straight jet streaks, visible in Fig. 8.10[127] might indicate traces of turbojet-powered, high-speed Ar 234 interventions,

---

[123] See Kay. German Jet Engine, p. 101.

[124] The odyssey of BMW's development team after its leave and little by little relocation from the Berlin-Spandau plant in response to heavy Allied air raids end of 1943, first 700 km in the south-west to Wittring/Saarbruecken, then mid-1944 back to an underground salt mine at Neu-Staßfurt, some 150 km west of Bln.-Spandau, and thereafter in four stations after war's end, finally in 1950 to Snecma, Paris-Villaroche will be described in a detailed portrait of Hermann Oestrich, Chap. 9.

[125] See Wikipedia, 'Kampfgeschwader 76' in English.

[126] See Wikipedia, 'Battle of the Bulge' in English.

[127] See Cole, The Ardennes, Ch. XXI, p. 527.

Fig. 8.10 *'Battle of the Bulge/Ardennes'*, Bastogne, 26 or 31 Dec. 1944, *'contrails'* of USAF aircrafts and Luftwaffe Ar 234s © US Army

and as such the first photo-documented encounter of turbojets and piston engines. The unit also flew reconnaissance missions over Antwerp's docks and airfields on 1 January 1945. The German air force caused serious damage to Allied air bases in north-west Europe then, but it sustained losses from which it could not recover. The Allied counterattack in early January succeeded in pushing the Germans back and by the end of the month the Allies had regained the positions they held six weeks earlier. British Prime Minister Winston Churchill said the Battle of the Bulge was *'undoubtedly the greatest American battle of the war'*; it was also one of the bloodiest. The Allies could offset these losses, but the German side had drained its manpower and material resources. The Allies resumed their advance and in early spring crossed into the heart of Germany—via an undestroyed bridge across the *Rhine* at Remagen.[128] Here, the most notable use of the Ar 234 in the bomber role was the attempt to destroy this *Ludendorf Bridge*. Between 7 March 1945, when it was captured by the Allies, and 17 March, when it finally collapsed, the bridge was continually attacked, most notably by Ar 234s of III./KG 76, carrying 1000 kg bombs; KG 76 reported high losses during this period, 9–13 March 1945. On 21 March their base at Achmer[129] was bombed, ten Ar 234s were lost and a further eight damaged. By 1 April 1945 the group had just eleven machines on strength, with seven serviceable and 27 pilots of which 16 were ready for action.

In summer 1945 the Allied air superiority over Western Europe had become back-breaking, and posed a considerable problem for the Luftwaffe. Two camps quickly developed, both demanding the immediate introduction of large numbers of jet fighter aircraft. One side, represented by General Adolf Galland, responsible for readiness,

---

[128] See Wikipedia, 'Battle of Remagen' in English.

[129] See Wikipedia, 'Achmer Aerodrome' in English.

training and tactics of the fighter force, reasoned that superior numbers had to be countered with superior technology, and demanded that all possible effort be put into increasing the production of the Me 262, even if that meant reducing production of other aircraft in the meantime. A second group pointed out, that the Me 262 alone would likely do little to over-come its inherent problems of unreliable powerplants and landing gears, against existing logistic deficits and the general fuel shortage. Instead, a new inexpensive design should be built, following the concept of a 'throwaway fighter'. Reichsmarschall Hermann Göring and Armaments Minister Albert Speer supported the idea, and finally got their way: A contract tender for a single-engine jet fighter that was suited for cheap and rapid mass production was established under the name *'Volksjäger (People's Fighter)'*.

The RLM design competition specified a single-seat fighter, largely made of wood and other non-strategic materials, powered by a single BMW 003 turbojet engine (since the new Heinkel HeS 011 was not ready in time) and, most importantly, could be produced and assembled by non-skilled labour. Specifications included MG armament, a weight of no more than 2000 kg, with maximum speed specified as 750 km/h at sea level, operational endurance at least half an hour, and a take-off distance of no more than 500 m. The requirement was issued on 10 September 1944, with basic designs to be returned within 10 days and readiness to start large-scale production by 1 January 1945; finally, informally already on 15 September 1944 the Heinkel entry He 162 was selected for production.[130] The He 162 A-2 first prototype with BMW 109-003 engine flew, piloted by Heinkel development chief pilot Gotthold Peter (1912–1944), 69 days after the production go-ahead for 1000 units in a planned *'Gewaltaktion* (act of violence)', already on 6 December 1944 from Luftwaffe airfield Schwechat-Heidfeld, Vienna—with a maximum flight speed of 840 km/h. The flight had to be stopped prematurely after the glued wood structure disintegrated in parts. On a second flight already on 10 December, again with Peter at the controls, in front of various officials, the glue again caused a structural failure; this allowed the aileron to separate from the wing, causing the plane to roll over and crash, killing Peter.

Main He 162 construction facilities were at Eger near Salzburg, the Hinterbrühl, a satellite of Mauthausen concentration camp and the Mittelwerk, factually a satellite of Buchenwald concentration camp; He 162 output was expected to be 1000 a month by April 1945. The giant Mittelwerk-*'Dora'* plant at Nordhausen,[131] 50 km east of Goettingen, was a key centre of the German military production towards the end of the war. It relied on slave labour, and was to have been a centre of V-1, V-2, He 162 and Junkers as well as BMW turbojet engine production. The Mittelwerk plant was expected to produce up to 2000 Me 262 and He 162 aircraft per month, but by the time production ended on 10 April 1945 with presumably just 25 Me 262 and 18 He 162 finished, of which the first of the Mittelwerke production block *'310'* with Work No. 31001 is illustrated in

---

[130] See Wikipedia, 'Heinkel He 162' in English.

[131] See 'Mittelbau-Dora Memorial', https://www.buchenwald.de/en/29/ in English.

Fig. 8.11 Heinkel He 162 with 1x BMW 003 E-1, Junkers Bernburg Airfield, ~ March 1945, Flugkapitän H. Steckhan on far left © R. Forsyth

Fig. 8.11—together with Junkers test pilot Hermann Steckhan.[132] On 27 March 1945, Steckhan planned to fly a He 162 demonstration for pilots of *Jagdgeschwader 1*[133] over Bernburg. But the Heinkel crashed upside down shortly after take-off, its wings and engine breaking off (similar to Peter's fatal accident during the He 162 second flight, 3.5 months earlier at Vienna-Schwechat) and its cockpit being crushed; somehow, Steckhan survived.[134]

8.1.3   The *Heinkel HeS 011* Development Programme

In 1945, nine years after Hans von Ohain's entry at the Ernst Heinkel Works, Rostock—there was still no turbojet engine in production. So, even by fully acknowledging the innovative design approach of the HeS 011, the core of this section in continuation of Sect. 6.2.2, the Heinkel and since April 1941 the Heinkel-Hirth turbojet development programme was a failure. The reasons, all the more depressing in view of the early pioneering, are obvious in hindsight. Over time, the quality of the recruited engineering team was too low, characterised by frequent leaves—as for M.A. Mueller and R. Friedrich, and their efforts were dispersed over a great many developments and projects. This situation changed somewhat in 1942, when by RLM order all focus had to be concentrated on developing a single turbojet, Helmut Schelp's new Class II *109-011* engine.[135] Now however, the new Heinkel-Hirth organisation at Stuttgart-Zuffenhausen appeared to be

---

[132] Details of Steckhan's CV and his early acquaintance with Helmut Schelp in 1935 have been described already in the context of Fig. 7.3.

[133] Then commanded by Oberst H. Ihlefeld (1914–1995),—see Wikipedia, 'Herbert Ihlefeld'in English.

[134] See Forsyth, He 162 Volksjäger Units, p. 60.

[135] See Wikipedia, 'Heinkel HeS 011' in English.

more geared for research than for developing an engine up to production status. Indirectly also reflected by a personal notion of Hans von Ohain, who repeatedly had indicated his preference of a leave back to Goettingen University. In the decisive years 1941–1943, neither the restless and erratically leading Ernst Heinkel nor the somewhat weak and not very energetic Hans von Ohain in his role as now solely responsible chief engineer were able to manage a sustainable turnaround. In addition, the layout chosen for the Heinkel HeS 011—under questionable circumstances on Schelp's personal insistence, proved to be difficult with unforeseen problem areas. Consequently at war's end, although this engine was the most powerful German turbojet then and claimed prone for production, it was by no means sufficiently developed.

Figure 8.12 provides an unique atmospheric insight into the leading ranks of this Heinkel development team in 1942. The main picture shows the Heinkel He 280, the first turbojet-powered fighter aircraft in the world, in July 1942 at Heinkel, Rostock-Marienehe, where on 27 August 1939 also the first turbojet-powered He 178 flight had taken place. The He 280 had a typical Heinkel fighter fuselage, elliptical wings, a twin-fin tailplane, and—for the first time—a tricycle undercarriage landing gear and was equipped with a compressed-air powered ejection seat (also to be used successfully on that aircraft in an emergency). From the beginning, Heinkel struggled with the tasks of developing aircraft and engine in parallel, Hans von Ohain's unchanged centrifugal engine concept, the HeS 8 with intended 700 kp take-off thrust was running into difficulties. Finally, on 30 March 1941 the first flight of He 280-V2 (test version #2) took place, piloted by Fritz Schaefer; however—*small cause, big effect*—due to leaks, they decided to remove the engine cowlings, as visible in the inset. The Heinkel development crept on up to 5 July 1942, when Schaefer also brought He 280-V3 to air. But the Messerschmitt competition was now already very close; Heinkel had squandered a project lead of at least one year, when Fritz Wendel took off the Me 262-V3, first time Jumo 004 turbojet-powered on 15 July 1942 from Messerschmitt's Leipheim airstrip. Heinkel, now under pressure, decided to go for a He 280 promotion movie, by re-enacting the first flight in a triumphant set-up—and with closed engine cowl. The picture—actually copied out of the movie sequence[136] of the first flight of the *second,* turbojet-powered He 280 V3 (GJ + CB)— shows behind the pilot Fritz Schaefer from—left-to-right—Hans Antz (1909–1981), together with Helmut Schelp responsible at RLM for the synchronised turbojet aircraft/ engine development in Germany, Dr. Hans von Ohain, Heinkel's young engine man, and to the right, the test pilots Gotthold Peter, Fig. 8.12, and Erich Warsitz.[137] However, all PR effort at Heinkel was in vain, when the He 280 was officially cancelled on 22 March 1943—and looking backward, as in Fig. 8.13, the spectacularly dragged-along, movie

---

[136] See minute 2:00 of 3:34 min YouTube video on He 280 re-enacted first flight: https://www. youtube.com/watch?v=IM0RhZeB49c

[137] See Sect. 6.2, entry section.

Fig. 8.12 *Rostock meets Hollywood:* He280 first flight re-enactment in July 1942: pilot Schäfer, behind—Antz (with hat), Ohain, Peter, Warsitz, l-t-r; original first flight on 30 March 1941 (inset) © V. Koos

starring engineers and their promising product remained an episode; the path for the Me 262, the first operational turbojet fighter, was free.

Already in May 1941, Schelp had alerted and prewarned Heinkel of a coming bomber specification, actually issued by RLM's Technical Office on 18 July 1941, to be powered by a new turboprop engine. The corresponding engine layout came from Schelp's department and consisted of two gas turbines, acting as hot gas producers to feed a third power turbine driving a variable-pith airscrew. Hans von Ohain's investigation of that scheme suggested that it would be necessary to design and develop first the basic turbojet engine before that could be later employed as one of the required gas producers for the turboprop. It is possible that Von Ohain started with an all-axial compressor design,[138] but before the end of 1941 Schelp must have *'asked'* Von Ohain to incorporate a diagonal impeller, a supposed compromise between axial and centrifugal flow, followed by three axial stages, which settled to a firm decision after a joint meeting on 26 August 1942.[139] As outlined already in the context of Fig. 7.5, the diagonal compressor idea resulted with all likelihood

---

[138] Known as *HeS 09*, of which however no drawings exist,—see Koos, Heinkel Raketen- und Strahlflugzeuge, p. 69.

[139] See Schlaifer, The development, p. 435, who comments this unusual intervention: '*The two cases in which the Ministry ultimately did dictate . . . the design of an engine were the mixed-flow Heinkel*

Fig. 8.13 The He 280 V2, first flight re-enactment (Fig. 8.12)—with complete, tractor-drawn He 280 V3 (GJ + CB) movie set-up

as a kind of late souvenir of Helmut Schelp's study stay at Stevens Institute of Technology in 1936.[140] The efficiency of the diagonal stage was certainly compromised to the axial, and meant an increase in engine diameter, but it was claimed of having robust blades, less susceptible to foreign object damage and icing, with advantages during start and better stall resistance than an axial compressor. The latter argument is supported by Mueller's 1939 patent description, where he believes that a diagonal front stage would allow a smoother blading curvature and thus a better, stall-free handling of stagnation pressure rises at engine inlet.In the same direction go Hans von Ohain's radial- and diagonal-flow considerations, Fig. 8.14,[141] which he presented on 31 January 1941 in a session of the German Academy of Aeronautical Research. His paper '*Special features of radial turbojet engines*',[142]

011, begun in 1942, and the turboprop version of that engine, the 021, begun by Daimler-Benz in 1943.'

[140] A more detailed discussion of diagonal compressor patents in Sect. 12.2 will list besides Birmann's US patent with priority of 1932, patents of M.A. Mueller's teacher at the Technical College Cologne,1927/1928, Dr.-Ing. Bruno Eck in 1937, by M.A. Mueller still at Junkers Dessau for a comparable turboprop project in 1939 (see also Sect. 8.3), and finally, for H. von Ohain and V. Vanicek at Heinkel-Hirth in 1942.

[141] See Eckardt, Detailed Flow Investigations, and—see Eckardt, Untersuchung der Strahl/Totwasser-Strömung.

[142] See Von Ohain, Besonderheiten, pp. 116–118.

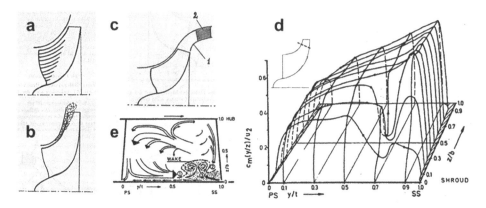

Fig. 8.14 Hans von Ohain, 1940, radial/diagonal compr. impeller design/flow considerations (**a–c**), high-speed centrifugal compressor mean (**d**) and secondary flow (**e**) laser measurements, 1972

immediately before his competitor at Heinkel, Max A. Mueller talked about '*Axial turbojet engines and motorjets*',[143] discussed and illustrated corresponding design consideration. Fig. 8.14a shows by parallel lines, attached to the impeller leading edge, regions of blade curvature, which reach far downstream (especially close to the casing). Von Ohain considers this an essential progress in comparison with foregoing, strictly radial impeller blading with an attached, short axial '*inducer*' upfront. Explaining Fig. 8.14b he states: '*The relatively large channel height of radial blowers is prone to flow separation in the area of curvature in the radial direction. Besides losses in rotor flow itself, large additional losses can occur by that non-uniform or undefined flow in the diffuser down-stream. These disadvantages can be tackled by defining a controlled streamline curvature.*' By differentiating between centrifugal forces on air mass particles in the meridional plane, towards the rotor hub due to streamline curvature, and radially outward due to impeller rotation, he postulates a controlled channel curvature so that the components of the two forces compensate each other. The diagonal impeller shape, Fig. 8.14c is a self-evident design solution here, but now the critical duct curvature—1—after the impeller is separation prone, thus distorting the axial diffuser—2—function downstream.

In fact, this diagonal-to-axial flow transition region, at the entry of a subsequent three-stage axial compressor, developed to a very critical flow region of the finally picked HeS 011 flow annulus, Fig. 8.16. Of course, Ohain could only have a premonition of the complex impeller flow field, which has been investigated in detail by advanced laser velocimetry only some 30 years later at the DLR Institute for Airbreathing Engines, Cologne-Porz. Fig. 8.14d shows the measured meridional velocity component c_m, in

---

[143] See Mueller, TL-Triebwerke. The order of presentations might have been arranged in favour of Von Ohain by Schelp and the organising RLM.

Fig. 8.15   Heinkel HeS 011 diagonal compressor rotor, solid Al milling by J.M. Voith, Heidenheim, ca. diameters: entry 0.48 m, exit 0.58 m, 6 March 1947 © ww2aircraft.net

reference to the rotor tip speed u_2 of ~300 m/s, for one impeller flow channel at ~80% meridional length, with PS—blade pressure side and SS—blade suction side, casing 'Shroud' in front and rotor hub in the rear. The flow pattern shows by and large a distribution close to theoretically predictable potential flow calculations, but the shroud/suction-side corner is severely distorted. The secondary flow pattern of Fig. 8.14e depicts here a pronounced, loss-concentrating '*wake*' area, so that the 3D configuration of the, in 1960 predicted, in the meantime '*classical*' Dean-Senoo '*Jet/Wake*' flow distribution at centrifugal impeller discharge had been confirmed.[144]

The first non-flight prototype of the HeS 011 V1 was built from steel castings and sheet metal, and no attempt was made to save weight or to meet aircraft design standards. The compressor set, drum-type rotor and turbine set were connected by a long tie-rod to take out axial imbalance, supported by three sets of anti-friction bearings. As illustrated already in Fig. 7.5 (r), the first bearing set comprised three ball races, in this respect comparable to Jumo 004 and BMW 003, positioned in front of the diagonal compressor. The compressor set of diagonal compressor rotor followed by a three-stage axial compressor, now—what will become important—explicitly designed with a degree of reaction R 0.5, was tested up to full speed of 11,000 rpm in December 1943 on a DVL test bed at MAN Augsburg.[145]

The diagonal stage of the compressor, Fig. 8.15, was milled out of a drum of forged aluminium alloy. This operation proved extremely difficult, but was finally achieved by the

---

[144] See Dean, Rotating Wakes.

[145] By the end of 1944 in addition, the production version of HeS 011 A-0 compressor, Fig. 8.16, was also tested as only full-size compressor at AVA Goettingen, for which W. Encke reported a rather disappointing efficiency of just 80%. The new test rig used a 4 MW DC electric motor with a step-up gearbox at each end of its shaft.

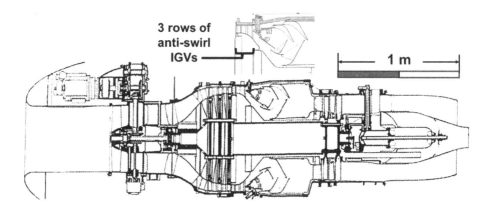

**3 rows of
anti-swirl
IGVs**

1 m

Fig. 8.16 Heinkel-Hirth HeS 011 A-0 turbojet engine, 1944—for a multitude of fighter and bomber projects © V. Koos

company J.M. Voith, Heidenheim, within alarming ~3000 man-hours. The two-stage axial turbine of the first prototype used solid blades, held in their discs by means of pinned V roots.

Figure 8.16[146] illustrates the corresponding series production standard Heinkel 109-011 A-0, which included a shortening of the engine by redesigning the combustion system and by supporting the rotor on two, instead of three bearing sets. The compressor and turbine were now bolted to the ends of a much simplified drum rotor, supported by the two bearings on each end, a ball race in front of the diagonal compressor wheel—and a single roller race behind the turbine.

The axial inducer on the quill shaft, upstream of the accessory tower shaft had 11 blades; little change was made to the diagonal compressor, now specified against the original integral approach with 12 inserted blades. All blades of the three-stage axial compressor, downstream thereof, were of hollow, steel sheet construction, riveted in their discs, while the stator blades were riveted into sheet metal shrouds. As a sign of clear aerodynamic emergency, where original engines had used a single row of combustor inlet guide vanes (IGV) after the last axial compressor stage, there were now three (!) rows of anti-swirl IGVs,[147] as illustrated in the enlarged inset of Fig. 8.16.[148] Interesting details of duplex fuel injectors as foreseen for the production engines, have been given elsewhere.[149] Air cooled,

---

[146] See Koos, Raketen- und Strahlflugzeuge, p. 72.

[147] On 27 Feb. 2019 Prof. V. Guemmer, TU Munich, commented this *'triple vane set-up'* as *'highly endangered by 3D wall BL separations, to be assessed by later published de Haller criteria'*.

[148] See Whittle, Gas Turbine Aero-Thermodynamics, p. 101, provides in hindsight an interesting, general comment in this context: *'The compression process in a gas turbine is performed in a centrifugal compressor or an axial flow compressor or a combination of both. If both types are used, the axial flow compressor invariably precedes the centrifugal type,'* so, in this respect the HeS 011 compressor arrangement represented a clear exemption.

[149] See Kay, German Jet Engine, p. 50.

hollow blades, designed by Heinkel's vibration specialist Max Bentele (1909–2006)[150] et al. have been manufactured as those for the BMW 003 at WMF Wuerttembergische Metallwarenfabrik, Geislingen an der Steige.[151] Finally at the engine back-end, a variable-area exhaust nozzle was introduced, with movement of the open-ended tail cone being by means of a hydraulic cylinder.

In a listing similar to that for Jumo 004/BMW 003, in foregoing Sects. 8.1.1 and 8.1.2, the characteristic axial-diagonal-axial compound compressor data for the Heinkel HeS 011 A-0 were:

• mass flow rate ṁ	29.1 kg/s
• total pressure ratio PR	4.2
• rotational speed n	11,000 rpm
• circumferential velocity $u_{1, \text{Tip}}$	334 m/s
• isentropic ad. efficiency η_C	0.80

supplemented by the following engine data:[152]

• combustor efficiency (burnout)	0.97
• turbine entry temperature TET	800 °C
• turbine isentropic ad. efficiency η_T	0.78
• thrust	1300 kp
• specific fuel consumption sfc	1.31 kg/(kgh)
• engine weight	950 kg (865 kg target)
• engine height/width	1.08/0.864 m
• engine length	3.455 m

After 1944, practically all German aircraft designers presented projects on the basis of the HeS 011,[153] but as mentioned already in the introduction to this section, disappointingly, the HeS 011 turbojet engine never reached series production; only 19 prototypes have been built in total, of which one had been planned to be mounted on Messerschmitt's Me P.1101 variable-sweep fighter aircraft, found in the *Oberammergau shops*[154] at war's end with adjustable wings, and which was later reproduced as *Bell X-5*[155] experimental

---

[150] See Wikipedia, 'Max Bentele' in English.

[151] The manufacturing of these *'Topfschaufeln* (pot blades)' has deen described in detail by H. Schubert in—see Hirschel, Aeronautical Research in Germany, pp. 444–450.

[152] See Kay, German Jet Engine, p. 51.

[153] A selection of planned applications has—see Kay, German Jet Engine, p. 51 f.

[154] Messerschmitt's evacuation quarters at war's end, 90 km south of the Augsburg-Haunstetten headquarter.

[155] See Wikipedia, 'Bell X-5' in English.

aircraft in the United States, with continuously variable, sweep adaptation during flight, Fig. 10.13.

To the strange, disorganised impression of the Heinkel-Hirth organisation after 1944 belong a subsequent double-move out of the Zuffenhausen headquarter. First the threat of Allied bombing necessitated the dispersal of factories in the Stuttgart area and the continuation of work at safer locations. Beginning in spring of 1944 a salt mine, called the *Staatliche Saline Friedrichshall,* at Kochendorf[156] was partially prepared for Heinkel-Hirth by *OT Organisation Todt,*[157] using SS prisoners and forced labour troops. Approximately one quarter of the 144,250 m^2 underground installation was allocated to (Ernst) Heinkel-Hirth, known with little camouflage as *'Ernst Werk'*. In August/September 1944 Heinkel-Hirth was ready to resume work on the 109-011, but without convincing explanation Hans von Ohain's team reappears shortly thereafter 300 km further east at Kolbermoor[158] under *'BMW wings'*, a move which certainly would not have occurred without agreement of the RLM (H. Schelp). Here HeS 011 production was scheduled a second time to begin in March 1945 in a factory and former cotton spinning mill, already being used as a dispersal of BMW's Munich works. On official request, BMW looked into opportunities for technical assistance to Heinkel-Hirth, and the first experimental manufacturing trials of the engine. In meetings on 15 and 19 February, continued in March 1945, Hans von Ohain's team met with Oestrich, Hagen and others from BMW for a kind of design review. A short film sequence was shown after an 8-h test, where the 109-011 combustion chamber was in bad condition, possibly by a thermodynamically poor match between compressor and turbine.

Thus, Heinkel-Hirth found themselves in a difficult position, but the story might have been very different, as the disempowered Ernst Heinkel and his Stuttgart antipode Professor Wunibald Kamm propagated openly, had RLM officialdom, meant was mainly Harald Wolff and Helmut Schelp, not *'prematurely killed Mueller's promising HeS 30'* turbojet engine. The fact that Hans von Ohain could feel relatively safe at Kolbermoor, virtually on *BMW territory*, might have been of special personal importance to him, as the following dramatic episode outlines.

The *'special documents collection'* of the Deutsches Museum, Munich contains a reference file for Ernst Heinkel—with HeS 011 documents for a 1-year period after the end of 1943, and an especially interesting document in the context above: The carbon copy of a six-page hearing of a witness at a court-martial on 25 November 1944.[159] Heard was 28 years-old Werner Hilgendorf, then since one month in basic training as armoured infantryman at Gnesen/Gniezno, but since July 1939, engineer at Heinkel, Rostock, and

---

[156] Kochendorf lay 50 km north of the Heinkel-Hirth headquarter at Stuttgart-Zuffenhausen.

[157] See Wikipedia, 'Organisation Todt' in English.

[158] Kolbermoor lies about half-way between Bad Aibling and Rosenheim, 50 km south-east of BMW's Munich-Milbertshofen headquarter.

[159] See Budrass, Hans Joachim Pabst von Ohain, and Deutsches Museum, DMM/ASD, FAA 001/0323.

mostly in the team of Hans von Ohain. Actually, the lives of Von Ohain and Hilgendorf were closer coupled than generally known. '*Physicist Pabst* (pope)' was Swabian name-jokingly one of a continuously rewarmed criticism of his tendency towards fundamental research—and against practical engineering solutions.[160] In line with these considerations Heinkel had sent Ohain in June 1939 to his *fraternity brother*[161] Prof. Wunibald Kamm, head of FKFS Stuttgart, and then apparently a strong supporter of NS ideology, to deepen his practical engineering expertise in a versatile design office environment. This opened the opportunity to Kamm to send Hilgendorf, who had entered FKFS in 1938, in exchange to the Heinkel Works at Rostock,[162] and to stress in continued personal consultation to Ernst Heinkel his superior, analytical knowledge of human nature, applied to the fundamentally different characters of Von Ohain and Max A. Mueller, mostly—as in a May 1941 letter to Ernst Heinkel—to bring the latter's unique talents to fruition in the Heinkel organisation, while '*Herr von Ohain's tasks could be fulfilled by any somewhat talented diploma engineer after a sufficient period of vocational adjustment.* '[163]

Hilgendorf's hearing was part of a preliminary investigation, initiated by the military superior of the new conscript, to whom he had indicated '*difficulties during turbojet development*', insinuating, Ohain might have delayed the military readiness of his engine due to scientific, but also ethical-religious reasons. Indirectly, of course, he put also Ohain's immediate superior since autumn 1941, Harald Wolff, into focus. He reported his observations, how Ohain and Wolff hampered continuously the project activities of Max Adolf Mueller's team, up to the point in 1942, when Ernst Heinkel recommended Mueller's leave. Ohain and Wolff wanted to keep their influence at Heinkel, at the same time with severe doubts about the outcome of the National-Socialistic War. Ohain had shown increasingly indifference towards his responsibility in turbojet development, and—supported by Wolff—just '*played around technically*'. Already in mid-1943 Ohain had stated that '*his turbo engine*' will not find application still during this war, while Wolff had remarked several times, that Germany will lose this war.

Besides this accusation of sabotage, Hilgendorf apparently wanted to get revenge for some frustrations which he might have had to endure in Ohain's team, but mostly for his

---

[160] Budrass quotes a letter from Heinkel to the EHAG board, dated 15 March 1944, characterising Ohain as '*excellent physicist, who requires the strong hand of a chief engineer, who stops Ohain's constantly observed notion to "*bricolage*" over the past six years*'.

[161] Both Heinkel (1888–1958) and Kamm (1893–1966) belonged to the '*Burschenschaft* (fraternity) *Ghibellina*'at TH Stuttgart,—see Wikipedia, 'Wunibald Kamm' in English. After the war Kamm met Hans von Ohain again at the Wright Patterson Airforce Base in Dayton, Oh., before he continued 1953–1956, 17 years after Schelp's stay, as professor at Stevens Institute of Technology, Hoboken, NJ.

[162] At Rostock Hilgendorf joined Mueller's team in July 1939, moved with this group back to Heinkel-Hirth, Stuttgart in mid-April 1941, and joined here after Mueller's leave the united engineering team under Von Ohain's sole leadership in autumn 1942.

[163] See Potthoff, Wunibald I.E. Kamm, p. 228.

conscription to the infantrymen, which reached Hilgendorf on 20 October 1944. The last point was not mentioned openly, but draft calls to experienced engineers in such important development projects were normally turned down by the employing companies. Hilgendorf experienced the bitter consequences of this decision. Gnesen was in the centre of the Soviet-Russian offensive in early 1945; he died on 22 January 1945 in a military hospital at Landsberg an der Warthe/Gorzów Wielkopolski, 130 km east of Berlin. Indirectly, Hilgendorf's documented arguments were supported by Ernst Heinkel, who addressed the core of the conflict in a note in preparation of a RLM meeting with Schelp on 10 November 1944, i.e. 14 days before Hilgendorf's hearing: *'Killing of HeS 30 was a mistake!'*

More specifically, at the *'core of the conflict'* was the selection of the right axial compressor degree of reaction. Von Ohain had already shown corresponding design flexibility and followed Mueller's HeS 30 example by picking R 0.5, also for the HeS 011 three-stage axial compressor, and the indirectly accused Schelp had reversed his decision in favour of the AVA compressor design concept for Jumo 004 and BMW 003 already in April 1941, when Brown Boveri Mannheim received study contracts for the *'Hermso I—III'* axial exchange compressors with R 0.5.[164] So Schelp must have lived all the time under the threat, that it was actually he, who should have gone court-martialled, not for intentionally doing wrong, but for imprudent over-speeding and trusting in AVA competence too easily. The AVA side of Encke, but especially Betz is different; since the mid-1920s and a serious economic crisis in Germany, Betz had planned a financial rush by successful patenting. After a showdown with Brown Boveri, CH, in 1932,[165] he was obviously under the impression that the R 0.5 area was no longer protectable by own patents, so—again by a certain laziness—they preserved their patented R 1 preference, pronounced the corresponding design advantages, but ignored considerably inherent performance drawbacks and—due to excessive axial thrust, as visualised in Fig. 8.2—the overall concept limitation to total pressure ratios PR < 5.

Budrass—after thoroughly reviewing and evaluating the facts—comes to the conclusion that Hilgendorf's accusations, especially against Wolff, might have been partially justified in principle. But—in hindsight—the most important contribution of Wolff's and Von Ohain's delay tactics was presumably that the planned, first HeS 011 production start in the Kochendorf underground facility, expectable under horrible circumstances, had been prevented in September 1944.

Schelp's whole approach is somewhat tragic, he tried to build a design solution kit for all eventualities by making the turbojet development programme as broad as possible; consequently, he tried to streamline/condense on the other side areas, where he took superior German know-how for granted, in the fluid mechanic responsibility range of Prandtl/Betz

---

[164] See Chap. 9, for Hermann Reuter.

[165] As discussed before in Sect. 4.3.3, and in the context of Fig. 4.49 in the *'Patent Arena'*.

and Encke; but actually here he was wrong and misled, and *turbomachinery* did not really belong to AVA's areas of excellence.

After more than 6000 manufactured Jumo 004 turbojet engines up to War's end, of some 500 BMW 003, and insignificant numbers of HeS 011, Schelp's final view is more than astonishing: '*In evaluating the three turbojet engines under development and in production, Schelp considers the HeS 011 to be the best and most promising long range project. Second in standing is the BMW 003 while Junkers would come last, … as a production shop. The initial plan was that Junkers would produce the 004 as it was at that time while BMW would develop the 003 as an interchangeable power unit. When the 003 had reached a satisfactory stage of development, it was to be put in production and the 004 withdrawn and Junkers given the opportunity to develop their engine.*'[166]

8.1.4   Summary of German Turbojet Engine Developments

While research and development for turbojet engines was widely organised—as described above—in the companies Junkers, BMW and Heinkel, which placed wherever required specific subcontracts to Institutes of the large research organisations like AVA Goettingen or DVL Berlin-Adlershof, it had been questioned for a long time if the large research organisations could fulfil—not the least in view of the required secrecy—independently meaningful tasks at all.[167] British post-war expertise knows in this context (without further proof): '*In a notable direction however, Schelp failed to mobilise available resources. The research establishments, throughout, resolutely refused to join in the effort on gas turbines. Most of the research workers either felt that the gas turbine was not ripe for practical development or else wished to conduct the research themselves.*'[168]

It is relatively recently that areas of '*Gemeinschaftsforschung (community research)*' have been identified, where scientists could follow their research activities under wartime conditions, also focussed to military relevant projects. Besides '*material research*',[169] there are now the activities of a '*Sonderausschuss Windkanaele (Special committee wind tunnels)*' as a second, equally structured area of aeronautical research.[170] This committee

---

[166] See C.I.O.S., Interrogation, p. 7/8,—or '*o si tacuisses, philosophus mancisses*', (lt., Boethius or Bible, Job/ Hiob 13.5, and in English: 'If you had remained silent, you would have continued to be a philosopher'.

[167] Helmuth Trischler, leading German expert on the history of aeronautical research, addressed the apparent problem, that '*the NS State defined no concrete, specifically designed research programme for the needs of air armament*',—see Trischler, Luft- und Raumfahrtforschung, p. 255.

[168] See Postan, Design and Development of Weapons.

[169] See Maier, Forschung als Waffe, p. 772.

[170] See Schmaltz, Vom Nutzen und Nachteil, and—see Kay, German Jet Engine, p. 220, where also is mentioned that the BMW 003 nacelle, investigated at AVA Goettingen as a 1/5th scale wind- and water tunnel model, was result of these activities.

Wing-mounted turbojet engines		Fuselage-mounted turbojet engine	
2 engines on straight wings		Radial engine in a/c nose	
2 engines on swept wings		Tail engine w. front air intake	
4 engines on straight wings		Tail engine w. top air intake	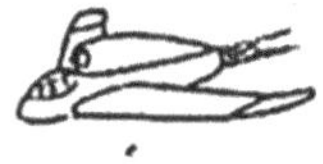
2 engines in flying wing a/c		Tail engine w. wing air intake	
		Tail engine w. air intake ring	
		4 engines at wing roots	

Fig. 8.17 Too Late—AVA Goettingen community research results, 1943–1945: Turbojet installation examples © F. Schmaltz

was led by Dietrich Kuechemann (1911–1976),[171] in 1943, still scientist at the AVA Institute for Theoretical Aerodynamics, and later, head of the AVA Institute for Wind Tunnel Research. He defined a working programme with special focus on turbojet engine installations and nacelle aerodynamics. He suggested a principle project structure as illustrated in Fig. 8.17,[172] by basically differentiating between '*wing- and fuselage-mounted turbojet engines*'.

Besides the aerodynamic engine characteristics, there should the interactions of the engine with other fuselage elements be investigated. Here, Kuechemann differentiated between *interference effects* of single components with significant effect on the engine performance, and an engine impact on fuselage features, as e.g. the nacelle's distorting influence on wing flows, or jet/empennage interaction on fuselage flight stability; on the other hand, also engine inlet distortions as a result of wing interference had to be investigated.

---

[171] See Wikipedia, 'Dietrich Küchemann' in English.

[172] Graphically adapted,—see Kuechemann, Stroemungsvorgaenge, pp. 8–10.

In June 1944, Hans Georg Lemme of Kuechemann's Institute presented a comprehensive internal report on wind tunnel model measurements of a *Heinkel P.1065* turbojet-powered bomber, where for the first time two- and four-engine configurations were simulated by the installation of small, e-powered turbofans—with and without jet interference.

In July 1944, Kuechemann presented a second, extended research programme to which Hermann Schlichting (1907–1982),[173] head of the Institute for Aerodynamics at TH Brunswick, made research suggestions for improving *'flight behaviour'*. He pointed out, that so far the nacelle installation had been mostly looked at in its effect on flight performance, a view, which had to be shifted more and more in the direction of high speed flight behaviour with unknown phenomena like premature shock waves, and the effect of flow separations on flight stability.

Schlichting suggested large-scale surface pressure measurements for a better physical understanding of wing/nacelle interactions.[174]

One year after the war, Kuechemann moved to England and started work at RAE Farnborough. In 1953 he and his mathematics and aerodynamics colleague, Johanna Weber (1910–2014),[175] published the still-standard work on the topic, *Aerodynamics of Propulsion,*[176] based on their work at the AVA Goettingen from 1940 to 1945. Kuechemann continued his work on high-speed flight, and was part of the team involved in the development of the *delta wing* in England, which eventually led to the advanced wing shape used on *Concorde* (1976–2003).[177]

Another area of German community aero-research commenced already in March 1944, but flourished only after the war, that of flow investigations within axial turbomachinery cascades. Interestingly, it was H. Schlichting, who originally came from the now precluded area of aircraft aerodynamics, who occupied the experimental area of compressible, viscous cascade wind tunnel investigations, first by erecting a high-speed cascade wind tunnel (HGK Hochgeschwindigkeits-Gitterwindkanal) in 1956 at DFL Brunswick. This unique facility had been projected by *Norbert Scholz (1921–1974)*, and allowed for independent Mach and Reynolds number variations.[178] Further names out of the renowned German post-war cascade-flow community on transonic axial-flow compressor research are Leonhard Fottner (1938–2002), Wolfgang Heilmann (1934–1988), Heinz Hoheisel

---

[173] See Wikipedia, 'Hermann Schlichting' in English.

[174] Partially as a result of these research activities on aircraft nacelle aerodynamics, the lack of large high-speed wind tunnel capacity became imminent and was compensated by the large 76 MW wind tunnel facility at Oetztal. This, at war's end unfinished project of LFM Munich, has been described by—see Hirschel, Aeronautical Research in Germany, p. 193 f.

[175] See Wikipedia, 'Johanna Weber' in English.

[176] See Kuechemann, Aerodynamics of Propulsion.

[177] See Wikipedia, 'Concorde' in English.

[178] See Hirschel, Aeronautical Research in Germany, p. 412.

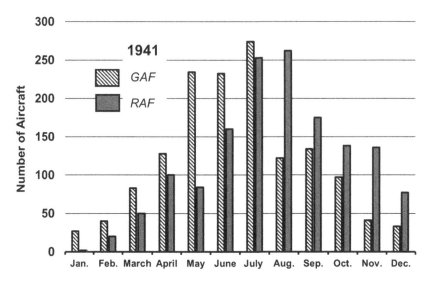

Fig. 8.18 Monthly Luftwaffe (GAF) and Royal Air Force (RAF) aircraft losses during 1941 © *Flight,* Jan. 1942

(1932–2018), Reinhard Kiock, Hanns-Juergen Lichtfuss, Hans Starken and Heinrich Weyer.

With respect to turbojet engine development and the possible involvement of large research organisations it has been criticised, that there was especially never a concerted effort to solve the combustion problem, or to collect fundamental know-how in this area to the advantage of all players. Correspondingly, a number of people were working on turbine blade cooling, but the work was largely uncoordinated and led in many cases to duplicating results. Broadly speaking, it seemed that whilst RLM were rigid in their control of the firms, they had—presumably with the rare exemption of *'nacelle aerodynamics'* above— too little to say in the work programmes at the research establishments.

Of the many possible charts to illustrate the German/British air warfare, Fig. 8.18 deals for 1941 with German and British aircraft losses in the Northern Area of the war, with two—after the well-known Luftwaffe defeat in the 1940 *'Battle of Britain'* and continuing British air superiority—remarkably different halves of the following year.[179] Fig. 8.19[180] illustrates the German down-turn towards war's end, especially by the Luftwaffe attrition, indicated by the early steady decline of flight training hours per pilot, in comparison to now combined RAF and AAF forces and corresponding material superiority after the official US

---

[179] See *Flight,* 8 Jan. 1942, p. 25. The publication of these, for the British side then rather disturbing data without any delay, is most noteworthy. The figures include the episode of a RAF Wing in Russia, but exclude the Middle East, and Luftwaffe aircraft shot down by the Royal Navy.

[180] See Schabel, Die Illusion, p. 198, from USSBS, The Defeat of the German Air Force (European Report, No. 59), p. 19, Fig. 8 (redrawn).

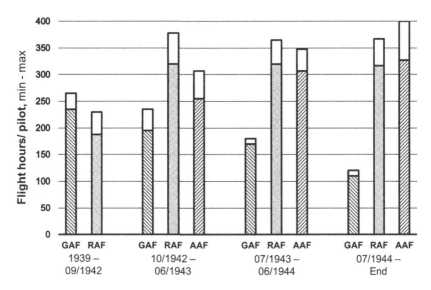

Fig. 8.19 Flight training of Luftwaffe (GAF), Royal Air Force (RAF) and US Army Air Force (AAF)

entry into World War II on 11 December 1941,[181] with predictable horrible consequences, which were only attenuated towards war's end by the inescapable fuel shortages.

The development of the innovative turbojet engines and their late appearance in the war theatre has put them in the emotional category of '*Wunderwaffen* (wonder weapons)', which in the perspective of some insiders have been thwarted by individual failure. Most critical were Luftwaffe engineers, who stated in their '*1944 Rechlin Memorandum*': '*If it would be possible to write a true story, an objective history of Luftwaffe technology since 1934, already present casual bystanders, not to mention later successors had to put the whole in the category of an insane phantasy or made-up satire. Nobody would seriously believe, that the reality kept so much shortcoming, bungling, obscurity, wrong empowerment, misjudgement of objective truth, and wilful ignorance of reasonable solutions.*'[182]

---

[181] See Flight, 16 March 1944, p. 289–291; correspondingly grew the '*total military AAF personnel*' between 1940–43,118 to 1944–2,385,000, the '*labour force*' in US engine plants between 1940–25,000 to Jan. 1944–361,000, and in US airframe plants between 1940–86,000 to Jan.1944–1,022,000.

[182] See Budrass, Die Mobilisierung, p. 299, requoting 'Command of Luftwaffe Test Centres, Side notes on the organisation of Luftwaffe technology, 20 May 1944, Bundesarchiv Militärarchiv (BA-MA) Freiburg, ZA 3/25. In German '*Würde man imstande sein, eine wahrheitsgetreue Darstellung, eine objektiv richtige Geschichte der Technik in der Luftwaffe seit 1934 zu schreiben, dann würde schon heute ein Unbeteiligter oder erst recht ein Nachfahre das Ganze für eine mit krankhafter Fantasie erfundene oder ersonnene Satire halten. Denn es würde niemand im Ernst annehmen, dass es in Wirklichkeit so viel Unzulänglichkeit, Pfuscherei, Verworrenheit, Macht am*

Latest after 1943 all activities of RLM and their responsible officers were highly focussed on the organisation of production capacity, production processes and a desperate fight for the accordingly required manpower, often through slave labour. Typical for the increasingly chaotic German procurement system was that each branch of the services organised its own needed output. Chapter 7 shows Helmut Schelp already in August 1943 at the Sachsenhausen-Oranienburg concentration camp to organise some forced labour production support—in the wake of the responsible General Engineer Franz Mahnke.

After Ernst Udet's suicide in November 1941, Erhard Milch became additionally *'Inspector General'* of the Luftwaffe, a post in which he was in charge of all aircraft production and supply during most of World War II. Between spring 1942 and summer 1944, he established at the RLM's large conference hall a system of regular Tuesday- and Friday-meetings with 70–80 attendants from the Ministry and industry. There were 190 meeting sessions of on average 4 ½ h length, of which 121 were recorded by professional, former Reichstag (parliament) stenographers, and in the meantime evaluated, especially with respect to the forced labour issue.[183] The minutes were widely distributed up to autumn 1942,[184] thereafter with severe restrictions only.

Milch had also set the tone in this documented struggle for work force by his indecisive comment on Tue 14 September 1943: *'Either KZ* (concentration camp inmates) *or proper Germans, ...whoever.'* And Franz Mahnke added: *'The best solution is and remains* "concentration camp"*; if we managed that, we would have certainty, that the installed machinery would be also occupied.'*[185] Speakers from industry copied and kept this verbal aggressiveness, surprisingly early, as *Auto-Union* director and influential head of industrial *'main engine committee'* Dr. William Werner (1893–1970)[186] on 18 August 1942: *'We help ourselves, catch people and machines, and make this happen with brutal force.'*

---

falschen Platz, Verkennung der objektiven Wahrheit und Vorbeilaufen an den vernünftigen Dingen insgesamt geben kann.'

[183] See Schmunk, Entweder KZ.

[184] On 31 Aug. 1942,—see Wikipedia, 'Harro Schulze-Boysen' in English, was arrested in his RLM office; he had led an espionage ring for the Soviet-Union, later to be known as *'Red Orchestra (Rote Kapelle)'.* He had had access to the Milch protocols, and as Milch jokingly stated in the minutes's continuation on 20 Oct. *1942* '... *had mixed up Luftwaffe Required and Actual Strength figures, so that my office appeared in Russia, but also in England and Amerika, more positive than in reality. Quite comfortable though, otherwise, I would have been ashamed by these laughable low numbers.'*

[185] In German *'Entweder KZ oder ordentliche Deutsche ...'*(Milch), and *'Die richtige Lösung ist und bleibt Konzentrationslager. ... Wenn wir das fertigbringen, hätten wir die Gewissheit, dass man die Maschinen dort aufbauen kann und an den Maschinen ein paar Leute stehen hat'* (Mahnke).

[186] See Wikipedia, 'William Werner' in German: *'Wir helfen uns selber, fangen uns die Menschen und Maschinen und hauen das durch.'*

Fig. 8.20 Police/Pölitz hydrogenation plant: Ruin (2007, l), Allied air raid (29 May 1944, m), orientation map (r)

On much lower level, the author heard his father Hans Eckardt (1921–2014) talking later about his *'adventures'* as young Luftwaffe Flak-Lieutenant on the retreating Italian front, the *'Gothic line'*[187] near Rimini, in November 1944. Their standard anti-aircraft gun *'Flak 18'*, in short the *'Eight-eight'*, of calibre 88 mm had the tendency to wear out the inner *'soul barrel'* already after some 900 shots. Wearing was most pronounced in the rear section of the 4.94 m long barrel, close to the breech, so that it was normally sufficient to exchange just this first ~1.65 m long section. Since ordnance supplies were close to breakdown, it was decided to organise the procurement in typical Luftwaffe self-help on battery level—by sending the author's father by train to the home base of Flak-Reg. 33 at Halle-Leuna.[188]

In May 1944, Allied air forces started with massive bombing raids on German fuel refineries and hydrogenation plants—which soon brought any meaningful Luftwaffe operation to a halt. On 12 May 1944, the eighth US Airforce flew attacks with 935 bombers to the largest plant at Leuna in mid-Germany and four other places, followed in the morning hours of 29 May 1944 by attacks on the second largest plant at Pölitz near the Baltic coast, Fig. 8.20, and five other sites. The Pölitz synthetic fuel plant was attacked by 224 Boeing B-24 s, of which eight were downed of combined Flak impact and during a subsequent aerial battle with Luftwaffe *Zerstoerergeschwader ZG 26* (destroyer squadron)[189] over the island

---

[187] See Wikipedia, 'Gothic Line' in English.

[188] Part of the successfully accomplished *'adventure'* was an unauthorised extension of the trip over the weekend, some 350 km further north, first by train in a permanently *'Occupied'* WC to Stettin/ Szczecin, followed by a 30 km march through the Pölitz/ Police ruins, and then by sailing across river Oder in a fisher boat to the completely surprised family, with the new-born son waiting, at Ganserin/ Gasierzyno, Fig. 8.20 (r). On the way back several barrel exchange sections were duly picked up at the regiment headquarters at Halle-Leuna, and carried back to the Italian front.

[189] See Wikipedia, 'Zerstörergeschwader 26' in English.

of Wollin, beyond the northern rim of the Fig. 8.20 (r) orientation map; these separated B-24 s were presumably on their way to neutral Sweden.[190]

Repeated air raids brought the production to a practical standstill within six months. Lack of fuel not only hampered the mobility of German ground forces, but especially dramatic were the operational restrictions for the new Me 262 turbojet fighters, so that the fuel industry was additionally stripped of substantial air protection. Consequently, the air defense of these critical sites lay mostly on the shoulders of heavy Flak units. For example, the 14th Flak Division, responsible for protecting Leuna had 28,000 troops—with 20,000 supporting, auxiliary personnel from '*RAD Reichsarbeitsdienst (Reich work service)*'. As the most heavily defended industrial target in Europe, a total of 6552 bomber sorties dropped 18,328 tons of bombs on Leuna; vice versa, Leuna bombing in 1944/1945 cost the eighth USAF1,280 airmen and in total 119 planes.

During the now rapid downturn of German warfare economy, the Luftwaffe '*Kriegstagebuch (War diaries)*'[191] tried to document unimpressed also the overwhelming figures of US arms output, e.g. in the week of 21 to 28 January 1945: '*Total number of aircraft produced in 1944: ~96,400, for January 1945: ~7000, including 1300–1350 four-engined a/c. Total number of B-29 'Super Fortress' up to end of January 1945: 2100. Aero engine production in January 1945: ~27,000, including approximately 400–500 turbojet engines. As estimated end of 1944, the total US aircraft production 1945 will presumably achieve 87,000 a/c ... For the English turbojet fighter a/c, there are only turbojet engine figures available for the output of RR Derby in November 1944 ~20–30, rising in December to ~50.*'

In power generation the severe shortage of coal supplies in 1944–1945 had unexpected, often overlooked consequences. Ever larger sections of the German grid were operated as frequency blocks, until in 1944 the number was reduced to two, namely, the *Central German* and the *Western German* frequency blocks. Towards the end of the war, frequency reduction had to be resorted to. The minimum reached was 43.3 Hz (c/s) in the Central German Block and about 41 Hz in the Western German block. Consequently, three phase electrical motors in industry ran at a lower speed, as this is directly related to line frequency and internal motor construction.

The final apocalypse cannot be easily put in words—nor visualised; perhaps Giffard's doctor father, and one of Britain's leading historians at present, David Edgerton,[192] found in a similar situation a kind of answer, when he picked for final illustration in his 2011

---

[190] By a strange coincidence, 2 h before the air raid culminated over Pölitz/Police on 29 May 1944, 8.30 am, the author was born in Ganserin, a small village of 500 inhabitants on the eastern shore of the Oder estuary, Fig. 8.20, orientation map (r). At ~9 am the author's grandfather cycled some 5 km to the next post office, to inform father Hans via telegram of the happy childbirth, when he was hit by a shell splinter—unhurt though, just the bicycle clip on the left trouser leg was sharply cut off.

[191] See https://www.cdvandt.org/ktb-chef-tlr.htm in German, with comments in English, covering in eight parts the period of 13 Dec. 1944 to 4 April 1945.

[192] See Wikipedia, 'David Edgerton (historian)' in English.

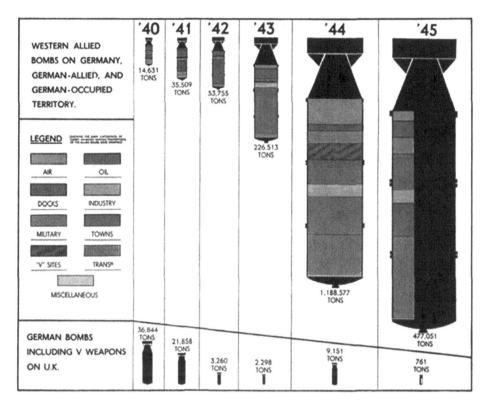

Fig. 8.21 *'Relative bombing weights'* © D. Edgerton

'Britain's War Machine',[193] a requote from Lord Tedder's 1947 'Air power in War', Fig. 8.21.

For a summary of this Sect. 8.1 *'German turbojet engine developments and related activities'*, 1940–1945, the author retreats to the position of a professional German historian, Lutz Budrass, and to a few corresponding foreign views, mostly put down shortly after war's end. Budrass closed his review of Hermione Giffard's book *'Making jet engines in World War II'*, which has been discussed at length in Sect. 8.1.1, by stating: *'German historians are not easily moved to . . . praise German science, technology and industry, not because Germany lost the war, but because those have been contaminated to their very core by the crimes of the Nazi regime. To always remind of the crimes whenever achievements are pointed out has become a specific art of German historians which has resulted in a number of methodological and theoretical innovations. By applying these innovations German historians have at least succeeded in doing away with those reminiscences of contemporaries like Albert Speer, Erhard Milch, Ernst Heinkel and others which have far too long dominated the discussion. We have also created a history*

---

[193] See Edgerton, Britain's War Machine.

of German science, technology and industry which is less shining and less glamorous, sometimes of a German thoroughness, but cautiously depicted and balanced. [194]

The foreign voices highlight the fact that the adoption of the new technology in Germany happened most remarkably under enforcing and precarious circumstances of war, while the plenty resources of the United States, and in its wake Great Britain and USSR shifted this change in the post-war period, ending before the Korean War. The chorus shall be opened by Joseph Ermenc[195] with his foreword to the 1965 interview with Helmut Schelp: *'The German fighter jet plane, the Messerschmitt 262, which appeared in German skies toward the end of 1944 was a fearful weapon. Its speed was over 500 miles per hour or around 100 miles per hour greater than the fastest U.S. or British fighter. One Me 262 on the average was able to shoot down 5 B-17 bombers. Indeed by February 1945 General Eisenhower was informed that if the ground forces didn't conquer Germany by June 1945, the production of jet and rocket planes by the Germans would make it impossible for the daytime bombing of Germany without unacceptable losses.'* [196]

Sir Roy Fedden (1885–1973) cast his thoughts after the noteworthy *'Fedden Mission'*[197] to Germany, June/July 1945, in the following short summary: *'Series production of jet engines in large quantities was undoubtedly in a more advanced state in Germany than in Britain and the USA, and had the war continued and had their factories not been overrun, they would have producing several thousand jet engines per months by this autumn [1945]. In the middle of 1946, the output would have been at a rate of 100,000 jet engines per annum, at least.'* [198]

The American technical historian Edward W. Constant, II, who coined for the developments of basically four pioneers, Frank Whittle of England, and the Germans Hans von Ohain, Herbert Wagner and Helmut Schelp, in the decisive 15 years 1928–1943 the term *'Turbojet Revolution'*, wrote in view of the German achievements: *'During the last desperate years of the war, the Germans put more different types and a larger number of turbojet-powered aircraft into service than any other nation. Germany also probably had the most comprehensive program for advanced turbojet development, both engine and airframe, of any of the powers.'* [199]

And finally, the ground-breaking, and as such, one of the first turbojet historians, Robert O. Schlaifer (1914–1994)[200] put his conclusions in view of an US post-war trauma in the sentences: *'The most serious inferiority in American aeronautical development which appeared during the Second World War was in the field of jet propulsion. Had the Germans*

---

[194] See Budrass, Review. The hun, p. 187.

[195] Joseph J. Ermenc (1912–2005), Professor Emeritus, Dartmouth College, NH.

[196] See Ermenc, Interviews, p. 97.

[197] See Sect. 10.1, and see—Wikipedia, 'Fedden Mission' in English.

[198] See Christopher, The race, p. 76.

[199] See Constant, The Origins, p. 208.

[200] See Wikipedia, 'Robert Schlaifer' in English.

Fig. 8.22 London Docks, 7 September 1940: general war impingement,—also of turbojet developments

put their jet fighters in production a year sooner, as they were technically perfectly able to do, or had the Allied campaign in Europe come a year later, the use of jet fighters by the Germans might have had a most serious effect on the course of the war. [201]

8.2 British Turbojet Engine Developments

8.2.1 A *Phoney War* on Turbojet Propulsion

While the real war reached England in full in autumn 1940—with massive *Luftwaffe* bombing raids, Fig. 8.22, and general deprivations, but specifically also negative impact on turbojet engine developments, the *phoney war* [202] continued on the publication front, on the German side with all likelihood by a misunderstanding.

Back in 1935, before Hans von Ohain joined Heinkel, he pushed the patenting of his ideas on a private basis with the support of a Berlin patent attorney. [203] Sole contact, reviewer, sounding board and sometimes even obstinate critic on behalf of the Reichspatentamt (Reich patent office) Berlin was all the time, up to the issuing of *'secret patents'* on Ohain's insistence in 1938, *Regierungsrat* (senior civil servant) Gerhard Gohlke. The patenting process had many unforeseen implications which led to Ohain's post-war confession, that presumably both his and Whittle's earlier turbojet patent might

---

[201] See Schlaifer, The Development, p. 321.

[202] See Wikipedia, 'Phoney War' in English.

[203] Patent attorneys Ernst and Carl Wiegand, for details, see Sect. 12.1.1.

have been not justified at all, had the patent searches on both sides been carried out more carefully, e.g. by producing Maxime Guillaume's FR534,801 on *'Propulseur par réaction sur l'air* (Air reaction propulsor)' with priority 3 May 1921. But, of course, the German patent office had to be blamed more seriously for missing Whittle's more recent GB347,206 *'Improvements relating the propulsion of aircraft and other vehicles'* with priority 16 January 1930.[204] It appears that Gohlke was conscience-smitten thereafter, which caused him to write up all relevant patented inventions in due course, which naturally excluded Ohain's *'secret patents'*. Under the uncommon German header *'Heizluftstrahltriebwerke* (Thermal-air jet engines)', he outlined this material in a series of four sequential articles of 21 pages in the internationally distributed magazine *'Flugsport* (Aero sports)' between 4 January and 15 February 1939. One can speculate that Gohlke wanted to address both an engineering and patenting audience, in Germany and abroad—mostly to finish the general lack of mutual information, and definitively not to claim a kind of German superiority in war technology.[205]

It appears that G. Geoffrey Smith, then editorial director of *FLIGHT (and Aircraft Engineer)*,[206] London's weekly aeronautical magazine, replied indirectly to Gohlke by a series of four articles under the headers *'Possibilities of Jet Propulsion'*, and *'Jet Propulsion of Aircraft, Parts II-IV'*[207] between 28 August 1941 and 9 October 1941, in which he repeated and commented 2/3 of Gohlke's patent collection, but added also in the concluding Part IV a comprehensive, well-founded 12-point *'Summary of Advantages of Jet Propulsion'*, and in the famous FLIGHT style a 3D graph of the new revolutionary propulsion system, Fig. 8.23 (l).[208] Keeping in mind, that Whittle' s radial *W.1* engine had its successful maiden flight on 15 May 1941 in the *Gloster E.28/39* aircraft, Fig. 8.24, it is remarkable how close the FLIGHT graph came to the German reality of that time.[209]

The sketched engine has some similarity to the all-axial Jumo 004 turbojet engine; the set-up of which was never published from the German side during the war, so that the FLIGHT graph may have contained as extra-feedback-irritation also some British intelligence information. After an eight(?)stage *'axial flow compressor'*—with provisions for *'turbine air cooling'*, the combustor consists of *'cylindrical combustion chambers'* (six for

---

[204] A German secret report 'Sondertriebwerke des Auslands (Foreign jet engines)', signed by Colonel-Engineer Dietrich Schwencke, Fig. 8.3, issued on 15 March 1944, lists three Whittle patents, his first one of 1930 (GB347,206), the dual-flow radial impeller of 1935 (CH188,758), and the twin-compound engine of 1936 (CH195,823). For the rermainder the German side assumed unpublished 'secret patents'.

[205] Actually the international demand for Gohlke's summary information was so high, that an English version of his article was launched immediately in Feb. 1942,—see Gohlke, Heizluftstrahltriebwerke.

[206] In 2019, now—*FlightGlobal* celebrated the 110th anniversary of *FLIGHT* magazine, still located at *Dorset House*, in the vicinity of Regent's Park.

[207] See FLIGHT, Vol. 40, 1941, pp. 115–117, (II) pp. 155–158, (III) pp. 190–201, (IV) pp. 239–242.

[208] See Smith, Gas Turbines, 3rd ed., p. 10.

[209] Heinkel He 280 turbojet aircraft, first flight, 30 March 1941, and Junkers Jumo 109–004 A turbojet engine, with a series of first test runs between 11 Oct. 1940 and end of Jan. 1941.

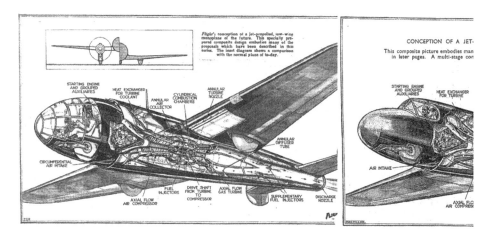

Fig. 8.23 *FLIGHT,* first turbojet illustrations, 9 Oct. 1941 by *MM* (l) and Smith's book, 3rd ed., April 1944, by *Max Millar* (r) ©FlightGlobal

Jumo 004), and also the Jumo 004 typical, sectioned *'drive shaft from turbine to compressor'* is visible. The depicted *'axial flow gas turbine'* has—different—clearly more than just the one, air-cooled Jumo 004 stage. But already the subsequent plug-cone nozzle and the row of *'supplementary fuel injectors'* homes in again to Junkers design considerations. The aircraft is characterised by the typical German framed cockpit design (e.g. of the *Me 264* long-range bomber), the *'circumferential air intake'* reappeared at war's end in the Messerschmitt *Me 1110/II* fighter project, Fig. 8.35, and elliptic wing tips and the twin-disk tail configuration could have been imagined at the Heinkel design offices. The inset along the upper rim of Fig. 8.23 (l) illustrates the specific turbojet advantage, the absence of an airscrew, which enables the aircraft to be of low built.[210]

Since December 1942, G.G. Smith used to summarise his special *Flight* publications on *'Gas Turbines and Jet Propulsion for Aircraft'* in a small, correspondingly entitled booklet,[211] which was already updated by a second edition in June 1943, followed by a third in April 1944. This 1944 edition contained also for the first time a copy of the 3D turbojet aircraft, but now for obvious reasons in an adapted, modernised and more English version. Fig. 8.23 (r) shows as most significant change a *Meteor*-type cockpit, now with asymmetric air intake at the bottom of the fuselage. In addition, the formerly half-hidden landing gear in the wings disappears now completely. The function of the *'heat*

---

[210] Schwencke's secret report of 1944 repeats Fig. 8.23 (l) with the predictable interpretation, that here with all likelihood a look alike of the actually coming GB/US turbojet aircrafts has been shown; in addition, the *'squirt'* engine of Fig. 8.24 (l) has been reproduced, with the expectation that these new aircrafts will be seen in the European war theatre not before autumn 1944. The announcement of 6 Jan. 1944 (below) of joint UK/ US turbojet developments was equally confirmed.

[211] See Smith, Gas Turbines.

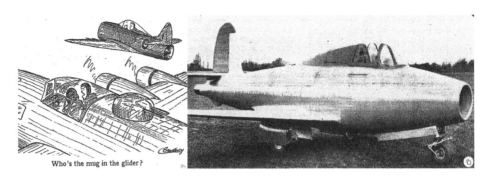

Who's the mug in the glider?

Fig. 8.24 *FLIGHT*, first English turbojet illustrations: *'squirt'* cartoon, 27 Jan. 1944 (l), *Gloster E.28/39*, 5 Oct. 1944 (r) ©FlightGlobal

exchanger' is no longer necessary for turbine air-cooling, and the *'supplementary fuel injectors'* for an indicated afterburner have been eliminated, as well as the *Heinkel* association of the tail configuration, replaced by a conventional central fin arrangement. And, not the least, famous *Flight* artist *Max Millar* dared again to put his full name in the lower left corner of the frame, where in 1941 stood just his initials *MM*.

On 13 January 1944, *Flight* opened under the header *'The Whittle-Carter Combination'*: *'The most jealously guarded secret in modern aviation history, and at the same time the best-known, was suddenly revealed on January sixth when a joint statement was issued by the War Department, Washington, on behalf of the Royal Air Force and the U.S. Army Air Forces, mentioning for the first time the existence of the jet-propelled Gloster monoplane... That the announcement should be made just now, and that it should be a joint statement by America and Great Britain, is just and proper. The development work has been to a considerable extent a co-operative affair, and the interchange of information gained on both sides of the Atlantic cannot but hasten the development of this form of propulsion. But it is equally just and proper that it should be fully realised that British inventiveness and genius, on the one hand, and British official appreciation of the possibilities, on the other, should be given the credit for the pioneer work done by Great Britain in this field of endeavour, which is so fraught with startling progress ... The two men to whom, first of all, should go the credit are Group Captain Frank Whittle, who began work on his schemes in 1933, and Mr. W.G. Carter, Gloster's chief designer, who produced the aircraft for the Whittle power unit. Both were successful in their way, and the time taken to get the aircraft into the air was remarkably short, considering all the novel features which had to be studied.'*

Shortly thereafter on 27 January 1944, the Gloster *'Squirt'*, what became officially the *E.28/39*, was released—first as a cartoon, Fig. 8.24 (l), and only the *Flight* issue of 5 October 1944 could present the one and only official photograph of that aircraft, Fig. 8.24 (r)—with the caption: *'OUT OF ITS INFANCY: The first British jet-propelled aircraft which first flew as long ago as May 15th, 1941. The jet power unit was designed by Group Captain (now Air Commodore) Whittle and built by Power Jets Ltd. Design and construction of the airframe was by the Gloster Aircraft Co. Ltd, and their chief test pilot, the late Flt.-Lt. P.E.G. Sayer, carried out all the original flying tests.'*

Six weeks after the unveiling of the Gloster E.28/39 turbojet aircraft, *Flight* was also in a position to inform their readers, disappointingly, *'that jet propulsion was* (not) *an exclusively British invention which had put the Allies a very big jump ahead of the enemy. It must have come as something of a shock at many a British and American breakfast table, therefore, when news of German jet-propelled fighters in action on the Western front was first mentioned some little time ago.'*[212] Together with the shadow silhouette and 3D artist's impressions of the dual-turbojet-powered Messerschmitt Me 262, also the similar Heinkel He 280 and the short range, rocket-powered interceptor Messerschmitt Me 163 were presented as *'three different types of German "jet fighters", which made their appearance during the past summer ... Actually there were several German jet-propelled prototypes flying as long ago as 1942, and it is now believed that the famous German "ace" of the 1914–1918 war, Ernst Udet, was test-flying a "squirt job" when he met his death in 1941.'*[213]

During wartime, *Flight's* reporting was tightly constrained by the UK Ministry of Information, which hindered especially the ability to cover the development and manufacture of new combat aircraft. But also trivial *'human weaknesses'* prevented many a journalistic scoop. As the 1940s drew to a close, turbine technology was central to one of the decade's biggest aviation stories—the maiden flight of the world's first jet airliner on 27 July 1949. But the momentous event was missed by all London air correspondents, including *Flight's* editor. The de Havilland press officer, unaware of the imminent first flight, sent the gathered press corps home from Hatfield early.[214]

8.2.2   RAE: From Metrovick to Rolls-Royce and Armstrong Siddeley

The turning of the decade from 1939/1940, as described in continuation of Sect. 6.3.3, marked a technological break for the turbojet engine developments at RAE Farnborough. For a long time, Alan Griffith and the RAE team led by Hayne Constant failed to see the turbojet's capacity to use excess air ingestion to increase mass flow and thrust, and to reduce temperature, which would overcome material temperature limitations, and more significantly, would boost propulsive efficiency, especially at high aircraft speeds. Apparently, their strive for turbo-prop performance in comparison to piston engines limited their scope to established parameters like shaft output, aircraft speed, engine weight and specific

---

[212] See *Flight*, 16 Nov. 1944, pp. 526–528.

[213] See Wikipedia, 'Ernst Udet' in English. Udet, factually Chief of Procurement and Supply for the *Luftwaffe*, committed suicide on 17 Nov. 1941 by shooting himself; causes might have been the newly started *'Operation Barbarossa'* against the Soviet Union in summer 1941, and related issues with the *Luftwaffe's* needs for equipment outstripping Germany's production capacity.

[214] After FlightGlobal, 24 May 2019, *'The 1940s—Flight goes to war.'*

fuel consumption. Therefore, they did not recognise that a turbojet could be an efficient propulsor even though individual compressor and turbine component efficiencies were well below those necessary for a practicable turboprop engine.

In mid-1939 the indicated *'break'* at RAE coincided with A.A. Griffith's leave to Rolls-Royce, which had become interested in aero gas turbines some time before, but had no one in its staff available for GT studies, until Griffith was headhunted from Farnborough to Derby. He then resumed studies of his counter-rotating contra-flow engine, which steadily had been phased out at RAE, and which will be further reviewed in the context of Fig. 8.28. The first RR intention then was to use it for a ducted fan, though in principle the basic unit was capable of being used both as turbojet and turboprop. Griffith's first 1½ years at Rolls-Royce had been used for preliminary calculations and some combustion testing. After starting with an extremely small workforce, all in a sudden in spring 1941 the project received extraordinarily high priority up to October 1941, when the rig was run on compressed air. However, running under own power was delayed until 1943, when a suitable combustion system was finally available.

As outlined already in more detail in Sect. 6.3.1, David R. Pye, Fig. 6.31 (r), Director of Scientific Research at the Air Ministry, visited Power Jets's new premises at Ladywood on Fri, 30 June 1939, and witnessed a test run of the engine up to 16,000 rpm. Pye, who had been rather sceptical before, was now convinced that Whittle *'had the basis of an aero-engine'*,—and it was clearly recommendable to follow Whittle's turbojet example,—and to give up RAE's turboprop developments. Since Power Jets had no capacity, therefore, by July 1939 Metropolitan-Vickers Electrical Company, in short *MV* or *Metrovick,* at Old Trafford, Manchester, had become involved with the RAE's turbojet project instead.[215]

It is exactly then, that RAE's axial compressor designs made the decisive strides to leap-frog the initial compressor design basis laid by *Brown Boveri.* In continuation of Fig. 4.24, and the plot of average *Stage Pressure Ratios SPR* over time, Fig. 8.25 indicates now this basis of Brown Boveri's *industrial* compressor design experience by the elliptic area with $SPR < 1.1$. The 21-stage industrial compressor delivered to RAE,[216] and seriously tested at Farnborough after 1 August 1941 confirmed this technology level with $SPR \sim 1.075$— which at that time was clearly shared by RAE's nine-stage *'B.10, Betty'* and the following 17-stage *'D.11, Doris'*, in Fig. 8.25 marked as #7 and #9.

As the RAE was not equipped for large-scale manufacture, it was arranged that detail design and manufacture should be carried out at Metrovick. The responsible MV chief engineer was Karl Baumann, who appointed David Smith to lead the MV design, development and manufacturing team; work started at the company the following year under an Air Ministry contract. The experimental non-flight engine *B.10* proved successful when tested

---

[215] As outlined in the context of Fig. 6.38, a foregoing contact between RAE and Metrovick on axial superchargers had been initiated by Pye as early as 3 June 1937,—with a serious working start not later than 13 Dec. 1937.

[216] See Fig. 4.24, #9.

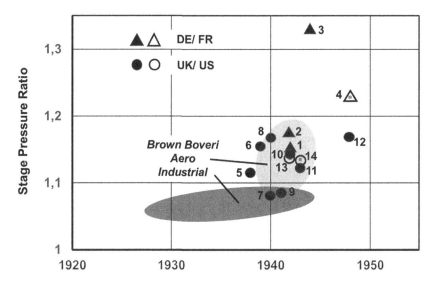

Fig. 8.25 Development History of Aero Axial Compressor Technology: Stage pressure ratio over time—for 14 DE/FR and UK/US prototypes, as tabulated:

(A) German turbojet engines (DE)

1. 8-stage, Junkers Jumo 004-A, 1942

2. 7-stage, BMW 003 A-1, 1942

3. 4 + 1 diagonal stage, Heinkel HeS 011 A-0, 1944

(B) French turbojet engines (FR, German derivative)

4. 7-stage, Snecma ATAR 101 V1, 1948

(C) British turbojet engines (UK)

5. 8-stage, 'Anne', RAE exp. compressor, 1938

6. 6-stage, 'Ruth', RAE exp. compressor, 1939/40

7. 9-stage, 'B.10, Betty', RAE exp. compressor, 1940

8. 9-stage, 'Freda', RAE exp. compressor, 1940

9. 17-stage, 'D.11, Doris', RAE exp. compressor, 1941

10. 9-stage, 'F.2', RAE/Metrovick *F.2* compressor, 1942

11. 14-stage, 'Sarah', RAE/Metrovick compressor, 1943

12. 12-stage, 'Avon', RR *Avon* compressor, 1948

(D) US turbojet engines (US)

13. 14-stage, 'TG-100', GE compressor (based on BBC data), 1942

14. 11-stage, 'TG-180', GE *J35* compressor, 1943

at MV in October 1940, first tested as separate compressor and turbine sections using steam to power them. In October 1940 they were run as a single complete engine for the first time. During testing it was decided that water cooling was not needed, and was replaced by an air cooling system, and the turbine was allowed to run red hot at 675 °C. Experiments with *Betty* convinced the team that any sort of piping between sections led to unacceptable losses, so the *'dispersed compound engine'* concept, Fig. 6.41, was left as too inefficient. At the same time, it was decided that pressure ratios on the order of PR 5 would be sufficient for near-term engines, so it was decided to abandon the two-spool approach for the time being. During construction, several high-temperature alloys had become available with creep strength up to 700 °C, and Constant demonstrated in a new report,[217] that these materials would allow to produce a turboprop that would outperform existing piston engines. The new *'D.11 Doris'* design with 2000 hp turboprop target power, consisted of an enlarged *B.10-like* 17-stage compressor, driven by an eight-stage turbine section, and a mechanically separate fivre-stage lp propeller turbine. However, in view of Whittle's centrifugal turbojet successes of the time, the axial *Doris* turboprop concept was soon to be considered as outdated, and slow progress delayed a difficult testing phase up to 1941.

In the meantime, RAE inaugurated the new cooperation with Metrovick, which had accelerated the turn to the *'pure turbojet'*, by designing a new nine-stage axial compressor designated *'Freda'*, for a mass flow of 22.7 kg/s, a total pressure ratio PR 4 at 7390 rpm, and a tip diameter of 0.563 m.[218] In December 1939 the Freda compressor was fitted with a turbine section to become the first self-running axial turbojet in England, the *F.1* with 975 kp thrust, immediately followed by a slightly larger *F.1A* with 25% more mass flow and 1220 kp thrust. In a strive for a near-term production design, the F.1A was turned over to Metrovick in July 1940, and a production effort started there as the *F.2*. After some intermediate steps the next salient programme milestone was achieved when the first all-axial flight engine Metro-politan-Vickers F.2 ran in a test cell during December 1941. In Fig. 8.25, the black UK bullets of #8 for 'Freda' and #10 for the F.2 engine compressor clearly mark the achieved, considerable jump in stage pressure ratios, which nevertheless remain in the shaded range of Brown Boveri's *aero* compressor developments.

With respect to the achieved performances of the various RAE compressors, a late confession from Hayne Constant in 1958 is remarkable: *'In the early days, following the lead given by Griffith, a number of compressors had been designed relying on a combination of pure theory and divine inspiration. This seemed to work well (Betty) until we strayed into the region of compressibility, when one or the other of the ingredients let us down. We had learnt quite early on the vital importance of the cascade test for providing fundamental information on which all compressor designs have come to be founded.'*[219] But, as Bailey highlights correctly, *'It was not until 1944 that a high-speed wind tunnel was available for studying compressibility effects in compressor cascades.'*[220] This lack of appreciation of an important development tool of the late 1920s, the cascade wind tunnel, can be blamed on both competing sides in Germany and England; it was a major setback to the development of the axial compressor in this early critical period.

---

[217] See Constant, The Internal Combustion Turbine.

[218] See Kay, Turbojet, Vol. 1, p. 19. As a kind of forerunner to higher stage pressure ratios (but still unsatisfactory performance) as finally achieved with *'Freda'*; *'Ruth'* #6 in Fig. 8.25, was designed with a higher pressure rise per stage in 1939, and tested in 1940 only.

[219] See Constant, Pyestock's contribution.

[220] See Bailey, The Early Development, p. 13.

Fig. 8.26 Karl Baumann (l), David M. Smith (m) and Robert R. Whyte (r), with Metrovick F.2 design team, ~ summer 1945 ©SIM Archive

The successful development of the Metrovick F.2 turbojet engine continued by a Gloster F.9/40 Meteor prototype aircraft test with two F.2 underslung nacelle installations at RAE on 13 November 1943. Development of the F.2 ended in 1944,[221] but continued in a number of follow-up versions, which finally led to the considerably larger *F.9 Sapphire* at ASM Armstrong Siddeley Motors. Presumably, Fig. 8.26 was taken as a kind of farewell greeting on the occasion of the third F.2 prototype being transferred from MV premises to Farnborough in September 1945, after 210 running hours, of which 106 h were accomplished successfully in an *Avro 683 Lancaster* flying test bed.[222]

Before leaving Fig. 8.25, it is worthwhile to observe the three German turbojet engines #1—#3 (depicted by black triangles), shown there for comparison, where the stage pressure ratios of both Jumo 004 and BMW 003 represent very well the then state-of-the-art of axial

---

[221] See Schlaifer, Development of Aircraft Engines, p. 430/431, has interesting information on the achieved technology standard: '*In November 1943, about the time when the design of the* (Jumo) *004B was fixed for production, the F-2 weighed 1510 lb and was cleared for experimental flight at 1800 lb thrust. The specific weight of the F-2 was thus 0.84 lb/lbf, or a shade more than the 0.83 of the 004B.*' and '. . . *it must be remembered that the design of the compressor of the 004 was virtually fixed before the end of 1941 . . . For this . . . reason it is scarcely fair to compare the 004 compressor with the axial compressor of the F2, which delivered a pressure ratio of 4:1 with 90% efficiency by the end of 1943.*' Comparable values were only achieved by H. Reuter's '*Hermso*' compressors out of BBC Mannheim in 1944–1945, see Sect. 9.1.

[222] See Kay, Turbojet, Vol. 1, p. 47.

compressor design, while the *'4 + 1 diagonal'* stages of the Heinkel HeS 011 reach by far the best value in justification of the complex concept selection (besides the outlined drawbacks). For French turbojet engines of the time, just the impressive seven-stage Snecma ATAR 101 V1 compressor of 1948 has been marked—with an open triangle #4, in association to its BMW heritage. The two *'open US bullets'* #13 and #14 stand for the two General Electric compressors of that wartime period. Finally, the good though not outstanding 1948 value of #12 for the 12-stage Rolls-Royce Avon compressor appears to reflect—especially in comparison to the 1940 achievement for #8 *'Freda'*—some of the conservative principles of A.A. Griffith as responsible designer.

David M. Smith (1900–1986) had already been portrayed in Fig. 6.40 (r); although a steam engine engineer the achievement for which Smith will be best remembered will be his key role in the development of the first British axial flow aircraft jet engine. Though close contacts continued between Metropolitan Vickers and the RAE, from 1940 onwards MV produced its own schemes, rather than relying on RAE ideas alone. During 1937–1940 Smith had built up an excellent team to design, develop and test gas turbine engines, Fig. 8.26. His leadership then and for the next seven years entitles him, according to his Royal Society biographer Sir Owen Saunders, *'to a place of high honour in the early development of axial flow jet propulsion engines in Britain.'*

British technical historians appear to be of split opinion who merits the crown for opening the path to the simple, pure axial turbojet configuration, classically represented by the Metrovick F.2. For Antony Kay, F.2 *'was developed under overall responsibility and the technical direction of Karl Baumann'*,[223] but for Peter Lloyd in the RAeS obituary for Hayne Constant in April 1968, the F.2 was *'the first axial engine which he had inspired and was responsible for.'*[224] Perhaps Hayne Constant, who certainly had lifelong interest in the laurels of F.2 design as well, could have the last word. In 1942, when the F.2 achieved a successful 24 h endurance test run, surprisingly, he was still in favour of Griffith's contra-flow concept: *'Both Griffith and myself have believed all along that the contra-flow represented the ideal form of gas turbine so far as we could visualise it. Anything else we embarked on was no more than an insurance or temporary expedient while the contra-flow was under development'.*[225]

During 1941 Metrovick began work on a turbofan version of the F.2 engine, with the aim of creating an engine more economical for slower, long-range flight. Karl Baumann, Figs. 6.40 (m) and Fig. 8.26, consulted with the RAE and then commenced the design work on this so-called thrust *Augmentor*, designated *F.3*. Baumann's F.3 scheme was to take an F.2/2[226] turbojet and add, behind the two-stage turbine, another four-stage, contra-rotating

---

[223] See Kay, Turbojet, Vol. 1, p. 46.

[224] See Lloyd, Hayne Constant.

[225] See Constant, Influence of the R.A.E. Reference and quotation, courtesy of Andrew Nahum, 12 Oct. 2021. The continuation of Griffith's contra-flow design story at Rolls-Royce will be addressed in the next Sect. 8.2.3.

[226] F.2/2 was a redesigned F.2, 1090 kp thrust version of August 1942.

turbine, which was connected to a two-stage, contra-rotating fan. The result was to produce a total jet efflux having a much larger mass flow, and much lower velocity than the original jet; the engine had a much higher propulsive efficiency at medium flight speeds, and fuel consumption was greatly reduced. Though it had a good performance, the F.3 was before its time and did not catch on. Since 3 October, 1941, business-related Baumann belonged to the Gas Turbine Collaboration Committee (GTCC), which Harold Roxbee Cox, since 1965 Baron Kings Norton, had founded and chaired, helping to pool industrial ideas and experience amongst what had become eleven participating companies in the meantime.[227]

Apart from the MV Gas Turbine Department at Barton, north of Trafford Park, work on compressors was undertaken at the Trafford Works, where the necessary great power and speed variation was provided by geared steam turbine drives. Of great importance was the work carried by MV in collaboration with *High Duty Alloys* from Slough in the west of London, from 1940 in developing a process to mass-produce precision-forged, light alloy compressor blades. By 1943 this process had been perfected and a major step towards the mass production of the turbojet had been taken. Responsible on behalf of Metrovick had been Robert R. Whyte, Fig. 8.26, an engineer who was with the Company during the whole time that they made gas turbines, and was responsible for building them. He became Superintendent of the Gas Turbine Department in the late 1940s, and eventually became responsible for building turbines of all types at Trafford Park. In addition, he was a member of the Manufacturing Sub-Committee of the Gas Turbine Collaboration Committee, set up in 1941 by the Ministry of Aircraft Production.

Besides Metrovick, *Armstrong Siddeley Motors (ASM)* at Coventry became a second tier British aero-engine firm, interested to participate in the newly emerging field of aero gas turbine engines. The early technical history of Armstrong Siddeley Motors has already been described in Sect. 6.1.1, and as part of Fritz Heppner's CV up to the sudden release of the famous car and piston-engine designer Stewart Tresilian in the personal excursion, Chap. 5. On 3 October 1941, ASM agreed with the Ministry of Aircraft Production (MAP) that it would abandon its current development programmes of piston engines, especially after a series of German air raids on the ASM Parkside plant had destroyed the corresponding *Deerhound* and *Wolfhound* development and test facilities in the months before. In due course, MAP counselled the firm to focus its efforts instead on the new turbojet engines, and recommended for starting a collaboration with Metropolitan-Vickers, and the planned Metrovick F.2 turbojet engine series production. The logic of MAP was that ASM needed to sharpen up its development practices. ASM would learn about the new turbojet engine concept and MV benefit from ASM's expertise in lighter weight structures

---

[227]After the nationalisation of Whittle's Power Jets in 1944, making Roxbee both Chairman and Managing Director, Karl Baumann participated from 28 April 1944 onwards in GTTACC, the Gas Turbine Technical Advisory and Co-Ordinating Committee. His patent record comprises 81(!) grants for assignees like British Westinghouse, Westinghouse Electric, Vickers Electric, and in his own name.

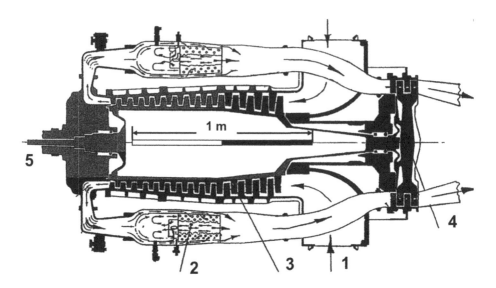

Fig. 8.27 Armstrong Siddeley *ASX* turbojet scheme, 1270 kp, 1944: 1 air inlet, 2 combustion chambers (11), 3 14-stage axial compressor, 4 two-stage turbine, 5 turboprop drive (optional)

and production techniques necessary in the aero industry; in this way it was hoped the development of the F.2 could be accelerated, as it was promising but overweight.

Up to this point, there appears to be agreement, thereafter commences *'mined territory'* with dominant perspectives of interest groups. In addition, an ASM-internal management conflict between old piston-engine rule with a moderate approach to future turbojet designs and a rather radical switch to the new technology overshadowed the *'transition year'* after mid-1941. To simplify the indicated conflict narrative, the personalised story of chief engineer Stewart Tresilian and the German *'newcomer'* Fritz/Fred Heppner has been factored out for the moment, and will be newly resumed in Sect. 8.2.4.

The strategic F.2-collaboration with Metrovick did not work out as ASM management and MAP might have hoped, when the actual work shares were negotiated. Metrovick simply insisted to keep the position of a *'design lead company'*, so that ASM was not allowed *'to undertake any redesign or modification work'*.

The RAE continued working on axial compressor design after the F.2 success. The original *'Freda'* compressor was later enlarged into 'Sarah', #11 in Fig. 8.25, with the addition of a further five low-pressure stages as part of a collaboration with Armstrong Siddeley Motors (ASM), and eventually became the *ASX,* Fig. 8.27. It took until mid-1942 for the MAP to persuade ASM to begin with a simple axial flow engine. The RAE volunteered to help ASM design an engine similar to the F.2 but incorporating their latest thinking. However, none of the RAE designs would go on to be a success on their own. The F.2 design was not put into production, although an enlarged version F.9 was very successful as the ASM Sapphire.

The engine's eleven combustion chambers were designed and manufactured by Joseph Lucas Ltd, but used a different system of vaporising fuel burners, designed by Sidney Allen (1909–1973), rather than the atomising fuel burners used by other British engine manufacturers, an unique ASM design feature.[228] In order to accommodate particularly long combustion chambers, the firm chose a reverse-flow arrangement for their compressor, Fig. 8.27, so that the air flowed through the compressor from back to front—an arrangement that caused serious aerodynamic losses in the engine. In October 1942, perhaps after lobbying from ASM engineers, including Pat Lindsey who was a key figure in the firm's turbojet work, the RAE approved the company's engine design and suggested to MAP that an order for six jet engines and spares be placed with ASM on the condition that *'technical supervision of the detail design by RAE should be stipulated'*; the contract was granted on 7 November 1942. The first ASX unit ran on a converted Deerhound test bed on 22 April 1943—a impressive record. In late summer 1943, MAP gave ASM a second contract to begin research on the *ASH*, a Heppner aero gas turbine design in a considerably higher power class, carrying his ennobling name letter, discussed in Sect. 8.2.4, and from MAP's perspective with many *'untried features'*.

Instead of pursuing a completely new engine design after the single, experimental ASX that it built, ASM decided to use the turbojet unit as the basis for a turboprop, the *ASP*. By 20 April 1945, the company had completed 22 h of test running on the new engine. The ASP was developed into the *Python* under Ministry contract, and in October 1945, it recorded the highest performance of any turboprop in Britain: 3656 shp plus an additional 500 kp of jet thrust. Unlike the Metrovick F.2, the Python was put into limited military service after the war, a striking success for a firm that had had no experience with axial gas turbine engine design before 1942.

8.2.3 Alan Griffith at Rolls-Royce: A Mediocre Balance Up To 1945

'A.A. Griffith and Early British Activities Towards Aero Gas Turbines' had already been addressed at length in Sect. 4.2. In 1929, much encouraged by the test results from his turbo-compressor rig, Griffith carried out a more extensive assessment of the gas turbine as an aircraft engine.[229] He had shown that high blade performance could be obtainable at the *'design point'* of the machine, where the blade angles and airflow are designed to be mutually matched. He now went on to consider how an engine with a multi-stage axial

---

[228] There are inherent problems with high-pressure atomising, but with the ASM system the fuel is sprayed at low pressure into a *'walking stick'* 180 degree curved tube where it vaporises, to give near-perfect burning at all fuel flows, which can vary considerably between sea-level take off and high altitude flight idle. This system was later applied to the Viper and Sapphire series of engines and was licensed to Westinghouse and Curtiss-Wright in the United States. It was used on the RR Pegasus, M.45 and Viper, and was applied to overcome the smoke problem on the Olympus 593 for Concorde.

[229] See Griffith, The Present Position.

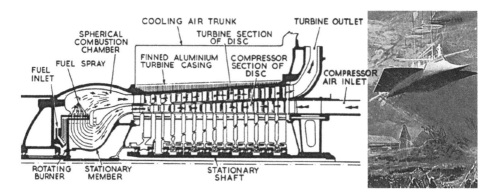

Fig. 8.28   Alan A. Griffith: Rolls-Royce *C.R. 1* contra-flow engine, 14-stage hp compressor/turbine section ~1943 (l), George Griffith: air ship *'Ithuriel'*, illustration by Fred T. Jane, 1893 (r)

compressor would behave at *'off design conditions'*—as the engine speed was dynamically varied, for example. Faced with an uncertainty like this, Griffith felt fundamentally challenged and bent his mind to inventing a way round.[230] This was his *Contra-Flow* engine concept, of which Fig. 8.28 (l) illustrates the 14-stage high-pressure section of the Rolls-Royce C.R.1 engine of the early 1940s. In Fig. 4.28 (r) already one test wheel of a nine-stage rig had been depicted, which was tested by Baxter and Smith at RAE Farnborough in 1940.[231] This experimental evidence clearly showed that this originally considered *'ingenious scheme of Griffith'* was by no means the right way to go. [232] *'There were severe difficulties in minimising leakage across the many running seals between the stages, and the contra-flow arrangement awkwardly constrained the engine layout. Overall, Griffith's solution turned out to be worse than the problem! But he did continue to put his faith in it for a long time,'* as Frank Armstrong summarised this turbojet development episode.[233] Indeed, what had been brilliant and surprising in 1929, must have been recognised as a folly dead-end in 1942, which points to a considerable lack of flexibility and a certain hardening in Griffith's character.

---

[230] As described in the context of Fig. 4.28, Griffith's 1929 patent of the typical, radially stapled compressor/turbine blade had forerunners from Stolze-Barkow, 1905 and Mérigoux, 1910.

[231] See Baxter, Contra-Flow Turbo-Compressor Tests. Acc. to—see Hodgson, Armstrong Siddeley, p. 6, ASM received from RAE the manufacturing order for this CR rig under the designation *'C.6* (some-times C.5)' in 1938 to produce the *'equivalent'* of 200 hp. Supervised by M. Cutler, the rig was ready for RAE testing end of 1939, and is now exhibited at Science Museum, London. Information, courtesy to A. Nahum, 11 Oct. 2021. Foregoing experiences with this contra-flow design might have alleviated Heppner's corresponding design adaptations as in the ASH engine concept, Fig. 8.31.

[232] As said already in Sect. 4.2.2, Baxter and Smith described the fundamental sealing problems between turbine and adjacent compressor rows only *'between the lines'*, with nevertheless a positive general outlook on their master's concept.

[233] See Armstrong, Farnborough, p. 25–26.

There can be at least three deep crises identified in Griffith's vita, first—when he had to give up certainly with some teeth-grinding his early scientific field of *'fracture mechanics'* after the *'soap film affair'* with Prandtl, Sect. 4.2.1, which resulted in a certain doggedness in following the Goettingen axial compressor tests by Betz in the late 1920s—in officially *'unpublished internal RAE reports'*, instead in the form of a free and open scientific exchange.

The second, apparently severe defeat from Griffith's perspective must have been the assessment of the ARC Engine Sub-Committee under Tizard in April 1930, that Griffith's input had not proven *'the superiority of the* (gas) *turbine with respect to the reciprocating engine'*, after Griffith optimistically had suggested the study of a 500 hp engine, using the contra-flow principle, with superior fuel economy, weight and power at altitude. Importantly, he also proposed the construction of an experimental 14-stage contra-flow unit similar to Fig. 8.28 (l), which could later be used as part of a complete engine. However, though the panel had not recommended a major commitment of embarking on an engine development programme, they by no means rejected Griffith's ideas entirely but made positive recommendations for experimental research at RAE. *'It is therefore surprising that although Griffith returned to RAE in 1931 and became responsible for engine research, these recommendations were not acted upon and no further RAE experimental research on gas turbine powerplants was done for more than six years!'* According to Armstrong, *'it remains uncertain as to whether Griffith's personal disappointment was the decisive reason for the lack of action, or whether other factors were also involved.'* [234]

Thirdly, it appears that Griffith—though officially appointed as Head of the RAE Engine Department in 1938,—left already on 1 June 1939 to Rolls-Royce, Derby, somewhat frustrated that Hayne Constant, now responsible for all RAE gas turbine work, had pushed CR flow concepts more and more aside. Griffith had been hired by the Chairman of Rolls-Royce Ltd. Ernest W. Hives (1886–1965) [235] as simple *'research engineer'* with the sole task *'to go on thinking'*, directly reporting to Hives and working mainly on the initiation and prosecution of aero engine research. To provide ideal conditions for quiet thinking, Griffith was given a *'pleasant room'* in the Rolls-Royce guest house, in vicinity of Hives private home *'Hazeldene'* at Duffield, a village some 8 km from Derby.

At Rolls-Royce Griffith apparently used the given freedom of research to devote himself to a continuation of his contra-flow compressor studies, and induced Rolls-Royce in 1942, according to Rubbra to build the 14-stage hp compressor/turbine stages, Fig. 8.28 (l), of a huge *C.R.1 contra-flow engine*, which would have been followed by six low pressure stages driving ducted fans. [236] The stages rotated freely on a fixed shaft, each in the opposite

---

[234] See Armstrong, Farnborough, p. 28.

[235] Thanks to Hives, no less than a hundred and sixty thousand *RR Merlins* were produced by 1945.

[236] In preparation, the giant *C.R.1* engine had been shown also as 3D perspective during a *'secret'* visit of King George VI and Queen Elizabeth at the Rolls-Royce Derby Plant, Nightingale Road, on 8 Aug. 1940; this took place at the height of the *Battle of Britain* and the Rolls-Royce production lines were working flat-out to supply *Merlin* engines powering the *Spitfires* and *Hurricanes*,—see Eyre, 50 years.

direction to its neighbours; there were no stator rows and therefore the rotational speed was low. Radial and thrust bearings were provided for each stage disk which carried at its periphery a ring of compressor blades. A shroud ring was formed in segments at the root and tip of the compressor blades; on the outer shroud, each blade carried an integral turbine blade. Hence, each compressor stage was driven by the turbine blades which it carried at its tip.[237]

The first test run of the 14-stage high-pressure unit was on 3rd March 1942. The outcome was very similar to that documented by Baxter and Smith for the 1940 RAE tests: The investigated 14-stage rig proved difficult to start, and test running showed that a very great deal of development would be needed to approach the design performance. The fact that the turbine and compressor blades were indivisible would add to the cost of testing alternatives and methods of blade manufacture were in their early stages. Accurate profiling of aerodynamic shapes had not been developed and the blade form in the test engines had been seriously modified to suit available milling cutters, thus jeopardising their performance and inducing premature surging. The compressor had a much lower efficiency than Griffith had predicted on the basis of accurate blade profiling and effective interstage sealing. Work was continued for a while, but eventually the success being achieved elsewhere with axial-flow compressors of conventional layout caused the ARC to recommend the discontinuance of these RR research activities. Before this disruption actually occurred, already in 1943 after an early cooperation between Rolls-Royce and Power Jets on the WR.1 engine was cancelled, Derby began work on an engine with three concentric shafts, designated *RCA3*, for *Rolls-Royce Compound Axial three-spool* engine. This three-spool engine concept certainly reflected to a moderate extent Griffith's idea of diminishing the danger of compressor stall by individually self-adjusting rotor speeds—with high-, intermediate- and low-pressure axial compressors on independent shafts and driven by separate turbines. This move prepared not only the end of the C.R.1 concept, but laid also the ground to Rolls-Royce characteristic, exclusively used three-spool engine configurations up to the present *RR Trent* engine.

Although Griffith had stuck to his contra-flow concept until about 1944,[238] he slowly came around and submitted in early 1945, at a time when Heppner had left ASM to join RR Derby, a slightly less complex engine scheme, the C.R.2. The new design gave up the contra-flow feature, but nevertheless called for a four-stage ducted fan and a 20-row counter-rotating axial compressor, driven by an eight-stage turbine.[239] After Heppner had followed Griffith in design complexity for a while, it appears that Heppner's patented idea of a counter-rotating *'high-speed propulsion plant'* (patent priority 26 March 1941)

---

[237] See Rubbra, Alan Arnold Griffith, p. 126.

[238] With all likelihood Griffith was confirmed in contact with Fritz Heppner, who propagated similar contra-flow, counter-rotating design principles of integral compressor/turbine rotors as head of the ASM Advanced Projects team, as outlined in Sect. 8.2.4.

[239] See Giffard, Making Jet Engines, p. 92.

had gained now Griffith's favour. As outlined in the context of Fig. 5.11, the German side had abandoned the corresponding BMW 109-002 concept already latest in 1941.

Finally, underlining Griffith tenacity in maintaining his positions, reviews within Rolls-Royce and possibly a side-view especially to Metrovick led to a different approach. In mid-1945 he produced a proposal, called the *AJ.65*, that was to lead to the first Rolls-Royce axial-flow jet engine—the *RR Avon*.[240] In 1946 he came forward with a scheme for a larger and more sophisticated jet engine incorporating the *'bypass'* principle whereby, a portion of the air entering the engine after passing the entry compressor stages, was bypassed around the high-pressure central *'core'*, leading to a—two compressors in series—arrangement for the core flow, subsequently used in world's first turbofan engine, the *RR Conway*.

In the late 1940s, Griffith carried out pioneering studies in *VTOL vertical take-off and landing* technology, such as controlling in the hover using air jets. The resulting *TMR Thrust-Measuring Rig*[241] was a pioneering VTOL aircraft, which brought the distinction of being *'the first jet-lift aircraft to fly anywhere in the world'* to Rolls-Royce in the 1950s. TMR was powered by a pair of *RR Nene* turbojet engines, which were mounted back-to-back horizontally within a steel framework; in turn, this framework was raised upon four legs fitted with casters for wheels. The TMR lacked any lifting surfaces, such as wings; instead, lift was generated purely by the thrust being directed downwards. Due to its unconventional appearance, it was understandably nick-named the *'Flying Bedstead'*.

Characterised by his in the meantime legendary tenacity to apparently or really outdated research and development topics, in combination with a certain crankiness, latest this nickname pushed attention of some technical historians back in the direction of Alan's father, George Griffith (1857–1906)—and a VTOL air ship, which associated many since their youth with that author's name in the spirit of Jules Verne, Fig. 8.28 (r): *'Alan Arnold Griffith was born on 13 June 1893, the eldest of three children of George Chetwynd Griffith and his wife Elizabeth Brierly. George Griffith was a colourful, rumbunctious, larger-than-life character, a prolific science fiction writer and a buccaneering explorer, poet, schoolmaster and journalist, who barnstormed his way around the world with strong-minded if somewhat incoherent views, anti-monarchist, anti-republican, socialist and communist, yet fiercely pro-British.'*[242] The late Victorian writer and noted explorer began writing so-called *marvel tales* in the style of Jules Verne, fantasy stories dealing besides the mentioned VTOL airship, with *'heavier-than-air flying machines, air to surface missiles, compressed air guns, submarines ... spectacular aerial, land or undersea combat'*, and enjoyed tremendous success in Britain. Published in 1893, simultaneously the year of Alan

---

[240] The *Avon* was the first Rolls-Royce axial compressor turbojet to enter production. Used in both civil and military aircraft, over 11,000 were built from 1947–1974. Applications covered 12 types, including the first *Comet* and *Caravelle* airliners; industrial versions of the Avon also remain in service today.

[241] See Wikipedia, 'Rolls-Royce Thrust Measuring Rig' in English.

[242] See Cantor, The Equation of Materials, p. 253, and—see Wikipedia, 'George Griffith' in English.

Arnold[243] Griffith's birth, his debut novel and most celebrated work, *'The Angel of the Revolution. A Tale of the Coming Terror'* was the first best-selling *'scientific romance'* and Griffith's success paved the way for subsequent authors of the genre, notably H.G. Wells.

The following excerpt from *'The Angel of the Revolution'*, end of chapter 46 *'Victory'*, provides an impression of George Griffith's writing, here in describing the hovering moment of Fig. 8.28 (r): *'... Millions of eyes were turned up at once, and beheld a vision which no one who saw it forgot to the day of his death:* Ithuriel, *..., her silvery hull bathed in a flood of light from ... electric lamps. In her bow, robed in glistening white fur, stood Natasha, transfigured in the full blaze of the concentrated searchlights. A silence of wonder and expectation fell upon the millions at her feet, and in the midst of it she began to sing the Hymn of Freedom. It was like the voice of an angel singing in the night of peace after strife. Men of every nation in Europe listened to her entranced, as she changed from language to language; and when at last the triumphant strains of the Song of the Revolution came floating down from her lips through the still night air, an irresistible impulse ran through the listening millions, and with one accord they took up the refrain in all the languages of Europe, and a mighty flood of exultant song rolled up in wave after wave from earth to heaven,—a song at once of victory and thanksgiving, for the last battle of the world-war had been lost and won, and the valour and genius of Anglo-Saxondom had triumphed over the last of the despotisms of Europe.'*

The aeronautical relevance of George Griffith's works is not only through his son Alan Arnold, and the latter's somewhat illusionary, contra-flow engine design, but also by his illustrator Fred T. Jane (1865–1916), who created in 1909 what is published regularly up today—*'Jane's–All the World's Aircraft'*.[244] With all likelihood the VTOL propulsion concept of airship *'Ithuriel'*,[245] Fig. 8.28(r), was imagined by Fred Jane; Arthur Rubbra specified it as follows: *'Forward propulsion was provided by a group of three airscrews on horizontal shafts at the rear and lift (presumably to augment the wings for vertical take-off) was provided by five airscrews on vertical shafts above the fuselage.'*[246]

As an explorer of the real world, George Griffith shattered the existing record for voyaging around the world, completing his journey as *'the real Phileas Fogg'* in just 65 days. He also helped discover the source of the Amazon River. Beginning in 1900 he suffered from increasing health problems, so that he decided to move the whole family to the better climate of the Isle of Man. There was only a short relief; George Griffith died on 4 June 1906 at Port Erin, Isle of Man of liver cirrhosis at the age of only 48; young Alan,

---

[243] Certainly not only coincidentally, *Arnold* is also the main protagonist in GG's key work, inventor of the revolutionary, global peace-enforcing *Aeronef* (airship, in French), in romance with *'The Angel of the Revolution'* Natasha, the daughter of *Natas* (sic! Satan in reverse), a Russian Jew and leader of the *'Brotherhood of Freedom'*.

[244] See Wikipedia, 'Fred T. Jane' in English.

[245] As a story in the story, *Ithuriel* is the name of an angel in the epic poem *'Paradise lost'* of John Milton (1608–1674),—see Wikipedia, 'Ithuriel' in English.

[246] See Rubbra, Alan Arnold Griffith, p. 117.

13, was then on the spot together with his mother—looking forward, according to Rubbra, to a *'rather poor and unsettled future'*.

In a kind of résumé, Edward Constant described the British status of gas turbine aero-engines at War's end: *'Certainly neither Griffith's own turboprop nor the turbojet that grew out of the axial-flow work Griffith spawned at the RAE saw war service, but Griffith is so intimately connected with the turbojet revolution that he does deserve considerable credit in its creation. Griffith was the first man in any country to recognize that the high-efficiency components he had designed meant that a gas turbine aero-engine could be practicable. Yet whatever accolades Griffith might deserve, it was Frank Whittle who created the first English turbojet.'*[247]

And, on 1st October 1941 Whittle's *W1X* from the E.28/39 was flown to the US with Power Jets personnel, carrying a complete set of drawings. These were handed over to *General Electric*, who rapidly had one of them manufactured and on test. On 3rd June 1942 Whittle flies out to GE to assist them;[248] he returns on 14 August. On 2nd October 1942, *Bell P-59 Airacomet* made its first flight powered by two *GE I-A* engines, the GE version of W1. This was a fruitful cooperation with improvements flowing both ways.

8.2.4   Fritz Heppner at ASM: The Rise and Fall of a German *Wunder*-Engineer

Fritz Heppner's career at Armstrong Siddeley up to the sudden dismissal of the foregoing chief engineer Stewart Tresilian on Mon 19 January 1942[249] has already been described in Chap. 5. Formally, Henry S. Rowell (57) was named as Tresilian's successor,[250] but the dramatic circumstances made it clear that the supposed stand-alone turbojet expert from Germany, Fritz Heppner (38) dominated the terrain. Seven years after his flight from Nazi-Germany into the darkness of a completely uncertain professional future, he and his family were *'arrived'*.

In comparison to Heppner, Tresilian never had the backing of parent Hawker Siddeley, or of Siddeley's General Manager, D.Sc. Henry S. Rowell. After his post-graduate studies at the National Physical Laboratory, Rowell had studied two years at the Universities at Goettingen and Berlin, and alone therefore might have been more inclined to Heppner's engineering ideas than others. Sopwith, Spriggs and Rowell instructed Tresilian to back F.A.M. Heppner's ASH gas-turbine contra-flow rotating shell engine design, Fig. 8.31 (l), a

---

[247] See Constant, The Origins, p. 218.

[248] Whittle, then 34, made it a *'secret stay'* by checking in to downtown 'Hotel Statler Boston', today—see Wikipedia, 'Boston Park Plaza' in English, as *'Mr. Whitely'*, before he demanded a phone installed in his room not connected to the main switchboard, etc. (GE Aviation Blog, 7 Feb. 2019).

[249] See R. Hodgson's website, http://www.designchambers.com/wolfhound/index.htm#Heppner

[250] Rowell had already hoped to become ASM's chief engineer in May 1939, when actually Tresilian got the job,—see R. Hodgson's website.

derivative of his 1942 turbofan patent, Fig. 5.13 (r), which was closer to Griffith's CR engines by adopting the latter's *'staged turbine (top)/compressor (bottom) blading'*, Fig. 8.31 (r). Tresilian refused, wishing instead, in late 1941, to back very close collaboration with Metrovick on the F.2 and look further into new advanced axial gas-turbine proposals, which his newly hired ex-Rolls Royce engineering freshman Thomas Pitt de Paravicini,[251] assisted by Brian Slatter, then a bright post-graduate research student,[252] had suggested in an internal ASM report *'Turbine aero engines'* in September 1941.[253] This report included an *UDF Un-Ducted Fan*[254] unit and various *HBPR High Bypass Ratio* engine configurations, Figs. 8.32 and 8.33. Frankly, Tresilian told the Directors Sopwith and Spriggs, representing parent Hawker Siddeley's upper management, that Heppner was *'making monkeys out of the lot of them'*. As a result, he was fired and given 2 h to leave.[255] Within the next two years the ASH in its various proposed guises harmed ASM's reputation with the Ministry and—though continued even after Heppner's leave in early 1945 by a geared ASH version, was never completed, and finally ended in 1946.[256]

Nevertheless, Heppner's image as an *'experienced turbojet engineer'*[257] could not have been completely unfounded; Fig. 8.29 tries to provide some insight, into otherwise *'unchartered ground'*. Here the activity timelines of six salient German and two British

---

[251] Cambridge graduate T.P. de Paravicini, since 1935 former research assistant of legendary turbocharger designer James P. Ellor at Rolls-Royce Derby, left ASM shortly after Tresilian's firing to Bristol, followed six months later by J. Bucher et al., who—according to Hodgson, somewhat sibylline—*'worked during the war period for H.M.G. (the British Government) in Europe, including Germany'.*

[252] B. Slatter (1920–2016) joined—after a stay at RAE Farnborough—Rolls Royce Derby, ending his career as Technical Director of RR Gas Turbines; his *Olympus* engines powered most of the Royal Navy fleet in the *Falklands War* of 1982.

[253] This report was followed in December 1941 by *'An Introduction to Gas Turbine and Jet Propulsion Theory'*—possibly one of the UK earliest treatises on the subject—with strong effect in favour of axials,—see Hodgson, Armstrong Siddeley, p. 10. Both reports should be at RRHT Derby Archives—to re-open in late 2022; information, courtesy of Neil Chattle, RRHT Derby, 22 Oct. 2021.

[254] This preceded the Metrovick *F5 UDF* by at least 2.5 years, and GE's counter-rotating, open propfan of the early 1980s,—see Wikipedia, 'General Electric GE 36' in English, by more than 40 years.

[255] See Hodgson, Armstrong Siddeley, p. 17: *'Heppner had demanded an immediate apology, which he did not receive. On Monday, 19 January 1941, Tresilian was asked by Rowell to apologise in writing to Heppner. He refused. Rowell said that if he did not, Heppner would leave immediately. He still refused. Rowell then telephoned Spriggs and/or Sopwith in London. A few minutes later, Tresilian announced to some of his design staff that:* "I've upset the apple cart and have to go". *He was given 2 h to leave. As a parting present, Spriggs gave him 12 days' pay (he had no contract) and ensured that no-one in the industry would employ him. He remained unemployed for some six months until he became the RAF-USAAF engine co-ordinator.'*

[256] The end of ASH development documents T.B. 23 in the *'Heppner* (file) *Box'* at RRHT Derby Archives: *'The A.S.H. Mk IV'*, dated 16 Feb. 1946.

[257] See Giffard, Making Jet Engines, p. 107, an expression which appears not to be justified by facts.

turbojet engine designers are shown in the time frame 1930 (Griffith) to 1945, differentiated for studies, development and production activities, and all embedding the scarce information known about Heppner's whereabouts.

The top bar for Hans *von Ohain* (1911–1998) commences in 1934 with his preliminary experimental studies at *'Becker's Garage'*, Goettingen, and implies over time his various Heinkel(-Hirth) turbojet engine developments, as e.g. *HeS 3B, HeS 08 and HeS 011*, of which the latter saw some production close to War's end. *M.A. Müller* (1901–1962) was responsible for Heinkel's unfinished *HeS 030*, a development which he had started in the wake of Professor Herbert *Wagner* (1900–1982) at Junkers Magdeburg in 1936, with a project under the designation *Junkers RT0*. Anselm *Franz* (1900–1994) developed the *Junkers Jumo 004* turbojet to become the most successful German turbojet programme with more than 6000 produced units. The name of Hermann *Oestrich* (1903–1973) is closely connected to the *BMW 003* turbojet development, more or less in parallel to Franz's Jumo 004, though Oestrich familiarised with aero engine thermodynamic studies already at DVL Berlin as early as ~1933. As a kind of *'lonely wolf'* Helmuth *Weinrich* (1909–1988)—with all likelihood in early acquaintance with Fritz *Heppner* (1904–1982)—started a self-made mechanic's shop keeper career at Chemnitz as early as 1931, propagating (after Heppner's leave in January 1935)—long-time unheard—a counter-rotating turbojet engine concept (with small demonstrator units). Finally, this turbojet engine nucleus made it with the support from RLM's Hans Mauch under the wings of the Bramo/BMW Berlin-Spandau team, where it dominated after 1939 the BMW-side of the RLM sponsored development programme as *BMW 002* for another two years, while the later successful *BMW 003* was considered then only as a simple, preparatory test vehicle. Fritz Heppner had—at best as Weinrich's discussion partner—the possibility to observe and indirectly influence these early developments only as long as he was in Germany. But, he had the additional benefit, that this very programme remained on the rise, while he was already in England. Finally, the illustration sketches roughly the corresponding works of Frank *Whittle* (1907–1996) and his various Power Jets radial engine products, with—in comparison—relatively early commitment for series production. In addition, the early, in hindsight fruitful study beginnings of Alan A. *Griffith* (1893–1963) are indicated for 1927, with the known lengthy break of all activities, and a continuation of engine developments at RAE and Rolls-Royce, but without production realisation up to 1945.

Heppner in England might have had some short contact with turbomachinery issues while working for the *Roditi* (engineering and trade) *Agency*, London in 1935/1936, though most of his work was rather in the automotive field of hydro-mechanical gearboxes and torque converters, as indicated by his patent applications during that period.[258]

---

[258] As indicated already in Chap. 5, Heppner might have used this *'turbojet development gap'* also for broadening his aerodynamic (patent) knowledge at the London Patent Office, in view e.g. for his inventions/ activities in the context of Figs. 8.34 and 8.36.

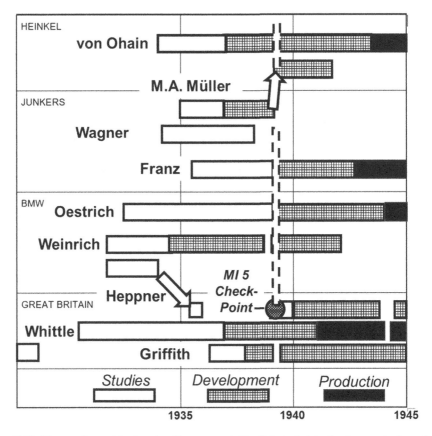

Fig. 8.29 Heppner's positioning in German and English turbojet development chronology, 1930–1945

The decisive setting of points towards his career as Armstrong Siddeley's de-facto chief engineer after 1941 must have happened at the marked '*MI 5 Check-Point*', when young John B. Bucher (26) started his first '*verification mission*' to Germany on the occasion of the '*Lilienthal Convention Berlin*', 12–15 October 1938, as described already in Sect. 6. 1.3. There is certainly some speculation, but as the vertical dashed marker in Fig. 8.29 illustrates, the period of '*end of 1938*' was certainly best suited to bring back a positive impression of the BMW 109-002 turbojet development in focus—to evaluate Heppner's further importance in England. Assuming then, that corresponding positive feedback reached not only MI 5 and the assigning Ministry, but also the ASM upper management—and Heppner, some of the following developments are then perhaps easier to understand.

It is Richard Hodgson, who introduces in this context a completely new aspect for interpreting Heppner's increasingly self-conscious behaviour during 1941, and his corresponding backing by ASM's upper management; in his words: '*What none of the*

ASM engineers could understand was why their serious reservations (against Heppner) *were always summarily dismissed by Dr. Rowell and the Board. The answer may be simple and purely commercial. Heppner's schemes were at times quite brilliant reverse engineering of already existing complex schemes that he had seen—despite formally not being allowed to do so—but were sufficiently different to attract their own patent protection. Heppner held or was named in over 50 patents[259] during this time. Anything he came up with might have been about the only engine ... over which the State would <u>not</u> have had some major control. A very interesting post-war prospect. And what he was promising—in a very plausible manner—was a unit of such performance that it would have been a one horse race. The obsession by all of the aero-engine companies with post-war rights during 1940/1/2 is quite remarkable, and caused Hayne Constant at the RAE [apparently a former card-carrying member of the party][260] to ask on several occasions:* "Why were the interests of the aero-engine companies higher than the national interest?" *This is recorded in memos at the Public Record Office'.[261]*

So, given ASM's stalled piston-engine programme, without realistic chances to participate with the new turbojet engine technology in any profitable wartime business, the described attitude might have spread, and should certainly be taken into account by evaluating further developments around the German *wunder*-engineer.

On 6 November 1939 Heppner's patent files show for the first time during his stay in England *turbomachinery* activities:

- this is the priority date of his GB536,238, as illustrated in Fig. 8.30, a gas turbine design, clearly still without any ambition towards a flight application, and
- at approximately the same date, he sends the patent specification of his '*High speed propulsion plant*', Fig. 5.11 (bottom), before official application to the Air Ministry's Directorate of Scientific Research, led by Major George Bulman.[262]

Instead of grasping this *great opportunity* from Heppner's perspective, the Ministry influenced by H. Constant turned this proposal down—for being overly complex. Besides the application of two more patents, nothing happens up to the internment of the Heppner family on Isle of Man—between end of May, 1940 and mid-March, 1941. Thereafter, he decides immediately as a first step to file for the rejected proposal a patent application,—what

---

[259] This number implies multi-national applications; the number of Heppner's unique patented innovations is rather close to 20 ($15\times$ turbomachinery-, and $5\times$ automotive-related).

[260] Insinuating Constant's membership in the Communist Party, which the Author was unable to verify.

[261] Hodgson, Armstrong Siddeley, p. 15.

[262] See Chap. 5, in the context of Fig. 5.11 (bottom). The official, negative reply must have reached Heppner still in Dec. 1939.

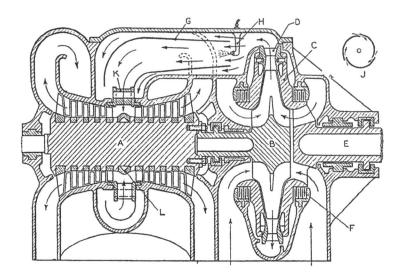

Fig. 8.30 Heppner's stationary gas turbine plant, with pneumatic power transmission to drive shaft E, ~ 1941 © *Flight* 1942

became—US2,360,130 'High speed propulsion plant', with priority 26 March 1941.[263] In addition, his stationary gas turbine proposal, was officially made public as GB536,238 on 9 May 1941, and in due course discussed in *Flight* at length by G.G. Smith.[264]

A leading feature of this design, Fig. 8.30, is that the output shaft E is completely separate from the rotor shaft and power is transmitted pneumatically by the air flow from the compressor. As a result a useful stage of speed reduction between turbine rotor and output shaft can be obtained without mechanical gearing. Consequently, the design differs radically from the usual construction in which the output torque is taken directly from a common rotor shaft, and represents the difference between the torque exerted by the turbine and that absorbed by the compressor. To avoid axial thrust, a twin, steam-turbine influenced axial-flow turbine is employed, of which rotor A is directly attached to the balanced compressor impeller B. Coaxial with the impeller is a *'cage rotor'* C, carrying a ring of radial-flow blades D; this rotor C is integral with or attached to the output shaft E, sealed against the compressor casing. By design, the output shaft is arranged to rotate at about one-third of the turbine rotor speed. The torque delivered by the turbine is necessarily transmitted wholly to the compressor impeller and, as no stationary reaction members intervene, the same torque will be passed to the output rotor.

---

[263] This priority date is only 10 days after Heppner's return from Isle of Man internment. Apparently, he was highly committed to proceed.

[264] See Smith, Gas Turbines, 3rd ed., pp. 77–79, but a publication in 2nd edition is likely. as well, in addition to *Flight*, 1942.

Thereafter, the air enters a conical combustor tube G, into which fuel is injected by nozzle H. The chamber is built up of overlapping shrouds, detail J, thereby furnishing longitudinal, tangentially directed air inlets. A stunning similarity with ABB's well-known 'EV burner principle'—some 40 years ahead of time.[265] Excess air flows between the chamber walls and the engine casing, and generates a diffusing, swirling flow mixture with the combustion gases, the temperature of which is thereby considerably reduced up to turbine inlet. The gases enter the turbine through a stationary ring of air-cooled guide vanes K; at turbine centre the radially entering gases are split and guided in opposite axial directions by a suitably shaped flange L of heat-resisting steel.

The same turbine/compressor torque-balancing principle was already applied consistently to Heppner's large turbofan engine designs, starting from the patented US2,416,389, actually entitled '*Torque Balancing of Jet Propulsion Turbine Plant*', Fig. 5.13 (bottom, right) with priority 17 June 1942, up to what became the ASH engine, Fig. 8.31.[266]

Though the ASH turbofan has obvious similarities with Griffith's C.R.1 contra-flow engine concept, Fig. 8.28,[267] it is also closely related to Heppner's foregoing large turbofan, as depicted as '*T.B. engine*' in Fig. 5.13, if only the counter-rotating turbo-components are replaced by the '*staged, turbine (top)/compressor (bottom) blading*'. In fact a closer look reveals, that Heppner combined here both principles, as visible along the non-rotating inner shaft, where Heppner—deviating from Griffith—keeps only every second disc in rotation. Not much is known about the development history of this concept; that there must have been personal contacts between Heppner and Griffith in this context is evident. That they had a lot in common (and to talk about) after Heppner's recent stay of more than nine months on Isle of Man, and especially at Port Erin, was certainly helpful.

After Heppner's counter-rotating turbojet, Fig. 5.11, had been turned down by H. Constant in the name of the Air Ministry as '*overly complex*', the second proposal, the '*T.B. engine*' of Fig. 5.13 (bottom, right), presented at the Ministry on 5 May 1942 was even more challenging—with now three sets of compressors and a contra-rotating turbine; and the denial came promptly already in May 1942. Surprisingly after these drawbacks, the third approach, the ASH concept of Fig. 8.31, which had been presented to the Ministry apparently as early as 7 December 1942,[268] was finally accepted on 28 August 1943. One

---

[265] See Eckardt, Gas Turbine Powerhouse, p. 308 f.

[266] It is likely that the designation of the complete set of 22 *T.B: reports*, T.B. 1 to T.B. 23 (with T.B. 11 missing) of the 'Heppner Box' at RRHT Derby Archive, stands actually for 'Torque Balancing'.

[267] Besides Heppner's turbofan set-up with two fan stages at flow entry, both concepts correspond with inner compressor and outer turbine locations, and the rotating fuel injection, see also Fig. 8.37 (r) for a related design solution of Heppner's German '*partner*' H. Weinrich. Unchanged, also ASH had the disadvantages of a long thin shaft and numerous bearings with difficult lubrication.

[268] The date is remarkable since ASM indicated by the decision to present ASH instead of ASX to the Ministry, a clear preference for Heppner's engine concept. The ASH base patent US2,428,330 'Assembly of Multi-Stage Internal-Combustion Turbines Embodying Contrarotating Bladed Members' for F. Heppner alone, had a priority date of 15 Jan. 1943.

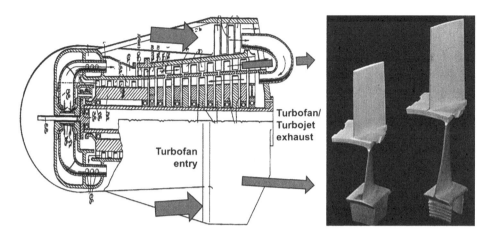

Fig. 8.31 *ASH* turbofan engine, 5000 kp thrust, September 1942 (l), *ASH Mk III* prototype, manufactured turbine (top)/compressor (bottom) blading, April 1944 (r) © RRHT Derby, r

can speculate about the reasons for this decision, but after Griffith's C.R.1 investigations at Rolls-Royce were already supported, the same could fairly not be denied to the—in the sense of the continuous Ministerial demands—simpler ASH adaptation, and it brought as a side effect the buy-in of the rewarded ASM management to ASH—and ASX.

Interestingly a comment out of Heppner's 'Project Section' with respect to the ASH design survived, dated 24 February 1943: '*No further basic development is expected for the time being, and the writer feels convinced that a clear energetic development of these schemes will lead to that type of light and efficient aero engine for which we have been hoping.*' The fundamental TB reports[269] (for the TB and the ASH engines) put Heppner in the position of a '*theoretician*', underlined by the fact that his applied axial compressor design tools were absolutely up-to-date.[270]

After piston-engine development work had been ended at ASM in July 1941, this opened the opportunity for Thomas P. de Paravicini and his assistant Brian H. Slatter to investigate the possible uses of gas turbine engines as a means of aircraft propulsion. Both started literally from first principles, but their first report, produced within two months time-lapse, contained a bunch of innovative ideas for turbojet and turbofan units: '*The fan diameter was 48 inches with a total of 24 blades. Also suggested was a twin unducted fan,*

---

[269] From RRHT Derby Archive: TB 01 'Note on the Thermodynamic Basis of the T.B. Engine' (19 Nov. 1942), TB 02 'Detailed Theoretical Notes on the T.B. Engine', and TB 03 'A Review of Alternative Designs Working on the Thermodynamic Cycle Already Discussed' (16 Oct. 1942). The pictures of manufactured 'staged' turbine (top)/ compressor (bottom) blading, Fig. 8.31 (r) are part of TB 16 'The A.S.H. Mark III', dated 21 April 1944.

[270] In 1942/1943, Heppner apparently used Howell's 1942 reports,—see Howell, The Present Basis of Axial Compressor Design, Parts 1 and 2.

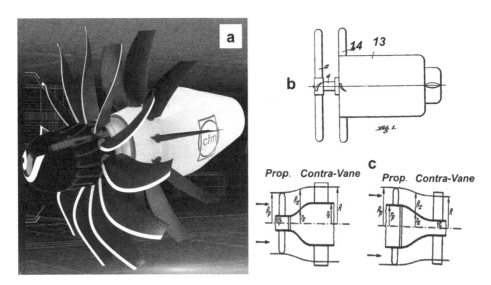

Fig. 8.32 Visionary Heppner I, 1941 [copy of de Paravicini/Slatter]: (**a**) *CFM Open Fan* concept, 2021, (**b**) F. Heppner *UDF* patent, 1941, (**c**) A. Betz, Propeller with stationary *'contra-vanes'*, 1938

contra-rotating unit of the same diameter having two sets of 12 blades. It was expected to produce between 2500 hp to 3500 hp with anticipated flying speeds up to 500 mph.'[271]

While the two Paravicini/Slatter reports of *'Tresilian's men'* are still missing, Heppner's occupation and/or transformation[272] of their innovative ideas into a patent, exclusively in his name, is still visible in US2,404,767 *'Jet propulsion plant'*—with GB priority on 28 October 1941, Figs. 8.32 and 8.33. This application, some four weeks only after the first Paravicini/Slatter report *'Turbine aero engines'* had been issued, occurred apparently in a hurry by a highly self-confident Heppner, still with Tresilian in charge as ASM chief engineer.

The circumstances of patenting aside, the ideas were remarkable for the time,—and Heppner may be called *'visionary'* solely for preserving the memory thereof, by patenting. Figure 8.32a shows the June 2021 *CFM International*[273] presentation of a new *'Open Fan'* propulsor, with a rotating propfan of 3.5–4 m diameter, followed by a row of stationary, case-mounted de-swirler vanes, both adjustable and from composites. This may push the engine bypass ratio up to BPR ~30+, with a corresponding reduction of thrust specific fuel

---

[271] See Lawton, Parkside, pp. 68–69.

[272] There is some probability, that Heppner used the de Paravicini/ Slatter propulsors, and adapted them with his typical counter-rotating core engine.

[273] The engine manufacturer association of *General Electric Aviation* and *Safran Aircraft Engine*, which commonly investigate future environmentally-friendly and fuel-efficient engine concepts in a technology programme *RISE—Revolutionary Innovation for Sustainable Engines*.

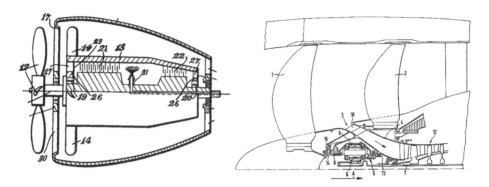

Fig. 8.33 Visionary Heppner II, 1941 [copy of de Paravicini/Slatter]: Heppner's patented *'semi-CRISP'*, 1941 (l), *MTU* typical *CRISP* patent (Inventor H. Klingels), 1998 (r)

consumption by −20% in comparison to present advanced CFM LEAP engines, of BPR 11.

Though Heppner's patent, Fig. 8.32b, appears to have some similarity with the new CFM concept, it actually is—due to the second row of fan blades being mounted on a *rotating casing 13*—a forerunner of another advanced open rotor configuration of already the 1980s, GE's UDF Un-Ducted Fan. In addition, Fig. 8.32c proves that also the idea of stationary de-swirling 'Contra-Vanes' dates back to a 1938 publication of Albert Betz, AVA Goettingen.[274] This Betz report was considered as so important that it was available in English as NACA Technical Memorandum within a record time of 10 months, stating: *'On heavily loaded ship propellers and on airplane propellers with very high pitch, the spiral motion of the slip-stream contains considerable kinetic energy which can be converted into effective energy to a large extent by contra-vanes.'* In principle, also the two ASM researcher could have used it as inspiration for their report, and Heppner, who certainly knew the complete background due to his comprehensive survey of patent publications, might have had less qualms under these circumstances. Apparently, good ideas are and were rare—and tend to be copied; no problem, as long as the real inventors will be honourably cited.

Heppner's 1941 propulsor patent covers a second, two-stage partially-ducted CR-fan version, Fig. 8.33 (l), which—named here as *'semi-CRISP'*—has some similarity with the *'CRISP Counter-Rotating Integrated Shrouded Propfan'* concept of MTU Aero Engines, illustrated in Fig. 8.33 (r) by a patented, special design with the air intake to the core engine placed in-between the two propfan rotors, invented by Hermann Klingels, MTU.[275]

---

[274] See Betz, Zur Theorie der Leitapparate fuer Propeller, also as NACA TM No. 909 (Sep. 1939).

[275] MTU patent DE19,828,562 'Triebwerk mit gegenläufig drehenden Rotoren (Counter-Rotating Engine)'with priority 26 June 1998. An earlier MTU patent ES2,023,020 'Turbofan', invented by H. Grieb and H. Geidel had been filed with priority 11 April 1988,—see Geidel, Gearless CRISP. The 'Geared CRISP' base version was not patentable, due to two premature publications of 1986,—see Grieb, Turbofan and Propfan.

Over more than 50 years this 1941 Heppner patent was used up to the 1998 Klingels patent in 41 cases as refence invention, and belongs herewith certainly to the most quoted roots of advanced aero engine technology. Foundations to the MTU CRISP engine concept were laid in a comprehensive technology programme 1988–1993, supported by BMFT (Federal Ministry for Research and Technology), and in close cooperation with DLR Deutsches Zentrum für Luft- und Raumfahrt (German Aerospace Centre), especially with the DLR–Institut für Antriebstechnik (Propulsion Technology) under Prof. Heinrich Weyer. These technology development and demonstration activities were carried out in close partnership with Pratt & Whitney, East Hartford, CT, which developed after a production start in 2016 in the meantime to the very successful 'GTF Geared TurboFan' programme, predominantly for the Airbus A320neo and aircraft derivatives, in the thrust class 15–33 klbf.[276]

Still in 1941, and only two months after the foregoing propulsor inventions of Figs. 8.32 and 8.33, Heppner filed another fundamental invention now in the wider area of aircraft-related propulsion, Fig. 8.34 (inset).[277] Following the basic patent idea, a future application of this 'Propulsive Fuselage Concept' is illustrated in Fig. 8.34, a 2020 study result of Bauhaus Luftfahrt, Munich as part of EU research project CENTRELINE.

It has long been known in the field of marine propulsion[278] that the propulsive efficiency is improved when fluid from the wake of a ship hull (or aircraft fuselage) is used as part or all of the propulsive stream. Betz explained this beneficial effect of reduced entry momentum,[279] and patented it together with Ackeret in general, as early as 1923.[280] He points out that with wake ingestion the power expended can actually be less than the product of the forward speed and craft drag; consequently, wake-adapted propulsors are commonplace for torpedo and other marine applications. For aircraft propulsion, similar to the torpedo, wake ingestion has shown clearly beneficial already e.g. for cruise missiles, where the concentric aft-located single engine propulsor can capture most of the body wake. Tests confirmed here a 7% power reduction due to wake ingestion,[281] a value in the expected order of magnitude of 5% sfc reduction for future aircraft with aft-fan installation.[282]

---

[276] See Wikipedia, 'Pratt & Whitney PW1000G' in English.

[277] Patent GB577,950 'Improvements Relating to Jet-Propelled Aircraft', with priority 31 Dec. 1941.

[278] See Froude, Description (1883). This Robert E. 'Froude, the Younger' had generated a related formula, which was renewed and brought to public attention by Sir Geoffrey I. Taylor (1886–1975),—see also the 'Griffith-Taylor-Prandtl scandal' in Sect. 4.2.1.

[279] See Betz, Tragflügel, p. 281 (1927, in German), and—see Betz, Interference, p. 218 (1966, in English).

[280] Patent DE513,116 'Verfahren zur Verminderung des Widerstands eines Körpers in Flüssigkeiten oder Gasen (Drag reduction of a body in fluids and gases)', with priority 5 Sep. 1923,—see Sect. 12. 1.2 and Fig. 12.7. Heppner's 1941 patent addresses a tail-mounted aft-fan as aircraft-specific variant thereof.

[281] See Smith, Wake ingestion.

[282] See Seitz, Concept validation, and—see Wiart, Exploration of the Airbus 'Nautilus'. The Heinkel He 211B, jet-powered 20+ passenger aircraft project of 1961 (design S. Günter) had a tail-mounted turbojet installation very close to Heppner's 1941 patent, with an accordingly projected sfc improvement of 10%.

Fig. 8.34 Visionary Heppner III: *'Propulsive Fuselage Concept'*, Heppner's GB patent drawing, 1941 (inset), *PFC* study aircraft EU Project *CENTRELINE*, 2020 © *Bauhaus Luftfahrt*

A closer look to *BLI* (boundary layer ingestion) on aircraft fuselages reveals that Heppner was not alone with his aft-fan placement. Interestingly, Frank Whittle patented explicitly a mid-fuselage BLI suction installation as early as 1938, Fig. 8.35 (top), with the wording: *'The invention consists primarily in a propulsive device including a combustion engine, which aspirates all or parts of its air from the airstream passing close to the surfaces of the aircraft, preferably from those regions where the relative airflow has been retarded relative to the aircraft, commonly referred to as boundary layer air.'*[283] Indicating a remarkable change of mind, since two years earlier, he apparently wanted to get rid of the inlet boundary layer to improve propulsive efficiency, by stating: *'It may be found possible to enhance the propulsive efficiency by deriving some or all of the air flow from the boundary layer at or over any desired part of the structure.'*[284]

In addition, Fig. 8.35 (bottom) illustrates another mid-fuselage BLI application for the projected Messerschmitt P.1110/II single-seater fighter aircraft with 40 deg swept-back wings and a V-tail unit.[285] This project with a maximum speed of 1006 km/h at 7 km altitude was an effort to establish the optimum airframe location for jet engines and other equipment in the quest for maximum aerodynamic efficiency. Consequently, fuselage cross-sectional area was one of the parameters looked at most closely, as the Messerschmitt design team led by the Austrian engineer Otto Frenzl (1909–1996) anticipated by about

---

[283] Whittle pat. GB512,064, 'Improvements relating to the propulsion of aircraft', priority 25 Feb. 1938. As discussed by—see Smith, Gas Turbines, 3rd ed., p. 97, Whittle stated that the forward-facing *'air scoops'* of the shown patent drawing had radially exaggerated dimensions.

[284] Whittle patent GB471,368 (same title as GB512,064), priority 4 March 1936.

[285] See Masters, German Jet Genesis, p. 131.

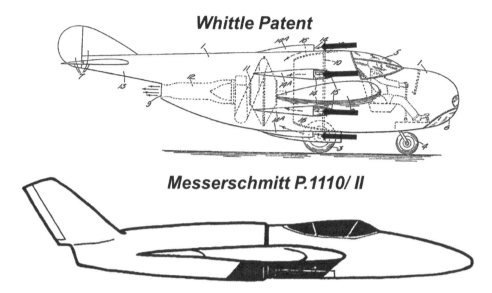

Fig. 8.35 Contemporary *Boundary Layer Ingestion* attempts: Whittle GB patent drawing, 1938 (top), Messerschmitt fighter aircraft design, 1945 (bottom)

10 years the work of the American R. Whitcomb on the *'transonic area rule'*.[286] The air intake was an annular slot running round the whole circumference of the fuselage. Behind the pressurised cabin (with ejection seat) was a self-sealing fuel tank, with the HeS 011 turbojet engine mounted in the tail. The armament of three to five 30 mm MK108 cannon[287] was housed in the nose.

There is some probability that Heppner used for the first time the term *'Boundary layer propulsion'* in 1944,[288] but fundamental investigations on wing boundary layer suction[289] go back to a first introductory presentation of Jakob Ackeret (1898–1981), then in 1926 at AVA, first experimental investigations of Oskar Schrenk (1901–1990)[290] by using the AVA high-lift aircraft *Junkers AF-1* in the early 1930s, after 1936 as modified *Fieseler 'Storch' AF-2* ('suction-stork'), and the Swiss Werner Pfenninger (1913–2003).[291]

---

[286] See Wikipedia, 'Area rule' in English.

[287] See Wikipedia, 'MK 108 cannon' in English.

[288] Fritz Heppner, 'The general theory of boundary layer propulsion', ASM (internal) T.B. Report No. 18, dated 3rd Nov. 1944, quoting O. Schrenk's NACA TM 974 of April 1941.

[289] See Wikipedia, 'Boundary layer suction' in English, and for an overview—see Gerber, Untersuchungen, in German, a 1938 doctoral thesis at Ackeret's ETHZ Institut für Aerodynamik.

[290] See Schrenk, Grenzschichtabsaugung, NACA TM No. 974.

[291] See Historisches Lexikon der Schweiz HLS, 'Werner Pfenninger' in German. Pfenninger laid ground in the area of *'suction-type laminar flow control'*,—see Braslow, A History, p. 6.

Before proceeding with Heppner's final report, it is worthwhile to review in short two documented, financial transactions which illustrate his special relationship with Armstrong Siddeley's upper management:

- His name appears in the preserved ASM Director's Minute Book, p. 199, in the context of a meeting held on Fri 21 May 1943: *'The secretary reported that on 4th July 1942 a loan of £1590.6.6 had been made to an employee, Mr F. M. Heppner. The loan is secured by a mortgage on a free-hold property and the capital sum is repayable by monthly instalments of £20. Interest on the loan is payable at the rate of 4% per annum. The Board confirmed that the granting of this loan on the terms mentioned was in order,* '[292]
- on 11 May 1943 Fritz Heppner assigned in a contract all his, then 35 turbo-related patents of 1941–1943, and all future patents to ASM, in exchange for a 3% royalty up to a maximum of 10,000 £ per annum.[293]

Heppner's final T.B. Report No. 18 on *'The General Theory of Boundary Layer Propulsion'*, is of special relevance as his last, personally signed engineering document under the date of 4 November 1944, Fig. 8.36, and in view of his thereafter beginning severe medical case history, resulting in a diagnosis of insanity (schizophrenia),[294] which brought him to closed mental asylum from end of 1945, up to his death in 1982. The subject of the report combines boundary layer propulsion for an aircraft with corresponding fuselage and wing technology applications. The first one was realised by the known aft-fan installation, Fig. 8.34, while the wing BL removal is by means of two *'suction slots'* per wing area, as indicated in Fig. 8.36. In a certain way Heppner's *'Theory'* appears to be planned as a kind of answer to O. Schrenk's April 1941 NACA TM, in which he pointed out in his Introduction: *'. . . a theory of boundary layer removal by suction, which can be mathematically carried out in individual examples, would constitute a great relief and help for experimental labor. The number of variables in suction experiments is always very great and might be considerably reduced by appropriate calculations.'*

For the study Heppner assumes a 100 t aircraft *'with a proposed range of 4,000 miles'*, to be powered by five ASH engines of 3000–3400 kp TO thrust each. Following apparently

---

[292] Following an interpretation of Peter Barnes, Librarian of the Coventry & Ansty Branch of RRHT of 17 March 2017, the loan was issued three days after Heppner had bought the house at Leamington Spa, Fig. 5.7 (r), with some likelihood per cheque by Tom Sopwith himself. The 1600 £ of 1942 would correspond to 80,000 £, or 92,000 € today (11/2022), sufficient for the house in 1942.

[293] Acc. to an *Espacenet* check, this might have been ~15 basic turbo-related ideas, times multi-country affiliations; the upper limit of 10,000 £/a would correspond to presently ~600,000 €/a. Details of the royalty reference calculation basis are not known.

[294] The word *schizophrenia* translates as 'splitting of the mind' and is Modern Latin from the Greek words *schizein* (σχίζειν, 'to split') and *phrēn*, (φρήν, 'mind'). Its use was intended to describe the separation of function between personality, thinking, memory, and perception. Classification subtypes of schizophrenia as *paranoid, disorganized, catatonic, undifferentiated*, and *residual type* were difficult to distinguish between and are no longer recognised as separate conditions. Instead, *stress* in all forms, which Heppner had more than enough, is considered a main cause.

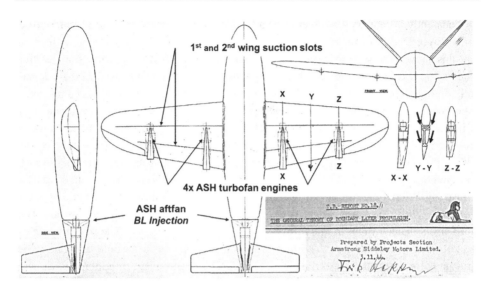

Fig. 8.36 Visionary Heppner IV, 1944: '*Augmenter*' Boundary Layer Propulsion aircraft concept, with wing and fuselage BL suction

Schrenk's example, who investigated drag reduction by suction on 40% thick wing profiles, he also applies very thick profiles, in which the four ASH bypass engines are partially buried, together with the suction slots ducting, visible in the enlarged wing section Y-Y of Fig. 8.36.[295] The effect of the combined boundary layer propulsion is expected by a −30 % reduction in mission fuel. Some criticism caused Heppner's slot positioning which possibly would have been better arranged closer to the wing trailing edge, but in general his wording appears to be concise and precise, without signs of a mental distortion. Possible exemption, the entry to his '*Conclusions*' (p. 8) which sounds somewhat constructed: '*The above report has been specially asked for by a representative of M.A.P. to ascertain the writer's views on future development. It appears from the question that the possibilities of the ducted fan in conjunction with boundary layer intake have so far not been fully realised. Thorough investigation on the lines indicated above will ensure a rapid engine and plane development,*' and (p. 11) '*It appears therefore that cargo can be delivered over 4000 miles at less than 1 lb. fuel per lb. cargo. ... The aircraft could be welded from stainless steel as has been done already, so that its life would be independent of weather conditions etc. (sic!)*[296] *The life of the engine will be dependent mainly on the careful development of the bearings. ...*'

---

[295] Not mentioned are possible inlet distortions, which came into focus only later.

[296] The Author feels not in a position to judge, if the foregoing remark is '*sound*' under the precarious wartime conditions end of 1944.

Apparently in January 1945 there was a last job change, Heppner left ASM and joined Rolls-Royce Derby for his remaining eight months in freedom.[297] There is no indication that this move was supported by Griffith, but during the final stages the story of this relationship takes an unexpected, interesting twist. Without knowing an exact date, apparently during Heppner's stay at Rolls-Royce in the first half of 1945, Griffith commenced experiments with suction to control the boundary layer on wings, the development of turbojets having made this more viable. This effort formed part of RAE Farnborough's programme of empirical research, the intention being to use boundary-layer suction on the upper surface of the wing to remove slower-moving and turbulent air from the wing surface and thereby reduce drag.[298] Eventually a full-size wing section was manufactured by Armstrong Whitworth Aircraft in readiness for a flight trials programme, and a *Hawker Hurricane IIb* Z3687 was selected to serve as the platform for this test panel. The Hurricane's maiden flight in this form was made on 23 March 1945. The sole purpose of the tests was to find out how far back across the wing the airflow would continue to be laminar before transition to turbulent occurred. The results were disappointing, it was found that on the new section the airflow transition/separation point had moved aft only by a small margin of the chord.

Here, in an attempt to produce something better, Griffith suggested taking the wing design further by introducing *suction* to modify the boundary layer flow. The aircraft chosen to assess the *Griffith suction wing* in flight was Gloster Meteor III EE445, the original outer wings being removed and replaced by new sections. Suction was provided by a modified blower normally used to cool the rear bearings of the Meteor's Rolls-Royce Derwent I engines. Rolls-Royce's Barnoldswick experimental department produced three modified engines, designated Derwent III, with the new capability to draw boundary layer air away via suction slots, positioned in both wing surfaces at 75% chord to take away the boundary layer air that was stuck to the wing's surface. Additional supporting data was recorded using a 1/7.5-scale model in RAE Farnborough's high-speed wind tunnel, during November 1945 and May 1946, but the actual flight transition tests were delayed up to early spring of 1948. The Griffith suction wing assessment lasted until December 1948;[299] overall the results were a disappointment and fell short—and remained so since then—of what had been hoped also by Heppner for a low drag, laminar aircraft wing development.

A last, scientifically bizarre episode in Fritz Heppner's life was reported by his son Chris (then 11),—for 8 December 1944, when Heppner received the letter of his election as

---

[297] Mail of Dave Piggott, RRHT Derby, 8 April 2013: '*In my many conversations with Geoffrey Wilde, he talked about Heppner's time in Derby just after the war. After he fell out with AS, he joined RR for a short time but was not a success at Derby either and soon left. Wilde maintained that his technical logic was not sound and some time after leaving RR he had mental health problems.*'

[298] Tony Butler, 'Griffith Wing Meteor', 11 July 2019: https://www.key.aero/article/griffith-wing-meteor

[299] See Brown, Wings on my sleeves, describing a quite dramatic Meteor test flight with Griffith's suction wing on 1st Oct. 1948.

Fellow of the Institute of Physics (IOP), enthusiastically greeted by his family when returning home to Leamington Spa from *Parkside*. IOP is a scientific charity that works to advance physics education, research and application; at present it has a worldwide membership of over 50,000, supporting physics in education, research and industry.[300] However, as was nearly to be expected, regularly repeated mail contacts between 2015 and 2018, directly to IOP or via present-day IOP members to find out in their Archive more about Heppner and his afterwards potentially *'erased'* election as a Fellow of the Institute of Physics were not successful.[301]

After a *'collapse'* presumably in August 1945, Heppner was brought to the mental hospital Hatton,[302] 4 km west of his Leamington residence, then to Scotland, and after the Scottish hospital became too expensive, to Essex; he finally moved with his wife to a modest house in Harrow, London, near his sister who lived nearby in Wembley, where GEC had its main site; he died in Harrow from a massive stroke on 3 March 1982.

Next reappeared *'friend'* Hellmut Weinrich in the post-war patent records, after he returned in February 1950 from the Soviet internment camp (former CC) Buchenwald, first for a short while to his hometown Chemnitz, before he moved to Oberhausen-Sterkrade, to work at *GHH Gutehoffnungshütte*.

This *'patent signal'*, more or less at the first possibility appears to recall a number of common ideas which he could have developed together with Fritz Heppner before 1935. In 1951 Weinrich filed a patent[303] for a *'Gasturbine für wechselnde Belastungen* (Gas turbine for changing loads) with five-stage axial lp compressor and three-stage axial lp turbine, characteristically interconnected by a free-wheeling(!), dual-flow radial compressor/turbine hp wheel, Fig. 8.37b. The similarity of this device with Heppner's 1941 patent[304] of a compact dual-flow centrifugal impeller with free-wheeling diffusor, Fig. 8.37a, is stunning. In addition, this Weinrich patent has a rotating combustor feature, Fig. 8.37c, with fuel supply through the centre shaft, identical to Heppner's preferred solutions as visible for the *T.B. engine*, Fig. 5.13 (bottom, r, see central fuel feed on the right rim), and for the *ASH engine*, Fig. 8.31 (see central fuel feed on the left), and completely different e.g. in comparison to Griffith's *C.R. engine*, Fig. 8.28 (see off-centre fuel feed on the left). Two observations are

---

[300] Fellow of the Institute of Physics (FInstP) is the highest grade of membership, elected by making *'an outstanding contribution to the profession.'* The Fellow's gown follows the pattern of the Doctor's robes of Oxford University in black with 4″ cuffs in violet damask, the cuffs slightly gathered with red cords and violet buttons; Fellows wear a doctor's bonnet in black velvet with red tassels.

[301] In principle, Heppner's affiliation to 'Physics' followed a certain logic due to his 'Vordiplom' in *Theoretical Physics* at TH Dresden on 22 June 1929, Fig. 5.2 (m), while he graduated from TH Berlin-Charlottenburg in ~1932 as *Dipl.-Ing.*

[302] See Wikipedia, 'Central Hospital, Hatton'in English, stating that—as a kind of post-war trauma— *'the hospital was overcrowded for over 20 years between 1945 and the late 1960s.'*

[303] Weinrich patent DE899,298, with priority 9 Dec. 1951.

[304] Heppner patent US2,334,625, 'Turbomachine', with priority 26 March 1941.

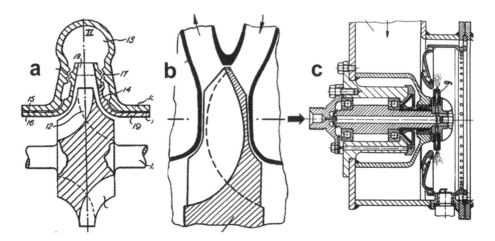

Fig. 8.37 Heppner/Weinrich patent symbiosis: (**a**) Heppner dual-flow radial impeller, 1941, Weinrich (**b**) dual-flow compressor/turbine, and (**c**) rotating fuel injection, both 1951

worthwhile to be mentioned in this context. Weinrich, as described in Chap. 5, with more than 50+ grants a kind of *'patent entrepreneur',* had none just for the counter-rotation principle, though he had proposed and was in charge for the BMW 109-002 design, Fig. 5.11 (top), and left this feature apparently *'open'* for Heppner's 1941 application, as illustrated by Fig. 5.11 (bottom, l). On the other hand it is remarkable, and perhaps a sign of early thoughtfulness, that Heppner in his 1941 patent filings never touched the combustor zone, thus *'leaving a door open'* for Weinrich in 1951.

In 1968, Christopher Heppner, Fritz Heppner's then 35 years old son, visited his father (64) in the Scottish sanatorium, remembering:[305] *'I tried to interest him in the modern jet in which I had just crossed the Atlantic, but he went into a long spiel about boundary layer control (he still retained some of his aeronautical physics!) and how much better things were in the "Universe" where he was chief engineer—compared to that, my stories of jets, were boring stuff.'*

8.3   BBC Aero Gas Turbine Activities and Derivatives

After a short positioning of the *BBC Aktiengesellschaft Brown Boveri & Cie.,* holding company with headquarters at Baden, CH and its relationship to the expanding, increasingly mighty and independently operating German daughter company *Brown Boveri & Cie. AG (BBC), Mannheim*–Käfertal during the War, Sect. 8.3.1, technically this chapter

---

[305] Mail to the Author, dated 24 Jan. 2014.

will deal first with the Swiss Brown Boveri's turbojet engine developments in Sect. 8.3.2, followed by BBC Mannheim's salient achievements in the area of altitude engine test facilities, Sect. 8.3.3, their tank gas turbine and derivative developments, Sect. 8.3.4, and related military recuperated aero gas turbine research activities at AVA/BBC, Sect. 8.3.5, all up to 1945. The turbojet engine developments for the *BMW 003C/D* on the German side of Hermann Reuter and his BBC Mannheim development department, founded as *TLUK Turbinen-LUftfahrt-Konstruktion* (Turbine aeronautical design) in April 1941, will be part of Chap. 9.

8.3.1   Brown Boveri: A Classic in Business Administration

Brown Boveri and the unique, historic relationship development to its daughter companies belongs to the few multi-national corporations with subsidiaries that were larger than the parent company. So it is not too surprising that at times the Swiss central business unit ran into difficulties to maintain managerial control over some of its larger subsidiaries, here especially BBC Mannheim in Germany.

Up to 1936 BBC Mannheim surpassed the parent house considerably, Fig. 8.38[306] (l), and on the basis of a joint-stock capital of 12 Mio Reichsmark (RM) achieved actual revenues in the order of 112 Mio RM; consequently, the stock capital had to be duplicated to 24 Mio RM in 1938. In reference to the starting values in 1933, the revenues had grown by 400% seven years later. Though BBC Mannheim was not directly involved in weapon production, its manufacturing of electric power generation equipment had of course considerable relevance in this respect. In the power regime BBC held the third rank in Germany in 1939 with 17% market share, behind Siemens-Schuckert and AEG, both with 40%. Soon the order intake exceeded the actual production capacities, so that e.g. customers for power plant equipment in Mannheim had to wait up to two years at the beginning of WW II. This gap between demand and actual production capability grew considerably during the war years, first throttled by Baden's opposition against further huge investments, after 1943 by significant destruction of the Mannheim manufacturing plants by allied bombing raids.[307]

---

[306] See Schubert, BBC-Mannheim.

[307] See Eckardt, Gas Turbine Powerhouse, pp. 18.21, and—see Wikipedia, 'Bombing of Mannheim in WW II' in English. Before the Yalta conference, 4–11 Feb.1945, US-President Roosevelt instructed Lt. General L. Groves as the responsible US Army authority, that if the atomic bombs were ready before the war with Germany ended, he should be ready to drop them on Germany; besides *Berlin*, the area *Mannheim-Ludwigshafen* was considered as alternative target. Considerations, which became obsolete after Germany's unconditional surrender, effective 8 May 1945,—see Wikipedia, 'Manhattan Project' in English.

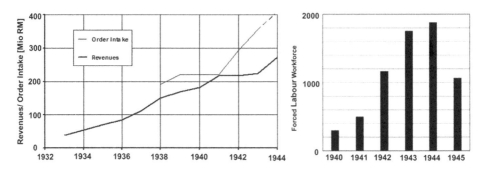

Fig. 8.38 Brown Boveri Mannheim 1932–1945: Revenues and order intake (l), forced labour workforce (r)

Since 1937, BBC Mannheim had become a dedicated *'marine plant'*, first by delivering steam turbine propulsion units up to 200,000 shp for several battleships,[308] Fig. 8.39 (m), then as prime source for German submarine electric drives. As *'state-economically important enterprise'* the number of BBC employees drafted to military services was limited to a maximum of 10%; consequently, the forced labour workforce at Mannheim was relatively modest and limited to less than 1900 in 1944, Fig. 8.38 (r), which represented approximately 30% of the employed manpower in total.

The other strong support out of Mannheim for the German Air Force since 1941 came from the above-mentioned, dedicated department for aircraft turbo propulsion *TLUK*. A third fascinating area of *'dual-use technology'* was wind tunnels for scientific research as well as for immediate military purposes. After the first installation of that kind, Ackeret's unique supersonic wind tunnel at ETH Zurich, 1932–1934, BBC was active in this field throughout Europe.[309] While Baden specialised in these high-speed facilities for aerodynamic testing, Mannheim became practically single source for all kinds of high-altitude test facilities for piston and jet engines. Further details on this widely unknown area of BBC's technology and product history have been collected in the following Sect. 8.3.3. During wartime, naval steam turbines, the altitude engine test equipment and the turbojet engine developments fell under the responsibility of BBC Mannheim's Director and Chief Engineer for Thermal Machinery, Ulrich Senger (1900–1973), Fig. 8.39 (r). In early 1948, Senger, who had joined BBC in 1922, founded the Institute of Aeronautical Propulsion

---

[308] Besides the shown—see Wikipedia, 'German battleship *Tirpitz*', commissioned in Feb. 1941 with 163,000 shp, the not completed—see Wikipedia, 'German aircraft carrier *Graf Zeppelin'* was equipped with a set of four BBC steam turbines of in total 200,000 shp; these deliveries were supplemented by huge helical gear drives, coming out of BBC Baden, CH.

[309] For a complete survey of BBC-designed and -manufactured high-speed wind tunnels,—see Eckardt, Gas Turbine Powerhouse, pp. 126–151.

Fig. 8.39 Brown Boveri Mannheim—in WW II: Hans L. Hammerbacher, CFO/CEO (l), battleship *'Tirpitz'* w. BBC steam turbine drives, 1941 (m), chief engineer Ulrich Senger (r)

Systems (ILA) at TH Stuttgart, and was full professor and director of the engineering laboratory up to his retirement in 1968.[310]

As said, BBC Mannheim managed relatively early to shake off any significant commercial control from BBC headquarters in Switzerland; on the technical side there was some cooperation between Mannheim and Baden in the 1920/1930 time period in the area of the Holzwarth constant-volume gas turbine and the subsequent Velox boiler activities, which must have included also the early beginnings of axial compressor design. But, immediately before and during World War II technical contacts were largely non-existent,[311] and did also not re-strengthen immediately after the War, when Baden might have significantly profited by German wartime turbojet development experiences in their finally failed effort to establish the Company as turbojet engine manufacturer through their French subsidiaries. Under Senger's directorate, departments for land turbines (Head Heini Meyer) and turbo-compressors (Head Max Schattschneider) were established on 6 December 1935, followed in April 1941 by the department for turbojet developments TLUK, led by the then 30-year-old Hermann Reuter.[312]

---

[310] In this period he managed the erection of a new high-altitude test facility, which in the meantime under his successors Profs. Wolfgang Braig and—since 2001—Stephan Staudacher became an internationally renowned test facility, regularly occupied by MTU Aero Engines, Rolls-Royce Germany, etc.

[311] As a confirmation,—see Senger, Die Betriebskennlinien mehrstufiger Verdichter, which may be understood as an independent position paper in 1941 in this important turbomachinery area; it was part of Senger's planned doctoral thesis which however, he was not able to finish during wartime; in 1948 he became Professor for Thermal Fluidmachinery and Steam Power Plants at the TH Stuttgart—without the normally requested engineering doctorate, but with an extraordinary broad range of practical technical expertise.

[312] Hermann Reuter (1911–1981) worked from 1935 to 1967 for BBC Mannheim; in 1948 he became successor of U. Senger; for details of his turbomachinery design activities,—see the next Sect. 8.3.4 and Chap. 9. Reuter finished his professional life as Professor for Mechanical Engineering at TH Karlruhe from 1967 to 1976.

During the Nazi regime General Director Karl Schnetzler (1876–1950) and his right hand, the influential solicitor and Chief Financial Officer Hans Leonhard Hammerbacher (1893–1964),[313] Fig. 8.39 (l), followed the Party line on the surface, but were no Party members and tried to keep the Company as far as possible out of the ideological and political implications of the time. Both were married to women with Jewish descent, which might explain these reserves.

As the figures for annual revenues and order intake illustrate, Fig. 8.38 (l), both men were convicted to conservative German-nationalistic ideas and were thus also fully committed to exploit the wartime business boom for their Company, even against repetitive warnings from the Swiss parent company. This attitude changed after 1943, when the factory production at Mannheim-Käfertal was significantly hampered by severe bombing damages, and the administration had to be transferred to the *'safe'* Heidelberg, 20 km in the south-east, and the production to several underground sites.[314] The Mannheim-Ludwigshafen area (BASF, BBC) belonged to the most heavily attacked areas by Allied *'carpet bombing'* raids[315] in Germany during WW II—with the consequence that 2/3 of BBC's production and engineering sites at Mannheim-Kaefertal were destroyed at war's end. Consequently, also the gas turbine history of BBC in Germany could be reconstructed indirectly only by means of secondary sources.

---

[313] Hammerbacher followed Schnetzler as CEO in 1945, and kept this position up to 1958. He had prepared—see Wikipedia, 'Kurt Lotz' in English, as his successor, apparently also with the target to unite the Baden and Mannheim businesses, now with a certain, in view of size, natural German dominance. Consequently, Lotz led both Companies alternately for two days from Baden and Mannheim, until he realised—what he called a certain 'anti-German opposition'—see Spiegel, Schweizer Spitze in German. Lotz left BBC in 1968 to Volkswagen in succession of Heinrich Nordhoff, where he launched the *'Golf'* as mid-term replacement of Ferdinand Porsche's then still popular *'Beetle'*.

[314] See Ruch, Geschäfte und Zwangsarbeit, and—see Schubert, BBC-Mannheim. At War's end 2/3 of the Käfertal production was in ruins. The move of the BBC administration was accelerated after a RAF 554 bomber raid in the night of 5/6 Sep.1943 which destroyed Mannheim nearly completely. Heidelberg's bombing safety can be deduced from operetta and films—see Wikipedia, *'The Student Prince'* in English, based on the play *'Old Heidelberg'*, with 608 performances *en suite* the longest-running Broadway show of the 1920s, revived there in 1943, before intensified Allied bombing started. The BBC industrial gas turbine and turbomachinery department under Max Schattschneider was dispersed to the *Robert Bunsen School* at Neuenheim—across River Neckar from Heidelberg,—see Kay, German Jet Engine, p. 196.

[315] See Wikipedia, 'Bombing of Mannheim in World War II' in English. Between 16 Dec. 1940 and 2nd March 1945 more than 150 air raids with a total bomb load of 25,181 t were recorded, killing 1700 people. These were 0.6% of the 284,000 inhabitants only due to an extraordinarily high bomb shelter capacity. Ludwigshafen-Mannheim had been *'earmarked'* as a potential target for nuclear bombing within the *US Manhattan Project*, a disastrous plan not executed due to the German capitulation on 8 May 1945. As a result of war 1300 BBC employees had lost their lives and in 1950 still 600 were reported missing.

8.3.2   Swiss Brown Boveri Turboprop Design Exercises

Immediately after the realisation of the stationary, power generating gas turbine in 1939, Brown Boveri looked for mobile GT applications, first in BBC's invention of a gas turbine locomotive, and practically at the same time also in flight applications.[316] The aero ambitions of Brown Boveri in Switzerland, which first had been documented in the extensive patent conflict with AVA Goettingen 1932–1936, Sect. 4.3.3, under the header *'In the patent arena'*, re-appeared clearly in 1939 again in a patent filing for an axial turboprop engine configuration, now officially under the name of Claude Seippel as inventor, Fig. 8.40.[317]

A closer look to the patent claims reveals that here a clear focus is put on combustion stability; the compressor blading is designed to steadily reduce the axial compressor flow velocity on the way to the combustor, and to re-accelerate the flow correspondingly thereafter in the turbine. The actual combustion chamber has a thoroughly shaped, diffusing inner contour and in this respect has a high similarity with the *'Venturi chambers'* of Hans von Ohain's first secret patent of 1938.[318]

Actually this patent received attention on both sides of the *'Channel'*, first—as mentioned—in a *Flight* issue of 15 October 1942, and correspondingly in the collected *Flight* material of G.G. Smith's book,[319] and in a secret report which D. Schwencke, Fig. 8.3 (l), issued in a limited batch of 50 copies on 15 March 1944 only for the higher Luftwaffe echelon.[320] Though Schwencke played the material as being the result of a kind of secret

---

[316] For the BBC gas turbine locomotive,—see Eckardt, Gas Turbine Powerhouse, pp. 196–203, with obvious consequences to the derivative tank gas turbine design, Sect. 8.3.4, and in due course the 'stand-alone' locomotive combustor stimulated also ideas for aero combustor designs and fuselage-mounted gas generator aircraft applications, described in GT Powerhouse, pp. 208–209, which led also to a BBC patent CH221,503 with priority 9 Dec. 1940.

[317] Inventor Claude Seippel, 'Gas turbine plant', CH214,256 and US2,326,072, with priority in Germany: 28 June 1939.

[318] Hans von Ohain's first secret patent No. 317/38 and the corresponding background of the diffusing *'Venturi chamber'* combustors will be discussed in detail in Sect. 12.1.1.

[319] See Smith, Gas Turbines, 3rd ed., April 1944, pp. 75–76, '... *Only 20 to 30 per cent of the air delivered by the compressor is required for the combustion of fuel; the remainder is employed to lower the temperature of the combustion gases,'* and *'Flame forms at or beyond the* (Venturi chamber) *grid bars* (20), *which furnish a number of friction surfaces along which the boundary layers of the mixture move at lower velocity than the ignition velocity. Small flames are formed locally which preheat and ignite the main mass of the mixture and thus maintain stable conditions.'*

[320] See Schwencke, Strahltriebwerke. The distribution list of the report comprises a.o. RLM State Secr. E. Milch; Chief of Luftwaffe (Lw) Planning Office—see Wikipedia, 'Ulrich Diesing' in English; RLM GL/C-E3 H. Schelp; Chief of Lw Gen. Staff—see Wikipedia, 'Günther Korten' in English; Chief of Lw *Führungsstab* (Operational Staff)—see Wikipedia 'Eckhard Christian' in English, General Fighter A/C A. Galland, Gen. Bomber A/C W. Marienfeld, Gen. Reconnaissance A/C G. Lohmann, and Lw HQ—see Wikipedia, 'Lager Robinson' in German, and—see Wikipedia, 'Hans Jeschonnek' (death) in English.

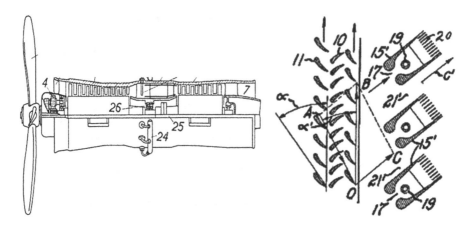

Fig. 8.40 Swiss Brown Boveri turboprop engine patent, 1939: Geared all-axial design with propeller and thrust nozzle (l), annular combustor arrangement with '*Venturi chambers*' (r)

service operation, it was for the BBC patent actually based on the 1942 *Flight* article,[321] and correspondingly, he was misled in his conclusion, that '*the Swiss Brown Boveri designed, presumably in close cooperation with the English, a well-configured, multi-stage, all-axial turbojet engine with an especially heat-shielded annular combustor*'.

BBC of Switzerland followed then already ideas of a post-war turbojet development and production together with their French subsidiary *C.E.M. Compagnie Électro-Mécanique*, and their aero daughter company at Paris-Le Bourget since ~1940, *SOCEMA Societé de Construction et d'Équipements Mécaniques pour l'Aviation*, which will be outlined in detail in Sect. 10.3.2. From 29 to 31 July 1941 Claude Seippel, BBC's chief engineer for gas turbines travelled from Baden, CH, via Geneva to Lyon and Grenoble in the then unoccupied *Vichy Zone* of France, where he met the chief engineers of C.E.M. Georges Darrieus (1888–1979),[322] well-known and good friend to Claude Seippel, and of SOCEMA, Paul Destival, whom Seippel then met for the first time to discuss and support future aero engine projects, unknown to the German occupants.[323] The circumstances of this trip are noteworthy—after the armistice on 25 June 1940, France was split in an

---

[321] This is also a strong hint, that the mutual patent information system was only of limited use during wartime.

[322] Darrieus's contributions to Brown Boveri's early axial compressor developments have been outlined in—see Eckardt, Gas Turbine Powerhouse, pp. 98–101.

[323] Details of this trip originate from Cl. Seippel's diary 1941, which reads in this context in French: '*29/31 VII—Lyon, Darrieus, Destival; Grenoble Maue. pénurie!*', i.e. after the aero gas turbine meeting at Lyon, Seippel continued to Grenoble to contact BBC's hydro and electro machinery plant with the recorded observation of a '*mauvaise pénurie*', a serious lack of food. Information, with special thanks to Cl. Seippel's son Olivier (1926–2012), also employed in various functions at BBC, who helped to revive personal memories of his father, especially by providing insight into the BBC part of Cl. Seippel's diary notes.

occupied north zone and a so-called '*zone libre*' of the Vichy regime to which belonged Lyon and Grenoble. On 11 November 1942 whole France was occupied by the Germans, so the Lyon meeting under consideration took place still during the Vichy period.

It was actually in that very year 1941, that a team of technicians and specialists in steam and industrial gas turbines at SOCEMA began looking at the possibilities of developing the gas turbine for powering aircraft. Initially this team, led by Paul Destival, head of the technical department working on gas turbines at C.E.M., was completely isolated from the engineering world by the German occupation. It is said that they were unaware of the existence of actual German and British[324] work in the aeronautical gas turbine field, although they were aware of patents such as those of Frank Whittle and Brown Boveri. In fact, the SOCEMA engineers had considered aeronautical use of the gas turbine well before war broke out and, in particular, a turboprop engine was seen as the most likely way forward.

With respect to travels out of Switzerland, there are two subsequent events worthwhile to note. From 14 to 19 September 1941, a mixed Swiss university/industry delegation consisting of Prof. J. Ackeret, ETH Zurich; Cl. Seippel, BBC Baden; C. Keller, Escher-Wyss Zurich, and the Ackeret co-workers P. de Haller and G. Dätwyler met Professors L. Prandtl, A. Betz, A. Busemann and O. Walchner at AVA Goettingen to discuss orders for two Ackeret-designed Mach 4 high-speed wind tunnels from BBC to Germany.[325] And on Fri 4 December 1942, Professor J. Ackeret alone followed Prandtl's invitation to lecture about '*Towards the gas turbine*' at the annual session of the German Academy for Aeronautical Research at Berlin. In principle Ackeret repeated on 2½ pages the status of the power generation gas turbine development which Brown Boveri had achieved with the Neuenburg plant in 1939. More remarkable is an attached, printed discussion note by the then 67-years-old Prandtl, in which he congratulated '*our Swiss colleague to this exceptional success of a very special gas turbine application. Given Stodola's long-time preparatory work, it may not be purely accidental that this happened actually in Switzerland, and with all modesty we may remind that Ackeret—after his stay at Goettingen—has beneficially introduced there also flight technical design aspects, etc.*' Taking the certainly existing, strict confidentiality rules into account, there is nevertheless not the slightest hint to assume that neither he nor Ackeret were informed—5 months after the Me 262 first flight—about the actually achieved status of turbojet developments.

Turboprop designs were and remained the prime aero gas turbine applications for BBC Switzerland up to War's end—and beyond (Sect. 10.3). Hans Pfenninger proposed in 1944 a detailed design of a 2500 hp all-axial turboprop flight engine, Fig. 8.41 (l), for which he

---

[324] See Destival, SOCEMA and—see Destival, French Turbo-Propeller; given the described, close contacts between RAE Farnborough and BBC Baden on the one side, and the unchanged close cooperation between BBC and its French subsidiary C.E.M., this statement surprises.

[325] See Eckardt, Gas Turbine Powerhouse, pp. 131–135, '*The AVA/ LFM wind tunnel projects*'.

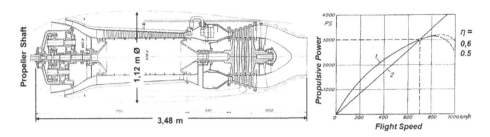

Fig. 8.41 Brown Boveri 2500 shp turboprop proposal—with 16-stage axial compressor and 5-stage axial turbine, using journal bearings (l), turboprop *1* and turbojet *2* propulsive power vs. flight speed (r), by H. Pfenninger, Baden 1944 © ABB Archive

carried out also a correct performance analysis, with considerable propulsive power advantages of the turboprop configuration below 700 km/h flight speed, Fig. 8.41 (r).[326]

The Swiss industrial policy between Allies and Third Reich during the War caused also some criticism afterwards; perhaps Churchill's conciliatory post-war evaluation came close to the truth: '*Of all the neutrals Switzerland has the greatest right to distinction. She has been the sole international force linking the hideously sundered nations and ourselves. What does it matter whether she has been able to give us the commercial advantages we desire or has given too many to the Germans, to keep herself alive? She has been a democratic State, standing for freedom in self-defence among her mountains, and in thought, in spite of race, largely on our side.* '[327]

8.3.3   Altitude Test Facilities from Brown Boveri Mannheim

Preparatory steps towards the introduction of altitude test facilities at the end of WW I at BBC Mannheim have already been described at the end of Sect. 4.1.2, and shall be repeated here in short. First aero engine altitude test facilities had been used by the engine company *Maybach* on Mt. Wendelstein (1838 m) in the Bavarian Alps as early as 1916/17, Fig. 4.16. On 17 October 1918 Sanford Moss, GE carried out turbo-supercharged *Liberty* engine tests on the highest accessible mountain-top in the United States—Pikes Peak, 14,109 ft

---

[326]See Pfenninger, Die Gasturbinenabteilung bei BBC. See also Fig. 10.20, where these developments are brought in the context of SOCEMA's TGA 1 turboprop. To the rather uncommon tasks of that BBC department belonged e.g. the development of an accelerating drive for the landing wheels of Swiss Airforce aircrafts to prevent the excessive rubber wear after touch-down at landing speeds up to 150 km/h; the solution was realised by means of an oil drive with a screw pump/ motor combination.

[327]See Churchill, Triumph and Tragedy (The Second World War, Vol. 6), p. 712, often interpreted as a neutralising comment towards Stalin's hefty critique against Switzerland.

(4300 m) in the Rocky Mountains.[328] In the 1920s the Italian *Aeronautica Militare* had founded a renowned, long-time reference facility at the Guidonia National Aeronautical Test Centre near Rome, which allowed for a complete simulation of air-cooled motor operation of max. 500 hp power up to 4900 m.

In August 1933 the German Aerospace Centre DVL at Berlin-Adlershof decided to build an engine altitude test facility[329] with partial simulation capability up to 24,000 m. Main contractor was BBC Mannheim, which received also the right to build copies of this DVL standard facility N_I at the military 'E-Stellen' (E—Erprobung—test facilities), at a number of research institutions and at the various engine companies. Between 1933 and 1943 BBC delivered 14 such N_I facilities[330]—with mass flows between 1.2 and 3.0 kg/s.

As early as 1941 the opinion spread amongst the various operators and BBC that a complete altitude simulation of flight conditions within a *'climate chamber'* was inevitable—with turbomachinery and cooling equipment for mass flows up to 20 kg/s. In total, six of these N_{II} *facilities* were ordered from BBC Mannheim between 1940 and 1944, of which only the one[331] for BMW Munich-Milbertshofen could be put into operation in mid-1944 by testing a BMW 003 turbojet. The heart of the plant was a test chamber of 3.8 m diameter

---

[328] Katharine Lee Bates' lyrics of *'America the Beautiful'* are said to be owed to the stunning vistas from the summit of Pikes Peak. Besides the exceptional scenery, there is a hidden technical relationship to Switzerland, and Brown Boveri's early railway past, Fig. 4.2. Since 1964 the Manitou and Pikes Peak Railway, the highest rack railway in the world, uses equipment from SLM Swiss Locomotive & Machine Works, Winterthur,—see Wikipedia, 'Pikes Peak Cog Railway' in English. M&PPR uses the *Abt rack system*, limited to maximum grades of 25%. Many rack railroads like the one on Mt. Rigi, 15 km east of Lucerne, CH, use the *Riggenbach system*, also called *'ladder rack'*. The steepest cog railway in the world is the Mt. Pilatus Railway, 5 km south of Lucerne. It uses the *Locher rack system* to climb grades of 48%. The steepest cog locomotive track along the Panama Canal has short grades of even fifty percent, it uses also the Riggenbach system.

[329] The idea was decisively boosted when Asmus Hansen, who had submitted in 1930 at TH Dresden his doctoral thesis *'Die thermodynamischen Grundlagen des Höhenflugmotors'* (The fundamental thermodynamics of the altitude flight motor), struggled with an engine fire while descending from a stratosphere flight up to 13 km with the *Junkers Ju 49* in 1932, Fig. 4.34—and was forced to an emergency landing with *'stehender Latte'* (seized propeller) at the Junkers airfield, Dessau.

[330] See Barth, Zur Entwicklungsgeschichte der deutschen Flugtriebwerks-Hoehenpruefstaende (with the following information on BBC altitude test unit sales provided by H. Koeckritz, BBC Mhm. in 1966), N_I units were delivered to DVL Bln.-Adlershof (4), E-Stelle Rechlin (2), BMW Munich, FKFS Stuttgart, DB S.-Untertuerkheim (2), Junkers-Motorenwerke Dessau (2), DFL Braunschweig, AVA Goettingen, Rheinmetall-Borsig Bln.-Tegel, Heinkel-Hirth S.-Zuffenhausen; in addition, Moscow CIAM etc. (5), up to June 1941, not completely delivered,—see Ruch, Geschäfte und Zwangsarbeit, p. 85.

[331] At least the equipment of six more N_{II} facilities was delivered by BBC, but not put into operation: DB S.-Untertuerkheim, DVL Bln.-Adlershof, E-Stelle Rechlin (2), BMW Munich, LFM Munich. The two N_{II}—sets for the 'E-Stelle Rechlin', 130 km in the north of Berlin, are said, were stored *'in a wood near the field'*—see C.I.O.S, Gas Turbine and Wind Tunnel Activity. The E-Stelle had at Rechlin-Laerz two parallel concrete runways, each 2400 m long and 50 m wide. BBC charged for one N_{II} altitude test equipment 6,300,000 RM with an extra one Million RM for the air conditioning plant.

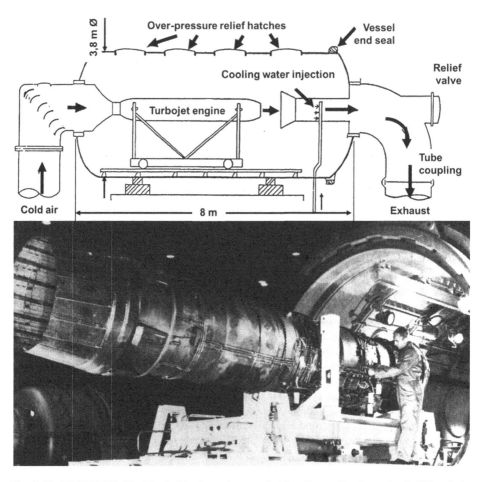

Fig. 8.42 BMW/BBC *'Herbitus'* altitude engine test facility: Test cell scheme (top), GE turbojet engine installation at *AEDC ETF* ~ 1955 (bottom)

and 8 m length, in which either a piston engine or turbojet engine was mounted, and where ambient temperatures could be varied from +55 °C to -70 °C, Fig. 8.42 (top). A three-stage BBC centrifugal compressor, Fig. 8.43 (r), with 3.3 MW drive motor and with a compression ratio PR 2.4 gave a maximum airflow of 25 kg. The air first passed through a heat exchanger, which could be used to heat the air if required, but normally the air was cooled by water sprays, passing thereafter through a Freon plant with seven-stage centrifugal compressor, which brought the temperature down to -15 °C, and was then expanded through a turbine doing useful work, which reduced the air temperature to -70 °C if required. By an adaptable bypass duct between compressor outlet and turbine inlet, any desired air temperature could be obtained. From the expansion turbine, the air passed to the engine within the test cell, giving the necessary ram pressure for simulating speeds up to 900 km/h. Altitude conditions were produced in the exhaust of the test cell by first cooling

Fig. 8.43 BMW/BBC *'Herbitus'* altitude engine test facility: Crate shipment from BMW Munich-Milbertshofen site to AEDC, 1946 (l), BBC 3-stage centr. compressor inspection at AEDC (r)

with water sprays, bringing the temperature down to 250 °C, and then through a cooler to reduce the temperature further to 40 °C. Two additional four-stage centrifugal compressors with 4.5 and 2.7 MW e-drives worked as vacuum pumps either in series or parallel, depending on the simulated altitude to produce the requisite low pressure in the exhaust. If the exhaust pumps worked in series, an intercooler was placed in-between. The pressure range at entry to the test cell was between 1.5 and 0.008 bar, corresponding for the latter value to an altitude of 16.5 km.[332] The test facility, code-named *'Herbitus*[333] Anlage (facility)'*, had a power requirement of in total 30 MW, so that the testing had to be limited

---

[332] Detailed facility descriptions are contained in—see Hirschel, Aeronautical Research in Germany, pp. 216–224, and—see Stoeckly, Bavarian Motor Works, in English, and in—see Hirschel, Luftfahrtforschung, pp. 201–209, and—see Barth, Zur Entwicklungsgeschichte, in German. In a letter, dated 1 Feb. 1971, from Herbert Gassert (1929–2011), BBC Mannheim's Chairman of the Board between 1980–1987, to MTU board member Dr.-Ing. Karl Ad. Müller, the BBC turbomachinery for *'Herbitus'* has been specified: two air compressors V1103 and V1102, three gas compressors V1604, V1104 and V704, including gearbox transmissions and e-drive motors, one air expansion turbine—with type identification, V standing for radial compressor, first three digits for impeller diameter in cm, and the fourth digit for the number of stages.

[333] Camouflaging, but for insiders nevertheless revealing project names were a speciality of educated German military; other examples will be *'Hetralin'* in Sect. 8.3.5, and *'ATAR'*, Chap. 9. The code name *'Herbitus'* of the BMW altitude engine test facility was demystified recently as follows: It was located ~100 m south-west of what is today the street corner of Lueneburger- and Konstanzer-Strasse in Munich-Milbertshofen, where then (and still today) the housing BMW Power Station, Bldg. 100 existed, Fig. 8.46. East thereof was the eponymous *'Oberwiesenfeld'*, site of F.-Z. Diemer's altitude flight record in 1919, Fig. 4.29, till 1939 Munich's only civil airport and what became the *'Olympia-Park'* in 1972. Ober-Wiese(n)-Feld has Wiese/meadow/grassland in the middle which in lat. is 'herba'; accordingly, 'herbescere' means 'to pullulate/to sprout' and the past participle thereof 'herbitus' something like 'it is pullulated/it is sprouted' or 'green overgrown'—with the additional benefit that the ending *US* could be interpreted as reference to the initials of Ulrich Senger as the responsible BBC chief engineer. In this context, special thanks go to Maria Kissler, Bochum for her kind support in Latin grammar.

to night hours to simulate test flight velocities up to 900 km/h and altitudes up to 13 km. A day-time crew of ten was used for maintenance, while night operation was carried out by another 10 men. Jet thrusts were measured mechanically, later by using load cells. In addition, a periscope in the control room enabled combustion conditions to be examined at two longitudinal positions.

Testing ran till the end of WW II on 8 May 1945, and continued immediately thereafter under US control, mostly with the German turbojets Jumo 004 and BMW 003 as test objects, but also with the Rolls-Royce *'Derwent'* and the de Havilland *'Goblin'* engine in autumn 1945.[334]

Beginning mid-1946 the 'Herbitus' facility was disassembled and shipped to the United States, Fig. 8.43 (l), where the now *ETF 'Engine Test Facility'*, re-constructed at AEDC Arnold Engineering Test Center in Tullahoma/Tennessee, had been put back in operation as the first of the transferred German facilities as early as 1954, Fig. 8.42 (bottom). The head of the BMW 'Herbitus' plant Christoph K. Soestmeyer (1910–1997)[335] was made to accompany the shipment and re-erection, and at the same time to carry out the planning of a considerable facility enlargement.[336]

In parallel to the transfer of the unique BMW altitude test facility, there were also all stored materials—mostly from BBC Mannheim—for the planned, first 1×1 m, Mach 10 hypersonic wind tunnel project[337] shipped from the site at Kochel am See, some 75 km south of the *'Herbitus'* installation, also to AEDC and put into operation as *'Tunnel A'*[338] in the late 1950s.

Figure 8.44 illustrates the combination of both facilities at AEDC Tullahoma in the X-15, Mach 7, experimental rocket-powered aircraft test programme. The versatility of the continuously modernised and updated, nevertheless in principle unchanged plant since the concept fixation in the early 1940s is demonstrated in Fig. 8.45, illustrating a 2020 ETF set-up in test cell J2 of the advanced Pratt & Whitney F135 afterburning turbofan for the Lockheed Martin F-35 Lightning II, USAF's most recent single engine strike fighter.

---

[334] See Bailey, The Early Development, p. 60.

[335] Ch. Soestmeyer came with practical experiences at the smaller DVL altitude test facility in 1941 to BMW Munich, where he together with U. Senger, BBC Mannheim and a number of responsible engineers from the altitude test facilities at Rechlin and Junkers Dessau designed the *'Herbitus'* concept in a rather rare cooperative approach. Soestmayer stayed several years as the AEDC ETF 'Manager of Operation',—see Hirschel, Aeronautical Research.

[336] One post-war addition to the Tullahoma turbomachinery comprised two large DEMAG centrifugal compressors with PR 2.5, V_1 114 m³/s entry volume flow and 14 MW max. power consumption,— see Barth, Zur Entwicklungsgeschichte, p. 21.

[337] See Eckardt, Gas Turbine Powerhouse, pp. 137–148, and—see Eckardt, The 1×1 m hypersonic wind tunnel. At War's end the turbomachinery and materials deliveries were ~80% complete on stock at Kochel. One of the decisive differences between Kochel planning and Tullahoma realisation was unlimited use of e-motor drive technology, so that the planned hydraulic turbine drives were no longer required for the compressor field.

[338] The designation *'Tunnel A'* dates back to planning of the Peenemuende team under W. von Braun in the late 1930s.

Fig. 8.44   Versatile AEDC *ETF Engine Test Facility*, 1958–1959: Reaction Motor 50 klbf liquid rocket test for X-15 aircraft (l), 1/18 scale X-15 model in 1×1 m AEDC *'Tunnel A'* (r) © USAF

Finally, the aerial view on BMW's present day headquarters and manufacturing plant at Munich-Milbertshofen, in immediate vicinity of the tent-architecture of the 1972 Olympic Stadium by G. Behnisch, F. Auer, and Frei Otto's early design application of FEM Finite Element Methods is an appropriate surrounding to remember the original 'Herbitus' location as another lasting piece of great engineering art, Fig. 8.46.[339]

8.3.4   BBC Support to Tank Gas Turbines and an Aero Derivative

If true, it was as early as mid-1943 that the notoriously self-centred Max Adolf Mueller,[340] formerly of the *Junkers Jumo* aircraft powerplant division in Dessau, and then *Heinkel-Hirth's* jet engine division at Stuttgart-Zuffenhausen, proposed the use of a gas turbine for

---

[339] The excellence of the German altitude test equipment may be also put in better perspective in comparison to the Farnborough altitude (motor) engine test chamber, for which two independent statements of UK visitors at Wright Field in September 1938 exist, as minutes of meetings,—see Douglas, The Secret Horsepower Race, p. 66 f. Rolls-Royce's Stanley Hooker *'was of the opinion, that the Farnborough altitude test chamber devoted most of its effort to making the equipment work and very little actual testing had been completed,'* and A.A. Griffith stated shortly thereafter more cautiously, that *'their air-cooling system, which is an air engine wherein the air is compressed, cooled and expanded, is not entirely satisfactory. While it is a compact arrangement, it is necessary to pass the air through a centrifuge to take out the snow. Even then under cruising conditions they have trouble with icing.'*

[340] See the foregoing Chap. 7, and—see Wikipedia, 'GT 101' in English. M.A. Mueller was then not only already long-time NSDAP party member, but had apparently also strong ties to Waffen-SS technical institutions (without knowing his actual rank).

Fig. 8.45 Versatile AEDC *ETF Engine Test Facility*, 2020: Pratt & Whitney F135, 48 klbf afterburning turbofan test (l), for Lockheed Martin F-35 Lightning II combat aircraft (r) © USAF

armoured vehicle engines. A gas turbine would be much lighter than the 600 hp-plus class, gasoline-fuelled reciprocating piston engines being used in the next-generation tanks, to that time primarily sourced from the *Maybach-Motorenbau GmbH*, Friedrichshafen for the *Wehrmacht*'s existing armoured fighting vehicle designs, that it would considerably improve their power-to-weight ratio and thereby improve cross-country performance, and potentially outright speed.

At that time, the only land vehicle which had been driven by a gas turbine was a Swiss Federal Railway (SFR) locomotive, built by Brown Boveri, Baden and put into operation on 1 September 1941, Fig. 8.47.[341,342] For BBC, the new GT locomotive was the highlight of the company's 50th anniversary celebrations on 29–30 September 1941, touring with guests between Zurich—Baden—Olten, and back again. With respect to fuel-economy the GT-powered locomotive had to find its place between steam and diesel drives. For stationary plants the condensation steam turbines were hard to beat, but with atmospheric exhaust, as required for locomotives, the specific fuel consumption (sfc) was already in favour of the GT locomotive. On the other hand, though the sfc comparison was clearly pro diesel locomotive with 180 g/PSh compared to 370 g/PSh for the GT configuration,[343] the considerable price difference between diesel fuel and heavy fuel oil (*'Masut'*) made the GT

---

[341] See Eckardt, Gas Turbine Powerhouse, pp. 196–204.

[342] Shown is a pre-stressed concrete bridge with Gotthard granite cover, re-built in 1945–6, after a foregoing steel lattice construction of 1876, was partially destroyed by an US air raid with 20 bombs on bridge and the neighbouring hamlet Rheinsfelden, killing three,—see Wikipedia, 'Aerial incidents in Switzerland in World War II' in English.

[343] This GT sfc value applies for 50% load, and the integral use of a heat exchanger in the GT locomotive engine. In view of the GT 102 tank engine (without heat exchanger), the resulting part load sfc should have been between 500 and 600 g/PSh,—information, courtesy of Norbert Lang, 13 Dec. 2021.

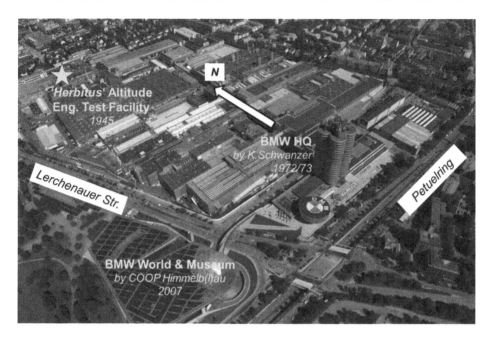

Fig. 8.46 Location of *'Herbitus'* Altitude Engine Test Facility, 1945, in aerial view of present BMW Munich plant site, 2021 © BMW

concept appear economically attractive at first. Consequently, one of the first GT locomotive modifications in December 1942 was the adaptation to cheaper fuel oil; the negative consequences were dark smoke emissions. Thereafter, test trips took place on the route Basel—Zurich—Chur with trailing loads of up to 500 t. The demonstrated maximum power was 2800 hp, the peak velocity reached 128 km/h. From 3 June 1943 until 18 July 1944, the GT locomotive was in continuous daily service for 297 days on the 150 km long, secondary line from Winterthur to Stein/Bad Säckingen, where the elevated railway tracks follow the Rhine river for more than 100 km, in immediate vicinity to the border and in good visibility from the German side, Fig. 8.47 (r).[344]

Prime reason for initiating such a radical project like the GT-powered tank was—what BBC had already used as argument for the GT locomotive—that the gas turbine could operate under a much lower grade fuel than the internal combustion engines then in service, and such a consideration was of great importance in view of the Reich's critical fuel position. Another essential reason for interest in the gas turbine was that it offered more power for a given size and weight, i.e. the 1150 hp of a GT tank engine vs. the 620 hp net power of the Maybach. Of particular interest was the specific power output as the ratio of

---

[344] BBC historian and railway expert Norbert Lang expressed reservations against this insinuated GT locomotive *'show case'*, since the secondary non-electrified line Winterthur—Basel with the BBC Muenchenstein assembly plant at the Basel end represented a classical test route to omit the bottleneck Zurich and the blocking of the critical *Boezberg* Jura crossing; courtesy of NL, 10 Dec. 2021.

Fig. 8.47 The daily exhibition along the German frontier, 1943–1944: Brown Boveri 2000 hp GT locomotive Am 4/61101 (l), *Glatt Bridge*/Rheinsfelden, CH (r) en route © ABB Archive (l)

engine power to vehicle weight in hp/t, which the proposed GT *Panther* tank would offer, especially in comparison to the Russian *T.34* tank.[345] The higher the power-to-weight ratio the more manoeuvrability and acceleration the vehicle would possess, with consequent tactical advantages.

Tank types	Weight t	Spec. power output hp/t	Max. speed km/h
Russian T.34	26	20	53
German Panther	46	13.5	54
German GT-Panther	*~43*	*27*	*54*
American Sherman	30	13	39
British Churchill	38	10	26

Overall responsibility for the GT-powered tank was in the hands of Dipl.-Ing. Otto Zadnik (1887—died after 1956) of Porsche KG, though Ferdinand Porsche himself participated in tank testing, as illustrated by Fig. 8.48.[346] Here, Porsche can be seen

---

[345] In 1943 the *Panther*, a MAN design with effective 620 hp, 12 cyl. V, 23 l, liquid-cooled Maybach HL 230 P30 gasoline piston engine was introduced against the *T.34*, with 50,000 units production record holder during WW II, and which had surprisingly appeared on the 1941 Russian battlefields. Subsequent tabulation adapted—see Kay, German Jet Engine, p. 161.

[346] Porsche's honorary Dr.-Ing. title, actually his second after TH Vienna in 1916, was granted by TH Stuttgart for winning the *1924 Targa Florio* on places #1—#3 with his *Mercedes SSK,* and 126 hp, 2 l, 4-cyl. compressor engine; 6 years later the title became part of the *'Dr.-Ing. h.c. F. Porsche GmbH'*. Initially, this 'Porsche KG' had only a small staff of thirteen, but in 1945 this had risen to forty and some 120 workers, by which time the office had been dispersed from Stuttgart to Gmuend and Rheinau, Austria,—see Wikipedia, 'Porsche' in English.

Fig. 8.48 *Porsche* tank development: Testing of *VK30.01 (P)* concept for *Tiger* tank, ~1942, O. Zadnik (ctr.) and F. Porsche (r) in provis. turret (l), Dr.-Ing. h.c. Ferdinand Porsche (1875–1951, r)

standing in the provisional turret of VK30.01 (P)—a 1940 heavy tank prototype[347]—to the right with his characteristic hat, while Zadnik with beret is seen in the centre. He was primarily an electric transmission specialist and together with his small team, had the task of designing the transmission gear and overlooking the GT installation in the vehicle. The electric drive of the Swiss GT locomotive and its compact installation of the powerplant must have looked attractive for the German engineers. Besides these technical associations, even more intriguing is the location of the small Porsche design office at Rheinau(weg) in Bregenz-Hoechst, literally *on* the Austrian-Swiss border.[348] The office site is only some 60 m north of the border, represented here by the right bank of the *'Old Rhine'* river, some 10 km before its confluence into Lake Constance. Since Zadnik grew up in nearby Bregenz, he may simply have favoured his home town in his choice of office location. However, the location in open space is surprising, especially since similar German facilities were usually underground at that time. The most reasonable assumption is that the vicinity of the Swiss border either promised some shelter against Allied air raids, or alleviated cross border *'communications'*.

Since 1943 the GT tank installation planning followed several possible options under time pressure, but finally on 25 September 1944 the 45 t *Panther* tank was selected as optimum choice, Fig. 8.49 (l); of the few preserved original installation drawings, Fig. 8.49 (r) shows the chosen GT 102 gas turbine concept in the rear compartment of a *Panther II* version.[349] Zadnik's task as overall project leader was to ensure that the complete

---

[347] VK30.01 (P) stands for 'VollKetten (fully tracked)' vehicle, of 30 t target weight, and (P)orsche's first design,—see Wikipedia, 'VK 3001 (P)' in English.

[348] In Google Maps, search for < Rheinauweg, Höchst, Bregenz, Austria >.

[349] See Wikipedia, 'Panther tank'in English, and for Panther II installation drawing,—see B.I.O.S.: 'Report on German Development of Gas Turbines for Armoured Fighting Vehicles'.

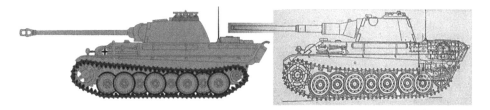

Fig. 8.49 *Panther* tank, 45 t (l), *Panther II* with heavier armour, 50 t, and GT 102 installation (upgrade not realised, r)

installation would meet the specified requirements and the work coordination of several specialists concerned with the component parts and the GT control equipment. The principal specialist was Dr.-Ing. Alfred Mueller,[350] who had design responsibility of the GT power units in association with the '*Kraftfahr-Technische Lehr-* and/or *Versuchsanstalt KTL/KVA der Waffen-SS* (Motor Technical Educational and/or Research Establishment of Waffen-SS)'[351] at St. Aegyd, Niederdonau, approximately 50 km south of St. Poelten, which was founded from KTL Vienna in July 1944 on order of SS-Führungshauptamt (Leadership Main Office)[352] for the intended transfer of turbojet engine technology to tank propulsion.[353] As a kind of project kick-off, in a meeting at the *HWA Heereswaffenamt* (Army ordnance office) at Dresden on 30 June 1944 an administrative task description was handed out to Alfred Mueller and his assistant Dipl.-Ing. Paul Kolb. Thereafter, a taskforce

---

[350]Reference literature, e.g.—see Wikipedia, 'GT 101' in English, often mixes actions and responsibilities of Max Adolf Müller (1901–1962),—see Chap. 7, and Dr. Alfred (Hermann) Müller (1907–1979). In general, the author follows the corresponding assessments of—see Kay, German Jet Engine, p.156 f. A further complication was a name change/ extension of the latter in 1969 to Alfred H. Müller-*Berner* by adding his wife's name to end the life-long naming confusion. Dr.-Ing. A. Mueller, aircraft engine designer at BMW Munich since 1933, received his academic degree for a thesis on '*The inner cooling of turbocharger turbine blades*' at Kamm's FKFS in 1941. In Jan. 1944 he finished designing the excellent turbo-supercharger for the BMW type 801 J altitude aero piston engine, before he took over the tank GT development task. Finally, in Feb. 1945, he was replaced by Max Adolf Mueller. Since 1957, after an early stay at C.A. Parsons & Co., where he led a small 1000 hp tank GT development team, and a continuation at Leyland Motors in England as Chief Development and Research Engineer for engines, trucks and busses since 1952, A. Müller worked in the piston engine development at Daimler Benz Stuttgart under Fritz Nallinger, and ended his career in 1972 as Director of the DB utility truck development under Hans Scherenberg,—see ATZ (1972), p. 456.

[351]Both designations—KTL and KVA—are in use in this context.

[352]See Wikipedia, 'SS *Führungshauptamt*'in English. Since 1943 the SS-FHA was led by Hans Jüttner; an encounter of Helmut Schelp with Jüttner on '*forced labour*' issues in the context of turbojet engine production in mid-August 1943 has been described in Sect. 7.4.

[353]See Rabl, Das KZ-Aussenlager St. Aegyd, p. 14 and p. 24. Beginning 2nd Nov. 1944 existed there also for five months a '*KZ-Aussenlager* (dependence)' with some 300 prisoners from KZ Mauthausen, apparently to build tank gas turbine test facilities. The project was not realised, nevertheless 46 inmates lost their lives by murder, suicide and cruel working conditions.

'*Gruppe Versuchsbau der Waffen-SS* (Group test construction of Waffen-SS)' was founded at KTL Vienna,[354] and known under the code name '*Alfred*'. Apparently increased Allied bombing threat caused the dislocation of KTL Vienna to the small 2000 inhabitant village St. Aegyd am Neuwalde, some 80 km south-east of KTL Vienna, as early as August 1944, when the SS confiscated 42,000 m^2 of land of the later concentration camp area, the '*Caritas House*', Fig. 8.52, and the present elementary school, all Catholic Church property.

Between July 1944 and February 1945 Alfred Mueller and his small team investigated carefully all possible tank gas turbine configurations, which have been traced and documented in detail by Antony Kay.[355] The first scheme comprised a diagonal front compressor, adopted from the Stuttgart-based Heinkel-Hirth 109-011 A turbojet engine, followed by a five-stage axial compressor coupled direct to a two-stage turbine to provide hot gases for a third power turbine, mounted coaxially, but without mechanical connections to drive the vehicle gears. This design, also known as *GT 101* in a variant with a pure nine-stage axial compressor from Hermann Reuter's Brown Boveri design team, suffered from the drawback that when the load was removed, e.g. during a gear change, the third power/ working turbine would dangerously overspeed. So after some unavailing trials this development path had to be given up, and was finally replaced in December 1944 by an earlier concept with a secondary combustion power branch, Fig. 8.50a,[356] which became known as *GT 102*. Now the third, power turbine was no longer coaxially behind the compressor turbine, but removed as a self-contained unit. Therefore, the GT 102 consisted of a nine-stage axial compressor group driven by three turbine stages, and a separate '*power branch*' consisting of a second combustion chamber and a two-stage axial working turbine which connected with the tank transmission. This separate working turbine could be easily prevented from over-speeding by a slide valve control in the air supply ducting from the compressor to the secondary combustion chamber. The long, hollow turbine stator and rotor aerofoils from sheet metal clearly show A. Mueller's foregoing turbocharger design practice at BMW. Cooling air was taken from the compressor diffuser and passed through the centre bearing housing into the hollow drum-type rotor. As illustrated by Fig. 8.50b both stator vanes and rotor blades had a weld fixing—1 to the aerofoil foot section. The air-cooled turbine blades had a special cooling insert with dimpled surface—3, formed

---

[354] Located in the newly erected barracks of 1940 at the south side of Schönbrunn palace garden, today Maria-Theresien-Kaserne, and immediately after the war used by British (armoured tank) troops as 'Schönbrunn Barracks, H.Q. Vienna Grn.' which may explain that a substantial part of the KTL engineering team (Müller, Kolb, Hryniszak, Zadnik) found occupation at C.A. Parsons & Co., Newcastle,—see patent of a 'Gas turbine plant', CH275,240 with priority 7 March 1949, a design derivative of GT 102,—and Leyland Motors (Müller) up to ~1956, the latter company later known for their truck gas turbine project.

[355] See Kay, German Jet Engine, pp. 156–173.

[356] A rare original drawing of Federal Military Archive BA-MA Freiburg, RS/5, 459: 'Tätigkeitsbericht der kraftfahrtechnischen Lehranstalt der Waffen-SS, Aussenstelle "Alfred" (KTL Report, field office "Alfred")', 1 Feb. 1945.

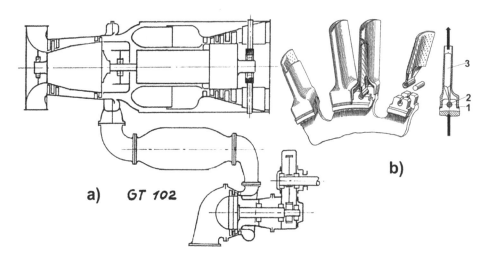

Fig. 8.50 GT 102 Gas turbine tank engine: (**a**) Concept drawing with secondary combustion power branch, (**b**) Air-cooled turbine blading © A. Kay, mod

from 2 mm thick sheet, through which was pushed a pin—2, and both were then inserted into the allotted hole in the disc edge.

Although the compressor had to deal with the full airflow of 10 kg/s, the compressor turbine was charged with only 70% thereof, since approximately 30% went immediately after the compressor in the secondary power branch. This opened up the possibility of shortening the exhaust diffuser, a most desirable aspect in view of the limited installation space with the axis of the compressor unit at right angles to the Panther tank's longitudinal axis, as illustrated in Fig. 8.51, with rear (l) and side view (r). Air entered the Panther's engine compartment through heavy grilles in the port side of the hull's upper rear surface. The compressor was of the nine-stage axial-type utilising BBC's standard 50% reaction, and was designed by the BBC Mannheim engineers Hermann Reuter, Karl Waldmann and Waldemar Hryniszak.[357] After the compressor diffuser, the annular combustion chamber of the main GT engine was provided, with a similar layout to that of the BMW 109-003 A turbojet. In order to take advantage of this experience, all dimensions remained largely unaltered, but the combustor diameter was made smaller and the number of burners reduced from 16 to 14.

Finally, the main shaft ended with the three-stage compressor turbine, for the design of its high-pressure blading special care was taken, since this component determined also to a large extent the overall efficiency of the power unit. Secondary combustion chamber and the working turbine it supplied were mounted below the main GT shaft side by side and

---

[357] According to—see Kay, German Jet Engine, p. 166: *'Brown Boveri & Cie., Mannheim having been chosen for the task because of the excellent results it had been obtained in designing a new axial compressor for the BMW 109–003* (C) *turbojet'* as part of the RLM-issued *'Hermso projects',*—see Chap. 9.

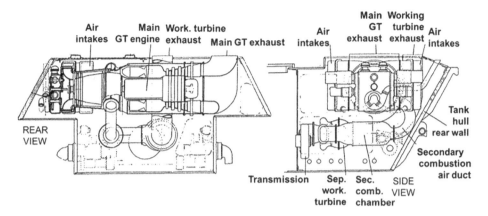

Fig. 8.51 *GT 102* gas turbine installation in *Panther* tank, 1944: Rear view (l), side view (r) © A. Kay, mod

with their axis at right angles to the compressor axis. The working turbine shaft could be connected to a *Zahnradfabrik Friedrichshafen ZF Type 305* transmission, consisting of a two-speed epicyclic gear operated by two stationary electromagnetic, multi-plate clutches.

The main data for the GT 102 installation were:

• Output shaft power	1150 hp
• speed, main shaft max.	14,000 rpm
• secondary shaft max.	20,000 rpm
• compressor pressure ratio PR	4.5
• combustion gas temperatures	800 °C
• fuel consumption	370 g/PSh
• overall thermal efficiency	16%

The fuel consumption for GT 102 was estimated to be little better than for GT 101, and about double that for the standard 700 hp Maybach gasoline piston engine. For further fuel savings, the planned *GT 103* unit was simply the *GT 102* with the addition of a regenerative heat exchanger,[358] for which improved sfc values between 150 and 230 g/PSh, and an overall thermal efficiency of 28.8% had been predicted.[359]

---

[358] The GT 103 heat exchanger with slowly rotating ceramic drum was within the responsibility of the BBC specialist W. Hryniszak, who also participated in the regenerator design of a long-range *'Amerika bomber'* turbojet engine, which is dealt with in the next Sect. 8.3.5. This project like that for the tank GT was configured as a *'last minute War effort'*—not only for the utilisation of low-quality liquid fuels, but also solid coal dust/chips were seriously investigated,—see Sect. 8.3.5.

[359] The overall thermal efficiencies of 16 percent without heat recovery, and of 28.8% with heat exchanger have to be put in perspective to the about 35 percent for a good internal combustion piston engine of the time.

The GT 102 hardware production was addressed in the aforementioned KTL Report, presumably authored by Alfred Müller: '*The manufacturing drawings for GT 102 (which implies GT 101 as well) are nearly finished. Some single parts are already in production, others are due to follow immediately. Manufacturing is mainly planned to occur at Mittelwerke G.m.b.H* (Harz Mountains, V-2 rocket, main production site). *Based on present scheduling the GT 102 unit will first be started during June* (1945). *At that time the compressor might already have been fine-tested, since its test rig is presently already relocated from BBC Mannheim to Mittelwerke, ready for operation at end of February. The design of the GT 102 working turbine requires still some more 4–6 weeks. Consequently, the complete GT 102 unit will be ready to test latest at end of July 1945.* '[360]

The criticality of the GT 102 overall length, accommodated at right angles to the Panther longitudinal axis and by taking the ends of the unit up to the sloping hull sides, which originally were occupied by fuel tanks, triggered predictably some SS-internal criticism from the camp of Max Adolf Mueller. He had instrumented his former Junkers/Heinkel compressor designer Rudolf Friedrich, who now after leaving Heinkel-Hirth before Mueller already in May/June 1941, was at the dedicated navy company Brückner-Kanis at Dresden. For the tank gas turbine design activities in 1944, Friedrich executed again, what he successfully had carried out already 3 years earlier at Heinkel Rostock for the HeS 030 turbojet engine, a considerable reduction in GT 102 length by shrinking the compressor stage count from nine to seven, leading to '*GT 102 Ausführung* (version) *2*' with the apparent consequence that the relative tip Mach numbers at compressor inlet were pushed at or beyond critical limits. Since KTL had no own expertise to decide the emerging conflict between their established engineering members, Dr. Alfred Mueller and Max Adolf Mueller, SS-Brigadeführer (Major General of Waffen-SS) and Commander of KTL Vienna Walther Neblich (1895–1945)[361] invited to a secret consultation conference on Sat/Sun 26/27 August 1944 at St. Aegyd, to decide on these GT 102 design alternatives. By some lucky coincidence Bruno Eckert (1907–1983),[362] then assistant at Professor Kamm's FKFS Stuttgart, later after the War, member of the management board of the Daimler Benz AG, Director of DB gas turbine technology, launching managing director of MTU Aero Engines, Munich in 1969, and author of the most popular Springer textbook on '*Axial and Radial Flow Compressors*', wrote his personal memoirs in 1982, which contain also an account of this conference episode.[363]

---

[360] Quoted and translated after—see Rabl, Das KZ-Aussenlager St. Aegyd, p. 25.

[361] Neblich committed suicide on 3rd April, 1945 at Vienna.

[362] Information, courtesy of Helmut Schubert, 21 Dec. 2021.

[363] See Eckert, Erinnerungen, p. 64–67. Seven pages of Minutes of that Meeting are preserved in the personal estate of the participating Prof. E. Schmidt at Hist. Archive of TU Munich, HATUM NL7.2 Ernst Schmidt, then LFA Braunschweig. The MoM confirms, besides several SS ranks, the attendance of H. Schelp and W. Brisken, RLM TLR-E3 VII, Dr. A. Müller and M.A. Müller, L. Ritz (see next Sect. 8.3.5), K. Bammert, R. Friedrich, Dr. V. Vanicek and J. Kruschik. Information, courtesy of Dr. Chr. Rabl, 03 Jan. 2022.

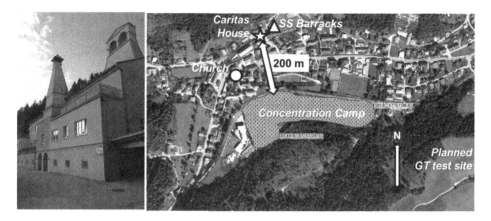

Fig. 8.52 St. Aegyd am Neuwalde, A, 1944: *'Caritas House'*, location of SS GT conference (l), survey map (r, 2000+)

Eckert remembered the extraordinary secrecy precautions upfront, which would not have allowed him even to inform his superior, Professor Kamm of FKFS; a requirement, which Eckert ignored. Then he describes the circumstances of being picked up by a Waffen-SS delegation at St. Poelten railway station, driving a cavalcade of heavy cars on narrow roads into the night with a *'frenzy speed'* up to nowhere, a *'strange monastery-type building'*, what was revealed to him the next morning as the SS-occupied *'Caritas House'* at St. Aegyd, Fig. 8.52 (l).[364] The relative comfort of his accommodation was by far outweighed by the irritating and disturbing sight of numerous KZ inmates in a neighbouring court/meadow in their typical blue-grey striped *'pyjamas'*, a sight—as he claims—then completely unfamiliar to him.[365] The aerial view of St. Aegyd, Fig. 8.52 (r) has a marked *Concentration Camp* area, of which approximately 1/3 to the east covered the actual barbed-wire fenced prisoner camp, *'Schutzhaftlager* (Protective custody camp)', in a distance of roughly 200–400 m to the expert conference at Caritas House. While apparently some underground manufacturing facilities had been planned in the slope south

---

[364] Founded in 1828, the house was run as cast iron work up to 1885, and presumably belonged to the industrial imperium of—see Wikipedia, 'Karl Wittgenstein' in English, until it was sold in ~1930 to the Catholic *Caritas* social service organisation, present address Berggasse 6. Karl's youngest son— see Wikipedia, 'Ludwig Wittgenstein' in English, had studied Mechanical Engineering at TH Berlin-Charlottenburg, from where he moved on in spring 1908 to study for a doctorate in aeronautics at Victoria University of Manchester. Before becoming the renowned philosopher, he filed there in 1910 a patent GB27,087 on *'Improvements in Propellers applicable for Aerial Machines'*—with small combustion chambers at the tip of the propeller blades.

[365] Eckert's observation caused some uncertainty in view of the confirmed conference date end of August 1944 and the official opening of the St. Aegyd satellite concentration camp on 2nd Nov. 1944. With some likelihood he referred to the *'SS Barracks'* under construction in the vicinity of *Caritas House,* Fig. 8.52 (r).

of the marked Concentration Camp area, there was also a field test site planned for the tank gas turbines under development on the open space south-east of the forest, and some 80 m above the CC area.[366]

To carry out his consultation job, Eckert received two voluminous design files after arrival at Caritas House, which as he quickly identified came from his FKFS-colleague Hermann Reuter[367] for a rather conventional nine-stage axial compressor, while the alternative design came from Rudolf Friedrich, then head of turbomachinery development at Brückner & Kanis, Dresden, and contained the seven-stage quasi-transonic axial. Apparently, Reuter's design corresponded more with Eckert's own experiences, while Friedrich's approach in the spirit of Max Adolf Müller's aggressiveness kept many unknowns to him. In hindsight, while writing his memoirs before 1982, Eckert jokingly mentioned the two contestants Reuter and Friedrich of 1944, now peacefully united as Professors for Mechanical Engineering at TH Karlsruhe.

After World War II, gas turbine tank propulsion had its high time in the 1980–1990s and is about to disappear again. The first gas turbine powered tank in series production was the Swedish turretless *Stridsvagn 103*, in service between 1967 and 1997, and (under)powered by a 490 hp Caterpillar 553 gas turbine.[368] The Russian *T-80*, with entry-into-service 1976, was the second main battle tank to be equipped with a 1000/1250 hp gas turbine, while the newest generation of Russian battle tanks *T-14 Armata* apparently relies again on 1500–2000 hp diesel power.[369] In the wake of the Russians, the GT-powered *M1 Abrams* entered service in 1980 and currently serves as the main battle tank of the US Army and Marine Corps. The 1500 hp, three-spool *Honeywell AGT1500* gas turbine drive was originally designed in the 1960s by the *Lycoming Turbine Engine Division,* and produced in the Stratford Army Engine Plant, CT. Consequently, it also received some German turbojet design flavour by Anselm Franz (1900–1994), chief designer of the *Junkers Jumo 004* turbojet engine, and during his final professional years vice-president at Avco Lycoming. However, very soon it became evident that the AGT1500 had a part-load fuel consumption problem, which initiated replacement studies in the mid-1980s. An *Advanced Integrated Propulsion System (AIPS)* programme was performed for the US Army Tank

---

[366] It is known that the tank GT prototypes were planned to be tested on the RLM-owned *'Hochdruckfeldprüfanlage* (High-pressure field test facility)' at Brueckner, Kanis & Co., Dresden, then led by Carlotto Martin, born 1903, FKFS employee and one of the last Dr.-Ing. students of Enno Heidebroek (1876–1955), Prof. for Mechanical Engineering since 1931 at TH Dresden, and its first post-war headmaster (Rektor),—see Wikipedia, 'Paul Kanis', and—see Wikipedia, 'Enno Heidebroek', both in German. In 1959 under Prof. U. Senger, C. Martin was responsible for the erection of the TH Stuttgart altitude engine test facility, technically closely related to the BMW *'Herbitus'* plant of Sect. 8.3.3,—see Hirschel, Aeronautical Research In Germany, p. 465.

[367] Reuter was employed by BBC Mannheim, and in 1944 a kind of guest researcher as FKFS,—see Chap. 9.

[368] See Wikipedia, 'Stridsvagn 103' in English.

[369] See Wikipedia, 'T-80' and—see Wikipedia, 'T-14 Armata', both in English.

Automotive Command (TACOM) to develop a *'lighter and smaller engine with rapid acceleration, quieter running and no visible exhaust'*, i.e. the *LV100* gas turbine engine, as team work among GE Aviation, Lynn, MA and Textron Lycoming,[370] Stratford, CT for engines and systems, and MTU Aero Engines, Munich and Friedrichshafen for GT low pressure turbine, recuperator and powerpack design.[371] Goals were an engine that offered replacement with not too many structural modifications, a reduced part count by 43%, improved reliability by over 400%, and a 50% reduction in fuel consumption at idle. This would have increased the tank's operating range considerably, while still offering 1500 hp to drive the M1 Abrams along at its accustomed lightning-fast clip—speeds over 100 km/h have been reported. However, with the demise of GT-powered *Crusader howitzer* the complete LV100 engine programme was shelved, and replaced after 2005 by a steady *TIGER* (sic!)[372] *Total InteGrated Engine Revitalization* program. Still, fuel consumption is a critical cost factor for the US Abrams M1 fleet, so predictably a diesel engine might be the most promising contender for a successor tank, and the engine presumably in common with the future Leopard-2 successor, the European MGCS Main Ground Combat System.

The KTL project to develop a gas turbine power plant for land traction had reached the described, transitional stage of detailed design and prototype manufacturing by the end of the War. In February 1945, Dr. Alfred Mueller was replaced by Max Adolf Mueller, with closer SS affiliation, who followed during the last remaining months the idea of an expendable Porsche 109-005 turbojet engine, as partial derivative of the tank engine project.

During the second half of 1944, the German side gradually lost their sites from which to launch the *Fi 103 (V-1)* flying bombs against London; forced back to central Holland the V-1 units could only attack targets such as Antwerp within the limited range of some 240 km. This situation generated at RLM Technisches Amt the design idea for a small expendable turbojet unit to increase the V-1 range limitations, but also to get rid of several operational short-comings of the pulsejet missile. Probably around October 1944, the official order for design studies was given, followed by the projects *BMW P.3307*[373] and

---

[370] Replaced for the subsequent *LV100-5* by Honeywell Aerospace, Phoenix, AZ.

[371] See Brockett, LV100 AIPS Technology. Further participating companies were GM's Allison Transmission Div., Indianapolis, IN, Donaldson Co. Inc., Minneapolis, MN, for air filtration system, and RCA, Burlington, MA, later replaced by Bendix, South Bend, IN, for digital engine control. MTU's share comprised a two-stage, variable geometry power turbine and a unique compact recuperator package with shock-absorbing U-tubes, where the 7 mm long elliptic *'lancet profiles'* were produced in a patented, continuous sheet metal folding/ welding process,—see H. Grieb and W. Schlosser, US4,766,953 with priority 24 March 1987, *'Shaped tube with elliptical cross section for tubular heat exchangers and a method for their manufacture'*.

[372] Possibly in association to Porsche's *Tiger I* tank of 1942, Vk45.01 (P), with 8.8 cm gun, arguably the most famous tank in WW II.

[373] See Kay, German Jet Engine, p. 139; except that the expendable BMW P.3307 was a competitor to the Porsche 109-005, not many details of design and programme achievements are known.

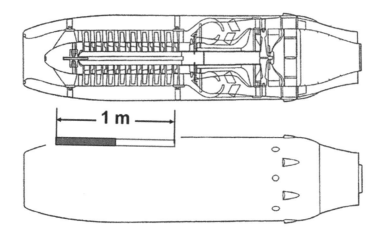

Fig. 8.53 *Porsche 109-005* expendable turbojet engine, 1945, cross section and outside view © H. Schubert

the *Porsche 109-005*,[374] of which apparently only the latter represented a kind of serious effort. In comparison to the *Argus 109-014* pulsejet with 350 kp static thrust, the alternative Porsche turbojet on the background design experience of M.A. Mueller, Fig. 8.53, produced approximately 500 kp and was planned to achieve a range of 700 km at top speed of 800 km/h, to escape the *Hawker Typhoon* interceptors over England. At War's end M.A. Müller was captured by US troops at what became US Camp Schlatt,[375] where he completed this 109-005 drawing for the Americans. For a number of reasons,[376] the compressor entry region was somewhat complicated for an expendable engine. A diagonal impeller was geared to run at 4500 rpm, while the following 8-stage axial compressor with

---

[374] See Wikipedia, 'Porsche 109-005' in English.

[375] Camp/ Lager Schlatt is 15 km north-east of Vöcklabruck, 200 km west of St. Aegyd,—see Uziel, Arming the Luftwaffe, pp. 75–76, who knows in this context: *'Müller, who was well connected to the Nazi Party ... approached the SS* (after leaving Heinkel) *with his design proposal. Initially the SS wanted to adapt his engine as a gas turbine for use in tanks, but while working for the SS, Müller also developed a new jet aero-engine, the TL-300* (Porsche 109–500), *which never proceeded beyond the design stage. In 1945 the KTL was evacuated to Sulzhayn in the Harz Mountains and then to Schlatt in the Austrian Tyrol.'* Meant is Sülzhayn/ Ellrich, approx. 9 km north-west of Mittelbau-Dora;—see also H. Reuter's moves together with KTL in the last months of War in Sect. 9.1.

[376] See Kay, German Jet Engine, p. 154; in addition, also anti-icing considerations were listed pro diagonal impeller.

50% reaction blading[377] rotated with 14,500 rpm, producing a relatively modest pressure ratio of 2.8, at 78% efficiency. After a dual-row, anti-swirl compressor exit vane arrangement followed eight individual combustion chambers with rotating (!) fuel injectors, enclosed in an annular shell, and a single stage air-cooled turbine, with blades similar to that of the GT 102 turbine, Fig. 8.50, according to BMW design principles. The turbine cooling air was not ducted from the highest pressure level at compressor exhaust, but was foreseen by Mueller to be sucked in by 'ram scoops' on the outer nacelle surface, Fig. 8.53 (bottom). Apparently the US reviewers were critical to some of Mueller's ideas, quite justly so, since this might have developed to a serious design flaw. Connecting the turbine cooling holes with the atmosphere would have resulted in hot air from the turbine escaping to the atmosphere—exactly the opposite of what is required.

8.3.5   Regenerated Aero Gas Turbine Developments

The project of an 'Amerikabomber (America bomber)' was raised by RLM as early as 1938,[378] to obtain a long-range strategic bomber for the Luftwaffe that would be capable of striking the United States, and specifically New York City from Germany, requiring a round-trip distance of about 11,600 km. However, the progress of several related projects was initially slow, and it was only after Germany had declared war on the United States four days after the Pearl Harbor attack by Imperial Japan, that the RLM started the programme more seriously in the spring of 1942, with the result that due to the lack of high-power piston engines a larger, six-engine aircraft was called for. Finally, after AVA regenerator studies of 1942 became known, which offered considerable range advantages for correspondingly modified PTL gas turbine engines, the RLM Technisches Amt launched design contracts[379] for a RGT Regenerated Gas Turbine turboprop engine for Me 264 application, of which in the following the Brown Boveri Mannheim/Heidelberg engine concept is presented in combination with the integral regenerative heat exchanger of AVA Goettingen. Nearing War's end with inherent fuel shortages, these joint BBC/AVA engineering activities were even expanded to solid hard-coal briquette burning.

---

[377] Though the Porsche 109-005 turbojet with Mueller's diagonal/ axial compressor combination achieved only PR 2.8, it had the same outer diameter of ~0.4 m as the GT 102 tank engine with an nine-stage axial compressor and PR 4.5, explanation: Mueller apparently choose a repetitive axial blading for cost reasons.

[378] Based on an otherwise unconfirmed Spiegel report 'Operation Pastorius. Hitler's Unfulfilled Dream of a New York in Flames', 16 Sep. 2010, a mock-up of the four-engined Me 264 had been presented to Hitler during a visit of the Messerschmitt Augsburg plant already in 1937.

[379] Besides BBC, H. Schelp launched a second engine design contract with AEG Berlin which is sketched in—see Kay, German Jet Engine, p. 223.

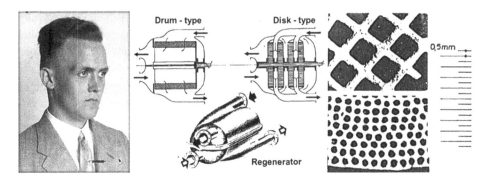

Fig. 8.54 AVA regenerative heat exchanger development, 1943–1945: L. Ritz, ~1940 (l); regenerator concepts (m); ceramic specs. (r)

The mastermind behind these disruptive aero-regenerator technology developments at AVA Goettingen was Dipl.-Ing. Ludolf Ritz (1908–1991),[380] Fig. 8.54 (l). He had studied aeronautics at TH Darmstadt, before he joined AVA Goettingen in January 1932.[381] Already in winter 1932 he found *'anti-icing'* as his future main field of research, which soon led to the general introduction of external and internal wing air-heating of all Luftwaffe aircraft.[382] In 1939 Ritz became head of the newly founded AVA Institute for Low Temperature Research,[383] and the regenerator activities were a spin-off of this de-icing research. The AVA was the sole establishment responsible for research on aircraft icing in Germany. In addition to the Goettingen laboratory facilities and icing tunnels, flight tests of de-icing equipment under natural conditions over the Alps were carried out

---

[380] Data (Sign. 28,002/1449), courtesy of Judith Käpplinger, KIT Karlsruhe Archive, 18 Jan. 2022.

[381] Quoted from Ritz's personnel file, courtesy of Dr. Jessika Wichner, DLR Archives, Goettingen, 8 April 2010.

[382] See Green, Wartime Aeronautical Research, p. 627–631, who describes the findings of Canadian scientists and engineers by touring German research facilities in late 1945, amongst these Ritz's AVA activities (de-icing, regenerative hex, coal burning). The author John J. Green was born and educated in aeronautics in England, he arrived in Canada in 1930, where he quickly became a leader in Canadian aeronautics. Between the foundation of *ICAS* in 1957,—see Wikipedia, 'International Council of the Aeronautical Sciences' in English, up to 1982, he was highly influential in this renowned aeronautical society, which he led as president between 1972 to 1978.

[383] In German: AVA Institut für Kälteforschung, the scientific staff comprised eight assistants then. After the War Ritz joined—like the St. Aegyd tank engine developers (A. Mueller, Kolb, Hryniszak, Zadnik)—C.A. Parsons & Co., Newcastle, where he contributed to the British nuclear reactor programme (Bradwell, Calder Hall, with Parsons steam turbines, and 30 related Ritz patents). In 1961, he returned to Germany, recommended by the German nuclear fission pioneer,—see Wikipedia, 'Otto Hahn' in English, as head of the newly founded *'IRB Institut für Reaktorbauelemente* (Reactor construction elements)' at the *'Kernforschungszentrum* (Nuclear research centre) *Karlsruhe'*. W. Hryniszak (1910–1996) stayed in England, and published in 1958,—see Hryniszak, Heat Exchangers.

from Munich Airport, in conjunction with the Luftwaffe. Meteorological flights were made from Prague and there was a cold weather experimental ground station in Norway where simulated icing tests of propellers were undertaken. In 1938 Ritz initiated the construction of a large, refrigerated and pressurised tunnel (*'Großer Kältekanal'*), for high altitude tests of turbojet engines and rocket motors, as well as for icing tests.[384] The open working section was 3×1.95 m and a speed of 130 m/s was obtainable, using four fans having a total driving power of 1640 kW. The most interesting feature was the absence of turning vanes, replaced by rotating cylinders at each of the four inner corners, which prevented boundary layer separation and thereby enabled practically very small aerodynamic corner losses. In addition to overcoming the icing problem associated with turning vanes, this scheme also permitted the use of very short diffuser length, giving a compact tunnel with reduced refrigeration requirements.[385] The refrigeration installation comprised a four-stage ammonia system with 2150 kW driving power, giving a minimum tunnel temperature of $-55\,°C$. The tunnel could be evacuated to 0.1 bar by means of three four-stage centrifugal compressors, which would also remove the combustion exhaust of turbojet tests. The concrete walls of this tunnel had a thickness of 1.5 m, covered by 0.3 m cork insulation. Unfortunately, as the tunnel was nearing completion in 1942, a spark from a welding torch ignited the cork insulation and the entire tunnel section was damaged beyond repair.

Ritz documented his *RGT Regenerator Gas Turbine* concept and investigations at AVA Goettingen immediately after the War in a comprehensive report,[386] from which the following results are taken. Performance studies showed early on, that a heat exchanger with 90–95% temperature exchange ratio (effectiveness) was possible with the counter-flow, rotating regenerator drum- or disk-type concept, Fig. 8.54, (m), due to the ~30-times higher exchange area density in comparison to conventional *'tube and shell'* heat exchangers (recuperators), so that also aeronautical applications came into reach. It was already accepted that, with a heat exchanger (*'hex'*) of only 80% effectiveness a gas turbine could equal the efficiency of a state-of-the-art piston engine. In Ritz's revolutionary innovation the heat transfer through-flow material matrix is alternately exposed to the heating and cooling gases, by rotating the matrix drum or disk(s) by small electric motors, typically with 30 rpm. Two arrangements of the heat exchanger rotors were experimented with, a radial or cross-flow 'drum-type' had a tubular, woven metal mesh structure, while

---

[384] See Trischler, Luft- und Raumfahrtforschung, p. 201. As an aside, the deep founding work for the 'Kältekanal' produced also an unexpected benefit for Goettingen. These works were contracted to specialised Italian miners, and incidently a brother of these miners opened in ~1940 Goettingen's first ice cream parlour, today *'Eisfieber'* at Groner Str. 37.

[385] A patented feature: DE723,718 with priority 14 June 1939, inventor L. Ritz, *'Anordnung zur Verminderung von Eckenverlusten in Kanälen und Rohrleitungen* (Arrangement for reducing corner losses in ducts and pipework)'.

[386] See Ritz, Abriß der Theorie. With priority 12 June 1952, Ritz filed several regenerator patents (US2,925,254; GB760,913; DE1,078,595; CH329,165) which actually appear to prove him being the first inventor of this revolutionary hex concept.

the axial 'disk-type' was fabricated from ceramics. The preference for ceramics was due to its high heat storage capacity at low specific weight, and the possibility of simple cleaning by continuously reversed through-flow direction. The correct choice of ceramic would give a reasonable resistance to thermal and mechanical shock,[387] and sufficiently high heat conduction into the matrix wall thicknesses between the holes without permitting too high a conduction between the hot and cold sections directly through the rotating element. Special attention was directed to the horizontal rotor sealing between hot and cold sections: Ritz claimed a feather-type labyrinth seal with leakages of less than 1% for compression ratios of up to 7:1. Various manufacturing techniques for the ceramic honeycomb matrix configurations were tested, Fig. 8.54 (r); the upper 1.5×1.5 mm square channels resulted from an extrusion process of the 'green' ceramic material before burning, while the circular, < 0.5 mm holes of the lower structure originated from embedded wool threads, which were later burnt out during ceramic hardening.

As said, Helmut Schelp issued on 9 September 1942 an order to the team BBC/AVA for designing a turboprop engine with Ritz regenerative heat exchanger for the Messerschmitt long-range bomber Me 264.[388] While AVA coordinated their part of the joint project activities under the neutral designation '*K 13*', Brown Boveri Mannheim used for keeping the secret—at first inexplicable—the term '*Hetralin* project' instead; a possible explanation came completely unexpected—comparable to the riddle of the code-named 'Herbitus' facility of Sect. 8.3.3—but here from a medical corner.[389]

A layout of the 1943, joint 5000 hp project[390] of a Brown Boveri turboprop (PTL) using Ritz's twin drum-type rotating regenerators from AVA Goettingen is shown in Fig. 8.55, with the sophisticated, BBC-invented operational scheme on the left. Besides *regeneration*, the biggest asset of this arrangement is the addition of *one reheat stage*,[391] which—as

---

[387] An aspect especially important for W. Hryszniak's design of the GT 103 tank heat exchanger.

[388] Acc. to Kay, Turbojet, Vol. 1, p. 176—Walter Brisken and Emil Waldmann, key members of Schelp's RLM department, carried out many studies of turboprops with heat exchangers in pursuit of maximum economy over long ranges. Date of RLM recuperated turboprop order from—L. Ritz 'Aktenvermerk, Betr. Hetralin K13 (note for the files)', 4 p., 12. Jan. 1943, DLR Archive Goettingen.

[389] '*Hetralin*', chemically a Dioxybenzol-hexa-methylene-tetramine contains 60% of Hexamethyl-tetramin, also known as *Urotropin*. At times of WW I—before sulphonamides and penicillin became available, 'Hetralin' (Urotropin was brought to market by Schering AG, Berlin in 1895)—in fact rather the latter content Hexamethylamin/Urotropin was—see Butler, Text-Book of Materia Medica—'*...used as a germicide in the urine. It is eliminated in the urine and broken down as formaldehyde. It is particularly valuable in cystitis, in gonorrhoea, in typhoid fever, and in all conditions in which it is desired to avoid urinary infection or to lessen its severity.*' This type of remedy was known in rough-joking trooper jargon also as '*tube cleaner*', an association actually not too far away, when a ceramic regenerator disk with minute passages has to be cleaned from coal flue gas congestions. To complete the story, in 1920 'Hetralin' usage has been patented for a '*New or improved process of brewing or preserving beer...*', GB143'506.

[390] The 5000 hp engine shaft power is specified for an altitude of 10 km.

[391] As an approximation to an iso-thermal change of condition.

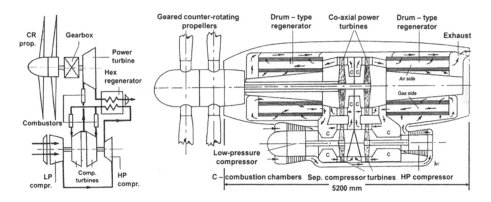

Fig. 8.55 BBC turboprop project *RGT* w. AVA twin heat exchangers, 1943: Schematic arrangement (l), engine layout (r) © A. Kay, mod

outlined by Ritz in detail—increases the mass-flow specific power for the selected total pressure ratio PR 7 from 345 hp/(kg/s) for a conventional GT cycle without reheat, to 430 hp/(kg/s), or by +25%, without any loss in efficiency.[392] The GT flow path starts with a—*LP compressor*, followed by a—*HP compressor* to produce the total pressure ratio of 7, both driven by separate—*Compressor turbines*. After passing the—*Hex regenerator*, the air is split up to be burnt in two—*Combustors*, before being expanded in the hp and lp compressor turbines. The re-merged turbine outflow is then led to the reheat—*Combustor*, before being expanded in the—*Power turbine,* and the remaining heat in the turbine outflow is used in the final—*Hex regenerator* passage.

The *'Hetralin K13'* project (in AVA terminology) stretched over a considerable time from the initial RLM order on 9 September 1942 up to the last documented test results at AVA on coal combustion in February 1945, Fig. 8.57. Astonishingly, Schelp revealed this project already on 4 November 1942 to a larger audience:[393] *'After project preparations, Dipl.Ing. Ritz, AVA Goettingen, has shown that a gas turbine with heat exchanger at appropriate altitudes may reach fuel efficiencies of 150 g/PS, or less. Pre-development activities for a 5000 PS engine at 10 km will be jointly undertaken by BBC Mannheim and AVA'.* Out of the initial project phase survived two *Minutes of Meetings*[394] in the DLR Archive Goettingen, which illustrate that the 'Forschungsführung' (L. Prandtl) was informed about this pre-stage of what became the 'America bomber' project. The goal

---

[392] For optimum cycle PR ~ 20,—not in technical reach in 1943—the gain in specific power by reheat addition would have been even in the order of +50%.

[393] Deutsche Akademie fuer Luftfahrtforschung, Berlin, 4 Nov. 1942, Session *'Triebwerksplanung* (Engine planning)', quoted in—see Gersdorff, Flugmotoren, p. 272.

[394] L. Ritz *'Aktenvermerk, Betr. Hetralin K13'*, dated 12 Jan.1943, and Horst Otterbach (1910-xx) *'Aktenvermerk 2. Besprechung am 31.3.43 zwischen Herrn Encke und Otterbach'*, dated 1st April 1943, from DLR Archive Goettingen.

was to develop the engine components separately, only adjusted by pressure ratio and selected mass flow, which with ṁ 6.5–8 kg/s was rather low. Originally the component work was seen on the AVA side, while BBC should carry out the integration and overall design. Remarkable, that RLM GL/C-E3 (Schelp) ordered in this context to inform BBC about the DVL activities on air-cooled turbine blading, while AVA provided supplementary information on turbine steam cooling. The project responsibility was shared between Hermann Reuter, Brown Boveri, Mannheim, Ludolf Ritz, and the *'scientific employee'* Horst Otterbach for AVA Goettingen. The latter authored also the second note on a meeting with W. Encke on 31 March 1943, about a required axial compressor design with 9600 rpm, 6.5 kg/s mass flow and a pressure ratio of 3.2 ata (bar). Encke indicated that a possible compressor design could be ready within one month, however, with surprising preconditions. *'He objected the compressor manufacturing at BBC, since this company competes with AVA e.g. at Junkers with unfair methods.*[395] *In addition, it has to be observed that BBC Mannheim is still obliged to report to the Swiss headquarter at Baden, from where the news will go to England.'* Vice versa, Otterbach indicates that he did not reveal the achievable efficiencies with heat exchanger against Encke's keen interest, and in due course, the discussed axial compressor design was passed to the BBC share.

According to the actual component arrangement of this BBC design in the responsibility of Hermann Reuter,[396] Fig. 8.55 (r), the overall unit length of this RGT engine measured 5.2 m, the vertical total depth with the underslung compressors and turbines was 2.42 m, of which the upper hex/gearbox cylinder had a diameter of 1.42 m, while the lower turbomachinery container measured 0.8 m in diameter.

As an example of the performance improvements to be expected with the RGT Regenerator Gas Turbine concept, a comparison was made on the basis of the *Messerschmitt Me 264* aircraft,[397] for which actually three piston-engine powered prototypes were built, of which Me 264 V1, Fig. 8.56a and photo, carried out its first flight from the Augsburg-Haunstetten airfield of the Messerschmitt Works on 23 December 1942. The aircraft had a wing span of 43 m, and was typically powered by 4x 1700 hp

---

[395] In German '. . . Konkurrenz mit nicht ganz einwandfreien Methoden'.

[396] H. Reuter (1911–1981), chief engineer and head of the BBC Mannheim aero gas turbine development group TLUK, and his flight GT developments will be presented in detail in Chap. 9.

[397] See Wikipedia, 'Messerschmitt Me 264', and for more details on the many development variants, http://www.luft46.com/prototyp/me264.html, and—see Griehl, Luftwaffe Over America, all in English.

Fig. 8.56 Messerschmitt Me 264 V1, long-range bomber/reconnaissance aircraft prototype 1943, (**a**) 4× 1700 hp *BMW 801 D*, (**b**) 2× 5000 hp *RGT* © W. Green, and A. Kay, mod. (l)

BMW 801 D piston engines, driving 2.7 m diameter, single-rotation propellers. For the same range of 18,000 km, the same payload of 5 t, and an assumed average cruising speed at 10 km altitude of 420 km/h, the mission fuel weight of the RGT concept would be reduced by approximately 60%, from 23 t (conventional) to 9 t (RGT) only. The corresponding gross weight reduction of the whole aircraft would change from 46 t (conventional) to 26 t (RGT), which would have reduced the wingspan for a RGT-powered Me 264 with 2,.9 m diameter, counter-rotating propellers to 35 m, Fig. 8.56b.

Coal and coal-dust combustion in industrial gas turbines were an issue at Brown Boveri since 1910.[398] In the late stages of Nazi-Germany the increasing lack of liquid fuel supplies put this on the scientific agenda again, for both cooperative projects between BBC Mannheim and AVA Goettingen: the regenerated gas turbine for long-range applications, and the tank gas turbine—with and without integral heat exchanger, as addressed in the foregoing Section.

Ludolf Ritz analysed coal combustion in gas turbines and came to the principal conclusion that conventional coal combustion operates roughly at an insufficient heat load level of $\sim$0.25 10^6 cal/m^{3}h, considerably below the typical gas turbine combustion load of 40–60 $\times$ 10^6 cal/m^{3}h. Consequently, the coal combustion had to be intensified by additional oxygen feeding to become an alternative to liquid fuel burning. In principle, this is possible at increased combustion pressures and velocities of the combustion air. Follow-

---

[398] See Eckardt, Gas Turbine Powerhouse, p. 298 f., and p. 180, where BBC coal combustion test facilities of 1943 are illustrated.

Fig. 8.57 Me 264 coal combustion as liquid fuel alternative, Feb. 1945: AVA/BBC coal combustion test chamber (l), pressed briquette burn-off (r) © DLR Archive

ing foregoing Russian research experience, Ritz achieved this at temperatures above 800 °C in the process laboratory test set-up of Fig. 8.57 (l), at combustion air velocities of 30 m/s.

9

Excursion IV: Hermann Reuter (1911–1981) and Hermann Oestrich (1903–1973)

This last *excursion* chapter deals with two leading turbojet engineers in Germany, Fig. 9.1, and their—in parts associated—war-time activities. Especially for Reuter, this will be a view on the quality of his and his team's early achievements without precedence,—but commonly in the *'rollercoaster'* transition period and thereafter,—a view on lasting post-war industrial impact, based on engineering versatility and adaptability to new circumstances.

Dipl.-Ing. Hermann Reuter, eight years younger than Dr.-Ing. Hermann Oestrich, and a highly talented mechanical engineer, became in 1941 BBC Mannheim's first department leader specialised on non-industrial gas turbines, modestly and tirelessly working for the Company's success during the War and thereafter. He qualified his small team to excellent and unique results in a number of widespread gas turbine applications in air, on land and sea.[1] Initiated by Schelp, the axial compressor design from BBC for the BMW 003 C/D versions was considered the best war time achievement in this context, which put—then controlled by Oestrich—Snecma's post-war *ATAR* engine developments in an excellent position for achieving the military success story of the *Mirage* fighter aircrafts.

In the time before, as BMW's turbojet chief engineer, Oestrich led not only the technology-wise most advanced German turbojet development project (with derivatives) between 1938–1945, but represented personally also a kind of development continuity into the post-war period as chief engineer of the French *SNECMA* and its successful *ATAR* turbojet engine family after 1950. Comparable to a *'soldier of fortune'*, he managed then to

---

[1] The indicated examples—for air, by the advanced axial compressor development for the BMW 109-003 C/D versions will be addressed in this chapter,—for land applications has been outlined already in the foregoing tank gas turbine Sect. 8.3.4, and—for sea, by illustrating Reuter's installation design of a Bristol Rolls-Royce Olympus TM1a on a 700 t gunboat of the Finnish Navy, also in the present chapter.

D. Eckardt, *Jet Web*,
https://doi.org/10.1007/978-3-658-38531-6_9

Fig. 9.1 Hermann Reuter, ~
1976 (l), Hermann Oestrich, ~
1955 (r), © U. Gutbrod (l)

exploit out of the circumstances the best for him personally (and—with some curtailment—
for his team), and to turn the *'wheel of fortune'* in his favour from a factual loser position at
War's end to a self-confident, acknowledged and wealthy *'Knight of the Legion of
Honour'*.[2]

9.1   Hermann Reuter, a Man of Hidden Qualities

Hermann Reuter was born at Wuppertal-Elberfeld, only 50 km south-east of Hermann
Oestrich's birthplace at Duisburg-Beeckerwerth. He went to the Gymnasium Elberfeld,
some 110 years after Friedrich Engels (1820–1895),[3] before he studied Mechanical
Engineering at TH Darmstadt up to 1932, and TH Munich, where he was graduated to
Dipl.-Ing. *'with excellence'* in 1934, Fig. 9.2.[4] He joined Brown Boveri Mannheim in
1935, then 24 years old, where he stayed for 32 years, climbing up from test engineer and
mechanical designer, to head of design department, to Technical Director, before he moved
on to *KIT*, the *Karlsruhe Institute of Technology* as Professor in Mechanical Engineering
between 1967 and 1976.

After a thorough in-house training under Eugen Senger, Fig. 8.39 (r), he was put in
charge of the newly founded BBC department *TLUK* for aeronautical turbomachinery in
April 1941. It can be assumed with some likelihood, that the establishment of this

---

[2] This special part of Oestrich's vita will be outlined in Sect. 10.4.

[3] See Wikipedia 'Friedrich Engels' in English. In 1936, the Gymnasium Elberfeld was named after
Wilhelm Dörpfeld (1853–1940), archaeologist and excavator of *Troja,* to prevent a threatening
naming by the NS-regime as *'Langemarck-Schule',*—see Wikipedia 'Langemark' in English. The
name of the scholar and social-revolutionist Engels certainly would have been a rightful candidate as
well, however, as communist he was not acceptable for the reigning *'national-socialists'* then.

[4] Data (Sign. 22008_461 and 21011_923), courtesy of Judith Käpplinger, KIT Archive, 17 Jan. 2022.

Fig. 9.2 Hermann Reuter's study places: Gymnasium Elberfeld 1921–1929 (l), TH Darmstadt 1932 (m), TH Muenchen 1934 (r)

department happened in the context of the—already described RLM request[5]—for a redesign of Encke's axial compressors for the Jumo 004 and BMW 003 prototypes, after serious doubts about the AVA design principle with a degree of reaction close to R 1 must have been ventilated, and at the same time news about Brown Boveri's successful realisation of industrial gas turbines—and more specifically, about an aero design collaboration with RAE in England spread—from Switzerland to Mannheim, or directly to Berlin. Further post-war information from BBC Mannheim[6] indicated that at that time the Jumo 109-004 A development struggled with vibration problems,[7] which after being overcome Junkers-internally concluded their compressor cooperation with Brown Boveri. Consequently, the further BBC compressor activities were solely directed to a potential BMW 109-003 retrofit programme. A seven stage variant, later dedicated as BMW 003 C version, was ready already during 1942, but then shelved, and somewhat surprisingly re-activated in 1944. At that time the corresponding compressor rig had been damaged in foregoing surge tests, and had to be restored first, which delayed the programme till War's end.

This delay might have even had some benefit for the then still unexperienced BBC team, which apparently used the extra-time for some performance improvements. The RLM-funded compressor design work was known as the '*HERMSO I—III*' projects, which reveals here Hermann Oestrich already very early and rather untypical for the time, as a kind of *personal* beneficiary. He, the responsible chief engineer of the BMW 003 turbojet engine development programme,—and as repeatedly proven later, a real believer in the Nazi fad of creating (double-) meaningful and funny project acronyms, shows up here easily to identify, as a combination of *HERM*ann *Oe*Strich (with final

---

[5] See Sect. 7.4, most remarkably the request was not initiated out of H. Schelp's responsibility, but followed a suggestion from the RLM Technical Office's airframe section GL/C-E2.

[6] Letter of late BBC Chairman of the Board, Herbert Gassert (1929–2011)—see Wikipedia, 'Herbert Gassert' in German, to MTU board member Karl-Adolf Müller, dated 1 Feb.1971, which Gassert forwarded to the author on 8 Dec. 2008.

[7] See Sect. 8.1.1 in the context of Jumo 004 A.

capitals reversed).[8] The *Hermso project* responsibility at BBC was with Hermann Reuter (30) of the newly founded department TLUK/ Ve,[9] and his personnel (with ages in 1945), including Hermann Schneider (39), Joseph Krauss (43) and Karl Waldmann (45). As outlined, the BBC compressor design in comparison to Encke's design approach, had 50% reaction blading, sharing the pressure rise equally between rotor blades and stator vanes. The object was to achieve greater efficiency, air mass flow and pressure ratio while still being interchangeable with the AVA compressor design. The *Hermso I,* seven-stage compressor achieved this target with excellent performance:

• mass flow rate ṁ	20.0 kg/s
• total pressure ratio PR	3.4
• rotational speed n	9800 rpm
• isentropic ad. efficiency η_C	0.84

supplemented by the following (predicted) BMW 109-003 C engine data[10]:

• thrust, static	900 kp
• specific fuel consumption sfc	1.27 kg/(kph)
• engine weight	610 kg
• engine diameter	0.69 m
• engine length	3.415 m

These results were gained in runs with the compressor only, which was housed in a specially built casing with multiple static pressure reading points, a vertical outlet duct and a rear extension shaft for the external drive.

Latest after the issuance of the *Hermso* compressor studies by RLM to Reuter's group at BBC Mannheim after April 1941, one can assume that a kind of back-up support by the

---

[8] See 'Herbitus', Sect. 8.3.3, and 'Hetralin', Sect. 8.3.5; later Oestrich's corresponding talents re-appeared in 'Hermos S.A.', and especially in the term 'ATAR'.

[9] TLUK/ Ve—with some probability standing for 'Turbine-Luftfahrt-Konstruktion/ Verdichter (Turbine-aviation-construction/ compressor)'. Other department members were identified in—see Wikipedia, 'Operation Lusty' in English, standing for 'LUftwaffe Secret TechnologY', post-war personnel tabulations: Willy Horni(n)g (35), Lohmeir or Lohmeyer, Hermann Henke (21), Gunter Casper (34), and in addition Karl Fickert, Waldemar Hryniszak and Ivo Dane (33), of which the latter Dipl.-Ing. I. Dane achieved some prominence with his Wilhelmshaven engineering office, and in ~1982 as co-founder of *DGW Deutsche Gesellschaft für Windenergie* (German society for wind energy). Hans Roskopf, though originally member of the BBC team, headed very soon Oestrich's group of mechanical designers, first at BMW, and thereafter also at SNECMA. Head of BBC Mannheim's industrial compressor department was Max Schattschneider, mostly known for the corresponding equipment for engine altitude test facilities, and wind tunnels like—see Eckardt, Gas Turbine Powerhouse, p. 122 f., and—see Eckardt, The 1x1 m hypersonic wind tunnel Kochel/ Tullahoma 1940–1960.

[10] See Kay, German Jet Engine, p. 125.

aero turbomachinery specialists of the *FKFS Forschungsinstitut für Kraftfahrwesen und Fahrzeugmotoren Stuttgart* (Research Institute of Automotive Engineering and Vehicle Engines Stuttgart) had been agreed, either internally between Ulrich Senger (1900–1973), Fig. 8.39 (r), BBC Mannheim's chief engineer for steam turbines and turbomachinery, and Professor Wunibald Kamm (1893–1966), the mighty head of the politically well-connected research institution,[11] or even with the explicit consent of Helmut Schelp, since he and the Air Ministry could not afford another AVA design aberration, mostly the result of an isolated, academic ivory-tower attitude and of non-communication. This brought Hermann Reuter and his team in immediate working contact with the FKFS specialists Dr.-Ing. Bruno Eckert (1907–1983) and Dr.-Ing. Fritz Weinig (1900–1970), Fig. 9.3, first sporadically in design reviews, but after September 1943, and intensified bombing raids on Mannheim and Stuttgart, in a joint office community at the safe FKFS outpost at Unterlenningen, some 30 km south-east of Stuttgart. This cotton spinning mill of C.A. Leuze, Fig. 9.3 (m), founded in 1861, struggled then economically due to cotton import/ valuta difficulties,[12] and had to host and shelter several war-related activities, like the displaced FKFS team.

Eckert studied between 1927 and 1934 with excellent results at TH Stuttgart—and an unusually broad curriculum in mechanical, electrical and aeronautical engineering. Thereafter, he came first to Kamm's THS institute, and later to FKFS as department head for fluid mechanics, where he stayed up to war's end. The project of an institute wind tunnel was his first opportunity to train his further specialisation in axial blower and compressor design. During the war he was responsible for the air cooling system of the Daimler-Benz *T-34* copy *VK30.01(DB) 'Panther' tank* version with 1200 hp *MB 507*, 12 cyl. water-cooled diesel engine. Already in 1940 Eckert had finished his doctoral thesis on automotive cooling blowers, of which he was then obliged to write two additional versions thereof (for DKF and VDI),[13] since the university original was declared *'secret'*.[14]

When he started his axial compressor design studies, the overall know-how was still rather limited. There were some theoretical cascade studies, but the theory was limited to inviscid, incompressible flows. Numerous practical design questions with respect to loss coefficients, component efficiencies and acceptable blade loadings could still not be answered. Consequently, many early compressors, like e.g. those for the Brown Boveri

---

[11] Ulrich Senger should have established contacts to TH Stuttgart and the renowned FKFS under Prof. Wunibald Kamm latest in 1936, when BBC delivered—as mentioned before—the 3rd N_I altitude test facility to this prospering Institute,—with then up to 650 employees.

[12] Since 1941 C.A. Leuze supported the search for alternative cotton sources, especially from Ukraine and Crimea.

[13] *DKF* Deutsche Kraftfahr Forschung (German automotive research), *VDI* Verein Deutscher Ingenieure (Assoc. of German Engineers), here both as publishers.

[14] See Eckert, Das Kuehlgeblaese (1940), with the DKF and VDI thesis extra versions. On top there appeared several short versions,—see Eckert, Kuehlgeblaese (MTZ, 1940) and—Kraftfahrzeug-Kuehlgeblaese (1942–1943).

Fig. 9.3 Reuter's compressor design *'boot camp'*: Dr.-Ing. Bruno Eckert, ~1972 (l), FKFS outpost 1943–1945 at C.A. Leuze, Unterlenningen (m), Dr.-Ing. Fritz Weinig, ~1960 (r) © Leuze (m)

industrial gas turbines, were too lightly loaded and required—unacceptable for competitive turbojet engine designs—too many stages for a certain pressure ratio. Considerable credit goes to Eckert for carrying out the required systematic investigations at FKFS, and for developing—based on Weinig's numerical methods—a reliable and practical compressor flow calculation tool-kit. This opened the door to effective and useful axial compressor designs for turbojet engines, which however—with the exemption of Reuter's designs— found only post-war applications. In addition this comprehensive work is documented in Eckert's, together with Erwin Schnell, fundamental textbook, which was published first in 1953, followed by a 2nd edition in 1961.[15]

Fritz Weinig (1900–1970) belonged—after a study start in 1920 at TH Karlsruhe— already one year later to Hermann Foettinger's students at TH Danzig, and commenced his first employment after the Dipl.-Ing. examen on 2 June 1926, as one of the first employees at Foettinger's new Institute for Fluid Mechanics and Turbomachinery at TH Berlin-Charlottenburg.[16] The mentioned numerical method for complex turbomachinery flow calculations was co-authored by Eckert and Weinig.[17] Between 1931 and 1945, he was

---

[15] See Eckert, Axialkompressoren.

[16] Today, HFI Hermann-Foettinger-Institute for Fluid Mechanics at TU Berlin. Weinig's first professional task was to optimise the shape of torpedo hulls by means of *Foettinger's Vector Integrator* on behalf of TVA (TorpedoVersuchsAnstalt/torpedo test facility) Eckernfoerde, where Weinig was responsible for the numerical calculations. On 7 July 1929 he earned his doctorate with a thesis 'Über die graphische Berechnung der Strömungsverhältnisse und der Leistungsaufnahme in einem gegebenen Turbinenrad (On the graphical determination of flow conditions and power consumption in a turbine wheel)',—followed by his habilitation;—see Hager, Hydraulicians (Weinig).

[17] Under the latter aspect, F. Weinig belongs to the earliest pioneers of this discipline in Germany, summarised for the first time in his 1935 book,—see Weinig, Die Stroemung um die Schaufeln, followed e.g. by W. Traupel's doctoral thesis of 1942 (dedicated to Joh. Seb. Bach)—see Traupel, Neue allgemeine Theorie. It appears that the foundations—also here—were laid in England by the

Professor of Fluid Mechanics at the Technical Universities of Berlin and Stuttgart. During the period from 1934 to 1939, he was the chief of the DVL general aerodynamics section in Berlin. From 1939 to 1945, he acted as consultant scientist for aero engines, compressors and turbines, at FKFS, where he was especially prominent in introduction and completion of new applications of *'profile cascade/ lattice theories'.*[18]

Reuter's team also built another compressor intended for the improved 109-003 turbojet engine. This was the *Hermso II* 10-stage axial compressor, Fig. 9.4,[19] which was predicted to achieve an isentropic adiabatic efficiency of even 90%, higher than any other German compressor of the time. During 1944, BMW were requested to redesign the 109-003 to give greater thrust and fuel economy while still retaining the same installed volume. The particular applications in mind were long-range reconnaissance versions of the *Arado 234*, Fig. 8.7 (r), which would operate at ceilings up to 17 km, and benefit considerably by using engines with such decreased specific fuel consumption. By the end of the war, no prototype of this 109-003 D had been built, though considerable headway was made with the initial 003 C/D compressor development. A test compressor, with all likelihood tested at GHH Oberhausen with a reduced number of stages only, was running at an efficiency of 89% to give a PR of 3.2, and the indications were that the requisite PR 4.95 could be achieved with an efficiency of 85%. Consequently, the excellent BMW 109-003 D compressor data were:

Momentum or *Actuator Disk Theory,* connected with W.J.M. Rankine (1865), A.G. Greenhill (1888), and R.E. Froude (1889), and supplemented by *BET Blade Element Theory* of W. Froude (1878), D.W. Taylor (1893) and the Polish inventor S. Drzewiecki, otherwise known for building the first e-battery-powered submarine in 1884. The author suggests a corresponding historic sketch of the early numerical techniques for fluid mechanics, turbomachinery flow calculations and their interrelations, as a kind of scientific pedigree to comprehend, teach and apply advanced present-day methods.

[18] Mainly based on the spreading reputation of his translated propeller book,—see Weinig, Aerodynamik der Luftschraube, Weinig came to work in the USA, first up to 1951 for the Air Material Command, Wright Field, Oh, thereafter for GE Aircraft Engines at Cincinnati, Oh, where he instructed designers to use his theory derived from thin circular-arc profiles with small camber up to 1958. The late Leroy H. Smith, jr. wrote in a private communication in Nov. 2008: *'By the time I arrived at GE in 1954, Dr. Weinig's 'deviation angle method'*—see Smith, Axial Compressor Aerodesign Evolution—*was no longer being used. Personally, he was set in his analytic ways and hard to communicate with. Our design methods had moved on, and he had trouble accepting that.'* The communication problem meant he was not very effective as a consultant. In ~1960 he left the USA, returning to relatives at Leinsweiler, 30 km west of Karlsruhe. He built a house with a grand view to near—see Wikipedia, 'Trifels Castle' in English, known as the site where *Richard the Lionheart,* King of England (36) was imprisoned in 1193. Information on Weinig's last years, courtesy of Maja Theisinger and Thomas Stuebinger, Leinsweiler, 24. Jan. 2022.

[19] The faded BBC drawing original TLUK-Ve, K10021 was found on the wall behind U. Senger's desk at TH Stuttgart, with the hand-written inset, containing the date of issue 8 Oct. 1942, forwarded to the author,—courtesy of Senger's former assistant H. Gassert, 8 Dec. 2008. In contact with (H. Reuter?) BBC Mannheim, Antony Kay must have received a copy which was used for a re-drawing with English labels in—see Kay, German Jet Engine, p. 126, Fig. 2.100.

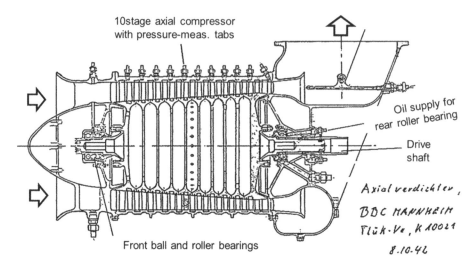

Fig. 9.4 BBC experimental axial compressor for BMW 109-003 D, 1944: $\dot{m}$ 25 kg/s, PR 4.95, n 10,000 rpm (l), hand-written inset (r) of 1942 drawing © A. Kay mod

• mass flow rate $\dot{m}$	25.0 kg/s
• total pressure ratio PR	4.95
• rotational speed n	10,000 rpm
• isentropic ad. efficiency η_C	0.85

Supplemented by the following BMW 109-003 D engine data:[20]

• thrust, static	1100 kp
• specific fuel consumption sfc	1.1 kg/(kph)
• engine weight	620 kg
• engine diameter	0.70 m
• engine length	3.656 m

Rumours of Reuter's compressor design masterpiece reached already in August 1946 also the Rolls-Royce turbojet engineering headquarters at Derby, Fig. 9.5, and consequently might have had its share in influencing Rolls-Royce's wholehearted jump to the axial engine concept, after initial moves by Lionel Haworth's *RCA3* in 1945, and Alan Griffith's subsequent proposals towards *Avon* and *Conway*.[21]

The distribution list on top of this RR-internal circulation note, Fig. 9.5, dated 27 August 1946, reads like a *Who's who* in aero engine design:

---

[20] See Kay, German Jet Engine, p. 126.
[21] See Sect. 8.2.3.

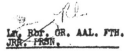

Lm, Rbr, GR, AAL, FTH,   E.7/HP,27,8,46,
JRR, PRSN,

c, Hs,

AXIAL COMPRESSOR DESIGN,

        I am circulating herewith a copy of an interrogation
report of Hermann Reuter on Axial Compressor Design, which
refers particularly to the results obtained on three
compressors built by B.B.C. Mannheim,

        It states that two of these three were tested in the
laboratory and curves indicate that 86% efficiency was
obtained at a compression ratio of approx,.3,5,

        Reuter discusses the principles used in the design
for the blading of these compressors,

Fig. 9.5 Rolls-Royce Derby, Internal note, 27 August 1946: *'Axial Compressor Design'* by BBC Mannheim © RRHT Derby

Rbr	Arthur A. Rubbra (1903–1982)	*Kestrel, Buzzard, Merlin, Griffon, Welland,* hand-written, forwarded to
AEl	Albert G. Elliott (1889–1978)	*Eagle to Kestrel,* 1937 Chief Eng. Aero Eng. Div., 1945 Director
GR	Alan A. Griffith (1893–1963)	integral in first design and the creation of Britain's first operational axial-flow turbojet engine
AAL	Adrian A. Lombard (1915–1967)	W.2B production, Derwent, Avon, Conway, 1949 Chief Des., 1958 Director Aero Eng. Div.
FTH	Frederick Taylor Hinkley (1908–?)	1944 Chief Des. Aero Eng., 1960 Comm. Director, 1969 OBE
JRR	John R. Reid	
PRSN	Harry Pearson (1913–1995)	1949 Chief Research Eng., 1960 Director Aero Eng. Div.
c. Hs	Ernest W. Hives (1st Baron Hives, 1886–1965)	1950–1957 RR Chairman

The reference *'Interrogation report of Hermann Reuter'* was based on a 4 h interview carried out by US Navy Technician Kenneth Campbell, Wright Aeronautical Corporation on Fri afternoon/Sat morning 26/27 November 1945 at Mannheim,[22] who had also met—as he mentions—Reuter's *mentor* on design methods Bruno Eckert for *six days*[23] in September 1945 at Unterlenningen, where Eckert actually stayed up to mid-1946 for one year before he decided to proceed with a better working offer at the French *Turboméca*.[24] The reason for the lengthy interrogation of Reuter in person was actually to get his comment on the remarkable compressor performance which in the meantime already had got so much attention as '*the best German axial compressor*'.

Reuter admitted frankly the benefit of the Eckert-Weinig education lessons in aero-thermodynamic theory, but the reader feels his impressive independence as born mechanical designer to pick from theory selectively, several times he expresses his suspicion that application of too sophisticated theory might be overdone, especially after the interviewer tried to make a point about some flaws in radial flow equilibrium and forced vs. free vortex design principles. However, the most important compromise which Reuter made to his design for practical reasons was the use of only one blade form for all stages (eight stages for Hermso 2) of the rotor, one blade form for all stator stages, except the inlet and the last, and of course, two special blade forms for these.

Challenged by Campbell on these simplifications, '*Reuter argued that the proof of the pudding was the eating, and he still got 88% efficiency. He suggested that, like many other theoretical considerations in axial flow blading design, it is probable that the effect of tip losses and boundary layer root and tip so distorts the radial distribution of flow that the truth may lie between the design which ignores the above consideration and the design which completely believes in it.*' An assessment, which his BBC colleagues back in Switzerland might have underlined at the same time, based on their own, independently gained experiences—and a similar post-war discussion with Theodore von Kármán and his team at Baden.[25] And '*The writer concluded that Mr. Reuter gave out all the information*

---

[22] See US Navy Dept, Interrogation Report; the printed, first date 22 Oct. 1945 in para 1 of this report is apparently a typing error,—and should read 26 Oct. 1945, also Cam(p)bell's typed signature is misspelled.

[23] See Eckert, Erinnerungen, p. 75 f., who at that time acc. to his memoirs ran an engineering office and consequently, was fully paid for one year by the US Navy, issuing e.g.—see Eckert, Series of Articles (containing ~100 measured compressor maps), including extras like a *Volkswagen 'Kübel'* (jeep) to bridge the daily 10 km transfer from Kirchheim/Teck, where he lived, to the Unterlenningen FKFS outpost. To Eckert's amusement, the Navy all-round carefree package for the (air-cooled) Volkswagen comprised several hundred litres of gas, and 100 l water-glycol cooling liquid (!).

[24] In 1952, Eckert returned to Germany and commenced work at Daimler-Benz AG Stuttgart-Untertuerkheim, where he became Director, Head of the Department of Fluidmachinery—with responsibility, mainly for (automotive) gas turbines, and after 1969 for the newly founded MTU Motoren- und Turbinen-Union Muenchen GmbH (today MTU Aero Engines), a merger of the post-war turbojet activities of DB, BMW and MAN. After his retiring he wrote his 'Erinnerungen' (biography) which represent a rather rare witness report especially of the early German turbojet developments from an engineer's perspective.

[25] See Eckardt, Gas Turbine Powerhouse, p. 146 f.

which he had in his head applicable to this subject. Reuter seemed to be slightly surprised at the interest shown in his work and was very frank about having merely applied the work of others to the particular assignment which had been given him. It was quite evident that he is an experienced engineer in the sense that he holds a high regard for the structural, machining and other practical considerations, and is thoroughly aware that going to the ultimate in the meticulous application of theory is not always justified in view of lack of research knowledge supporting it.'

Besides the aerodynamic loading characteristics, the BBC concepts carried distinctive design features as *'hand-writing'* of the mechanical chief designer H. Reuter. One speciality is the light-weight drum-type compressor rotor of Fig. 9.4, which contained considerable Brown Boveri manufacturing technology of their industrial gas turbines, which will be discussed further in the context of Figs. 9.15 and 9.16, as foundation of the identical *'tambour rotor'* of the ATAR V-1 turbojet engine.

Another valuable technology platform for Oestrich's post-war turbojet engine developments in France represented the BMW 109-018, the only *Class IV* turbojet in Helmut Schelp's classification. The new engine was aimed at giving a static thrust of 3400 kp, and was intended for operation at altitudes up to between 15 and 18 km, and for bringing a twin-engine, short range *Junkers Ju 287* bomber up to the critical speed of Mach 0.82; it had forward swept wings, to avoid interception by outrunning enemy fighters.[26] Development of the 109-018 was officially stopped end of 1944, but work continued on Oestrich's far-sighted initiative, Fig. 9.6 (l). The compressor was of the 12-stage axial type, designed in the responsibility of Reuter's group engineers Löffler and Karl Fickert. One complete compressor was built, but to avoid its capture, it was intentionally destroyed at Neu-Stassfurt before it could be tested as planned at either Dresden or Oberhausen, Fig. 9.6 (r). An explanation for this otherwise rather unique measure for keeping last-minute secrecy may represent the 109-018 compressor main data:[27]

• mass flow rate $\dot{m}$	83.0 kg/s
• total pressure ratio PR	~ 7.0
• rotational speed n	5000 rpm
• isentropic ad. efficiency η_C	0.79

A light construction was used for the 109-018 compressor, the first five disks being of *Dural* and the remainder of steel. But, different to the in this respect more advanced BMW 009–003 C/D with its drum-rotor design, Fig. 9.4, the larger 109-018 was closer to the pre-stressed, tie-bolt rotor construction of the 109-003 A (and similarly the Jumo 109-004 B). Much of the strength of the sheet-metal compressor casing was derived from the

---

[26] See Wikipedia, 'Junkers Ju 287' in English.

[27] See Kay, German Jet Engine, p.131 f.

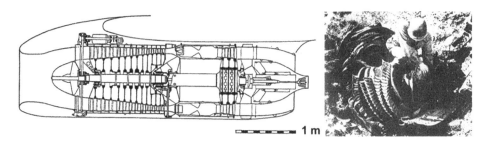

Fig. 9.6 *Oestrich's secret*, May 1945, Stassfurt: *BMW 109-018* turbojet, 3400 kp (l), GI checks destroyed 12-stage BBC compressor, PR 7 with *Dural* blades/disks (r) © H.Schubert (l), A. Kay (r)

substantial channels, formed on the outside of the compressor stator supporting rings, which in turn were covered and stiffened by a steel skin, Fig. 9.6 (l).[28,29]

One can assume that in 1945 the principles of BBC and BMW design concepts were largely integrated so that e.g. a further Class I engine growth to 1700 kp take-off thrust as reflected in the project *P.3306*, Fig. 9.19, could not further be differentiated between these design origins. These statements are nevertheless worth to be kept in view of Hermann Oestrich's post-war activities: his starting designs were the configurations which had been prepared jointly together with the Brown Boveri turbojet engineering team till the end of war.

After war's end rumours spread about Reuter's BBC aero GT design team *TLUK/Ve* and its achievements, so the Allies became somewhat nervous of not knowing on his whereabouts. This was especially justified in view of the fact that the Russians had occupied BMW's underground engine development facilities at Neu-Stassfurt, 20 km south of Magdeburg in the meantime, where the associated BBC experts were presumed. Finally, on 26 June 1945 morning under a radiant sky with 27.5 °C recorded temperature, BBC's upper management was ordered to line up for reporting by the responsible US Navy Lieut. S.P. Robinson, according to the not so deliberate *'Interview with Directorate at Heidelberg'*:[30]

- Hans L. Hammerbacher—President and Managing Director,
- Herr Deichman[n]—Director of Engineering,
- Dr. Caspari—Director of Sales,
- Dr. Saenger [Senger]—Chief Engineer, Turbine Design,
- Herr Kockeritz [Koeckritz]—Sales Engineer contacting RLM,

---

[28] This typical double-wall structure is also visible on the related design of—see Wikipedia 'Pirna 014' in English, especially at the depicted exhibition model at Deutsches Museum Munich.

[29] See Kay, German Jet Engine, p.135, Fig. 2.106—with the caption: *'An allied technician examining the* (destroyed) *compressor, apparently exploded with an internal charge which split open the casing and stator rings'.*

[30] See C.I.O.S, Gas Turbine and Wind Tunnel Activity Brown Boveri(e) Cie. Name corrections in [..].

Herr Lesch, in charge of design of supersonic tunnels was not present. Reuter's team was dispersed from Heidelberg to Sulzheim on March 28 (sic), 1945 and left there for an unknown destination in Southern Germany on April 6 (sic), 1945.

The situation was apparently still highly unclear, Reuter was actually reported *'missing'*, Senger had just returned on 16 June, and Caspari the day thereafter, but to get Reuter and his team[31] was important now, more important than Brown Boveri's whole lined-up directorate. After some tense moments, obviously Senger handled the situation, as quoted:

> Discussion with Saenger cleared up a few details regarding Reuter's whereabouts and the move to Sulzheim. BBC and RLM had the greatest difficulty in keeping Reuter and his staff out of the army and sometime in 1944 it appeared that the only way this could be done was to assign them to the Kraftfahrtechnischen Lehranstalt der Waffen-SS, located then at St. Aegyd, north of Vienna. This transfer is reported to have been but a formal one to insure deferment of technical personnel. The men were registered by the Waffen-SS but did not do any military service or wear a uniform. Sometime around the first of the year the KTL was dispersed from Vienna to Sulzheim and it was there that Reuter and his staff went on March 26, 1945. It was further learned that the whole of the organization left Sulzheim in fifty trucks with all their records and equipment on April 8. Their exact destination is unknown but a BMW dispersal point, with the code name "Kahla" near Rosenheim, was mentioned.[32]

---

[31] The C.I.O.S. report lists: Schneider, Waldmann, Krauss, Hornig, Henke, Dane, Casper; not mentioned Hraniczek, responsible for the GT103 regenerator, who was also part of the KTL tank team.

[32] There is some mis-information: As described already in Sect. 8.3.4, the KTL evacuation in early 1945 went from St. Aegyd to Sülzhayn/ Ellrich, with living quarters for SS personnel, approximately 9 km north-west of the huge underground Mittelbau-Dora production facilities (and not as wrongly spelled 'Sulzheim'). Of mixed-up dates, Mon 26 March, and Fri 6 April, 1945 are more likely (than the also quoted 28 March and 8 April). The target of the 50 truck KTL convoy was *'Camp Schlatt'* in Austria, some 60 km south-east of Passau, where also Max Adolf Mueller went to US captivity, and not the BMW dispersal point Kolbermoor near Rosenheim, where also Hans von Ohain was found at war's end,—see Conner, Hans von Ohain, p. 144. *'Kahla'*—the name of a Thuringian porcelain since 1844—as code name for Kolbermoor near Rosenheim—had apparently been picked with special subtlety, per se a brandmark of Oestrich, since it could be easily mixed-up with 'KALAG', code name of Oestrich's related underground facilities of former BMW Berlin-Spandau at Neu-Stassfurt, 500 km north of Kolbermoor, Sect. 9.2, and named after an unrelated salt mine 'Kaliwerk Löderburg AG'. In the 1940s Kahla had several manufacturing sites spread all-over Germany, one of these small plants was at the 'Alte Spinnerei' (Old cotton spinning mill) building at Kolbermoor, today practically a suburb in the west of Rosenheim, some 50 km south-east of Munich. Even today there is *Mahlwerck Porzellan GmbH*, so there is high probability that Senger's *'Kahla'* near Rosenheim, refers to this site. When finally in spring 1946 Reuter after a—what he called in his CV—*involuntary 6 months stay* at Parsons, Newcastle (presumably with the remainder of the KTL group) returned to the Brown Boveri headquarters at Heidelberg, his actual route back home was not reconstructed, so that the question from which of the interrogation camps he was finally released (*Kolbermoor, Schlatt or Newcastle*), has to remain unanswered.

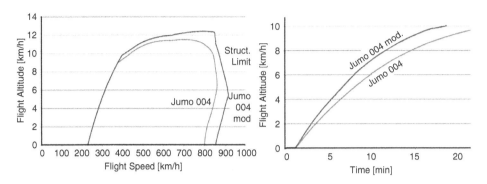

Fig. 9.7 Performance assessment Jumo 004 mod. (BBC) vs. Jumo 004: Me 262 flight envelope (l), climb time (r) © ALR/EADS

After the numerous detailed discussions in the foregoing on engine performance, especially in view of the *exotic* AVA Goettingen design with a degree of reaction R 1 axial compressor blading, as implemented in the series production engines Jumo 109-004 B and BMW 109-003 A, it was worthwhile to carry out a thought experiment to determine the potential effect of a modified Jumo compressor—with unchanged stage number, blade count and dimensions, however with a degree of reaction of R 0.5 and an average performance advantage of the BBC designs of seven percentage points in isentropic compressor efficiency. Fig. 9.7, highlights corresponding study results which have been carried out in 2009 for an accordingly modified Messerschmitt Me 262 fighter aircraft without (standard) and with 'Jumo 004 mod.' turbojet engine.[33]

As seen, there would have been considerable performance improvements: The *Jumo 004 mod.* has +8% in flight altitude, or a + 7% higher speed, Fig. 9.7 (l), its climb time up to 8 km would have been reduced by −22% (11 min instead of 13.5 min), Fig. 9.7 (r). And, the compressor efficiency advantage alone would have increased the mission range by +16%.

In 1948 Hermann Reuter became successor to Ulrich Senger as BBC Mannheim's Director and Chief Engineer for Thermal Machinery. In this period fell his last engineering project on ship propulsion, before he left the company in 1967 to become a Professor for General Mechanical Engineering[34] at IPEK Institute for Product Engineering at the KIT Karlsruhe Institute of Technology. The activity for Reuter's last, extraordinarily challenging project started in 1961, when the German MoD ordered a unit consisting of a Bristol-

---

[33] These investigations have been carried out by Drs.sc.techn.ETH—see Wikipedia, 'Georges Bernard Bridel' in German, and Marc Immer, ALR Aerospace Project Development Group Zurich, —see Wikipedia, 'ALR Piranha' in English.

[34] In German: Reuter had the 'Lehrstuhl für Allgemeine Maschinenkonstruktionslehre' at IPEK between 1967 and 1976.

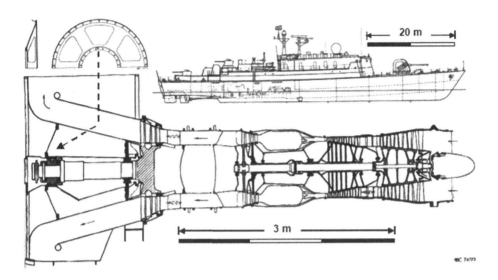

Fig. 9.8 H. Reuter as naval architect, 1965: RR *Olympus TM1a* w. BBC free power turbine for 16 MW pump-jet drive (bottom), of FNS 'Turunmaa' 700 t gunboat, 37 kn, 1969–2002 (top inset, right) © ABB (bottom), Wärtsilä (top)

Siddeley Engines Ltd. (BSEL) marinised Olympus 200 series engine[35] as a gas generator, together with a Brown Boveri Mannheim two-stage long-life marine free power turbine. Figure 9.8 illustrates the set-up of the Olympus two-spool gas turbine and the early version of the BBC two-stage free power turbine.[36] Especially the sophisticated bearing shock absorption structure indicates Reuter's familiarity with similar design tasks for tank GT suspension; modern tank GT's are specified for and undergo acceptance barge shock tests up to 50 g accordingly. Correspondingly, the bearing structure of the free power turbine is a mechanical design highlight: The spokes of the shown rear disk shell are extraordinarily thin, so that they act as flexible springs and heat flow restrictors, but the thrust and buckling stiffness is high enough to absorb considerable bearing forces in case of a shock impact.

In 1965 the Finnish Navy commenced a re-equipment programme with ships designed to suit the needs of protecting their shallow coastal waters in the Baltic Sea, especially for anti-submarine warfare and trade protection roles; internationally these vessels were

---

[35] The *Olympus Mk 200* had a 16,000 lbf take-off thrust and was used for the first *Avro Vulcan B2* bomber. The initial design of this second generation '*Olympus 6*' began in 1952, a major redesign with five lp and seven hp compressor stages and a '*canullar combustor*' with eight interconnected flame tubes. In spite of a much greater mass flow, the size and weight was little different to earlier models. Best known was the Rolls-Royce/ SNECMA *Olympus 593*, a reheated version of the Olympus which powered the supersonic airliner *Concorde* that was started in 1964. BSEL and Snecma Moteurs of France were to share the project. Acquiring BSEL in 1966, Rolls-Royce continued as the British partner. See Wikipedia, 'Rolls-Royce/Snecma Olympus 593' in English.

[36] See BBN, Schiffs-Gasturbine (author H. Reuter).

labelled as *corvettes*. The result was the order for the 700 t gunboats '*FNS* (Finnish Navy Ship) *Turunmaa*', Fig. 9.8, and '*FNS Karjala*' from Wärtsilä's Hietalahti shipyard in Helsinki, both of which had been handed over to the Navy on schedule in autumn 1968, not least due to the fact that BBC Mannheim had accomplished the machinery work in time for ship integration.[37]

The original requirement was for a compact vessel with 70 m length at the water line, 7.8 m breadth and 2.7 m depth at the standard displacement, capable of carrying the new *Bofors TAK 120* (Torn Automatisk Kanon, 120 mm) turret automatic gun as the main armament. In 1985–86 both ships were refitted, amongst other alterations the original Mercedes-Benz high-speed diesel engines were replaced by $3 \times$ MTU diesel engines of 2200 kW, each driving a propeller up to maximum cruising speeds of 17 kn within a range of 5000 nm.[38] The GT-powered speed was 37 kn, corresponding to 68.5 km/h.

9.2   Hermann Oestrich, a Clever Engineer, Always Ready in Time

Hermann Oestrich's official CV is easily accessible,[39] but for details it requires '*deeper digging*'—in direct meaning. His birthplace Beeckerwerth, an area of 13 km² in a big bend of River Rhine north of Duisburg, was bought up nearly completely by the '*Ruhr-Tycoon*' August Thyssen (1842–1926)[40] for a 600 m deep coal mine,—in immediate vicinity to the 250 m wide river a considerable technical challenge. The sudden change of the living circumstances for Oestrich's family are not known in detail—but solely the growth of local population from 200 farmers to 4000 miners must have been a tremendous, uprooting change.

Hermann Oestrich attended the newly built (Prussian) '*Realgymnasium*' of 1911 at Muelheim an der Ruhr, Fig. 9.9, of which already the building appears to reflect the spirit of the time. This *classical* gymnasium had Latin as first, English as second language, and optional either Old Greek or French, of which young Hermann decided farsightedly for the latter,—shortly after the Great War a surprising tendency towards language and country, which Oestrich preserved during his lifetime. Confronted with a necessary decision

---

[37] Beginning its sea trials in early 1968, the '*Turunmaa*' was the first Olympus-powered warship to enter service, some 6 months before '*HMS Exmouth*', the first British ship which had been refitted to trial the propulsion system for the Royal Navy and in due course, the first warship entirely propelled by gas turbines.

[38] See Wikipedia, '*Turunmaa*-class gunboat' in English.

[39] See Wikipedia, 'Hermann Oestrich', and—see 'Beeckerwerth' for Oestrich's birthplace, both in English.

[40] See Wikipedia 'August Thyssen' in English. Actually in 1906, '*Oestrichshof*' was acquired completely by Thyssen's mining organization—see Wikipedia, 'Gewerkschaft Deutscher Kaiser' in German, and Oestrich's father decided thereafter to terminate as farmer—apparently under satisfying terms; information, courtesy of Michael Kanther, Stadtarchiv Duisburg, 16 Feb, 2022.

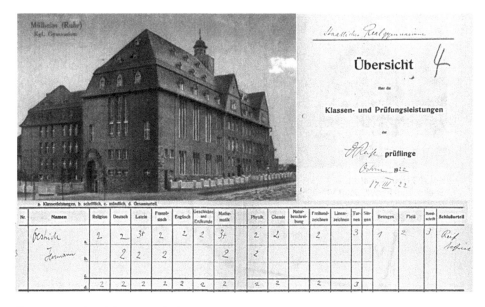

Fig. 9.9 H. Oestrich at Muelheim an der Ruhr, 1922: Staatl. Realgymnasium, 1911; and *'Abiturzeugnis'* (Certificate). © Stadtarchiv Muelheim a.d. Ruhr

amongst the Allied countries in 1945, his French preconditioning must have played a considerable role. For the rest his 'Abiturzeugnis', Fig. 9.9 (bottom), shows a consistent *'2' (Good)* across all disciplines.[41]

Between 1923 and 1927 he studied Mechanical Engineering at Hannover and Berlin, before he entered *DVL Berlin-Adlershof, Institut für Motoren, Arbeitsprozesse und Thermodynamik* (Engines, working processes and thermodynamics), since 1933 led by Professor Dr.-Ing. Fritz A.F. Schmidt, where Oestrich had the responsibility of the thermodynamics group. In 1935 he joined *Siemens Flugmotorenwerke* (flight engine works) at Berlin-Spandau, which became *Bramo Brandenburgische Motorenwerke*, as a separate Siemens subsidiary in 1936, and finally *BMW Flugmotorenbau, Entwicklungswerk Spandau* in the summer of 1939. Most of Oestrich's engineering activities during his BMW years has already been described in Sects. 6.2.4 and 8.1.2, except for the last 1-year period up to War's end, and his years at SNECMA in France up to his retirement in 1960, which will be addressed in the following. He became chief engineer of BMW Flugmotoren Spandau after earning a doctorate on *'Der Gaswechselvorgang in Hoehenmotoren* (The

---

[41] Final certificate for university qualification, courtesy of Yvonne Kurzeja, Stadtarchiv Muelheim a.d. Ruhr, 19 Aug. 1921; deviating was only his *'Turnen'* (gymnastics) and *'Handschrift'* (hand writing); the right column *'Schlußurteil'* (final comment) carries the entry *'Reif befreit'*, so that he had not to undergo an oral examination. Today the gymnasium carries the name *'Otto-Pankok-Schule'*, after a local painter.

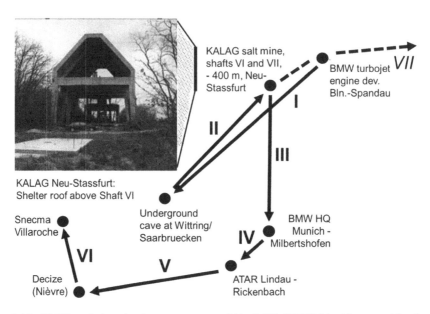

Fig. 9.10 BMW turbojet development team, 1943–1946/ SNECMA *'Groupe O'* odyssey, 1945–1950

cyclic gas exchange of high altitude flight engines)' at the TH Berlin-Charlottenburg in 1937.

When the Bramo company was absorbed by BMW, the intention was that all jet-engine development should eventually be centred at Spandau, and to a large extent this plan was realised. However, partly because of the diversification in power plant types and, later, through dispersal of plants and sub-contracting, various aspects of the turbojet work were conducted in various places. Here we follow mainly H. Oestrich and his development team, which carried out between 1943 and 1946—later under the French designation *'Groupe O'*—quite an odyssey, Fig. 9.10.

In December 1942 appeared a first directive of the *Reichsministerium für Rüstung und Kriegsproduktion* (Reich Ministry for armament and war production) under Albert Speer about *'The dislocation of unique armaments productions, e.g. ball bearings'*. However, move of the turbojet development work was only triggered after the beginning of heavy bombing of Berlin in autumn 1943, which caused a large part of the work to be dispersed to caves at Wittring,[42] 30 km south of Saarbruecken, *arrow I* in Fig. 9.10. Following a direct

---

[42] The limestone mine Wittring(en) had two code named underground areas *'Kalk* (chalk) I', 35,000 m^2, in reserve for army purposes, and *'Kalk II'*, 30,000 m^2 for *BMW 003-0* and *-A* series production preparation; in addition,—see Kay, German Jet Engine, p. 97, there were also turbojet development departments allocated there. In Oestrich's memories, however,—see Mouton, Hermann Oestrich—the development team stayed together at Spandau up to the end of 1944, before the transfer to Neu-Stassfurt was accomplished—questionably late in Feb. 1945.

RLM order to evacuate the Spandau Works on 24 May 1944, and the Allied landing in Normandy on 6 June 1944, both Spandau and the Wittring location in France, close to the German border had to be given up, and finally—with short interruptions—ended up in a disused salt mine at Neu-Stassfurt, where shafts VI and VII were used 400 m underground, under the already explained code name *KALAG, arrow II*.[43] The allocated BMW team comprised some 1800 engineers and workers, which explains the daily, lengthy and cumbersome personnel change process by means of limited elevator capacity.[44]

The double shaft mine Neu-Stassfurt was about 35 km south of Magdeburg, where BMW-Kalag occupied the complete -400 m floor level, though a connection between both shafts, which were 900 m apart, had first to be opened; the linear stretch of the BMW underground space was ~2.3 km. Individual rooms for 'Quality control' and the 'Design office' measured typically 46×25 m, with an average height of 10 m. The total working space comprised approximately 15 rooms (with solid remaining pillars in between), or an equivalent total floor space of 12,000 m^2. The floors were covered with concrete, the rooms illuminated, and electricity, water and compressed air supplies installed; nets under the ceiling prevented rock fall. The underground facilities were prepared for a workforce of 2000, which would have required a (planned) ventilation of 220,000 m^3/h.

However, moving underground proved to be anything but a simple operation, as outlined by Daniel Uziel.[45] Mines, limited by size and capacity of their elevators, were particularly troublesome. The small size of many elevators and their slowness seriously delayed traffic in and out of the factory, and limited the size of machine tools that could be brought in; these problems were especially apparent in the move of BMW's factories into the Stassfurt mines. Very often, due to the limited capacity of the elevator, it took around three hours to change the 12 h shift of 2000 workers. The elevators and their prominent towers were also vulnerable to air attacks, which for Kalag Neu-Stassfurt led to the mounting of a concrete shelter roof above shaft VI, as visible in the inset of Fig. 9.10; close to this must have been also the sheds with engine test facilities.

In the vicinity of shaft VI there was also a satellite camp of KZ Buchenwald, with a capacity of 2000 prisoners, actually occupied by some 500 prisoners mostly of French origin, which must have additionally alleviated the contacts with Oestrich's team at war's end. After Allied bombing in autumn 1944 had destroyed the Spandau works completely, *Kalag* remained the only operational BMW turbojet engine development facility, with focus on the building of three 109-003 C test engines, while the 109-018 had less priority. Last traces of the BBC compressors in Germany for the BMW turbojet versions—003 C and D have been found in an—otherwise not to be commented—'*Endzeit* (End times)' document:

---

[43] *Kalag* was the designation of the BMW section, while the complete industrial compound at Neu-Stassfurt was code-named '*Reh* (Deer)'.

[44] See Nouzille, La folle histoire.

[45] See Uziel, Arming the Luftwaffe, p. 131.

"Secret": Minutes of 003 development meeting at Stassfurt on Sun(!), 10 March 1945

Attendants:	Part time:	cc.
H. Fl. Stabsing. Schelp	*H. Buske*	*H. Dir. Bruckmann 2x*
H. Fl. Haupting. Brisken	*H. Hagen*	*H. Dir. Dr. Ulsamer*
H. Fl. Obering. Klemmer	*H. Prestel (*Karl Prestel, BMW chief designer missed the evacuation of Oestrich's team from Stassfurt to the south at end of April 1945; he had tried to return shortly to the Berlin-Spandau headquarters and was cut-off by advancing Russian troops	*H. Bev. Wolff [He-Hi, AK Head., -Pres.]*
H. Dir. Dr. Oestrich	*H. Fischer*	

. . .

Ref. –003 C

- *target thrust—900–950 kp*
- *start of series production—October 45*
     with compressor version BBC-Hermso, *alternatively compressor blading from Brueckner, Kanis & Co.,*
- *BMW guarantees that the serial mechanical design of the BBC compressor will be finished till 30 March 45,*
- *H. Brisken states, BBC compressor testing shall commence end of March on the premises of 'Mittelwerke'; necessary test equipment will be forwarded from Mhm. to 'Mittelwerke',*[46]
- *BMW (Hagen) agrees to start combustor testing one week after engine arrival at DVL*
- *Moosburg test site.*

---

[46] This comment appears to underline that the meeting participants were not too familiar with the BBC Mannheim preconditions and difficulties at that time, there is even doubt that BBC had a compressor facility of the required power ready for tests at all. It appears that all known BBC compressor tests for the *Hermso* projects were run or planned to be run at *GHH GuteHoffnungsHuette* Oberhausen. *Mittelwerke G.m.b.H* is as an SS industry somewhat identical with *Mittelbau-Dora* near Nordhausen, Harz Mts., where the underground production of A4/V-2 rockets was concentrated since Aug. 1943, maintained by 60,000 prisoners from the Buchenwald Concentration Camp, of which est. 20,000 lost their lives. According to J.-Ch. Wagner, present head of the Buchenwald Memorial, the compressor test facility cannot be located, however, one can presume that the described activities were part of a work relocation to Mittelwerke underground production for the *He 162 'Volksjaeger'* with *BMW-003 A* engine. For the same reason, i.e. to stabilise production against bombing impact on industrial surface sites, the manufacturing of the so-called *'Rudermaschine (steering unit)'* for the A4/V-2 rocket had been transferred before already from BBC Saarbruecken to Mittelwerke as well. See Wagner, Produktion des Todes, p. 205, and Allen, The Business of Genocide.

Ref. –003D
- *target thrust—1100 kp*
- *compressor operational altitude—16,000–18,000 m*
- *compressor blading BMW, parallel designs Brückner-Kanis and BBC, max. diam. 700 mm*
- *target dates for development:*
  - *project drawings due—15 April 45*
  - *termination of mechanical design—1 July 45*
  - *completion of first unit—1 November 45 [!]*
      This latter date is especially uncertain, since it is fixed without knowing of the Munich production possibilities.'

Eight weeks before the actual end of WW II, the *'business as usual'* attitude of these minutes is remarkable. At the time of the meeting on 10 March 1945, the Russian front had come to an intermediate standstill at River Oder, some 200 km east of Stassfurt, while in the west the reference date, 7 March 1945, coincides with the US crossing of the *'Bridge of Remagen'*, approximately 320 km away. Stassfurt lay (still) in the US occupation zone, when US and Russian troops met at Torgau on River Elbe on 25 April 1945. On 1 July 1945, the US troops left the area according to the inter-allied agreement, such considerably expanding the Russian occupation zone to what became after 7 October 1949 *GDR German Democratic Republic.*

On 12 April 1945 armoured tank units of the 83rd Division of the 9th US Army reached Stassfurt and the BMW facility at Neu-Stassfurt. Already one day later, a ten-man American technical team, consisting mainly of Pratt & Whitney engineers, arrived in Stassfurt to interrogate Oestrich and others. They were immediately impressed by the advanced *BMW 018*, Fig. 9.6 (r), and the *BMW 003 C*, of which a test run was demanded—and fulfilled. Before, Oestrich had hidden in a theatre-like orchestration those of his technical records which were supposed to be intentionally found, i.e. mainly those about the standard BMW–003A version—on the churchyard. After the detection of these materials by US troops, which had received the decisive hint from an observing Belgian *DP* (displaced person), he behaved very cooperative so that further searches e.g. to detect the *Hermso I—III project drawings* of the alternative, considerably more advanced BBC compressors were omitted. In interrogations Oestrich stated that the aerodynamic design of the BMW 003 had been fixed nearly four years ago, and by a redesign a nearly 50% in output could be obtained. He claimed that in the development of this engine the firm had received no significant help from national research organisations whose work in this field was considered ineffective. The firm had done some research on its own, but the main emphasis, especially in the last two years, had been on the immediate development problems.

Between 16 April and 10 May the site was carefully stripped by the US troops from finished or partially assembled engines, test equipment, drawings and calculation files. On 11 May 1945, after nearly a month of US interrogations and a threatening Russian occupation of the area, Hermann Oestrich and Hans-Georg Muenzberg (1916–2000)[47] were flown to Munich for urgent interrogation, while the *'baggage'* followed after some time by truck convoy to Munich-Milbertshofen, Fig. 9.10, arrow III.[48]

In July 1945 at Munich, Sir Roy Fedden[49] on behalf of British authorities asked Oestrich and part of the BMW team to work out a PTL engine project for a turboprop transport aircraft. During that period Oestrich was approached by French Secret Service DGER[50] at Munich and—together with Dipl.-Ing. Kurt Donath,[51] responsible for BMW-003A pre-production, flown to Paris in August, for what became negotiations with the Air Ministry about hiring a 100 men team with families. This contact was a kind of natural consequence of the developments over the past five years, during which the German aero-engine industry had contracted several French firms to produce German piston-engines under license. Prominent among them was BMW, whose chairman Popp had suggested as early as November 1940 while visiting Paris, the formation of a business cooperation with Gnôme et Rhône (G&R), the largest French aero engine producer and prime ancestor company of Snecma.[52] BMW basically sought to free capacity for the production of the

---

[47] See Wikipedia, 'Hans-Georg Münzberg' in English, was a member of Oestrich's design team till 1956, thereafter in parallel to his continuing SNECMA occupation Professor for Turbojet Propulsion at TU Berlin, after 1964 full-time professor at TU Munich.

[48] See Kay, Turbojet, Vol. II, p. 182; Oestrich in his memoirs—see Mouton, Hermann Oestrich, — remembers that 3–4 autobuses were rented from German entrepreneurs(?) for the transport *'en autoroute'* to Munich, which under the given circumstances is rather unlikely, and may be owed to the fact that he wrote these memoirs in ~1960. Acc. to a 3rd version, in a meeting on 3 June 1945 Oestrich pointed out to US officers the immediate threat of a Russian occupation of the area, and that a safe evacuation of his core team with families and some materials would require four trucks only. See Werner, Kriegswirtschaft, p. 353, and BMW Archive UA 704/1.

[49] As Oestrich remembered—see Mouton, Hermann Oestrich, Fedden, accompanied by Dr. Stern as translator,—see Chap. 2, <Sep. 1920>—had met him already on Tue 19 June 1945 still at the Neu-Stassfurt underground facility.

[50] Present—see Wikipedia, 'Directorate-General for External Security' in English, abbreviated DGSE, dates back to the 1944 *'Direction Générale des études et recherches' DGER*, which in itself had to be substantially reformed end of 1945, when 8300 of 10,000 full-time intelligence workers were fired.

[51] Up to 1957, Kurt Donath (1902–1971) was Technical Director of BMW AG; Franz Popp (1886–1954), who had been with BMW since the beginning, had departed in 1942 and was replaced by Kurt Donath at that time. Donath had come to BMW through Bramo, and he would be tasked with leading BMW out of the post-war woods. At the time of the Paris trip together with Oestrich he represented BMW alone, and besides providing support by encouraging Oestrich and renewing old Gnome & Rhone connections, he might have had—in the end futile—ideas to get also BMW involved into the coming *'Groupe O'* project.

[52] On Popp's initiative a copy of the Blvd. Kellermann *'Blockhaus'* was planned at the BMW Munich-Allach plant (today MTU Aero Engines) after that visit,—see Fig. 9.14 (l). However, the realisation of that two-storey bunker with 32,000 m² floor space (17 m high, 160 × 35 m, 2 m wall thickness) under the code name 'Walnuss (walnut)' was delayed up to 1945, when only 65% of that building were finished,—see Werner, Kriegswirtschaft, pp. 173–177.

new *BMW 801* radial engine by outsourcing the production of the older *BMW 132* to G&R. A total of 5000 labourers worked in France for BMW, but BMW management was said never really happy with this enterprise.[53] The company became infamous for slow production, building only 8500 engines by May 1944, when the Germans had been estimating 25,000. An air raid of the RAF completely destroyed the original Paris-Gennevilliers factories on 9/10 May 1944. With the end of the war, the company was in no condition to continue in the aero-engine business, and what was left was nationalised on 29 May 1945, creating the *Société Nationale d'Étude et de Construction de Moteurs d'Aviation (SNECMA)*.

Upon his return to Munich, Oestrich had to fulfil his British (interrogation) obligations first between 25 August to 5 September 1945 at London-Wimbledon, in the company of 24 other high-ranking German specialists, like Wernher von Braun and Helmut Schelp. Returning to Munich, he and a team of (only) 12 was offered a rather humble 6-months contract by the Americans, which confirmed his first—and after his 1920 decision at the Gymnasium Muelheim a.d.R. for studying French—renewed long-time impulse in favour of the generous French offer.

Of course, Oestrich's perspective on the case was not decisive. He and his team were part of an Allied race for German technology residues, and the French (Communist) post-war government wanted their substantial share, not only to limit potential advantages of their *Great Power* competitors.[54] In general, each of the Allied countries followed different technology acquisition and transfer patterns. Similar to the Russians, the French tried to get/ hire complete, from the very beginning fully functional expert groups, while the American interest was focussed more on individual *top shots* in the business.[55] There are estimations that France acquired comparable to the Soviet Union in total some 800 German experts in military technology in the post-war period. Consequently, there are numerous reports of conflicts in this French-American acquisition process.[56]

In September 1945 the French representatives at Munich organised the transfer of Hermann Oestrich and his BMW team to the French occupation zone, risking and causing

---

[53] See Uziel, Arming the Luftwaffe, p. 46.

[54] See Albrecht, Rüstungsfragen, p. 98.

[55] In this context, England was completely disinterested and believed strongly in its own resources. Consequently, the USAAF, relying on General Arnold's US-UK technology acquisition agreement followed only half-heartedly. Only the US Navy, in permanent competition with USAAF, started at Heinkel-Hirth, Stuttgart-Zuffenhausen, a small-scale series production of HeS 011 turbojet engines, for which Hans von Ohain was brought back from Kolbermoor.

[56] See Ebert, Willy Messerschmitt, p. 293, reporting that the advancing 1st French Army led by General Jean de Lattre de Tassigny violated several times the agreed US zone, and captured near Wertach/ Allgäu, ~ 40 km in the west of Messerschmitt's Oberammergau site, a complete set of drawings for Messerschmitt's swept wing fighter prototype P.1101, and in addition the project documentation for P.1102—P.1112. Materials which had to be handed back to the Americans on 9 June 1945 only, see also Sect. 10.2.2.

a real '*Eklat* (scandal)' with the local US administration. Following the official ATAR story,[57] a technical mission consisting of the Chief Engineer and Director of STAé (Service Technique Aéronautique) Guy du Merle (1908–1993),[58] accompanied by the head of the STAè motor section Daum, and the coming Chief Engineer of the Rickenbach facilities Delbègue came then fully committed to Munich to fix in discussions with Oestrich the framework of a collaboration agreement,[59] started already at Paris in last August, and details of his transfer to the French zone. In October 1945 the transfer was carried out '*par des engages volontaires français de l'armée américaine* (by hired French volunteers of the US Army)',[60] which means that approximately 10 '*Dodge 1.5 t flatbed*' US Army trucks[61] with Moroccan drivers in US uniforms were instrumental for the successful '*coup*'. Gimbel reported, that the French from US perspective had more or less '*kidnapped*' the BMW team: '*They spirited away twelve specialists who had already been selected and cleared for eventual evacuation to the United States and whom the U.S. Airforce employed temporarily at Bayerische Motorenwerke (BMW) in Munich while contractual details were being worked out.*'[62] Oestrich and his core team were transferred over 190 km from Munich-Milbertshofen to a former Dornier plant at Lindau-Rickenbach on the eastern shore of Lake Constance, where in the meantime the French Army had established a car repair shop, see arrow IV in Fig. 9.10. There, under French directory, and in a relatively short time, Oestrich collected a team of 185 specialists, largely from former *BMW Triebwerksbau*, but also from *Junkers, Daimler Benz, Heinkel-Hirth, VDM, Arado* and *Dornier*. This such formed unit receives the designation: *Atelier Aéronautique de Rickenbach*, of which the first capital letters formed officially the abbreviation '*ATAR*' as name for the new engine family.

Oestrich's inscrutable personality may at best be illustrated by the obviously highly cultivated game of inventing project code names with hidden meanings, as shown by other name inventors on the German side already for '*Hetralin*' and '*Herbitus*'. Oestrich must have been a master in this discipline. Compressor project designations *Hermso I–III* (for design studies carried out by H. Reuter and his BBC team for BMW's turbojet developments under H. Oestrich) can be traced quite easily to a combination of his shortened first name and initial letters in his family name—reversed. Some insiders might have found it strange already then, that an official deal between two large companies was put to such a personalised level. When in 1950 the need occurred to name a small

---

[57] See Bodemer, L'ATAR, p. 63.

[58] See Wikipedia, 'Guy du Merle' in English.

[59] The final agreement between the French Government and Hermann Oestrich, installing him as Technical Director for five years, was signed on 25 April 1946.

[60] See Bodemer, L'ATAR, p. 63, footnote 1.

[61] For a rare photo of the convoy during daylight,—see André, Les Turboréacteurs, p. 4. H. Oestrich in a typed ~1960/1961 account in German describes a dangerous border crossing to the French Zone during '*night and fog*'. Report, courtesy of P. Mouton, 31 Jan. 2022.

[62] See Gimbel, Science, p. 31.

company which Oestrich had received from the Air Ministry to park for 10 years till his retirement the royalty income for investing his *'personally-owned technology'*[63] into ATAR, the idea was easily borrowed from the past—and *Hermos S.A.*[64] was born.

Certainly, Oestrich's creativity was also involved to name *ATAR*, the newly emerging jet engine family, which according to traditional translation stands as abbreviation for *'ATelier Aéronautique de Rickenbach'*,[65] keeping the memory to this small suburb/village in the east of Lindau on the shore of Lake Constance, where the engineering nucleus of 120 had started in 1946. Besides this surface meaning, however, ATAR has in good German tradition of code name inventions a multi-facetted meaning. It was e.g. the name of an Iranian *'God of fire'* who shows up already in the holy script of *Zoroaster*, 700–600 BC, the *Avesta*. Zoroaster himself is also called the *'laughing prophet'*, and still further below, *ATAR*[66] stands in old-Persian language also for *'the holy fire'* and *'the man, annealed in fire'*. Possibly, altogether interpretations with which Hermann Oestrich, 1922 scholar of the Humanistic/Classical Gymnasium (specialised in old languages) at Muelheim a.d.R. and certainly knowing his *Nietzsche*,[67] might have liked to be associated with. Albrecht[68] knows that *'ATAR'* was actually the self-determined group name (of sophisticated name creators) in a kind of protest against the blunt code name *'Groupe O'*—and its possible end-reading as Null/Zero.

In regard of the locally celebrated *All Saints' Day* on Thu 1 November 1945, the design work of *Groupe O* started in earnest on the following Friday, when Oestrich received at Rickenbach in written form *'diréctives générales* (general directions)' as a 'Note #1' on what was expected from his team, more precisely defined after his studies, and what was agreed again at length in written form in 'Note #2', dated Sat 3 November 1945 under the

---

[63] The correct terminology of this phenomenon—out of Oestrich's perspective—is called *'reproduction rights'*, according to his *'contrat de licence et de collaboration technique'* which H. Oestrich, in person, attributes on the basis of his patents to the French Air Ministry for the ATAR production, and for which he is compensated by royalties, which largely remain deposited in the company *Hermos S. A.*, Casablanca, Morocco, up to his leave in 1960,—see Sect. 10.4, and—see Bodemer, L'ATAR, p. 78.

[64] Hermos S.A. (Société Anonyme) had been founded at Casablanca still under *Sherifien legislation* in 1950, which officially represented an aero engine repair shop with some 170 employees in 1959.

[65] This is apparently the official *SNECMA* version—see Bodemer, L'ATAR, p. 63, while Kay, Turbojet, Vol. 2, p. 182 has wrongly *'Atelier Technique Aéronautique Rickenbach'*.

[66] For the complexity of ATAR interpretations,—see Wikipedia, 'Atar' in English, especially the sub-section on 'The cult of fire'; in the present Persian-Zoroastrian calendar both the 9th day and the 9th month carry the name 'adar', and the 5th Aryan godhood after Sky, Water, Sun, and Moon—is Fire/Atar.

[67] See Wikipedia, 'Friedrich Nietzsche', 'Philosophy of Friedrich Nietzsche', *'Thus Spoke Zarathustra'*, and 'Zoroastrianism', all in English.

[68] See Albrecht, Rüstungsfragen, p. 107.

header of the local *STAe Office* at Kressbronn, 13 km west of Lindau-Rickenbach.[69] This important meeting on Fri 2 November 1945 attended besides Oestrich, the well-known trio du Merle—Daum—Delbègue, fixing

- the *'turbo-réacteur 3306'* as prime study object with 1700 kp take-off thrust, which Oestrich promises to investigate in view of a possible turboprop version within 4 weeks, and
- *'turbo-réacteur 018 et turbo-propulseur 028'*, as not less urgent, though Oestrich claims for the extra-propeller and -gearbox work an answer within 6 weeks only.

Correspondingly, the ATAR team defines after a short thermodynamic study the following target characteristics, what would become ATAR 101 V ('Versuch'/prototype):

- take-off thrust—1700 kp
- speed—8500 rpm
- mass flow—46 kg/s
- compr. pressure ratio 4.2
- weight—850 kg
- maximum diameter—0.89 m
- compressor efficiency—0.82
- turbine efficiency—0.78
- turbine entry temp.—650 °C
- seven-stage axial compressor with drum-type ('tambour') rotor from light metal
- rotor blades from light metal, mounted in annular rotor grooves
- rotor vanes, similarly from light metal, mounted in annular casing grooves
- annular combustion chamber with 20 burners, and 20 centred fuel injectors
- single stage turbine with air-cooled Ni-Cr blading *('Sirius HT')*, and with cooling inserts.

---

[69] Kressbronn was officially the naval base of the *'Fottille du Lac de Constance'*, which comprised some 40 mostly smaller units (*'vedettes'*), manned by 9 officers, 45 sergeants, 159 privates and sailors, and 19 gendarmes. The history of one boat is well documented: In 1938 the *Fishing Control Boat Lachs* (Salmon) was registered at Ziegenort/Trzebież, in Fig. 8.20 (r) on the west bank of River Oder opposite of Gasierzyno/Ganserin, in 1940 it was stationed as *Motorboat C 109* of the German Navy at Calais in preparation of *Operation Sea Lion,* in 1942 it was attached to the Fishery Office Bregenz on Lake Constance, where in 1945 it was confiscated by the French Navy, and again registered as *Vedette Héron (B4),* in 1950 handed over to the *Water Police Lindau* as *Zander,* and in 1960 finally sold to a private owner. From US perspective, Kressbronn has a special technical landmark by a small, 72 m long suspension bridge across the creek *Argen,* built in 1896/7, and— unconfirmed—young Othmar Ammann (1879–1965) on the spot, later known as designer of New York's, 20-times longer *GWB George-Washington-Bridge* (1931), see also Chap. 7.

9.3   Groupe O: *en marche*

After 15 June 1946 the *Ministère de l'Air* decided to attach and integrate an increasing number of French engineers to the German core team. Consequently, Rickenbach in the occupied zone was given up, and Oestrich's design office, then some 160 persons, was transferred to heartland France, to Decize on River Loire, in Departement Nièvre.[70] In total 272 waggons filled with materials, removed machine tools and equipment, office furniture, instruments, documents and paperwork left Lindau railway station in the coming months, *peu à peu* up to the end of 1946.[71] A smiling Hermann Oestrich and the first group of 120 German and Austrian technicians arrived on Sat 27 July 1946, Fig. 9.11 (l). The transfer and installation of the Germans was not easy though, material difficulties were accompanied by considerable psychological reservations, how the locals would react about the arrival of some 120 German families. Here the presence and support by Communists in the town hall was a considerable advantage. For Decize a fine example of thorough dialectic training has been passed on, where Communist leader Jacques Duclos[72] stated in a discussion in August 1946: *'Those who are criticising the installation of the Germans in Decize, cannot be anything other than* (former) *collaborators!'*[73]

The whole complex, called *'Cité Voisin'*, working and living quarters of personnel and attached families were allocated in an area of empty barracks buildings, erected in 1939 for the *'Gardes Mobiles'*, and during the war occupied by the German *Wehrmacht* (Army). The barracks were situated somewhat elevated at the outskirts of Decize, and dominated there the Loire valley.[74] The garages were used as offices, and along the *'main road'* was a camp of German PoWs,[75] separated from the *'Cité'* by a barbed wire fence.

---

[70] This is the official French explanation; in addition the—see Wikipedia, 'Allied Control Council' in English, at Berlin, the governing body of the Allied Occupation Zones in Germany and Allied-occupied Austria after the end of WW II decreed in April 1946 that all armament-related studies or productions in Germany had to be brought to an end. Consequently, preparations for leaving Rickenbach started in May 1946.

[71] The Rickenbach plant was handed back to *Dornier* in 1948—empty, in this respect presumably copying the retreating Wehrmacht. The German-French exchange of military booty is best document-ed for—see Wikipedia, 'Greif cannon' in English, which was set-up at Ehrenbreitstein Fortress on the River Rhine, opposite Koblenz, in 1524. In 1799 the French brought it to Metz, and in 1866 to the Musée de l'Armée in the Hôtel des Invalides, Paris. In 1940 the Germans sent it back to Koblenz, where the French removed it again in 1946. It was only as part of the Franco-German reconciliation, that French President François Mitterand signed in 1984 with German Chancellor Helmut Kohl an agreement for permanently *'lending'* it to the fortress Ehrenbreitstein.

[72] See Wikipedia, 'Jacques Duclos' in English.

[73] In French: *'Ceux qui critiqueraient l'installation des Allemands à Decize ne pouvaient être que des collaborateurs,'*—see Carlier, Les débuts.

[74] Today, Gendarmerie Nationale, 105 Av. de Verdun, Decize.

[75] See Eggers, Professor Gerhard Eggers (1912–1998). PoW—prisoner of war.

Fig. 9.11 *Groupe O*—from occupied Germany to France: At Decize, H. Oestrich arriving on 27 July 1946 (l), 120 engineers with families at *Cité Voisin,* ~1950 (r)

Up to 1947, Dr. H. Oestrich as Technical Director had built a team of highly qualified and experienced specialists, most of those with BMW background still in their former positions:

- Deputy Director and Regulation—Dr. A. Stieglitz
- Advanced Projects—Dr. H. Triebnigg
- Mechanical Design—H. Rosskopf
- Thermodynamics—Dr. H.G. Münzberg
- Mechanical Integrity and Materials—Prof. H. Wiegand
- Engine Controls—S. Decher (Junkers)
- Aerodynamic Studies—Prof. A.W. Quick (Junkers)
- Engine Installation—G. Eggers (Junkers)
- Turbocomponent Aerodynamics—K. Fickert (BBC)
- Test Department—C. Jouanneau (Rateau)
- Engine Tests (Villaroche)—L. Menz
- Flight Tests (Villaroche)—H. Borsdorff (Rechlin)
- Technical Documentation—K. von Gersdorff (RLM)
- Mechanical Workshop—H. Pouyaud
- Assistant to Technical Director—R. Constant (Rateau)

For the transition phase from Rickenbach to Decize in approximately July 1946 exists a rare original document,[76] the blading plan of the first axial compressor of the ATAR 101 V, Fig. 9.12. Characteristically, the seven-stage axial compressor blading of in total 869 aerofoils is split in three groups (Gruppe) I–III, which indicates blades of the same profile shape, but different blade length and adjusted incidence angles.[77] In general the

---

[76] See Bodemer, L'ATAR, p. 65.

[77] See Kruschik, Die Gasturbine, 2nd ed., p. 158.

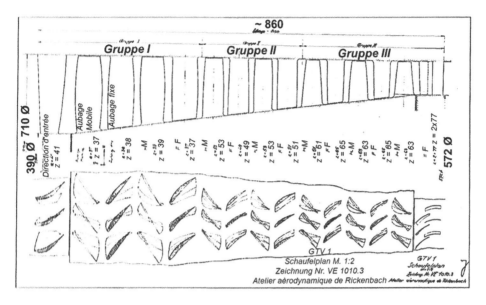

Fig. 9.12 *GTV Groupe Technique Voisin*, 1946: © A. Bodemer *ATAR 101 V*, first drawing of axial compressor blading, with blade count, dimensions in mm

blading is clearly of the 50% reaction type, with the rotor blades designated as '*Aubage Mobile or M*', and the stator vanes correspondingly as '*Aubage fixe or F*'. The number of aerofoils per row is designated by 'z'. The compressor has—like the BMW 109-003 and the Jumo 109-004—a row of inlet guide vanes ('*Direction d'entré*'), and at the compressor backend a double-row of exit guide vanes to remove the remaining exit swirl at combustor entry.

The—additionally typed—drawing label in the lower right-hand corner reads, top—down:

- *GTV 1* (Groupe Technique Voisin) No. 1,
- *Schaufelplan M. 1:2* (Blading plan, scale 1:2),
- *Zeichnung Nr. VE 1010.3* (Drawing No. VE.. Verdichter-Entwurf/compressor design),
- *Atelier aèrodynamique de Rickenbach* (with aèro . . ., instead of aéro . . ., a tiny flaw after short time practising French).

The advanced status of the compressor design with PR 4.2, corresponding to an average Stage Pressure Ratio of 1.227 has already been highlighted by point '4' in Fig. 8.25.

The new camouflaging designation '*GTV Groupe Technique Voisin*' insinuated that the unit belonged to the traditional French aircraft manufacturer *SA des Aéroplanes G. Voisin*, which actually stopped the aircraft production after WW I, continuing in the automotive sector together with *Citroën* up to 1939, when *G&R* acquired the remainder, so that in 1945

Fig. 9.13 *'Groupe O'*—arriving at Melun-Villaroche: Reception and office building, 1950 (l), *ATAR 101 V* on test bench, 1948 (r) © Safran Archive

SNECMA, capturing G&R, succeeded here more or less automatically (Fig. 9.13).

Personally, Hermann Oestrich steadily integrated to the new tasks:

- 1948—French citizenship,[78]
- 1950—SNECMA's Technical Director Gas Turbines,
- 1952—Knight of the Order of the Légion d'Honneur,[79]
- 1960—retirement, and consultant to SNECMA, up to his death in 1973.

Oestrich's appointment in 1950 went along with the official integration of *GTV* into *SNECMA* on 15 June 1950,[80] and correspondingly, with new contracts for some German and Austrian engineers, while others left France, either back to their home countries,[81] or in search of other turbojet development offers, e.g. in Egypt, South Africa, USA and Argentine. Michel Garnier (1916–1989), long-time his deputy, followed Hermann Oestrich as SNECMA's Technical Director in 1960. His name is connected with the *ATAR 9 K* for the *Mirage IV*, the *Larzac*, the *Olympus* for the *Concorde*, and nearly 80 own, patented inventions.[82]

Finally, in 1952/3 arrived the last German members of the former Groupe O at the new working place at SNECMA, Villaroche, and privately settling mostly in the neighbouring Dammarie-les-Lys, Département Seine et Marne, with then 5000 inhabitants.

---

[78] One can assume that this unusually early acceptance of Oestrich's request for French citizenship was a precondition for the favourable agreement on his patent royalties/ production rights settlement scheme with the French government,—see Sect. 10.4.

[79] See Wikipedia, 'Legion of Honor' in English.

[80] See Bodemer, L'ATAR, p. 78.

[81] However, returning engineers had to change their field of work, since Allied restrictions prevented aeronautical activities in Germany up to 1955.

[82] See Sect. 10.4 for comparison. Information, courtesy of Air & Cosmos, No. 1267, 13 Jan. 1990.

Fig. 9.14 *Gnome et Rhone/Snecma,* Paris—a disputed inheritance: *Blvd. Kellermann* production works with *'Blockhaus'* (arrow), drawing 1938 (l), *l'Humanité* caricature, 1950 (r) © Safran Arch. l

Already in June 1946 the first drawing sets for prototype construction reached the former *Gnome et Rhone* fabrication plant at Boulevard Kellermann, Fig. 9.14 (l), which now belonged—renamed—to *SNECMA*. In 1947 commenced the first component tests—for combustion chambers, compressors, air-cooled turbines and the complex new ATAR regulation system, as derived from the Jumo 004,[83] mostly at the 300 km distant Melun Villaroche Aérodrome,[84] where the set-up of a small test centre had been started. Fig. 9.13 (l) shows the combined (civil airport) reception and test centre office building, while the test facilities, as shown typically in Fig. 9.13 (r), were located nearby. Snecma, charged with design, testing and production, delivered the first ATAR 101 V1 prototype engine in early March 1948 from the Kellermann works to the Melun-Villaroche test centre, and the first test runs, Fig. 9.13 (r), took place there between 26 March and 5 April 1948. During a 1.5 h test, the contractually agreed Phase A target of 1680 kp thrust at 7500 rpm was successfully achieved. Already on 21 May 1948, the Phase B target of 2200 kp thrust was surpassed with the second built turbojet engine *101 V2*, and finally in January 1950 and after more than 1000 test hours, a record thrust of 2700 kp, three-times that of the (1945 planned) *BMW 109-003C* was demonstrated.[85] Over a period of 25 years the ATAR family of military turbojet engines was produced after 1950 in 14 systematically expanded versions in a thrust range from 2400 to 7200 kp, so that it belonged with more than 5000

---

[83] See the following Sect. 9.4.

[84] See Wikipedia, 'Melun Villaroche Aérodrome'in English. The airport with 1975 and 1300 m long runways, situated 35 km south-east of Paris city centre, saw over time with the various occupants numerous aircraft types, first at the beginning of WW II, e.g. heavy French fighters Potez 631—with 2× 700 hp Gnome et Rhone GR 14M piston engines, thereafter Luftwaffe bombers like Junkers Ju 88A, Heinkel He111H, and Dornier Do217E, followed by the 9th USAF Douglas A-26 and A-20 bombers, which after war's end were more and more replaced by DC-3/C-47 transport aircrafts.

[85] See Bodemer, L'ATAR, p. 79.

built units to the most successful in that category world-wide; ATAR engines powered among others the French fighter aircrafts *Ouragan, Mystère, Super Mystère, Etendard, Vautour*, and *Mirage III*.[86]

While the technical post-war integration of the German engineers at SNECMA went smoothly, the political acceptance was at times highly disputed, and caused some disturbances. Symbolically the *Usine Kellermann*, G&R's impressive main production plant on Boulevard Kellermann in the 13th arrondissement of Paris, Fig. 9.14 (l), represented the *genius loci* of this remarkable piece of French-German rapprochement, while the central figure in this context was Charles Tillon (1897–1993), a French metal worker, Communist, trade union leader, politician and leader of the French Resistance during World War II.[87]

The Boulevard was built in 1861, today a 1.2 km section in the south of the in total 35 km long *BP Boulevard Péripherique,* named after François-Christophe Kellermann (1735–1820), 1st Duke of Valmy.[88] The plant covered a triangular area of ~22,000 m^2 along the north face of Bd Kellermann close to Porte de Chantilly,[89] and was steadily expanded to Europe's largest aero engine production site for some 14,250 workers, after it had been severely damaged during the German retreat at the end of WW I.[90] This was mainly the work of one business leader from the age of 29, Paul-Louis Weiller (1893–1993),[91] who from 1922 to 1940 developed here the most important airplane engine factory in Europe. In 1938 he erected along the western rim of the plant area a huge bomb-

---

[86] The 5000 built ATAR engines correspond roughly to 1500 built fighter aircraft from Dassault.

[87] See Wikipedia, 'Charles Tillon' in English.

[88] See Wikipedia, 'François Christophe de Kellermann' in English. Kellermann's family came from Saxony; the—see Wikipedia, 'Battle of Valmy' in English, on 20 Sep. 1792 was a strategically rather insignificant battle between Prussian coalition troops and the newly formed French revolutionary army under Kellermann, best known in German curriculum by Goethe's attendance as visiting *'battle stroller* (Schlachtenbummler)', and his pathetic after-remark to Prussian officers *'From here and today a new era has begun, and you can say, you were part of it.'*

[89] In the drawing of Fig. 9.14 (l), the present *Rue Cacheux* branches off to the north in the lower left corner. During WW I some 25,000 G&R radial engines were produced here, and another 75,000 from G&R licensees. Also Manfred von Richthofen, in the Red Baron's famous tri-decker *Fokker Dr.I*, flew with the *Oberursel Ur.II* an—in 1916 un-licensed—clone of the *Le Rhone 9J*, a 110 hp rotary engine.

[90] On 11 April 1918, after the battery—positioned between Fourdray and Crépy-en-Laonnois in the Forêt de Compiègne, 120 km in the north-east had been set-up since 23 March 1918, one (of three) Krupp 37 m long barrel—see Wikipedia, 'Paris-Geschütz' in German, sent a calibre 21 cm, 106 kg grenade over a distance of ~120 km which hit the *Usine Kellermann*; with a muzzle velocity of up to 1645 m/s, the flight time was about 3 min, so that an allowance of ~60 m for earth rotation had to be considered. The gun was manned by 60–80 soldiers of the German Navy, familiar with the big calibre ammunition.

[91] See Wikipedia, 'Paul-Louis Weiller' in English. Weiller, himself a flight ace during WW I, was born to a rich Jewish Alsation family, who rose to prominence in business, finance, and politics during the nineteenth century. In 1908, his mother Alice Weiller-Javal (1869–1943) was the second passenger woman on board of Wilbur Wright's double decker at Le Mans; she was murdered at Auschwitz.

proof, windowless *'Blockhaus'*[92] as factory building, as marked by the arrow in the Fig. 9.14 (l) drawing, 130 m long, 30 m wide, with five storeys above ground, and two sub-terranean levels, all connected with huge elevators.

In August 1944 General De Gaulle, head of the French government in exile, proposed Charles Tillon as the coming new *Ministre de L'Air* in the hope to catch the French working class for the idea of a renewed French aviation. On the engine side followed a double-strategy to close the existing gaps as a result of the German occupation e.g. by encouraging Hispano-Suiza France into license production of the offered Rolls-Royce *Nene* engine, while *Gnome et Rhone*, the by far largest French engine manufacturer was accused of a too close collaboration in the foregoing period, and was consequently nationalised as decreed on 28 August 1945.[93] This was SNECMA's starting point, though without corresponding experience, so that the decision pro nationalisation meant also at the same time a decision in favour of the full acquisition of German turbojet engine technology as offered by Oestrich and his team; in fact, Tillon was Air Minister under de Gaulle from 10 September 1944 to 21 November 1945, and was as such formally responsible for the hiring of *'Groupe O'*. Following the war he was again elected a deputy, and between 1944 and 1946 after his turn as Minister of Air, also Minister of Armaments and Minister of Reconstruction and Town Planning. After the communist ministers were dismissed in 1947, he was made partially responsible for the PCF's (Communist Party) military policy, then a *'Fight for Peace'* policy which simply followed pro-Soviet Union interests, and fought strictly a French re-armament. Consequently, when in 1950 the reorganisation of SNECMA, with Hermann Oestrich as Technical Director was made official, there was strong opposition from PCF and associated trade unions, strongly voiced in their central organ *l' Humanité,* Fig. 9.14 (r).[94] On 12 May 1950, in an article in the PCF central newspaper l'*Humanité,* the communists accused the French government *'that they installed former Nazis to preside now French factories, like that Ostreich* (sic) *who has been placed at the head of a big Paris plant',*[95] underlined by the attached caricature of Fig. 9.14 (r), entitled *'La réorganisation de la SNECMA'.*[96]

---

[92] The German *'Blockhaus'* is used in French synonymously for bunker/ casemate; interestingly at Dresden exists still today the—see Wikipedia, 'Blockhaus (Dresden)' in German, across the River Elbe, and opposite of the *'Golden Rider'* monument of *August, the Strong* on horseback; this cubic building replaced in 1732–1737, planned by the French architect Zacharias Longuelune, a foregoing wooden toll station.

[93] See Werner, Kriegswirtschaft, p. 287. Constanze Werner talks about a *'trustful relationship without technology exchange'* between BMW and G&R. For her, it is striking that only G&R was accused for *'collaboration'*, while all other law suits at the *Cour de Justice de la Seine*, including the *'Case Renault'* ended with a *'non-lieu* (suspension)'; the corresponding files are protected under French archive law with a 100 year *'blocking period'*.

[94] See Chouat-Hugonnet, Le Comité d'entreprise, para 24, footnote 15.

[95] In French: *'... présider les usines françaises par des nazis mal blanchis comme cet Ostreich (!) qu'il met à la tête de la grande usine parisienne.'*

[96] On the left pillar, the placard reads 'Bureau d'em boche', with a certain double-meaning with 'boche—German',—see the comprehensive Wikipedia, 'List of terms used for Germans' in English, and 'empocher—on the gravy train ("absahnen" in German)'.

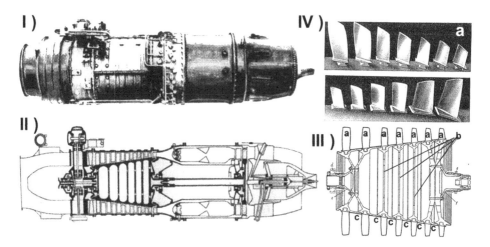

Fig. 9.15 ATAR 101 V mechanical design features, 1948: (I) Horizontally split compressor casing, (II) general arrangement, (III) compressor—welded tambour rotor (101 B1), IV) light alloy blades (a) and vanes © A. Bodemer

9.4   ATAR: And Its Technology Ancestors

This final section of a comparative design(er) review of *Reuter vs. Oestrich* will highlight the most essential mechanical design features of the ATAR 101 V—and trace their origins in a comparison of Oestrich's BMW 109-003 basic design concept and typical gas turbine design characteristics from Brown Boveri.

The most significant change, the selected degree of reaction for the compressor blading aerodynamics—from AVA's R 1 to BBC's R 0.5—has already been outlined in detail in the context of Fig. 9.12. The corresponding isentropic efficiency benefit of the BMW 003 C as basis for the seven-stage ATAR 101 V axial-flow compressor should have been, as deduced before, in the order of five to seven points. In addition, Fig. 9.15 shows in condensed form the main mechanical design differences:

The outside view of the engine, I), illustrates point-blank the horizontally split compressor casing, not to be found in Oestrich's 003 design, Figs. 6.29 and 8.9, so that a BBC origin via the *Hermso* studies becomes likely.[97] Though already Kay had made the point for the 109-003 A-1/A-2 compressor casing: '*The compressor casing.., cast in magnesium alloy (Elektron),*[98] *... was not split longitudinally and thus differed from the Junkers*

---

[97] The ATAR engine family kept this feature roughly up to the ~1970 M53 series, or the capability of sustained supersonic flight, when the higher thermal loads prevented the further use of casing half-shells and its inherent '*omega-deformation*'.

[98] The cast ATAR 101 V split casing was either from *NF A-U4N (France)* or Mg-Zr alloy *ZRE 1*,—see Kruschik, Die Gasturbine, 2nd ed., p. 157.

109-004',[99] but as Fig. 6.17 confirms, already Wagner's foregoing *RTO* engine at Junkers Magdeburg carried this characteristic design mark, and followed thus Wagner's now classic *'I knew the stationary gas turbine concept'*, to which belonged somewhat naturally Brown Boveri's *'split casing'* design, Fig. 6.15.

Further, the ATAR 101 V general arrangement, II) in Fig. 9.15, highlights at first glance the unconventional new *'tambour* (drum-type)' construction of the compressor rotor, obviously different to Oestrich's foregoing preloaded disc arrangement of the BMW 003, Fig. 8.9, but identical to Hermann Reuter's revolutionary light-weight compressor design concept for the experimental 109-003 D version, Fig. 9.4. Details of the applicable rotor manufacturing process will be discussed in the following by means of the enlarged rotor depiction III)—of the 1950 ATAR 101 B1 version. The complete compressor consisting of discs, blades (a) and vanes, IV), was like the BMW 003 and 018 from—in parts forged—aluminium alloy, for the BMW turbojet engines from *'Dural'*,[100] and for the early ATAR series from RR 58.[101] The blading, with *'hammer heads'* at the blade roots and correspondingly *'dovetail fixations'* for the vanes, was attached in simple circumferential grooves, a considerable factor for cost and assembly time saving. However, the apparently greatest advantage of the split casing configuration resulted from the possibility to balance and subsequently mount the rotor as one complete unit.

After this proof that the ATAR rotor design actually originated from Brown Boveri, Baden, CH, it is worthwhile to look at its inherent manufacturing technology from the beginnings of steam and gas turbine technology in more detail, Fig. 9.16. Brown Boveri had decided to publish its *'welded rotor technology'* at the forum of the 2nd WPC World Power Conference at Berlin, 14–26 June 1930.[102] The presentation was carefully orchestrated together with the internationally renowned Professor Aurel Stodola, similar to the certification run of the world's first power generation gas turbine under Stodola's supervision, nine years later. Stodola[103] presented the innovative manufacturing approach for a *'highest speed rotor drum of rim-welded forged steel discs without centre bore'* by

---

[99] See Kay, German Jet Engine, p. 108.

[100] Information, courtesy of Robert Marmelic, 1 Nov. 2016.

[101] See Wikipedia, 'Hiduminium' in English, a pre-WW II Rolls-Royce development, with the abbreviation for HIgh DUty AluMINIUM.

[102] The conference with nearly 4000 attendants took place at the State Opera House (*'Krolloper'*), opposite of the *'Reichstag'* (parliament building); information, courtesy of E. Blocher, Siemens Comm., 9 Feb. 2022. It was hosted by *Siemens*; Albert Einstein (1879–1955) lectured on *'The Space, Field and Ether Problem in Physics'*. Sir Arthur Eddington (1882–1944)—a distinguished astrophysicist who first explained Einstein's *Theory of Relativity* in English and led the first expedition to confirm it—said in his address that, in the future, *'subatomic energy would provide the plain diet for engines previously pampered with delicacies like coal and oil'*. As a society highlight the *'Weltkraftfest'* was celebrated in Berlin's *'Sportpalast'* on 18 June evening, 650 waiters served 500 kg turtles (as soup), flushed by 2000 bottles of champagne, and 4000 of exquisite whites and reds.

[103] See Stodola, Technisch-wirtschaftliche Fortschritte. In addition, explicitly mentioned were the broadened disk rims and the forged shaft stubs on the outer discs.

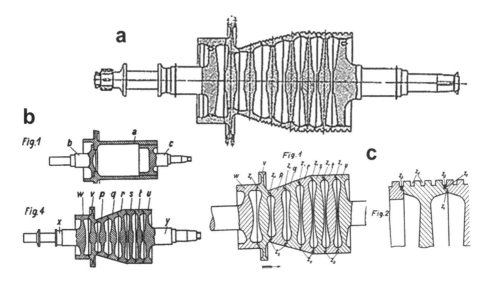

Fig. 9.16 Gas turbine *'drum rotor'* manufacturing technology, 1930: (**a**) Stodola WPC Berlin paper, June 1930; BBC patents (**b**) GB347,374; (**c**) GB362,813 (German priority 2 May 1930)

Fig. 9.16a, thus quoting the essential patent features, which were then released with an effective priority just six weeks earlier for Brown Boveri, Baden, CH, Figs. 9.16b, c.

Professor Heinrich Triebnigg,[104] member of Oestrich's engineering team, described the ATAR 101 rotor assembly, such that individual discs were brought together by a *'joining technology'*[105] to a disc package, to which the rotor stubs were flanged at both ends. This 'joining' was actually a shrinking process in which an—in final operation colder—disc was warmed up and then shrunk to a protruding rim neck of the neighbouring—in operation warmer disc, as illustrated in Fig. 9.16c, for a turbine rotor, for which the operational temperatures decrease from left to right.[106] Thereafter, the circumferential grooves for blade fixation were inserted and the drum rotor as a whole was dressed over. In addition the disc connection had to be strengthened by an appropriate welding process, suited for the light metal disc materials. This required technology for joining Al-based materials had been invented in time in 1941 as *GTAW Gas Tungsten Arc Welding*[107] by the inventor Russell Meredith at Northrop Aircraft,[108] who used a Tungsten electrode and a *laminar* Helium gas

---

[104] Prof. H. Triebnigg (1896–1969): 1949–1955 at TU Berlin, Institute for machine elements, 1955–1967 TH Darmstadt, chair for combustion engines and flight motors.

[105] In German *'Fügetechnik'*.

[106] Information, as described in the relevant BBC patents, courtesy of the late Fredy Häusermann (16 May 1942—16 Nov. 2016), 31 Oct. 2016.

[107] See Wikipedia, 'Gas tungsten arc welding', and 'Gas metal arc welding', both in English.

[108] For the close connection between Northrop's chief engineer V. Pavlecka with Brown Boveri's A. Meyer—see Sect. 6.4.2 and Fig. 6.52 (r); patent US2,274,631 'Welding torch' had a priority date of 4 Jan. 1941.

shielding. Northrop called the process *Heliarc,* and sold patent and trademark rights to the Linde Division of UCC Union Carbide Corp., an US subsidiary of Linde AG Munich, which had been confiscated in 1917 and merged to UCC thereafter.[109] In addition, Linde, New York, filed a correspondingly necessary patent for superimposing a high-frequent AC to the welding current.[110]

Since the 1930s German industry including BMW (automotive and aero engines) was using an early form of GTAW called *'Arcatom',*[111] where the arc was between tungsten rods, cooled by hydrogen. France had since 1909 its own, specialised welding institution *'La Société de Soudure Autogène Française (SAF)',*[112] which mastered all details of this German atomic-H_2 welding process after WW II,[113] so that the availability and application of suitable welding processes for the ATAR 101 aluminium rotor manufacturing can be taken for granted.

A nice engineering example how a (German) combustor design drawback was turned into an advantage in advanced regulation technology over time at SNECMA has been illustrated in Figure 9.17, handwritten notes taken by young Pierre Mouton at ETAVA in 1959.[114] In the left column he characterises the *'English* single can *combustor technology',* mainly based on the inventions of chief engineer Dr. J. Stanley Clarke of Joseph Lucas Ltd, with

- *big pilot flame, but large pressure drop for the single can combustor,*
- *very good flame stability and good combustion efficiency, but Δp is important, $\Delta p = 10\%$ of the inlet dynamic pressure,*
- *total pressure = static pressure + dynamic pressure.*

---

[109] This development was reversed in 2018, when on initiative of—see Wikipedia, 'Wolfgang Reitzle' the German Linde AG and US company Praxair Inc. were merged to Linde plc., Dublin.

[110] Patent GB593,536 'Improvements in and relating to welding' by Wilber B. Miller for Linde Air Products Co., NY, with priority 14 June 1944, required to disperse/ reduce the isolating alumina layers. Information, courtesy of Peter Adam, 7 Nov. 2016.

[111] See Wikipedia, 'Atomic hydrogen welding' in English, replaced after WW II by *'GMAW Gas metal-arc welding',* mainly because of the availability of inexpensive inert gases.

[112] *SAF,* majority-owned since 1929 by *Air Liquide.*

[113] It must remain open, if based on own developments or by technology transfer, most entertaining would have been by a 1943 GE training movie, https://www.youtube.com/watch?v=uZwYMyHlWXk

[114] P. Mouton, born in 1939, was a student at ETAVA École Téchnique Aéronautique de Ville d'Avray (today IUT Institut Universitaire de Technologie), 8 km west of Eiffel Tower, from 1954 to 1959. The school was created in 1946 for staffing the renewal of the French aeronautical industry; information, courtesy of P. Mouton, 7 Feb. 2022. See Mouton, Junkers Jumo 004, p. 27.

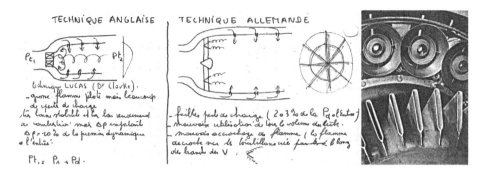

Fig. 9.17 Combustor analysis (P. Mouton, ETAVA notes, 1959): English can combustor (l), German annular combustor (m) and BMW 003 combustion chamber, rear view (r) © P. Mouton

Comparable considerations (m) for the '*German* annular *combustor technology*' with

- *small pressure loss (2 to 3% of the inlet dynamic pressure),*
- *poor usage of the total flame tube volume,*
- *poor flame holding, the flame is detached by vortexes behind the V arms* [within the 16 burners, hidden by the 2½ conic burner shields, visible in Fig. 9.17 (r)].[115]

The described deficit of the BMW combustor led to a poor combustor efficiency, so that 12% of the injected fuel burnt still while passing the turbine, impossible to operate with a fixed exhaust nozzle. Under development pressure the German engineers found no solution for the combustor problem, but installed a pilot-regulated, mechanically adaptable nozzle instead. The longitudinal movement of the exhaust '*bullet*', Fig. 8.9, was by means of an electric motor which drove shafts leading to bevel gears. An electrical switch in the aircraft cockpit was used to select three positions of the bullet, and hence the exit area:

Switch position	(condition)	Nozzle exit area
A Anlassen	engine start and idling	Maximum
S Starten	take-off and climbing	Minimum
H Horizontal	horizontal flying	Intermediate

Since this combustor problem applied also to the Jumo 004, one can generally state that both German turbojet production engines required as a considerable drawback the installation of variable nozzles to compensate for the '*parasitic*' post-combustion after the turbine. And, since there was no time for a complete restart, also Oestrich's team continued with these flaws for the ATAR 101, and the hydromechanical compensation system. In the late

---

[115] In Fig. 9.17 (r), the 2½ visible burners in the upper part lie ~25 cm behind the plane of the air feed pockets, visible in the lower part; compare also Fig. 8.9.

1950s, when supersonic flight speeds of the Mirage III and IV revealed increasing deficits in the regulation accuracy of the established hydromechanical system, Albert Stieglitz, Oestrich's deputy and head of SNECMA's regulation department initiated a remedy which in due course led to a patent for an electric temperature *'trim correction'*,[116] as a starting point for further analogue and then digital improvements of the ATAR control unit. Of the many subsequent regulation *'firsts'* at SNECMA, the fully modulated afterburning system which could even be ignited during the acceleration to maximum power has to be mentioned, as well as the acceleration/deceleration control by limiting the fuel flow/compressor discharge pressure ratio; ideas still in practical use today, and associated with the names of Lothar Menz and Wilhelm Jurisch.[117]

Finally, the close relationship of the basic BMW 109-003 turbine design with that of the ATAR 101 up to ~1950 has been described already by Hermann Oestrich himself in a post-mortem German publication on BMW 003 development, issued in 1989,[118] Fig. 9.18.

This hollow turbine blade was, according to Oestrich, the result of *'the most comprehensive, the most difficult and the most interesting development'*, and the non-availability of suitable materials requested the air-cooled configuration. The first version was a derivative of BMW's in-house turbocharger turbines, patented already in 1938 for Gustav Zellbeck and Alfred Müller.[119] However, the required welding of two milled profile halves caused insurmountable cracking problems, so that a second, very expensive, nonetheless unsatisfactory variant was developed together with Leistritz, before a third, sheet metal

---

[116]Patents for A. Stieglitz et al.: FR1,391,769 and US3,312,057, *'Regulator device for gas-turbine engines and rotary like units'*, with priority 29 Nov. 1963, where the idea of a temperature-based burner control had been already part of a foregoing Junkers patent DE898,699 for S. Decher et al., with priority 6 March 1943. Most remarkably, the Jumo 004 speed controller (later in ATAR engines) contained a *'Fliehpendel'* as a stand-alone feature, the 1:1 implementation of Watt-Boulton's *'fly-ball governor'* of 1788. In the 1960s, Decher, who belonged to the Jumo turbojet team of A. Franz between 1938–1945, thereafter to the 'Groupe O', had joined again A. Franz, now at Avco Lycoming.

[117]Information, courtesy of Pierre Mouton, 21 and 27 Jan. 2022. W. Jurisch returned to Daimler Benz, Stuttgart in ~1959, filing there some 25 verifiable patents in his name, before he moved on to MTU Munich in ~1970, to contribute to the newly emerging controls department in the following decade.

[118]See Oestrich, Entwurf und Ausreifung, p. 109 f. The paper, dating from 1946, was presented in 1989 in Munich at the DGLR-Symposium *'50 years of turbojet flight'*, presumably administered by K. von Gersdorff, in *'Groupe O'* responsible for *Technical Documentation*. The 15 p. presentation is illustrated by nine slides (#22–25, and #29–33), of which Fig. 9.18 (l, m) depicts #32, Fig. 9.18 (r)—see Bodemer, L'ATAR p. 72, indicating the switch to a larger original volume flow; all slides carry the *ATAR* logo, remarkably with bi-lingual captions in German and Spanish.

[119]Patent US2,297,446 *'Hollow blade for exhaust gas turbine rotors'*, with priority 3 Dec. 1938. Gustav Zellbeck (1913–2011) was the uncle of Prof. em. Hans Zellbeck, MTU Friedrichshafen and TU Dresden; Alfred Mueller was addressed at length in Sect. 8.3.4. Interestingly, this design was heavily criticised by H. Schelp after the war,—see C.I.O.S., Interrogation, especially the *'root fastening, which puts a high concentrated load on the neck above the root which has resulted in a great number of blade failures.'*

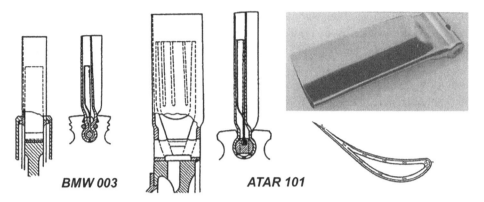

Fig. 9.18 BMW turbine cooling technology transfer: *BMW 109-003* final serial rotor blading (l),
SNECMA ATAR 101 rotor, production configuration up to ~1950 (m, r). © Schubert, Bodemer r

configuration[120] finally represented the successful solution, Fig. 9.18 (l), which was
accordingly adapted to the larger ATAR 101 turbine blade. A sheet metal insert, visible
in Fig. 9.18 (r) reduced already for the BMW 003 turbine blade the cooling air through-
flow area,[121] intensified thus the heat transfer and reduced the overall cooling air mass flow,
blown off after the fourth compressor stage, to only 2%. Together with the turbine vane
cooling, these measures secured an average turbine entry temperature TET < 1050 K, for
the late BMW 003 versions in preparation of series production, indirectly (pyrometer-)
controlled by a turbine blade surface temperature limit of max. 970 K.[122]

The early ATAR 101 prototype development with air-cooled turbines was also
overshadowed with difficulties and failures, however, by the beginning of 1950—soon
after the 500 h endurance run was achieved,[123] a nickel-chrome alloy became available[124]
that permitted a change to solid turbine blades. Thus, costly air cooling of these blades was
dispensed with and blades with an efficient twist could be made, resulting in a better turbine
performance and an improved specific fuel consumption. In summary, the development
process up to 1950 resulted in the following essential design changes of the ATAR 101 V
with 1700 kp thrust, in comparison to the BMW 003 A base engine (800 kp):

---

[120] Manufactured at BMW on the basis of conically rolled plates from P. Leistritz GmbH, Nuremberg.

[121] See Schubert, Turbine, p. 249, the cooling air insert could be traced back to a BMW employee
Sonntag; for the circumstances of the later patenting of this and other ideas, here for H. Oestrich and
H. Rosskopf in US2,559,131, with priority 22 April 1948, see also Sect. 10.4.

[122] See Schubert, Turbine, p. 250.

[123] As a development milestone, the 500 h test was accomplished with a single stage, air-cooled
turbine with 53 blades with inserts (2% compressor mass flow), limited to ~800 °C, 0.797 m diameter
(340 m/s tip speed), Fig. 9.18 (r), from 14% Ni and 16% Cr alloy *SIRIUS HT* of *Sociètè Holzer*,—see
Bodemer, L'ATAR, p. 73.

[124] Solid turbine blades from *PER 2* of *Société Aubert & Duval*, similar to Nimonic 75,—see
Bodemer, L'ATAR, p. 85.

Turbojet	BMW	- 003 A1	- 003 C	- 003 D		ATAR 101 V
Project		P.3302	Hermso I	Hermso II	P.3306	Groupe 'O'
Thrust F	kp	800	900	1,150	1,700	1,700
Speed	rpm	9,500	9,800	10,000	8,700	8,050
Weight W	kg	570	610	620	900	850
Comp.Stages	-	7	7	10	7	7
PR	-	2.7 - 3.1	3.4 - 4.0	4.95	4.2	4.2
Deg.Reaction	-	0.9	0.5	0.5	0.5	0.5
Comp. Eff.	-	0.78	0.84 - 0.88	0.85 - 0.90	0.85	0.82
Sfc	kg/kph	1.35	1.27	1.10	1.18	1.30
Mass flow	kg/s	19.3	20.0	25.0	40.0	46.0
F / W	-	1.33	1.47	1.80	1.88	2.00
Eng. Diam.	m	0.690	0.690	0.700	0.850	0.886
Eng. Length	m	3.565	3.415	3.656	3.200	3.660

Fig. 9.19 Turbojet engine development, 1942–1950: From *BMW 103-003 A1* to *Snecma ATAR 101 V*

- new seven stage compressor with drum-rotor, stators in split casing (*BBC Hermso I*),
- enlarged combustion chamber with new mixers.
- new set of turbine inlet guide vanes with improved cooling,
- new turbine rotor blading with improved cooling,
- new bearing arrangement on three (instead of four) roller bearings,
- improved regulation of (unchanged) system *Junkers* (fabrication *Bronzavia*),
- simplified variable, bullet-type exhaust nozzle.

The pedigree of the French ATAR turbojet family has been outlined in the foregoing—with the conclusion that it represented a natural post-war continuation of the German BMW 003 developments up to 1945. Figure 9.19 underlines this impression by a stepwise, chronological tabulation of the essential engine parameters—from the base version BMW -003 A1 via three intermediate study versions for -003 C, -003 D and P.3306—to the final ATAR 101 V.

As an extension, Fig. 9.20 compares for that period the specific thrust figures of early (British) radial and axial turbojet engines. Most remarkable—in both Figures—again is the significant negative influence on compressor efficiency, of the fateful AVA decision to use as single exemption for their axials Jumo 004 B and BMW 003 A1 a degree of reaction R 1.0 for the compressor blading. And, in a second view to Fig. 9.19, that Oestrich's ATAR 101 V with R 0.5 and an isentropic compressor efficiency of 0.82 missed to exploit the full potential of this design approach, visible in the neighbouring column for the BMW project P.3306 with η 0.85.

Power plant	Compressor type	W Weight kg	F Thrust kp	F / W kp / kg	F / A kp/ m²
De Havilland Goblin II	Single sided, radial	703	1,410	2.00	1,080
Rolls-Royce Derwent V	Double sided, radial	582	1,590	3.60	1,760
Rolls-Royce Nene R.N. 1	Double sided, radial	726	2,270	3.13	1,820
Jumo 004 B	8 stage, axial	720	900	1.25	1,980
BMW 003 A1	7 stage, axial	570	800	1.40	2,140
ATAR 101 A0	7 stage, axial	910	2,200	2.40	3,530

Fig. 9.20 Comparison of radial vs. axial turbojet engines, 1942–1950: Specific thrusts F/W and F/A

The impact of the German design flaw is also visible in the specific thrust figures of Fig. 9.20, where only ATAR 101 A0 indicates with thrust/ frontal area F/A 3.530 the clear superiority of the axial configurations, under this—especially for further developments—decisive aspect.[125] Here, the advantage of the German turbojet engines is—for obvious, already discussed reasons—not very pronounced in comparison to the radial Rolls-Royce engines; the relative leap frog of the ATAR in comparison to the BMW ancestor was mainly the result of short-term improvements of the compressor aero-thermodynamics. The apparently small difference between BMW 003 and Jumo 004 has of course to consider the compressor, seven to eight stage count, and the additional development challenge of the BMW annular combustor. However, both features were from the beginning implied in Schelp's—for BMW technologically more challenging—design script, and thus indirectly, advantageous for the subsequent development activities on the French side.[126] The secret of this astonishing thrust jump is based on an inherent growth in engine mass flow, which Münzberg structures in three typical steps:

- The first axial compressor design approaches were characterised by reservations against critical Mach numbers, so that even the relative compressor tip flow should be clearly subsonic.
- Next is to impose by the compressor inlet guide vanes (IGV) a c_u/r = const., *'solid body'* swirl distribution, which for a given mean axial flow component at entry induces a

---

[125] See Münzberg, Konzeptionsentscheidungen, p. 69 f.

[126] A fact, which Münzberg admitted,—even appreciated, certainly also out of Oestrich's perspective,—see Münzberg, Konzeptionsentscheidungen, p. 71.

smaller c_{ax} in the tip region, and larger values near the hub. This 'solid body' swirl approach allows for a higher mass flows, even by observing upper relative Mach number limitations; this was applied for the early ATAR 101 versions with IGVs.[127]

- Finally, by removing the row of IGVs, and a corresponding '*free vortex*', c_u r = const. compressor inlet flow, the compressor tip section moves in the transonic, or even supersonic (relative flow) regime. This—aerodynamically challenging—design path was introduced to the ATAR family in the early 1960s (e.g. *ATAR 9 K series*, 6000+ kp thrust with afterburner, *K* for *Kellermann*).

After nearly 50 intensive, and commercially successful development and production years, the ATAR manufacturing ended in February 1994;[128] besides some negative, in hindsight questionable character traits of Hermann Oestrich, which will be commented in Sect. 10.4, the trustful French-German engineering teamwork immediately after war's end exemplified by practical work the outlook for a prospering and peaceful Europe.

---

[127] Already ATAR 101 A with a new set of IGVs, and correspondingly improved compressor front loading; for 101 B an increased number of compressor vanes allowed to increase pressure ratio and mass flow. Finally, 101 C and D had higher rotational speeds with IGV modifications, and an all new last compressor stage.The introduction of an eight-stage compressor configuration in 101 E (1954) improved mass flow and pressure ratio by ~15 percent,—see Bodemer, L'ATAR, p. 98.

[128] See Gersdorff, Flugmotoren, p. 329.

The Post-War Period 1946–1955

10

After a short '*orientation phase*' immediately after War's end which the Allies consistently used to search and get informed, to analyse and wherever appropriate to get hold of German aeronautical technology, the industrial transfer of this revolutionary wartime technology towards a new generation of civil and military turbojet high-speed flight commenced very rapidly, and happened especially in the UK and USA in a breadth and speed, considerably beyond the scope of these final sections. Therefore, the following Sect. 10.1 focusses primarily to Great Britain's successful path to the world's first commercial jet airliner, the De Havilland DH.106 *Comet*—with substantial, though indirect German impact on aircraft aerodynamics. Correspondingly, Sect. 10.2 will address in short mainly the German aero engine technology transfer to the Soviet Union, and the transfer of the swept-wing aircraft design principle to the United States—in-time for the Korean war fighter applications. Section 10.3 will highlight the short period up to 1950 of Brown Boveri's, in the end in vain, post-war aero engine activities in France, and together with its French subsidiaries C.E.M. and SOCEMA. Finally, Sect. 10.4 outlines Snecma's chief engineer Hermann Oestrich as a strange '*soldier of fortune*', in a kind of financial privatisation of German turbojet engine design know-how transfer, in collaboration with the French secret service.

10.1 Great Britain: An Industrial Head Start, Still Going Radial

Over time some rumours about certain performance deficits of the German turbojet engines, resulting from AVA's unusual and unfortunate pick of a degree of reaction 1.0 for their axial compressors in both Jumo 004 and BMW 003, might have reached Great Britain already before War's end. In hindsight there were several occasions for such information leaks:

D. Eckardt, *Jet Web*,
https://doi.org/10.1007/978-3-658-38531-6_10

- Already before 1939 Brown Boveri in Baden, Switzerland and Mannheim, Germany must have realised the weak point in the AVA design concept during the described patent conflict, Figs. 4.48 and 4.49,
- thereafter, semi-officially confirmed in April 1941 by Schelp's order to H. Reuter's group at BBC Mannheim for the alternative compressors with R 0.5, Chap. 7,
- and latest, after the first Jumo 004 hardware was inspected at Pyestock in 1944, as described by Bill Bailey:[1] *'What was seen left a feeling of deflation. Instead of a new approach to a high efficiency aircraft gas turbine compressor, all that appeared was blading for about 0/100% reaction, with a large part of the stator made from shaped steel plate. It was very disappointing. ... In this time the Aerodynamics Section never bothered to make any analysis of the blading design.'*

Given reactions from Rolls-Royce's upper management on 27 August 1946 in the context of H. Reuter's compressor design achievements, Fig. 9.5, these judgements might have been somewhat premature, but up till then premonitions of the earlier kind were quickly confirmed still during the remainder of 1945:

- On 1 June 1945 appeared an—before apparently withheld—article of R. Howell in the Proceedings of IMechE[2] with the accordingly classifying key statement: *'The prime interest in impulse blading (100% reaction) ... is its use in the earliest Parsons compressors at the beginning of this century and its more recent use in the compressors of the German Jumo 004 and B.M.W. 003 jet propulsion engines. Most other known compressors have 50 per cent reaction blading.'*

  Followed by the verdict: *'It will be noted that 90 per cent reaction shows up as very inferior to 50 per cent reaction, and for the Jumo 004 mean stage temperature rise of 16.5 deg. C, the former gives only 81 per cent stage efficiency compared with 88 per cent for the latter. The former efficiency, though only calculated, is comparable with those obtained from actual running of the Jumo 004 engines,'*
- and finally, after *Power Jets R&D* had had a closer look to the Jumo 004 impulse blading in December 1945[3]: *'As expected, the* (compressor*) performance is not particularly good and from engine test measurements the efficiency appears to be between 75 and 80 per cent.'*

After these preliminaries it is not surprising—especially from the British perspective—that a first group of US/UK *'technology inspectors'* arrived already on 9 May 1945 jointly at *LFA Luftfahrtforschungsanstalt,*[4] Brunswick/Braunschweig-Völkenrode, where Professor Adolf Busemann (1901–1986)[5] had contributed to the development of swept-wing, high-

---

[1] See Bailey, The Early Development, p. 56.
[2] See Howell, Design of Axial Compressors.
[3] See Power Jets R&D, The Junkers Jumo 004, p. 343.
[4] See Wikipedia, 'Luftfahrtforschungsanstalt' in English.
[5] See Wikipedia, 'Adolf Busemann' in English.

speed aircraft aerodynamics[6] in his LFA Institute for Gasdynamics. Fig. 10.1 illustrates this group of scientific VIPs with, from left to right:

- Hugh L. Dryden (1898–1965),[7] then NACA advisor to the US Air Force, and coming NACA Director, 1946–1958, here as member of the von Kármán-led *SAG Strategic Advisory G*roup, established by General Arnold in October 1944,
- (Sir) Ben Lockspeiser (1891–1990),[8] then since 1945 director-general of the UK Ministry of Aircraft Production (MAP), here in his role as host *'Mr. MAP Völkenrode'*,
- Theodore von Kármán (1881–1963),[9] world-famous aeronautical scientist, here in his position as *SAG* team leader,[10] and
- Albert P. Rowe (1898–1976),[11] a British physicist and senior research administrator, who played a major role in the development of *Radar* before and during World War II.

Besides the mentioned *SAG*, the Americans entertained additional scientist and technology collection programmes—*'Operation Paperclip'*[12] and *'Operation Lusty'*,[13] specialised on *Luftwaffe Secret Technology*, complemented by the English *'Operation Surgeon'*,[14] and the SU 'Operation Ossoawiakim'.[15]

According to Professor Peter Hamel, between 1971 and 2001 director of the Institute of Flight Systems at DLR Braunschweig, and local aero historian,[16] the British *'Operation Surgeon'* under Ben Lockspeiser had two phases:

- *'Scientific analysis and evaluation'*, July 1945–June 1946, based on interrogations of leading LFA experts, followed by in-depth reporting (180 German experts wrote 252 detailed Technical Monographs),

---

[6] See Wikipedia, 'Swept Wing' in English.

[7] See Wikipedia, 'Hugh Latimer Dryden' in English,—see also Fig. 6.12.

[8] Lockspeiser in reality, other than—see Wikipedia, *'The dam busters* (film)' in English, suggested the essential *'spotlights altimeter'* for the accomplishment of the 1943 RAF's 617 Squadron attack on the Möhne, Eder, and Sorpe dams in Germany with Barnes Wallis's *'bouncing bomb'*; a technique in use by RAF Coastal Command aircraft for some time.

[9] See Wikipedia, 'Theodore von Kármán' in English.

[10] Besides H. Dryden belonged to von Kármán's travel group also the Chinese Caltech aerospace engineer—see Wikipedia, 'Qian Xuesen' (also transcribed as *Tsien*), the American wind tunnel specialist—see Wikipedia, 'Frank Wattendorf', and Boeing's chief of aerodynamics—see Wikipedia, 'George S. Schairer', who also took the Fig. 10.1 photo; all foregoing Wikipedia articles in English.

[11] See Wikipedia, 'Albert Rowe (physicist)' in English.

[12] See Wikipedia, 'Operation Paperclip' in English.

[13] See Wikipedia, 'Operation Lusty' in English.

[14] See Wikipedia, 'Operation Surgeon' in English.

[15] See Wikipedia, 'Aktion Ossawakim' in German.

[16] See Hamel, Birth of Sweepback.

Fig. 10.1 UK/US mission to LFA Braunschweig-Völkenrode, 9 May 1945: H. Dryden, B. Lockspeiser, Th. von Kármán, A.P. Rowe, l-t-r; Photo G.S. Schairer

- *'Dismantling of facilities'*, October 1946—July 1947, when 14,000 t of wind tunnel hardware and test equipment were shipped to UK aero research and test facilities at Bedford, Farnborough and Cranfield, in conjunction with 4900 *'liberated'* documents and scientific files, stripped from LFA library stock.

After World War II both West Germany and East Germany were obliged to pay war reparations to the Allied governments, according to the *Potsdam Conference Agreement*, dated 2nd August 1945.[17] While the Soviet Union removed mostly industrial installations from their occupied zone, what became *GDR*, the Western Allies confiscated predominantly large amounts of German patents, copyrights and trademarks, worth some estimated 10 billion (1948) US dollars—and wind tunnels:

- The US booty comprised the BMW *'Herbitus'* altitude engine test facility, Figs. 8.42–8.46, the equally revolutionary Kochel 1 × 1 m Mach 10 hypersonic wind tunnel[18] with complete, stored equipment deliveries from Brown Boveri Mannheim, both re-erected at AEDC Tullahoma, TN, where they are still in use.

---

[17] See Wikipedia, 'World War II reparations' in English.
[18] See Eckardt, The 1 × 1 m hypersonic wind tunnel, and—see Eckardt, Gas Turbine Powerhouse: The hypersonic wind tunnel for Kochel/Tullahoma 1943–1968, pp. 137–146.

In addition, a 0.4 × 0.4 m supersonic wind tunnel delivery from BBC Switzerland was picked up at LFM Munich-Ottobrunn, still uncrated, and flown(!) end of June 1945 in a B-17 on the initiative of Drs. Tsien, Wattendorf and Dryden, SAG from Paris-Orly to Wright Field, where the further whereabouts got lost.[19]

- France removed from their occupation zone at Oetztal, Austria a huge, high-speed wind tunnel under construction since 1942, which also belonged to LFM Munich. The Oetztal wind tunnel was supposed to have an 8 m diameter nozzle with atmospheric operation up to high subsonic Mach numbers. This—the biggest wind tunnel in the world at the time—was designed for 76 MW driving power, provided directly by means of Pelton water turbines to two counter-rotating 15 m diameter axial-fan wheels. The hydraulic power was to be furnished by a flow of water of 18 m^3/s with a head of 530 m. Between December 1945 and June 1946 French authorities ordered to dismantle all movable Oetztal material, and transported it in 13 freight trains to Modane-Avrieux in the French Alps, where the set-up was reconstructed and finished. This unique facility has been operational since 1952 as the ONERA wind tunnel *S1 MA*,[20] indispensable to all recent Airbus aircraft developments.

- Finally, the British stripped all wind tunnels of the LFA Braunschweig site in their Lower Saxony occupation zone.[21] Typically, Fig. 10.2[22] shows the machine hall for two 0.94 × 0.94 m, Mach 1.1–1.6 transonic wind tunnels *A-9 a, b* in 1944 and 1947, originally belonging to Busemann's LFA Institute for Gasdynamics, and re-erected at *National Aeronautical Establishment* (NAE, then under construction) at Bedford as 3 × 3 ft *Supersonic Wind Tunnel (SWT),* in operation up to 2003.[23]

These were exciting days at Braunschweig; already on 10 May 1945, one day after he had taken the historic photo of Fig. 10.1, George S. Schairer (1913–2004),[24] Boeing's chief of aerodynamics informed in a seven-page letter to his then superior Ben Cohn at Seattle: '*The Germans have been doing extensive work on high-speed aerodynamics. This has led to one very important discovery. Sweepback and sweepforward have a very large effect on critical*

---

[19] See Eckardt, Gas Turbine Powerhouse: The AVA/ LFM wind tunnel projects, pp. 131–135. Von Kármán's inspection team had found the wind tunnel in early May 1945; in addition, several LFM-visits of Ch. Lindbergh are documented in his '*wartime journals*' during that period.

[20] See Thiel, Oetztal, and—see Tsien, Technical Intelligence Supplement, p. 96.

[21] See Uttley, Operation 'Surgeon'.

[22] Two GHH centrifugal compressors of 195 m^3/s volume flow and 1800 rpm were powered by two 5 MW AEG e-motors of 600 rpm, coupled by means of a 1:3 gearbox.

[23] Thereafter the facility was scrapped in 2004. DLR Braunschweig (P. Hamel) organised the back-transport of the A-9b compressor '*Druckspirale*' casing (exhibited near Braunschweig airport), and of a Zeiss Schlieren parabolic reflector (preserved at Deutsches Museum Munich).

[24] See Wikipedia, 'George S. Schairer' in English.

Fig. 10.2 Busemann's *A-9* windtunnel, LFA Brunswick, 1944 and 1947: 2 × (5 MW eDrive with 195 m³/s radial compressor) l; work done, r © P. Hamel

Mach No.'[25] He also told Cohn to distribute the letter to other companies as well, although only Boeing and North American made immediate use of it. Boeing was then in the midst of designing the B-47 *Stratojet*, and the initial model was a straight-wing design. By September 1945, the Braunschweig data had been worked into the design, which re-emerged as a larger six-engine design with more robust wings swept at 35 deg—with a first flight date of the Boeing B-47A[26] on 17 December 1947, Fig. 10.3.

Similarly, Ben Lockspeiser had his *'Damascus experience'* at Völkenrode, and within days he also wrote a report back to his home base about a possible exploitation of the stunning facilities: '... *we should bridge over the gap of some five to ten years which I see no means of doing by any other method. ... The equipment ... is such that we cannot expect to be able to build its parallel in a number of years and the knowledge possessed by its scientists is such that it will fill* (our) *gaps. ... It would, in our view, be difficult to exaggerate the importance to this country of exploiting these facilities to the full.*'[27]

Lockspeiser's time at Farnborough had convinced him of the need to explore the unknowns of supersonic flight and he set to work on this in 1943. So he chose, with the consent of Sir Stafford Cripps, then his Minister, the *Miles Aircraft Company* to design and

---

[25] See Wikipedia, 'Swept wing' in English, and—see Heinzerling, Flügelpfeilung; Heinzerling, in reference to a L. Bölkow publication, dates the letter rather improbably late on 5 Oct. or Nov. 1945.

[26] See Wikipedia, 'Boeing B-47 Stratojet' in English.

[27] See Nahum, "I believe the Americans have not yet taken them all", p. 103, also valuable for its detailed insight into the British 'Operation Surgeon', then led by Major George P. Bulman, Chap. 5, Sects. 6.1.1 and 6.1.3, who at the same time was also responsible for building NAE Bedford. Information, courtesy of A. Nahum to the author, 28 May 2020.

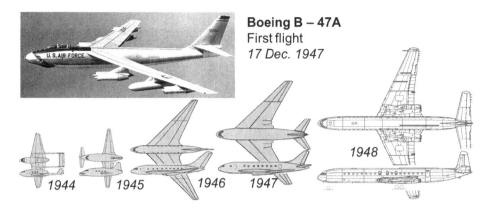

Fig. 10.3 Rapid spread of swept-wing aircraft design: *Boeing B-47A*, 1947 and *De Havilland DH.106 Comet*, 1944–1948

build Britain's first experimental all-metal supersonic aircraft, the *M52*, to be powered with a thrust-augmented Whittle W2 jet engine—to achieve a maximum speed of 1000 mph. During the LFA visit he realised quickly the vital importance of swept-back wings for supersonic flight; this led him to cancel the straight-wing M.52,[28] a decision for which he was much criticised, especially after the rocket-powered Bell X-1,[29] the first manned supersonic aircraft broke the *sound barrier* in level flight on 14 October 1947 up to Mach 1.06, piloted by Charles E. Yeager (1923–2020)[30]—with a straight wing design.[31] Afterwards Lockspeiser was harshly criticised for his caution in refusing to press on more urgently with British manned supersonics. Instead he had decided for *SWT tests* and radio-controlled rocket-powered models, but eventually one did fly at over 900 mph, thus proving that the basic aerodynamic design was completely satisfactory.

At one glance the hesitant British approach to high-speed flight, and the sudden release of these reservations in 1946 after the LFA results had become known, is illustrated in the 1944–1948 timeline of the first turbojet-powered commercial airliner *De Havilland DH.106 Comet*,[32] Fig. 10.3. The Comet 1 prototype first flew on 27 July 1949; it featured an aerodynamically clean design with four *de Havilland Ghost* engines buried in the wing roots, a pressurised cabin, and large square windows, Fig. 10.5a, b.

Still during the War, the de Havilland company had been successful with their famous DH.98 Mosquito[33] twin-engined aircraft, particularly outstanding in many roles. In January

---

[28] On 31 Jan. 1946,—see Hamel, The birth of sweepback, p. 12.

[29] See Wikipedia, 'Bell X-1' in English.

[30] See Wikipedia, 'Chuck Yeager' in English.

[31] Bell apparently evaluated also released M.52 test data before the first flight of Yeager's X-1,—see Hamel, The birth of sweepback, p. 12.

[32] See Wikipedia, 'de Havilland Comet' in English.

[33] See Wikipedia, 'de Havilland Mosquito' in English.

1941 the ARC, at the instigation of its chairman Sir Henry Tizard, suggested de Havilland to design and build a jet fighter (with own engine), giving them a free hand since no specification was issued. After studying the work of Whittle, the RAE and Gloster, Major Frank B. Halford (1894–1955) decided that their first turbojet *H.1*[34] should have a thrust of 1360 kp, and have a single-sided centrifugal compressor, to prevent lengthy axial compressor development, and 16 separate, straight-through can-type combustion chambers. The turbine was to be a single-stage type and the whole engine was to be kept as simple as possible. In due course, Gloster's first prototype of the F.9/40 fighter, fitted with two H.1 turbojets, each rated at a static thrust of 907 kp, made its first flight on 5 March 1943 from Cranwell, piloted by Michael Daunt.

The *H.1/Goblin* turbojet lent itself readily to redesign, leading to development of the more powerful *Ghost* turbojet, especially for the emerging civil market, giving as early as 1946 a designed thrust of 2270 kp for the new *Ghost 50*. The seventeen impeller blades were gently curved at their central intake portion, Fig. 10.4; it was only 30 years later when Professor John Moore (1943–2020) of Cambridge's Whittle Lab identified the impeller's elliptic blade geometry of a German turbomachinery research programme, going back to the same design principles as the *Ghost*.[35]

As the design of the Comet proceeded, many changes were made. The sweep-back of the wing was approximately halved to 20 deg on the leading edge, and the fuselage was lengthened to accommodate up to 32 passengers, Fig. 10.3. Two Ghost 50 engines were buried in each wing root, each being fed from a plain, oval air intake and exhausting via a long jet pipe that extended beyond the wing's trailing edge, Fig. 10.5a.

Within a year of entering airline service, problems started to emerge, three Comets being lost within 12 months in highly publicised accidents, after suffering catastrophic in-flight break-ups. Two of these were found to be caused by structural failure resulting from metal fatigue in the airframe, a phenomenon not fully understood at the time; the other was due to overstressing of the airframe during flight through severe weather. The Comet was withdrawn from service and extensively tested. Design and construction flaws, including improper riveting and dangerous concentrations of stress around some of the square windows, Fig. 10.5b, were ultimately identified. As a result, the Comet was extensively redesigned, with oval windows, structural reinforcements and other changes. Although sales never fully recovered, the Comet improvements culminated in the redesigned Comet 4 series which debuted in 1958, and remained in commercial service until 1981. From the Comet 2 onwards, the Ghost engines were replaced by the newer and more powerful 3180 kp *Rolls-Royce Avon AJ.65* engines. To achieve optimum efficiency with the new powerplants, the air intakes were enlarged to increase air mass flow, Fig. 10.5c.

---

[34] De Havilland bought Halford's firm and reformed it as the de Havilland Engine Company, renaming the H.1 and H.2 as the *Goblin* and *Ghost* respectively.

[35] See Moore, Eckardt's impeller; laser-measured velocity distributions of this impeller have already been presented as Fig. 8.14d, e; in due course John and Joan Moore's ideas, the availability of measured high-speed data and of the rotor geometry assisted considerably to the ground-breaking improvements of 3D Navier-Stokes turbomachinery flow calculation methods in the early 1990s.

Fig. 10.4 De Havilland Engine Co. *Ghost 50*, 2270 kp TO thrust, 1949, with PR 4.6 radial compressor, shown as smaller-scale *Goblin*

Upgraded Avon engines were introduced on the Comet 3, and the Avon-powered Comet 4 was highly praised for its take-off performance from high-altitude locations such as Mexico City. The Comet 4 was considered the definitive series, having a longer range, higher cruising speed and higher maximum take-off weight. These improvements were possible largely because of Avon engines, with twice the thrust of the Comet 1's Ghosts. Deliveries to *BOAC British Overseas Airways Corporation* began on 30 September 1958 with two 48-seat aircraft, which were used to initiate the first scheduled transatlantic services, and—much later than originally expected, the introduction of all-axial aero gas turbines in commercial service.

Soon after the '*booty period*' of reparations had ended in the early-1950s, also the pre-war partnership between Rolls-Royce and BBC Switzerland was re-established, as shown in Fig. 10.6. Especially in the area of high-altitude engine test facilities, where the BMW '*Herbitus*' plant had set new standards,[36] there was need for action at the outset of the successful RB 211 turbofan development programme. Rolls-Royce chairman and managing director Lord Ernest Hives had decided that a dedicated *Altitude Test Facility (ATF)* should be built at RR Derby's *Sinfin A* site, which was officially opened by Prime

---

[36] Altitude test facilities from BBC Mannheim had been described already in—see Sect. 8.3.3. For a comprehensive survey of newly built facilities in NATO countries up to the 1970s,—see Krengel, Air-Breathing Engine Test Facilities Register; there, RR Derby Sinfin A site on pp. 6–19. In general, this survey illustrates in a nutshell the new weight distribution within (NATO) industries, expressed in turbojet engine altitude test facilities, differentiated in [national/industrial] ownership: Canada [1/−], France [7/−], Germany [1/−], Italy [−/−], UK [5/3], USA [25/16].

Fig. 10.5 *De Havilland Comet,* 1948—1958: (**a**) *Comet 1* with *DH Ghost* intakes; (**b**) *Comet 1* with square windows, Hatfield, Oct. 1949, (**c**) *Comet 4* with *RR Avon* intakes

Minister Harold Macmillan in 1958. The ATF was built at a cost of around £5.5 m[37] and enabled engines to be tested in environments that they would experience up to about 24 km (80,000 ft) flight altitude. The ATF was a large, complex installation requiring cooling and refrigeration plant and special exhauster stations. The amount of electrical power required was such that it needed its own 11 kV power supply intake. Of the four BBC main extractor sets in the rear of Fig. 10.6, each one had a 16.2 MW synchronous e-motor, a gear train and an axial-flow compressor; in the foreground is the supplementary exhauster set. The ATF inlet conditions allowed for mass flows up to 180 kg/s, inlet pressures up to 5 bar, and an inlet temperature variation between 193 and 453 K; correspondingly, the exhaust capacity for mass flows was up to 272 kg/s, at exhaust pressures down to 0.0275 bar, and an upper temperature limit of 2100 K. Most impressive was the full transient capability for GT testing, from idling to maximum *MIL rating* within 5 s (up to stability 10 s).

10.2   German Aeronautical Technology Transfer to the Allies

10.2.1 Building on German Achievements in the Soviet-Union

The know-how transfer about German aeronautical, turbojet engine and swept-wing technology towards the SU commenced apparently long before War's end; A. Kay reported about the capture of Franz Warmbrünn, then a Wehrmacht soldier, on 1 Januar 1944 in

---

[37] Equivalent to approximately 160 MEUR in 2020.

Fig. 10.6 BBC turbomachinery equipment for *ATF Altitude Test Facility* at Rolls-Royce Derby, UK, 1957 © ABB Archive

Belarus, who had worked in Heinkel's von Ohain team before, and was claimed of having had information about Messerschmitt's turbojet aircraft activities as well.[38] This resulted in announcements from the State Defence Committee (GKO) to A.M. Lyulka to proceed with the development and construction of a turbojet of 1250 kp static thrust, ready for test runs within 9 months on 1 March 1945, and accordingly to S.A. Lavochkin, to design and build the experimental fighter for Lyulka's engine. By March 1945 the first pieces of Junkers Jumo 004 turbojets had reached CIAM, the Central Institute of Aviation Motors at Moscow. The following month crates of new BMW 003 turbojets were found at Heinkel Rostock, and soon after, Jumo 004 engines and Me 262 jet fighter spares were discovered at Warnemuende airfield nearby. On 28 April 1945 the production of a small batch of Jumo 004B turbojet engines, designated as RD-10, was ordered at no.26 GAZ (state aviation factory) at Ufa.[39] Here, chief designer V.Ya. Klimov and his deputy N.D. Kuznetsov,[40] Fig. 10.7, both of whom were to become famous in Soviet turbojet development, were tasked with producing to Soviet standards the detailed drawings for the Junkers 109-004B turbojet and for its assembly at no.1 Scientific Test Institute of the Red Air Force (NII) at Moscow-Zhukovsky by 1 July 1945. To assist the production of this 109-004B

---

[38] See Kay, Turbojet, Vol. 2, p. 17, and earlier comments in the context of Fig. 7.16.

[39] Ufa is located 1200 km east of Moscow, 100 km west of Ural mountains.

[40] See Wikipedia, 'Nikolai Dmitriyevich Kuznetsov' in English.

Fig. 10.7 Russian-German turbojet engine relationships: General Designer Nikolai D. Kuznetsov (1911–1995), ~1975, (l) HBPR engine workshop, Kuybyshev, 3rd April 1992, l–t–r: R. Walther, G. Brines, G. Schill, N.D. Kuznetsov, D. Eckardt, V.S. Anissimov (r) © Kuznetsov (l), Eckardt (r)

demonstration batch at Ufa, the Austrian designer of the (widely failed) Jumo 222 piston-engine(!) development programme[41] and ex-technical director of the Junkers engine plant in Dessau, Ferdinand Brandner (1903–1986)[42] was released from a PoW camp, and brought in to support the Klimov-Kuznetsov team. The effort was however delayed, and it was not until the beginning of 1946 that a few RD-10 engines had been built.[43] At the same time assembly of Jumo 109-004 turbojets was also continued, presumably in the captured underground factory at Muldenstein near Dessau, by using still available component stock. From here N.D. Kuznetsov organised presumably also make-up supplies for Ufa.

At Zhukovsky, also the first flight of a Messerschmitt Me 262 took place on 15 August 1945, piloted by A.G. Kochetkov; up to November 1945, he made 18 flights in this aircraft, followed by other pilots since October 1945. However, on 17 September 1946 the aircraft crashed, killing the test pilot F.F. Demida. Correspondingly, on 24 April 1946 a MiG-9 aircraft took off from Moscow's Tushino airfield for the first time, powered by two BMW 109-003 (in SU designation M-003) turbojet engines, and piloted by TsAGI[44] test pilot, engineer, writer and air force Col. Mark L. Gallai (1914–1998).

---

[41] See Wikipedia, 'Junkers Jumo 222' in English.

[42] See Wikipedia, 'Ferdinand Brandner' in English.

[43] Acc. to—see Kotelnikov, Early Russian Jet Engines, p. 29, the total of RD-10s completed in 1946 was 59, another 447 were assembled in 1947, and 833 in 1948.

[44] For TsAGI at Moscow-Zhukovsky,—see Wikipedia, 'Central Aerohydrodynamic Institute' in English.

In March 1946, a restart of turbojet activities in the Russian occupied zone of Germany came in reach, organised in *Special Design Bureaux (OKBs)*. Since—latest after the Potsdam Conference—the organisation of these OKBs in Germany was in contravention of agreements between the victorious Allies, their work—like also that of the 'Groupe O' at Rickenbach—had to be carried out in great secrecy. In fact, when the Soviet press made a big fuss about US Navy encouragement of post-war Heinkel-Hirth manufacture of the 109-011 turbojet in 1946, the programme under Hans von Ohain at Stuttgart-Zuffenhausen was very quickly shut down. In direct consequence it was predictable that both France and the Soviet-Union would be forced to remove their ambitious turbojet development projects from German ground, latest in autumn 1946.

For the time being Russian activities were split in two units, *OKB-1*, in the former Junkers plant at Dessau was to develop the *Jumo 004E* turbojet as afterburning version with 1200 kp static thrust under ex-Junkers engineer Dr. Alfred Scheibe (b.1896), while *OKB-2*, led by ex-BMW engineer Karl Prestel (1906–1990) in the former BMW underground plant in Neu-Stassfurt looked to a 1050 kp, thrust-augmented 109-003C, and to a restart of the 109-018 activities for a 3400 kp turbojet version. Typical for the Soviet management system, and unnoticed from the German side, especially *OKB-1* at Dessau stood in immediate competition with the Klimov-Kuznetsov group at the Ufa no.26 GAZ, and in fact the Soviet side produced superior results in time. In May 1946 the first Junkers 109-004E was ready at OKB-1, but a disappointing thrust of only 1050 kp was obtained on the test bench. Contrary to the Junkers teams under A. Franz and now A. Scheibe, which consistently failed to demonstrate a significant thrust increase by afterburning, the Russian design approach went apparently in the right direction, by demonstrating 1240 kp thrust in the first successful afterburner flight of a Lavochkin La-150F on 10 April 1947, achieving 950 km/h in horizontal flight at 5 km altitude.

The breakthrough happened by combining in a *RD-10F* an unchanged Jumo 004B gas generator in the front, and an enlarged rear afterburner/nozzle section, Fig. 10.8,[45] as suggested by Igor A. Merkulov (1913–1991),[46] who therefore should be acknowledged for the first successful afterburner demonstration. The first Soviet swept wing test aircraft *La-160 'Strelka'* (arrow) was developed in due course also at the Lavochkin Design Bureau. The La-160 was like the US Bell X-5 the Soviet equivalent to the Messerschmitt P.1101, Fig. 10.13, and had a thin wing with 35 deg. sweep along wing leading edge, Fig. 10.9. On 24 June 1947 test pilot and air force Col. Ivan E. Fedorov flew the '*Strelka*', powered by the afterburning RD-10F in horizontal flight at an altitude of 5.7 km to a new record speed of 1050 km/h.

This was also in view of fuselage aerodynamics a first independent success for the Soviet engineers. During design, captured German swept wing data had been available, but further, indispensable and extensive wind tunnel studies were required from TsAGI. Here,

---

[45] Adapted drawing; original—see Müller, Junkers Flugtriebwerke, p. 264.

[46] Since 1933, Merkulov had worked as development engineer for *subsonic* ramjets.

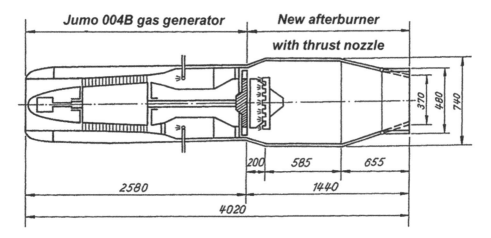

Fig. 10.8 *Klimov/Kuznetsov RD-10F*, 1240 kp static thrust with *OKB Merkulov* afterburner, measures in mm © R. Müller mod

the necessity of upper-wing boundary layer fences at subsonic speeds was revealed, and consequent-military mission to Oberammergau in the US zone, Sect. 10.2.2, it is more than likely that Soviet intelligence managed to get access to data for the Emergency Fighter Competition (EFC) at Berlin, either at RLM premises and/or at DVL Berlin-Adlershof.[47] It is known that in autumn 1944, a first proposal with 40 deg swept wings similar to the P.1101 was submitted to EFC, but not accepted. Already five months after the La-160, the SU's first operational fighter Mikoyan-Gurevich MiG-15 took off for its first flight on 27 November 1947, Fig. 10.15.

Nikolai Dmitriyevich Kuznetsov was born in Aqtöbe/Aktyubinsk, Kasachstan in 1911, some 600 km south of Ufa. He had trained in the aero-engine department of the Moscow Aviation Technical College in 1930, and was then a fitter at no.26[48] aero-engine GAZ until being en-rolled into the aviation technical department of the VVA Air Force Academy Zhukovsky.

Antony Kay provides the following comment: '*His later creative and original engine work was put to good use solving problems on many M-type piston engines at Klimov's OKB. Kuznetsov was a modest but very hard-working man with exceptional scientific and technical talents. He was also an outstanding organiser and his persistence and insights were ideally suited to solving difficult engine developmental problems.*'[49] During 1946

---

[47] See Green, Warplanes, p. 665—a 6.56 ft-span model of the P.1101 was exhaustively tested in the DVL Adlershof wind tunnel as part of W. Voigt's swept wing design activities at Messerschmitt between July 1942 and Sep. 1944, the official start of the EFP programme, Sect. 10.2.2.

[48] See Kay, Turbojet, Vol. 2. p. 26, refers here to *no.24* GAZ, if not a typing error (no.26 GAZ?), the location of no.24 GAZ could not be identified.

[49] See Kay, Turbojet, Vol.2, p. 26, and—see Kotelnikov, Early Russian Jet Engines, p.31, acc. to this source: '*N.D. Kuznetsov was one of the world's greatest enthusiasts for the Merlin piston engine; he had commanded Hawker Hurricane units during the war.*'

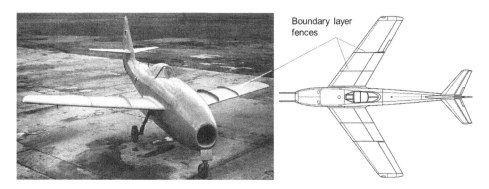

Fig. 10.9 *Lavochkin La-160 'Strelka'* swept-wing fighter prototype, with RD-10F turbojet engine, 24 June 1947, 1050 km/h

Lt. Col. Kuznetsov was solely appointed to the post of chief designer at no.26 GAZ in Ufa, while most other designers joined Klimov's OKB at Leningrad. Kuznetsov proceeded to develop a 3000 kp thrust turbojet RD-12—with centrifugal compressor. During tests, this impeller disintegrated, injuring Kuznetsov and another engineer. This accident delayed the development of not only the RD-12, but also a less powerful RD-14 version, intended for a 3-engine fighter. Presumably in this context, no.26 GAZ was disbanded at the end of 1948, and Kuznetsov found himself unemployed, until he was ordered to take up a new post at *experimental OKB.2* near Kuybyshev (Samara), where the work of the German engine engineers under the lead of Dipl.-Ing. F. Brandner continued, Fig. 10.10 (l).

Kuznetsov strove to get the best out of his organisation; he eliminated the long-standing competition between former BMW and Junkers personnel, and reorganised them into new teams for greater cooperation—not the least by adjusting unbalanced salaries. Later, he organised a technical education programme in which the Germans gave lectures. Those Soviet specialists who passed the German language examination received a 20% bonus. He preserved a lifelong respect and appreciation for German engineering, still clearly visible and expressed when in 1992 a new generation of German engineers appeared at his design office as part of a German government initiative, to discuss future projects on fuel-efficient, civil HBPR (high-bypass ratio) engine concepts, an area which his company had started with the *NK-93*[50] in time. Fig. 10.7 (r), illustrates a corresponding workshop snapshot of the *'glavnyy konstruktor'* N.D. Kuznetsov (then 81), and his head of design office V.S. Anissimov, amidst a mixed Pratt & Whitney, East Hartford, CT (G. *Jerry* Brines) and MTU Aero Engines, Munich delegation (Rainer Walther; Günter Schill, translator from czaristic-founded Harbin, PRC,—speaking admired high-Russian; and the Author).

---

[50] See Wikipedia, 'Kuznetsov NK-93' in English.

Fig. 10.10 '*Upra*'—German village life 2700 km from home, 1947: F. Brandner—in front of his house (l), main road with apartment blocks, erected by German PoWs (r)

Back to the post-war period. In the early morning hours of Tue 22 October 1946[51] the Soviet Military Administration in Germany (SMAD) launched the secret '*Operation Ossoawiachim*'[52] in their east-German occupation zone (later GDR), in which 2500 hand-picked German scientists and engineers together with 4000 accompanying family members were brought to the Soviet Union, thus violating strongly inter-allied agreements. These specialists contributed considerably to a know-how transfer in rocketry, aviation and turbojet engines, telecommunication, optics and navigation, etc., and, the latest group returned to Germany only in 1958. To illustrate the extent of the '*operation*', saying is that all other railway traffic in the area had been interrupted for five days after 22 October. The train for the selected 530 aircraft and engine specialists left from the railway station at Dessau centre, '*behind the* (ominous) "*Zuckerfabrik/Zuckerraffinerie*"', Fig. 4.31; they were given mere 4 h to collect their belongings and pile into trucks waiting in the streets. In most cases they were allowed to take their families and furniture,[53] but for others these would have to follow later. The specialists came from the engine factories Junkers, at and

---

[51] On 9 July 1946 Stalin ordered the beginning of all dismantling activities on 22 Oct. 1946, after the last and only democratic elections in East Germany up to 1990 had been fixed for Sun 20 Oct. 1946, and a negative impact of these Soviet measures on the election results should be prevented.

[52] Originally in the 1930s/1940s, *OSSOAWIACHIM* stood for a Soviet '*Society for the promotion of defense, aviation and chemistry*', namely to generate prospective candidates for the Red Air Force.

[53] German observers were impressed by the '*Red Army technology*' to let turned tables simply rattle down stairways—on the table plate. A few cases were reported, in which NKVD had loaded girlfriends and weekend families of German specialists independently in the same train, who met then each other for the first time on the first Russian platform.

near Dessau, BMW Neu-Stassfurt,[54] and Heinkel Rostock, and from all aircraft factories on this Soviet controlled territory. After a lengthy train journey, the aircraft specialists were unloaded in a small village, 120 km north of Moscow,[55] while some 250 engine specialists travelled on another 900 km eastward to Upravlencheskiy village—in short *'Upra'*—near Kuybyshev (now Samara), which was to become known as no. 2 experimental GAZ. Design documentation had been loaded separately with the luggage, but unfortunately it was discovered after arrival, that during the reloading of the train at Brest-Litovsk, necessary because of the change in the track gauge, the safe containing the design documentation had been lost. The early working conditions of the engineers reflected this situation—a few privately owned *'Dubbel. Handbook for mechanical engineering'*, represented common German engineering standard, and marked the starting point for the next six years.

Soon after the German attack of the Soviet Union on 22 June 1941, Kuybyshev—outside the range of German air raids—grew in political and strategic importance, when already in October the government (except Stalin) and the associated military administration, the Ministry of Foreign Affairs and the complete diplomatic corps (20 embassies and missions) were evacuated here, followed by Moscow's *Bolshoi Theatre* and numerous renowned artists. Rather by chance Kuybyshev had the large PoW camp no. 234, inmates of which were sent to the 20 km distant 'Upra' in the north, where they erected in record time several of the apartment block houses, visible in Fig. 10.10 (r) along the main road, using standard four-room apartment layouts of German residential building co-operatives of the time. When the families of the German specialists finally arrived end of October 1946, they found their name plates at the doors.[56]

By 1947 the intensification of the Cold War brought the need for an intercontinental bomber on the Soviet side to rival emerging US developments (B-36, B-52); Stalin demanded a bomber that could reach the USA, and fly back—like Hitler's *'Amerikabomber'* project since 1938, Sect. 8.3.5. The responsible *Tupolev* design office decided that such an aircraft—what would become the *Tu-95 (Russian Bear)*,[57] Fig. 10.12,—should be of about 200 t take-off weight,[58] and to obtain the required performance and range, four turboprops of between 12,000 and 15,000 shp were deemed necessary from the beginning. The German teams had brought with them the projected turboprops from Junkers, the 109-022 with ~ 9500 eshp (equivalent shaft hp) at sea level, and from BMW, the 109-028 with ~12,600 eshp, shaft and jet combined at sea level and

---

[54] The BMW team under K. Prestel was designated OKB-2 after Soviet occupation of the Neu-Stassfurt area, 1–4 July 1945, after the leave of H. Oestrich and his team in the wake of the retreating US troops; the move of the Prestel group to the east is indicated by the *dashed arrow VII* in Fig. 9.10.

[55] For the account of German rocket engineers and their families,—see Albring, Gorodomlya Island.

[56] See Hartlepp, Erinnerungen an Samara.

[57] See Wikipedia, 'Tupolev Tu-95' in English.

[58] Actual MTOW: Tu-95, 188 t; B-52, 221 t.

800 km/h, which had their basis in the Junkers 109-012 (2780 kp TO thrust) and the BMW 109-018 (3400 kp TO thrust) turbojets respectively.

The first flight of the Tu-95, Fig. 10.12, powered by four 11,000 shp *NK-12* turboprop engines took place on 16 February 1955, at a time when the last group of 200 German specialists had already returned on 5 July 1954, most of them to Dresden, GDR, where the government launched with some enthusiasm a civil airliner project *Baade 152*[59] in competition to the western aero industries, a competition which was finally lost in 1960 after aircraft prototype crashes, and a subsequent east-west specialists' *'brain drain'* still before the erection of the Berlin Wall on 13 August 1961.

The NK-12 turboprop engine, still in production and still a class of its own, especially in view of the gearbox transmission power, has an annular, conical magnesium alloy air intake casting to support the compact gearbox, Fig. 10.11 (l), and the front main bearing. The annular inlet to the 14-stage axial compressor had variable IGVs from steel. The compressor casing was a four-piece welding construction of two mm steel sheet stiffened by U-section steel rings, a clear borrowing of the BMW 109-018 turbojet design,[60] Fig. 9.6,—and apparently only marginally visible in accessible NK-12 drawing, like Fig. 10.11.[61] Following Kay's detailed description[62]: *'These rings also supported the rotor blade casing rings and the stator blade outer rings. The stator blades were inserted in and welded to their rings. At the inner end of the stator blades were other U-section support rings. This complicated stator assembly was annealed as a whole after welding and then separated longitudinally into two halves.'*

The engine has a 14-stage axial-flow compressor[63] of a maximum speed of 9250 rpm and 86–88% efficiency, producing pressure ratios between 9:1 and 13:1 depending on altitude, at a mass flow of 65 kg/s, with variable inlet guide vanes and blow-off valves for

---

[59] See Wikipedia, '152 (Flugzeug)' in German, and—see 7 min YouTube video https://www. youtube.com/watch?v=gIl58IUOx4k The *Baade 152* aircraft was powered by four—see Wikipedia 'Pirna 014' in English, turbojet engines of 3300 kp TO thrust, to a large extent designed and manufactured by those specialists, who had already put the NK-12 into air. On 1st February 1991, just 4 months after Germany's official re-unification date, a *Pirna 014* turbojet engine became permanently exhibited at the Deutsches Museum Munich, with some 30 of the former SU specialists' team still attending.

[60] Alternatively, an influence of the similarly structured Junkers 109-012 compressor casing is possible.

[61] This information is also missing e.g. in the NK-12 drawing of—see Kay, Turbojet, Vol. 2, p.37.

[62] See Kay, Turbojet, Vol. 2, p. 36.

[63] The NK-12 compressor design responsibility had Dr.-Ing. Hans-Joachim Schroeder (b.1906) and Dipl.-Ing. Hans Deinhard (b.1913), the combustion system relied on ideas of Schelp assistant Dipl.-Ing. Emil Waldmann (1911–1975),—see Sects. 7.1 and 7.3. The turbine design came from Dr.-Ing. Gerhard Cordes (b.1912), who was later with DB Stuttgart. The Waldmann family lived at *Upra* in one of the 4-storey stone houses, Fig. 10.10; E. Waldmann's further stations: VEB Strömungsmaschinen Dresden—DB Stuttgart—MTU Munich (together with Schroeder, Deinhard, Prestel); information—courtesy of Hans Waldmann, 2nd Nov. 2018.

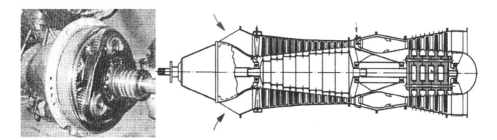

Fig. 10.11 Kuznetsov 15,000 shp NK-12M turboprop engine, 1957, with H. Bockermann's planetary CR gearbox,(l) © Kay (l), Hartlepp (r)

engine operability. The combustion system used is a *cannular-type*, actually a combination of an annular BMW combustor and the Junkers single can principle: each of the 12 flame tubes is centrally mounted on a downstream injector that ends in an annular secondary region. The contra-rotating propellers and compressor are driven by the five-stage axial turbine with small clearances (and correspondingly claimed 93% efficiency), and a turbine inlet temperature of 1250 K.

The highlight of the NK-12 design is shown in Fig. 10.11 (l), the massive planetary and spur gearbox between the two coaxial, counter-rotating airscrew shafts and the compressor shaft with a transmission ratio of approximately 1:11,[64] and a still record high power transmission of more than 15,000 hp.[65] This engineering masterpiece was designed by Dipl.-Ing. Herbert Bockermann (1907–1996), who had joined Junkers Dessau as gearbox specialist in 1936, then responsible for worm-gear adjusted, fully automatic variable pitch propeller developments. At '*Upra*' he designed the still unmatched NK-12 gearbox, and at post-war VEB Flugzeugwerke Dresden as part of the *Baade 152* turbojet aircraft developments, he led the department of gearbox design. As said, the NK-12 turboprop had its roots in the Junkers 109-022, and the BMW 109-028, but its specific fuel consumption was only about half(!), and its weight only about 65%, that of its German forerunner engines, some 5–8 years before.[66]

In the course of Tu-95 development, Tupolev requested 15,000 hp which were eventually provided in 1957 in the shape of the NK-12M, Fig. 10.11 (r). The aircraft range is in the order of 13,000 km at a maximum speed of 910 km/h. The Tu-95 with the NATO designation *Bear* is the only turboprop bomber in the world to enter front line service in

---

[64] Typically at cruise the NK-12 engines run at 8300 rpm, and the airscrews at constant 750 rpm, while power being varied by fuel flow and airscrew pitch.

[65] The next best aero gearbox of the 1960s is that of the—see Wikipedia, 'Rolls-Royce Tyne' in English, turboprop engine with 6100 eshp and a power-to-weight ratio of 2.55 hp/lb, slightly better than that of the NK-12 with 2.3 hp/lb, compare Fig. 2.3.

[66] Of considerable influence were of course Soviet material improvements over time, not the least after the surprising Rolls-Royce *Nene* and *Derwent* engine technology transfers.—see Kotelnikox, Early Russian Jet Engines, pp. 61–84.

Fig. 10.12 *Tupolev Tu-95 MS 'Bear'* cruise missile carrier, 2020, powered by 4× 15,000+ hp *NK-12MPM* turboprop engines, with 5.6 m Ø *AV-60 T* CR propellers

the late 1950s, which due to its versatility stayed in production and operation up to today, Fig. 10.12.

10.2.2 Building on German Achievements in the West

The advanced aircraft technology transfer from Germany 1945 to the USA 1950, more or less in parallel to a comprehensive collection and engagement of top scientists and engineers, is well documented,[67] and shall be here repeated only punctually. As a follow-on project to Messerschmitt's successful, Jumo 109-004B-powered Me 262, the *Messerschmitt P.1101,*[68] after the Me 262 another brainchild of Woldemar Voigt[69] (1907–1980), was a single-seat, single-jet fighter aircraft project developed as part of the *EFP Emergency Fighter Programme (Jägernotprogramm, 3rd July 1944), and* as such a reaction to Luftwaffe's High Command expressed need to strengthen defence against Allied bombing raids.

Adolf Busemann (1901–1986)[70] had proposed swept wings as early as 1935 in his ground-breaking paper[71] at the 5th Volta Congress, Rome; Willy Messerschmitt researched

---

[67] See Hamel, Birth of Sweepback, and—see Gimbel, Science, Technology and Reparations.

[68] See Wikipedia, 'Messerschmitt P.1101' in English.

[69] See Wikipedia, 'Woldemar Voigt (Ingenieur)' in German.

[70] See Wikipedia, 'Adolf Busemann' in English.

[71] Ad. Busemann, 'Aerodynamischer Auftrieb bei Überschallgeschwindigkeit (Aerodynamic lift at supersonic speed)', 5th Volta Congress, Farnesina, Rome 30 Sep.–6 Oct. 1935.

the topic from 1940. In April 1941, Busemann proposed fitting a 35 deg swept wing (*Pfeilflügel II*, literally 'arrow wing II') to the Me 262,[72] the same wing-sweep angle later used on both the North American F-86 Sabre and Soviet Mikoyan-Gurevich MiG-15 fighter jets. Though this was not implemented,[73] W. Messerschmitt continued with the projected HG II and HG III (*HochGeschwindigkeit*, high-speed) derivatives in 1944, designed with a 35 and 45 deg wing sweep, respectively.

The months June/July 1944 were dedicated to high-speed testing at the Messerschmitt premises, to sound out Me 262 dive and horizontal speed limits:

- On Sun 25 June 1944, Luftwaffe-Sgt. Karl-Heinz Herlitzius[74] pushed his pre-series Messerschmitt Me 262 S-2, V-11 with serial canopy, powered by 900 kp Jumo 004 B-1 engines, out of an altitude of 7 km into a 35 deg. nose-dive (*'Steiler Vollgassturz'*) over Leipheim, achieving a camera-documented, maximum speed of 1004 km/h, thus breaking the 1000 km/h barrier,
- on Thu 6 July 1944 an unknown pilot accelerated a modified pre-series Me 262 S-3, V-12 with '*racing canopy*', powered by 1000 kp Jumo 004 C/D engines, in horizontal flight over Leipheim up to 998.5 km/h, and finally on the same day
- Messerschmitt test pilot, flight captain, and glider world champion (1937) Heini Dittmar[75] achieved in his 1600 kp rocket-powered Me 163B V-18 in horizontal flight over Lager Lechfeld, 20 km south of Augsburg, a top speed of 1130 km/h.

While the Me 262 *Hochgeschwindigkeit I* (HG I) flight-tested in 1944 had only small changes like the mentioned '*racing canopy*', the HG II and HG III designs were far more radical. The projected HG II combined the low-drag canopy with a 35 deg wing sweep and a V-tail. The HG III had a conventional tail, but a 45 deg wing sweep, and *Heinkel HeS 011* turbojet engines embedded in the wing roots. The dive tests determined that the Me 262 went out of control at Mach 0.86, and that higher Mach numbers would cause a nose-down trim that the pilot could not counter. Messerschmitt believed the HG series of Me 262 derivatives was capable of reaching transonic Mach numbers in level flight, with the top speed of the HG III being projected as Mach 0.96 at 6000 m altitude. After the war,

---

[72] See Radinger, Me 262, p. 75, and—see Rizzi, Aircraft Aerodynamic Design.

[73] The Me 262 is often referred to as a swept-wing design as the production aircraft had a small, but significant leading edge sweep of 18.5 deg, which likely provided an advantage by increasing the critical Mach number. Sweep, uncommon at the time, was added after the initial design of the aircraft. The engines proved heavier than originally expected, and the sweep was added primarily to position the centre of lift properly relative to the centre of mass.

[74] Herlitzius was then on recovery leave from his *Me 110*, III. Group of ZG 26 (Zerstörer/destroyer Group). On 20 Feb. 1944 *(Black Sunday)* 11 out of 16 Me110 of III./ZG 26 were downed after daytime mission by USAF *P-47 Thunderbolts* near Wunstorf base (20 km west of Hanover), with 8 of 11 crews killed, Pilot Herlitzius and Radio Op. Röder injured; Herlitzius died in a 1947 car accident.

[75] See Wikipedia, 'Heini Dittmar' in English.

the RAE Farnborough re-tested the Me 262 to help with British attempts at exceeding Mach 1, achieving speeds of up to Mach 0.84 and thus confirmed the results from the dive tests.[76]

Basically the P.1101 prototype was a test vehicle for investigating swept-wing aerodynamics with the possibility to adjust the wing sweep angle prior to flight from 30, 40–45 deg. The wings incorporated automatic landing edge slots as well as trailing edge flaps. Though on 28 February 1945 the RLM settled on a competing design, the Focke-Wulf Ta 183 as EFP winner, the RLM decided to continue reduced funding in order for Messerschmitt to carry out experimental flights, testing the swept-back wing at anticipated speeds up to Mach 1.

On 29 April 1945 the US Army occupied the town of Oberammergau,[77] discovered the Oberammergau barracks complex and found there the prototype P.1101 nearing completion, together with a Me 109 in Building 615, Fig. 10.13. This building was then part of Oberammergau's 'Hötzendorf Barracks',[78] comprising some 20 new buildings, inaugurated in October 1937, Fig. 10.13 (l), and in military use up to August 1943. At that time the Army was ordered out of the Barracks and the Messerschmitt Design Bureau, based in Augsburg-Haunstetten, and now referred to as 'Upper Bavarian Research Institute' for security reasons, moved in one month later, triggered by severe Augsburg air raids on 25 February 1944. Professor Willy Messerschmitt (1898–1978) was based in Oberammergau from November 1944, working on the development of the P.1101 jet fighter. After the design existed for a while rather fluent in three versions, wind tunnel tests were carried out for a number of wing and fuselage profiles,[79] a decision was made to construct a full-scale test aircraft, which finalised the design with the fourth version, and on 4 December 1944 the selection of production materials began, Fig. 10.14. Under increasing war pressure it was decided to launch prototype production and detailed design with load calculations in parallel, while existing components such as Me 262 wing sections, an extended landing gear originating from Me 109, and flight components were utilised where feasible. The prototype was fitted with a Heinkel HeS 011 engine mock-up, but alternatively also a Jumo 109-004 B could have been used for the planned first flight in June 1945.

After visits by military intelligence, technical exploitation teams and rather casual 'battle strollers' like Henry Kissinger and his life-long friend Captain Fritz Kraemer,[80]

---

[76] See Wikipedia, 'Messerschmitt Me 262' in English.

[77] Oberammergau, some 90 km south of Munich, is best known for—see Wikipedia, 'Oberammergau Passion Play' in English, since 1634, and actually in 2022.

[78] The Hötzendorf Barracks was named after Field Marshal,—see Wikipedia, 'Franz Conrad von Hötzendorf' in English, (1852–1925), who was the Chief of Staff of the Austrian-Hungarian armed forces at the outbreak of the First World War; today 'NSO NATO School Oberammergau'.

[79] Finally, different NACA profiles were picked for the inner and outer wing sections.

[80] See Wikipedia, 'Henry Kissinger', and—see Wikipedia, 'Fritz G.A. Kraemer', both in English.

Fig. 10.13 Oberammergau's passion for high-speed flight (I): Inauguration of *'Hötzendorf Barracks'*, Oct. 1937 (l), *Messerschmitt P.1101* swept-wing fighter prototype with *HeS 011* engine mock-up, May 1945 © NSO (l)

Fig. 10.14 (right inset), the P.1101 prototype was shipped first to Wright Patterson AFB,[81] then to the Bell Aircraft Works in Buffalo, NY in 1948, where still an 3300 kp Allison J35[82] turbojet engine was installed. However, damage found after more thorough inspection, mostly as a result of the post-war visitor enthusiasm at Oberammergau and rough handling during the subsequent transit, ruled out any possibility for repair. Therefore, most P.1101 design features including the Allison engine option were integrated to the new *Bell X-5* experimental aircraft, which became the first aircraft capable of varying its wing geometry while in flight.[83]

A kind of *'northern aerodynamic saga'* had it, that using P.1101 technology brought by a former Messerschmitt employee still in February 1945 to Switzerland,[84] there picked up by Saab's Frid Wänström (1905–1988) in late 1945,[85] the first Western post-war swept-wing fighter aircraft emerged, the *Saab 29 'Tunnan'* with a first flight on 1st September 1948, powered by the 2300 kp, *de Havilland Ghost* turbojet, which determined the

---

[81] At WPAFB *'the P.1101 was generally considered by U.S. aircraft designers as something of an example of German development irrationality'*,—see Green Warplanes, p. 666.

[82] See Wikipedia, 'Allison J35' in English, originally developed as General Electric TG180.

[83] See Wikipedia, 'Bell X-5' in English; the X-5 wing actuation was carried out by electric motor drives.

[84] A less-risky form of technology transfer is described in this context by—see Mettler, Die schweizerische Flugzeugindustrie, acc. to which Swiss Customs intercepted in April 1945 a declared *'newspaper package'* in transit from Germany to Spain, containing drawings of a swept-wing turbojet fighter, which was forwarded to EFW Eidgenössische Flugzeugwerke (Swiss Federal Aircraft Works) Emmen for analysis/ copying, and returned on the way to Spain; at the same time with some certainty this incidence may mark the introduction of wing-sweep to EFW's own N-20 swept-wing fighter project with four, wing-buried turbojet engines,—see Wikipedia, 'EFW N-20' in English.

[85] See Wikipedia, 'Frid Wänström' in English. Documents preserved by *'Saab Veterans'*, Linköping clearly show that Wänström picked up information on Me 262 II and III as well as Me 163, but nothing on P.1101; information, courtesy of Art Ricci, KTH Stockholm, via Ernst Hirschel, 29 March 2022.

Fig. 10.14 Oberammergau's passion for high-speed flight (II): *P.1101* front view (at Wright Field) and wing plan, May 1945; Henry Kissinger and Capt. Fritz Kraemer (inset, r) © NSO, inset r

characteristic *'flying barrel (flygande tunnan)'* aircraft shape.[86] After 1951, Lippisch's delta wing specialist Hermann Behrbohm,[87] who had been also part of the Messerschmitt P.1101 development team at Oberammergau, joined Saab AB at Linköping, S, and influenced the subsequent supersonic delta-wing fighter developments *Saab 35 Draken* and *Saab 37 Viggen* in the 1950/1960s.[88]

British transonic aircraft designs of the 1950s such as the *Hawker Hunter* fighter,[89] the *Vickers Valiant*[90] and the *Handley Page HP70 Victor*[91] bombers also utilised the swept-wing technology, developed at Goettingen, Braunschweig and Oberammergau.

Produced by *North American Aviation,* Los Angeles, CA, the *F-86 Sabre*[92] is best known as the United States' first swept-wing, transonic fighter that could counter the

---

[86] See Wikipedia, 'Saab 29 Tunnan' in English; since 1952, also with the first British afterburner turbojet engine, the 2800 kp Ghost RM 2A.

[87] See Wikipedia, 'Hermann Behrbohm' in English.

[88] See Wikipedia, 'Saab 35', powered by Svenska Flygmotor (SFA, later Volvo Aero) RM6B, modified licensed RR Avon Mk48 with 4890 kp dry, and 6535 kp wet thrust,—see Wikipedia, 'Rolls-Royce Avon'; and—see Wikipedia, 'Saab 37', powered since 1962 by SFA RM8B,—see Wikipedia, 'Volvo RM8', modified licensed P&W low-bypass turbofan engine JT8D-1, with 7350 kp dry, and 12,700 kp wet thrust.

[89] See Wikipedia, 'Hawker Hunter' in English.

[90] See Wikipedia, 'Vickers Valiant' in English.

[91] See Wikipedia, 'Handley Page Victor' in English. In 1951 the swept-wing appeared first on the HP 88, an experimental aircraft intended to forward the development of the HP 70 Victor.

[92] See Wikipedia, 'North American F-86 Sabre' in English.

Fig. 10.15 Korean War—West meets eastern swept-wing application, 1950: North American F-86 Sabre (l), Mikoyan-Gurevich MiG-15 (r), Chino Air Show, CA, May 2009 © Brian Lockett

swept-wing Soviet MiG-15[93] in high-speed dogfights in the skies of the Korean War (1950–1953), fighting some of the earliest jet-to-jet battles in history, Fig. 10.15. The F-86 incorporated much German research into its design, employing a 35 deg swept wing and automatic leading edge slots. Flown for the first time on 1st October 1947, the Sabre survived many initial teething problems to become the premier USAF fighter of the Korean War. The North American F-86 Sabre was the first American aircraft to take advantage of flight research data seized from Nazi-Germany at the end of World War II. These data showed that a thin, swept wing could greatly reduce drag and delay compressibility problems that had bedevilled fighters such as the *Lockheed P-38 Lightning* when approaching the speed of sound. By 1944, as shown before, German engineers and designers had established the benefits of swept wings based on experimental designs dating back to 1940. A study of the P.1101 data showed that a swept wing would solve the early Sabre's speed problem, while also a *slat* on the wing's leading edge that extended at low speeds would enhance low-speed stability.

Less than a decade after German capitulation, its advanced military aero technologies had spread during Cold War and reached the super-powers. The MiG-15 was one of the first successful jet fighters to incorporate swept wings to achieve high transonic speeds. In aerial combat during the Korean War, it outclassed straight-winged jet fighters, which were largely relegated to ground-attack roles. In response to the MiG-15's appearance and in order to counter it, the United States Air Force rushed the North American F-86 Sabre to

---

[93] See Wikipedia, 'Mikoyan-Gurevich MiG-15' in English.

Korea. The XP-86 prototype, which led to the F-86 Sabre, was rolled out on 8 August 1947; the first flight occurred on 1 October 1947 with George Welch (1918–1954) at the controls, flying from Muroc Dry Lake, now Edwards AFB, CA.

By the end of the Korean hostilities, an often cited US claim was that the F-86 had shot down 792 MiGs, with a loss of only 76 Sabres, a victory ratio of 10:1. After the XP-86 testing on the basis of a GE 35 C3 turbojet of 1800 kp static thrust, the first models to see combat, the F-86A, were powered by a 2400 kp static thrust General Electric J47 engine; later F-86H models were more powerful and used, both for air-to-air and ground support, GE J73-GE-3 engines of 4200 kp static thrust.

The MiG-15 is believed to have been one of the most produced jet aircraft with more than 13,000 manufactured; licensed foreign production may have raised the production total to almost 18,000. In comparison, the Sabre production is considerably lower, but by far the most produced Western jet fighter, with a total production of all variants at 9860 units.

10.3   Switzerland: In Search of Its Post-War Possibilities

It appears that the Swiss BBC at Baden, beginning with the early contacts with British aero engine developments before WW II, continuing with Seippel's stand-alone turboprop patent of 1939 and in due course an intensified collaboration on this subject with the French daughter company C.E.M./SOCEMA, must have matured plans for a potential engagement in this newly emerging business after war's end. This was not the least quickened by official policies of political and military institutions in Switzerland after the war towards autonomous aircraft and aero engine developments. This latter aspect will be dealt with in the following Sect. 10.3.1, while the French-Swiss turbojet development activities under BBC-C.E.M.-SOCEMA company label will be described in Sect. 10.3.2.

10.3.1 Swiss Turbojet Engine and Aircraft Awakening

The *'jet age'* started in Switzerland on Wed 25 April 1945, precisely 8.46 h in the morning, when a brand new Messerschmitt Me 262[94] jet landed at Duebendorf, 20 km in the east of Zurich. The aircraft was piloted by OFR[95] Hans G. Mutke (1921–2004), who after an

---

[94] The aircraft, a Me 262 A-1a/R1, Werks-Nr. 500,071, amongst *'aficionados'* known as the *'White 3'*, belonged to the 9th echelon of JG 7 (JagdGeschwader), at that time stationed at Munich-Riem and/or Bad Aibling.

[95] *OFR*—abbreviation for OberFaehnRich, officer candidate.

alarm start from Fuerstenfeldbruck airstrip, 20 km north-west of Munich, recognised the Me low on fuel and decided for an emergency landing in neutral Switzerland.[96]

The new jet alerted the Swiss aeronautics community:

- Claude Seippel wrote about his eye-witness-experience in his diary on 18 May 1945, *'Duebendorf visite d'un Duesenjaeger'*;
- Jakob Ackeret filed in his ETHZ office under the date of 28 May 1945 comparative flight performance calculations/graphs between Me 262 and Gloster Meteor jet-powered aircrafts,
- and finally, on 23 August and 14 September 1945 the pilot Hans G. Mutke was interviewed officially by an expert team, comprising besides military personnel also Professor J. Ackeret, ETH Zurich and Dr. C. Keller, Director of Escher Wyss Research Department.

At that time the Jumo 004 B engine had been on test already at the EFW Emmen[97] engine test facility, but the frailty of the German series production engine (especially due to the lack of Cr, Ni, Mo, alloying elements) limited the total test time clocked up till 1948 to only 16 h. Actually, the Me 262 handbook had given 30 h $\pm$ 20% as expectable operation time. Finally, on 30 August 1957 this aircraft was returned from Switzerland for permanent display at the Deutsches Museum in Munich.

Amidst the Me 262 hype at the end of April 1945, the *KTA KriegsTechnische Abteilung* (Ordnance Dept. for War Technology) of the *Militärdepartement* (Mil. Dept.) had launched a national *RfP Request for Proposal*[98] for new jet propulsion systems with participation of

- BBC Brown Boveri & Cie., Baden,
- Escher Wyss AG, Zurich, and
- Gebr. Sulzer AG, Winterthur,

and for the following technical demands:

---

[96] Typical Mutke, this is not the only story in this context,—see Wikipedia, 'Hans Guido Mutke' in English, where also his disputed claim of breaking the sound barrier on 9 April 1945 before—see Wikipedia, 'Chuck Yeager' in English, on his Bell X-1 flight on 14 Oct. 1947, is outlined in detail.

[97] EFW Eidgenössische Flugzeugwerke (Swiss federal aircraft works) at Emmen, 5 km north of Lucerne.

[98] See Munzinger, Duesentriebwerke, comprising also a detailed documentation of the various company proposals, and—see Kay, Turbojet, Vol. 2, p. 234 f. Already on 21 Dec. 1944 a high-ranking BBC delegation had visited EFW Emmen, comprising Claude Seippel, Director GT developmt. Baden, Georges de Diesbach, Director C.E.M./ BBC Paris and Räto Gilly, head compressor. sales, BBC Baden. O. Seippel (son of Cl. S.) put this in context to the emerging N-20 programme and a suggested Swiss-French collaboration, which obviously was denied in view of these exclusively Swiss activities.

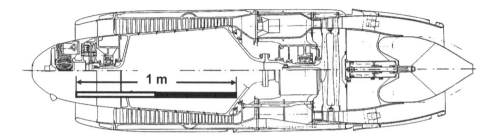

Fig. 10.16 Swiss ordnance *KTA* request for proposal, April 1945: Brown Boveri 1075 kp turbojet engine © Munzinger mod

- thrust/power 1500 hp,
- large over-powering capability for take-off and climb,
- low fuel consumption at base load,
- upscaling of the proposals to 3000 and 4500 hp.

The results of this design competition were impressive, but it may be sufficient here to depict just one BBC achievement for a conventional turbojet engine in comparison to the reference Junkers Jumo 004 B-1 engine, Fig. 10.16. Going from left to right, the BBC design relied on the proven *Riedel starter* (piston) *engine* located in the engine entry spinner, on an axial compressor of 13 stages with a degree of reaction R 0.5 with BBC's characteristic welded, drum-type rotor construction,[99] on six circularly arranged combustion chambers (like the Jumo original), on an uncooled two-stage axial turbine of 700 °C inlet temperature, followed by an afterburner duct for reheat and an equally Jumo-adapted, variable exhaust nozzle.

The *'preliminary'* Brown Boveri proposal of 30 July 1945 comprised a turbojet engine with 1075 kp TO thrust, and in comparison to the 900 kp production version of the Jumo 004 B-1, besides the indicated thrust gain of 24%,—a 23% weight saving and a 26% length reduction, respectively a 21% smaller diameter.

In addition, on 11 September 1945 Brown Boveri contributed a 2650 kp/1500 shp single-rotor (SR) turboprop engine to the KTA contest, in continuation of its foregoing turboprop design exercises during wartime, Sect. 8.3.2. In comparison to the turbojet, Fig. 10.16, with a calculated engine thermal efficiency of 14.8%, this turboprop was predicted with 22%.[100] Most remarkably, Escher Wyss Zurich[101] suggested a hereto related

---

[99] A preserved, unchanged feature up to the 2200 kp SOCEMA TGAR 1008 turbojet engine of 1948, Fig. 10.21.

[100] See Munzinger, Duesentriebwerke, p. 21.

[101] Between 1936-1938 EW developed the first reversible, variable pitch propeller, also for braking purposes during landing,—see Kay, Turbojet, Vol.2, p.235. This department was headed by Ackeret's assistant Curt Keller,—see Sect. 4.1.3, who worked out design proposals for turbojet engines and Swiss heavy-water nuclear reactors (based on German technology),—see Eckardt, Gas

2500 kp/1700 shp SR-turboprop, but with 1350 kg weight considerably heavier than the 790 kg of the Brown Boveri design. With a statorless, counter-rotating co-axial compressor/turbine core-flow arrangement, comparable to the Weinrich/Heppner CR turbojet engine, Fig. 5.11, the E-W concept was 10% shorter, and featured an additional '*double planetary gearbox*' for the SR propeller drive. Escher Wyss explained the design intent to expand the effective turboprop applicability by reducing the turboprop rpm at higher flight speeds, while for take-off (ignoring the noise aspect) the turboprop speed could be correspondingly increased, and all this without compromising speeds and thermal efficiency of the core engine.

The following tabulation is based on a 3.2 m propeller diameter:

	Propeller speed rpm	Prop. tip speed m/s	Core rotor rpm	Core '*stator*' rpm
Cruise	1310	220	10,600	10,600
Take-Off	1790	300	10,500	9400

Taken all available information together, it appears that BBC followed the KTA Request for Proposal with considerable attention. Based on competitive technology levels perhaps still during 1945, but certainly early in 1946 something must have happened—BBC's interest slackened and finally the Sulzer Brothers AG, Winterthur was the only national company which followed wholeheartedly the engine development programme towards a Swiss fighter aircraft *EFW N-20 'Aiguillon'* (Stinger). A detailed analysis of the BBC strategy change is not available yet, the BBC board might have simply realised that the industrial platform for such an undertaking was too small—and turned its interests rather in the direction of a potentially larger military market in France.

As said, between 1948 and 1952 EFW Emmen developed an indigenous jet fighter, particularly tailored to the country's needs, the *N-20* with top speeds in the high subsonic regime (1200 km/h), Fig. 10.19. In May 1948, and after Brown Boveri had turned its attention to their French turbojet engine development projects,[102] Sulzer Brothers AG, Winterthur were contracted to develop a *Swiss Mamba SM-01* turbojet engine[103] to power the N-20, Fig. 10.17. Besides the attainment of a critical Mach number, there were design requirements for a good climb rate of 2 min from 1 to 10 km, and a high combat ceiling of 15 km. Because of the mountainous nature of the country, the take-off and landing was to be made in a fraction of the length of the available runways: instead of the then usual

---

Turbine Powerhouse, AK engine and FN 207, p. 289. The author could not find out, if the Heppner/ Weinrich CR turbojet engine concept had reached Curt Keller at Escher-Wyss still up to 1945 e.g. via BMW/ Bramo Berlin-Spandau, or after the War in close cooperation with Armstrong Siddeley Coventry; the responsible F + W SM-01 engine designer W. Spillmann came from the Escher Wyss Research Dept. under C. Keller.

[102] See next Sect. 10.3.2

[103] SM-01 turbojet of 635 kp static thrust, to be deduced from the Armstrong Siddeley Motor's 1500 shp 'Mamba' turboprop engine,—see Wikipedia, 'Armstrong Siddeley Mamba' in English.

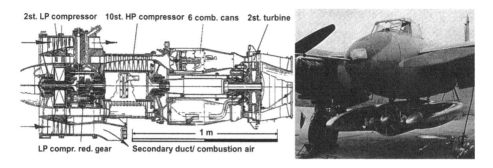

Fig. 10.17 *EFW Swiss Mamba* 635 kp turbofan engine: (l) SM-01, 1st run 12 Jan. 1950 (l), DH.98 Mosquito flying test-bed with centre-mounted SM-01, 1st turbofan flight 7 Oct. 1952 (r) © A. Kay mod

600–900 m, take-off was to be made within 200–350 m, and landing within 250–300 m.[104]

In May 1948 the *Militärdepartement*, Berne[105] issued a military specification, which from todays perspective sounds utopic, and impossible to fulfil: The new weapon system should be suited for aerial defence as well as ground combat, and contained ad-hoc numerous, innovative *'firsts'* like *'cranked'*, swept-back wings in two stages, four new turbojet engines embedded in a through-flow wing with afterburning, high-lift and thrust reversing capabilities, landing parachute, automatically adjustable wing leading edges,[106] an interchangeable weapon bay as streamlined ventral container, a pressurised cabin—to be completely blown off with separate pilot and cabin parachute (to rescue the instrumentation, including head-up display). The weaponry comprised two 20 mm cannons, unguided rockets and up to 800 kg of bombs for short range missions.

All this in view meant, that neither the military contractors nor the executing engineers foresaw the trend of the 1950s towards complex weapon systems, with air-to-air and air-to-ground guided missiles, nor the increasing need to integrate combat aircrafts technically in ground-based guidance and control systems. In hindsight it is therefore not surprising that the N-20 aircraft and SM-01 engine development programmes were stopped during 1952,[107] though the 4-year-development-programme had provided some astonishing

---

[104] See Kay, Turbojet, Vol. 2, p. 239, and—see Bridel SVFW-Lecture, 23 Nov. 2007: *'Fünf Pioniere'*.

[105] Since 1979 EMD Eidgenössisches Militärdepartement (Federal Military Department).

[106] A feature, presumably deduced during the joint Swiss-Swedish analysis of the Me 262 HG II and III information package, which arrived in April 1945, Sect. 10.2.2.

[107] For the continuing/prolongating developments towards the FFA P-16 ground attack fighter, powered by a 5000 kp Armstrong Siddeley *Sapphire* turbojet engine,—see Bridel, Schweizerische Strahlflugzeuge und Strahltriebwerke, and—see Wikipedia, 'FFA P-16' in English.

results, and though involved engineers were left with a strong, long-lasting feeling of deprivation.

Notwithstanding the outcome, it is worthwhile to highlight a few programme achievements. At a KMF meeting[108] on 29 June 1948, and after the government had dropped its support for Sulzer's complex 1800 kp *DZ-45* turbofan engine, EFW proposed that Armstrong Siddeley's 1135 eshp *Mamba* turboprop engine[109] to be developed as a bypass turbojet under the designation *EFW Swiss-Mamba SM-01* for the four-engine, swept wing N-20 fighter aircraft project.[110] This approach required considerable modifications and more detailed work was needed to enable the engine to be slid on rollers into the N-20 wings. Consequently, the combustors had to be re-arranged for a smaller diameter, and the turboprop gear drive replaced accordingly by a new reduction gearing between the 15,000 rpm hp spool and the new two-stage lp compressor of 1.58 pressure ratio, Fig. 10.17 (l). The remainder with 10-stage axial compressor, six combustion chambers and a two-stage turbine reflected the original turboprop design for the *Douglas DC-3* aircraft propulsion in 1949. The secondary (bypass) airflow was fed to combustor ducts along both sides of the engines, where either additional fuel could be burnt in reheat operation, or the cold air was flap-deflected to wing slots to act as lift augmentation or reverse thrust.[111]

On 7 October 1952 engine flight trials began at Emmen, using a *De Havilland DH.98 Mk IV Mosquito* (recce) that had landed at Duebendorf safely after an engine failure on 30 September 1944, and been impounded since then. As shown in Fig. 10.17 (r), the SM-01 was suspended beneath the Mosquito's fuselage, flanked by aerodynamic fairings; flight tests were made up to 8 km altitude without problems, actually assumed to be the first flight tests with a two-circuit *turbofan* engine.

In line with the innovative, experimental character of the N-20 project, it is not surprising that the influential long-time team of Jakob Ackeret and Curt Keller had also prepared an advanced aero propulsion version of their 1935 invention of a closed cycle gas

---

[108] *KMF Kommission für militärische Flugzeugbeschaffung* (Military aircraft acquisition commission), in 1948 with four mil. Members, and Dir. Buri, F + W Emmen, and Profs. Ackeret and Amstutz, ETHZ.

[109] See Wikipedia, 'Armstrong Siddeley Mamba' in English.

[110] The EFW SM-01 had a design bypass ratio of ~2, and a take-off static thrust of initially only 500 kp installed, later raised to 750 kp. Of the six engines built, SM-05 was brought to 1660 kp thrust, and elevated 1000 °C TIT, compared to the foregoing 800 °C. Intermediately, SM-03 had an improved BBC lp compressor, while SM-05 was equipped with contra-rotating one-stage hp and two-stage lp turbines from Brown Boveri—without intermediate gears, so that the lp compressor blade tips ran in the supersonic regime.

[111] See Spillmann, A ducted fan engine, p. 18; report and information, courtesy of G. Bridel, 16 Nov. 2016, and—see Kay, Turbojet, Vol. 2, p. 241.

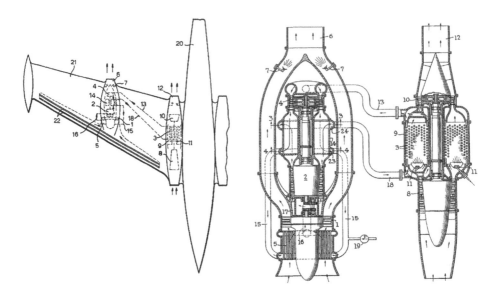

Fig. 10.18 Ackeret-Keller closed-cycle engine patent for *EFW N-20*, 1955: Aircraft installation with wing leading-edge suction (l), twin-turbojet engine arrangement with closed cycle core 2-4 (m, r)

turbine for the N-20, which was finally patented in 1955, Fig. 10.18.[112] The *'dual unit'* concept comprised the wing-embedded turbojet engine—in the patent called *'jet propulsion unit'*—with closed-cycle core between stations 2 and 4, in the centre of Fig. 10.18, here with the extra feature of a boundary layer suction slot along the wing leading edge. In addition, the *'thermal power engine'* at the wing root/fuselage transition, of which the combustion chamber is part of a larger heat exchanger, was the energy source for the closed cycle circulation. The inventors claimed as the main advantage, the constantly high thermal efficiency, independent of the operation altitude, so that just the fuel flow to the *'thermal power engine'* and the closed-cycle pressure level had to be regulated; a clear drawback of this concept—up till today—is and was the difficult, light-weight realisation of the required heat exchangers. Nevertheless, the concept saw already a first realisation effort in the late 1950s as part of the 1 BUSD USAAF development programme for *ANP Aircraft Nuclear Propulsion*,[113] where the *'Indirect Air Cycle'* with a small nuclear reactor instead of the heat source of the original AK *'thermal power engine'* was then assigned to Pratt & Whitney; however, the developments never came close to flight-ready hardware. In the meantime the possible application scenario of the closed-cycle propulsion concept was

---

[112] Jakob Ackeret and Curt Keller, US2,820,599 *'Dual unit jet propulsion plant for aircraft'* with priority 3rd Feb. 1955. The original invention of the closed-cycle engine, US2,172,910 *'Power plant'* with priority 12 July 1935 had C. Keller as sole inventor,—see Eckardt, Gas Turbine Powerhouse, p. 287 f.

[113] See Wikipedia, 'Aircraft Nuclear Propulsion' in English.

switched to space, where e.g. a spacecraft powered by a *NTR Nuclear Thermal Rocket*[114] could potentially get to *Mars* in just 3–4 months, experts say—about half the time required using traditional chemical rockets.

As said before, Adolf Busemann, who together with Albert Betz pioneered and obtained a German secret patent for the swept wing for high-speed flight in 1942,[115] suggested a wing sweep of 35 deg for the Me 262 in April 1941, but severe pitch-up stability problems must have been encountered at high-speed wind tunnel tests, and the actual Me 262 sweep built was 18 deg only, more to ensure stability by accommodating the position of the centre of gravity than to delay transonic drag rise. Immediately after the War flight testing slowly brought reality to the short-term euphoria when drawbacks like tip stall, pitch-up instability and loss of aileron control became evident, and required remedying; in Boeing's G. Schairer's words *'Applying sweepback and jet engines to jet aircraft proved to require very extensive development of* new concepts *for many design details. Each jet aircraft has required special and different tailoring of details.'* The intensity of competing swept wing research activities is illustrated by the following tabulation of first flights:

* 1st October 1947   North American Aircraft F-86 *Sabre,*
* 27 November 1947   Mikoyan-Gurevich MiG-15,
* 17 December 1947   Boeing B-47 *Stratojet,*
* 17 April 1948   EFW N-20, 0.6 scale model,
* 1st September 1946   Saab J-29 *Tunnan.*[116]

Figure 10.19(l) illustrates the first flight of the towed 0.6 scale model of the EFW N-20, piloted by Walter Läderach, behind a C-3603 ground attack aircraft.[117] The *'cranked'* shape of the N-20 wing leading edge is clearly visible; possibly the inboard sweep of 25 deg, identical to the Saab J-29 Tunnan, was fixed relatively cautiously as a result of joint

---

[114] See Wikipedia, 'Nuclear Thermal Rocket' in English.

[115] German secret patent No. 732/42 for A. Betz and A. Busemann, with priority since 9 Sep. 1939. Also in 1939 Hubert Ludwieg (1912–2001) demonstrated for the first time the principle superiority of swept wings in high subsonic airflows in the 11×11 cm high-speed wind tunnel of AVA Goettingen,—see Straßl, Verringerung, and for a comprehensive survey—see Meier, German Development of the Swept Wing 1935–1945.

[116] See Rizzi, Aircraft Aerodynamic Design.

[117] The K + W C-36 from the Eidgen. Konstruktionswerkstätten Thun (Swiss Federal Construction Works) was flown by Ernst Wyss; foregoing manned wind tunnel tests of this model aircraft in the large Emmen subsonic wind tunnel are shown in https://www.youtube.com/watch?v=-tb4Gmr9NLo, a 8:28 min video clip, between minutes 1:00 and 2:00, followed by rocket- and turbojet-powered flights with 4×100 kp Turboméca Piméné engines, all scenes piloted by W. Läderach.

Fig. 10.19 *EFW N-20 Aiguillon* all-wing Swiss jet interceptor, 1946–1952: Tow test of 0.6 scale, piloted, wood model, 17 April 1948 (l), N-20.10 taxi test, with 4x SM-01 engines, 8 April 1952 (r)

Swiss-Swedish studies, while the *'daring 40 deg'* of the N-20 outboard wing sections was within the sole responsibility of the EFW aerodynamicists.[118]

The full scale N-20 aircraft, Fig. 10.19 (r), was estimated to have a maximum speed of 1095 km/h, but the converted Swiss Mamba SM-01 turbofan, first time test-flown under a De Havilland D.98 Mosquito in 1948, Fig. 10.17(r), did not generate adequate thrust. Considerable further work was still required for the definitive two-shaft SM-05 engine, which was meant to generate 1500+ kp TO thrust. The prototype was completed in 1952 and, fitted with four SM-01 engines, flew briefly during a taxi test on 8 April 1952, but both engine and aircraft developments were cancelled soon afterwards.[119]

10.3.2 BBC/C.E.M./SOCEMA: Going Aero

As already described in Sect. 8.3.2, first contacts between Brown Boveri, Baden, CH, represented by its renowned head of gas turbine department Claude Seippel and the then still undercover aero engine activities of C.E.M's daughter-company SOCEMA, represented by their chief designer Paul Destival, took place at Lyon on 29 July 1941. At that time and with support from the *STA Service Technique du Ministère de l'Air*, then in Vichy, SOCEMA was encouraged to pursue a project for a 2500 shp turboprop engine. In order to conceal the work's purpose from the Germans this engine was designated the *TGA*, which could stand for Turbo-Groupe d'Autorail, but secretly stood for *Turbo-Groupe d'Air*; to make the link with railways more credible, the contract for the work was issued via the French Railway SNCF.

---

[118] Actually A. Betz added two supplementary ideas to the swept wing secret base patent,—in Nov.1939, a wing with increasing or decreasing sweep angle over span width, practically the later 'cranked' N-20 wing shape, and—in Dec.1939, a mechanism for in-flight sweep angle variation, as realised in principle in the Bell X-5 prototype.

[119] See Wikipedia, 'EFW N-20' in English.

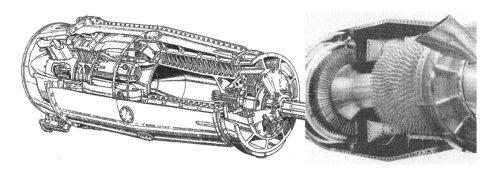

Fig. 10.20 SOCEMA C.E.M. TGA 1bis, 2500 shp turboprop engine, 1947: 3D general arrangement (l), TGA 1 mock-up with welded rotor (r) © A. Kay mod

The TGA 1 turboprop was based on a 1939 patent of Brown Boveri,[120] Fig. 8.40. As described in this context, this was also the BBC turboprop concept which a German Secret Service Report of 1944 had erroneously interpreted as part of a Swiss/British collaboration. Based on the agreement of dimensions and performance data, there is also some probability that the fully designed 2500 shp turboprop configuration of Fig. 8.41, that Hans Pfenninger[121] had carried out at BBC Baden in 1944 without providing any further information on the background, was actually planned as an alternative/reference proposal for the TGA 1.

The TGA 1 design,[122] Fig. 10.20, was initially laid out with axial compressor, annular combustion chamber and axial turbines, a surprisingly modern layout adopted by a *'team in isolation'*. In addition, simplicity and ease of manufacture were the prime TGA design targets: the rotor was carried on only two bearing points, a BBC design concept from early on. The compressor with PR 3.6 was of the 15-stage axial type,[123] with its blades fitted to a welded, drum-type rotor, again a characteristic BBC design feature as already could be seen at the Snecma ATAR 101-V, Fig. 9.15 (r). The first compressor stages had steel blades, the rest was in alloy. Because the SOCEMA engineers had some experience of compressors with 50% reaction, they chose to design a reaction type axial compressor.[124]

Beginning in 1943, several examples of the TGA 1 turboprop were run and improvements were introduced in 1945 with the TGA 1bis. The compressor gave little trouble and had 85% efficiency; however, as was usual in turbojet development, the combustion chamber caused many problems. A vaporising fuel system was developed,

---

[120] Patent US2,326,072 'Gas turbine plant' for Cl. Seippel with German priority on 28 June 1939.

[121] See Pfenninger, Die Gasturbinenabteilung bei BBC, p. 686.

[122] See Destival, SOCEMA, and—see Kay, Turbojet, Vol.2, p. 204 f.

[123] See Fig. 4.24, reference point 12.

[124] Actually, the first row had 100% reaction, gradually—typically within three stages—reduced to 50%, following here patent FR781,182 on behalf of BBC Baden with priority 3rd Feb. 1934, which was already part of the AVA/ BBC patent conflict, Sect. 4.3.3: *In the Patent Arena.*

but finally was replaced in the prototypes by ten flame tubes. Two prototypes were built, the first running on the bench at the end of 1947, but no engine was run with an airscrew.[125] The prototypes were not test flown and after concept and working cycle were being assessed as superseded, the project was abandoned.

Immediately after War's end, Claude Seippel renewed his contacts with C.E.M./ SOCEMA during a trip to Paris on 18 June 1945, as marked in his diary,[126] together with his nomination as successor of Adolf Meyer as Director of the BBC Thermal Department on 22 June 1945.[127] A longer stay at Paris was between 6 and 10 October 1945, with Adolf Meyer, Georges de Diesbach and Paul Destival attending.[128]

Learning in January 1944 from a British broadcasting[129] of jet-powered aircraft flying without propellers, and after a first *Junkers 109-004B* turbojet had been found after *'Libération'* in August 1944 in a machine-gunned and burnt-out truck near Orléans, pure turbojet engines got highest French development priority. In due course, French aircraft manufcturer *SNCASE Société nationale des constructions aéronautiques du Sud-Est*, Toulouse,[130] in short *Sud-Est*, asked SOCEMA to design a turbojet in the Jumo 004 class for a ground attack bomber under consideration, what became the *SNCASE 'Grognard'* (French: Grumbler) in 1946, Fig. 10.22.[131]

SOCEMA's new turbojet TGAR 1008 was conservatively designed for speed of manufacturing and reliability, Fig. 10.21.[132] In designing the eight-stage axial compressor

---

[125] The exhibited TGA 1bis mock-up, Fig. 10.20 (r), shows a planned three-bladed propeller of 4.24 m diam. from Société Rateau.

[126] The corresponding entry has for the evening also his attendance of a performance of *'Antigone'* at the—see Wikipedia, 'Théâtre du Gymnase Marie Bell' in English.

[127] At that time Seippel also gave up his function as head of the GT department. The actual transfer of the thermal directorate took place on 1 April 1946.

[128] Seippel's diary record has a stay at Geneva on 3rd Oct. 1945, a first meeting of the group at Lyon on 4/5 Oct. 1945, and then a 2-days car drive to Paris on 5/6 Oct. 1945, which Seippel apparently criticised by a sceptic comment: *(Coste!)*

[129] After the official radio announcement on 7 Jan.1944, C.B.C. Canadian and B.B.C. British Broadcasting Corporation invited Geoffrey Smith, managing editor of *'Flight'*, to give talks over the radio from London, on what jet propulsion meant and its implications. The broadcast over the Canadian network was made at 6.15 pm on the same day, and a longer talk followed a few days later on the B.B.C. Overseas and Home programmes. In a further broadcast talk on 28 March 1944, Smith dealt with the subject on a somewhat wider basis. The author integrated abstracts from the scripts in his standard reference book of that time,—see Smith, Gas Turbines and Jet Propulsion for Aircraft, p. 112 f.

[130] See Wikipedia, 'SNCASE' in English.

[131] See Wikipedia, 'Sud-Est Grognard' in English.

[132] See Destival, French Turbo-Propeller, and—see Kay, Turbojet, Vol. 2, p. 206. The Destival paper had been presented at RAeS London on 11 Nov. 1948, initiated after a visit of H. Roxbee-Cox at C.E.M. in May 1948. The documented discussion has contributions of RAeS President Roxbee Cox; G.S. Moult, Chief Eng. De Havilland Engine Co.; A.G. Smith NGTE; W.H. Lindsey, ASM.

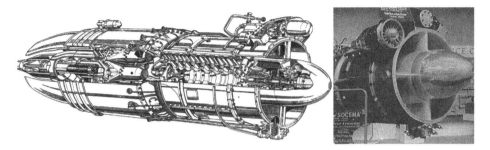

Fig. 10.21 SOCEMA/C.E.M. *TGAR 1008*, 2200 kp turbojet, 1949: 3D Gen. Arrangement (l), *Salon de l'Aéronautique*, Paris, 1949 (r) © Bodemer mod

for this engine, it was decided to copy the Junkers concept as best short term solution,[133] but the development of an *cannular* combustion chamber with atomising fuel system and 20 separate, cylindrical flame tubes (arranged in two rows to adjust the TET profile), was undertaken as a specific French solution. The highlight of this engine was however the single stage turbine. The turbine inlet guide vanes were film-air-cooled, presumably on the basis of BBC patents which had been applied for the first time by Brown Boveri in 1938. According to Kay's description *'the air entered each hollow vane at the inner end and then out through slots near the leading edge to cool the under surfaces, and also out at the tip to cool the outer surface wall. This method developed by M. Darrieus, Chief Engineer at CEM, allowed the vanes to work at a temperature 250 °C lower than the transiting hot gases.'*[134] Also the engine speed governor was based on the Junkers hydromechanical control unit with elastic, *'isodrome'* feedback.

---

[133] This statement from—see Kay, Turbojet, Vol. 2, p. 206, implied also a degree of reaction R ~ 1 (!) compressor blading like for Jumo 004B, including the typical (3× roller) thrust bearing package, as confirmed by—see Destival, French Turbo-Propeller; this reflects the long-time communication interruption between BBC/C.E.M. specialists. Nevertheless, the TGAR 1008 compressor had a PR 3.7 (and an ad. efficiency of 85%), which is represented in Fig. 4.24 by Point 16, comparable to the best military compressor designs from BBC Mannheim.

[134] See Kay, Turbojet, Vol. 2, p. 206; the principle of using compressor air for turbine disk and straight through-flow, turbine blade cooling was first applied in the *Lorenzen turbocharger*, based on a BBC license of the 1920s, Fig. 8.4 I), and—see Eckardt, Gas Turbine Powerhouse, p. 56 f. and p. 399 f. From Lorenzen there is a direct connection via DVL to the application of that relatively simple turbine cooling principle in German turbojets Jumo 004B and BMW 003. Turbine aerofoil film cooling was first patented by C.E.M.'s Georges Darrieus, US2,149,510 for profile stagnation line/ leading edge ejection, with priority as early as 29 Jan. 1934, and for various turbine showerhead/ film cooling configurations by the BBC engineering managers Adolf Meyer, DE710289, with priority since 8 Feb. 1938 and by Paul Faber, US2,236,426, with priority since 27 July 1938. Acc. to Kay, the Darrieus patent found an application in TGAR 1008. Turbine blade film cooling was first practically applied by BBC for the *VTF 225 turbocharger*, developed since 1938 for RR aero piston engines, see Fig. 6.37.

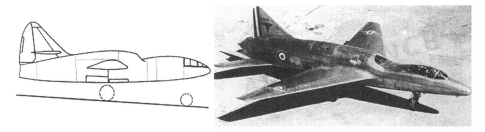

Fig. 10.22 SNCASE SE.2400 *'Grognard'* ground attack bomber, 1945–1954: Single seater w. straight wings +2× TGAR 1008, Jan. 1946 (l), 2 seater w. swept wings +2x RR Nene, 1st flight 14 Feb.1951 (r)

A TGAR 1008 prototype successfully achieved a two hour test run at 1500 kp thrust at 5500 rpm on 30 April 1948 at Le Bourget; steadily by 1949 the engine was developing a thrust of 2200 kp at 6600 rpm, accomplishing even a successful 150 h type test. In the end the TGAR had with 45 kg/s about twice the design mass flow of the original Jumo 004 B, at a considerably better compressor efficiency of 84% at PR 3.7, and a cooled turbine efficiency of 88% in combination with a turbine entry temperature TET of 550–600 °C.

The *SNCASE Grognard* aircraft design was originally released as a single seater aircraft with straight wings in January 1946, to be powered by two *'stacked'* SOCEMA TGAR 1008 turbo-jet engines, Fig. 10.22(l). However, in the time frame 1947–1950 SOCEMA's progress stalled somewhat, so that Rolls-Royce engines *Nene*, with 2× 2240 kp TO thrust, and *Tay (Nene with afterburner)*, with 2× 2835 kp TO thrust came into consideration, to be produced in France by Hispano-Suiza. Figure 10.22 (r) illustrates the two seater version *SE.2415 Grognard-II* with stretched fuselage and 32 deg swept wings; two boundary layer fences, installed on the outer wings after the initial SE.2415 tests, are visible in the outer wing sections. In the end substantial specification changes eliminated the *attack* category completely, which jeopardised the future development of *Grognard's* unpressurised airframe, and resulted to a programme stop in the early 1950s.

The SOCEMA gas turbine development history nevertheless found a reconciliatory end *in France.*[135] In October 1952, the visitors of the *Paris Auto Show* would discover an astonishing car with its shapely body hiding a gas turbine, Fig. 10.23: The idea to design a very compact gas turbine and to install it into a car had been born. The automobile part of this project was entrusted to a French engineer-designer Jean-Albert Grégoire (1899–1992).[136] The prototype was made of a cast *Hotchkiss-Grégoire* aluminium chassis with Al plate body. Given the fact, that a gas turbine does not generate any braking effect upon deceleration, an electromagnetic brake system was added to the transmission. The

---

[135]The cancellation of the SOCEMA aero engine development activities implied obviously also BBC's reaction to abandon—after their earlier retreat in Switzerland, now also in France—any plans for future company commitments to the emerging turbojet/aero gas turbine business.

[136]See Wikipedia, 'Jean-Albert Grégoire' in English

Fig. 10.23   SOCEMA Grégoire 1st GT-powered car, 100 hp, 200 km/h, 1952

100 shp SOCEMA gas turbine *TGV 1 Cematurbo* of 130 kg weight used in this prototype had a maximum gas generator speed of 45,000 rpm,[137] correspondingly a two-stage axial lp turbine drive speed of 25,000 rpm, which was further reduced, and—in theory—was designed to accelerate the 1300 kg vehicle with a drag coefficient $c_d \sim 0.19$ up to a speed of 200 km/h. Predictably, the final development of the SOCEMA Grégoire was far from being accomplished in 1952; however, this (single prototype) reminiscence to the early French/Swiss aero GT history can still be seen as an inspiring display at the *Museum of the 24 Hours of Le Mans*.

10.4   Hermann Oestrich: Taking Care of His People and Himself

As far as it is possible to reconstruct a 50+ year old personal history, it appears that Hermann Oestrich appreciated a high recognition amongst most of his team members and their families, largely for arranging the *'big deal'* with the French authorities, and the possibility of work for all team members in a critical period, where e.g. representatives of the United States would have allowed him to be joined by a handful of collaborators only, if he had decided in favour of the American offer. However, *'sotto voce'*, and very quietly over time emerged once in a while the portrayal of Oestrich's authoritarian character, of a certain wiliness combined with dedicated selfish attitudes.

Fortunately, there exist still reliable French sources which alleviate the sensitive task of a corresponding assessment. First, there is the *'ATAR Bible'* of Alfred Bodemer and Robert Laugier,[138] which describes in detail the construction of a business strawman *'La société Hermos'* in the early 1950s, as a precondition of Oestrich's later *'happiness'*. Careful readers certainly will have immediately detected Oestrich's fab over decades, to generate

---

[137] The *Cematurbo* gas generator consisted of a single stage centrifugal compressor of PR 3.5, and a single stage axial turbine; for further data,—see Kruschik, Die Gasturbine, p. 746 f.

[138] See Bodemer, L'ATAR, p. 78.

project names just by playing around with his name—from the *Hermso projects* of 1941,[139] to the new company name *Hermos S.A.,* now 10 years later. This 'Hermos S.A.', a *Société Anonyme*, was founded in 1950 at Casablanca, Maroc, by Oestrich on the initiative of the French Air Ministry, and a capital investment of then one million FFrs,[140] as an aero (piston) engine repair shop—with 170 employees in 1959. The target was

- to invest here a considerable (80%) amount of the license fees which Oestrich received during ATAR production from the French government,
- to allow for a SNECMA participation in Hermos stock between 34 and 49%, and
- to keep these arrangements up to 26 May 1960.

Officially the deal was designed as to look like a promotion programme in aeronautics for a needy north-African enterprise, unofficially Hermos S.A. was the vehicle to compensate Oestrich for his personally owned, intellectual property used for ATAR production. Actually the question was, who defined, documented and applied this technology under consideration,—in the end it was all in Hermann Oestrich's hands. His naming talents have already been highlighted several times; more of the category *'laughing prophet'* (for Zarathustra/ Zoroaster) was the trick how Oestrich licensed his ATAR technology to France. In his former engineering life up to 1945, apparently he was not a great inventor. *Espacenet*, the European patent database, has up to 1945 only two inventions, one in his name alone and another, filed in partnership;[141] in Fig. 10.24 marked as *'BMW'* applications for 1945. This figure shows the number of annual patent application in *Oestrich's name* over the corresponding year of patent priority: Beginning 1947, all in a sudden he develops apparently extraordinary creative talents; in the following three years he files in his name and receives personally corresponding French patent grants thereafter of not less than fifty-one (51) patents, of which approximately one quarter (12) came from members of his team,—all in the royalty-prone field of turbojet engine technology. Alone in the years 1947 and 1948, he files not less than 16 patents annually. One can assume that these patents were at the same time precondition and result for the contractual license agreement with the government,[142]—and under the given circumstances, it is clear that most of these now filed ideas in his name had been generated—and mostly applied—originally already before 1945. State-owned possession in *Intellectual Property Rights (IPR)* before War's end had been elegantly privatised thereafter.

As an example that the described patenting process could be perceived differently, and more benevolently in favour of H. Oestrich, especially by Snecma's upper management,

---

[139] See Sect. 9.1, and Fig. 9.19.

[140] Corresponding to approx. 17,000 EUR in today's money.

[141] Oestrich patent DE1,020,491: 'Combustor gas mixer'with priority 24 Jan. 1941, and Hagen-Oestrich patent DE1,020,492: 'Propulsor (Motorjet)' with priority 19 Jan. 1942.

[142] Officially the *'contrat de licence et de collaboration téchnique'* regulated the license fees and the *'reproduction rights'* of the French government.

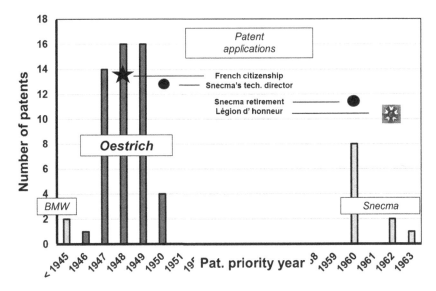

Fig. 10.24 Hermann Oestrich and his patented inventions 1941–1963: Applications in the name of *BMW*, *Oestrich* and *Snecma*

there is the corresponding account of Pierre André, former Snecma chief engineer, about Oestrich:[143] *'Himself holder of numerous patents, he (Oestrich) took the task to evaluate the patents (or parts thereof) of his close collaborators to be compensated, and to make sure that necessary patents were maintained.'*[144] First—as seen—his personal patent harvest before 1945 was rather meagre, and truly the 25% patent share of his team in the patented German technology block for the decisive years 1946–1950 was certainly compensated when Oestrich acquired their rights, but then Oestrich alone was the *'profiteur'* of the later ATAR turbojet engine mass production, and the corresponding flow of royalties.

As an essential administrative precondition for the *'deal'*, Oestrich received extraordinarily early the French citizenship already in 1948, and in 1950 he was appointed as SNECMA's technical director.[145] Thereafter the excessive invention creativity was gone again and remained so up to his premature retirement in 1960 with 57, in the same year when his personal technology royalties became fully accessible—in total some 14 MEUR

---

[143] See André, Les turboréacteur de grande puissance, p. 4.

[144] In French, 'Détenteur lui-même de nombreux brevets, il prit en charge la valorisation des brevets (ou parts de brevets) de ses proches collaborateurs contre rétribution par les interessés, en vue d'assurer le maintien de leur validité.'

[145] The appointment meant that his first five-year-contract of 26 April 1946 was prematurely prolonged, remarkably—without a French competition clause. In addition, Oestrich's fortune grew considerably after 1955, when SNECMA was appointed single source for French military turbojet engines, ending the Hispano-Suiza license agreement with Rolls-Royce.

Fig. 10.25 Dassault/SNECMA success story: 1500 aircraft + 5000 engines Dassault *Mirage IIIS* (Suisse) with 6000 kp SNECMA *ATAR* 09C3 © P. Steehouwer

in present value.[146] After a short flaring up of Oestrich's patent activities for *'SNECMA'* with eight filings in the context of VTOL airstrips in 1960, these activities calmed down quickly, 10 years before his death on 2 April 1973, aged 69, at Paris.

Before in 1962, Hermann Oestrich was knighted as Officer of the Légion d'Honneur for his merits in *ATAR* development. Under the described settings Dassault's *Mirage* aircraft and Snecma's *ATAR* engines merged to a successful product combination in the post-war era. More than 1500 aircraft and 5000 engines were sold, not the least also in the 1960s to the *Swiss Air Force* after ambitions for national developments had been given up, Fig. 10.25. Perhaps, in such IPR-prone environment, and if the initial intellectual property contents from Brown Boveri would have been known, a good rebate could have been negotiated ...

Neither in B.I.O.S./C.I.O.S. interviews immediately after the War nor in later publications, Hermann Oestrich mentioned e.g. the BMW 003 C/D engine developments, and especially the contributions of Hermann Reuter's group of BBC compressor designs in the various *'Hermso I–III'* studies for the later success of the ATAR turbojet engines.

---

[146] See Nouzille, La folle histoire du Dr. Oestrich; this l'Express article lists 800 million francs in royalty payments during the 1950s, corresponding to 90 million new francs in 1999, the year of the publication. The author speculated that in the 1950s a certain share of that money might have been reserved to support Oestrich's former Nazi acquaintances *'on the run'*, but could not provide any substance to this allegation.

Summary

Though in the end clearly a spin-off of military warfare during WWII, the turbojet engine started its lasting triumph in commercial aviation only after the 1960s—as turbofan.[1] Since then, the World Bank lists monotonously growing numbers of passengers transported per year worldwide, with a preliminary all-time high in 2015 of 3.44 billion passengers. Likewise, the number of registered carrier departures worldwide has reached a peak in 2015 with almost 33 million take-offs.[2] According to a recent forecast,[3] the production value of all gas turbines worldwide in 2021 was 61.6 BUSD (61,600,000,000 USD, 100%), of which civil aviation covered 68%, military aviation correspondingly 15.6%, and the remainder of roughly 10 BUSD or 16.4% of the total annual GT market was reserved for non-aviation, in BUSD numbers 7.5–2–0.5, for electrical power generation—mechanical drives—and marine applications. Nearly undisturbed the aviation gas turbine/turbojet engine segment showed a steady monotonic growth of 2.75% annually over the past 20 years, not the least simply driven by the annual growth of the world population by presently more than 1%.

Over now more than 60 years, turbojet engineering has sustainably supported this trend, Fig. 11.1,[4] by increasing its performance expressed in *propulsive efficiency*[5] steadily along

---

[1] See Sect. 10.1 and Figs. 10.3–10.5 for the early civil transport developments around the De Havilland DH.106 *Comet*.

[2] See Wikipedia, 'Civil aviation' in English.

[3] Forecast International, Newton, CT, quoted in Lee Langston, 'Gaining Altitude', Mechanical Engineering, Vol. 144, No. 4, June/July 2022, pp. 40–45.

[4] The *Figure* was presented in a MTU-internal lecture series on 17 March 1993: D. Eckardt *'From RTF to APS, advanced design for future engine projects'*.

[5] See Wikipedia, 'Propulsive efficiency' in English.

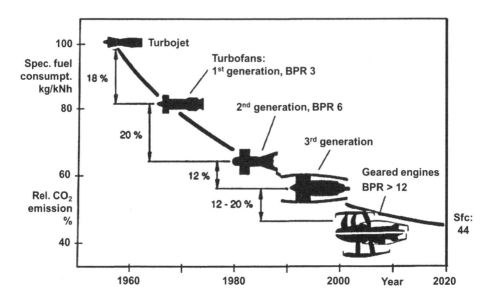

Fig. 11.1 Turbojet engine generations, 1960–2020: Sfc performance for *Cruise* conditions DE, *mtu* 1993

just one design parameter, the *By-Pass Ratio BPR*, visualised by the growing turbofan engine diameters over time. BPR is defined as a mass flow ratio *'cold/ hot'*, e.g. for BPR 10 the bypassing (outer), 'cold'/unheated turbofan air flow is ten-times bigger than the (inner) mass flow through the driving turbojet/core engine. Correspondingly, the thrust-*specific fuel consumption sfc*[6] for the pure 'single cycle turbojet' engine (1960, BPR 0) varied to the present 4th *'geared engines'* generation (2016, BPR 12.5) and beyond, by decreasing sfc from the shown turbojet starting value of 100 kg/h per a thrust unit of 1 kN (equivalent to 102 kp or 224.4 lbf) in 1960 to an expected value of sfc ~44 kg/kNh, as shown for the year 2020, for the new, still coming generation of geared turbofan engines with a bypass ratio increase from BPR 12.5 to the new target of BPR ~17.[7]

In line with the indicated turbojet/turbofan sfc values shrink the relative CO_2 emissions,[8] here indicated alternatively in percentage points. In addition the engine

---

[6] Given for defined *Cruise* conditions of one turbo-engine, flying at 11 km altitude, at Mach 0.8.

[7] The engine concept had been visualised in this *Figure* by the then (~1990) investigated *CRISP* Counter-Rotation Integrated Shrouded Propfan (top) vs. the *ADP* Advanced Ducted Prop mid-fan concept (bottom). In 2016, approx. 25 years after the corresponding, conclusive PW/ MTU joint design studies and technology demonstrations the first *GTF Geared Turbofan Engine* was introduced as *PW1100G* with BPR 12.5 for the *Airbus A320neo* aircraft,—see Wikipedia, 'Pratt & Whitney PW1000G' in English. The reasons for this environmentally disastrous delay were the deliberately floatable oil and kerosene prices, and general industrial-political hesitance.

[8] Another highly favourable improvement is a noise reduction of up to 31 dB (~75%) in comparison to the foregoing V2500 turbofan engine level.

nitrogen oxide emissions are environmentally critical, and even engine-exhausted water vapour condensation at high altitudes, the effect of the so-called *'engine contrails'*, Fig. 8.10, on global warming has to be taken into account in future aviation emission reduction initiatives, on the way to the targeted *'zero emission'* flight.

A big asset in favour of the gas turbines' future usage is its inherent *multi-fuel capability*, which played already a role in the German wartime decision pro aero gas turbine, in view of increasingly shrinking resources of highly refined hydrocarbon fuels. Key aspects in view of future environmentally acceptable aviation will be—besides corresponding engine/aircraft integration studies e.g. on tank capacity, wing-body aircraft configurations, and the provision of reliable and secure airport infrastructures—the usage of sustainably produced *'green hydrogen'*, and/or the combustion of system-optimised SAF Sustainable Aviation Fuels.[9] For the industry in general, this conversion represents a huge challenge, but in view of possible *'technicalities'*, there appear to be no fundamental hurdles to adapt the GT combustion technology accordingly.

With respect to the *'invention'* of the turbofan idea, Frank Whittle in his 1981 book[10] leaves little doubt that he in 1936 had the decisive idea *'that means to "gear down" the jet were needed, i.e. to aim for a high mass low velocity jet instead of the low mass high velocity jet of the simple jet engine'* were required to increase the *propulsive efficiency*. Different to a turbo-prop solution, *'the writer … remained convinced that the propeller was not the answer for speeds of 500 m.p.h. (800 km/h) and above, so turned his attention to the concept of using part of the gas generator output energy to drive a high mass flow low pressure compressor or "fan"'*.

So far Whittle's conclusive wording; regretfully, his then suggested design solution in his patent GB471,368 with priority 4 March 1936 has no similarity to present-day turbofans, in his own words: *'The primary problem was the mechanical arrangement, i.e. how to connect the fan to its driving turbine without using the two spool arrangement which, at that time, seemed to present formidable bearing, coupling and assembly problems.'*

The situation on the German side is somewhat different. The first turbofan-/ZTL-Patent[11] DE767,258 goes with priority of 12 September 1939 to Hans von Ohain. His compressor drive arrangement follows the ZTL-concept, Fig. 7.10, of Schelp's 1937

---

[9] The topic *SAF* comprises contents and solutions of—see Wikipedia, 'Synthetic fuels', 'Aviation biofuel' and 'Power-to-X', all in English, under the reservation that necessary renewable energies can be provided in sufficient quantities and at acceptable cost. At the centre of the *Synfuel* concept is the *FT Fischer-Tropsch* 'indirect conversion' synthetic fuel manufacturing, of which Fig. 8.20 (l) had already shown the (ruins of the) Police/Pölitz hydrogenation plant; by early 1944 German synthetic-fuel production had reached more than 124,000 barrels per day from 25 plants.

[10] See Whittle, Gas Turbine Aero-Thermodynamics, p. 217.

[11] The term *ZTL* for *'Zwei-Kreis-Turbinen-Luftstrahltriebwerk* (Two-cycle turbo-air-jet engine)'* is apparently documented in H. Schelp's 1937 study for the first time, see Sect. 7.2 and Figs. 7.10 and 7.11. At that time alternative designations like *'Doppelstrahltriebwerk* (Double turbojet engine)'* at

'*Flugbaumeister*' study at DVL Berlin-Adlershof—with one turbine drive for front fan and hp compressor combined. Consequently, it is also very probable that the corresponding four turbojet concepts of Fig. 7.10 originate out of this DVL context in 1936–1937, with possible contributions of Heinrich Kuehl and Karl Leist, who was later responsible for the design of the first realised turbofan engine *DB 109-007* at Daimler-Benz Stuttgart, Figs. 5.12 and 5.13.[12] For the time being it must be assumed that Schelp's 1937 DVL study provided the first comprehensive turbojet/turbofan performance survey, and thus introduced also the corresponding design and terminology guidance in Germany.

In summary, *Jet Web* outlines from early beginnings the important developments in the history of the air-breathing turbojet engine. Although the most common type, the gas turbine powered jet engine, was certainly a product of WWII, many of the needed advances in theory and technology leading to this invention were made well before this time, as described along the '*Survey*' timeline of Chap. 2. Interdependent '*connections*' of various early European research and industrial gas turbine development efforts, dating back in parts still before WWI, lead to the specific role of the Swiss Brown Boveri & Cie. in this web-type information and component know-how exchange process, Chaps. 4 and 6.

Finally, the jet engine was clearly an idea whose time had come. Frank Whittle submitted his first patent in 1930, and dominated the further British developments. On the German side, Hans von Ohain, though distinguished by the first turbojet-powered flight on 27 August 1939, did not achieve a comparable position. Stunningly, here a young self-made RLM -'*Flugbaumeister*' Helmut Schelp set-up and coordinated three German turbojet engine industrial development and series production strings, which culminated 1944 in the introduction of the first operational turbojet fighter aircraft, the *Messerschmitt Me 262*, powered by the *Junkers Jumo 004* turbojet engine, of which more than 6000 were built, Chap. 8. For some time after the War, British designs dominated, but by the 1950s there were many competitors in the emerging civil market, particularly in the US with its huge, supportive arms-buying programmes, Chap. 10.

Symbolically, it is apparently more than coincidental, that in 2021 a military contract was awarded in which the U.S. Air Force's fleet of 76 *Boeing B-52 Stratofortress* strategic bombers will receive new engines, replacing Pratt & Whitney's original *TF33* engines. The eight-engine B-52 were entered into service in 1955, so according to plan, this backbone of the former U.S. Strategic Air Command (SAC)[13] might become the first 100-year-old flying jet-powered aircraft family. The new 17,000 lbf turbofan engine, the *F130*, with a 1.27 m diameter fan (BPR 4.1) is the military version of the *BR700-725 A1* business jet engine, now in use for the *Gulfstream G650*, Fig. 11.2. The engine is derived from the *Rolls-Royce BR700* family of turbofans, developed under the engineering leadership of

---

Heinkel or '*Manteltriebwerk* (Shrouded turbojet engine) at AVA were in use,—see Koos, Heinkel Raketen- und Strahlflugzeuge, p. 70.

[12] DB 109-007 on test: 27 May 1943, with a calculated sfc of 107 kg/kNh in 12 km altitude, to compare with Fig. 11.1.

[13] See Wikipedia, 'Strategic Air Command', now USAF Global Strike Command (GSC).

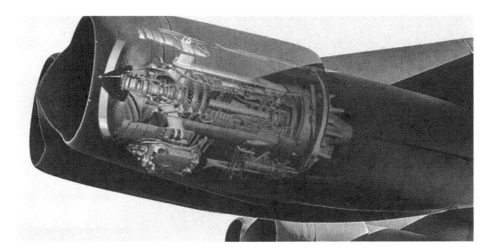

Fig. 11.2 Sustainable life prolongation for Boeing B-52: Rolls-Royce *F 130 (BR 725)* turbofan engine retrofit, $8\times$ 17 klbf © RR

Prof. Günter Kappler[14] by a joint venture between BMW and Rolls-Royce in 1990.[15] As *Jet Web* outlines in detail, BMW Flugmotorenbau together with Junkers Flugzeug- und Motorenwerke started the series production of the first turbojet engines, this means the new B-52 engines will have a joint British-German heritage, dating back to the described exciting times of still unchartered engineering territory.

---

[14] See Wikipedia, 'Günter Kappler' in German.

[15] The German mastermind behind this idea—see Wikipedia, 'Eberhard von Kuenheim' in English, outlined his motivation during LH450 flight FRA-LAX on Mon 3 April 1989, sitting on the armrest besides two keenly listening MTU engineers on their way to what was the first *GTF* promotion tour, simply by pointing to the BMW logo, a white/blue quartered circle in the Bavarian colours—'*We have to do it*'. Perhaps, also a good motto, for new generations of dedicated, innovative and inspired young GT engineers. (Special thanks for this information to the H.-A. Geidel archives, 28 June 2022).

Attachments

12

12.1 On Aero Propulsion Patents 1930–1950

12.1.1 Hans von Ohain's Secret Turbojet Patents

<u>His Early Patent History up to 1939</u> Hans Joachim Pabst von Ohain (1911–1998) started serious patent search at the RPA Reichspatentamt (Reich patent office) Berlin as part of his invention of a *'light microphone'*[1] in 1934, which became part of his doctoral thesis in physics at the Goettingen University.[2] Since his family lived still in Berlin, it was rather convenient for him to browse through patent filings in the large RPA reading and research room at Gitschiner Strasse 100,[3] just 1 km off the offices of his patent attorney Dr. Ernst Wiegand (born 1905),[4] sharing office with (presumably his brother) Dr. Carl Wiegand.[5]

---

[1] Inventor Hans von Ohain, patent CH184,920 'Verfahren und Vorrichtung zum Umwandeln von Schwingungen in Lichtschwankungen', with priority 2 June 1934, and GB443,184 'Improvements in or relating to Methods and Apparatus for the Conversion of Mechanical or Electrical Oscillations into Light Variations'. Siemens bought the patent still in 1934 for 3500 RM.

[2] See von Ohain, Ein Interferenzlichtrelais; von Ohain quoted the official translated title in—see Ermenc, Interviews, p. 12, as *'The Application of Zero Order Optical Interference for Translation of Sound Waves Directly into Electrical Impulses'*.

[3] The RPA was a remarkable institution with 1600 employees in 1938, working in 900 rooms (24,000 m^2) distributed on 7 floors,—see Spencer, The German Patent Office.

[4] The technical qualification of this young patent attorney is difficult to assess. In written communication with Hans von Ohain and his turbojet engine proposals, he uses consistently terms like *Strahlapparat (jet apparatus), Strahlungsapparat (radiation apparatus),* and even *Bestrahlungsapparat (radiation treatment apparatus).*

[5] In 1935 the patent law firm C. and E. Wiegand was registered under the address Tempelhofer Ufer 10, Berlin. In the mid-1920s the elder C. Wiegand was also the patent attorney of Hugo Junkers, and

D. Eckardt, *Jet Web*,
https://doi.org/10.1007/978-3-658-38531-6_12

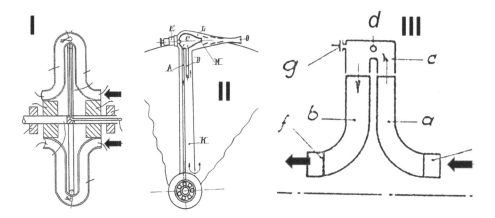

Fig. 12.1 Hans von Ohain's first secret patent #318/38, 1935–1939, I) pat. application, 15 May 1935, II) Serrell's *'Venturi'* burner, 9 Aug. 1927, III) #318/38 secret patent, 1939 © V. Koos mod. I)

Hans von Ohain finished his thesis examination at Goettingen in early 1935, and was thereafter focused on the principle test preparation of his *'Garage turbojet engine model'* in the court yard of Professor Pohl's Institute in June 1935[6]—and his first corresponding patent application in this context, which was filed in his name already on 15 May 1935, Fig. 12.1 (I).[7]

This *'rotating U-tube configuration'* as a compact gas turbine, has in itself a curious background history, normally known as *Nernst-Turbine* or *Nernst-Rotor,* after the 1920 Nobel Prize winner in chemistry and Professor for Physical Chemistry at the Berlin University Walther Nernst (1864–1941), based on a first publication of his idea in Aurel Stodola's famous book, but again without providing an actual publication source.[8] Finally, there exists a hint[9] from the mid-1960s, that Nernst's idea was claimed of having been actually patented in the 1906–1908 time frame, only to realise in hindsight that a principal

---

in due course of Bauhaus Dessau, by mediation of Bauhaus designer—see Wikipedia, 'Marcel Breuer' in English, and then ongoing disputes on the protection of tubular furniture designs (e.g. Breuer's *'Wassily Chair'*),—see Neurauter, Das Bauhaus, p. 287, and information, courtesy of Kathleen Neubert, Stadtarchiv Dessau-Rosslau, 17 May 2022.

[6] See RLM permission letter, dated 12 June 1935, reproduced in *Spectrum der Wissenschaft,* 2/95, pp. 10–13, http://webdoc.sub.gwdg.de/pub/phys/2013/spektrum-1995-2-p10-13.pdf

[7] Hans von Ohain, patent application O. 21 822 'Verfahren zur Herstellung von Luftströmungen insbes. zum Antrieb für Luftfahrzeuge (Method to generate air jets especially for aircraft propulsion)'.

[8] See Stodola, Dampf- und Gas-Turbinen, p. 1010 f. In addition—see Whittle, Gas Turbine Aero-Thermodynamics, p. 103.

[9] See Jost, The First 45 Years, p. 9. A patent search showed no such patent application by Nernst. Already Stodola mentioned the weak point of the concept, that a realistic circumferential speed of 400 m/s would provide only an insufficient compression ratio of 2.3.

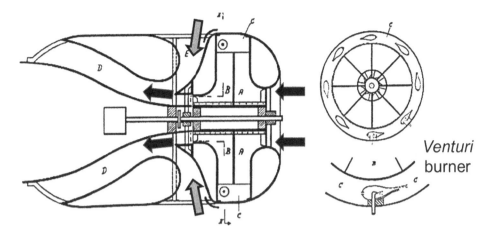

Fig. 12.2 Hans von Ohain's second secret patent # 317/38, 1935–1939: General arrangement (l), cross section w. *'Venturi'* burners (r) © V. Koos mod

mistake prevented a practical usage of this concept. Though, there is no confirmation that Hans von Ohain had been informed about this background, it is not unlikely that he got the news of the failed Nernst patent during the last year of his studies of physics at Berlin,[10] before he moved on to Goettingen for his PhD doctorate. This was an area of his immediate interest, and one could imagine that he felt challenged to correct Nernst's fault.[11]

On 9 November 1935 Hans von Ohain's application for a second turbojet engine patent was sent to the Reichspatentamt; more or less a copy of the first principle application of 15 May 1935, now with more technical substance, Fig. 12.2.[12] After several corrections, supplementary remarks and renewed applications of both patents, a first official statement of the Reichspatentamt appears in the files on 28 February 1936, as a letter from Regierungsrat (senior civil servant) Gerhard Gohlke. Here, he refers in view of both von Ohain patent applications mainly to a technical description of the *'Huettner turbine'*,[13] where there is stated that *'the different centrifugal forces of an unheated and an heated arm of a rotating U-tube can generate a flow'*. He announces a deadline up to 3rd May 1936 for both applications, to make appropriate corrections, otherwise prospects for a patent grant

---

[10] Today *HU Humboldt University*, up to 1945 *Friedrich-Wilhelm University*, Berlin.

[11] Von Ohain's HeS 3B engine, in principle in line with the Nernst concept, achieved a compressor pressure ratio of 2.8—close to Stodola's prediction, while Whittle's WU compressor was close to 4.0.

[12] Hans von Ohain, patent application O. 22 *104*, 'Verfahren und Vorrichtung zum Erzeugen von Gasströmungen (Method and apparatus to generate gas flows)'.

[13] See Huettner, Die Huettner-Turbine, and patent DE640,558; in addition Gohlke mentions the patents of Marconnet (FR412,478), Campini (FR741,858) and Sama (GB343,942), which exploit the compression effect of centrifugal forces on air flows or air/fuel mixtures. However, nothing is said about the then already granted patents of Whittle (GB347,206, 1930) and Guillaume (FR534,801, 1921).

would be poor. The period is used to cancel the application .. *882*, Fig. 12.1
(l) completely,[14] to modify application ... *104*, Fig. 12.2, to fulfil Gohlke's expectations,
and to deduce thereof accordingly a new principle application, Fig. 12.1 (r).[15] On
10 February 1937 patent attorney E. Wiegand confirms the reception of a von Ohain letter
from the foregoing day—including patent descriptions GB275,677 and FR523,427,
promising *'to carry out all necessary adaptations'*.

Not only the fact is unusual, that a patent attorney receives information about third party
inventions from his client, the more are the contents. Fig. 12.1 (II) illustrates the core idea
of GB275,677 from J.P. Serrell, an US citizen employed at *Farman* Paris, for a '*reaction-
driven propeller'*.[16] It is obvious that von Ohain wanted to direct his attorney's attention to
the *Serrell burner* with large similarity to von Ohain's '*Venturi burner'* of Fig. 12.2, and
thus a threat in view of the patentability of his idea. And Wiegand had confirmed to carry
out necessary adaptations of the application text to prevent a patent conflict in this context,
even though Gohlke had not indicated any suspicions yet.

Interestingly, the '*Venturi burner'* interfered not only with Hans von Ohain's patent
planning, but also with Frank Whittle's 1930 patent GB347,206, Fig. 12.3, however with
completely different consequences. As revealed in his 1945 *Clayton Lecture*,[17] Whittle
stated: '*I applied for my first patent in January 1930. The principal drawing of the patent
specification'* shows '*that I tried to include the* (thermo-) *propulsive duct, or "athodyd" as
it has since been called, but this had been anticipated at least twice, so the drawing* (of the
'thermo-duct/ Venturi burner', Fig. 12.3 (r)), *and relevant descriptive matter had to be
deleted from the specification.'* A clear evidence that the thoroughness of patent search was
then significantly different in Great Britain in comparison to the German Reich, not von
Ohain's fault, but certainly to his benefit.[18]

---

[14] Officially RPA-rejected on 1 Sep. 1936.

[15] Hans von Ohain, patent application O. 22 *938* (O.22104 derivative).

[16] GB275,677 with priority 9 Aug. 1927; FR523,427 with priority 19 Jan. 1920 from H.-F. Melot
addresses the diffusing guide vanes, visible inside Serrell's '*Venturi-type burner'*, Fig. 12.1 (II).

[17] See Whittle, The Early History.

[18] In 1945 Whittle was not in a position to comment on this duplicity with von Ohain's patent, which
was only partially revealed for the first time in 2001—see Conner, Hans von Ohain, p. 125 f.—and in
full to a limited circle in 2011. Regretfully, since then, the responsible German Patent Office DPMA
Munich was not able to close the still existing information gap and provide basic public information
on this salient twentieth century invention; instead, the '*DPMA Erfindergalerie* (Inventors Gallery)'
perpetuates the Whittle—von Ohain patent saga without mentioning the '*secret patent'* complication
at all,—see https://www.dpma.de/ponline/erfindergalerie/e_bio_ohain.html in English. The original
von Ohain papers, presumably with his patent copies, have been donated to the Smithsonian National
Air and Space Museum, Washington D.C. in 1995 (collection number 1995-0059); information,
courtesy of J. Anderson, Wright State Univ., Dayton Oh., 29 April 2013. Copies were filed at WSU,
Special Collections and Archives, Collection Inventory, Series V: Patents, 1907–1997, Box 8/Folder
8: First Patent w/Corresp., 1935–1947; information, courtesy of Hanns-Juergen Lichtfuss,
15 Oct. 2011.

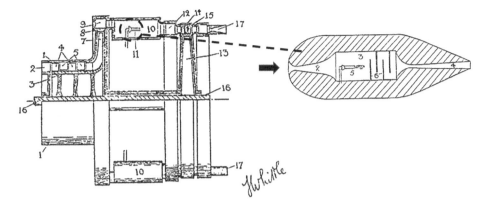

Fig. 12.3 Frank Whittle patent drawing GB347,206, 16 January 1930: General arrangement (l), *'Venturi'* burner (*'thermo-prop. duct'*) deleted from Whittle's specification (r) © F. Whittle mod

Hans von Ohain's patent story up to the outbreak of War in 1939 is finished in short. On 7 June 1937 he informed E. Wiegand that his still pending patent applications, Fig. 12.1 (III) and 12.2, had been transferred to Ernst Heinkel, future contacts should go through Josef Bucher, the head of EHF's patent office.[19] As far as visible in the filed records, internal discussions addressed both the possibility of a *'secret patent'* as well as international patent applications. An argument which especially Ernst Heinkel was inclined to follow, after Herbert Wagner in person received official RLM permission to file his turboprop invention as GB 495,469 with priority 8 February 1936, Fig. 6.16 (r)—and also Frank Whittle became visible on the English side; Ernst Heinkel's ego wanted to be part of this emerging, internationally prestigious turbojet patent contest. These discussions appear of having ended in a strangely balanced manner, when

- guessed in the first half of 1938, both von Ohain patent applications .. 104, and ... 938 must have been decided within RLM (!)[20] to become *'secret patents'* #317/38,[21] Fig. 12.2, and #318/38, Fig. 12.1 (r), and surprisingly

---

[19] In the 'Heinkel files' at Deutsches Museum Munich, the patent transfer is actually dated 1 month earlier, on 7 May 1936: Hans von Ohain is compensated by 6000 RM (5 monthly salaries), of which he is paid (a) 2000 RM immediately, and—typical Ernst Heinkel—the rest split in (b) for the achievement of a HeS 3B weight reduction below 300 kg at 800 kp thrust (both not achieved), and (c) for the first flight, which actually happened on 27 Aug. 1939.

[20] This was not completely unexpected, since in spring 1936 Hans von Ohain, still at Goettingen, had informed the RLM in a letter about his pending patent applications ... *822* and ... *104*, and asked, if there were official reservations against international applications. (No answer found).

[21] Courtesy to Evelyn Benke, DPMA, who on 10 Aug. 2011 informed *'Patent application O. 22 104 (XI/62b) was transferred to Reich Minister for Aviation and Supreme Luftwaffe Commander on 9 Sep. 1938 to become #317 in the list of secret patents 1938, therefore also quoted as #317/38.'* Both patents received the uniform title 'Strömungserzeuger für gasförmige Mittel (Flow generator for

• Max Hahn, Hans von Ohain's assistant of the early days at the Goettingen garage of Bartels & Becker filed—presumably shortly thereafter—now as Heinkel employee under the (German) priority date of 27 May 1938, what became later known as patent US2,256,198: 'Aircraft power plant'.[22]

The exact events in these early months of 1938 remain in the dark, but it is obvious that the actual decisions must have been made in the Air Ministry (RLM), and not in the Reich Patent Office—by G. Gohlke. On 6 December 1938[23] Ernst Heinkel also asked officially Ernst Udet for the permission to make foreign patent applications for turbojet inventions in the name of EHF Ernst-Heinkel-Flugzeugwerke. On 13 February 1939 the RLM (no name) declares the von Ohain turbojet project with the targeted first flight a few months ahead, as a whole as *state secret* ('*Staatsgeheimnis*'), and as a direct consequence the further information flow is restricted; on 1st August 1939 General Engineer Dipl.-Ing. Roluf Lucht (1901–1945) and as General Staff Engineer, head of all General Engineers,[24] suggests that Heinkel should only send communications to General Engineer Gottfried Reidenbach (1899–1977) with aircraft responsibility, General Engineer Dipl.-Ing. Wolfram Eisenlohr (1893–1991) with engine responsibility, and Major Dipl.-Ing. Uvo Pauls (1902–1989) with responsibility for the Luftwaffe test centre Peenmuende-West, who would decide about the further distribution in their areas. In this perspective there is some likelihood that the '*secret patent*' decision for the two von Ohain applications in 1938 had been made by Eisenlohr and/ or Lucht.

The immediately following sequence of events came absolutely unexpected, but one may observe that they fell after the Munich Conference, 29/30 Sep. 1938, in a short, three-months period of *political détente*, which also had inspired e.g. the Lilienthal Convention,

---

gaseous media), #317/38 (former application ... *104*) with priority 10 Nov. 1935, #318/38 (former ... *938*) correspondingly on 5 June 1936.

[22] The corresponding US patent grant dated from 16 Sep. 1941. The patent drawing of Max Hahn's '*reverse annular combustor*' concept, widely in line with the HeS 3B flight engine, has already been shown as Fig. 6.23 (l). Actually Hahn's *German* patent application was never processed. Within EHF there was confusion, when Hahn's invention was ear-marked as '*secret patent*' first, and then the claim was not realised. The paperwork of Hahn's invention was found at War's end as reference H 155,928 '*Rückstoßantriebsvorrichtung* (Reactive propulsion unit) amongst a body of 146,000 other unprocessed applications '*in limbo*'; information, courtesy of E. Benke, DPMA, 9 July 2013. To get German patent activities newly started after war, these remnants of the Nazi regime had to be documented on microfilm by an Allied *F.I.A.T.* team: H155,928 is stored on film reel 129 G, S. 7442 ff., available at DPMA. Schelp in – see Ermenc, Interviews, p. 112 f., called it '*unfortunately*', that Hahn's '*name was used on several patents which are actually Ohain's patents.*' And, same source, p. 41, Hans von Ohain saw Hahn's patent '*justified*' due to the implied combustor innovations, and '*I have really forgotten about this secret phase; I don't think it accomplished much since I am sure that English Intelligence knew everything.*'

[23] This date and all others in this context are backed by corresponding documents in the 'Ernst Heinkel files' in the Deutsches Museum Archive, Munich.

[24] Actual appointments in 1940,—see Wikipedia, 'General-Ingenieur' in German.

Berlin, 11–15 October 1938, which has been discussed at length in Sect. 6.1.3. The list of *'selected participants'* of that international meeting has been attached in the following as Sect. 12.2, where *'RegR G. Gohlke'* can be also found as representative of the Reich Patent Office.

After the apparently premature end of the patenting process of von Ohain's inventions by the RLM declaration of putting them into the status of *'secret patents'*, one can imagine that this decision left the main protagonists with *'mixed feelings'*:

- Hans von Ohain might have been relieved that the incriminatory investigations of potentially violating other intellectual property rights ended, he could refocus to his demanding daily work up to the first turbojet flight, still with the certainty in mind that his role as inventor of his apparatus was permanently granted,
- Ernst Heinkel might have overcome his short-phased disappointment that his early investments in the new turbojet technology were not justly acknowledged, by accepting the alternative patents for Max Hahn instead, and
- Gerhard Gohlke, who could not have been satisfied by his fragile and shaky role in the foregoing patent clinch with von Ohain's patent attorney Ernst Wiegand, decided to launch a comprehensive survey of the available patent literature in the prosperous area of *'Heizluftstrahltriebwerke* (Thermal-air jet propulsion)', which was published first in German in several issues of the magazine 'Flugsport' between January/February 1939, and translated into English in early 1942.[25]

Roughly there were 30 patents addressed in Gohlke's collection, but most were subsonic ramjet-type heated ducts and pulsejets, and there were only six turbojet engine concepts, as shown in Fig. 12.4, together with inventor's name, patent number and year of priority. This kind of early *'Hall of Fame'* lists Maxime Guillaume[26] as first turbojet engine inventor in 1921, followed by Frank Whittle, who is represented by three patents in the years 1930,[27] 1935 and 1936. Alf Lysholm filed for the Swedish AB Milo Company in 1933, a jet engine

---

[25] Hans von Ohain mentioned Gohlke in his *Foreword* to—see Mattingly, Elements, p. xxiii: *'So the early jet engine concepts were forgotten for a long time. They were unknown to Sir Frank Whittle, to me, and to the British and German patent offices. In 1939, however, the retired patent examiner Gohlke found out about the early jet patents and published them in a synoptic review.'* Ohain's statement about Gohlke's retirement status is questionable, since he was still promoted to Oberregierungsrat in 1940, and finished only at RPA in 1944, when the S-Bahn line had been destroyed during air raids, making his daily 8 km trip from his home in Berlin-Friedenau to RPA at Berlin-Kreuzberg otherwise too cumbersome. It appears that also Ernst Wiegand (33) enjoyed the premature end of the patent dispute. On 22 Oct. 1938 he was registered #91 in tourist class on board of Norddeutscher Lloyd SS *'Bremen'* en route from Bremerhaven to New York.

[26] See Wikipedia, 'Maxime Guillaume' in French: Born in 1888, his year and place of death are still missing. His last known French address in 1970 was 380, Route de Coursegoules, 06620 Gréolines.

[27] On Whittle's GB347,206 patent perception in Germany, there is known *'The Whittle patent was registered at the Berlin Patent Office on 14th August 1931. Subsequently, it was circulated to all*

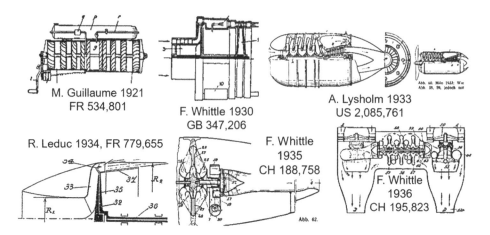

Fig. 12.4 Gohlke's Turbojet Engine Patent Collection, *Flugsport* 1939

in various forms—with axial and radial turbines, as turbojet and turboprop. Finally, Gohlke might have signalled with René Leduc's 1934 patent of a *'turbine-type jet generator'* that there were in the meantime better possibilities to object against von Ohain's *'Nernst-Rotor'* than just by referring to the *Huettner-Turbine*, as actually used in Gohlke's official reply. It is striking that Herbert Wagner's turboprop invention, Fig. 6.16 (r), patented as GB495,469 with priority 8 February 1936 (in Germany) has not yet been listed here; the publication date of 11 November 1938 might have been too late for Gohlke's *Flugsport* publication.

The Secret Patents at War's End—and Beyond Since the end of 1943 the RPA Berlin was forced to take precautionary measures against the increasing threat of Allied air raids; ten patent divisions with 180,000 files were evacuated in trains of barges[28] via the *Oder-Spree Canal* to Goerlitz, Helmut Schelp's hometown, and from there further by trucks to the former monastery of Striegau/ (today) Strzegom, Pl, then used as jail, and to the county court building at Jauer/Jawor nearby, some 290 km south-east of Berlin, and 50 km west of Breslau/Wroclaw, the capital of Silesia.[29] In early 1944 further stocks, mainly the RPA library of 250,000 volumes and five million patent files were brought by prisoners to a bomb-proof office space, 600 m below ground of a potash mine at Heringen on River Werra/Hesse, half-way between Kassel and Erfurt. And, with the eastern front nearing Silesia, all secret materials were brought back to Berlin in January 1945, and forwarded from there, apparently including the originals of the ~12,000 'secret patents', in February 1945 also to Heringen. On 30 March 1945, a few hours before US troops occupied the area,

---

aeronautical establishments by technical journal—about November 1931. The AVA at Göttingen was one recipient.' Information, courtesy of Ian Whittle, 23 Oct. 2014.

[28] No railway capacity could be freed up.

[29] See https://www.dpma.de/english/our_office/about_us/history/140yearsofthepatentoffice/1941-1950/index.html

the Heringen mine shaft was officially given up by RPA personnel, except for a small group under the command of a Regierungsrat z.b.V.[30] *Franke* and party member, who stayed to execute a secret destruction order from the 'Reichsverteidigungskommissar' in charge.[31]

Espacenet, the European patent search tool, shows presently 30 patents for 'Ohain Hans', of which nine have a priority of 1945 or earlier. Of these, five under the designation '*Jet engine* (Strahltriebwerk)', refer to the invention of a '*pressure wave exchanger*' for aircraft propulsion, comprising one main patent with priority 14 January 1944, and four supplementary patent derivatives. Except for the aero application, Hans von Ohain's patent has great similarity with Claude Seippel's patent DE724,998 'Pressure Exchanger' with priority 7 December 1940.[32] In addition, Hans von Ohain's patent of a '*light microphone*' is preserved in a German and Swiss version. Finally, there are two turbomachinery related patents in the name of Hans von Ohain:

(i) DE767,258, 'Strahltriebwerk insbesondere für Luftfahrzeuge (Jet engine especially for aircraft)', a turbofan/bypass engine configuration as originally filed on 12 Sep. 1939, and

(ii) DE767,808, 'Verdichteranlage für Heizluftstrahltriebwerke (Compressor unit for heated-air jet engine)', a diagonal compressor impeller as used in the Heinkel HeS 011, with priority 23 May1942, Fig. 12.5 (r).

Though the official patent paperwork had been largely destroyed, the new German Patent Office, operative again since October 1949, used the opportunity based on a '*first transitional (patent) law*'[33] to reinstall, prolongate and publish patents on demand of the patent owners, if these participated in this process by providing official records in their ownership.

---

[30] Z.b.V., '*zur besonderen Verwendung* (under special assignment)'.

[31] See Wikipedia, 'Reich Defense Commissioner' in English, with all likelihood—see Wikipedia, 'Fritz Sauckel' in English, in charge for 'Wehrkreis IX Kassel'. At the International Military Tribunal Nuremberg, Sauckel was found guilty of war crimes and crimes against humanity, and was hanged on 16 Oct. 1946. The *Franke* team used 5×1 l thermos with liquified air as fire accelerants, which incinerated the pile of secret patent residue-free; post-war, this brought RegR Franke to prison, as documented by the Office of the U.S. High Commissioner for Germany, Background Information No. 68, 13 Sep. 1951. The *Franke story* has been documented as No.5 in a 25 piece serial article '*Was wurde aus den deutschen Patenten?* (What happened to the German patents?)' published by R. Wilder in the Nov.1945–Dec.1966 German magazine 'Der Kurier', Berlin, Aug./Sep.1958, No.188-212.

[32] See Kay, German Jet Engine, p. 52 f. Here, under the designation '*Tuttlingen Engine*', the whole concept is presented as originating from Hans von Ohain, without any reference to earlier corresponding publications. Seippel had mentioned '*Gas turbines*', and '*Supercharging for heat engines*' as potential applications of his patent idea, which was practically realised in the BBC 'Comprex' turbocharger; for a detailed comparison of Seippel's and Von Ohain's corresponding patents—see Eckardt, Gas Turbine Powerhouse, pp. 201–206.

[33] Erstes Überleitungsgesetz, dated 8 July 1949.

With respect to *'secret patents'*, one had to observe the fundamental difference between the US system, based on *'secret applications'* only, which the government maintains *'secret'* as long as needed e.g. for 50 years, before the standard patent phase follows thereafter, and the German *'secret patent'* system over the standard patent lifetime of 20 years. These re-established German patents after 1949 are marked in the database by a front page imprint *'Granted on the basis of § 30, clause 5 patent law, in the meantime cancelled.'*[34] Accordingly, information on von Ohain's two turbomachinery patents above were released on his ~1952 demand for the turbofan engine concept, (i) on 23 May 1953, with a remaining validity up to 12 September 1959, and for the diagonal compressor wheel, (ii) on 10 September 1953, with a remaining protection time up to 23 May 1962. Regretfully, he did not demand then the reconstruction and publication of his essential *'secret patent #317/38'*, Fig. 12.2, with priority 10 November 1935,—either because the remaining validity up to 1955 was too short, or because a publication, which would have brought the patent text permanently into the public domain was not in his interest.[35]

As illustrated in Fig. 12.5, the diagonal compressor concept had a short, and remarkable German patenting history, after H. Schelp might have learnt about R. Birmann's design presumably as early as 1936, as described in the context of Fig. 7.5. The first patent application for DE920,090, Fig. 12.5 (l), was filed by Bruno Eck[36] with priority of 8 December 1937, followed by Eck's scholar Max A. Mueller, who was granted DE767,969 with a priority of 18 January 1939, Fig. 12.5 (m), and finally by the responsible HeS 011 designer Hans von Ohain, who filed what became DE767,808, Fig. 12.5 (r), together with Dr. Viktor Vanicek with a priority of 23 May 1942.[37] Surprisingly for the date,[38] more than 10 years after original issuance, this (and only this) Ohain patent text has after the *'Claims'* a *'Search Report'*, which points out that for technical differentiation also GB347,206 had been considered, actually Frank Whittle's first patent *'Improvement relating the propulsion of aircraft and other vehicles'* with priority of 16 January 1930!

---

[34] In German 'Erteilt auf Grund des inzwischen aufgehobenen § 30, Absatz 5 Patentgesetz.'

[35] Another such opportunity was missed in 1989 when Hans von Ohain was celebrated at Munich on 26/27 October 1989 by a key note speech on the occasion of the 50th anniversary of the first turbojet flight,—see DGLR Publ., 50 Jahre Turbostrahlflug, p. 38. Here his References indicated his secret patent No. 317/38, together with other secret patents and important contract material, etc. in the possession of the Deutsches Museum Munich; none of these claims could be verified.

[36] Dr.-Ing. Bruno Eck was long-time head of the fluid mechanics laboratories at the FH Fachhochschule (Engineering Academy) Cologne, author of the well-known textbook 'Introduction to Technical Fluid Mechanics (Einführung in die Technische Strömungslehre), Springer, 1935—and in the years 1927–1928 influential teacher to his student Max Adolf Mueller,—see Sect. 7.5.

[37] In addition, there was DE873,190 with priority 7 Dec. 1943, for the inventors H. Jacobs (also with DE866,878 responsible for the HeS 011 annular combustor) and V. Vanicek, who suggested a diagonal turbine to complement a diagonal compressor, a configuration which acc. to—see Koos, Heinkel Raketen- und Strahlflugzeuge, p. 71, had already been rejected by RLM (H. Schelp) in July 1942.

[38] Patent publication on 13 April 1953, presumably the *'Search Report'* is a post-war addition, signalling extreme thoroughness in patent search on the German side—now.

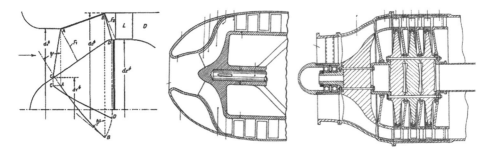

Fig. 12.5 German diagonal compressor patent pedigree: B. Eck 1937 (l), M.A. Mueller 1939 (m), H.v. Ohain et al. (1942, r)

Ohain - in Conner, Elegance in Flight, p. 34 - speculates, if at times patent reviewers in England and Germany had detected the first turbojet patent in 1921: *'Actually, turbojet patents of the Frenchman Guillaume (French physicist, Charles Edouard Guillaume) should have been the cause for the rejection of practically all further turbojet patents. Why the German patent examiner didn't find Guillaume's patent is nebulous to me. I learned that the English patent examiner didn't find it either, so he granted Whittle a patent far too broad, which should not have been granted to Whittle'.*

Nice story, if not referred to the wrong *Guillaume*: Charles E. Guillaume (1861–1938), was a Swiss physician from Neuenburg/ Neuchâtel,[39] who received the 1920 Nobel Prize in Physics for inventing and investigating between 1886 and 1919 *'Invar/ Elinvar'*, a Fe-Ni alloy of extraordinarily small thermal expansion coefficient and temperature-independent modulus of elasticity. These inventions had considerable influence on the advancement of metrology, and specifically to Swiss precision watch making. The French agricultural engineer and turbojet inventor Maxime Guillaume, born 1888, received in 1921—then resident in the Département Meuse/Maas[40]—the first patent FR534,801 for a *'Reactive propulsor on air,*[41] using in principle a rudimentary gas turbine to power an aircraft, as shown in Fig. 12.4.

Somehow fitting to Hans von Ohain's foregoing, non-transparent *'secret patent'* story, the whole post-war reception of turbojet engine historiography culminated in a unique, industrial *'coup'*, what H. Giffard called *'the birth of the dual inventor narrative in the*

---

[39] See Wikipedia, 'Charles Èdouard Guillaume' in English; as the son of a Swiss horologist the *'Guillaume balance'* in horology is named after him. Coincidentally, the first 4 MW power generation gas turbine from Brown Boveri, Baden CH was installed at the Neuchâtel utility works in 1939,—see Eckardt, Gas Turbine Powerhouse, p. 181 f.

[40] See Wikipedia 'Maxime Guillaume' in French, for the most comprehensive collection of his CV data, though year and place of his death—as mentioned already in FN 26—are still unknown.

[41] In French: 'Propulseur par réaction sur l'air'.

United States'.[42] During the early 1970s a new generation of powerful turbofan engines was introduced by the *'big three OEMs'*—for General Electric the *CF6*, for Pratt & Whitney the *JT9D*, and for Rolls-Royce the RB211. For GE Aviation, then managed by the self-assured and charismatic Gerhard Neumann (1917–1997),[43] the repetitive remembrance to single-source wartime turbojet kick-off support by the British, and especially in the person of Frank Whittle, became annoying. Also, ever-lasting discussions on patent priorities, suspicion of *'patent foul-play'*, and the long-time meaning of in reality sometimes very short advantages in the parallel turbojet engine development schedule between Whittle in England, and von Ohain in Germany, led de-facto to a kind of mutual *'co-inventor acknowledgement'*, that both developments occurred completely independent, without significant background/ cross communication, and with by and large equivalent results.[44]

In January 1966, both turbojet inventors (presumably) met for the first time, when Hans von Ohain received the prestigious AIAA's Goddard Award 1966,[45] which Sir Frank Whittle had already received as first recipient on Tue 26 January 1965 for *'his imagination, skill, persistence, and courage in pioneering the gas turbine as a jet propulsion aircraft engine, thus revolutionizing military and commercial aviation for all time.*'[46] Here, in the USA Hans von Ohain and Frank Whittle were first seen as co-inventors of the turbojet engines, and this status became more and more established in the late 1970s, when both men, now living in the United States, visited jointly numerous public events to commemorate turbojet development history.[47] As an example, Fig. 12.6 (l) illustrates the second day

---

[42] See Giffard, Making Jet Engines, p. 222. About a possible motivation she speculates: *'After having collaborated enthusiastically with Britain during the war, American authorities were eager to make full use of German expertise after the war (bolstered no doubt in part by officials who needed a compelling case for importing German engineers).*

[43] See Wikipedia, 'Gerhard Neumann' in English.

[44] In the 2000s, after the death of both 'co-inventors', this 'agreement' was several times publicly called into question by a bravely fighting Ian Whittle, Sir Frank's son and professional B 747 pilot, causing apparently some frowning, if not 'effective hits' amongst the addressed.

[45] The Goddard Award 1966 went jointly to Hans J.P. von Ohain, for *'his contributions to the achievement in 1939 of the first successful application of turbojet propulsion to aircraft'*; and to A.W. Blackman, UAC and G.D. Lewis P&WA for *'their contributions to the understanding of the phenomenon of combustion instability ...'*

[46] Apparently, the information that Whittle and von Ohain received the Goddard Award 1966 jointly, as stated in—see Conner, Hans von Ohain, p. 251 is wrong (again); the 1965 presentation to Sir Frank alone, has been documented by a *British Pathé* video on the internet. Possibly, Whittle might have attended von Ohain's 1966 ceremony as a guest ...

[47] See Giffard, Making Jet Engines, pp. 222–225. Here she gives the Smithsonian Institution's National Air and Space Museum's new exhibit on jet aviation, opened in July 1981, a special role for distributing the dual-inventor narrative: *'Its display case on the "Pioneers of Jet Propulsion" declared that Whittle and von Ohain are "rightly regarded as the inventors of the first practical jet engines", although the jet engine "reflected the efforts of many individuals from many nations". ... The authoritative gallery neatly juxtaposed American jet history with its foreign origin stories.'*

of a meeting on 3–4 May 1978, here as a GE-sponsored event moderated by Neumann's successor as Senior Vice President, GE Aircraft Engines (officially since October 1979) Brian H. Rowe (1931–2007)[48] at the National Museum of the US Air Force at Dayton, OH, which was documented in a common booklet.[49] The two-days meeting had been hosted by the Wright-Patterson Air Force Propulsion Laboratory,[50] of which Hans von Ohain, naturalised US citizen since 1951, had become ARL Chief Scientist in 1963, with main responsibilities for energy transformation and air-breathing jet propulsion.[51]

With emerging interest in the turbojet engine history in Germany, marked by the publication of the first issue of Kyrill von Gersdorff's et al., afterwards so successful book *'Flugmotoren und Strahltriebwerke (Flight motors and jet engines)'*[52] in 1980, the German Museum in Munich launched an initiative to expand and complete its collections into the range of turbojet engines. Von Ohain had provided information[53] that both the historical first flight aircraft *Heinkel He178*, as well as the first corresponding turbojet engine *Heinkel HeS 3B* had been destroyed during a heavy Allied air raid on the *'Deutsche Luftfahrtsammlung (German aviation collections)'*, Berlin-Moabit, in the night of 22/23 November 1943.[54] Consequently, this led to the decision to commission the building of two 1:1 scale replicas of the HeS 3B, for which Hans von Ohain and Wilhelm Gundermann provided reconstructed drawings. The task was neatly carried out by MTU Aero Engines's apprentice shop, Fig. 6.23, and after a last approval on site by Hans von Ohain, Fig. 12.6 (r), the HeS 3B engine replica was handed over in Ohain's presence to the Deutsches Museum on 15 May 1981.[55] In due course the second HeS 3B engine copy was presented at the Smithsonian Institution's National Air and Space Museum, in the newly opened *Jet Aviation Gallery* in July 1981, where it was positioned besides the original *Whittle W1X* turbojet engine, which Power Jets (R&D) had gifted to the Institution already in 1949.

---

[48] See Wikipedia, 'Brian H. Rowe' in English.

[49] See Whittle, An Encounter, mainly the documentation of an illustrated, public joint interview of both inventors.

[50] See Wikipedia, Wright-Patterson Air Force Base, in English.

[51] See Ebert, Pabst von Ohain.

[52] See Gersdorff, Flugmotoren.

[53] See Conner, Hans von Ohain, p. 254.

[54] For details about the *'Deutsche Luftfahrtsammlung'*, and its present whereabouts in Poland,—see Eckardt, Gas Turbine Powerhouse, p. 60.

[55] See Conner, Hans von Ohain, p. 254. Deutsches Museum published von Ohain's speech together with a short introduction by their responsible curator Walther Rathjen in DM's *Kultur & Technik*, 1981, Issue 2 'Aufbruch in den Überschallflug. Das erste Strahltriebwerk (Towards supersonic flight. The first jet engine)', in which the achievement of placing this exhibit near term besides Whittle's engine at SI NASP Museum was already proudly announced.

Fig. 12.6 Making of < *Independent Jet Engine Inventors* >: *'The Encounter'*, 4 May 1978, F. Whittle, B. Rowe (GE), H. von Ohain (l), HeS 3B turbojet engine replica presentation, May 1981, H. von Ohain, W. Hansen (MTU) (r) © GE (l), MTU (r)

12.1.2 Other Patents

Besides Heppner's patented, specific *BLI boundary-layer injection*, aft-fan configuration of 1941, as discussed already in detail in Fig. 8.34, *Betz and Ackeret* patented the positive effect of boundary layer re-energising by *blowing*, Fig. 12.7 (l),—or *suction*, Fig. 12.7 (r) in more general form in DE513,116, as claimed as an elegant method to diminish the flow drag of a body with a wake already in 1923, requiring only minimum description:

Method for drag reduction of a flow-embedded body, so that by suction or blowing the excessive accumulation of boundary layer material will be prevented.'[56]

The versatile and always on the go-engineer Max Adolf Mueller (1901–1962) acquired over time an impressive patent portfolio of 163 granted patents; other than e.g. H. Oestrich, Fig. 10.24, his output of inventions has also a substantial content before and during the war, Fig. 12.8 (l). Most remarkable among Mueller's early inventions in the 1935–1940 time-frame (30) is e.g. DE768,072 'Einrichtung zur Beeinflussung der Stroemungsgrenzschicht bei Flugzeugen mit einem ein Axialgeblaese mit Zwischenentnahme enthaltenden Rueckstossantrieb (Aircraft wing boundary layer modifications by means of an axial blower as part of a reaction-type propulsor)' with priority of 22 October 1938, filed during his time as Junkers employee at Magdeburg. This concept has some similarity with

---

[56] Betz-Ackeret patent claim in German: *'Verfahren zur Verminderung des Widerstandes eines nicht als Quertriebskörper dienenden Körpers in relativ zum Körper bewegten Medien, dadurch gekenn-zeichnet, daß durch Absaugen oder Abblasen eine übermäßige Ansammlung der Grenzschicht verhindert wird.'*

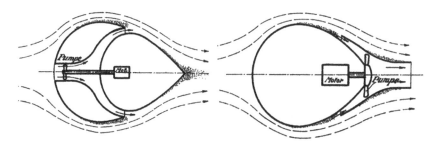

Fig. 12.7 Betz-Ackeret drag-reduction patent DE513,116, with priority 5 Sep. 1923

Heppner's later *'Augmentor'* Boundary Layer Propulsion aircraft concept with wing BL suction, Fig. 8.36, and uses H. Wagner's patented all-axial turboprop engine GB495,469, with priority 8 February 1936, as reference wing-integrated turboprop engine, Fig. 6.16 (r).

However post-war, after a *'low'* during and immediately after the war, Mueller's remarkable *'high'* with 53 filed patent applications between 1951 and 1955 shows also a principle similarity to Oestrich's considerations and activities. Employed now at the machine technology company *'Goetzewerke Burscheid'*,[57] founded in 1887, and known over time for their heat-resistant metal seals and piston rings, he filed plenty of ideas together with *Goetze* (45), which he must have observed and rather coincidently picked up during wartime. Now, to stay in the same category of machinery—applications were filed to be applied for heavy (agricultural) tractors, what before were heavy armoured vehicles and tanks. A typical example of Mueller's corresponding, wide-spread patent haul is the *'Startwagen fuer fahrgestellose Flugzeuge (Take-off carrier for aircraft without bogie)'*, DE926,769 with priority 16 January 1954, which found broader applications during the *Arado Ar 234* bomber test phase with *BMW 109-003* turbojet engines in 1944.[58]

Another remarkable example of Mueller's creativity and engineering art is a perfectly designed patent application *'Mechanism for Adjusting Turbomachinery Guide Vanes'*, with priority 17 March 1942, Fig. 12.8 (r), representing first time patent protection of *VIGV Variable Inlet Guide Vanes* arrangements of axial compressors,[59] more than 10 years ahead of Gerhard Neumann's claim for being the first, who had patented this important design feature in his US2,933,234 'Compressor Stator Assembly' with priority

---

[57] Burscheid is located 25 km north-east of Cologne.

[58] See e.g. Kay, German Jet Engine, p. 106; there are no indications that Mueller was part of these developments, with all likelihood he had observed *'Startwagen'*-experiments at the *Rechlin* test centre.

[59] The supplementary VIGV patent for centrifugal compressors and turbochargers acquired Alfred Mueller, BMW already with DE703,364, and with priority 24 Dec. 1938.

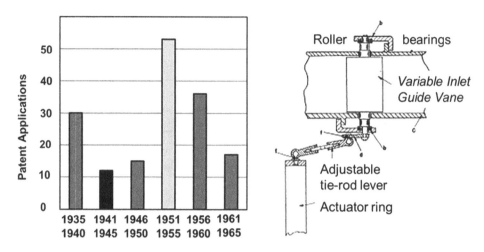

Fig. 12.8 Max Adolf Mueller (1901–1962) patent history: Patent applications (163) between 1935 and 1965 (l), VIGV patent (CDG-759) with priority 17 March 1942 (r)

28 December 1954 for the then introduced series of GE civil (CF6) and military (J79) turbojet engines.[60]

Finally, the counter-rotating axial turbojet concept of Fritz Heppner (1904–1982), patented as US2,360,130 with priority of 26 March 1941, and illustrated in Fig. 5.11, was at that time a not so isolated idea as the sole comparison to the DB 109-007 turbofan engine, Fig. 5.13, might suggest. Actually, there are at least two other comparable inventions known, one from Metrovick's Karl Baumann (1884–1971), GB586,558, with priority of 12 January 1942, and another filed by the Swedish *STAL Finspong*, GB620,721, with priority of 13 October 1945, to drive a CR turboprop engine, without a conventional gearbox.

12.2 Convention of the Lilienthal Society for Aeronautical Research e.V., Berlin 12–15 October 1938, List of Selected Participants

After the Royal Aeronautical Society RAeS, founded already in 1866, the corresponding German research organisation was established in 1912, however the name changed during its more than 100 years of existence:

---

[60] The patent description was found, and apparently removed by US troops at Hirth Motoren Stuttgart as *'Einrichtung zum Verstellen von drehbaren Leitschaufeln für Turbomaschinen'*, translated and archived as CDG-759 (Captured German Documents) at the National Archives, College Park, Maryland, from where the author received the solely surviving English translation, courtesy of Nathaniel Patch, 23 Jan. 2012.

1912–1913	WGF	Wissenschaftliche Gesellschaft fuer Flugtechnik e.V.
		Scientific Society for Aeronautics e.V.
1914–1936	WGL	Wissenschaftliche Gesellschaft fuer Luftfahrt e.V
		Scientific Society for Aviation e.V.
1936–1945		**Lilienthal Gesellschaft fuer Luftfahrtforschung e.V.**
		Lilienthal Society for Aeronautical Research e.V.
1952–1962	WGL	Wissenschaftliche Gesellschaft fuer Luftfahrt e.V
		Scientific Society for Aviation e.V.
1962–1967	WGLR	Wissenschaftliche Gesellschaft fuer Luft- und Raumfahrt e.V
		Scientific Society for Aero- and Astronautics e.V.
1967–	DGLR	Deutsche Gesellschaft fuer Luft- und Raumfahrt e.V.
		German Aerospace Society e.V.

The Lilienthal Society of 1936–1945 organised three annual conventions 1936–1938 at Berlin, Munich and again at Berlin. The last convention took place at the Ufa-Palast am Zoo, a cinema complex and convention centre, from 12 to 15 October 1938—with 3000 registered, international participants. During the opening ceremony on Wed 12 October 1938 the RAeS President (Sir) Roy Fedden was honoured by the Lilienthal Ring Award. The Otto-Lilienthal Museum Anklam[61] preserves not only a number of original memorabilia to this German aviation pioneer, but collects also items in connexion with the name of Lilienthal and the Lilienthal Society. Amongst these collected museum items is e.g. besides the original Lilienthal Ring, an 182 page booklet of the 1938 Berlin Convention, containing approximately 2700 names of said in total ~ 3000 participants, which the Author gratefully had the opportunity to evaluate, especially in view of Sect. 6.1, dealing with *'International information exchange/ transfer'*.

The following tabulation contains in alphabetic order 58 names out of the larger number of 2700 of personalities somewhere addressed in the text with information in four categories, (1) the participant's name, (2) his first name, (3) the listed professional title, and finally (4) the related institution. Some of the original information is incomplete which was supplemented by the Author wherever possible by inserting the missing information in [brackets]. In this respect the tabulation is a specific *Index (of Names)*, which is additionally and comprehensively provided at book's end, but focussed here solely to the group of convention attendees, which at least in our imagination had the opportunity for three days in October 1938 to meet, to discuss and, perhaps, to understand each other better.

---

[61] Anklam, Otto Lilienthal's home town where he got his first flight lessons as a boy by observing gliding storks over the meadows and marsh areas around the small town of some 12,000 inhabitants, lies in the German federal state of Mecklenburg-Vorpommern approx. 20 km off the Baltic shore and 180 km north of Berlin.

SURNAME	First Name	Title	Institution
ACKERET	[Jakob]	Dr.-Ing.	Prof. at TH Zurich
BAEUMKER	Adolf	MinDirig.	Pres. Lilienthal-Soc.
BARTH	Hermann	Dr.-Ing.	[DVL Bln.-Adlershof]
BECKER	Karl Emil	Dr.-Ing.	Head of HWA Army Ordnance Office
BETZ	Joh. Albert	Dr.	Head of AVA Aerod. Test Est. Goettingen
BRANDENBURG	[Ernst]	Dr.-Ing. E.h. MDir.	RVM Reich Ministry of Transport
BRANDNER	[Ferdinand]	Dipl.-Ing.	[Junkers] Dessau
BRUCKMANN	[Bruno]	Dipl.-Ing.	[Bramo] Bln.-Spandau
BUCHANAN	[John]	Deputy Dir. Tech. Dev.	Air Ministry London
BUCHER	John B.	Dipl.-Ing.	Dir. and Chief Engineer ASM Coventry
BUSEMANN	[Adolf]	Dr.-Ing.	Prof. [LFA] Braunschw.
ECKENER	[Hugo]	Dr. Dr.-Ing. E.h.	Director Zeppelin F'hfn
ECKERT	Bruno	Dipl.-Ing.	[FKFS] Stuttgart
EISENLOHR	[Wolfram]	[Dipl.-Ing.]	RLM Fl.-Hauptstabsing.
ELLOR	[James E.]		Rolls-Royce, Derby
ENCKE	[Walter]	Dipl.-Ing.	[AVA] Goettingen
FEDDEN	[Roy]		Pres. RAeS Bristol
FRERICHS		Dipl.-Ing.	[Bramo], Bln.-Spandau
FRIEDRICH	Rudolf	Dipl.-Ing.	[Junkers] Magdeburg
FRIZ	[Max]		Dir. [BMW] Eisenach
GEORGII	[Walter]	Prof. Dr.	Head DFS Darmstadt
GOHLKE	[Gerhard]	RegR	Reich Pat. Office, Bln.
GREY	[Charles Grey]		Ed., *Aeroplane* London
HEINKEL	[Ernst]	Prof.	EHF Warnemuende
JACOBS	Hans	Engineer	[DFS] Darmstadt
KAMM	[Wunibald]	Prof. Dr.-Ing.	FKFS Stuttgart
KAMMHUBER	[Josef]	Oberst-Lt.	Head Org.-Staff
KNOERNSCHILD	[Eugen]	Dipl.-Ing.	[DVL Bln.-Adlershof]
KRUSE		Fl.-Stabsing.	Patent Att. Berlin
KÜHL	Heinrich	Dr.-Ing.	[DVL Bln.-Adlershof]
LINDBERGH	Charles	Oberst	
LIPPISCH	[Alexander]	Engineer	[DFS] Darmstadt
MADELUNG	[Georg]	Prof. Dr.-Ing.	Head Flugtechn. Institute Stuttgart
MADER	[Otto]	Prof. Dr.-Ing.	[Junkers] Dessau
MARQUARD	[Ernst]	Fl.-Stabsing.	RLM Office Head

(continued)

SURNAME	First Name	Title	Institution
MAYBACH	[Karl]	Dr.-Ing. E.h.	Dir. Maybach-M.
MESSERSCHMITT	[Willy]	Prof. Dipl.-Ing.	Head Supervis. Bd. Messerschmitt Augsb.
MILCH	[Erhard]	General	State Secretary
MÜLLER	Alfred	Engineer	[Bramo] Bln.-Spandau
MÜLLER	Max Adolf		[Junkers Magdeburg]
MUTTRAY	Horst	Dipl.-Ing.	[DFS] Darmstadt
MUTTRAY	Justus	Dipl.-Ing.	[Junkers] Dessau
NÄGEL	[Adolph]	Prof. Dr.-Ing.	Dir. Masch.-Lab. THD
OESTRICH	[Hermann]	Dr.-Ing.	[Bramo] Bln.-Spandau
PRANDTL	[Ludwig]	Prof. Dr.	Pres. Lilienthal-Soc.
RITZ	[Ludolf]	Dipl.-Ing.	[AVA] Goettingen
RUDEN	[Paul]	Dr.-Ing.	[TH] Hannover
SACHSE	[Helmuth]	Fl.-Haupt-stabsing.	Dir. BMW Munich
[Forbes-] SEMPILL	[William]	19th Lord Sempill	[Peer, Air Pioneer, Spy]
SCHELP	[Helmut]	Dipl.-Ing.	[RLM] Flugbauführer
SCHMIDT	Friedrich A.F.	Dr.-Ing.	[DVL Bln.-Adlershof]
SHENSTONE	[Beverley S.]	Engineer	Air Ministry London
TRIEBNIGG	[Heinrich]	Prof. Dr.-Ing.	TH Bln.-Charlottenburg
TWEEDIE	[W. Lawrence]		Air Ministry London
UDET	[Ernst]	Col.-General	RLM Head Techn. Off.
VACHELL	[John L.]	Grp. Captain	Brit. Embassy Bln. Air Att.
WAGNER	Herbert	Prof. Dr.-Ing.	Head Av. Institute THB [Board Junkers Dessau]
WALTER	Hellmuth	Engineer	[Hellmuth Walter KG] Kiel
WEINIG	[Fritz]	Prof. Dr.-Ing.	[DVL Bln.-Adlershof]

Bibliography

Ackeret, Jakob: 'Euler's Arbeiten über Turbinen und Pumpen (Euler's Works about turbines and pumps)', Introduction to Vol. II 15 of Euler's Works, special print to 'Euleri Opera Omnia', Series II Vol. 15, Orell Füssli, Zurich, 1957, 61 p.

Ackeret, Jakob: 'Auf dem Weg zur Gasturbine (Towards the gas turbine)', Schriften der DAL Deutsche Akademie der Luftfahrtforschung, special print from Vol. 7b, Issue 3, 1943, pp. 72-75, presentation to the DAL Berlin on 4 Dec. 1942

Ackroyd, John A.D.: 'The United Kingdom's contributions to the development of aeronautics', Part 1. From antiquity to the era of the Wrights, The Aeronautical Journal (AJ), Jan. 2000, pp. 9-30; Part 2. The development of the practical aeroplane (1900-1920), AJ Dec. 2000, pp. 569-59; Part 3. The development of the streamlined monoplane (the 1920s-1940s), AJ May 2002, pp. 217-268; Part 4. The origins of the jet age, AJ Jan.2003, pp. 1-47

Ackroyd, John A.D. and Riley, Norman: 'Hermann Glauert FRS, FRAeS (1892-1934)', J. of Aeronautical History, Vol.1, Paper No. 2011/02, 52 p. https://www.aerosociety.com/publications/jah-hermann-glauert-frs-fraes-1892-1934/

Ackroyd, John A.D.: 'The Spitfire Wing Planform: A Suggestion', J. of Aeronautical History, Paper No. 2013/02, 15 p. https://www.aerosociety.com/media/4843/the-spitfire-wing-planform-a-suggestion.pdf

Ackroyd, John A.D.: 'The Aerodynamics of the Spitfire', J. of Aeronautical History, Paper No. 2016/03, 28 p. https://www.aerosociety.com/media/4953/the-aerodynamics-of-the-spitfire.pdf

Ackroyd, John A.D.: 'Aerodynamics as the Basis of Aviation: How well did it do?', J. of Aeronautical History, Paper No. 2018/01, 62 p. https://www.aerosociety.com/media/8042/aerodynamics-as-the-basis-of-aviation-how-well-did-it-do.pdf

ACS American Chemical Society: 'The Houdry Process. A National Historic Chemical Landmark, 13 April 1996', commemorative booklet, 8 p. pdf, https://www.acs.org/content/acs/en/education/whatischemistry/landmarks/houdry.html

AEG Allg. Elektrizitaets-Ges.: 'Wir stellen vor (We present): Dir. Prof. Dr.-Ing. E.A. Kraft', Spannung (Voltage), Vol.1, Issue 8, pp. 247-248

Albrecht, Ulrich: 'Rüstungsfragen im deutsch-französischen Verhältnis (Armament issues in the German-French relationship)', pp. 97-133, in Engler, W. (ed.): 'Frankreich an der Freien Universität: Geschichte und Aktualität (France at the Free University: History and presence)', Franz Steiner Vlg., 1997, 292 p.

Albring, Werner: 'Angewandte Strömungslehre (Applied fluid mechanics)', De Gruyter, 5th issue, 1978, 420 p.

Albring, Werner: 'Gorodomlia. Deutsche Raketenforscher in Russland', Luchterhand, 1991, 249 p., and 'Gorodomlya Island: German Rocket Scientists in Russia', Books on Demand, Norderstedt, 2016, 248 p.

Allen, Michael Thad: 'The Business of Genocide: The SS, Slave Labour and the Concentration Camps', The University of North Carolina Press, 2003, 402 p.

André, Pierre: 'Les turboréacteurs de grande puissance en France et la contribution des ingénieurs allemands après la guerre', Revue Pégase, No. 111, 4th trim. 2003, 13 p.

Anon.: 'Die industrielle und kommerzielle Schweiz beim Eintritt ins 20. Jahrhundert (The industrial and commercial Switzerland entering the 20th century) in Schweizerische Groß-Industrie, Zurich 1903, pp. 721-728

Armstrong, Frank W.: 'The aero engine and its progress – fifty years after Griffith', Aeronautical Journal, Dec. 1976, pp. 499-520

Armstrong, Frank W.: 'Farnborough and the Beginnings of Gas Turbine Propulsion', J. of Aeronautical History, Paper 2020/02, pp. 16-37

ATZ:'Zum 65. Geburtstag von Dr. A. H. Müller-Berner (On the 65th birthday ..)', ATZ Automobiltechnische Zeitschrift, Vol. 74, Issue 11, 1972, p. 456

Aumann, Philipp and Köhler, Thomas: 'Vernichtender Fortschritt: Serienfertigung und Kriegseinsatz der Peenemünder "Vergeltungswaffen" (Devastating progress: Series production and war deployment of the Peenemuende "V-weapons"), Ch. Links Vlg., 2018, 198 p.

Bacon, Francis: 'The Works of Francis Bacon. Vol. 4', Cambridge University Press, 2011, 512 p.

Bailey, Bill: 'The Early Development of the Aircraft Jet Engine', unpubl. thesis, held by the Royal Aeron. Society, 1995, edited and supplemented by Ian Whittle, 2004, 71 p.

Bailey, Gavin: 'The Narrow Margin of Criticality: The Question of the Supply of 100-Octane Fuel in the Battle of Britain', The English Historical Review, Vol. CXXIII, No. 501, April 2008, pp. 394-411

Bakken, Lars E., Jordal, Kristin, Syverud, Elisabet and Veer, Timot: 'Centenary of the First Gas Turbine to Give Net Power Output : A Tribute to Ægidius Elling', ASME Turbo Expo Vienna, A, 14-17 June 2004, Vol.2, Paper No. GT2004-53211, pp. 83-88

Barth, Hermann: 'Zur Entwicklungsgeschichte der deutschen Flugtriebwerks-Höhenprüfstände (Development history of German flight engine altitude test facilities)', DLR paper for the 75th birthday of Adolf Baeumker, 1966, 35 p.

Bauerfeind, Klaus: 'Steuerung und Regelung der Turboflugtriebwerke (Jet engine steering and control)', Birkhaeuser 1999, 313 p.

Bauersfeld, Walther: 'Die Grundlagen zur Berechnung schnelllaufender Kreiselräder' [Basic evaluation of high-speed (axial) impellers], ZVDI, Vol. 66, No. 19, 13 May 1922, pp. 461-465 and No. 21, 27 May 1922, pp. 514-517

Baumann, Alexander: 'Motorensysteme (Engine systems)', Lecture at the 'II. Hauptversammlung der Wiss. Gesellschaft für Flugtechnik (II. General assembly of the Scientific Soc. for Aeronautics)', June 1913, in ZFM Zeitschrift für Flugtechnik und Motorluftschiffahrt, 1913, pp. 309-310

Baxter, A.D. and Smith, C.W.R.: 'Contra-Flow Turbo-Compressor Tests', A.R.C. Technical Report R & M No. 2607, June 1942, His Majesty's Stationary Office London 1951, 43 p.

Baxter, A.D: 'Professional Aero Engineer, Novice Civil Servant', The Book Guild Ltd, Lewes Sussex, 1988, 188 p.

BBC Brown, Boveri & Cie.: 'Axialgeblaese für Veloxkessel' (Axial blower for Velox boiler), TF Test Programme and Report No. 963/I, 22 Dec. 1932 – 8 May 1933, 11 p.

BBC Brown, Boveri Cie.: 'Axialgebläse für Rolls-Royce, Bestell. 95840 V1 (Axial blower for Rolls-Royce, Order 95840 V1)', TFVL Test Programme and Report No. 1240, dated 12 Jan. 1938 – 8 May 1941, 13 p.

BBC Brown, Boveri & Cie.: 'Flugzeug-Aufladegruppe VTF 225 mit Schaufelkühlung nach Darrieus für 12 Zylinder-*Kestrel*, 4-Takt Vergasermotor der Firma Rolls-Royce in Derby (Aircraft turbo-supercharger VTF 225 for 12 cyl.-*Kestrel* 4stroke carburettor engine of RR Derby), TFVL Test Programme and Report No. 1204, dated 29 Nov. 1938 – 1 Dec. 1939, 17 p.

BBM Brown Boveri Mitteilungen: 'Rueckblick auf die Entwicklung der Brown Boveri-Konstruktionen im Jahre 1933', Vol. XXI, No. ½, Jan./Feb. 1935, pp. 25-39

BBN Brown Boveri Nachrichten: 'Schiffs-Gasturbine mit Strahltriebwerk als Treibgaserzeuger (Ship gas turbine with turbojet engine as aero jet expander)', (H. Reuter), Vol. XXXXVII, No.12, Dec. 1965, Mannheim, pp. 19-27

Beisel, Dieter: 'Bomben-Stimmung. Ein deutscher Flugzeug-Konstrukteur wollte beim Bau der Atombombe den Amerikanern den Rang ablaufen', ('German aircraft designer wanted to challenge the Americans building the atomic bomb'), Kultur & Technik, German Museum Munich, 4, 1990, pp. 11-14

Benecke, Theodor (ed.): 'History of German guided missiles development', AGARDograph 20 (Proceedings of the first guided missiles seminar, Munich, 23-27 April 1956), Appelhans Brunswick 1957, 419 p.

Benecke, Theodor, Hedwig, Karl-Heinz and Hermann, Joachim: 'Flugkörper und Lenkraketen (Guided Missiles)', Bernard & Graefe Koblenz, 1999, 377 p.

Bernhardt, Hannelore: 'Zu Biographie, zu Werk und Wirken von Gerhard Harig (1902-1966), (Biography, work and action of . . .)', Leibniz online, 2/2006, 23 p. http://leibnizsozietaet.de/wp-content/uploads/2012/11/04-HBernhardt.pdf

Betz, Albert: 'Das Maximum der theoretisch möglichen Ausnützung des Windes durch Windmotoren' ('The maximum of the theoretically possible exploitation of wind by means of a wind motor'), Zeitschrift für das gesamte Turbinenwesen (Journal for turbine technology), 1920, 26, pp. 307-309

Betz, Albert: 'Tragflügel und hydraulische Maschinen (Aerofoils and Hydraulic Machinery)', in Handbuch der Physik 1927, Vol. VII, 4, pp. 215-288

Betz, Albert: 'Axiallader', Jahrbuch 1938 der Deutschen Luftfahrtforschung, pp. II 183 – II 185, additionally secret report 090/008 of Lilienthal-Gesellschaft fuer Luftfahrtforschung Fachgruppe fuer Flugmotorenforschung, Session 28 Feb. and 1 March 1938, Berlin – with discussion, pp. 3-8, in German, and 'Axial Superchargers', NACA TM No. 1073, Washington 1944, 12 p. https://ntrs.nasa.gov/citations/19930094393

Betz, Albert: 'Zur Theorie der Leitapparate fuer Propeller', Ingenieur-Archiv, Vol. IX, No. 6, Dec. 1938, pp. 435-452, in German, and 'The Theory of Contra-Vanes Applied to the Propeller', NACA TM No. 909, Washington, Sep. 1939, 28 p. https://ntrs.nasa.gov/citations/19930094507

Betz, Albert: 'Ackeret in Göttingen' ('Ackeret at Goettingen'), ZAMP March 1958, Vol.9, Issue 5-6, pp. 34-36

Betz, Albert: 'Interference Between Propeller and Vehicle: The Ducted Propeller', pp. 215-217 (1966), in Einführung in die Theorie der Strömungsmaschinen' ('Introduction to the Theory of Flow Machines'), G. Braun Karlsruhe, 1959, 272 p. in German, and Pergamon Press (1966) and Elsevier (2014), 236 p. in English

Biezeno, Cornelis B. and Burgers, Johannes M.: 'Proceedings of the first International Congress for Applied Mechanics, Delft 1924', J. Waltman jr. 1925, 460 p.

B.I.O.S.: 'Report on German Development of Gas Turbines for Armoured Fighting Vehicles', British Intelligence Objectives Sub-Committee, Final Report No. 98, 1945, 12 p.

B.I.O.S.: 'Vorkauf Rotating Boiler (Drehkessel) and Rotating Boiler Gas Turbine (Drehkessel Turbine), British Intelligence Objectives Sub-Committee, Final Report No. 931, Item No. 29, 1947, 26 p.

B.I.O.S.: 'German Gas Turbine Developments during the period 1939-1945' by J.W. Adderley, British Intelligence Objectives Sub-Committee Overall Report No. 12, 1949, 47 p.

Birkenfeld, Wolfgang: 'Der synthetische Treibstoff 1933-1945', Musterschmidt Vlg. Goettingen, 1964, 280 p., Vol. 8 of 'Studien und Dokumente zur Geschichte des Zweiten Weltkrieges'

Bloor, David: 'The Enigma of the Aerofoil', The University of Chicago Press, Chicago-London, 2011, 547 p.

Bodemer, Alfred and Laugier, Robert: 'L'ATAR et tous les moteurs à réaction français (ATAR and all other French turbojet engines)', Editions J.D. Reber, 1996, 336 p.

Boening. Renate: 'Ich wundere mich nur, dass wir alle mitgemacht haben. Erinnerungen an mein Leben in der DDR' (I just wonder, why we all participated. Recollections to my life in GDR), Vlg. B. Budrich, 2016, 165 p.

Bolter, John R.: 'Sir Charles Parsons and electrical power generation – a turbine designer's perspective', 1994 Parsons Memorial Lecture, Proc. Inst. Mech. Engrs., Vol. 208, Part A : J. of Power and Energy, 1994, pp. 159-176

Borck, Cornelius: 'Das künstliche Auge (The artificial eye)', pp. 159-176, in: Orland, Barbara (ed.): 'Artifizielle Körper – Lebendige Technik (Artificial bodies – living technology)', Chronos, Zurich, 2005, 288 p.

Boyne, Walter J. and Lopez, Donald S. (Eds.): 'The Jet Age. Forty years of jet aviation', Smithsonian Books, Washington DC, 1979, 300 p.

Boyne, Walter J., ed.: 'Air Warfare, An International Encyclopedia', Vol. 1, ABC-CLIO, 2002, 771 p.

Brandner, Ferdinand: 'Die Propellerturbinen-Entwicklung in der Sowjetunion : Vortrag (Turboprop development in the Soviet Union : presentation)', Schweizerische Bauzeitung, Vol. 75, Issue 32, 1957, pp. 508-511, https://doi.org/10.5169/seals-63403 and pp. 520-524, https://doi.org/10.5169/seals-63405

Brandner, Ferdinand: 'Ein Leben zwischen den Fronten (A life between the lines)', Welsermühl, 1973, 380 p.

Braslow, Albert L.: 'A History of Suction-Type Laminar-Flow Control with Emphasis on Flight Research', NASA Monographs in Aerospace History, No. 13, 1999, 79 p. https://www.nasa.gov/centers/dryden/pdf/88792main_Laminar.pdf

Bridel, Georges: 'Schweizerische Strahlflugzeuge und Strahltriebwerke (Swiss turbojet aircrafts and engines), Verlag Verkehrshaus der Schweiz Luzern, Special print No. 2, 1975, 103 p.

Brockett, Walter and Koschier, Angelo: 'LV100 AIPS Technology – for Future Army Propulsion', ASME 1992 Turbo Expo, Cologne, 1-4 June 1992, ASME paper 92-GT-391, 9 p., https://doi.org/10.1115/92-GT-391

Brown, Eric: 'Wings on My Sleeve: The World's Greatest Test Pilot tells his story', Hachette UK, 2008, 304 p. (Google books)

Bruckmann, Bruno W.: 'Unexpected – an aviation engineer's views', own publication, Phoenix, Az, 1986, 160 p.

Buckingham, Edgar: 'Jet propulsion for airplanes', NACA-TR-159, 1924, 18 p., https://ntrs.nasa.gov/search.jsp?R=19930091225

Budrass, Lutz: 'Flugzeugindustrie und Luftruestung in Deutschland 1918-1945' (Aviation industry and air armament in Germany 1918-1945), Droste, Duesseldorf, 1998, 976 p.

Budrass, Lutz: 'Hans Joachim Pabst von Ohain. Neue Erkenntnisse zu seiner Rolle in der nationalsozialistischen Rüstung' ('... New findings on his role within the Nazi arms industry'). In: Friedrich-Ebert-Stiftung, Landesbüro Mecklenburg-Vorpommern (Publ.): Technikgeschichte kontrovers. Zur Geschichte des Fliegens und des Flugzeugbaus in Mecklenburg-Vorpommern (= Beiträge zur Geschichte Mecklenburg-Vorpommern, Band 13), Schwerin 2007, pp. 52–69

Budrass, Lutz: 'Ideology and Business Strategy: Assessing Nazi Germany's Different Approaches to the Supply of Light Metals for the Luftwaffe', pp. 37-61 in Hans Otto Frøland et al. (ed.): 'Industrial Collaboration in Nazi-Occupied Europe. Norway in Context', Palgrave Macmillan, London, 2016, 465 p.

Budrass, Lutz: 'Die Mobilisierung von Forschung und Entwicklung in der deutschen Luftfahrtindustrie 1933-1945 (The mobilisation of research and development in the German aero industry 1933-1945)', pp. 295-325, in Flachowsky, Sören et al. (eds.): 'Ressourcenmobilisierung. Wissenschaftspolitik und Forschungspraxis im NS-Herrschaftssystem', Wallstein, Göttingen 2017, 632 p.

Budrass, Lutz: 'Review. *"The hun is not always ahead of us in secret weapons"* Some remarks on a new book on the history of the turbojet: Hermione Giffard, Making Jet Engines in World War II. Britain, Germany and the United States, Chicago, London 2016' in Technikgeschichte, Vol. 85, 2018, pp. 173-187

Buechi, Alfred: 'Ueber Verbrennungskraftmaschinen (On combustion engines)', Zeitschrift fuer das gesamte Turbinenwesen (Journal of complete turbomachinery), 1909, p. 313

Buechi, Alfred: 'Geschichtliches ueber den Ursprung der Idee, einige grundlegende Patente und die ersten kommerziellen Anwendungen der Buechi-Abgasturboaufladung an Brennkraftmaschinen (Historical origin, a few fundamental patents and the first commercial applications of Buechi exhaust gas turbocharging on combustion engines)', MTZ Motortechnische Zeitschrift, Vol. 18, No. 6, June 1957, pp. 171-175

Butler, George Frank: 'A Text-Book of Materia Medica, Therapeutics and Pharmacology', Palala Press, New York, NY, 2015, 696 p.

Cantor, Brian: 'The Equation of Materials', Oxford University Press, 2020, 288 p.

Carlier, Claude: 'Les débuts de la coopération aéronautique franco-allemande: Le 'Groupe O' 1945-1960 (The origins of French-German aeronautical cooperation..)', in Wilkens, Andreas (ed.) : 'Die Deutsch-Französischen Wirtschaftsbeziehungen 1945-1960 (The German-French economic relations ..)', Jan Thorbecke Vlg., 1997, 360 p. https://perspectivia.net/servlets/MCRFileNodeServlet/ploneimport_derivate_00009846/carlier_debuts.pdf

Caspary, Adolf: 'Wirtschaftsstrategie und Kriegsführung' (Strategy of Economics and Warfare), E.S. Mittler & Sohn, Berlin 1932, 166 p.

Catrina, Werner: 'BBC Glanz-Krise-Fusion, 1891-1991 Von Brown Boveri to ABB' (BBC Splendour-Crisis-Merger, 1891-1991 From Brown Boveri to ABB), Orell Fuessli Zurich, 2nd ed. 1991, 320 p.

Chouat-Hugonnet, Nicole: 'Le Comité d'entreprise de la SNECMA sous la IVe République ou l'hégémonie communiste à l'épreuve (SNECMA's workers council under the IVth Republic or communiste hegemony on test)', pp. 279-290, in Girault, Jacques: 'Des Communistes en France (Communists in France)', Histoire de la France XIXe-XXe, Publications de la Sorbonne, 525 p. https://books.openedition.org/psorbonne/60342?lang=de

Christiani, Karl: 'Experimentelle Untersuchung eines Tragflügelprofils bei Gitteranordnung' ('Experimental Investigation of an Aerofoil Cascade'), Luftfahrtforschung, Vol.2, Heft 4, 27 Aug. 1928, pp. 91-110, and doctoral thesis at Goettingen University with same title, Mitteilungen aus der Aerodynamischen Versuchsanstalt in der "Luftfahrtforschung", Goettingen, Vol. 2, No. 4, 1926, 91 p.

Christopher, John: 'The Race for Hitler's X-Planes: Britain's 1945 Mission To Capture Secret Luftwaffe Technology', The History Press, Oct. 2012, 240 p.

Churchill, Sir Winston: 'Triumph and Tragedy (The Second World War)', Houghton Mifflin (Trade), Boston, 1953, 800 p.

C.I.O.S.: 'Description of Junkers .004 (203) Jet Propulsion Engines', Combined Intelligence Objectives Sub-Committee, London, File No. XI-6, XII-9, and XIV-4, 1944-1946. 33 p. http://www.cdvandt.org/CIOS-XI-6-XII-9-XIV-4.pdf#page=30&zoom=auto,-155,577

C.I.O.S.: 'Gas Turbine Development by B.M.W.', Combined Intelligence Objectives Sub-Committee, London, File No. XXVI-30, Item No. 5, (F/ Lt. P.R. Price, R.A.F., M.A.P.), 25 April – 30 May 1945a, 37 p. https://www.cdvandt.org/CIOS-XXVI-30.pdf

C.I.O.S.: 'Gas Turbine and Wind Tunnel Activity Brown Boveri(e) Cie', Combined Intelligence Objectives Sub-Committee, London, File No. XXVII-22, Item No. 5, (Interview with Directorate reported by Lieut. S.P. Robinson USNR, F/ Lt. R.S. Sproule RCAF), June 1945b, 6 p., from Deutsches Museum Archive

C.I.O.S.: 'Interrogation of Dipl. Ing. Helmut Schelp', Combined Intelligence Objectives Sub-Committee, London, File No. XXXII-46, Item No. 5 & 26, (Reported by Lieut. S.P. Robinson USNR), August 1945c, 13 p., from Niedersächs. Landesarchiv Hannover

C.I.O.S.: 'Gas Turbine Development B.M.W., Junkers, Daimler Benz', Combined Intelligence Objectives Sub-Committee, London, File No. XXIV-6, Item No. 5 & 26, (S/ Ldr. P. Lloyd, M. A.P.), ~1946, 33 p., from ETHZ Library

Cole, Hugh M.: 'The Ardennes: Battle of the Bulge', St. John's Press, 2016, 750 p. https://history.army.mil/books/wwii/7-8/7-8_Cont.htm#toc

Cole, Lance: 'Secrets of the Spitfire: The Story of Beverly Shenstone, the Man Who Perfected the Elliptical Wing', Casemate Publ., 2012, 272p.

Cole, Lance: 'Secret Wings of World War II. Nazi Technology and the Allied Arms Race', Pen & Sword Aviation, Barnsley UK, 2015, 256 p.

Conner, Margaret: 'Hans von Ohain: Elegance in Flight', AIAA American Institute for Aeronautics & Astronautics, 2002, 300 p.

Constant, Edward W., II: 'The Origins of the Turbojet Revolution', Johns Hopkins University Press, Baltimore, MD, 1980, 328 p.

Constant, Hayne: 'The Internal Combustion Turbine as a Powerplant for Aircraft', RAE Note E 3546, March 1937 (unpublished)

Constant, Hayne: 'Influence of the R.A.E. on the early history of the Gas Turbine', RAE Report, Eng/2038.R/HC/21, dated 10 Sept. 1942 (unpublished), from MOSI Museum of Science and Industry (Manchester),

Constant, Hayne: 'The Early History of the Axial Type of Gas Turbine Engine', Proc. IMechE, Vol. 153, Issue 1, 1 June 1945, pp. 411-426

Constant, Hayne: 'Gas Turbines and Their Problems', Todd Publ. Group, London, 1948, 158 p.

Constant, Hayne: 'Pyestock's Contribution to Propulsion', The Journal of the Royal Aeronautical Society, Vol. 62, Issue 568, 1958, pp. 257-267

Corum, James S.: 'The Other Richthofen', World War II Magazine, Vol. 23, No. 3, Aug./ Sept. 2008, https://www.historynet.com/the-other-richthofen.htm

Darrieus, Georges: 'Contribution au tracé des aubes radiales des turbines' [Contribution to the determination of the profile of radial turbine blades (but axial flow)], *Festschrift* to Prof. Dr. A. Stodola's 70th birthday, Orell Fuessli Zurich, 1929, pp. 92-95

Davidson, Ivor M.: 'On the Development in Germany of The Supersonic Axial Flow Compressor', NGTE Memorandum No. M.16, April 1947a, 13 p.

Davidson, Ivor M.: 'Some Data Pertaining to the Supersonic Axial-flow Compressor', A.R.C. Technical Report R. & M. No. 2554, May 1947b

Dean, Robert C. and Senoo, Yasutoshi: 'Rotating Wakes in Vaneless Diffusers', Trans. ASME, J. of Basic Engineering, Sep. 1960, pp. 563-574

Decher, Siegfried: 'Die Entwicklung des Triebwerkes Jumo 004 bei den Junkers Flugzeug- und Motorenwerken in Dessau (Development of Jumo 004 at Junkers Dessau)', Flugwelt III; Issue 2, 1951, pp. 41-46

Deist, Wilhelm: 'Die Reichswehr und der Krieg der Zukunft' (The Reichswehr and the Future War), MGM Militärgeschichtliche Mitteilungen, Oldenbourg, 1, 1989, pp. 81-92

Denton, John: 'The Evolution of Turbomachinery Design (Methods)', PCA's 20th anniversary lecture 21 May 2009, 46 p. https://www.pcaeng.co.uk/library/PCA_20th_JDDenton.pdf

Destival, Pierre: 'SOCEMA Aircraft Turbines. The Record of a French Development Programme', Flight Magazine, 18 Nov. 1948, p. 608

Destival, Pierre: 'French Turbo-Propeller and Turbo-Reaction Engines', The Journal of the Royal Aeronautical Society, Vol. 53, Issue 458, Feb. 1949, pp. 111-136

Deutsche Luftwacht, Luftwissen: 'Hauptversammlung 1938 der Lilienthal-Gesellschaft für Luftfahrtforschung (General assembly 1938 of the Lilienthal-Society for aeronautical research)', Mittler & Sohn, Vol. 5, No. 11, 1938, pp. 389-424

DGLR Publ.: 'Herbert Wagner. Dokumentation zu Leben und Werk' ('Documentation of his Life and Works'), DGLR Deutsche Gesellschaft fuer Luft- und Raumfahrt, Bonn 1986, 159 p.

DGLR Publ.: '50 Jahre Turbostrahlflug (50 years of turbojet flight)', DGLR Report 89-05, Bonn 1989, 430 p.

Douglas, Calum E.: 'The Secret Horsepower Race: Western Front Fighter Engine Development', Harper Tempest, 2020, 480 p.

Duffy, James: 'Target: America: Hitler's Plan to Attack the United States', Praeger, 2004, 192 p., pdf

Dunham, John: 'A.R. Howell – The Father of The British Axial Compressor', ASME paper 2000-GT-0008, 9 p.

Ebert, Hans J.; Kaiser, Johann B. and Peters, Klaus: 'Willy Messerschmitt – Pionier der Luftfahrt und des Leichtbaus (Pioneer of aeronautics and light weight constructions)', Bernard & Graefe, Bonn, 1992, 430 p.

Ebert, Hans J.: 'Pabst von Ohain, Hans', in Neue Deutsche Biographie, 19, 1999, pp. 742-743 (online version)

Eck, Bruno: 'Ventilatoren (Ventilators)', Springer Berlin 1937, 197 p. (1st ed.), - 1952, 304 p. (2nd ed.) and - 1957, 493 p. (3rd ed.)

Eckardt, Dietrich: 'Detailed Flow Investigations Within a High-Speed Centrifugal Compressor Impeller', ASME J. of Fluids Engg., 98, 1976, pp. 390-402

Eckardt, Dietrich: 'Untersuchung der Strahl-/Totwasser-Strömung hinter einem hochbelasteten Radialverdichterlaufrad', DFVLR Cologne, doctoral thesis RWTH Aachen, DLR FB 77-32, 18 July 1977, 227 p. (in German); 'Investigation of the Jet-Wake Flow of a Highly-Loaded Centrifugal Compressor Impeller', NASA TM-75232, Jan. 1978, 195 p. (in English) https://ntrs.nasa.gov/citations/19780008108

Eckardt, Dietrich: 'Future Engine Design Trade-Offs', 10th ISABE Int'l. Symposium on Air Breathing Engines, Nottingham UK, 1-6 Sep. 1991, ISABE invited paper, 10 p.

Eckardt, Dietrich and Brines, Gerald: 'Technology Readiness for Advanced Ducted Engines', Moscow Aero & Industry Engine 1992, Propulsion Seminar, 8 p.

Eckardt, Dietrich and Rufli, Peter: 'Advanced Gas Turbine Technology - ABB/BBC Historical Firsts', ASME J. Eng. Gas Turb. Power, 124, 2002, pp. 542-549

Eckardt, Dietrich: 'The 1x1 m hypersonic wind tunnel Kochel/ Tullahoma 1940-1960', CEAS Space Journal, 7, 2005, pp. 23-36, and in German: https://www.dglr.de/publikationen/2015/340001.pdf

Eckardt, Dietrich: 'Gas Turbine Powerhouse. The Development of the Power Generation Gas Turbine at BBC – ABB – Alstom', De Gruyter Oldenbourg Munich, 2nd ed. 2014, 500 p.

Eckert, Bruno: 'Das Kuehlgeblaese des Kraftfahrzeugs und sein betriebliches Verhalten (The auto-motive cooling blower and its operational behaviour)', Dr.-Ing. thesis, TH Stuttgart, 1940, DKF Deutsche Kraftfahr-Forschung No. 51, 1941; and VDI publ., Berlin.

Eckert, Bruno: 'Kuehlgeblaese fuer Verbrennungsmotoren (Cooling blowers for combustion engines)', MTZ Motortechnische Zeitschrift, special print, 1940, 12 p.

Eckert, Bruno and Weinig, Fritz: 'Berechnung eines siebenstufigen Axialladers (Design computation of a seven stage axial supercharger)', FKFS Report 395 (BMW), 3 April 1941, 17 p.

Eckert, Bruno and Kobel, Willy: 'Berechnung eines sechsstufigen Axialladers (Design computation of a six stage axial supercharger)', FKFS Report 399 (BMW), 18 July 1941, 10 p. and Captured German Aeronautical Documents (Microfilm) CGD 276, Part C, Smithsonian National Air and Space Museum, 2001

Eckert, Bruno: 'Kraftfahrzeug-Kuehlgeblaese (Automotive cooling blowers)', in Neue Kraftfahrer Zeitung (New driver newspaper), 1942-1943, 26 p.

Eckert, Bruno: 'Series of Articles on Compressor and Fan Design Written by German Engineers: The Influence of physical dimensions ..and flow conditions ..on compressor characteristics', US Navy Dept. of Ships, 1946, 146 p. (Google books)

Eckert, Bruno and Schnell, Erwin: 'Axialkompressoren und Radialkompressoren. Anwendung/Theorie/Berechnung' (Axial and Radial Compressors. Application/Theory/Calculation), Springer Berlin, 2nd ed., 1961, 528 p.

Eckert, Bruno: 'Erinnerungen (Memories)', auto-biography, published by the author, Stuttgart, 1982, University Archive Stuttgart UASt 72/18, 244 p.

Eckert, Michael: 'Strategic Internationalism and the Transfer of Technical Knowledge: The United States, Germany and Aerodynamics after World War I', Technology and Culture, Vol. 46, No. 1, January 2005, pp. 104-131

Eckert, Michael: 'The Dawn of Fluid Dynamics: A Discipline Between Science and Technology', John Wiley & Sons, 2006, 296 p.

Eckert, Michael: 'Ludwig Prandtl: A Life for Fluid Mechanics and Aeronautical Research', Springer, 2019, 368 p.

Edgerton, David: 'Britain's War Machine: Weapons, Resources and Experts in the Second World War', Penguin UK, 2011, 456 p.

Edgerton, David: 'England and the Aeroplane: Militarism, Modernity and Machines', Penguin UK, 2013, 288 p.

Eggers, Karsten: 'Professor Gerhard Eggers (1912-1998)', pp. 285-302, in Pophanken, Hartmut; Schalipp, Klaus and Kuckuk, Peter: 'Ein Jahrhundert Luft- und Raumfahrt in Bremen: Von den frühesten Flugversuchen zum Airbus und zur Ariane. Sonderausgabe (A century of aerospace at Bremen: From early flight tests to Airbus and Ariane. Special print)', BoD – Books on Demand, 2018, 368 p.

Eiffel, A. Gustave: 'La résistance de l'air et l'aviation' (Air resistance and aviation), Imprimerie de la Cour d'Appel, Paris, 1911 and 'Les nouvelles recherches expérimentales sur la résistance de l'air et l'aviation faites au Laboratoire d'Auteul (New experimental investigations on air resistance and aviation at the Auteul Lab), H.Dunod and E.Pinat, Paris, 1914, 2 vol.

Encke, Walter: 'Untersuchungen an Modellraedern von Axialgeblaesen', Zentrale fuer Wiss. Berichtswesen (ZWB), UM-Mr. 3135, 1940, pp. 1-28 and 'Investigations on Experimental Impellers for Axial Blowers', NACA TM No.1123, April 1947, 28 p https://ntrs.nasa.gov/api/citations/20030064213/downloads/20030064213.pdf

Energy Transfer: 'Marcus Hook Industrial Complex', leaflet, 2019, 2 p. https://cms.energytransfer.com/wp-content/uploads/2020/04/Marcus_Hook_FactsSheet_Updated_2019.pdf

Ermenc, Joseph J.: 'Interviews with German Contributors to Aviation History', Meckler Corp. 1990, 185 p.

Eyermann, Wilhelm and Schulz, Bruno: 'Die Gasturbinen, ihre geschichtliche Entwicklung, Theorie und Bauart', ('The gas turbines, historical development, theory and design'), M. Krayn Berlin 1920, 2nd ed., 310 p.

Eyre, Donald: '50 Years with Rolls-Royce: My Reminiscences', Rolls-Royce Heritage Trust, 2005, 152 p.

Fedden, A. H. Roy: 'The Development of the Mono-Sleeve Valve for Aero Engines (Die Entwicklung der Einschiebersteuerung bei Flugmotoren)', special print, Lilienthal Society, 1938, 29 p. https://lilienthal-museum.museumnet.eu/sites/lilienthal-museum.museumnet.eu/files/archivalie/digitalisatepublic/hauptvers1938.pdf

Fedden, A. H. Roy: 'Next Decade's Aero Engines Will Be Advanced But Not Radical', SAE Technical Paper 330049, 1933, 25 p.

Ferris, John Robert: 'Intelligence and Strategy. Selected Essays', Routledge, London and New York, 2007, 408 p.

Flachowsky, Soeren: '"Das größte Geheimnis der deutschen Technik" Die Entwicklung des Stratosphärenflugzeugs Ju 49 im Spannungsfeld von Wissenschaft, Industrie und Militär' (The Biggest Secret of German Technology), Dresdener Beiträge zur Geschichte der Technikwissenschaften, No. 32, 2008a, pp. 3-32 http://tud.qucosa.de/api/qucosa%3A25368/attachment/ATT-0/

Flachowsky, Sören: 'Von der Notgemeinschaft zum Reichsforschungsrat. Wissenschaftspolitik im Kontext von Autarkie, Aufrüstung und Krieg (Science policy in reference to autarchy, armament and war)', Steiner, Stuttgart, 2008b, 545 p. – in English

Forsgren, Jan: 'Messerschmitt Bf 109: The Design and Operational History', Fonthill Media, 2017, 272 p.

Forsyth, Robert: 'Luftwaffe Mistel Composite Bombers', Bloomsbury Publ., 2015, 96 p.

Forsyth, Robert: 'Messerschmitt Me 264 Amerika Bomber', Bloomsbury Publ., 2016a, 80 p.

Forsyth, Robert: 'He 162 Volksjäger Units', Osprey Publ., 2016b, 96 p.

Franz, Anselm: 'Der Weg zum ersten Großserien-Strahltriebwerk Junkers Jumo 004 und spätere Entwicklungen in den USA (The way to the first mass-produced turbojet engine Junkers Jumo 004 and later US developments)', in '50 Jahre Turbostrahlflug', DGLR Munich, 26/27 Oct. 1989, DGLR Report 89-05, paper 89-218, 30 p.

Freund, Gerald: 'Unholy Alliance. Russian-German Relations From the Treaty of Brest-Litovsk to the Treaty of Berlin', Harcourt, Brace and Co., New York, 1957, 283 p.

Friedrich, Helmut: 'Erinnerungen von Rudolf Friedrich 1909-1998 (Recollections)', Self-publication, Baden-Baden, 2009, 129 p.

Friedrich, Rudolf: 'Vom Wagner/ Mueller RT0 Versuchsgeraet zum Heinkel-Strahltriebwerk He S30' (From the Wagner/ Mueller RT0 test engine to the Heinkel turbojet engine He S30), in '50 Jahre Turbostrahlflug', DGLR Munich 26/27 Oct 1989, DGLR Report 89-05, paper 89-217, 14 p.

Friedrich, Rudolf: 'Dokumente zur Erfindung der heutigen Gasturbine - vor 118 Jahren', (Documents to the invention of the modern gas turbine – 118 years ago'), VGB Essen Germany, VGB-B 100, 1991, 99 p.

Froude, Robert E.: 'Description of a Method of Investigation of Screw-Propeller Efficiency', Transactions of the Institution of Naval Architects, Vol. 24, 1883, p. 231

Geidel, Helmut-Arndt and Eckardt, Dietrich: 'Gearless CRISP – the Logical Step to Economic Engines of High Thrust', ISABE paper 89-7116, Proc. of the ISABE 9th Int'l. Symposium on Air Breathing Engines, Athens, 4-9 Sep. 1989, pp. 1088-1098

Génie civil, Le: 'Les perspectives de la propulsion à vapeur des avions géants (Outlook on steam propulsion for giant aircrafts)', Paris, 31 Dec. 1938, v. 113, no. 27, pp. 571-572

Gerber, Alfred: 'Untersuchungen über Grenzschichtabsaugung (Investigations of boundary layer suction)', Doctoral thesis, ETH Zurich Library, Research Collection, 1938, 72 p.

Gersdorff, Kyrill von, Schubert, Helmut and Ebert, Stefan: 'Flugmotoren und Strahltriebwerke (Flight Motors and Jet Engines)', 4th ed., Bernard & Graefe, Bonn, 2007, 558 p.

Giffard, Hermione: 'Engines of Desperation: Jet Engines, Production and New Weapons in the Third Reich', Journal of Contemporary History, 48, 4, 2013, pp. 821-844

Giffard, Hermione: 'Making Jet Engines in World War II: Britain, Germany, and the United States', University of Chicago Press, 2016, 319 p.

Giffard, Hermione: 'Engine of innovation: the Royal Aircraft Establishment, state design and the coming of the gas turbine aero-engine in Britain', Contemporary British History, Vol. 34, Issue 2, pp. 165-178, online 17 July 2019

Gilles, August; Hopf, Ludwig and von Kármán, Theodore: 'Vorträge aus dem Gebiete der Aerodynamik und verwandter Gebiete (Lectures on aerodynamics and related areas)', Conference proceedings Aachen, 26-29 June 1929, Springer 1930, 221 p.

Gimbel, John: 'Science, Technology and Reparations. Exploitation and Plunder in Postwar Germany', Stanford University Press, 1990, 304 p.

Goepfert, Rainer: 'Mikojan/ Gurewitsch I-250/Mig 13', Flieger Revue 05/2021, pp. 50-55

Gohlke, Gerhard: 'Heizluftstrahltriebwerke (Thermal-Air Jet Engines)', Flugsport, Vol. 31, No.1, 4 Jan. 1939, pp. 1-5; No.2, 18 Jan.1939, pp. 31-37; No.3, 1 Feb.1939, pp. 70-75; No.4, 15 Feb.1939, pp. 100-104 in German, and 'Thermal-Air Jet-Propulsion', Aircraft Engineering and Aerospace Technology, Vol. 14, Issue 2, 1 Feb. 1942, pp. 32-39, in English.

Golley, John, in assoc. with Sir F. Whittle: 'Whittle – The True Story', Airlife London 1987, 272 p.

Golley, John: 'Genesis of the Jet. Frank Whittle and the Invention of the Jet Engine', Airlife Publishing Ltd. UK, 1996, 272 p. version

Green, John J. et al.: 'Wartime Aeronautical Research & Development in Germany', The Engineering J., Oct. 1948, pp. 531-538 and 545; Nov. 1948, pp. 584-589; Jan. 1949, pp. 19-25

Green, William: 'Warplanes of the Third Reich', Macdonald, London, 2nd ed., 1972, 672 p.

Grieb, Hubert and Eckardt, Dietrich: 'Turbofan and Propfan as Basis for Future Economic Propulsion Concepts', AIAA 22nd Joint Propulsion Conference, Huntsville Al., 16-18 June 1986a, 12 p.

Grieb, Hubert and Eckardt, Dietrich: 'Turbofan and Propfan – Antagonism or Synthesis', 15th ICAS Congress, London, 7-12 Sep. 1986b, Proceedings ICAS-86-3.8.2, pp. 1099-1110

Griehl, Manfred: 'Luftwaffe Over America. The Secret Plans to Bomb the United States in World War II', Pen and Sword, Barnsley UK, 2016, 256 p.

Griffith, Alan A. and Taylor, Geoffrey I.: 'The Use of Soap Films in Solving Torsion Problems', Proceedings of the Institution of Mechanical Engineers, 93, 1, 1917, pp. 755-809

Griffith, Alan A.: 'The Phenomena of Rupture and Flow in Solids', Phil. Transactions of the Royal Society A, Vol. 221, Issue 582-593, 1921, pp. 163-198

Griffith, Alan A.: 'An Aerodynamic Theory of Turbine Design', RAE Report H.1111, 7 July 1926, 17 p., 4 Figs. with Appendix H.1111 'Application to the Gas Turbine Problem', 5 p. and 'Proposed Test Rig for Verifying the New Turbine Theory', Supplement to H.1111, 13 Sept. 1926, 3 p., 1 Fig. [Unpublished]

Griffith, Alan A.: 'The Present Position of the Internal Combustion Turbine as a Power Plant for Aircraft', Air Ministry Laboratory (AML) Report 1050A, Nov. 1929 [Unpublished]

Griffiths, Richard: 'Fellow Travellers of the Right: British Enthusiasts for Nazi Germany, 1933-1939', Faber & Faber, 2015, 414 p.

Gunston, Bill: 'Fedden – the life of Sir Roy Fedden', Rolls-Royce Heritage Trust, Historical Series No. 26, 1999, 352 p.

Hafer, A.A.: 'Gas-Turbine Progress Report – Materials, Cooling and Fuels', Trans. ASME, 75, 1953, pp. 127-136

Hagen, Hermann: 'Zur Geschichte des Strahltriebwerkes. Die Entwicklung bei BMW in den Jahren 1938 bis 1945' (On the history of the turbojet engine – BMW developments between 1938-1945),

DGLR paper 81-075 (MTU internal print), presented at DGLR Annual Convention, Aachen, 11-14 May 1981

Hager, Willy: 'Hydraulicians in Europe 1800-2000', CRC Press, 2014, Vol. 2, 994 p.

Hallett, George E.A.: 'Airplane Motors. A course of practical instruction in their care and overhauling, for the use of military aviators', Gibson Bros. Inc., Washington 1917, 79 p.

Hallett, George E.A.: 'Superchargers and Super-Charging Engines', SAE Transactions, Vol. 15, Part I (1920), pp. 218-231

Hamel, Peter G.: 'Birth of Sweepback: Related Research at Luftfahrtforschungsanstalt – Germany', AIAA J. of Aircraft, Vol. 42, No. 4, July-Aug. 2005, pp. 801-813

Harris, R.G. and Fairthorne, R.A.: 'Wind Tunnel Experiments With Infinite Cascades of Aerofoils', RAE report no. B.A.763, Sept. 1928, 40 p.

Harrison, Mark: 'The Political Economy of a Soviet Military R&D Failure: Steam Power for Aviation 1932-1939', The Journal of Economic History, 63,1, March 2003, pp. 178-212

Hartlepp, Heinz: 'Erinnerungen an Samara – Deutsche Luftfahrtspezialisten von Junkers, BMW und Askania in der Sowjetunion von 1946 bis 1954 und die Zeit danach (Samara memories – German aviation specialists from Junkers, BMW and Askania in the Soviet Union 1946-1954 and thereafter)', Aviatic Vlg. 2005, 176 p.

Hawthorne, Sir William: 'The Early History of the Aircraft Gas Turbine in Britain', revised version in '50 Years of Jet-Powered Flight', DGLR Report 92-05, Vol. II, pp. 48-98 and Notes and Records of the Royal Society, London, 1991, Vol. 45, pp. 79–108

Heinkel, Ernst, ed. Thorwald, Juergen: 'Stormy Life: Memoirs of a Pioneer of the Air Age', Dutton, New York, 1956, 256 p., and in German: 'Stürmisches Leben', Mundus, Stuttgart, 1953, 1st ed., 561 p.

Heinzerling, Werner: 'Flügelpfeilung und Flächenregel, zwei grundlegende deutsche Patente der Flugzeugaerodynamik (Wing sweep and area ruling, two fundamental German patents on aircraft aerodynamics)', Arbeitskreis Luftverkehr der TU Darmstadt, Kolloquium Luftverkehr, Darmstadt, 2002, pp. 1–49

Henschke, Ulrich K. and Mauch, Hans A.: 'How man controls', in: The Surgeon General US Air Force (ed.): German Air Force Medicine World War II, Washington 1950, Vol. I, pp. 83-91

Heppner, Christopher: 'Reading Blake's Designs', Cambridge University Press, 2009, 320 p.

Hilburn, Earl D.: 'Steam Will Power Tomorrow's Planes!', Modern Mechanix and Innovations, August 1932, pp. 112, 172-173

Hirschel, Ernst H., Prem, Horst and Madelung, Gero: 'Aeronautical Research in Germany. From Lilienthal until Today', Springer Berlin Heidelberg, 2004, 694 p.

Hirschel, Ernst H., Rizzi, Arthur, Breitsamter, Christian and Staudacher, Werner: 'Separated and Vortical Flow in Aircraft Wing Aerodynamics', Springer, 2021, ~ 320 p.

Hodgson, Richard: 'Armstrong Siddeley and Early Gas Turbines', Text of a talk at Rolls-Royce Heritage Trust, Bristol, Feb.1998 (also RRHT Coventry, Oct.1993; RRHT Derby, Jan.1996 and Pratt & Whitney, E.Hartford, Ct., Aug.1994, 26 p., (unpublished)

Hoff, Wilhelm, von Dewitz, Ottfried and Madelung, Georg: 'Jahrbuch 1928 der deutschen Versuchsanstalt fuer Luftfahrt e.V., Berlin-Adlershof' (Yearbook 1928 of the German test institute for aviation, reg. soc., Berlin-Adlershof), reprint, De Gruyter, Berlin 2019, 287 p.

Holley, Irving B.: 'Ideas and Weapons', Yale University Press, New York 1953, 222 p.

Horlock, John H.: 'Axial Flow Compressors. Fluid Mechanics and Thermodynamics', Butterworths Scientific Publications, London 1958, 187 p.

Howell, A. Raymond: 'The Present Basis of Axial Flow Compressor Design. Part 1 Cascade Theory and Performance', H.M. Stationary Office, June 1942, ARC R&M No. 2095 (RAE Report No. E 3946), and 'Part 2 Compressor Theory and Performance', Dec. 1942, RAE Report No. E 3961

Howell, A. Raymond: 'Design of Axial Compressors', Proc. IMechE, 1st June 1945, 153, pp. 452-462

Howell, A. Raymond: 'Griffith's Early Ideas on Turbomachinery Aerodynamics', Aeronautical Journal, Dec. 1976, pp. 521-529

Hryniszak, Waldemar: 'Heat Exchangers. Applications to Gas Turbines', Butterworths Scientific Publ., London, and Academic Press Inc., New York, 1958, 355 p.

Huebner, Walter: 'Der Deutschlandflug 1934, ein Wettbewerb um die beste Gemeinschaftsleistung' (Flight over Germany 1934, a competition about the best team performance), Luftwelt, No.12, 1934, 229 p.

Huemer, Fabian: 'Der "Volksjäger" Heinkel He 162. Forcierte Ressourcenmobilisierung im Angesicht der Niederlage', Master thesis, University Vienna, 2013, 172 p.

Huettner, Fritz: 'Die Huettner-Turbine (The Huettner Turbine)', ETZ Elektrotechnische Zeitschrift, Issue A, 1934, pp. 742-744

Huston, John W. (ed.): 'American Air Power Comes of Age, General Henry H. "Hap" Arnold's World War II Diaries', Vol. I, Air University Press, Maxwell Airforce Base, AL, Jan. 2002, 569 p., internet pdf.

Irving, David: 'The Rise and Fall of the Luftwaffe. The Life of Field Marshal Erhard Milch', e-book Parforce UK Ltd., 2002, identical with 1973 first ed. print version, 471 p.

Jantzen, Eilhard and Maier, Knut: 'Betriebsstoffe in der deutschen Luftfahrt', Bernard & Graefe, 2016, 488 p.

Jenny, Ernst: 'The BBC Turbocharger, a Swiss Success Story', Birkhaeuser Basel-Boston-Berlin 1993, 300 p.

Johnson, Dag and Mowill, R. Jan: 'Ægidius Elling – a Norwegian gas-turbine pioneer', Norw. Tech. Museum Oslo, March 1968, presented at the 1968 ASME GT Products Show, Wash. DC and VGB Kraftwetkstechnik (in German), Vol. 52, No. 2, 1972

Jones, Glyn: 'The Jet Pioneers: The Birth of Jet-Powered Flight', Methuen London, 2009, 228 p.

Jost, Wilhelm: 'The First 45 Years of Physical Chemistry in Germany', Annual Review of Physical Chemistry, Vol. 17, 1966, pp. 1-16

Joule, James P.: 'On the Air-Engine', read 19 June 1851, pp. 331-356, in Philosophical Trans. 1852, Part I, p. 65, in 'The scientific papers of James Prescott Joule', Physical Soc. (Great Britain), 1884, Vol. 1, 702 p., https://archive.org/stream/scientificpapers01joul#page/n3/mode/2up

Kalkmann, Ulrich: 'Die Technische Hochschule Aachen im Dritten Reich (1933-1945), (The technical high-school Aachen in the Third Reich)', Vlg. Mainz, 2003, 602 p.

Kantorowicz, Otto: 'Zur Leitfähigkeit gepresster Metallpulver (On the electrical conductivity of pressed metal powders)', J.A. Barth, Leipzig and Ann. Physik, Vol. 12, 1932, 51 p.

Kármán, Theodore von: 'Aerodynamics. Selected Topics in the Light of their Historical Development', Cornell University Press, Ithaca, NY, 1954, 203 p.

Kay, Antony L.: 'German Jet Engine and Gas Turbine Development 1930-1945', Airlife Publishing Ltd. UK, 2002, 296 p.

Kay, Antony L.: 'Turbojet. History and Development 1930-1960, Vol. 1 Great Britain and Germany', The Crowood Press Ltd. UK, 2007a, 272 p.

Kay, Antony L.: 'Turbojet. History and Development 1930-1960, Vol. 2 USSR, USA, Japan, France, Canada, Sweden, Switzerland, Italy, Czechoslovakia and Hungary', The Crowood Press Ltd. UK, 2007b, 269 p.

Keller, Curt: 'Axialgeblaese vom Standpunkt der Tragfluegeltheorie', (Axial fans in view of aerofoil theory), Dr. sc. techn. dissertation at ETH Zurich, Leemann Zurich 1934, 190 p. https://www.research-collection.ethz.ch/handle/20.500.11850/138814 – and Keller, Curt: 'The Theory and Performance of Axial-Flow Fans: Adapted for the Use of Fan Designers by Lionel S. Marks with the assistance of John R. Weske', McGraw-Hill Book Co., New York 1937, a selective translation and an abstract of the author's dissertation, 140 p.

Knoernschild, Eugen M.: 'Dampftriebwerke fuer Flugzeuge, (Steam engines for aircraft)', Luftwissen, v. 8, no. 12, 1941, pp. 366-373

Knothe, Herbert: 'Herbert Wagner 1900-1982', A biographic documentation, 2004, 22 p. https://ritstaalman.files.wordpress.com/2015/02/herbertwagnerkknothe.pdf

Koeltzsch, Peter: 'Was bleiben wird (What will remain)', Lecture to W. Albring's 100th birthday, Proc. of the Inst. f. Appl. Fluid Mech., TU Dresden, Vol. 10, 2014, pp. 3-48 https://www.researchgate.net/publication/270281151 , and www.albring.info

Koenig, Max: 'Gas Turbines', Trans. North-East Coast Institution of Engineers and Shipbuilders, Vol. XLL – 1924/1925, pp. 347-415.

Koenig, Max: 'Ueber eine Naeherungsmethode zur Ermittlung der Schwingungsperioden profilierter Kreisscheiben' (An Approximate Method for Determining the Vibration Modes of Profiled Circular Disks), ETHZ doctoral thesis, Print Keller&Co. Lucerne 1927, 128 p. https://www.research-collection.ethz.ch/handle/20.500.11850/136517

Koos, Volker: 'Heinkel He 176 - Dichtung und Wahrheit (Poetry and Truth)', Jet+Prop Nr.1, 1994, extended March 2019, 11 p. https://adl-luftfahrthistorik.de/dok/heinkel-he-176-raketenflugzeug.pdf

Koos, Volker: 'Ernst Heinkel – Vom Doppeldecker zum Strahltriebwerk' (. . . - From Double-Decker Towards Turbojet Engine), Delius Klasing Bielefeld 2007, 224 p.

Koos, Volker: 'Heinkel Raketen- und Strahlflugzeuge (Rocket and Jet Aircraft)', Aviatic, Oberhaching 2008, 288 p.

Koos, Volker: 'Heinkel He 70 "Blitz" – Synonym für das Schnellverkehrsflugzeug', Sep. 2016, 13 p. https://docplayer.org/78331015-Heinkel-he-70-blitz-synonym-fuer-das-schnellverkehrsflugzeug.html

Koppenberg, Heinrich: 'Das Wagner-Triebwerk (The Wagner engine)', Mitteilungsblatt der Vereinigung ehemaliger Angehoeriger der Junkers Flugzeug- und Motorenwerke (Notes of the union of former members of Junkers airframe and engine works), 1955

Kotelnikov, Vladimir and Buttler, Tony: 'Early Russian Jet Engines – the Nene and Derwent in the Soviet Union, and the evolution of the VK-1', Rolls-Royce Heritage Trust, Historical Series No.33, 2003, 120 p.

Kotze, Hildegard von, and Krausnick, Helmut (eds.): '"Es spricht der Fuehrer". 7 exemplarische Hitler-Reden ("The Fuehrer speaks". 7 exemplary Hitler-speeches)', S. Mohn, Guetersloh, 1966, 379 p

Krengel, Joachim H.: 'Air-Breathing Engine Test Facilities Register', AGARDograph No. 269, July 1981, 126 p. https://www.sto.nato.int/publications/AGARD/AGARD-AG-269/AGARD-AG-269.pdf

Kruschik, Julius: 'Die Gasturbine' (The Gas Turbine), Springer Wien 1960, 2nd ed., 873 p.

Kuechemann, Dietrich and Weber, Johanna: 'Stroemungsvorgaenge am Triebwerk (Flow mechanisms at turbojet engines)', in Betz, Albert: 'Monographien ueber Fortschritte der Luftfahrtforschung seit 1939 (Monographs on progress in aeronautical research since 1939)', K3, AVA Goettingen, 1946, total – 72 authors, ~7,000 p.

Kuechemann, Dietrich and Weber, Johanna: 'Installation of Jet Engines', pp. 205-209, in 'Aerodynamics of Propulsion', McGraw Hill, 1st ed., New York, 1953, 340 p.

Kuehl, Heinrich: 'Die Dissoziation von Verbrennungsgasen und ihr Einfluss auf den Wirkungsgrad von Vergasermaschinen (The dissociation of combustion gases and their influence on the efficiency of carburettor engines)', Dissertation TU Munich (Profs. Nusselt and Loschge), 10/358, 1935, 31 p.

Kuehl, Heinrich and Schmidt, Fritz A.F.: 'Die Eignung verschiedener motorischer Arbeitsverfahren für Höhen- und Weitflug (The qualification of various motor working principles for altitude and distance flight)', DVL Jahrbuch (annual book) 1937, pp. 433-437

Kuehl, Heinrich: 'Grundlagen der Regelung von Gasturbinentriebwerken fuer Flugzeuge (Basics of gas turbine engine controls for aircraft)', DVL FB 1796/1-3, Berlin 1943. Part 1 Calculation standardization of aircraft GT engine controls, NACA-TM-1143-Pt-1, April 1947; Part 2 Principles of Control Common to Jet, Turbine-Propeller Jet, and Ducted-Fan Jet Power Plants, NACA-TM-1143-Pt-2, April 1947; Part III Control of jet engines, NACA TM-1166-Pt-3, April 1947.

Kuehl, Heinrich: 'Die Brennkammer fuer Gasturbinen (The gas turbine combustion chamber)', BWK Brennstoff-Waerme-Kraft (Fuel-Heat-Power), Vol. 4, Issue 7, 1952, p 217.

Kuehl, Heinrich: 'Zum 65. Geburtstag von Dr. B. Eckert', MTZ Motortechnische Zeitschrift, Vol. 33, Issue 4, 1972, p. 184.

Lancaster, Otis E., ed.: 'Jet Propulsion Engines', Vol. XII of High Speed Aerodynamics and Jet Propulsion series, Princeton NJ, Princeton University Press 1959, 799 p.

Lang, Norbert: 'Charles E.L. Brown und Walter Boveri: Gründer eines Weltunternehmens (... Founders of a world enterprise), Verein für wirtschafts-historische Studien (in German), 1992, 100 p.

Langston, Lee S.: 'Visiting the Museum of the World's First Gas Turbine Powerplant ', Global Gas Turbine News – a Supplement to Mechanical Engineering Magazine, April 2010, p. 51

Lawton, Roy: 'Parkside: Armstrong Siddeley to Rolls-Royce, 1939-1994', The Rolls-Royce Heritage Trust, Historical Series No. 39, 2008, 240 p.

Lehmann, Joerg and Morselli, Francesca: 'Science and Technology in the First World War', CENDARI Collaborative European Digital Archive Infrastructure 2016, 26 p.

Leutz, Achim: 'Schiff Nr. 294 der Stettiner Maschinenbau-Actien-Gesellschaft Vulcan "Föttinger Transformator"', Hermann-Föttinger-Archiv 2012, 18 p., in German https://hermann-foettinger.de/preprints/versuchsschiff.pdf

Leutz, Achim: 'Föttinger und der Kaiser-Wilhelm-Tunnel (Cochemer Tunnel)', Internet-Paper in German http://hermann-foettinger.de/preprints/akwt.pdf , 2016, 9 p.

Lichtfuss, Hanns-Juergen and Schubert, Helmut: '75 Jahre Turbostrahlflug. Der Beginn eines neuen Zeitalters der Luftfahrt (75 years turbojet flight. A new era in aviation begins)', DGLR paper 340021, 2014, 50 p. https://publikationen.dglr.de/?tx_dglrpublications_pi1[document_id]=340021

Lindbergh, Charles A.: 'The Wartime Journals of Charles A. Lindbergh', Harcourt Brace Jovanovich Inc. New York, 1970, 1,038 p.

Lippisch, Alexander: 'Erinnerungen (Memories)', Luftfahrtverlag Axel Zuerl, Steinebach, 1982, 264 p.

Lloyd, Peter: 'Hayne Constant, CB, CBE, MA, FRS, Fellow 1904-1968', The Aeronautical Journal of the Royal Aeronautical Society, Vol. 72, April 1968, pp. 285-286

Lorence, Christopher: 'Why the Time for Open Fan is Now', GE Aviation Blog, 12 Nov. 2021 https://blog.geaviation.com/product/why-the-time-for-open-fan-is-now/

Lowell, William O.: 'Gas Turbine Design', SAE Technical Paper 400064, 1940, 14 p.

Maas, Ad and Hooijmaijers, Hans: 'Scientific Research in World War II: What scientists did in the war', Chapter 9 by Florian Schmalz: 'Aerodynamic Research at the Nationaal Luchtvaartlaboratorium (NLL) in Amsterdam under German Occupation during World War II', Routledge, 2016, 256 p.

Mach, Ernst and Salcher, Peter: 'Photographische Fixirung der durch Projectile in der Luft eingeleiteten Vorgänge', (Photographic visualization of projectile-induced phenomena in air),

Proc. Kaiserl. Akad. Wiss., Vienna, Math.-Naturwiss. Cl. (in German). 95 (Abt. II), 1887, pp. 764–780

Madelung, Georg: 'Beitrag zur Theorie der Treibschrauben (Propeller Theory)', in Jahrbuch 1928 der deutschen Versuchsanstalt für Luftfahrt e.V. (DVL annual book 1928), Berlin-Adlershof, De Gruyter 2019 (reprint), pp. 27-62

Maier, Helmut: 'Forschung als Waffe (Science as weapon). Rüstungsforschung in der Kaiser-Wilhelm-Gesellschaft und das Kaiser-Wilhelm-Institut für Metallforschung 1900-1945/48', Wallstein, Goettingen, 2007, 1,235 p.

Massie, Robert K.: 'Dreadnought. Britain, Germany and the Coming of the Great War', Random House, New York, 1991, 1,007 p.

Masters, David: 'German Jet Genesis', Jane's Publ. Co. London, 1982, 142 p.

Matthaeus, Juergen and Bajohr, Frank: 'The Political Diary of Alfred Rosenberg and the Onset of the Holocaust', Rowman & Littlefield, 2015, 536 p.

Mattingly, Jack D.: 'Elements of Gas Turbine Propulsion (Forword by Hans von Ohain, German Inventor of the Jet Engine)', McGraw-Hill India, 2005, (1996 1st ed.), 960 p.

McFarland, Stephen L. and Newton, Wesley Ph.: 'To Command the Sky: The Battle for Air Superiority Over Germany 1942-1944', Smithsonian History of Aviation paperback, Univ. Alabama Press, 2006, 344 p.

Meher-Homji, Cyrus B. and Prisell, Erik: 'Pioneering Turbojet Developments of Dr. Hans von Ohain – From The HeS 1 To The HeS 011', ASME paper 99-GT-228, 13 p.

Meher-Homji, Cyrus B.: 'The historical evolution of turbomachinery', ASME Proceedings of the 29th turbomachinery symposium, Sept. 2000, pp. 281-321

Meher-Homji, Cyrus B.: 'Enabling the Turbojet Revolution – The Bramson Report', ASME Global Gas Turbine News, Vol. 42, No. 1, 2002, pp. 16-20

Meier, Hans-Ulrich: 'German Development of the Swept Wing 1935-1945', AIAA pdf-publication 2010, 735 p.

Mettler, Eduard: 'Die schweizerische Flugzeugindustrie von den Anfängen bis 1961 (The Swiss aircraft industry from start up to 1961)', Polygraphischer Verlag, 2006, 140 p.

Meyer, Adolf: 'The Velox Steam Generator, its Possibilities as Applied to Land and Sea', Mechanical Engineering, Aug. 1935, pp. 469-478

Meyer, Adolf: 'The Combustion Gas Turbine: Its History, Development and Prospects', in Brown Boveri Review, Vol. XXVI, No. 6, June 1939, pp. 127-140 and, Proceedings of the Institution of Mechanical Engineers, London, Vol.141, Issue 1, 1 June 1939, with discussion, pp. 197-222

Middlebrook, Martin and Everitt, Chris: 'The Bomber Command War Diaries: An Operational Reference Book 1939-1945', Viking, January 1985, 804 p.

Mommsen, Hans and Grieger, Manfred: 'Das Volkswagenwerk und seine Arbeiter im Dritten Reich (The Volkswagen works and its labour force in the Third Reich)', Econ, 1996, 1055 p.

Moore, John: 'Eckardt's Impeller: A Ghost from Ages past', Cambridge Univ. (England), Dept. of Engineering, Report N77-28442/0/XAB; CUED/A-TURBO/TR-83, 1976, 22 p.

Moss, Sanford A.: 'The Gas Turbine, An "Internal Combustion" Prime-Mover', Ph.D. thesis Cornell University, May 1903, 48 p.

Moss, Sanford A.: 'Gas Turbines and Turbosuperchargers', Trans. ASME 66, 1944, pp. 349-350

Moss, Sanford A.: 'Turbo-superchargers in Europe', J. 'Archive', RR Heritage Trust, Derby & Hucknall Branch, No.81, Vol.27, Issue 2, Aug. 2009, pp. 20-28

Mouton, Pierre: 'Junkers Jumo 004 et BMW 003' and 'Débuts de l'électronique à Snecma (Electronic beginnings at . . .)', Prendre l'air, no. 3, Dec. 2019, pp. 26-30

Mouton, Pierre: 'Hermann Oestrich et son groupe «O»', RAP 1228 – Histoire de l'aéronautique (unpublished paper), 19 Jan. 2022, 17 p.

MTU Munich: 'Erinnerungen, 1934-1984, Flugtriebwerkbau in München (Memories 1934-1984, aero engines from Munich)', MTU Motoren- und Turbinen-Union München GmbH, May 1984, 144 p.

Mueller, Max Adolf: 'TL-Triebwerke in axialer Bauweise und Motor-Luftstrahltriebwerk (Axial turbojet engines and motorjet), DAL Deutsche Akademie der Luftfahrtforschung (German academy of aeronautical research), 5th scientific session, Berlin, 2 July 1943, pp. 123-130.

Mueller, Reinhard: 'Junkers Flugtriebwerke. Benzinmotoren, Flugdiesel, Strahlturbinen' (Junkers Aero-Engines. Petrol Engines, Aero Diesel, Turbojet Engines), Aviatic Oberhaching 2006, 382 p.

Mueller, Rolf-Dieter: 'Die deutschen Gaskriegsvorbereitungen 1919-1945. Mit Giftgas zur Weltmacht?' (The German Preparations for Gas Warfare 1919–1945. World Power by Means of Poison Gas?), MGM Militärgeschichtliche Mitteilungen, Oldenbourg, 1, 1980a, pp. 25-54

Mueller, Thomas J.: 'On the Historical Development of Apparatus and Techniques for Smoke Visualization of Subsonic and Supersonic Flows', AIAA Paper 80-420, 1980b

Mueller-Romminger, Fredric: 'Geflogene Vergangenheit – der Flughafen Ainring' (Flown past - the Ainring airport), Das Salzfass (The salt shaker), Heimatkundliche Zeitschrift des Historischen Vereins Rupertiwinkel (Local historic journal of the historical club Rupertiwinkel), 1996, Issue 2, pp. 118-123

Muenzberg, Hans-Georg: 'Konzeptionsentscheidungen bei der Turbostrahltriebwerksentwicklung (Conceptual decisions for turbojet engine development)', DGLR Report 89-05 *'50 Jahre Turbostrahlflug'* ('50 years of jet powered flight'), Munich 26/27 Oct. 1989, Paper 89-216, pp. 69 – 97

Munk, Max and Hueckel, Erich: 'Systematische Untersuchungen an Flügelprofilen' (Systematic Investigations of Wing Profiles), Technische Berichte der Flugzeugmeisterei der Inspektion der Fliegertruppen, Vol. I, 1917, 148 p.

Munk, Max and Hueckel, Erich: 'Weitere Goettinger Fluegelprofiluntersuchungen' (Additional Wing Profile Investigations at MVA Goettingen), Mitteilung Nr.17 in Technische Berichte der Flugzeugmeisterei der Inspektion der Fliegertruppen, Vol. II, Issue 3, 1918a, 407 p.

Munk, Max and Hueckel, Erich: 'Der Profilwiderstand von Tragflügeln' (The Profile Drag of Airfoil Wings), Technische Berichte der Flugzeugmeisterei der Inspektion der Fliegertruppen, Vol. II, 1918b, 451 p.

Munk, Max M.: 'Elements of the wing section theory and of the wing theory', NACA TR No. 191, 1924, 26 p.

Munk, Max: 'My early aerodynamic research. Thoughts and Memories', Annual Review of Fluid Mechanics, Vol.13, 1981, pp. 1-8 https://www.annualreviews.org/doi/abs/10.1146%2Fannurev.fl.13.010181.000245

Munzinger, Ernst: 'Duesentriebwerke – Entwicklungen und Beschaffungen für die schweizerische Militäraviatik (Jet engines – developments and acquisitions for the Swiss military aviation)', Baden-Verlag 1991, 226 p.

Murphy, Eugene: 'Hans Adolph Mauch 1906-1984', NAE National Academy of Sciences, Engineering, Medicine; The National Academies Press, Memorial Tributes: Vol. 3, 1989, 370 p, Chapter Hans Mauch, pp. 258-264, https://www.nap.edu/read/1384/chapter/46

Muttray, Horst W.: 'Untersuchungen ueber die Beeinflussung des Tragfluegels eines Tiefdeckers durch den Rumpf', Luftfahrtforschung, 11 June 1928, pp. 33-39 and 'Investigation of the Effect of the Fuselage on the Wing of a Low-Wing Monoplane', NACA TM No. 517, 27 p. https://ntrs.nasa.gov/archive/nasa/casi.ntrs.nasa.gov/19930090868.pdf

Muttray, Horst W.: 'Die aerodynamische Zusammenfuegung von Tragfluegel und Rumpf', Luftfahrtforschung, 1934, Vol.11, No. 5, pp.131-139 and 'The Aerodynamic Aspect of Wing-Fuselage Fillets', NACA TM No. 764, 21 p.

Nahum, Andrew: 'Two-Stroke or Turbine? The Aeronautical Research Committee and British Aero Engine Development in World War II', Technology and Culture, Vol. 38, No. 2, Apr. 1997, pp. 312-354

Nahum, Andrew: '"*I believe the Americans have not yet taken them all!*": the exploitation of German aeronautical science in postwar Britain', in Trischler, Helmuth and Zeilinger, Stefan (eds.): 'Tackling Transport', Science Museum, London 2003, pp. 99-138, https://pdfs.semanticscholar.org/7687/b66e87e0e97e1da40349c2bc3704b64a047f.pdf

NASA: 'Engines and Innovation: Lewis Laboratory and American Propulsion Technology', NASA SP-4306, 1991, 433 p. https://history.nasa.gov/SP-4306.pdf

Neal, Robert J.: 'A Technical & Operational History of the Liberty Engine. Tanks, Ships and Aircraft 1917-1960', Speciality Press, North Branch, MN, 2009, 616 p.

Neufeld, Michael J.: 'The Rocket and the Reich. Peenemünde and the Coming of the Ballistic Missile Era', Simon and Schuster, 1995, 368 p.

Neufeld, Michael J.: 'Rocket Aircraft and the "*Turbojet Revolution*"', Chapter 9, pp.207-234, in Roger D. Launius, ed. 'Innovation and the Development of Flight', Texas A&M University Press, 1999, 352 p.

Neurauter, Sebastian: 'Das Bauhaus und die Verwertungsrechte: Eine Untersuchung zur Praxis der Rechteverwertung am Bauhaus 1919-1933 (Bauhaus and patent rights: An investigation of actual patent exploitation at Bauhaus 1919-1933)', Mohr Siebeck, 2013, 528 p.

Nickelsen, Kärin; Hool, Alessandra and Grasshoff, Gerd: 'Theodore von Kármán. Flugzeuge für die Welt und eine Stiftung für Bern (Airplanes for the world and a charitable trust for Berne), Birkhaeuser (Springer), 2004, 260 p.

Nixon, Frank: 'Aircraft Engine Developments During the Past Half Century', The Aeronautical Journal, Vol. 70, Issue 661, Jan. 1966, pp. 150-167

Noack, Walter G.: 'Flugzeuggebläse', Rundschau ZVDI 1919, pp. 995-1026, and 'Airplane Superchargers', NACA TN-48, May 1921, 31 p.

Noack, Walter G.: 'Tests of the Daimler D-IVa engine at a high altitude test bench', (translated from Technische Berichte, 1918, Vol. III, Issue 1 by Paris Office NACA), NACA TN-15, 1920, 21 p.

Noack, Walter G.: 'Druckfeuerung von Dampfkesseln in Verbindung mit Gasturbinen', (Pressurized Firing of Steam Boilers in Connection with Gas Turbines), Z. VDI 1932, p. 1033 f.

Nockolds, Harold: 'The Magic of a Name', Foulis & Co., London, Nov. 1938, cited from Nov. 1959 reprint, 283 p.

Norris, Guy: 'GE at 100: Taking power to the skies', Aviation Week & Space Technology, 15-28 July 2019, pp. 22-23

Nouzille, Vincent: 'La folle histoire du Dr Oestrich. De BMW à la Snecma, l'itinéraire du père des moteurs de Mirage est un vrai roman d'espionnage (The crazy story of Dr Oestrich. The life of the father of the Mirage engine is a real spy novel)', L'Express (French weekly paper), Paris, 20 May 1999, 00:00 h, p. 59

Oestrich, Hermann: 'Aus der Entwicklung des Strahltriebwerkes ATAR 101 (On the development of the turbojet engine ATAR 101)', WGL Jahrbuch (yearbook) 1954, pp. 135-139

Oestrich, Hermann: 'Entwurf und Ausreifung des Triebwerkes BMW 003 von Dr. Hermann Oestrich † (Design and refinement of the BMW 003 turbojet engine, by the late Dr. H. Oestrich)', DGLR Report 92-05, '50 Jahre Turbostrahlflug' ('50 years of jet powered flight'), Vol. II, Munich 26/27 Oct. 1989, Paper 89-223, pp. 89-114

Ohain, Hans-Joachim Pabst von: 'Ein Interferenzlichtrelais für weißes Licht (An interference light relais for white light)', Annalen der Physik, Vol. 415, Issue 5, 1935, pp. 431-441

Ohain, Hans-Joachim Pabst von: 'Besonderheiten der Strahltriebwerke radialer Bauart (Specific Features of Centrifugal Turbojet Engines)', Deutsche Akademie der Luftfahrtforschung, Proceedings of session 'Strahltriebwerke' on 31 Jan. 1941, Berlin, pp. 109-120

Ohain, Hans-Joachim Pabst von: 'Turbostrahltriebwerke – ihre Anfaenge und Zukunftsmoeglichkeiten (Turbojet engines – their beginnings and future possibilities)', DGLR Report 89-05 '50 Jahre Turbostrahlflug' ('50 years of jet powered flight'), Munich 26/27 Oct. 1989, Paper 89-213, pp. 3–45

Olmsted, Merle C.: 'Turbo-Supercharger Development' in American Aviation Hist. Soc. J., Vol.12, 1967, pp. 99-111

Palucka, Tim: 'The Wizard of Octane: Eugene Houdry', Invention & Technology, Vol. 20, No. 3, Winter 2005, https://www.inventionandtech.com/content/wizard-octane-0

Parker Dawson, Virginia: 'Engines and Innovation: Lewis Laboratory and Americam Propulsion Technology', NASA Office of Management, Scientific and Technical Information Division, 1991, 276 p.

Pavelec, S. Michael: 'The Jet Race and the Second World War', Praeger Sec. International, 2007, 250 p.

Perry, Robert L.: 'Innovation and Military Requirements: A Comparative Study', Rand Corp., Sta. Monica, RM-5182-PR, 1967, 92 p.

Pfenninger, Hans: 'Die Gasturbinenabteilung bei BBC, Rueckblick und heutiger Stand', (The BBC Gas Turbine Department – in Retrospect and Today), Schweizerische Bauzeitung, Vol. 88, Issue 30, 23 July 1970, pp. 683-691

Postan, Sir Michael M., Hay, D. and Scott, J.D.: 'Design and Development of Weapons', HMSO (Part of History of the Second World War, UK Civil Series), 1968, 639 p.

Potthoff, Juergen and Schmid, Ingobert C.: 'Wunibald I.E. Kamm – Wegbereiter der modernen Kraftfahrtechnik' (Pioneer of Modern Automotive Technology), Springer 2011, 379 p.

Power Jets R&D: 'The Junkers Jumo 004 Jet Engine', Aircraft Engineering, Dec. 1945, pp. 342-347

Prandtl, Ludwig: 'Zur Torsion von prismatischen Staeben', (On the torsion of prismatic rods), Physikalische Zeitschrift, Vol. 4, 1903, pp. 758-770

Prandtl, Ludwig: 'Über Flüssigkeitsbewegungen bei sehr kleiner Reibung' (Fluid flow in very little friction), in Proceedings of the 3rd International Mathematical Congress Heidelberg 1904, publ. Leipzig, 1905

Prandtl, Ludwig: 'Tragfluegeltheorie, I. Mitteilung' (Aerofoil theory, 1st part), Nachrichten von der Gesellschaft der Wissenschaften zu Goettingen, Mathematisch-Physikalische Klasse (1918), News of the Scientific Soc. Goettingen, Mathem.-Physical Class (1918), Vol.1918, pp. 451-477

Prandtl, Ludwig et al. (eds.): 'Ergebnisse der Aerodynamischen Versuchsanstalt zu Goettingen', (Results of the Goettingen Aerodynamics Research Institute), I. Lieferung, R. Oldenbourg Muenchen, 1921, 140 p.

Prisell, Erik: 'The Beginning of the Jet Age', AIAA paper 2003-2721, 11 p.

Prommersberger, Juergen: 'Seeschlachten des 1. Weltkriegs: Der Kampf um die Ostsee' (Sea Battles of WW I, the Fight in the Baltic Sea), CreateSpace Ind. Publ. 2016, 678 p.

Pugh, Peter: 'The Magic of a Name: The Rolls-Royce Story, Part 1: The First Forty Years', Icon Books Ltd., 2015, 320 p.

Rabl, Christian: 'Das KZ-Aussenlager St. Aegyd am Neuwalde (KZ satellite camp)', Mauthausen Studies, Vol. 6, 2008, 164 p.

Radinger, Willy and Schick, Walter: 'Me 262: Entwicklung, Erprobung und Fertigung des ersten einsatzfähigen Düsenjägers der Welt (Me 262: development, testing and production of the world's first turbojet aircraft in operation)', Aviatic Verlag, Planegg, 1992, 111 p.

Rankin, Nicholas: 'Defending the Rock: How Gibraltar Defeated Hitler', Faber & Faber, 2017, 672 p.

Ritz, Ludolf: 'Abriß der Theorie und Konstruktion einer rationellen Gasturbine (Short theory and design of an economic gas turbine)', Parts I-VI and supplement 'Coal and RGT fuels', AVA Report No. 45K/5-11, Goettingen, July 1945, 159 p. in total

Rizzi, Arthur and Oppelstrup, Jesper: 'Aircraft Aerodynamic Design with Computational Software', Cambridge Univ. Press, 2021, 465 p.

Roe, Kenneth A.: 'Claude P. Seippel (1900-1986), National Academy of Engineering, Memorial tributes, Vol.4, 1991, 356 p.

Roginskij, Arsenij (ed.) et al.: 'Erschossen in Moskau...: Die deutschen Opfer des Stalinismus auf dem Moskauer Friedhof Donskoje 1950-1953 (Shot at Moscow...: The German victims of Stalinism at Donskoje cemetery, Moscow, 1950-1953)', Metropol, 2nd ed., 2005, 480 p.

Rosenthal, Jenny E.: 'On a Novel Form of Refrigerator', Journal of Applied Physics, Vol. 17, (1946), p.62

Rossmanith, H.P.: 'Fracture Research in Retrospect, an anniversary volume in the honour of G.R. Irwin's 90th birthday', A.A. Balkema Publ., Leiden NL, 1997, 580 p.

Roxbee Cox, Harold: 'British Aircraft Gas Turbines. The Ninth Wright Brothers Lecture, 17 Dec. 1945', J. of the Aeronautical Sciences, Vol. 13, No. 2, Feb. 1946, pp. 53-87 (with discussion)

Roxbee Cox ed., Sir Harold: 'Gas Turbine Principles and Practice', George Newnes Ltd, London, 1955, 960 p.

RSHA: 'Sonderfahndungsliste G.B. (Special search list Great Britain)', Berlin, 1940, 279 p. https://de.wikipedia.org/wiki/Sonderfahndungsliste_G.B.

Rubbra, Arthur A.: 'Alan Arnold Griffith 1893-1963', Biographical Memoirs of Fellows of the Royal Society, Vol. 10, Nov. 1964, pp. 117-136

Ruch, Christian et al.: 'Geschäfte und Zwangsarbeit : Schweizer Industrieunternehmen im "Dritten Reich" (Companies and forced labour: Swiss industrial enterprises in the "Third Reich")', ICE Vol. 6, Chronos, Zurich 2001, 384 p.

Ruden, Paul: 'Untersuchungen über einstufige Axialgebläse', Luftfahrt-Forschung (Oldenbourg), Vol. 14, Issue 7, 20 July 1937, pp. 325-346 and Issue 9, pp. 458-473 and 'Investigation of Single Stage Axial Fans', NACA TM No.1062, April 1944, 117 p.

Saunders, Owen A.: 'David Randall Pye 1886-1960', Biographical Memoirs of the Fellows of the Royal Society, Vol. 7, Nov. 1961, pp. 198-205

Saunders, Owen A.: 'David MacLeish Smith 1900-1986', Biographical Memoirs of Fellows of the Royal Society, Vol. 33, Dec. 1987, pp. 603-617

Schabel, Ralf: 'Die Illusion der Wunderwaffen. Die Rolle der Duesenflugzeuge und Flugabwehrraketen in der Ruestungspolitik des Dritten Reiches', (The illusion of wonder weapons. The role of jet aircrafts and anti-aircraft missiles in the 3rd Reich armament policy), Oldenbourg, Munich 1994, 316 p.

Schelp, Helmut: 'Alcohol-Blends in Gasoline', SAE Technical Paper 360058, 1936, 16 p., https://doi.org/10.4271/360058

Schelp, Helmut: 'Luftstrahltriebwerke für Schnellflugzeuge (Turbojet engines for high speed aircraft), in 'Strahltriebwerke (Turbojet engines)', Deutsche Akademie der Luftfahrtforschung Berlin, Arbeitstagung (working session) 31 Jan. 1941, pp. 15-37

Schelp, Helmut: 'Hochleistungstriebwerke (High performance power plants)', DAL Deutsche Akademie der Luftfahrtforschung (German academy of aeronautical research), 5th scientific session, Berlin, 2 July 1943, pp. 15-37; AEHS translation in (faulty) English, 23 p., - and German, 33 p.: http://www.enginehistory.org/Piston/Junkers/Hochleistungstriebwerke/index.html

Schelp, Helmut: 'Zur Geschichte des Strahltriebwerks – Die Entwicklung bis 1945, (Turbojet history – developments up to 1945)', Annual DGLR Conference, Presentation No. 81-073, Aachen, 11-14. May, 1981, DGLR Annual Book, IV(1981), pp. 073-1 - 073-17.

Schilhansl, Max: 'Näherungsweise Berechnung von Auftrieb und Druckverteilung in Flügelgittern', (Approximate calculation of lift and pressure distribution in airfoil cascades), Jahrbuch der Wissenschaftlichen Gesellschaft fuer Luftfahrt. 1927, pp. 151-167

Schlaifer, Robert: 'The Development of Aircraft Engines', Harvard University Press, Boston Ma., 1949, 545 p.

Schmaltz, Florian: 'Vom Nutzen und Nachteil der Luftfahrtforschung im NS-Regime. Die Aerodynamische Versuchsanstalt Göttingen und die Strahltriebwerksforschung im Zweiten Weltkrieg (Benefit and disadvantage of aeronautical research in the NS regime. The AVA Goettingen and turbojet engine research during WW II)', in: Christine Pieper/Frank Uekoetter (eds.), 'Vom Nutzen der Wissenschaft. Über eine prekäre Beziehung (The benefit of science. About a precarious relationship)', Stuttgart 2010, pp. 67-113.

Schmaltz, Florian: 'Kampfstoffforschung im Nationalsozialismus: Zur Kooperation von Kaiser-Wilhelm-Instituten, Militär und Industrie' (Chemical agent research in National Socialism: On the cooperation of Kaiser-Wilhelm-Institutes, military and industry), Wallstein, 2017, 676 p.

Schmidt, Fritz A.F.: 'Zündverzug und Klopfen im Motor (Ignition lag and engine knocking)', VDI FB 392, 1938, 24 p.

Schmidt, Fritz A.F.: 'Verbrennungskraftmaschinen: Thermodynamik und versuchsmäßige Grundlagen der Verbrennungsmotoren, Gasturbinen, Strahlantriebe und Raketen. Zündungs- und Reaktionsvorgänge im Temperaturbereich bis 4000°C' (Internal combustion engines: Thermodynamics and test-based combustion engine data, gas turbines, jet propulsion and rockets. Ignition and reaction processes up to 4,000°C), Springer, 2013, 649 p. (1st ed. 1939, 328 p.)

Schmidt, Fritz A.F.: 'The Internal Combustion Engine', Chapman & Hall, London, 1965, 579 p., enlarged version of the 3rd ed.

Schmunk, Stefan: '"Entweder KZ oder ordentliche Deutsche." Die Luftwaffe und der Arbeitseinsatz 1942-1944 ("Either KZ or Proper Germans". Luftwaffe and the deployment of labour 1942-1944), TU Darmstadt, Master's thesis 2005, 119 p. https://www.researchgate.net/publication/317570552_Entweder_KZ_oder_ordentliche_Deutsche_Die_Luftwaffe_und_der_Arbeitseinsatz_1942_-_1944

Schrenk, Martin: 'Probleme des Hoehenfluges', ('The problems of high altitude flight'), Zeitschrift fuer Flugtechnik und Motorluftschifffahrt, IXX, 1928, A 466

Schrenk, Oskar: 'Grenzschichtabsaugung', Luftwissen, Vol.7, No.12, Dec.1940, pp. 409-414, and 'Boundary Layer Removal by Suction', NACA TM No. 974, Washington, April 1941, 14 p. https://ntrs.nasa.gov/api/citations/19930094442/downloads/19930094442.pdf

Schubert, Helmut: 'Turbine – The hollow metal blade as solution for material shortage', pp. 244-252, in Hirschel, Ernst H., Prem, Horst and Madelung, Gero: 'Aeronautical Research in Germany. From Lilienthal until Today', Springer Berlin Heidelberg, 2004, 694 p.

Schubert, Joachim: 'BBC-Mannheim zwischen Weltwirtschaftskrise und Ende des Zweiten Weltkriegs – Ein geschichtlicher Abriss (BBC-Mannheim between the world economic crisis and the end of WW II)', Mannheim, 2019, 10 p. file:///C:/Users/eckar/Downloads/bbcchronik.pdf

Schwager, Otto: 'Der Weg zum Hoehenflugmotor', ('Towards the aircraft engine for high altitudes'), Flugwissen Vol. 5, Issue 5, 1943, pp. 136-144

Schwencke, Dietrich: 'Strahltriebwerke des Auslandes (Foreign turbojet engines)', G. Kdos. (Top secret), RLM Internal Rep. GL-C 14,937/44, 15 March 1944, 43 p.

Seippel, Claude: 'The Development of the Brown Boveri Axial Compressor', BBR Brown Boveri Review, Vol. XXVII, No. 5, May 1940, pp. 108-113

Seippel, Claude: 'From Stodola to Modern Turbine Engineering', The 17th Parsons Memorial Lecture, 31 Oct. 1952, to The North-East Coast Institution of Engineers and Shipbuilders, Bolbec Hall, Newcastle-on-Tyne, published in the Institution Transactions, Vol. 69, 1953

Seippel, Claude: 'The evolution of the compressor and turbine bladings in gas turbine design', Trans. ASME, J. Eng. for Power, 89, 1967, pp. 199-206

Seippel, Claude: 'Die Entstehungsgeschichte des vielstufigen Axialverdichters bei Brown Boveri' ('The development history of the multi-stage axial compressor at Brown Boveri'), Internal BBC report TT 7509, 30 July 1974, 13 p.

Seitz, Arne et al.: 'Concept validation study for fuselage wake-filling propulsion integration', ICAS 2018, 31st Congress Belo Horizonte, Brazil, 21 p., ICAS2018_0752_paper.pdf

Senger, Ulrich: 'Die Betriebskennlinien mehrstufiger Verdichter (The operation lines of multi-stage compressors)', Brown Boveri Nachrichten, Vol. XXIII, No.1-3, Jan.-March 1941, Mannheim, pp. 19-27

Serovy, George K.: 'A Method for the Prediction of the Off-Design Performance of Axial-Flow Compressors', PhD dissertation, Iowa State College, 1958, 74 p. http://lib.dr.iastate.edu/rtd/2265

Shelton, John K.: 'R.J. Mitchell at Supermarine: From Schneider Trophy to Spitfire', Standon books, 2nd ed., 2017, 372 p.

Sherbondy, Earl H. and G. Douglas Wardrop: 'Textbook of Aero Engines', Frederick A. Stokes Co, New York 1920, 363 p. https://archive.org/details/textbookaeroeng00shergoog/page/n4

Sietz, Henning: 'Es riecht nach Senf!' ('It smells of mustard!'), DIE ZEIT, 22 June 2006, https://www.zeit.de/2006/26/A-Tomka

Simms, Brendan: 'Hitler. Only the World Was Enough', Allen Lane 2019, 668 p.

Sinnette, John T., Schey, Oscar W. and King, J. Austin: 'Performance of NACA Eight-Stage Axial-Flow Compressor Designed on the Basis of Airfoil Theory, NACA Report No. 758, 1944, 19 p.

Smith, Apollo M.O. and Roberts, Howard E.: 'The Jet Airplane Utilizing Boundary Layer Air for Propulsion', Journal of Aeronautical Sciences, Vol.14, No.2, Feb. 1947, pp.97-109

Smith, G. Geoffrey: 'Gas Turbines and Jet Propulsion for Aircraft', Flight Publ. Co. Ltd. London, 1st ed. Dec. 1942, 2nd ed. June 1943, and 3rd ed. April 1944, 123 p.

Smith Jr, Leroy H.: 'Wake Ingestion Propulsion Benefit', Journal of Propulsion and Power, Vol.9, No.1, 1993, pp. 74-82

Smith Jr, Leroy H.: 'Axial Compressor Aerodesign Evolution at General Electric', ASME J. of Turbomachinery, July 2002, Vol. 124, pp. 321-330

Speidel, Helm: 'Reichswehr und Rote Armee' ('Reichswehr and Red Army'), Vierteljahreshefte fuer Zeitgeschichte, 1953, Issue 1, pp. 9-45 https://www.ifz-muenchen.de/heftarchiv/1953_1_2_speidel.pdf

Spencer, Richard: 'The German Patent Office', J. of the Patent Office Society, Vol. 31, Issue 2, 1949, pp. 79-87 https://www.dpma.de/english/our_office/about_us/history/140yearsofthepatentoffice/1941-1950/index.html

Spiegel, Mag.: 'Heinkel AG, Umwege über Vaduz (.., Detours via Vaduz), 30 Sep. 1958 https://www.spiegel.de/wirtschaft/umwege-ueber-vaduz-a-a7482ebd-0002-0001-0000-000041759242

Spiegel, Mag.: 'Schweizer Spitze (Swiss Peak)', 17 July 1966 https://www.spiegel.de/politik/schweizer-spitze-a-5f9d6cc3-0002-0001-0000-000046408025?context=issue

Spillmann, Werner: 'A ducted fan engine with bypass burning', EFW Emmen report no. FO-1491, 1956, 33 p.

Stern, William J.: 'The internal combustion turbine', Tech. Rep. of Advisory Committee for Aeronautics 1920-1921, Vol. 2, pp. 690-774 (Eng. Sub-Com. Report No. 54, Sept. 1920, 87 p.)

Stilla, Ernst: 'Die Luftwaffe im Kampf um die Luftherrschaft (The Luftwaffe's Fight for Air Superiority)', Ph D thesis, Uni Bonn, Bonn 2005, 315 p. https://bonndoc.ulb.uni-bonn.de/xmlui/handle/20.500.11811/2241

Stodola, Aurel: 'Dampf- und Gas-Turbinen', ('Steam and Gas Turbines'), J. Springer Berlin, 5th ed. 1922, 1,111 p.

Stodola, Aurel: 'Technisch-wirtschaftliche Fortschritte auf dem Gebiete des Dampfkraft-maschinenbaus in der Schweiz (Techno-economic progress of Swiss steam power technology),

Paper No. 213, pp. 3-77, in 'Gesamtbericht Zweite Weltkraftkonferenz (Complete report of the 2nd world power conference)', VDI Berlin, Vol. 5, 1930

Stoeckly, Eugene E.: 'Bavarian Motor Works Altitude-Test Facilities', Trans. ASME, Oct. 1946, pp. 743-750

Stoltzenberg, Dietrich: 'Fritz Haber: Chemist, Nobel laureate, German, Jew', Chemical Heritage Foundation, Philadelphia 2005, 336 p.

St. Peter, James: 'The History of Aircraft Gas Turbine Engine Development in the United States . . . A Tradition of Excellence', International Gas Turbine Institute of ASME, 1999, 591 p.

Straßl, Hans and Ludwieg, Hubert: 'Verringerung des Widerstandes von Tragflügeln bei hohen Geschwindigkeiten durch Pfeilform (Arrow wing drag reduction at high flight velocities)', AVA Report 39/H/18, 1939, pp. 1-5

Stüper, Josef: 'Flugerfahrungen und Messungen an zwei Absaugeflugzeugen', ZWB Berlin-Adlershof, FB1821, 1 July 1943, and 'Flight experiences and tests on two airplanes with suction slots', NACA TM No. 1232, Washington, Jan.1950, 104 p.

Szoelloesi-Janze, Margit: 'Fritz Haber 1868-1934. Eine Biographie', Ch. Beck, pocket book, 2nd ed., 2015, 928 p.

Thiel, Ernstfried: 'Von Oetztal nach Modane – aus der Geschichte des grossen Hochge-schwindigkeitskanals <Bauvorhaben 101> der Luftfahrtforschungsanstalt (LFM) München, spaeter Anlage S1 MA der Onera', DGLR-Jahrbuch 1986, Part II, pp.773-795

Thomson, George P. and Hall, Arnold A.: 'William Scott Farren, 1892-1970, Elected F.R.S. 1945', Biographical Memoirs of the Fellows of the Royal Society, Vol. 17, 1 Nov. 1971, pp. 215-241

Tilgenkamp, Erich et al.: 'Schweizer Luftfahrt' ('Swiss Aviation'), Vol. I (of 3), Aero Club Schweiz, Zurich 1941, 384 p.

Timoshenko, Stephen P.: 'History of Strength of Materials', McGraw-Hill, 1953, 452 p.

Tizard, Henry T. and Pye, David R.: 'The Character of Various Fuels for Internal Combustion Engines', Report of the Empire Motor Fuels Committee, Vol. XVIII, Part I, Publ. by the Inst. of Automobile Engineers, London, 1924

Toepser-Ziegert, Gabriele and Bohrmann, Hans: 'NS-Presseanweisungen der Vorkriegszeit, Anhang 1936 (NS press directives of the pre-war era, supplement 1936)', De Gruyter 2015, 1,906 p.

Tournaire, L.-M.: 'Sur des appareils à turbines multiples et à réactions successives pour utiliser le travail moteur que développent les fluides élastiques' (Multi-stage reaction turbomachinery to use the drive power of compressible fluids), memorandum to the Académie des Sciences de l'Institut de France, Paris, Vol. 36, 1853, p. 588 http://www.annales.org/archives/x/tournaire.html

Traupel, Walter: 'Neue allgemeine Theorie der mehrstufigen axialen Turbomaschine (New general theory of the multi-stage axial turbomachine)', ETHZ doctoral thesis, and AG Gebr. Leemann & Co., Zurich, 1942, 151 p.

Traupel, Walter: 'Thermische Turbomaschinen' ('Thermal Turbomachines'), Springer Berlin, 3rd ed. 1988, Vol.1, 579 p.

Trichet, Pierre: 'Paperclip, French Style', AIAA-paper 2009-962, 47th AIAA Aerospace Conf., 5-8 Jan. 2009, Orlando Fl., 31 p.

Trischler, Helmuth: 'Luft- und Raumfahrtforschung in Deutschland 1900-1970. Politische Geschichte einer Wissenschaft (Aerospatial research in Germany. Political history of a science)', Campus, Frankfurt, 1992, 542 p.

Trubowitz, Peter and Harris, Peter: 'When states appease: British appeasement in the 1930s', LSE Research Online, 2015, 50 p. http://eprints.lse.ac.uk/61659/

Tsien, Hsue-Shen et al.: 'Technical Intelligence Supplement: A Report Prepared for the AAF Scientific Advisory Group', US Headquarters Air Material Command, Wright Field, Dayton Oh., May 1946, 177 p. https://searchworks.stanford.edu/view/1689393

U.S. Navy Dept.: 'An Investigation of the Possibilities of the Gas Turbine for Marine Propulsion: Report submitted to the Secretary of the Navy', Bureau of Ships Techn. Bulletin No. 2, U.S. Gov. Printing Office, Jan. 1941, 45 p.

U.S. Navy Dept.: 'Interrogation Report Hermann Reuter on Axial Compressor Design', RTP/ TIB.3e, Ref. I/100 (Restricted), 9 Nov. 1945, (Kenneth Cambell, USN Technician), 7 p.

Uttley, Matthew: 'Operation "Surgeon" and Britain's post-war exploitation of Nazi German aeronautics', Intelligence & National Security, Vol.17, Issue 2, June 2002, pp. 1-26

Uziel, Daniel: 'Arming the Luftwaffe. The German Aviation Industry in World War II', McFarland & Co. Inc., Jefferson NC, USA and London, 2012, 303 p.

Vierhaus, Rudolf: 'DBZ Deutsche Biografische Enzyklopädie (Dictionary of German Biography)', Vol. 7 Menghin-Pötel, De Gruyter, Berlin, 2011, 927 p.

Vincent, Edward T.: 'Supercharging the Internal Combustion Engine', McGraw-Hill, New York, 1948, 315 p.

Vogel-Prandtl, Johanna: 'Ludwig Prandtl. 'Ein Lebensbild, Erinnerungen, Dokumente' (Ludwig Prandtl. Life, memories, documents'), Univ.Verlag Goettingen 2005, 256 p. http://webdoc.sub.gwdg.de/ebook/univerlag/2006/prandtl_book.pdf

Wagner, Herbert: 'Bericht ueber den Bau von Hoehenflugzeugen' ('Construction report about altitude aircraft'), Schriften der Deutschen Akademie der Luftfahrtforschung (Proceedings of the German Academy of Aviation Research), Heft 29, R. Oldenbourg, 1939, pp. 35-56

Wagner, Herbert and Kimm, Gotthold: 'Bauelemente des Flugzeugs (Construction elements of aircraft)', R. Oldenbourg, 1940, 296 p.

Wagner, Herbert: 'Meine Arbeiten am Strahltriebwerk' ('My turbojet engine activities'), Supplement to a letter to Kyrill von Gersdorff, dated 8 May 1980, pp. 126-127 in DGLR Publ.: 'Herbert Wagner. Dokumentation zu Leben und Werk' ('Documentation of his Life and Works'), Bonn 1986, 159 p.

Wagner, Jens-Christian: 'Produktion des Todes. Das KZ Mittelbau-Dora (Production of death. ..)', Wallstein (2nd ed., paperback in German/English), 2004, 688 p.

Wagner, R.: 'Praktische Ergebnisse mit Gegenpropellern (Practical results with counter-rotating propellers)', Jahrbuch der Schiffbautechn. Gesellschaft, 13th Vol. 1912, Springer Berlin, Section XV

Wallner, Lewis E. and Saari, Martin J.: 'Preliminary Results of an Altitude Wind-Tunnel Investigation of a TG-100A Gas Turbine – Propeller Engine, IV Compressor and Turbine Performance Characteristics', NACA RM No. E7J20, 13 Nov. 1947, 46 p.

Wark, Wesley K.: 'The Ultimate Enemy. British Intelligence and Nazi Germany, 1933-1939', Cornell University Press, 209, 304 p.

Wefeld, Hans-Joachim: '75 Jahre Akaflieg Berlin', Akademische Fliegergruppe, Berlin, 1995, 137 p.

Weinig, Fritz: 'Die Stroemung um die Schaufeln von Turbomaschinen. Beitrag zur Theorie axial durchstroemter Turbomaschinen (The flow around turbomachinery blading. A theoretical contribution to axial-flow turbomachines)', J.A. Barth Leipzig, 1935, 142 p.

Weinig, Fritz: 'Aerodynamik der Luftschraube', J. Springer, 1940, XVI + 484 p., and 'Aerodynamics of the Propeller', Air Documents Div., Intelligence Dept., Air Material Command, 1948, 501 p., (Google book)

Weise, Arthur: 'Überschallaxialverdichter', Lilienthal Ges. Report 171, Oct. 1943 (unpubl.) and translated as 'A Supersonic Axial Compressor', Information on Unconventional Compressors, U.S. Navy Dept., Buships 338, May 1946

Werner, Constanze: 'Kriegswirtschaft und Zwangsarbeit bei BMW (War economy and forced labour at BMW)', Oldenbourg, Munich, 2006, 447 p.

White, Roland J. and Antz, Hans M.: 'Tests on the Stress Distribution of Reinforced Panels', J. Aeronautical Sciences, Vol. 3, No. 6, April 1936, pp. 209-212

Whitfield, Jakob: 'Metropolitan Vickers, The Gas Turbine, and the State: A Socio-Technical History 1935-1960', PhD thesis, University of Manchester 2012, 203 p. https://www.escholar.manchester.ac.uk/uk-ac-man-scw:186390

Whittle, Frank: 'The Early History of the Whittle Jet Propulsion Gas Turbine', The 1st James Clayton Lecture 5 Oct. 1945, Proc. of IMechE, 152, 1, 1945, pp. 419-435

Whittle, Frank and von Ohain, Hans: 'An Encounter Between Jet Engine Inventors: Sir Frank Whittle and Dr. Hans von Ohain, 3-4 May 1978', Aeronautical Division Air Force Systems Command (1986), Paperback, 136 p.

Whittle, Frank: 'Gas Turbine Aero-Thermodynamics – With Special Reference to Aircraft Propulsion', Pergamon Press, 1981, 261 p.

Wiart, Ludovic and Negulescu, Camil: 'Exploration of the Airbus "Nautilus" engine integration concept', 31st Congress Belo Horizonte, Brazil, 12 p., ICAS2018_0135_paper.pdf

Wilkins, Mira: 'The history of foreign investment in the United States, 1914-1945', Harvard University Press, Cambridge Ma. And London, 2004, 981 p.

Wilson, David Gordon: 'Turbomachinery – From Paddle Wheels to Turbojets', Mechanical Engineering, Oct. 1982, pp. 28-40

Zacher, Hans: 'Otto Fuchs', DGLR Jahrbuch 1987, Vol. 2, Bonn, pp. 976-984

Zegenhagen, Evelyn: '"Schneidige deutsche Mädel". Fliegerinnen zwischen 1918 und 1945 (Dashing German gals. Air women between 1918 and 1945)', Wallstein, 2007, 504 p.

Ziemke, Earl F.: 'The U.S. Army in the Occupation of Germany, 1944-1946', Int'l. Law & Taxation Publ., 2005, 496 p. https://history.army.mil/books/wwii/Occ-GY/index.htm

Index